S0-DVF-422

Lifelong Human Development

Lifelong Human Development

Alison Clarke-Stewart
University of California at Irvine

Marion Perlmutter
University of Michigan

Susan Friedman

WILEY
John Wiley & Sons
New York Chichester Brisbane Toronto Singapore

Cover illustration: Roy Wieman
Interior and cover design: Dawn L. Stanley
Photo editor: Safra Nimrod
Copy editor: Gilda Stahl
Production supervisor: Lucille Buonocore
Technical Art and Illustration: Blaise Zito Associates, Inc.

Library of Congress Cataloging-in-Publication Data:

Clarke-Stewart, Alison, 1943–
Lifelong human development / Alison Clarke-Stewart, Marion Perlmutter, Susan Friedman.
p. cm.
Includes index.
ISBN 0-471-84723-2
1. Developmental psychology. I. Perlmutter, Marion.
II. Friedman, Susan, 1947– . III. Title.
[DNLM: 1. Human Development. 2. Personality Development.
BF 713 C611L]
BF713.C58 1988
155--dc19
DNLM/DLC 87-31701
for Library of Congress CIP

Printed in the United States of America

10 9 8 7 6 5 4 3 2 1

Preface

Lifelong Human Development is the product of a collaboration by Alison Clarke-Stewart, a specialist in the psychological development of children; Marion Perlmutter, a specialist in adult development; and Susan Friedman, a specialist in the written word. It has been our hope to present a balanced, up-to-date, and captivating picture of the psychological developments at each stage of the lifespan, from conception to death. It also has been our hope to make the classics in the field and the most recent research accessible to majors and nonmajors alike. To this end, we have included information about real children and adults. The people who appear in these slice-of-life portraits are all quite real, although they have been given fictional names to protect their privacy.

The book is organized chronologically, but teachers whose courses are arranged topically can easily assign chapters or parts of chapters to fit their needs. We have avoided overlap between chapters and have included cross-references where necessary. Each chapter includes relevant theory, classic and recent research, examples from real people's lives, key terms, summaries, suggested readings, illustrations, and boxed material on topics of particular interest. A glossary of terms apears at the end of the book.

The book as a whole is divided int eight major parts. Part One is devoted to the history, theories, and methods of lifespan developmental psychology. Part Two is devoted to heredity and prenatal development. Parts Three through Eight are devoted to successive stages of the lifespan. The chapters within each part are arranged in a repeating order: physical, cognitive, and then social and emotional development.

Acknowledgments

Textbooks themselves have a lifespan, and we are grateful to those who helped this one reach maturity. To the children and adults who allowed us glimpses of their lives, we owe special thanks. To the reviewers whose suggestions improved our manuscript, we are indeed grateful: Suzanne Getz, Pennsylvania State University; Anita Greene, Stanford University; Mary Harris, University of New Mexico; Carolyn Mebert, University of New Hampshire; Ann K. Mullis, North Dakota State University; and Gwendolyn Sorell, Texas Tech University.

The editorial staff at Wiley has been enormously helpful—as usual. We are grateful to our editors, Warren Abraham and Carol Luitjens, and to the production staff who have guided us into print: Gilda Stahl, copy editor; Dawn L. Stanley, designer; Lucille Buonocore, production supervisor.

To the members of our own families, for patience beyond their years and forbearance above and beyond the call of duty, double dessert and triple thanks.

Irvine, California — ALISON CLARKE-STEWART

Ann Arbor, Michigan — MARION PERLMUTTER

New Haven, Connecticut — SUSAN FRIEDMAN

Contents

PART ONE

Introduction

chapter one

History and theories

Developmental psychologists today work to describe, explain, and in some cases to modify the way that individuals behave from the beginning to the end of life. They work to understand how individuals are alike and how they are different. Historically, most developmental psychologists in the United States and Europe focused on children's development. They tended to see developmental change as leading in a single direction and toward a single point—maturity. They viewed childhood and adolescence as periods of growth and gain, adulthood as static, and old age as a period of loss and decline.

The lifespan perspective on human development

Today, developmental psychologists taking a **lifespan** perspective consider human development to involve simultaneous growth and decline throughout life, that is, from conception to death. Changes in different aspects of biological, psychological, and social function may have different starting points, end points, and developmental trajectories. What is constant—paradoxically—is change. Lifespan studies of the development of intelligence, memory, motivation, and personality have shed light on the ongoing change that characterizes human experience. For example, people seem to experience relatively minor changes in sensory memory as they age, but many experience more severe changes in word and name retrieval. Although to date, most of the cognitive changes studied among adults are declines, lifespan psychologists stress the potentials for growth during adulthood and old age (and conversely, for decline during childhood or adolescence). They suggest that no single period in life has a paramount influence on the course of development.

Precursors of the lifespan perspective

During the 1700s and 1800s, well before there was an intellectual focus on any specific period of development, such as childhood, some European philosophers proposed a lifespan view of human development. Johann Nicolaus Tetens (1736–1807) was a German philosopher who wrote about the entire lifespan, from conception to death. Tetens (1777) discussed aging as one component of lifelong development. He raised issues that remain important today: the relation between human nature and environment, the variation in differences between individuals, and the perfectibility of human development. Another German philosopher, Friedrich August Carus (1770–1808), took a lifespan approach to human development, describing the four periods of the lifespan as childhood, youth, adulthood, and senescence. He saw old age, for example, as a stage of both decline and as maturity and progression, the last and highest stage of development. In his 1808 work, *Psychology*, Carus described a science that would explain the developmental differences among individuals and the historical and other conditions under which such differences occur. Both Carus and Tetens wrote about issues of human development that remain vital today. Adolphe Quetelet (1796–1874), a Belgian statistician, also proposed a lifespan view of development. Quetelet's book, *A Treatise on Man and the Development of His Faculties* (1842), presented data on the entire lifespan and emphasized the importance of general laws of development as well as social and historical change.

The ideas of Tetens, Carus, and Quetelet lay dormant until early in this century. In 1902, a lone article (Sanford, 1902) appeared in the *American Journal of Psychology*, suggesting that development be treated as a continuous process from birth to death. Some years later, several textbooks (e.g., Buhler, 1929; Hollingworth, 1927; Pressey, Janney, and Kuhlen, 1939) appeared that took a lifespan perspective on human development. They focused on the processes within human development and their many dimensions, directions, and contexts. But it was not until the 1970s that the lifespan approach infiltrated the study of development among American psychologists.

The lifespan perspective today

According to the lifespan perspective, the individual, historical, and cultural conditions that individuals encounter during their lives modify the course of their psychological development. Thus what is learned about one set of individuals, in one culture, at one historical time may not generalize to other sets of individuals, in other cultures, at other times. For example, the older adult today is not only a product of many years of living but of early development in a less technological, less sophisticated world.

The lifespan perspective has demonstrated the importance of the historical contexts within which people develop. For example, children who grew up during the Great Depression, as adults, have been found to hold different values from those who were born in the complacent 1950s (Elder, 1974). Elementary school children who learn arithmetic on a computer probably will have different cognitive strategies from those who learned from *Jolly Numbers* workbooks. Lifespan developmental psychologists take major social, technological, and historical changes into account when they describe the development of individuals and groups over the lifespan. The lifespan perspective has broadened the horizons of developmental psychology.

The lifespan perspective has also helped to shape today's view of relatively distinct life stages. Today we largely take for granted the fact that infants and children look, speak, think, and play differently from older people. We believe that infancy and childhood unfold in gradual stages of development and that the standards for adults' and children's behavior differ greatly. "Don't be too hard on him," the doting grandparent insists, "He's only a child." Today children are sent to school and prohibited by law from working. Their families cherish the milestones of their development—the first steps, the first words, the early finger paintings and homemade cards—and tolerate many of their youthful displays of temper.

Only a few centuries ago, children as young as 6 were sent from home as servants or apprentice laborers. They were dressed as adults, expected to work like adults, and in many ways were treated like adults. Discipline was sometimes harsh, and children were expected to conform as early as possible to proper standards of behavior. Many children in the 16th, 17th, and early 18th centuries in the United States and England were treated in ways that today seem cruel and inhumane (Ariès, 1962; Piers, 1978). Parents punished their children severely, in the belief that a child's will had to be broken if society were not to be thrown into disorder, that only hard work and strict limits would lead to salvation. In extreme cases, children might be beaten or even killed for breaking rules. Of course, not all children were treated harshly. Many parents treated their children kindly and lovingly (Pollock, 1983). Brutal, battering parents were rare. Even in the 1500s, educated parents understood that children are different from adults, that they play, and need adults' protection, discipline, and education. Even so, today's notion of childhood as a unique period of innocence, joy, and gradual development is new.

How did we arrive at this point of seeing childhood as a distinct and important period of life? How did the study of development become an important area of research? To find out, we have to go back almost 300 years.

Philosophical roots: Locke and Rousseau

John Locke (1632–1704), a physician and philosopher, was perhaps the first scholar to recognize development as a fit subject of intellectual interest. He proposed that early experiences deeply influence later life. Children are not born as sinful creatures, he suggested. Instead, the infant is born as a blank slate, or *tabula rasa*, on which experiences in life write their story. Through interactions with people in the environment, each child develops his or her unique character and abilities. Because youth is a formative period, Locke suggested, it merits the careful attention of adults. By encouraging their children's curiosity and answering their questions, parents would encourage their children to develop into rational, attentive, and affectionate people. Such careful parents would not have to resort to punishment. Locke believed that individual children were born with differences in intelligence and temperament, but he emphasized the role of learning and the importance of the parent as a rational teacher of a receptive child. Locke rejected the idea, popular during his lifetime, that children are born with knowledge. In modern terms, Locke saw **nurture**, or external forces, as the driving force in development.

A second philosopher, Jean-Jacques Rousseau (1712–1778), challenged Locke's beliefs about development. Whereas Locke emphasized the influence of nurture on development, Rousseau emphasized the importance of **nature**, or internal forces. Rousseau argued that children are not simply inferior adults or ignorant students. Children, Rousseau believed, are born with their own individual natures, and adults must not expunge this individuality in a quest for reason or social order. Adult knowledge, he argued, cannot and should not be poured into a child like water into a pitcher. To Rousseau, children were active, testing, self-willed explorers of their world, who should be allowed to grow with little pressure from their parents. Through a process of natural unfolding, children would reach the point where they could reason logically. Only then, late in childhood, should adults begin to reason with and actively teach children.

These two philosophers can be considered the first direct ancestors of the psychology of human development. They were the first scholars to write about the importance and uniqueness of the period of childhood, and in their differences of opinion brought to the field of developmental psychology one of its central issues—the question of whether nature or nurture is primarily responsible for shaping development.

The scientific forefather: Darwin

The next ancestor of developmental psychology appeared a century after Rousseau, when Charles Darwin (1809–1882) upset accepted beliefs with his theory of evolution. With this theory Darwin radically changed the way that scientists thought about the development of species, of societies, and of human beings. The idea of development was central to Darwin's *Origin of Species*, published in 1859, and for a half century after this treatise was published, scientists looked for developmental changes in and across different species. Development over the human life cycle—from a watery existence before birth, to crawling and creeping, and into maturity—they believed, would recapture the evolution of the human species from water-dwelling forms, through lower animals, to the primates, and finally *homo sapiens*. It was said that **ontogeny**—the development of the individual—repeats **phylogeny**—the development of the species. Studying one would reveal truths about the other. Scientists began searching for these truths by documenting the genesis of behaviors in the development of children. As it turned out, they were mistaken in their belief that ontogeny repeats phylogeny. But even so, these scientists were responsible for making the study of development into a science.

Darwin himself kept a baby journal in which he recorded the visible changes in his eldest child, Doddy, from the time he was born. Thus, Darwin not only introduced a theoretical rationale for a scientific approach to studying development, he also introduced a useful technique for gathering data.

The early 20th century: developmental psychology as a fledgling science

G. Stanley Hall was the first psychologist to study children and adolescents.

G. Stanley Hall: psychological forefather

G. Stanley Hall (1844–1924) brought Darwin's wide-ranging curiosity and naturalist's outlook to the study of human behavior for its own sake. He was the first psychologist to take an interest in studying children and carried out many studies in the 1880s and 1890s. These were enthusiastic, spirited, if scientifically undisciplined attempts to fathom children's development. Hall administered more than 100 such studies.

By the beginning of the twentieth century, through the influence of Hall and his contemporaries, the study of children, or "child study," had been established as a legitimate discipline. A journal for research on children had been launched, societies to observe children had been formed, and a series of biographical studies focusing on childhood appeared. Hall also was the first psychologist interested in the study of adolscents. His huge two-volume *Adolescence* (1904) gave him the lasting title "father of adolescence." In it he integrated scientific, literary, and anthropological work on the psychology of adolescence. Although Hall is best known for his work on development in childhood and adolescence, after he retired he published *Senescence: The Last Half of Life* (1922).

Despite this great body of work, Hall's influence on other psychologists was not as strong as one might expect. Perhaps it was the breadth of Hall's interests, at a time when psychology was forming itself into firmly identifiable schools of thought, that limited the influence of this psychological forefather of developmental psychology. Hall was a generalist at a time when specialists took all the laurels. He was an intuitive, passionate enthusiast at a time when dispassionate, precise measurement and observation were the rule. His method was simply to ask adults about what they did as children or what their own children did. This survey method was, although unsystematic, an improvement over the loose narrative recording methods popular at that time. Hall was criticized for not having had a clear theory of development, for not having conducted experiments nor systematic observations.

What the field of developmental psychology needed were tools with which to gather objective and quantifiable data. Its only methods of study were descriptive observations, in which children were regarded rather like trees or plants, and questionnaires, which were used to call forth recollections of childhood from adults.

Binet and Simon: the first IQ test

It was in France soon thereafter that the study of development acquired an important scientific tool. Alfred Binet (1857–1911) and Théophile Simon (1873–1961) devised the first standardized test of intelligence. Binet and Simon's efforts were prompted by the French government's need to find and separately educate children of low intelligence. Until Binet and Simon devised their intelligence test, educators had unscientifically categorized children into fuzzy categories such as "moron," "idiot," or "imbecile." But Binet and Simon's intelligence test was more precise. It started with easy questions and ended with difficult ones, and in between it covered the range of mental abilities from perception to reasoning. After testing many children of all different ages in the same, standardized fashion, Binet and Simon calculated age **norms** for the test, that is, the number of questions that most children of a particular chronological age could answer correctly. Individual children's test scores then could be measured against clear, quantifiable standards. Thus, this test gave psychologists an objective tool for studying the development of intelligence.

Arnold Gesell: natural blooming

Arnold Lucius Gesell (1880–1961) also systematically observed individuals at different ages. He observed and photographed them unobtrusively, through a one-way window (see photo at the beginning of chapter 2), a method that developmental psychologists still use today. The information about development that Gesell collected through his observations also is used today. His timetables detailing when most children reach certain milestones of development, published in *An Atlas of Infant Behavior* (1934) and *Infant and Child in the Culture of Today* (1943), are still widely available and are referred to by parents and pediatricians to chart children's growth and progress. These observations and developmental timetables were used by Gesell himself, however, not just to get information about how adequately individual children were developing but to support his firm belief that development is **maturational**.

Like Rousseau, Gesell conceived of development as a process of natural unfolding on the stage of environment. Development, he suggested, is the result of inherited factors, rooted in anatomy and physiology. It requires no pressure from the external environment. Thus, for example, mechanical walkers or walking lessons are neither helpful nor harmful, for when children's bodies are ready, they begin to walk (see Figure 1.1). Similarly, when children are ready, they begin to talk, to tell stories, and to read. According to Gesell, children develop a bit like plants. Parents plant and tend them, but whether the plant turns out to be a hardy wildflower or a delicate rose is determined by the child's inheritance, not by the parents' care.

John Broadus Watson sought to make psychology an objective science by giving it one direction and one method, the observation of behavior.

John Watson: the birth of behaviorism

An entirely different view of development issued from Gesell's contemporary, John Broadus Watson (1878–1958). For Watson, raising children was more like building a house than planting a garden. Parents and others "construct" children out of learned behaviors, habits, skills, and feelings. To explain how they do so, Watson turned to the ideas of **classical conditioning** presented by Ivan Pavlov (1849–1936) in his work with dogs. Watson demonstrated that children could be taught to fear neutral objects by pairing them with a scary or aversive stimulus, just as Pavlov had gotten dogs to salivate when they heard a sound that had been paired repeatedly with the presentation of food. In a demonstration of this principle, Watson placed an infant on a table in front of him. Watson's assistant then took the baby's familiar tame rabbit out of its box and handed it to the baby.

> He starts to reach for it. But just as his hands touch it I bang the steel bar behind his head. He whimpers and cries and shows fear. Then I wait awhile. I give him his blocks to play with. He quiets down and soon becomes busy with them. Again my assistant shows him the rabbit. This

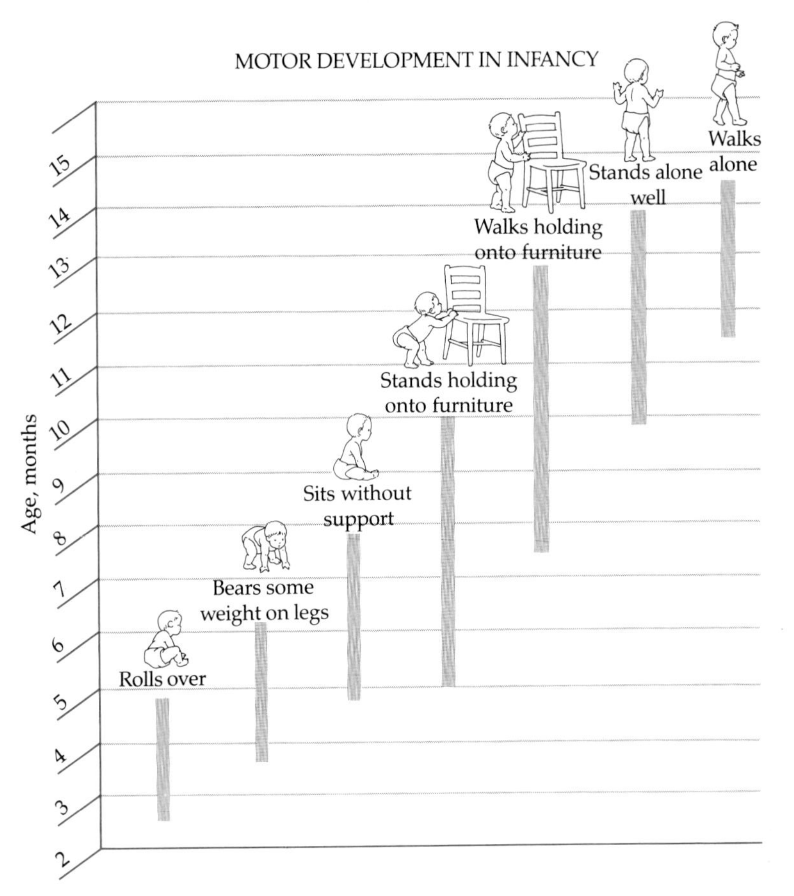

Figure 1.1 ***Motor development in infancy is*** **maturational.** ***Although different infants reach motor milestones at slightly different ages, all infants, regardless of their ethnicity, social class, or temperament, reach them in the same order. The bottom of each bar in the graph indicates the age at which 25 percent of the infants tested could perform the behavior. The top indicates the age at which 90 percent could perform it.***

> time he reacts to it quite slowly. He doesn't plunge his hands out as quickly and eagerly as before. Finally he does touch it gingerly. Again I strike the steel bar behind his head. Again I get a pronounced fear response. Then I let him quiet down. He plays with his blocks. Again the assistant brings in the rabbit. This time something new develops. No longer do I have to rap the steel bar behind his head to bring out fear. He shows fear at the sight of the rabbit. He makes the same reaction to it that he makes to the sound of the steel bar. He begins to cry and turn away the moment he sees it (Watson, 1928, pp. 52–53).

Watson's success in changing children's behavior led him to take a strong pro-nurture position on development. He asserted firmly that environmental conditioning brings about the child's learning and development and laid upon parents complete responsibility for their children's behavior:

> All we have to start with in building a human being is a lively squirming bit of flesh, capable of making a few simple responses such as movements of the hands and arms and fingers and toes, crying and smiling, making certain sounds with its throat. . . . Parents take this raw material and begin to fashion it in ways to suit themselves. This means that parents, whether they know it or not, start intensive training of the child at birth (Watson, 1928, pp. 45-46).

Watson's ideas about parents' responsibilities were unrealistic. Children are not just the products of their parents' training. Despite his controlled demonstrations, Watson's suggestions came more from his personal philosophy than from empirical study.

focus *John Watson: no coddling allowed*

John Watson had radical ideas about raising children according to behaviorist principles. One of his pet peeves was mothers who showed their children too much love.

> Loves are home made, built in. The child *sees* the mother's face when she pets it. Soon, *the mere sight of the mother's face* calls out the love response. . . .So with her footsteps, the sight of the mother's clothes, of her photograph. All too soon the child gets shot through with too many of these love reactions (Watson, 1928, p. 75).

The problem with "love reactions," he believed, was that they smothered a child's initiative and spontaneity. Watson asked mothers to

> . . . remember when you are tempted to pet your child that mother love is a dangerous instrument. An instrument which may inflict a never healing wound, a wound which may make infancy unhappy, adolescence a nightmare, an instrument which may wreck your adult son or daughter's vocational future and their chances for marital happiness (Watson, 1928, p. 87).

Parents, Watson argued, should behave like strict executives in an efficiently organized household. Like others of the day, Watson was concerned with "scientific management." The home, like the factory or the office, was to be managed efficiently, economically, and precisely. Like a model worker, an infant was to obey rules and to perform on time. Children were to be regulated and scheduled from their very first days. Feedings should be given the infant every four hours. As Watson wrote in a Children's Bureau pamphlet to mothers, *Are You Training Your Child to Be Happy?* (1930),

- *Begin when he is born.*
- Feed him at exactly the same hours every day.
- Let him sleep after each feeding.
- Do not feed him just because he cries.
- Let him wait until the right time.
- If you make him wait, his stomach will begin to wait.
- His mind will learn that he will not get things by crying.

Toilet training, Watson believed, should begin when the infant is 1 to 3 months old. (Today's parents usually are advised to wait until a child acts "ready" for toilet training, somewhere between the ages of 2 and 3 years.) By aggressively administering enemas and by scheduling the infant for sessions on the potty, mothers were to eliminate dirty diapers in the early months. Watson's advice sounds harsh to us today, although his goals are those of conscientious parents everywhere: to raise productive, independent, and responsible members of society.

But Watson did contribute an experimental method to the scientific study of children. Until Watson came on the scene, psychologists had done much of their research by asking their subjects to report on their subjective experiences, a method known as introspection. Watson criticized introspection as unverifiable and as clearly inappropriate for child or animal subjects. Wanting to establish psychology as a purely objective branch of natural science, Watson (1913) proposed that psychologists study only *observable* behavior and that their theoretical goal be simply the prediction and control of

Sigmund Freud dramatized the links between the physical pleasures of the infant and the sexual satisfactions of the adult.

behavior. This approach, which Watson aptly named **behaviorism**, was a far cry not only from the impressionistic work of the introspectionists but from the naturalistic research of Darwin, Hall, and Gesell.

Sigmund Freud: innocence lost

A third, very different theory of development was proposed by Sigmund Freud (1856–1939), a physician from Vienna. When Freud published his *Three Contributions to the Sexual Theory* (1905), he shocked his Victorian audience right down to their lace pantaloons. He claimed that children are born with the beginnings of sexual feelings, a belief that was not popular in the sexually repressive society of the day.

> Childhood was looked upon as ''innocent'' and free from the lusts of sex, and the fight with the ''demon of sensuality'' was not thought to begin until the troubled age of puberty. Such occasional sexual activities as it had been impossible to overlook in children were put down as signs of degeneracy and premature depravity or as a curious freak of nature (Freud, 1935, p. 58).

Freud, however, asserted that sexuality starts at the beginning of life, and because of it childhood is dramatic, complicated, and full of psychological conflict.

The sexual experiences of childhood, moreover, lay the foundation for adult behavior. Thus the infant's pleasure in nursing at the mother's breast foreshadows adult pleasure in kissing, sucking, and licking. Events connected with early oral pleasures (or deprivations) foreshadow adult habits like overeating, heavy drinking, or smoking. Freud was the first of developmental psychology's forebears to stress so clearly the connections between childhood and adult behavior. He saw a unifying thread to a person's development from birth to death. Early experiences, he believed, leave an unconscious mark in a person's memory. They are ''unconscious'' because a person forgets or represses them, and they are, therefore, unavailable to ordinary, conscious thought. Only through dreams, jokes, slips of the tongue, and other windows to the unconscious can a person retrieve unconscious ideas. Yet these unconscious ideas shape development.

In earliest infancy, the ideas and experiences that are most important center in the infant's mouth. In this **oral stage**, Freud believed, the infant focuses on the pleasures of sucking and taking in food with mouth and tongue. In this period, the mother must not frustrate the infant's impulses to suck and mouth objects. In the second year of life, the child focuses on the conflict between keeping things in and letting them go, between asserting his or her selfhood and will—''No, it's mine!''—and acting on generous impulses—''Here, Mommy.'' How the mother negotiates the conflicts of toilet training in this **anal stage**, Freud believed, shapes the child's concepts of self and others. During the **phallic stage**—in the next three years—the child develops an especially strong desire for the parent of the opposite sex. A boy wants to replace his father in his mother's affections, a conflict Freud called **Oedipal**, after Oedipus, the character in Greek myth who killed his father and married his mother. A girl wants to replace her mother in her father's affection, a desire called by Freud the **Electra conflict**, after another mythical Greek character. Said one not untypical 5-year-old girl, rather cruelly, to her mother, ''I hope that you die before Daddy does. Then I will marry him.'' At this stage, boys begin announcing, ''I'm going to marry Mommy,'' and children may try to maneuver themselves into their parents' bed. Children resolve their Oedipal and Electra conflicts by the end of this period, according to Freud, by identifying with their same-sex parent and thereby possessing their opposite-sex parent vicariously. After the phallic stage comes the quieter **latency period** of middle childhood, during which, Freud theorized, sexual feelings go underground and children immerse themselves in school and community activities. Finally, with puberty, the adult sexuality of the **genital stage** begins.

Freud suggested the existence of three mental structures within each individual. The first structure is the powerful, pleasure-seeking, instinctual **id**. It drives children—and adults—to want and desire, to yearn for satisfaction and pleasure. In infants' cries for food and attention—and smiles when parents deliver—and in the barrage of the toddler's ''I want cookie,'' ''Dat *my* bike!'' and ''No nap,'' we hear early echoes from the id. The **ego** is the mental structure that emerges with the rapid expansion of memory. It coexists, not always peacefully, with the id. Ego development helps children to control their impulses, to bury some ideas in the unconscious, and to allow others to bubble to the surface. Ego is realistic self-interest, and it governs the choice of realistic goals and realistic strategies for maximizing pleasure and coping with stress. In the preschooler's transparent attempts to get what he or she wants—''I love you, Nana. What did you bring me?''—we see evidence of the ego. The rules and inhibitions laid down by parents—''Brush your teeth before bedtime''—and culture—''Do unto others. . . .'' are internalized in the third mental structure, the **superego**. Informally called a ''conscience,'' the superego helps children to obey rules, follow ethical principles, and get along with others.

With his sweeping ideas about development, Freud influenced later thinking in psychology. Although not all his ideas have been accepted by the scientific community, some have been, and these have become part of the way we all think about people today. For example, we now believe that people have irrational and hidden motives, that children express sexual and emotional impulses, that their lust for pleasure often puts them at loggerheads with their parents' and others' needs for order, control, and reason, and that how children resolve internal and external conflicts affects their lives as adults. To a large extent, these beliefs started with Freud.

focus *Behavior modification*

Parents have long known that rewarding a child's behavior will cause it to recur. It doesn't take too many bedtime pleas for "one more glass of water, p'eeze" for parents to grump, "No more!" B.F. Skinner, an influential behaviorist, carefully raised his daughter according to principles of operant learning. Here Skinner describes his 9-month-old daughter sitting on his lap at nightfall:

> The room grew dark, and I turned on a table lamp beside the chair. She smiled brightly, and it occurred to me that I could use the light as a reinforcer. I turned it off and waited. When she lifted her left hand slightly, I quickly turned the light on and off. Almost immediately she lifted her hand again, and I turned the light on and off again. In a few moments she was lifting her arm in a wide arc "to turn on the light" (Skinner, 1979, p. 293).

Skinner later toilet trained his daughter with a potty chair that played a song whenever she urinated into it.

Operant learning techniques have been used to solve serious developmental problems, too. Children who are socially isolated and who barely speak have been trained to communicate. First they are rewarded for uttering any sound, then for words, then phrases, and finally for short sentences. Psychologists also have modified children's violent aggression with behavior modification techniques. They may reward the children, first, with tangible prizes like candies and stickers for cooperating and sharing. Later they may reward them with tokens like check marks or poker chips that are redeemable for toys, gum, or comic books. In addition, they may punish aggressive behavior with a brief time-out period such as five minutes alone in a room. After several weeks, children consistently given this treatment are usually cooperative, at least in the setting in which they have been reinforced. These are just a few examples of the many kinds of behavior problems that have been modified through reinforcement applied systematically, conscientiously, and consistently.

The middle 20th century: theories of development

Social learning theorists

Despite Freud's influence on the field of psychology, from 1930 to 1960 it was learning theory that dominated mainstream American psychology. American psychologists went beyond the principles of classical conditioning that Pavlov and Watson had explored. Their experiments showed that stimuli regularly associated with **primary (biological) drives** that are basic to human survival, including hunger and thirst, themselves became the objects of **secondary**, or **learned drives**. John Dollard, Neal Miller, and Robert Sears applied this principle of learning to children's development. For example, they conceived of an infant's attachment to the mother as a secondary drive, learned through association with the infant's primary hunger drive. They suggested that the infant learns to seek the mother's love and attention because she gives food and comfort and alleviates pain. With the principles of primary and secondary drives as their guide, Dollard, Miller, and Sears proposed a theory of development based on learning theory.

Operant learning theorists

Another group of psychologists also tried to explain development in terms of learning. Psychologists following the lead of B. F. Skinner, such as Sidney Bijou, Donald Baer, and Jacob Gewirtz, applied principles from **operant learning theory** to the study of children. The principles that they adapted were those of **operant conditioning**. Skinner had demonstrated that adults (and animals) keep acting in ways that are rewarded by pleasant consequences and stop acting in ways that are punished by unpleasant consequences. Bijou, Baer, and Gewirtz demonstrated that children's social behavior can be shaped by the **reinforcement** or **punishment** of their actions. Gewirtz (Gewirtz and Boyd, 1977b), for example, analyzed interactions between mothers and infants, showing that a mother's smiles and encouragement are reinforced by her infant's smiles, gurgles, and movements, while the latter, in turn, are strengthened by the smiles and movements of the mother. Operant learning theory had the virtue of being easy to demonstrate, especially in a tightly controlled setting. The effectiveness of reinforcement to change the frequency of children's behaviors—saying ''yes,'' paying attention, and so forth—was amply demonstrated in many **behavior modification** experiments in laboratories, schools, and homes.

Observational learning

Yet people seem to learn many things in the absence of obvious reinforcement. Children imitate their parents and others without getting tangible rewards. For example, the 3-year-old copies the way her mother's hands gesture as she speaks, and the 4-year-old stomps into his room and slams the door just like his older brother does. Why? A third group of learning theorists, including Albert Bandura, Richard Walters, and Walter Mischel, proposed a theory of observational learning to account for this kind of learning. They suggested that people watch and listen to how others behave and then sometimes imitate them. Children, according to this theory, need no obvious reward for copying, say, the way that parents push the shopping cart, brush their hair, hold their forks, or button their shirts, or for copying the way that other children speak, or blow bubbles with their drinking straws, or laugh at bathroom jokes.

Bandura conducted research in the laboratory to observe systematically the conditions under which imitation flourishes. There he found that children were most likely to imitate adults who were nurturant, powerful, affectionate, and who possessed things that the children wanted. At home, he suggested, children imitate their parents because their parents have these very qualities. According to observational learning theorists, the fact that parents are loving and powerful in their children's eyes is enough to motivate the children to imitate them. Parents need not resort to bribes of cuddly dolls or shiny red trucks to get the children to use toilets instead of diapers and spoons instead of fingers.

These three groups of learning theorists—espousing social learning, operant learning, and observational learning—kept alive Locke and Watson's pro-nurture view of development from the 1930s to the 1960s. They reflected the dominant American view, in psychology and in much of American society, of people's perfectibility, controllability, and changeability. But a number of other theories of development drew small but growing followings over this period as well.

The ethological approach

One such theory was **ethology**, the study of how animals behave in their natural habitats. Ethology stands in clear contrast to behaviorism, with its

research focus

Imitating a model

According to observational-learning theorists, children learn by imitating the behavior of parents, other children, film and television characters. To test the degree of children's willingness to imitate, Albert Bandura and his co-workers (Bandura, Ross, and Ross, 1963) showed 3- and 4-year-old children a film of a person acting aggressively and watched to see whether the aggressive acts would be imitated. The children were divided into four groups of 24; half of each group were boys, half girls.

The experimenters showed half of the children in the first group a man, the other half a woman, punching, hammering, hitting, and yelling at a large rubber clown. They showed another group a movie of the same adults carrying out their aggressive acts. They showed a third group of children a cartoon in which a cat punched, hit, and hammered the clown. The fourth group of children was exposed to no aggressive model at all. At this point, all of the children were allowed to play briefly with some attractive toys and then deliberately frustrated by having the toys removed. Finally, the children were allowed to play with a set of toys that included a large rubber clown. The question: would the children punch and hit the clown?

They certainly did. The children imitated the aggressive behavior that they had seen. Compared with the children who hadn't seen any aggression, the children who had seen it acted far more aggressively. Girls were more likely to imitate the female model, boys the male model. The cartoon prompted the greatest number of aggressive acts, but the filmed and the live adult model did their share as well of prompting aggression in the children.

These scenes from films of Bandura's research demonstrate how children imitate exactly the attacks of an adult model on a Bobo doll.

emphasis on how behavior may be modified by external forces in artificial environments. Ethology has much in common with the naturalistic approaches of Darwin and Gesell. One of the principles of animal behavior demonstrated by ethologists Niko Tinbergen and Konrad Lorenz, which later was picked up by developmental psychologists, was that animals are born with predispositions to learn particular forms of behavior during **critical periods** of their lives. Soon after they hatch, for example, ducklings begin to follow the nearest large moving object they see. This phenomenon is called **imprinting**. Under ordinary circumstances, ducklings imprint on their mothers and follow them about—a biological mechanism that helps keep the ducklings safely near their source of food and care. Under unusual circumstances, however, hatchlings might imprint on something else—a person or a large red ball, for example.

Influenced by the ethological approach, John Bowlby, a British psychoanalyst, explored the question of whether human children also become attached to their mothers or other people during a critical period in infancy. Bowlby analyzed the process through which infants develop deep and loving relationships with their mothers. Human infants, like hatchlings, need to stay close to their mothers to survive. The forms of behavior that infants are born with, like crying, grasping, clinging, and, later, following, help them to stay close to their mothers. From this physical closeness, children develop emotional bonds to their mothers. Bowlby stressed the dangers to children's development if their caretakers are numerous or unstable, emotionally or physically distant, insensitive or inattentive. His theory strongly influenced thinking about social and emotional development and was important in changing practices in orphanages and other residential institutions for children. It presented a very different view of the development of an infant's relationship with the mother from those proposed by learning theorists. For them, the infant is reinforced for interacting with the mother or develops a relationship with her because she satisfies the need for food. It is also different from that proposed by Freud, in which the relationship of mother and infant revolves around the sensual satisfaction of sucking. Bowlby studied the same issues as Freud and followed for some distance in his footsteps, but he was influenced by other theoretical approaches and other data and ultimately came up with his own unique theory of social development (further discussed in Chapter 7). The same is true of the next developmental theorist we discuss.

Erik Erikson: a psychosocial integration

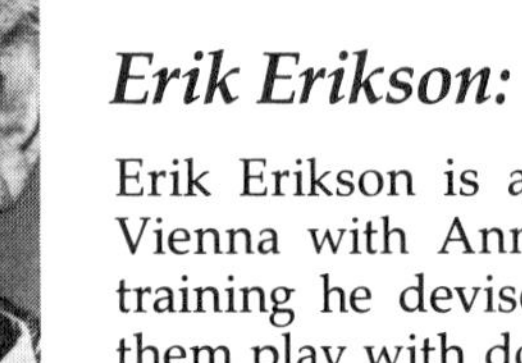

Erik Erikson has been a major contributor to our understanding of the psychological crises in childhood, adolescence, and adulthood.

Erik Erikson is a psychologist who first studied child psychotherapy in Vienna with Anna Freud, Sigmund Freud's eminent daughter. With this training he devised a way of interpreting children's behavior by watching them play with dolls and other toys rich in possible psychological meanings. Erikson then moved to the United States, where he did clinical interviews and therapy with such diverse groups as Northwest Indians, Harvard students, soldiers, and civil rights volunteers. Out of this varied background, Erikson expanded Freud's theory of **psychosexual development** and developed his own theory of **psychosocial development**.

Unlike Freud's, Erikson's theory of human development covered the whole life span (see Table 1.1). Erikson stressed social and cultural aspects of development, not just sexual ones, as Freud had done. Erikson proposed that the way in which individuals cope with their social experiences shapes their lives. Individuals, he suggested, cross eight different crisis points over the course of their lives. At each crisis, people are vulnerable to developing negative feelings like guilt, inferiority, or isolation. But they also stand poised to enlarge and deepen their personalities to encompass positive feelings like trust, intimacy, generativity, and integrity. In each crisis period, the "inner laws of development" create possibilities, which are shaped by the people and the social institutions—schools, church, judicial system, and so on—that fill or frustrate the person's needs. Although some parts of one's personality may be predetermined, the resolution of each phase of personality development is worked out within specific social situations.

In infancy, for example, the infant learns to trust or mistrust that his needs for food, warmth, and comfort will be met by the people around him. How reliably and responsively a mother attends to the infant's needs affects

Table 1.1
Developmental progression

Age	Freud's stages	Erikson's crises
1st year	*Oral Stage* Infants obtain pleasure through stimulation of the mouth, as they suck and bite.	*Trust versus Mistrust* Infants learn to trust, or mistrust, that their needs will be met by the world, especially by the mother.
2nd year	*Anal Stage* Children obtain pleasure through exercise of the anal musculature during elimination or retention.	*Autonomy versus Shame, Doubt* Children learn to exercise will, to make choices, to control themselves; or they become uncertain and doubt that they can do things by themselves.
3rd to 5th year	*Phallic (Oedipal) Stage* Children develop sexual curiosity and obtain pleasure through masturbation. They have sexual fantasies about the parent of the opposite sex and guilt about their fantasies.	*Initiative versus Guilt* Children learn to initiate activities and enjoy their accomplishments, acquiring direction and purpose. If they are not allowed initiative, they feel guilty for their attempts at independence.
6th year through puberty	*Latency Period* Children's sexual urges are submerged; they put their energies into acquiring cultural skills.	*Industry versus Inferiority* Children develop a sense of industry and curiosity and are eager to learn; or they feel inferior and lose interest in the tasks before them.
Adolescence	*Genital Stage* Adolescents have adult heterosexual desires and seek to satisfy them.	*Identity versus Role Confusion* Adolescents come to see themselves as unique and integrated people with an ideology; or they become confused about what they want out of life.
Early adulthood		*Intimacy versus Isolation* Young people become able to commit themselves to another person; or they develop a sense of isolation and feel they have no one in the world but themselves.
Middle age		*Generativity versus Stagnation* Adults are willing to have and care for children, to devote themselves to their work and the common good; or they become self-centered and inactive.
Old age		*Integrity versus Despair* Older people enter a period of reflection, becoming assured that their lives have been meaningful, and they grow ready to face death with acceptance and dignity; or they despair for their unaccomplished goals, failures, and ill-spent lives.

not only the relationship between the two but the more general attitude the infant forms toward the world. This attitude affects how well the infant deals with the next important crisis, when as a toddler he must deal with his earliest expressions of independence. Whether the toddler resolves this second crisis with self-control or with uncertainty and shame depends on how his parents respond to his attempts at independence—whether they respond with tolerance and support or with impatience and anger. The resolution of this second crisis in turn affects the child's ability to meet the next crisis, and so on.

Erikson, like Freud, saw development as occurring in distinct *stages* during a person's lifetime. He suggested that what happened in one stage was qualitatively different from what happened before or after, and developments in one stage built on those in earlier stages. All people progress through the stages in the same order, although not necessarily at the same rate. In normal development, people go forward through the stages, not backward.

This view of development as occurring in stages is different from the view held by learning theorists, who saw development as gradual, cumulative, and continuous. It is a view expressed strongly by the next theorist we discuss.

Jean Piaget: master theorist

Jean Piaget (1896–1980) was perhaps the one single theorist with the greatest effect on modern developmental psychology. Systematically and intensively observing children—his own three included—over his entire lifetime, Piaget developed and tested the fullest existing explanation of how human thought develops from infancy to adulthood. Trained in biology to observe expertly, Piaget watched, recorded, and sometimes intervened in the behavior of his young subjects. Here is an example of an observation he made on the crying of a newborn:

> Obs. 1: On the very night after his birth, T was wakened by the babies in the nearby cots and began to cry in chorus with them. At 0;0 (3)[1] he was

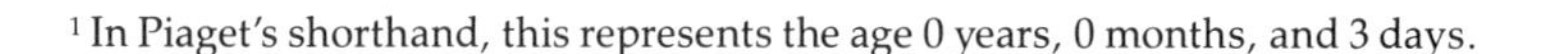

[1] In Piaget's shorthand, this represents the age 0 years, 0 months, and 3 days.

By observing children in their ordinary activities and by asking them probing questions, Swiss psychologist Jean Piaget was able to construct a theory of how children's minds develop.

> drowsy, but not actually asleep, when one of the other babies began to wail; he himself thereupon began to cry. At 0;0 (4) and 0;0 (6) he again began to whimper, and started to cry in earnest when I tried to imitate his interrupted whimpering. A mere whistle failed to produce any reaction (Piaget, 1962, p. 7).

Similar records were made as Piaget watched infants look for toys hidden under pillows, watched children pour water from one glass into another, asked children the rules to their games, and listened carefully as they revealed their reasoning. From such carefully recorded observations, Piaget wove hypotheses, tested them repeatedly, and then joined them into a comprehensive theory of the stages of cognitive development. In Piaget's theory, children are born ready to adapt to and to learn from the world. They do not have to be taught deliberately to crawl or to walk or, later, that objects obey certain physical laws and people obey moral rules. Whereas Locke, Watson, and the learning theorists believed that knowledge was imposed on the child from outside, and Rousseau and Gesell believed that abilities were inborn, Piaget set these two beliefs into balance. To him, children construct knowledge as they mentally organize information from the environment. Infants and children actively participate in their own development. They manipulate and explore their world, guided by their "mental structures" or mental representations of how things work. Qualitative changes in these mental structures occur for all children as they make shifts in their organization of knowledge in a particular order of stages (see Table 1.2).

Although Piaget's theory of cognitive development is in part already being replaced by other theories, Piaget's careful and systematic observations have provided the most vivid and detailed picture so far of the developing cognitive capacities of the child and adolescent. The notions he proposed offered important insights into children's thinking and reshaped how developmental psychologists thought about intellectual development, about the active way in which children approach learning, and about the interaction of nature and nurture. Piaget was a brilliant observer of children, and his work was instrumental in showing adults how to consider children's thought on *children's* terms. The broad distinctions he made among stages of development from infancy through adolescence continue to be useful for describing children's thought. There is no question that Piaget's work changed the way that people think about thinking and the ways in which they talk about the influence of nature and nurture.

Table 1.2
Piaget's stages of cognitive development

Stage	Activities and achievements
Sensorimotor Birth to 2 years	Infants discover aspects of the world through their sensory impressions, motor activities, and coordination of the two.
Preoperational 2 to 7 years	Children cannot yet think by operations, by manipulating and transforming information in basic and logical ways. They can think in images and symbols and form mental representations of objects and events.
Concrete Operational 7 to 11 years	Children can understand logical principles that apply to concrete, external objects.
Formal Operational Over 11 years	Adolescents and adults can think abstractly. Their thinking is no longer constrained by the givens of the immediate situation but can work in probabilities and possibilities.

Theories in perspective

Theories are important because they provide frameworks for ideas and information. They offer general rules and principles that help to organize empirical results. They give scientists something to hang their facts and findings on. Theories form the trunk and major branches; hypotheses and data, the smaller branches and twigs of scientific knowledge. In developmental psychology, each of the theories we have discussed serves this function. Each enriches our understanding of the complexities of human development. Each sheds light on the processes of development and helps to organize facts and hypotheses. Of course not all the theories we have presented can be correct. Which one is right? The truth is that no one theory is all right or all wrong. All the theories that we have discussed have been tried and found wanting in some respect, if only because they don't cover everything we know about development. Yet each also has contributed to our knowledge in long-lasting ways.

From Freud's psychosexual theory we have retained the ideas that unconscious motives affect behavior in subtle, complicated ways, that people's needs, impulses, and problems change as they get older, and that extreme experiences in the first few years of life can leave an indelible mark on later psychological development. From learning theory we have retained the idea that actions can be controlled by the manipulation of reward and punishment and by the actions of models. From Gesell's maturational theory we have retained the notion that there is a biological current underlying normal growth; development is not imposed totally from without. From Piaget's theory of cognitive development we have learned that infants and children are not passive receivers of experience but are, instead, active builders of their own knowledge. As the field of developmental psychology develops, it is encouraging to see that our explanations of development are growing less extreme and simplistic, more subtle and eclectic, as elements from a variety of theories—old and new—are absorbed and integrated.

New theoretical approaches

Developmental psychologists today are not limited to the theories we have described—maturational, psychosexual, psychosocial, learning, cognitive. Researchers in the 1980s are also influenced strongly by new theoretical approaches—information processing, dialectical, ecological, and sociobiological—and by new versions of the old theories—neobehavioristic, neo-Freudian, and neo-Piagetian.

Information processing

Information processing is an approach to understanding human behavior and development that came in with computers. People, like computers, take in information and remember it. How they do so is the central question in the information-processing approach. The roots of the information-processing approach are in the fields of computer science, communications theory, and linguistics. From computer science came the recognition that people and computers are both manipulators of symbols. From communications theory came the notions of information coding and channel capacity. From linguistics came information about how people build and use grammars. The information-processing approach—like computers—quickly became popular in the 1970s and now dominates the field of cognitive psychology.

Developmental psychologists who take an information-processing approach study how information is processed and remembered by people of different ages. Do 2-year-olds remember less efficiently than 8-year-olds? Do preschoolers tell more disorganized stories than older children? If so, why? Information processers suggest that young children do not think as well as older children because they have not yet learned strategies for deploying their mental processes efficiently and because they have accumulated less knowledge. The information-processing approach has proved useful in the study of cognitive development as researchers have uncovered both strengths and weaknesses in Piaget's theory. The results of their investigations will be discussed in later chapters on cognitive development.

Dialectical approach

To psychologists who take a **dialectical view**, human development is a process of continual change. All development evolves out of a state of imbalance. Developmental change is the product of a constant conflict between opposites. It proceeds through the resolution of existing conflicts and the inevitable emergence of new conflicts. In this view, contradictions produce conflicts, which produce crises and change. Upheaval and disequilibrium are utterly necessary. Stability and calm are merely temporary stepping-stones in the turbulent stream of human development, as a person achieves temporary syntheses between opposing positions. The child emerging from the tantrums and stubbornness of the toddler period with a degree of self-control and amiability, only to plunge headlong into vacillation between "Me a baby, Mommy" and "Me do it *myself*, Mommy. Me a big boy," illustrates this disequilibrium. Psychologists with a dialectical view do not ask how people achieve tranquility and find answers. These were the questions that Piaget asked. His stages of development represent cognitive equilibrium. Psychologists with a dialectical view ask how people create problems and raise questions (Riegel, 1976). The dialectical approach is opening new windows and producing new ideas for developmental psychologists interested in processes such as problem solving, communication, and language.

Ecological approach

Developmental psychologists who take an **ecological** approach consider how people accommodate to the environments in which they grow and live. They are interested in how settings and contexts influence behavior and development. According to Urie Bronfenbrenner, a prime spokesman for this approach, the settings or environments in which people live and grow can be viewed as systems, which exist simultaneously at various levels. The **microsystem** is composed of the network of ties between people and their immediate settings such as school or family. The **mesosystem** is composed of the network of ties among major settings in the person's life. For example, the mesosystem for an American 12-year-old might include ties among family, school, friends, church, and camp. The **macrosystem** is composed not of settings that directly affect the person, but of broad, general, institutional patterns in the culture such as the legal, political, social, educational, and economic systems.

Ecological psychologists systematically study these systems within which the person develops. Like the ecologist in the natural sciences who investigates the life and times of the porpoise by studying all the nearby sea creatures and the human boaters and hunters and water polluters who make up the porpoise's world, the ecological psychologist examines factors at many different levels in the human environment. For instance, he or she might investigate the social class of the children under study, their family's composition, and the type of day-care center or nursery school they were attending and then would look for complex patterns of interactions between development and these aspects of the environments.

Sociobiology

According to sociobiologists, behavior has a genetic basis. **Sociobiology** is a modern extension of the naturist positions of Rousseau, Darwin, Gesell, and Lorenz. Sociobiologists like Harvard's Edward Wilson, an entomologist, and Robert Trivers, a biologist, have suggested that social behavior, like physical traits, is the direct product of evolution and, furthermore, that the behavior that improves the chances of a person's genes being reproduced will be passed on genetically to the next generation. Some developmental psychologists have found that the sociobiological approach sheds light on important questions about development, such as why there are inborn differences between the sexes, how social groups are organized into dominance hierarchies, what the bases of the attachment between parents and children are, and how and why parents invest their time and energy in caring for their offspring. In the long run, how useful sociobiology will prove for understanding human development remains an open question, but it is a theory that is likely to influence the field of developmental psychology in the future and to enrich traditional psychological theories.

New versions of old theories

Erikson's psychosocial theory of development, as we pointed out, was a revision and extension of Freud's psychosexual theory. In the last decade, revisions of learning theory and of Piaget's cognitive developmental theory also have appeared. Bandura (1980), for example, incorporated cognitive elements into social learning theory and stressed the role of thinking in learning. People are more likely to imitate a model, for example, if they think that imitation is what is desired. Their actions also are influenced by their self-perceptions—their judgments about how effective they are likely to be in a situation. These self-perceptions are based on past successes, evaluations by other people, and observations of other people's performance. In another revision of social learning theory, Mark Lepper (1983) incorporated another cognitive element, people's understanding of why they are doing something. When children believe that external forces rather than their own internal motivations are making them do something, for instance, their intrinsic interest in doing it declines. These revisions of learning theory might be classified as **neobehavioristic**.

There are also **neo-Piagetian** theories—theories that are based on Piaget's structural theory of cognitive development but go beyond it. Theories by Juan Pascual-Leone (1980), Robbie Case (1985), and Kurt Fischer (1980) fall into this category. These theorists retain Piaget's idea that development occurs in stages, but they change the specifics. According to Pascual-Leone and Case, each higher stage is based on the person's ability to store more items in short-term working memory and to perform familiar cognitive activities automatically. According to Fischer, abilities develop in three tiers: sensorimotor skills, representational skills, and abstract skills. Each tier can be broken down into increasingly complex levels. As people master specific skills, they build upon and transfer skills from one domain to another. Development is gradual, and people attain relatively different degrees of proficiency within levels. Only in the skills that they are called on to perfect most consistently do people reach their highest possible level. The rule, not the exception, thus is for people to achieve unevenly within levels and to be at different levels for different skills.

These neo-behaviorists and neo-Piagetian theories represent developmental psychologists' best efforts to integrate new data into old theoretical frameworks. With all these new theoretical approaches—information processing, dialectical, ecological, sociobiological, neobehavioristic, **neo-Freudian**, and neo-Piagetian—our understanding of development undoubtedly will continue to grow and develop through the next decades as it has grown and developed since the beginning of the century.

The era of opposing theories and ''schools'' has faded. Psychologists still debate, but ideological barricades have been broken down. Today there is clearer recognition that many different theories can contribute to the understanding of lifelong development.

What developmental psychologists study

Developmental psychologists use theories and data to try to answer certain questions. What are these questions?

What is development?

The most basic underlying question for developmental psychologists, obviously, is ''What is development?'' The answer, though, is not as obvious as the question. Developmental psychologists must try to distinguish development—and more specifically, human development—from all other kinds of change, like learning, forgetting, brain damage, and drug effects. ''Development'' implies that the change is systematic, not random; that it is permanent, not temporary; that it is progressive, not regressive—it goes forward not backward; that it is steady, not fluctuating; that it occurs in regular, predictable order; that it occurs over some period of time within a person's lifetime, not in an hour or over two generations; that it occurs for all people, not just a few; and perhaps most centrally, that it is related to a person's increasing age and experience.

In childhood, adolescence, and into adulthood, development ordinarily means improvement and increasing complexity, competence, and efficiency. But what does development mean in adulthood? Is change in adulthood inevitably negative or regressive? Although many of the biological changes of adulthood are declines in the strength or frequency of responses, some adult development consists of improvement, progress, and accumulation. Is change in adulthood systematic, whether biologically, socially, emotionally, or cognitively? If so, which aspects of adult development are systematic? Developmental psychologists know, for example, about the sequence of reproductive changes for women from first heterosexual intercourse through pregnancy and menopause; about the sequence of the family life cycle. Adult development may well take place in terms of accumulated knowledge of the real world, deeper understanding of time and change, and opportunities to nurture others.

The nature of developmental change

The next obvious question for developmental psychologists is how abilities and behavior, language and emotions, moral understanding and gender roles, social skills and sympathy change as people get older. What can a 4-year-old do and say that a 2-year-old cannot? How is the moral reasoning of an adolescent different from that of a 10-year-old or a 30-year-old? An important question is whether the changes in any particular trait, ability, or behavior are continuous—increasing gradually and steadily over time, as learning theorists like Watson or Skinner suggested—or discontinuous—occurring in sudden spurts or stages, as Piaget proposed. To developmental psychologists, this issue is central. A related question is whether the rate and the pattern of development are the same for everyone. Do those who start out low on some ability remain at the bottom of their peer group, or do they later bob to the top, surpassing even their precocious peers? Do those who start out more competent than others stay that way, developing at a constant rate, or do they go through periods when their abilities develop less quickly and their age-mates are able to catch up?

To make judgments about continuity, developmental psychologists must face the problem of recognizing continuity when it happens. Psychological growth involves changes in *form*. The question for developmental psychologists then is how to recognize what different behaviors mean at different ages. Take crying, for example. Does an infant's crying mean the same as an adolescent's crying, because the overt behavior—tears, grimaces, sobs—is obviously similar, or does the infant's crying reflect a difficult temperament and mean the same as rulebreaking and rebelliousness in an adolescent? Developmental psychologists try to trace the threads of continuity and the plateaus of discontinuity in intellect, social actions, and personality as they answer questions about the nature of developmental change.

Domains of development

Another kind of question that interests developmental psychologists is how development in different areas of human functioning fits together. Typically, the field of developmental psychology is divided into three separate domains:

- physical development, including body changes and motor development;
- cognitive development, including thought and language; and
- social-emotional development, including personality and relations with other people.

But people are whole beings, and their behavior is not sharply divided into three neat categories. Development involves interconnected changes across all domains. As the infant eats and gains weight (physical domain), she smiles and coos at her parents (social-emotional domain), who then talk to her and stimulate her hearing (cognitive domain) and give her opportunities to develop skills like grasping a spoon (physical domain), while at the same time allowing her to explore the smells and textures of her food (cognitive domain) as they beam fondly at the mess she is making (social-emotional domain)—and so on. To study the development of the whole person, developmental psychologists ask questions that cross domains. For example, "How does feeling sad or having a physical handicap affect learning?" "How does knowing a second language affect adolescents' thinking and interacting with others?" "How do changing family responsibilities affect gender roles in middle age?"

What causes development?

The last general research question for developmental psychologists is "What causes development?" In the past, as we have seen, scientists interested in the causes of human development often cast their questions in extreme forms. Is development a product of inborn, internal forces—"nature"—*or* is it a product of environmental, external forces—"nurture"? Do people inherit intelligence, *or* does intelligence develop through personal experiences? Do people inherit their physical size and shape, their moral tendencies, their love of music, *or* do these qualities develop as a result of environmental influences? More recently, extreme positions of nature versus nurture have softened. Psychologists now ask more subtle questions about how internal and external forces work *together* to produce an individual's development. How are inherited aspects of a person's intelligence, morality, musicality, physical size and shape, they ask, influenced by environmental conditions? Which environmental conditions are most influential—family, divorce, income, occupation, community, nutrition, school, television, historical events, and so on? What are the limits of these environmental influences? How do inherited characteristics themselves influence the environment? How do the relative influences of heredity and environment change with age?

In even more subtle questions, developmental psychologists ask about the internal mechanisms of development. How do transitions between developmental levels of functioning actually take place? What makes the mind leap—or creep—to the realization that objects are permanent, that other people are worth pursuing, that words stand for things, or that you can't put square pegs into round holes?

The individual differences among people, which are greatest in old age, raise still another general question. What is the range of variation in height, intelligence, sociability, talent, coping styles, and other attributes? Human development is at once a universal process in life and a most individual and personal process. Much about every human being is predictable from knowing that being's sex, color, class, and other facts of birth. Although every human being is born, lives, and dies, no two of us are quite alike. And the further we develop, the more individual—the more like ourselves—we become. Developmental psychologists study these individual differences and try to balance their descriptions of the average, normative course of development with their descriptions of the individual paths taken.

Looking forward

Developmental psychology is a lively and important field of research and study. Developmental psychologists study questions that are intellectually challenging and pragmatically useful. They have much to teach—and much to learn.

Development is a complex journey, with winding paths, sharp curves, critical turning points, and many roads not taken. Developmental psychologists are now quite knowledgeable about many of the *facts* of development: they know what newborns can see and hear, what changes puberty and menopause bring, and the typical changes in memory from early to late adulthood. They are getting better at predicting the direction that a person will take on the developmental journey—whether he or she will do well in school, be active and fidgety, cheat on tests, or feel insecure in strange situations. They are learning more and more about the *processes* of development, about how people learn new habits, incorporate new information, form new relationships and mourn lost relationships, and about how these depend on inherited traits, past experiences, and present circumstances. Still, there is much to learn.

Future developmental psychologists will be less content to describe the developmental landscape and more concerned with explaining what underlies it. They will probe the connections between brain and behavior, hormones and health, inheritance and intelligence, culture and child care. They will search for evidence to support their theoretical constructs, and in their search for developmental continuities, they will integrate knowledge from different domains of development with knowledge gained from other areas of psychology as well as from fields like genetics, physiology, neurology, and anthropology.

Developmental psychology is a fascinating, fertile field, one that is well worth your cultivation. You will get your introduction to it in this book, as you go from the very beginning of human life to its end, seeing along the way the issues, facts, theories, and questions about how people develop. In this book, we have divided the life cycle into six stages—infancy, early and middle childhood, adolescence, adulthood, and late adulthood. In most people's experience, there are no sharp dividing lines between stages. But developmental psychologists tend to consider infancy the period from birth to age 1 and ½ or 2. Early childhood extends from infancy's end until age 5 or 6. Middle childhood, the elementary-school years, extends from roughly age 6 to 11. Adolescence begins with the physical changes of puberty and extends until the late teens or early twenties. Adulthood is made up of early adulthood, which extends through the thirties, and middle adulthood, which extends until the mid-fifties or mid-sixties. Late adulthood extends from the end of middle adulthood until death.

This book reflects our view of the state of the art in developmental psychology. Much space in the book is given to descriptions of abilities and behavior at different ages—what people can do at 2 months, 2 years, 12 years, 22, and 62 years. These are the fundamental data of developmental psychology and provide a necessary basis for understanding human development. But where possible we have also discussed theoretical explanations of what these abilities and behavior mean. There is no single theoretical framework around the different chapters of the book, because, as yet, there is no single theory around all the data in developmental psychology. Human development is extremely complex—more complex than the best of our theories. We have discussed aspects of different theories where they seem most important—either because, historically, they inspired the research or because, currently, they seem to explain it. We have in some places related data to traditional theories—behaviorism, Freudian, Piagetian—but the thrust of the book comes more from contemporary theoretical approaches—neobehaviorist, neo-Freudian, neo-Piagetian, ecological, information processing. Our view of theory is eclectic and integrative. Our view of development is also eclectic and integrative. We see development as interactive, characterized by both continuities and discontinuities, involving both overt behavior and underlying structures, subject to multiple internal and external forces which jointly give rise to a variety of developmental trajectories and outcomes.

Summary

1. The study of human development is a product of the 20th century, although its philosophical roots extend back to John Locke's and Jean Jacques Rousseau's ideas about human nature. G. Stanley Hall founded the study of development in the United States. Binet and Simon devised the first objective, standardized test of children's intelligence. Three powerful theories of development were proposed by Gesell (maturational), Watson (behaviorist), and Freud (psychoanalytic).
2. During the middle of the 20th century, American psychology was dominated by learning theory. Social learning theorists applied principles of primary and secondary drives to development. Operant learning theorists used a mechanistic learning theory to explain social behavior, proposing that it is a series of responses that can be increased by reinforce-

ment. Observational-learning psychologists suggested that children learn social behavior by observing and imitating others, without external reinforcement. Erik Erikson introduced his theory of human development as a series of crises over the whole life span. Jean Piaget elaborated an interactionist theory of children's cognitive development that remains influential today.

3. In the 1980s, new approaches likely to influence the study of child development are information processing, dialectical, ecological, sociobiological, neobehavioristic, neo-Freudian, and neo-Piagetian.

Key terms

lifespan
nurture
nature
ontogeny
phylogeny
norm
maturational
classical conditioning
behaviorism
oral stage
anal stage
phallic stage
Oedipal conflict
Electra conflict
latency period
genital stage
id
ego
superego
primary (biological) drive
learned (secondary) drive
operant learning theory
operant conditioning
reinforcement
punishment
behavior modification
ethology
critical period
imprinting
psychosexual development
psychosocial development
information processing
dialectical
ecological
microsystem
mesosystem
macrosystem
sociobiology
neobehavioristic
neo-Piagetian
neo-Freudian

Suggested readings

ARIÈS, PHILIPPE. *Centuries of Childhood*. New York: Random House (Vintage Books), 1962. The metamorphosis of the concept of childhood from the Middle Ages to the present view of childhood as a distinct phase of life, traced through paintings, diaries, school curricula, and the history of games.

GINSBURG, HERBERT, and OPPER, SYLVIA. *Piaget's Theory of Intellectual Development* (2nd ed.). Englewood Cliffs, N.J.: Prentice-Hall, 1979. An explanation of Piaget's theory and helpful examples of each stage of cognitive development, in words any undergraduate can understand.

HALL, CALVIN. *A Primer of Freudian Psychology*. New York: New American Library, 1979. Delivers what it promises, an understandable overview of Freud's enormous output, for those who are starting to examine his ideas.

KAGAN, JEROME. *The Nature of the Child*. New York: Basic Books, 1984. An elegantly written discussion of historical and contemporary views of the child which shows how ideas have changed over the years.

MILLER, PATRICIA. *Theories of Developmental Psychology*. New York: W. H. Freeman, 1983. A good overview of the major theories in development: Piagetian, Freudian, psychosocial, social learning, information processing, ethological.

chapter two
Methods

The scientific method

In Chapter 1, we traced the evolution of the study of development from a philosophy into a science. Today, the study of development is strictly scientific. The studies we now rely on are those that are based on the **scientific method**.

- In these studies, the researcher usually begins with a specific **hypothesis**. ''People have temperamental characteristics that are evident from the moment they are born.'' ''Prenatal growth can be hindered by mothers' smoking.'' A hypothesis is a hunch, guess, or prediction about the world. It may grow out of a chance observation, intuition, the findings of another study, or it may be based on a formal theory. Entire theories never can be tested in a single study; they are meant to be broken down into specific, testable hypotheses. The most valuable theories are those that lead to a coherent set of testable hypotheses. In areas with no existing theory or previous study, a researcher may begin with curiosity and a question: ''Do infants have different temperaments when they are born, and if so, how long do they persist?'' ''Does cigarette smoking during pregnancy damage a developing fetus?''
- The hypothesis or question then is tested against reality. Reality in scientific studies takes the form of objective facts and evidence—**data**—that are systematically collected. For the hypotheses stated, relevant data would include the regularity with which a number of different infants ate and slept and how much they cried; the occurrence of birth defects in infants of smoking and nonsmoking mothers.
- These data are then analyzed statistically so that the researcher can determine whether they support or disconfirm the original hypothesis.
- Finally, the study may be repeated, with different subjects, different procedures, or by different researchers. If subsequent studies produce results that confirm those from the original study, then the hypothesis and the theory from which it was derived are strengthened.

Research designs

To design a study that conforms to the scientific method, the researcher must make a series of choices. There is no single ''right'' research method. For one thing, there are different research designs to choose from. Some designs are more apt than others for particular purposes. But the choice that the researcher makes depends on more than the research question under investigation. It also depends on the resources available and on the researcher's stamina, taste, and judgment.

Longitudinal versus cross-sectional

To answer questions about the course of development, researchers must study people at different ages. But there are two basic ways of doing this. In a **longitudinal research design**, a sample is followed over an extended period, as they get older. In a **cross-sectional research design**, separate samples of different ages are studied. In the cross-sectional study, the subjects are chosen carefully so that the subjects in the different age samples are as nearly alike as possible in all ways other than age. They are of the same sex, social class, physical health, and so on. From the differences observed between the samples, then, the researchers can infer how behavior and abilities change as a result of increasing age. But they must make their inferences cautiously, because their samples may have differed in ways that they did not realize. The reason that a group of 9-year-olds is more advanced than a group of 6-year-olds, for example, may be that the 6-year-olds have all gone through a stressful infancy because war was declared soon after they were born or that the 9-year-olds all had participated in an enriching preschool program or have taken nutritional supplements. Having comparable samples at different ages is essential in cross-sectional research. If researchers are careful about selecting their samples, cross-sectional research has the advantage of producing at least a rough outline of developmental change more quickly than longitudinal research. Most of the research reported in this book is cross-sectional.

Longitudinal research takes the long view. It yields answers about individual patterns of change with age. It sheds light on continuities and discontinuities in development, showing, for instance, whether the development of independence from Mother takes place gradually and regularly or in spurts, dips, and plateaus, or whether bright 6-month-olds are still leading the pack of 2-year-olds, 6-year-olds, and college applicants on IQ tests. A longitudinal design is the only way of answering whether individuals' IQ test scores are stable over time, whether difficult babies become difficult children or law-breaking adults, whether underweight newborns are still small for their age years later—because the same people are studied at different ages.

But the longitudinal design also has weaknesses. Imagine that you want to study the development of intelligence from birth to maturity. The cross-sectional strategy is to choose a large sample of children, say 100 at each age from 6 months to 30 years, and give them all intelligence tests. The longitudinal strategy is to test the same small sample of, say, 100 children, repeatedly from 6 months until they turn 30. The longitudinal approach is not only slow—taking a full 30 years—but risky. Some of the subjects will move away, and some will drop out of the study. Those who drop out may differ in some way from those who remain in the study, and their absence may bias the study's results. The dropouts may be the least intelligent or the most intelligent. Also, after years of taking tests, the subjects in the longitudinal sample will show **practice effects**. They will have grown so used to taking intelligence tests that the test results will be affected, and so the researcher will have to compensate for that effect as well. Finally, even after making all these adjustments, the researcher still must be cautious in generalizing about the findings. Because only one **cohort** (people born in the same year) was followed, the researcher cannot generalize to people who grew up during a different historical epoch. Their experiences and consequently their patterns of development may have been unique. Longitudinal research has its drawbacks; it is laborious and time-consuming.

The best research design for assessing development includes elements from both longitudinal and cross-sectional strategies: the researcher studies several different-aged samples longitudinally over a period of years (see Figure 2.1). This **cohort sequential research design** has the advantages of being relatively short in duration (for subjects and researchers), of including more than a single historical cohort, and of charting individuals' development over time.

It can be used to differentiate among three kinds of influences on development (Baltes, 1983):

1. *Age-related influences*—the biological and environmental factors that are closely linked with chronological age, such as the development of language in 2-year-olds or secondary sex characteristics in 12-year-olds.
2. *History-related influences*—the biological and environmental factors that are closely linked with historical eras, such as the development of vac-

CROSS-SECTIONAL DESIGN

Grade 2

Grade 5

Grade 7

In a cross-sectional research design, the researcher studies groups of people who are of different ages, like the children in these three classes. This design reveals overall age-related changes, but not individual patterns of development.

LONGITUDINAL DESIGN

Age 15

In a longitudinal design, the researchers study the development of individuals over time. The main drawback is having to wait for the subjects to grow older. The main advantage is that the design reveals individual patterns of development, as seen in these three children.

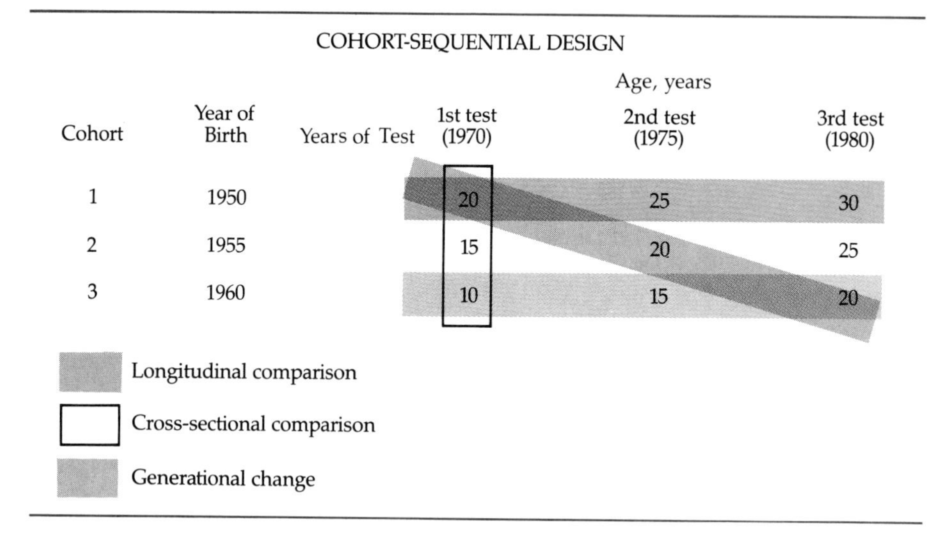

Figure 2.1 ***The cohort-sequential design combines elements of longitudinal and cross-sectional designs and also yields information about differences between generations or cohorts of subjects.***

cines against polio and measles, the Head Start program for early childhood education, or the Vietnam War.

3. *Non-normative influences*—the unique events that affect individuals in varying, often unpredictable sequences, such as winning a lottery, losing a job, or falling seriously ill.

Age-related influences are primarily important in the development of infants, children, and old people. In contrast, history-related and non-normative influences are primarily important during adulthood. Although each influences development, not every research design can or should include all of them.

Correlational versus experimental

Another choice in designing a study is between correlation and experiment. Researchers interested in finding out what causes development must choose one of these two designs.

In a **correlational design**, researchers look but don't touch. They measure relations among things without trying to intervene. If they are interested in studying the development of social skills, for example, they might observe children of mothers who are themselves extremely sociable and compare them with the children of mothers who are less sociable. In correlational research, researchers do not manipulate behavior. Instead, they systematically observe its occurrence in different groups of people or in people in different situations. Identical twins may be compared to fraternal twins; adopted children may be compared to those raised by their biological parents. People in one culture or social class may be compared to those in another. Researchers sometimes start out with a single, diverse group rather than two groups and look for differences within the group.

The correlational design yields valuable information about development in different environments. It is the design chosen by researchers who take an ecological approach to studying development. But this method of research has its limitations. One limitation is that a researcher often cannot measure or control all of the variables in the environment under study. People in different cultures, different social classes, or different families have experiences that differ in many ways, not just in the ways that researchers know about and measure. It is hard to determine which of these ways might be responsible for the differences observed in the subjects' behavior. Even more serious, it is impossible to separate cause and effect in a correlational study. The researcher cannot know, for example, whether parents are influencing children or children are influencing parents. Any observed association between the two only demonstrates that the behavior of parents and children is *related*. To solve this problem of cause and effect, researchers often intervene directly in the environment, providing "causes," of which they then measure the "effects." They do this by using an experiment.

In an **experimental design**, researchers do touch; they actually manipulate environments and look at the effects of their manipulations on behavior. An experimenter may read special books to children, reinforce children for playing with blocks, or show adults a movie. Those who receive the special treatment are the experimental group. The **control group** consists of people who are not treated specially. Subjects are assigned to one group or the other at *random*. The researchers then assess the effects of their manipulation on the experimental group's behavior by seeing how the behavior of people in this group changes in comparison with the behavior of people in the control group. Thus they know whether their manipulation caused a difference in the subjects' behavior.

Even with a control group, though, the experimental design has limitations. Often it is neither practical nor ethical to use human beings as experimental subjects. People must not be subjected to extreme changes in environment, for example. Researchers sometimes can use cats, rats, or primates as experimental subjects. There is always a danger in applying research findings on development in animals to human beings, of course, but sometimes the effort is worth making. In carefully constructed animal experiments, cause and effect can be shown precisely enough to justify tentative generalizations to human behavior. The work of Harry and Margaret Harlow and their associates at the University of Wisconsin (for example, Harlow and Griffin, 1965; Harlow and Zimmerman, 1959) provides a good example of experiments on animals to study development. These researchers performed many experiments with rhesus monkeys, depriving infant monkeys of social contacts for various periods of time, watching to see how mothers and infants reacted when they were separated, seeing how monkeys behaved when their only contact was with a wire mesh "mother" or a terrycloth "mother," and so on. These experiments with monkeys provided important hypotheses for researchers to follow in work on humans. Animal research is valuable because it allows for extreme and well-controlled manipulations of the environment.

Research on animal subjects allows investigators to set up conditions that would be unethical with human subjects. In this study, the young monkey hugs the terry cloth "mother" that he is being raised with and ignores a cold, wire "mother," even though she offers milk.

In experiments with people, researchers are limited not only in how extreme their manipulations may be but in how long their experiments may last. Researchers therefore often have measured only the immediate effects of relatively brief manipulations of behavior. In the observational-learning study by Bandura described earlier, for example, children in the experimental groups watched an adult or a cartoon character kick, punch, and throw a rubber Bobo doll and then were left alone with the Bobo doll, all within a brief period of time.

Recently researchers have expanded experimental designs beyond this short time frame. For example, they have trained parents to give children the experimental treatment. It is not ethical to ask parents to act aggressively toward, to hit, or to punish their children for the sake of an experiment. So what researchers have done instead is to train some parents—the experimental group—to be *less* aggressive toward their children, while other parents—the randomly chosen control group—continue to use their normal level of physical punishment with their children. Such training programs for parents represent a broader form of experiment, in which **interventions** are carried out over

Table 2.1
Pros and cons of research designs

Design	Pros	Cons
Longitudinal: Same sample observed at different ages.	Shows developmental curves for individuals or groups. Shows temporal sequences of events.	Expensive, time consuming. Subjects may drop out during course of study.
Cross-sectional: Different samples observed at different ages.	Gives view of average developmental changes with age.	Does not indicate individual growth curves. Does not indicate temporal relations.
Experimental: Controlled treatment given to subjects selected at random.	Effects of known, specified, controlled treatment can be determined.	Questionable whether findings apply to situations outside the often artificial one of the experiment. Treatment is usually short term.
Correlational: Subjects observed without researcher intervening.	Shows behavior and relationships as they occur in real life.	Cannot be used to determine cause and effect.

longer periods of time. They also take place not in the laboratory but in natural environments.

Table 2.1 summarizes the pros and cons of various research designs. Researchers who combine correlation and experiment—by first observing naturally occurring differences in behavior that are related to environmental factors and then trying to produce these differences in behavior by experimental manipulations—may capture the best of both designs.

Collecting data

Once researchers have chosen their research design, they must decide how they will collect the data they need to test their hypothesis or answer their research questions.

Naturalistic versus structured observations

Researchers may choose to focus on naturally occurring spontaneous behavior as subjects work or play freely at home, at school, or at work. Developmental psychologists taking an ethological approach would do this kind of **naturalistic observation**. So might a behaviorist who wanted to document naturally occurring reinforcement patterns. In contrast, a researcher might choose to observe behavior in constrained situations set up by the researcher—presenting the subject with a new task, an unfamiliar peer, or a tape recording of a baby crying, for example. These observations would be more **structured**. In a correlational research design, data often are collected in naturalistic observations. In experiments, data often are collected in structured observations. But structured observations also may be used in correlational research, just as naturalistic observations may be used in experiments.

In naturalistic observations, researchers try hard not to influence the behavior of the subjects they are studying. Some observations can be done without the subjects' knowledge and without any intrusion into their activities. These might be done in a public park or supermarket. But these observations tap only public behavior. Usually psychologists are interested in how people behave in their private lives. As we saw in Chapter 1, Arnold Gesell introduced the one-way window, through which the observer can see

In the naturalistic observation photo on the left, a child psychologist collects data on kindergartners. In the structured observation photo on the right, two observers record their observations as they watch adult subjects through a one-way window.

without being seen, as a means for observers to watch without interfering with people's activities. Unfortunately, not many homes, offices, or schools come equipped with one-way windows. Researchers doing naturalistic observation in these settings, therefore, must make every effort to fade inconspicuously into the woodwork. One way to do this is to take as little equipment as possible into the home or classroom. In most settings, cameras and lights and tripods would stand out, and so observers resort to notebooks and clipboards or small, hand-held video cameras. In a few studies, observers have hidden themselves and their equipment in small, portable booths that they took to the homes. But most often observers just sit quietly in a corner of the room out of the way. They are careful not to initiate interactions with the person they are observing.

To keep to a minimum the stress that being observed inevitably creates, observers try to put their subjects at ease before the observation begins. They may spend some hours or days in the setting to let the subjects get used to their presence. They describe their purpose to adults in the least threatening way possible, emphasizing the fact that they are not judges or experts making an evaluation, but that they are just gathering information about how people usually act.

Despite all these precautions, being observed inevitably affects the ''target,'' and when people are aware that they are being observed, they may act more stiffly, shyly, self-consciously, politely, and positively. Even so it is unlikely that they can or will disguise completely their usual ways of behaving just to put on a good show for the observer. After the first few visits from an observer, a person usually begins to act in his or her customary ways.

An alternative to observations by unfamiliar people is for parents, teachers, babysitters, and other familiar people to record children's activities or for adults to report their own activities. Charles Darwin, you recall, observed and wrote about his son Doddy's behavior. Darwin, of course, was a trained observer as well as a parent. Not all parents are so capable—or so objective. Nevertheless, most people can be guided and trained to make useful observations of their own or their children's actions. Researchers have used this strategy successfully in investigations of very young children's spontaneous expressions of sympathy (Zahn–Waxler and Radke–Yarrow, 1979; Zahn–Waxler, Radke–Yarrow, and King, 1979) and recall (Ashmead and Perlmutter, 1980), for example. Because these kinds of behavior crop up infrequently in young children and are most likely to occur in comfortable surroundings, an unfamiliar observer might have to spend hours waiting for a single instance. Having parents collect the data is the more efficient strategy for this kind of research. Parents might be less reliable and desirable as observers of their *own* actions, however. The temptation to consciously or unconsciously distort the facts would be stronger, and focusing attention on their own actions could lead parents to change their behavior. Researchers must be cautious about using this strategy.

One further strategy for collecting data naturalistically is to "bug" subjects' rooms with audio or video recorders. One mother who wanted to study her 2-year-old's private monologues after she had tucked him in for the night set up a tape recorder in his room (Weir, 1962). This method of tape-recording children's and their parents' talk has been used occasionally by other researchers, too. Barbara Tizard (Tizard, Carmichael, Hughes, and Pinkerton, 1980) sewed microphones into nursery school children's shirts or dresses to record their private conversations with other children and with teachers. In the future, video technology is likely to be used to observe people unobtrusively.

Only naturalistic observation can tell us about people's "real lives," "typical activities," and "normal encounters." This is their unique advantage. More structured observations have the advantages of giving researchers more control over the situation and of giving them a chance to look at how different people behave in a single, standard situation. It is all very well to know that children attending day-care centers act differently there from the way that children reared exclusively at home act at home, for example. But it is also important to find out whether day-care children act the same as home-reared children when they are in an unfamiliar room, given a set of unfamiliar toys, and introduced to an unfamiliar child or adult. Only then can developmental psychologists make general statements about how putting children in day care or keeping them at home affects their behavior.

Structured observations offer such standard situations. They also can be more efficient than naturalistic observation for producing particular kinds of data. Why wait around the schoolyard to observe adolescents' cooperation and competition, for instance, when you can structure games, races, and problems that will elicit these kinds of behavior in a half-hour session? Structured situations let researchers focus effectively on just the behavior they are interested in.

Like naturalistic observations, structured observations raise concerns about the intrusiveness of observers and their equipment. Although behavior in structured observations is more constrained than in naturalistic observations, the researcher still wants subjects to feel comfortable and not to act as if they are under a lens. But structured observations provoke other concerns as well. Researchers must try to make the setting itself comfortable and unthreatening, so they convert university classrooms and laboratories into "living rooms," furnished with rocking chairs and fireplaces, and camouflage their one-way windows behind stylish window blinds or partially drawn drapes.

Structured observations do not necessarily have to be done in laboratory settings, however. It is possible to structure observations of people's activities at home or at school—in their natural, everyday settings—by giving the people specific tasks. There may even be advantages to carrying out structured observations in these natural settings, where people are more comfortable.

Whether researchers choose to do their structured observations at home or in the laboratory, however, they must be careful to control all aspects of the environment. More than a few researchers have been dismayed to find that the toddlers they were observing ignored the attractive toys they had been provided with and spent all their time scrutinizing the nest of electric wires hooked up to the camera or studying the dustballs on the windowsill. Researchers using structured observations need to make the whole situation identical for all subjects and focused on the task at hand.

When these concerns are met, structured observations can yield important information. The assessments of infants' and toddlers' abilities, in which the child is shown blocks, a red ring, pictures, or a broken doll, represent one kind of structured observation that has been very revealing. Another is the so-called "strange situation," in which the relationships of infants and mothers are assessed with a series of carefully choreographed, standardized sequences of separations and reunions. Data from many such structured observations are included in our current knowledge of development.

One limitation of *both* naturalistic and structured observations is that they must focus on children's and adults' *overt* behavior—what can be seen, heard, recorded, and quantified. Researchers must take care not to project their own points of view and attitudes onto their research subjects, but record their observations of behavior systematically and objectively. The traditional method for finding out what people are thinking is to *ask* them, by using interviews and questionnaires.

Interviews and questionnaires

With interviews and questionnaires, researchers actively probe people's ideas, feelings, and motives. Their data come in the form of words rather than observed actions. Especially with children, interviewers have to work sensitively and skillfully in order to get true indications of their subjects' abilities without misinterpreting or distorting the responses they are given. Jean Piaget was one such skilled interviewer of children. Here he interviews a 6-year-old on his thoughts about lying.

PIAGET: Do you know what a lie is?
CLAI: It's when you say what isn't true.
PIAGET: 2 + 2 = 5. Is that a lie?
CLAI: Yes, it's a lie.
PIAGET: Why?
CLAI: Because it's not right.
PIAGET: The boy who said 2 + 2 = 5, did he know it was wrong, or did he just make a mistake?
CLAI: He made a mistake.
PIAGET: Then, if he made a mistake, did he tell a lie or not?
CLAI: Yes, he told a lie.
PIAGET: A bad one or not?
CLAI: Not very bad.
PIAGET: You see this gentleman? (Piaget points to a graduate-student assistant.)
CLAI: Yes.
PIAGET: How old do you think he is?

CLAI: 30 years old.
PIAGET: I would say 28. (The student says he is 36.)
PIAGET: Have we both told a lie?
CLAI: Yes, that's a lie.
PIAGET: A bad one?
CLAI: Not too bad.
PIAGET: Which is the worst, yours or mine, or are they both the same?
CLAI: Yours is the worst, because the difference is biggest.
PIAGET: Is it a lie, or did we just make a mistake?
CLAI: We made a mistake.
PIAGET: Is it still a lie, or not?
CLAI: Yes, it's a lie. (Piaget, 1932, p. 140).

Not all interviewers are so skillful or as patient as Piaget. They run the risk of asking leading questions, not probing responses deeply enough, not establishing rapport with their young subjects, and not interviewing all subjects in the same way. Interviewers can avoid some of these risks by using interviews and questionnaires that are prepared ahead of time. With practice, they can also learn to ask the same questions of all subjects in the same even-handed way. Prepared and practiced interviews have been used many times to probe people's attitudes and knowledge. Countless interviews and questionnaires, given to adults, parents, teachers, and children, have been used to collect data about development. With young children, however, unless the interview is very simple, short, and specific, interviewers often need the flexibility to follow the unpredictable, creative twists that their subjects introduce. In these cases a less structured interview is called for.

With older children and adults, researchers can often use paper and pencil questionnaires, which subjects fill out themselves. Compared to in-person interviews, this method has both advantages and disadvantages. Its advantages lie in its standardization of responses, its protection of subjects' anonymity, and its economy of administration (an entire group can fill out a carefully constructed questionnaire at one time under supervision). Its disadvantages are its lack of flexibility, its dependence on subjects' abilities to understand written questions, and its ineffectiveness for probing sensitive topics.

No matter how well prepared, sensitive, and flexible the interviewer is, no matter how carefully crafted the questionnaire, moreover, interviews and questionnaires have one limitation that resides not within the researcher or the instrument but within the subject. Whether intentionally or not, people may distort or misreport what they tell the interviewer or put down on the questionnaire. Young children may not remember events well or describe them accurately. Parents often misremember even such seemingly obvious facts as the age at which their children first walked or spoke. The possible solutions to this problem include asking only straightforward questions about objective happenings, not requiring or allowing reporters to make complicated subjective interpretations of their own or others' actions, and asking about things that happened in the recent, not the distant, past. There are also ways of wording questions so that subjects do not think that there is only one clearly desirable answer and feel impelled to give it rather than the truth. "Do you beat your child?" is *not* a subtle question. Researchers might do better with something more on these lines: "Children often are difficult to discipline, as I'm sure you realize only too well. Here are some stories about problems that children present. Please tell me how you might handle problems like these if they happened with your child. . . ."

One relatively easy way to get information about development is to administer standard tests. In the Peabody Picture Vocabulary Test, a child is asked to point to the appropriate pictures as the tester names objects. Part of many intelligence tests for adults involves spatial reasoning.

Tests

One quick way to collect data about development is to give standardized tests—IQ tests, personality tests, tests of perceptual abilities, and the like. These tests are widely available and offer researchers easily and objectively collected data to compare with the results of other researchers. When choosing a test, the researcher must be concerned with whether it is valid. **Validity** is the indication of how accurately a test measures what it is supposed to measure. If a thermometer registers 212 degrees Fahrenheit when it is placed in boiling water at sea level and registers 32 degrees Fahrenheit when the water is frozen, it is a valid instrument for measuring temperature. The measurement of psychological traits and abilities is more elusive.

True validity indicates how well a test conforms to the best scientific idea of the underlying trait or construct being measured—intelligence, self-esteem, social maturity, achievement motivation, and so on. But true validity is a theoretical concept, not empirically measurable. Therefore other measures of validity are used as approximations of true validity. A test may be considered valid, for one thing, if its items fairly represent the relevant body of knowledge. For example, the validity of a math test is high if the items have to do with arithmetic and algebra; the validity is low if the items have to do with the mating behavior of killer bees. Another measure of validity is a test's ability to predict success on a second variable supposedly related to the ability being tested. For example, if academic success is believed to depend on intelligence, we would expect people who get good grades to score higher on IQ tests. A third measure of validity is how well a test correlates with another established test. The validity of any newly devised intelligence test, for example, is usually determined by how highly individuals' scores on it correlate with their scores on established tests like the one devised by Binet and Simon. In choosing a test, a researcher must balance concerns about validity against the desire for innovation and the willingness to take risks. The best solution is for the researcher to use more than one test or measure of a trait.

The pros and cons of different methods of collecting data are summarized in Table 2.2.

Table 2.2
Pros and cons of data collection methods

Method	Pros	Cons
Structured Observation: Behavior observed in a standard, contrived situation.	Data from different subjects can be compared. Data not restricted to preestablished test responses.	Does not get at underlying attitudes. Not known whether observations can be generalized to other situations.
Naturalistic Observation: Spontaneous behavior observed in familiar environment.	Provides information about behavior in the real world. Offers description of activities, behavior, interactions.	Settings for different subjects are not comparable. Does not assess maximum performance possible. Some kinds of abilities may not be observable.
Questionnaire or Test: Same assessment given to all subjects.	Data from different subjects can be compared. Individual's performance can be compared to norms.	Data limited to preestablished responses.
Interview: Questions posed to children and adults.	Quick way to get information. Only way to assess conscious intentions and attitudes.	Interviewees are biased, not accurate reporters of past events or their own behavior.

Recording and coding the data

Once researchers have decided on their methods of collecting data, they must decide how to record and code the data. Researchers can choose from a wide variety of techniques for recording data—from writing a narrative account of interaction in a notebook, to pressing a computer key every time the subject does something, to making a clinical judgment about how warm a person is, to marking checks on a list at each occurrence of a kiss, smile, caress, or word of praise. Checking whether a subject passed or failed each item on a test is perhaps the easiest way of recording data. At the other extreme, during an interview or an observation, researchers may have to use cameras or tape recorders to catch everything that is happening. Mechanical devices for recording data, like electroencephalographs, electrocardiographs, spectrographs, and stabilimeters are also available for assessing brain waves, heart rates, vocalization patterns, and activity levels. Each of these techniques produces useful information. The choice of which one to use depends on the researcher's desire for detail and on the time and money available.

Turning this recorded information into data that can be analyzed, then, requires coding it into categories or numbers: how often a behavior was observed in two hours; how intense, or loud, or loving the behavior was; how many test answers were correct; and so on. Some of this coding can be done on the spot, if researchers are very experienced and use established checklists, tests, or rating scales. In other cases, observers turn detailed notes, taken during the observation or interview, into ratings, descriptions, or scales after the fact. When the observations are filmed or videotaped, researchers can go back and pore over their records to find each significant glance, raised eyebrow, and fleeting smile. Once again, whether the researcher codes every possible minute behavior or settles on broader ratings or categories is a personal choice. There are advantages and disadvantages to both methods. The first method gives more objective and detailed information, but it takes more time and may be more superficial. The second method gives more clinical and global information, but it may be more subjectively biased. (We discuss the problem of bias in the next section.)

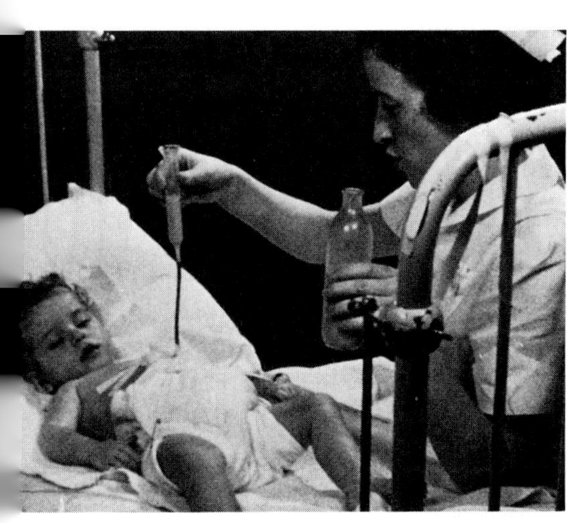
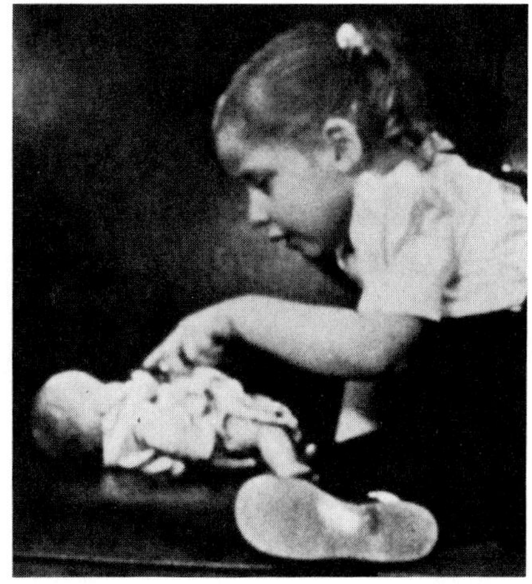
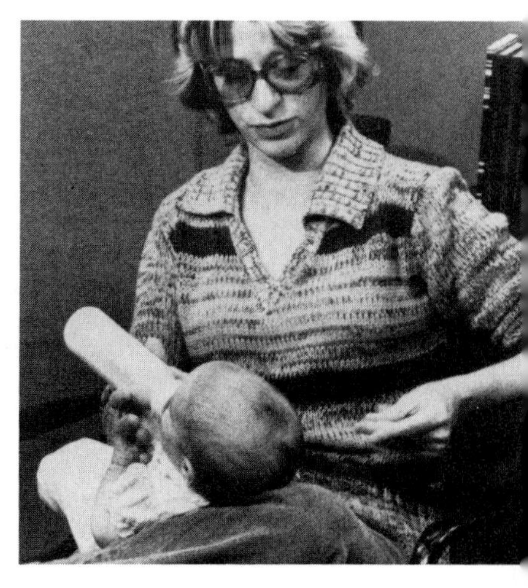

These photos illustrate the usefulness of a case study for finding out about human development, here the long-term effects of early experience. Monica was born with a malformation of the esophagus that required her to be fed by gastric fistula until she was 2 years old. She was not held in a caregiver's arms for feeding during this entire period. As a young child, Monica fed her dolls without holding them. Later, as a mother, Monica did not hold her own infants in her arms during feeding. (Engel, G.L., Reichsman, F., Harway, V.T., and Hess, D.W. Monica: Infant-feeding behavior of a mother gastric fistula-fed as an infant: A 30-year longitudinal study of enduring effects. In E.J. Anthony and G.H. Pollock (Eds.), **Parental influences in health and disease.** *Boston: Little Brown, 1985.)*

Problems in doing a study

Sampling

Not only do researchers have to decide how much detail they should record and how many acts of behavior they should code, but they also must decide how many subjects they should study. If they study too few—fewer than, say, 25—they run the risk that chance or individual variations among the subjects will bias their results. By limiting themselves to a few subjects, researchers also can detect only the strongest statistical relations among the variables they are studying. With only 15 subjects, for example, researchers have less than one chance in four of detecting a moderately strong correlation of 0.30. (Correlations are discussed later in this chapter.)

The nature of the research question and the chosen methods of collecting data serve as guidelines for determining the adequate number of subjects in a study. If the researcher collects extensive data on each subject, then there is some justification for including fewer subjects. Detailed **case studies** of one or a few subjects can give information that more superficial measures on a large sample of subjects do not. A case study of one girl, Genie, for example, who grew up in almost total isolation until she was rescued at age 13, has provided unique information about the limits of language development (Curtiss, 1977).

But usually, researchers do not encounter such cases. Instead, they study large numbers of subjects so that they can make statements about development *in general*. Researchers also try to choose subjects who will represent the entire population in which they are interested. A **representative sample** of American retirees, for example, would include retirees from all racial and religious groups and from all social classes, in the same proportions as the general population. If samples are not representative, the results obtained from them may not apply to the general population or be consistent with what other researchers have found. Researchers often find that a single group of subjects cannot aid them in answering all their questions. For that reason, having several groups of subjects is an advantage, and when the findings are the same for more than one group, the researchers can be confident that their findings are **robust**.

Researcher bias

Another problem that researchers must guard against as they collect data is that their own biases do not creep in and influence their results. This problem is most likely to crop up when the person collecting the data is aware of the hypothesis being tested and very much wants it to "work." Researchers' expectations then can color their observations and taint what should be neutral and bias-free. Even without knowingly intending to bias results, if they know what the hypothesis is, researchers can exert a subtle influence on their subjects' behavior. An example of how biases work is offered by an experiment in which a group of developmental psychologists (Goren, Sarty, and Wu, 1975) claimed to have demonstrated that infants prefer to look at pictures of human faces rather than at pictures of other, equally complex figures. Other researchers repeating the study did not get the same results. They suggested that the first group had unwittingly biased their results by holding the infants on their laps while they looked at the pictures and turning the infants slightly toward the face pictures. Such sources of bias must be eliminated from research.

To protect against bias, researchers collecting data must be "blind" to the subject's condition. They must not know to which group—experimental or control, middle class or lower class, day care or home reared—subjects belong. That way they will not consciously or unconsciously distort either the subjects' behavior or their own record of that behavior. In a **double-blind study**, neither subjects nor data collectors know who is in the experimental and control groups. To protect against researcher bias, in well-designed studies people other than the designer of the research act as observers, experimenters, or interviewers. Two or more independent observers also may observe the same subject, and this helps to cancel out personal biases. Observers also can be given instructions that are so detailed and specific that it would be hard for them to bias observations (short of outright cheating).

Reliability and replicability

One way that researchers demonstrate that their results are not biased is to show that their observations are **reliable** and their results **replicable**. Observations cannot be unique to one observer. Other people, looking at the same behavior, must agree that the person was smiling, or acting fearful, that the parent was loving or demanding, and so on. The observations must be, as far as possible, objective reflections of reality. For this reason, it is customary for more than one observer, rater, or tester to collect data in a study. Before the observers begin, they are trained until they agree on the meaning of the observational categories or interview responses. Then, from time to time, their agreement is checked. The degree to which they agree is one indication of reliability. Another indication is the extent to which several observations of a single subject are comparable. Researchers want to find out about subjects' typical activities, and so they must spend enough time observing at different times of day and on different days to be able to make general summaries of subjects' behavior. Tests, interviews, and questionnaires, too, like observations, are reliable if they yield essentially the same results when given to a person at different, closely spaced times.

When reliability is high, it is more likely that the results of the study can be replicated, if another researcher or the same researcher at a different time and with different subjects repeats the study. **Replicability** is another important criterion of scientific research.

Analyzing data

Once data have been collected and coded, they must be analyzed. For detailed case studies, a richly vivid clinical description may be enough. But for most studies, the analysis of data is done statistically—these days, usually with the help of a computer. Several different kinds of analysis are possible.

Descriptive analysis

If the researchers simply want to describe people's behavior, they can analyze the frequencies or levels of the behavior—the frequencies of smiles and sentences beginning with "I," incidents of helping, or levels of height, health, intelligence, or sociability. One very simple way to analyze such descriptive data is to sum up the occurrence of behaviors for the sample as a whole. When data are quantitiative and form an **interval scale**, researchers can calculate the **mean** or average frequency or level for the sample. "The average income for the families studied was $20,000 per year." "The average number of 'I' utterances made by each subject observed was 14." Even more informatively, researchers can calculate mean levels or frequencies for different groups in the study: men and women, blacks and whites, rich and poor, 28- and 76-year-olds. "The average IQ score for males was 100, for females, 102." "The average number of questions was 10 for 62-year-olds and 40 for 34-year-olds.""Half of the 46-year-olds and 70 percent of the 58-year-olds passed the test." Perhaps best of all for descriptive purposes, researchers can analyze their data graphically. They can plot a graph that represents the test scores of each person in the sample or the number of times each laughed, hugged, or poked. Usually, for a large sample, these scores and frequencies form a normal, or bell-shaped, curve (Figure 2.2). When data form a curve of this shape, they are said to form a **normal distribution**. At the tails of the curve are a relatively small number of very high and very low scores, and in the middle are a large number of intermediate scores. The same number of people have scores above the peak of the curve as below it. The bell-shaped curve illus-

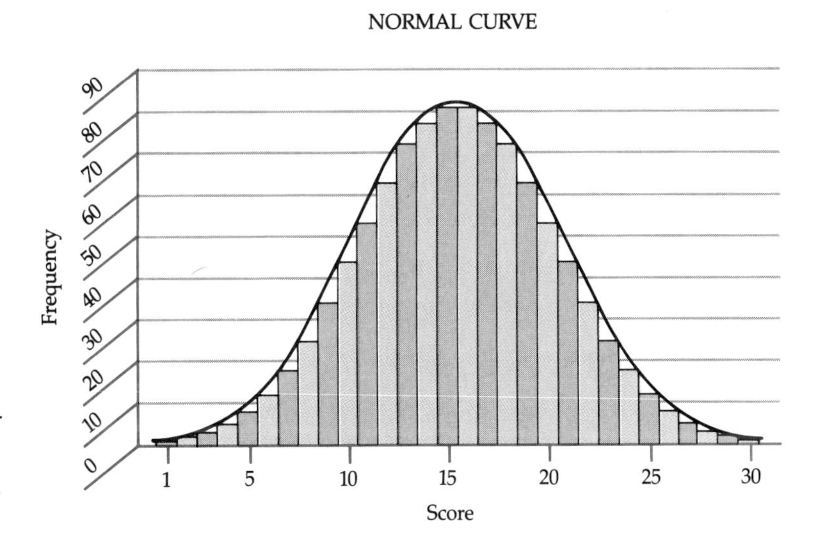

Figure 2.2 ***When the number of people receiving each score from 1 to 30 is plotted, it forms a symmetrical normal, or bell-shaped, curve.***

trates the fact that most people's scores—for height, weight, or intelligence, for example—cluster around the mean, or norm. Relatively few people are extremely short or tall, thin or fat, mentally retarded or gifted. On a normal curve of IQ scores (see Figure 12.1), for example, the mean is 100 points. The same number of people have scores between 90 and 100 and between 100 and 110, between 80 and 90 and between 110 and 120. But more people have scores between 90 and 110, a range of 20 IQ points, than between 70 and 90 or 110 and 130 points, which are also ranges of 20 IQ points.

Finding a normal curve in their data tells researchers that their sample is large enough and representative enough so that they may proceed with further statistical analyses, most of which require that the data to be analyzed be normally distributed. It also allows them to predict how tall, smart, or heavy people in the general population will be. For developmental psychologists, the more interesting graphic analysis shows how the levels or frequencies of behavior change with age. This analysis requires them to plot a **growth curve**, by recording mean scores of people at different ages. Growth curves take one of three basic shapes. An ability or behavior may increase in an S-shaped curve, in which the increase occurs rapidly over a certain period of time—the "formative period"—and then slows down and perhaps stops (Figure 2.3*a*). Height and verbal intelligence, for example, follow this general curve. Alternatively, an ability or behavior may keep increasing or decreasing steadily in a straight line (Figure 2.3*b*). This kind of growth is not likely to continue across the entire life span, however. Most abilities eventually reach a plateau. People do not keep getting bigger and better all their lives. Finding a straight line usually reflects the fact that researchers have included just one part of the life span—the verbal ability or height of 2- to 12-year-olds, for example. Some behaviors or abilities increase first, hit a peak, and then decrease as a person gets older (Figure 2.3*c*). Perceptual and motor abilities are two kinds of behavior that fit such an "inverted U"-shaped curve.

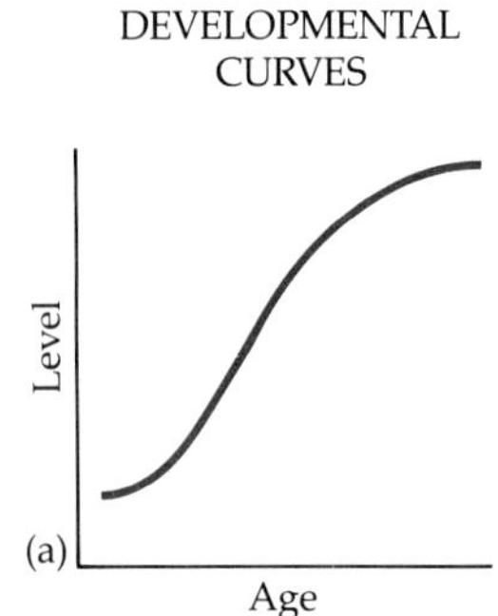

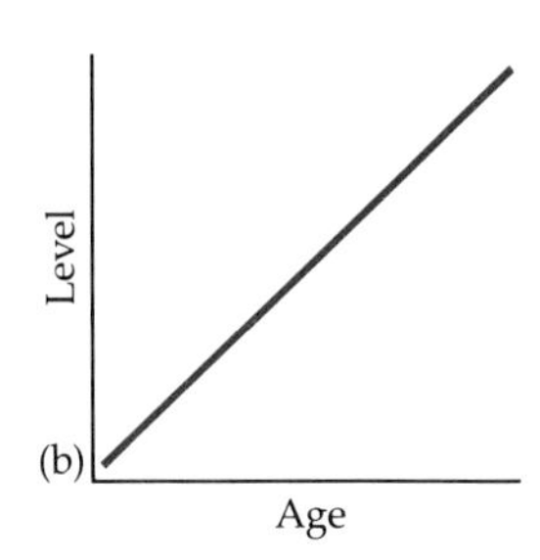

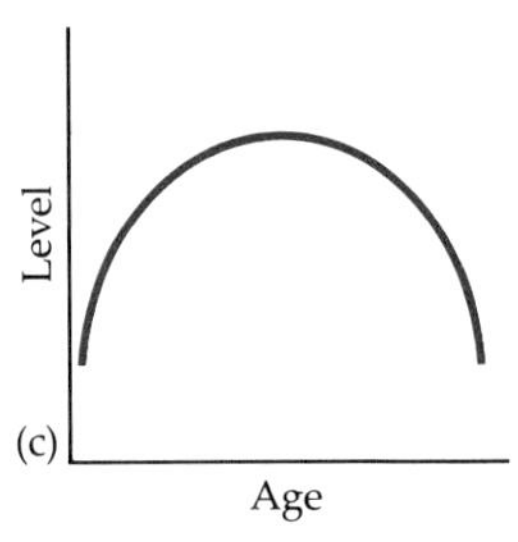

Figure 2.3 When growth or development over time is plotted, it usually takes one of these three shapes: **(a)** *an S-shaped curve, in which development increases gradually and then levels off;* **(b)** *a straight line, in which development continues to increase; or* **(c)** *an inverted U-shaped curve, in which development first increases and then declines.*

Significant differences

When subjects can be divided into different groups—experimental versus control group, 22-year-olds versus 62-year-olds, and so on—the researchers usually want to report not just what the mean or average scores for the groups are, but whether there are **statistically significant** differences between these mean scores. Differences are statistically significant when they are so large that they are not likely to have occurred by chance. Researchers usually set an acceptable probability level of something occurring by chance at 5 percent; that is, if an event has less than a 5 percent chance (one chance in 20) of having happened by sheer coincidence, it is considered statistically significant. There are standard statistical tests for determining how likely any given difference is to have occurred by chance. These tests take into consideration the number of subjects in the study, the size of the difference between the groups, and the variability among the subjects within the groups. Analyzing the data for significant differences tells researchers whether the differences they observed reflect *real* differences in human development and behavior or appear because the researchers happened to pick a sample with members who had extremes of ability or behavior.

Statistical relations

For some questions, researchers want to know not about differences between groups, but about how individual behavior relates to a particular background, age, experience, or ability. To learn about these relations, researchers usually use some form of the statistical analysis of **correlation**. A correlation reflects the degree to which two variables are related, that is, the degree to which two variables increase or decrease together. It is indicated by a number ranging from −1.0 to +1.0, or it can be plotted on a graph. In a perfect positive correlation (correlation coefficient = +1.0), the person who scores highest on one variable also scores highest on the other; the person who scores second highest on one variable also scores second highest on the other; and so on. A perfect correlation is illustrated in Figure 2.4*a*. In a perfect negative correlation (correlation coefficient = −1.0; see Figure 2.4*b*), the two variables are inversely related: the person scoring highest on one variable scores lowest on the other variable, the person scoring second highest on the first variable scores second lowest on the other, and so on.

In the real world, it is extremely unusual to find perfect correlations, positive or negative. Usually there is some degree of association, but it is not perfect. People who score above the mean in intelligence might be likely to score above the mean in sociability, for instance, or people whose parents are above average in intelligence will be likely to be above average in intelligence. Just how closely the rankings on the two variables match is reflected in correlation coefficients that range between 0.99 and −0.99 (see Figure 2.5).

There are statistical tests to establish whether the level of correlation observed is significantly greater than zero. Whether a particular correlation coefficient is statistically significant depends on how large it is and on how many subjects were included in the analysis. When there are only a few subjects in the sample, the correlation coefficient must be quite high (for example, for a sample of 15 subjects, a correlation coefficient must be larger than 10.50 to be significantly greater than zero). With a large sample, a much smaller coefficient is significant (for example, for a sample of 400, a coefficient of 0.10 is significant). The researcher can look up any correlation coefficient in a statistical table to see whether it is significant.

Interpreting results

After all the decisions about the design of the research have been made and the data have been collected and analyzed, there remains a final, critical task:

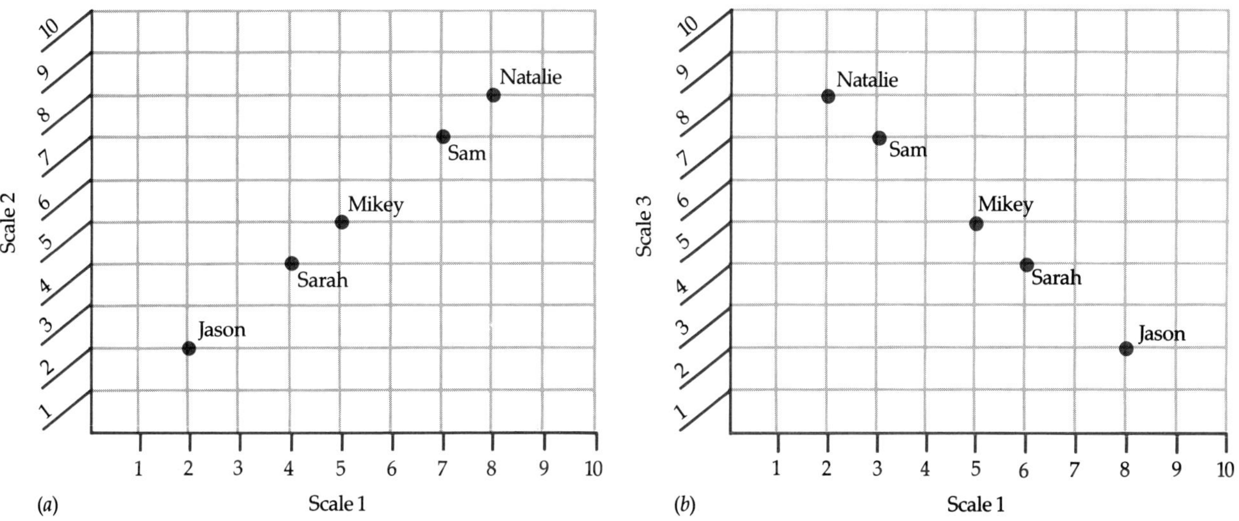

Figure 2.4 In a perfect (**a**) *positive or* (**b**) *negative correlation, subjects fall in exactly the same rank order on both of the scales.*

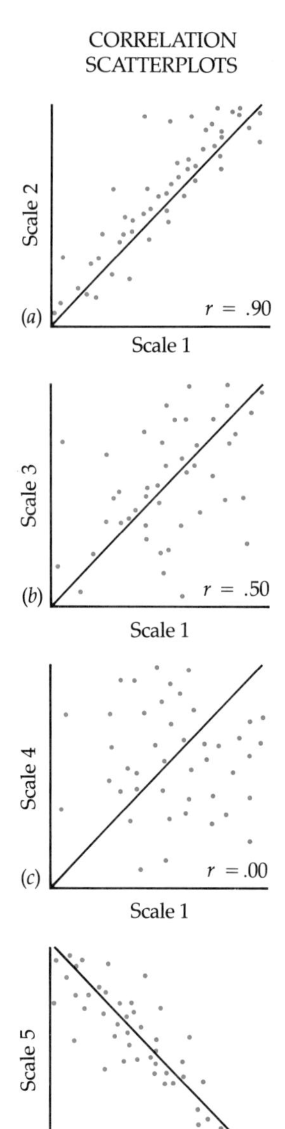

Figure 2.5 ***Scatter plots of four correlations, ranging from 0.90 to −0.90.***

the results must be interpreted. The researcher must explain just what the results *mean*. Interpreting results is no easy task. The relations discovered among the variables may point in different directions, or they may be confusing or inconsistent. They may not clearly confirm or disconfirm the researcher's hypotheses. Researchers must be careful not to overinterpret their findings or to draw more sweeping conclusions than are justified. They should not tout a correlation of 0.15 as exceptionally important. Even if that 0.15 is statistically significant, because 500 people were observed, 0.15 is still not impressive in absolute size. It does not explain much about what is happening. (To be precise, it explains only 2 percent of the variation observed in the other variable.) Similarly, researchers should not make headlines over an IQ difference of 5 points, which, with a large sample, is statistically significant but in real life is inconsequential. They should not lead people to think that their findings will predict the behavior of an individual. Statistical analyses apply only to groups and reflect only probabilities of occurrence.

Most important, researchers must not claim that one variable *caused* another when they have documented only that the two are *correlated*. No matter how high a correlation is, it does not mean that one variable is causing the other, only that the two are related. Finding a strong correlation between intelligence and sociability does not mean that people are sociable *because* they are intelligent. (Both may be a function of age, for instance.) Finding a high negative correlation between children's intelligence and their parents' punitiveness does not prove that punishing children *makes* them less intelligent. Less intelligent children may elicit more punishment from their parents, or less intelligent parents may be more punitive toward their children, for instance. Getting a highly significant correlation between parents' and children's IQ scores does not necessarily mean that intelligence is transmitted genetically. More intelligent parents may provide a more stimulating environment for their children, and this may cause their children to do better on IQ tests. Finally, finding *no* significant correlation between two variables does not necessarily mean that one variable is *not* influencing the other. One variable may cause another, but *more* of it does not cause *more* of the other. For example, some baby talk, or vitamin C, or play with toys may be necessary for infants to develop intellectually, but using *more* baby talk, vitamin C, or toys will not make babies more intelligent.

In the real world of complex human development, causes and effects are seldom clear-cut. Effects may operate in both directions at once, and causes may be many. By repeatedly measuring and seeing which variable changes first and which next, researchers can chart causes and effects more clearly. With more sophisticated statistical analyses, researchers can rule out some causal paths as implausible. But some causal questions are likely to remain unanswered, and researchers always must be cautious in interpreting their results.

Research ethics

As they design their studies and choose their research methods, researchers also must be concerned with the *ethics* of doing research. Clearly, researchers must not treat people cruelly or callously for the sake of scientific discovery. But what are the ethical limits on researchers? Is it ethical for a person to be shown a violent movie so that researchers can study the effects of filmed violence on aggression? Is it ethical to subject children to the stress of their mother's absence or of the sound of a baby crying helplessly to study the development of social relationships and empathy? Is it ethical to test old people's perception by having shapes loom down on them and then veer away only at the last moment? Clearly, researchers should not lie to subjects about the research or its possibly harmful effects.

Ethical questions about scientific research have been raised more and more insistently in recent years. University committees now must approve researchers' plans to make sure that subjects' rights are not violated. These committees must be satisfied that the research will do no harm to the subjects, that it will not subject them to events that are significantly different from those they might encounter in their daily lives, and that it offers potential benefits to science or society. They must be satisfied as well that subjects have given their "informed consent" to the procedures—that they know in advance what to expect and have consented to participate on the basis of their understanding. Because children are inexperienced and vulnerable, researchers must take extra care in using them as research subjects. Some investigations, done in the 1930s, '40s, and '50s, probably would not be approved by a human subjects committee today. In one investigation, for example, 11-day-old infants were submerged facedown in water to see whether they would swim (McGraw, 1939/1967). Some struggled, clutched for support, gasped for breath. The ethics of these research designs seem unacceptable to us now, in the 1980s.

Summary

1. Current knowledge about development has emerged from studies that follow the scientific method. Such studies begin with a specific hypothesis or question that is then tested against reality.
2. To study the course and continuity of human development, researchers study people at different ages. One way they do this is longitudinally—following a sample of subjects for an extended period; another way is cross-sectionally—observing separate samples of subjects of different ages. Researchers may overcome the limitations of these two research designs by studying a number of overlapping age samples longitudinally.
3. Researchers can choose between correlational and experimental research designs. In a correlational design, researchers do not manipulate behavior; they observe it systematically in different people and in different settings and look for relations between the behavior and other factors. In contrast, when they conduct an experiment, researchers actually manipulate environments and look at how these affect behavior. Research that combines correlation and experiment may offer the best of both designs.
4. Observations may take place in natural settings or in settings where the researchers can better control what happens. Different settings all may influence behavior in different ways. Naturalistic observation provides insight into people's "real lives" and typical activities. Structured observations give researchers the chance to see how different people behave in a single, standard setting.
5. Observations focus on overt behavior. Interviews, tests, and questionnaires probe what people think and feel—their feelings, thoughts, abilities, and motives.
6. Researchers must decide how much data to amass and how many subjects to study. Their aim is to study enough people and people who are representative of the entire population so that they can use their data to make general statements about development.

7. Researchers must take care that their own biases do not influence their results. Testing, observations, and interviews should be conducted by research assistants who do not know to which group subjects belong or the hypothesis of the study. Results must be objective, reliable, and replicable.
8. Researchers usually analyze average or mean differences between groups' behavior and test whether these differences are statistically significant, that is, whether they are large enough so that they are unlikely to have occurred by chance. Correlations are statistical measurements of how two variables increase or decrease in relation to one another.
9. In interpreting their data, researchers must take care not to infer too much, for example, claiming causes when they have documented only correlations.
10. Conducting research binds developmental psychologists to strict ethical standards. Research must not cause harm, people must give informed consent ahead of time, research must be confidential, and people's rights to refuse to participate must be respected.

Key terms

scientific method
hypothesis
data
longitudinal research design
cross-sectional research design
practice effects
cohort
cohort-sequential research design
correlational design
experimental design
control group
interventions
naturalistic observation
structured observation
validity
case study
representative sample
robust
double-blind study
reliability
replicability
interval scale
mean
normal distribution
growth curve
statistical significance
correlation

Suggested readings

COOK, R. and CAMPBELL, DONALD. *Quasi-experimentation*. Boston: Houghton-Mifflin, 1979. A discussion of research strategies on important social issues.

HOGARTH, ROBIN (Editor). *Question Framing and Response Consistency.* San Francisco: Jossey-Bass, 1982. A collection of articles by researchers in the behavioral and social sciences about research that uses interviews and questionnaires.

KIMMEL, ALLEN (Editor). *Ethics of Human Subject Research.* San Francisco: Jossey-Bass, 1981. An important and thorny issue—ethics of research—is discussed from a number of different perspectives.

REIS, HARRY (Editor). *Naturalistic Approaches to Studying Social Interaction.* San Francisco: Jossey-Bass, 1983. One in a series of paperback books on methodology in the social and behavioral sciences. This one describes six different observational methods.

RICHARZ, ANN. *Understanding Children through Observation.* St. Paul: West Publishing, 1980. A guide to practical observations of infants and children, with specific suggestions for research topics and recording forms.

SACKETT, GENE (Editor). *Observing Behavior* (Vols. 1 and 2). Baltimore: University Park Press, 1978. An advanced discussion of direct behavioral observation methods.

SILVERMAN, IRWIN (Editor). *Generalizing from Laboratory to Real Life.* San Francisco: Jossey-Bass, 1981. Another volume in the series on methodology in the social and behavioral sciences. This one discusses the scope of laboratory observations.

PART TWO

Foundations

chapter three

Heredity and environment

When I was pregnant with Sam, our first child, I believed that heredity contributed the basic physical elements—eye color, body type, the shape of nose and fingers, and the like. Environment, I thought, contributed the rest. Parents—the baby's environment—molded their malleable little creature to fit their patterns of life. It was parents who formed a baby's habits of sleeping and eating and who could make the baby an anxious, nervous wreck or a charming, outgoing pleasure. It was parents who created sleeping problems and feeding problems. I guess that I believed that heredity controlled development before birth and that then environment took over. Well, wake up and smell the coffee, Mother!

My pleasant fantasy lasted until about 60 seconds after Sam was born. He opened his eyes and looked around the delivery room with a calm, friendly, and curious expression on his face—not a malleable lump, but a fully formed little person.

In temperament, Sam has continued to be calm, friendly, curious, and from the first minute, exceptionally alert. This boy notices everything. He sleeps when *he* needs to. When he was two weeks old, he stayed awake for 11 hours straight with no signs of flagging. Sam nursed when *he* was hungry, not by any schedule we imposed—and I was going to be the mother who scheduled my baby's feedings so that the rest of life could go on undisturbed.

I know that we've had *some* influence on Sam's development. He's thriving because of the nourishment and love he gets from his family. But Sam certainly has cleared up my naïve assumptions about heredity and environment.

Nature–nurture interaction

As Sam taught his mother, we all are products of both heredity and environment. In times past, as we saw in Chapter 1, those who studied development usually thought that people developed according to the dictates of *either* environment—"nurture"—*or* heredity—"nature." But today, behavioral scientists know that nature and nurture, heredity and environment, both contribute continually and inseparably to a person's development. They are beginning to understand how very complex this process really is.

The **genotype** is all of a person's genetic inheritance. But the actual **expression** of the genotype as a person's visible characteristics and behavior is called the **phenotype**. The phenotype depends not only on the person's genetic inheritance but on all the environmental forces that affect the person from the moment of conception. Even if we knew a baby's genotype, we still would not be able to predict with certainty how the person would turn out. From the moment of conception, many possibilities are open to each unique human being.

Skin color is a good example of how genes and environment both contribute to development. Skin color depends on various genes—including those that produce melanin, a pigment that makes skin look dark or fair, and those that block the production of melanin, like the gene for albinism. The amount of sunlight affects how fully these genes are expressed. In regions with little sun, people born with genes for dark skin will have relatively light skin, because the genotype will not be fully expressed. But people with genes for dark skin who live in the tropics, where the sunlight is strong and continuous, will have dark skin. Their environment fosters the full expression of their skin color genotype.

Just as sunlight in the environment interacts with genes to determine skin color, stimulation in the environment interacts with genes to determine intelligence. For a child's genetically inherited intelligence to be expressed fully, the child needs loving interest and stimulating attention from adults. The amount of stimulation a child receives may actually alter the child's brain chemistry (Vandenberg, 1968). Thus the intelligence of a child who grows up in an enriched environment, with lots of toys and talk and teaching, is likely to be expressed more fully than the intelligence of a child who grows up in an impoverished environment.

In short, the expression of a person's genetic inheritance is influenced both by genotypic patterns and by factors in the environment. The genotype determines many of the person's basic characteristics and capacities, limits and potentials. The environment helps to determine the direction and speed of the person's development. Scientists have proposed several models to represent the complex relations between genotype and environment, between nature and nurture.

Models of gene–environment interaction

The **reaction range** model (Gottesman, 1963) shows how genotypes and phenotypes are related in different environments. In Figure 3.1, for example, Alice and Laura are twin sisters with identical genotypes who are adopted at birth into different families. Alice's adoptive parents dote on her, talk to and praise her, buy her books and puzzles, and send her to an excellent school. But Laura's adoptive parents ignore, punish, and thwart her, barely letting her attend school at all. Because the twins are reared in such different environments, their abilities also turn out to be different, despite their identical genotypes. At age 15, Alice scores 120 on an IQ test; Laura scores only 90. Thus the reaction range model shows how different phenotypes can develop from the same genotype in different environments. The model also illustrates the contrasting case: similar environments lead to more similar phenotypes

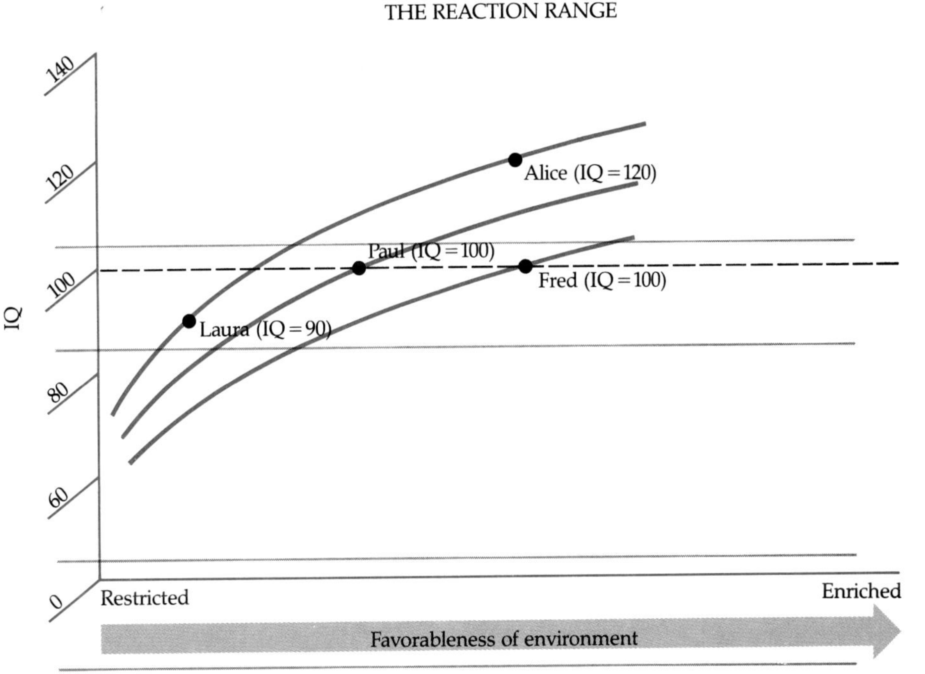

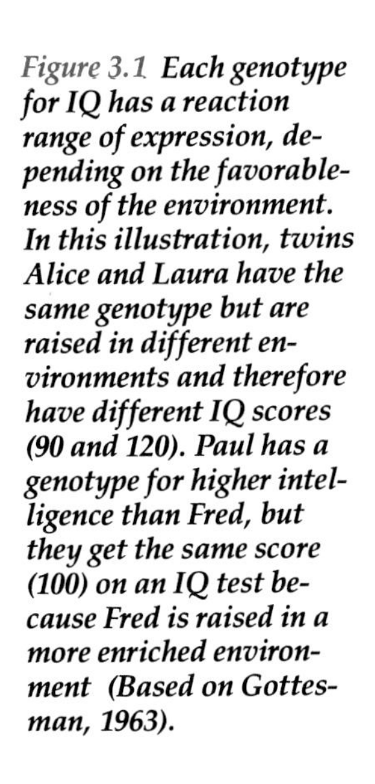
Figure 3.1 ***Each genotype for IQ has a reaction range of expression, depending on the favorableness of the environment. In this illustration, twins Alice and Laura have the same genotype but are raised in different environments and therefore have different IQ scores (90 and 120). Paul has a genotype for higher intelligence than Fred, but they get the same score (100) on an IQ test because Fred is raised in a more enriched environment (Based on Gottesman, 1963).***

among children with different genotypes. In Figure 3.1 Fred has a genotype for lower intelligence. Paul has one for higher intelligence. But Fred is given special attention, lessons, and schooling, whereas Paul is given just ordinary attention. Consequently, both turn out to have IQs of 100.

Not all human characteristics have the same reaction range as intelligence, however.

> Just after his third birthday, Sam visited the pediatrician for a routine checkup. As he chatted with Sam and Sam's father, the pediatrician charted Sam's height (41 inches) and weight (38 pounds). ''What are you feeding this child?'' the pediatrician joked. ''He's off the charts.''
>
> ''Iron filings,'' Sam's father replied. ''He eats more than I do.''
>
> ''You're tall yourself,'' the pediatrician said, ''so it stands to reason that Sam will be too.''

Since he was just a few weeks old, Sam's height has been in at least the 95th percentile for his age. Like his father and his grandfathers, Sam is going to be tall. Height is one characteristic that is under fairly strict genetic control. Other human characteristics—like intelligence and sociability—are more susceptible to variations in the environment and have larger reaction ranges (see Figure 3.2).

Characteristics under tight genetic control are said to be highly **''canalized.''** This term comes from a model of genetic-environmental interaction proposed by Conrad Waddington (1957) (see Figure 3.3). In this model, as a child develops, his or her genotype for a particular trait may be thought of as a ball rolling down the slopes and valleys of a three-dimensional landscape. If the valley floor is broad and its slopes are shallow, the ball can be thrown off course—derailed from the normal developmental path—by environmental conditions. The genotype for the trait is not well protected from environmental forces. But if the valley is deep, with a narrow floor and steep, protective walls, the ''winds'' of the environment are less likely to blow the ball off course. Even if the ball is displaced by a strong wind, it is likely to regain its course. This pathway along the deep valley is highly ''canalized'' and well protected from environmental forces. Even if some childhood illness interferes with Sam's growth for a time, for example, he is likely to make up the lost height when he recovers. But his youthful gregariousness—what his mother calls his ''animal high spirits''—may well be moderated under the demands of teachers and others who expect him to concentrate on studying and to blend with his classmates.

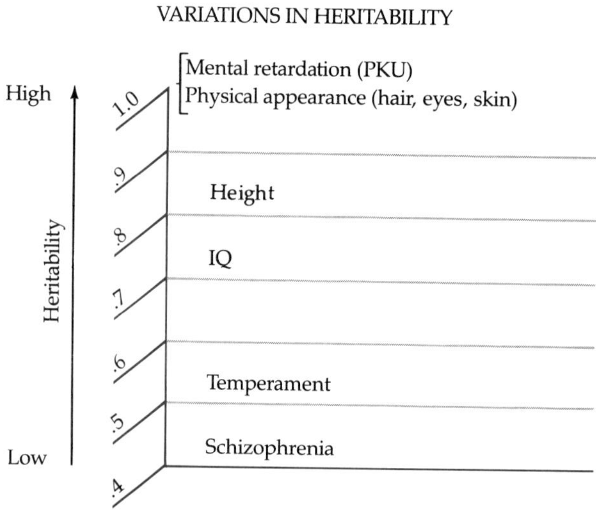

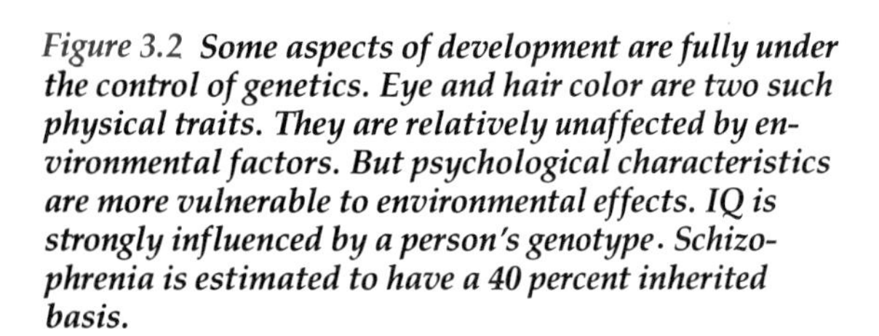

Figure 3.2 ***Some aspects of development are fully under the control of genetics. Eye and hair color are two such physical traits. They are relatively unaffected by environmental factors. But psychological characteristics are more vulnerable to environmental effects. IQ is strongly influenced by a person's genotype. Schizophrenia is estimated to have a 40 percent inherited basis.***

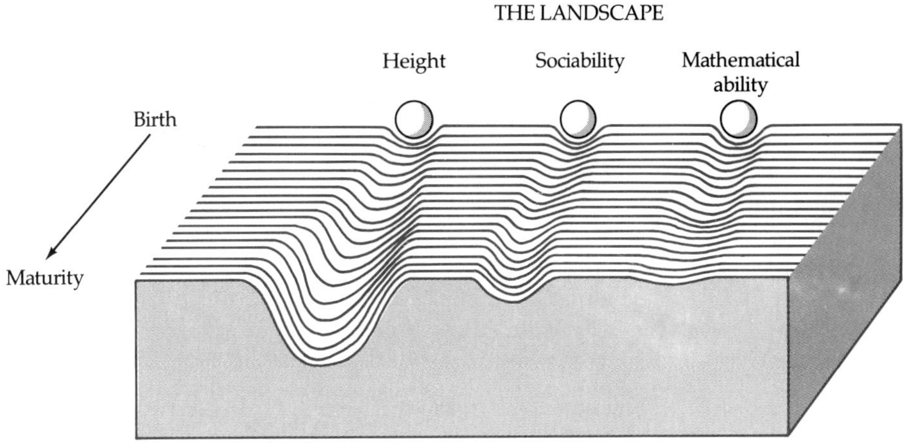

THE LANDSCAPE

Height Sociability Mathematical ability

Birth

Maturity

Figure 3.3 ***This adaptation of Waddington's (1962) landscape model of interaction between heredity and environment shows the hypothetical development of three different kinds of characteristics. Physical appearance, here height, is highly canalized—under strict genetic control. Social behavior, here sociability, is much less canalized and more susceptible to environmental influences. Intellectual development, here mathematical ability, is highly canalized in the early years but less so later.***

To add to the complexity of interaction between genes and environments, not only are there differences in canalization for different traits, but the degree of canalization for the same trait may change as the child gets older. In general, during a child's first few years, it seems, the valleys are deep and narrow, and development is highly canalized. Human infants follow the same path of development as they have for hundreds of generations. But later on, the degree of canalization of these characteristics diminishes and differences in environment grow increasingly influential.

> On a beautiful Sunday morning, 2½-year-old Sam and his parents joined friends and their 8-year-old son at the children's section of the Bronx Zoo. Timmy, the 8-year-old, was fascinated by the animals. He knew about snakes and owls from school books. He peppered his father with questions about the elephants. Timmy's teacher later reported that the zoo had made quite an impression on Timmy. He had done almost nothing but eat and sleep snakes ever since. At home, too, Timmy drew snakes; asked for books about snakes; and became an instant snake expert. But the zoo had made very little impression on Sam. The animals could have danced on their heads, and Sam wouldn't have noticed. The only thing Sam was interested in was Timmy, his idol.

The zoo trip may well have been formative for Timmy, influencing the course of his mental development. For Sam, the zoo was just another setting for watching Timmy. Children experience different environments differently at different ages.

Finally, there is one more wrinkle of complexity in the interaction of genes and environment. Not only does the environment affect the expression of a person's genotype, but the person's genotype affects the environment. Sandra Scarr and Kathleen McCartney (1983) proposed that in essence people make their own environments. They do so in two ways. First, their genotype evokes a particular response from the environment. Active, happy babies, for example, evoke more social reactions from their caregivers than expressionless babies. A bright, curious person evokes more interesting interactions than a dull, withdrawn one. Second, a person actively seeks the environment that he or she finds stimulating and compatible. The child who loves music learns to play an instrument. The girl who is gifted at languages studies anthropology or journalism. The boy with high visual-spatial ability becomes a pilot or a mechanic. This kind of active "niche picking" increases as children get older. Young children, who have less control over their environments, do less "niche picking" than adolescents. Adolescents make important choices on their own, choices that are in keeping with their own talents and inclinations, their genotypes.

It is clear that to understand how people develop, we need to examine both genetic and environmental influences and their interaction. In this chapter, we start at the beginning of that process, with the story of genetics.

research focus

Mothers' and children's IQs

"How do genes and the environment contribute to children's intellectual development?" was the question asked by a team of investigators from the Frank Porter Graham Center at the University of North Carolina (Yeates, MacPhee, Campbell, and Ramey, 1983). They followed 112 children from birth to age 4 and tested their IQs at intervals. They also observed the children's home environments to see the kinds of stimulation available there, and they gave IQ tests to the children's mothers to try to estimate the children's inherited intelligence levels.

Their results showed that as the children grew older, their IQ scores were increasingly well predicted by the combination of their mothers' IQ scores and the ratings of their home environments. As 2-year-olds, 11 percent of the variation in the children's IQs could be predicted by this combination of genetics and environment. As 4-year-olds, 29 percent could be predicted. But the relative importance of genetics and environment changed. For 2-year-olds, the mother's IQ was more highly related to the infant's IQ than the home environment was. Knowing how stimulating the home was added nothing to the researchers' ability to predict the infant's intelligence. But for 4-year-olds, the stimulation of the home environment was the main predictor. The mother's IQ added nothing to the prediction.

This study supports the view that after infancy, as the child begins to explore the world, a shift occurs in the relative importance of inheritance and environment for intellectual development. By the time the child is 4 years old, the study results suggest, home environment is as important as genotype in predicting IQ.

As children leave infancy and begin to explore the world, the stimulation in their environment becomes increasingly important in the development of their intelligence. By the time children are 4 years old, their environment weighs equally with their genotype in predicting their IQ scores.

Genetic building blocks

DNA and RNA

Each human cell contains 46 **chromosomes** in its nucleus. Each long and thin chromosome is made up of over 1000 **genes**, strung out like a chain. The genes are made of molecules of **deoxyribonucleic acid (DNA)**. The long DNA molecule is made of two strands of alternating sugar and phosphate molecules attached to nitrogen bases—adenine, cytosine, guanine, and thymine—which are twisted around each other and connected by rungs in the shape of a laddered spiral. During normal cell division, or **mitosis**, the DNA, which is packed tightly into the nucleus of the cell, "unzips," breaking its rungs and forming two separate strands. The nitrogen bases in each of these strands then pick up complementary nitrogen bases that are loose in the cell nucleus, making a new laddered spiral identical to the first. Thymine picks up adenine, adenine picks up thymine, guanine picks up cytosine, and cytosine picks up guanine. In this fashion all 46 chromosomes of the cell reproduce themselves. Then the two sets of 46 chromosomes move to different sides of the cell, a barrier forms between them, and the cell divides into two identical daughter cells.

The DNA in each human cell contains the chemical instructions for building a complete human being. It provides the instructions for creating a new member of the human species with basic human characteristics like limbs, head, and hair rather than fins or feathers; and it provides the instructions for creating a new individual with his or her own unique combination of characteristics like eye color, intelligence, height, and blood type. The particular order in which the molecules making up the DNA are arranged is the genetic code for these individual characteristics. A person's genotype, thus, is like a blueprint. To get from blueprint to human being, information in the DNA molecules is transcribed into molecules of **ribonucleic acid (RNA)**.These RNA molecules take the information from the DNA into the cytoplasm of the cell, which contains the raw material for making protein. Out of the proteins made according to the instructions from the RNA, the many functioning parts and capacities of each human being develop.

Meiosis

For a new human being to be created, a special kind of cell division, **meiosis**, must occur. Germ cells in the reproductive organs, the **testes** and **ovaries**, divide by meiosis to form **sperm** cells and **ova**. In this process, the chromosomes of the germ cells line up side by side in **homologous**, or parallel, pairs, one chromosome originally being from the person's mother and one from the person's father. While they are lined up, each chromosome doubles. At this stage, any of the four strands of a homologous pair may break. If two strands from different chromosomes break at the same point, **crossover** may take place, and pieces of the strands then join in a new combination. After the genetic material is exchanged, the paired and doubled chromosomes separate and separate again, as their cells divide and divide again, leaving sperm and ovum with only half the number of chromosomes that the other cells of the body contain. At **conception**, when the 23 chromosomes in the man's sperm cell join the 23 chromosomes in the woman's ovum, the number of chromosomes in the new cell, or **zygote**, is 46.

Chromosomes

Of the 46 chromosomes in each zygote, 44 (22 pairs) are **autosomes**, and it is these that carry most of the genetic code for the development of intelligence, height, eye color, and the like. The other pair of chromosomes determines whether the new human being is male or female. If the zygote contains two **X chromosomes**, a female develops. If the zygote contains one X and one **Y chromosome**, a male develops. Every zygote inherits an X chromosome from the mother, because all of her body's cells contain two X chromosomes, and so each of her ova contains one X chromosome. Because men's body cells contain an X and a Y chromosome, the father's sperm cells each contain *either* an X *or* a Y chromosome. Whether the infant turns out to be a boy or a girl therefore is determined by the sex chromosome in the sperm.

Inheritance of physical characteristics

Sam has his mother's head—her dark skin, white teeth, brown eyes, and shiny hair—and his father's body—his height, lankiness, and metabolism. "Every time he turns," his mother says, "I see a glimmer of myself, my sister, my parents, and my grandparents. The curve of his cheeks reminds me of Sam's father and grandfather. The shape of his eyes reminds his grandmother of her own mother's eyes. It's easy to lose yourself in the looking glass of your child."

Natalie is quiet and dark-haired. Plump, short, and slow-moving, like her father's mother, Natalie rarely runs at full tilt or jumps from high places. She has her father's dark, wavy hair and her mother's clear white skin. Everyone who sees her remarks on how much Natalie looks like her father when he was her age. In fact, when she saw a snapshot of her father as a toddler, she insisted that it was a picture of herself. Like both of her parents, Natalie has unusually beautiful light blue eyes. When her mother was pregnant, she knew one thing for sure about her developing child: he or she would have light blue eyes.

Although body weight is an inherited physical characteristic, it is influenced by environmental factors such as diet and exercise.

Hair and skin color, height and weight, eye color and shape are among the physical characteristics coded in genes located on the autosomal chromosomes. The probability of a person's inheriting curly hair or of growing to over six feet, for example, can be predicted with some certainty from knowing the hair type or height of the person's parents and grandparents. Natalie's mother was right in predicting Natalie's blue eyes. All the children born to her and her husband would inherit their striking blue eyes. Blue eyes are inherited through a simple pattern of genetic inheritance. Blue eye color is a **recessive** characteristic: A person has blue eyes only when both chromosomes in the pair carrying the gene for eye color carry the blue-eye gene. Brown eye color is a **dominant** characteristic: A person with brown eyes needs only one brown-eye gene in the pair. Thus, if both parents have blue eyes, they can contribute only blue-eye genes to their child, and that child will have blue eyes. If both parents have brown eyes, the chance that their child will have blue eyes cannot exceed 25 percent. This simple pattern of inheritance is followed not only for eye color, but for various other physical traits as well. For example, curly hair is dominant over straight hair, prominent noses over smaller noses, thick lips over thin, cleft chins over smooth, and shortness over tallness. But most physical characteristics follow more complicated patterns that involve more than one gene. They are **polygenic**. Predicting polygenic characteristics in a new human being, even knowing the parents' characteristics, is difficult.

Although these physical characteristics are inherited, extremes of environment—diet, health, living conditions, and the like—can affect their expression. As we have seen already, skin color is affected by the amount of available sunlight. Severe malnutrition gives African children thin reddish hair, even though they have inherited genes for thick dark hair. Body weight, especially in females, although an inherited characteristic, is influenced by diet. Second-generation Japanese-Americans, who eat the rich diet available in this country, are, on the average, taller than their parents who grew up in Japan.

Because of the complexities of inheritance patterns, the infinity of possible genetic combinations, and the fact that phenotypes are affected by the environment, we will never be able to predict with complete certainty before birth what any particular child will look like. True, we may know from observing the parents that a child is likely to be short and to have brown eyes, fair skin, and wavy hair. We may know from tests done prenatally that a fetus is male or female or carrying genes or chromosomes for certain specific disorders. But we still are far from being able to read the whole genetic code and even farther from knowing the extent of environmental effects. Genetic research is one of the most exciting fields in science today, and new discoveries about how genes work roll off the presses every month. But the more we find out, it seems, the more complex the picture becomes. Geneticists have found out, for example, that genes are not always static in their regulation of behavior. They may jump or split, creating complex changes in systems of genes. They may turn on and off at different points in development. Even though great strides are being made in the understanding of genetic transmission, some mystery in an individual's physical development will undoubtedly always remain.

Inheritance of behavioral traits

Behavioral characteristics are even more susceptible to environmental influences than physical characteristics are. They are also more difficult to dissect so neatly into identifiable pieces like noses, lips, and eyes. What is more, behavioral characteristics all seem to be polygenic. Geneticists have not yet identified a single gene that produces variation in a single behavioral trait. They do not even know of a complex behavior that is entirely under genetic influence. Only rarely can more than half of the variability of an observed behavioral trait be traced to genetic influence (Plomin, 1983). Thus, although the principles of genetics are extremely important to students of human development, it will be some time before we can specify how human chemistry affects complex human behavior. We may observe that a child like Natalie is calm, shy, and musical, but we are far from being able to identify the genetic bases for these traits. What we do know about the inheritance of behavioral characteristics we have learned through indirect methods like studying twins and family trees.

Methods of study

Twin studies

Sir Francis Galton, Charles Darwin's cousin, led scientists to one ingenious method for studying the heritability of behavioral characteristics. Galton realized that identical twins, who by definition have the same genotype, would give scientists a window through which to observe the contributions of environment and heredity. Since he made that suggestion, twin studies have been an important source of information about the relative contributions of environment and heredity to human behavior.

A pair of twins is either "identical," technically called **monozygotic (MZ)**—single-egg—twins or "fraternal," **dizygotic (DZ)**—double-egg—twins. Identical twins develop from one fertilized egg by mitosis and therefore have exactly the same genotype. Early in development, the contents of the single

Monozygotic twins (front), having identical genotypes, are identical in appearance and abilities. Dizygotic twins (back) can be as unalike as any two brothers or sisters.

egg divide into two embryos. In the uterus, the two embryos usually are enclosed within a single fetal membrane, although sometimes the identical embryos migrate separately into the uterus and attach at different points. Fraternal twins develop when two ova are released and fertilized by separate sperm at the same moment of conception. These two embryos have genotypes no more alike than those of any two brothers and sisters. They may look alike or not, be of the same sex or not.

Psychologists study twins by comparing the similarity, or **concordance**, of identical twins with the similarity of fraternal twins. Differences between identical twins can be laid directly to differences in environment. Differences between fraternal twins can be laid to differences in both environment and heredity. Comparing similarities between pairs of fraternal and identical twins suggests whether a characteristic is inherited.

Unfortunately, twin studies have limitations. They do not allow researchers to separate the effects of environment totally from the effects of heredity. For example, identical twins look so much alike that they may be treated more alike than fraternal twins are.

> My husband still has trouble telling our 13-year-old identical twin sons apart. Like most everybody else, he treats them like a single creature called, "You boys," as in "Will you boys please turn down that record player?"

Identical twins may also spend more time together, making their environments more nearly alike than those of fraternal twins. To overcome these limitations, psychologists have tried to find identical twins who have grown up in different families. The assumption is that such twins' similarities arise from heredity and their differences from factors in the environments in which they are raised. In fact, even before birth, the twins' environments may differ. They may have different access to their mother's blood supply; or one of them may suffer oxygen deprivation during delivery. Some twins are mirror images of each other. The consequences of this may be minor, such as differences in handedness, or they may be major, such as differences in dominant brain hemispheres with associated differences in general cognitive or personality styles. But even if these cases could be eliminated from the studies, the biggest problem with this approach is finding identical twins who have been raised apart. Fewer than 200 sets of identical twins raised in different families have been found and studied in the past half century. Of these, only three sets were reared apart in completely different environments and had no contact with each other from birth (Farber, 1981).

Family trees

Researchers also study the inheritance of characteristics by analyzing their presence or absence among members of entire families. One famous family-tree study focused on a family named by the investigator the ''Kallikaks'' (Goddard, 1912). ''Kallikak'' is a combination of the Greek words for ''good,'' *kalos*, and for ''bad,'' *kakos*. Martin Kallikak, a soldier in the Revolutionary War, so it was claimed, fathered an illegitimate son by a retarded tavern maid. Of the five generations and 480 descendants of the maid, many were drunks, prostitutes, thieves, felons, and ne'er-do-wells. Some 143 descendants were considered retarded. But Martin Kallikak also married a wife of ''good stock.'' Of the 496 descendants of this marriage, most were landowners, merchants, and teachers. One was reported to be ''sexually loose,'' and two had ''appetite for strong drink.'' Although this example may be more interesting than scientific, carefully done family-tree studies can yield important information about heredity. The open question, of course, is to what extent differences in rearing and environment interact with the differences in inheritance to produce lines of descendants with different characteristics.

In another type of family-tree study, investigators trace the relatives of individuals with a particular trait or disorder. If the incidence of the trait among first-degree relatives—parents, children, brothers and sisters—is higher than among second-degree relatives—aunts and uncles, nieces and nephews, cousins—the trait can be presumed to have at least some genetic basis. The limitation of this method is that the environments of first degree relatives also are likely to be more similar than the environments of second-degree relatives. Thus, although this method is useful for tracing the inheritance of traits *not* affected by the environment, it does not separate the contributions of environment and heredity for traits that *are* affected by the environment.

Adopted children

Another method of studying heredity involves adopted children. In one kind of adoption study, researchers compare the behavioral characteristics of adopted children with those of their biological and their adoptive parents. If the adopted children resemble their adoptive parents more than they resemble their biological parents, the resemblance is presumed to stem from environmental factors. Comparisons also can be made between the adopted children and their biological siblings who are raised by their natural parents. In one old study, Marie Skodak and Harold Skeels (1949) studied children of retarded mothers, who had stayed with their mothers or who had been adopted by parents with normal IQs. The average IQ of the adopted children was higher than the average IQ of the children who had stayed with their retarded mothers. This finding suggested that the home environment affects the development of children's intelligence. But the average IQ of the adopted children of retarded mothers was lower than the average IQ of adopted children whose mothers were not retarded. This finding suggested that genetic endowment also plays a role. This study clearly showed that both heredity and environment contribute to development.

More recently, efforts to study the contributions of heredity and environment in adopted and nonadopted children have gotten more complex. In the Colorado Adoption Project (Plomin, Loehlin, and DeFries, 1985), for example, researchers have compared the correlations between development and environmental stimulation for children in adoptive families with those of children in biological families. The correlations usually are higher in the biological families, leading the researchers to suggest that genetic factors not only contribute directly to children's development, but also mediate the effects of the environment on development.

Even though it has not been possible to trace behavior patterns directly to specific genes, these indirect methods—twin studies, family-tree studies, and adoption studies—have allowed developmental psychologists to determine whether heredity contributes to the development of particular abilities, behavior patterns, and behavior problems. In the future, researchers may use these methods to determine how and to what extent heredity actually contributes.

Experimental intervention

Twin studies, family trees, and adoption are methods used by behavior geneticists to study genetic bases of behavior. An alternative approach to studying nature-nurture interaction is to study the effect of the environment on development, by trying to change people's development experimentally. This is the method of experimental developmental psychology. The researcher assigns people to an experimental treatment—extra stimulation, increased discipline, more toys—or to a control condition in which they receive no extraordinary treatment. Then he or she assesses the effects of the experimental intervention. This method has been used most frequently by researchers to try to change people's intelligence levels. Unfortunately, this method, like the others, has limitations. There are both practical and ethical problems in intervening in people's lives in ways that are as extreme as one needs to test fully the limits of environmental effects on development.

Personality and temperament

Personality is the typical and consistent way in which a person acts and approaches the world. Sam is friendly, cheerful, and boisterous. His father calls him "sensitive but low-strung." Natalie is quiet, passive, and shy. Other children are aggressive, greedy, blundering, high-strung, demonstrative, tentative, or anxious. **Temperament** is an infant's natural disposition or style of activity—the infant's "personality." Sam's mother mentioned that in the hospital delivery room, Sam was already calm and curious. Natalie's parents found her disposed to be quiet and passive right from the start. In a longitudinal study of children's temperaments from infancy to adolescence, Alexander Thomas, Stella Chess, and Herbert Birch (1968) found that children's temperaments appeared early and seemed to be quite stable throughout the children's early years. They surmised from this that temperament has some genetic basis. One aspect of temperament that seems to be inherited is activity level and the behavioral characteristics associated with it—being fidgety, impatient, "on the go" or quiet, sedentary, and still. Although some people have criticized this study (we discuss these criticisms in Chapter 7), its findings are supported by evidence from twin studies, which also suggest that activity level has a genetic component: activity level is more concordant in identical than in fraternal twins.

Sociability is another aspect of temperament and personality that appears to be partly inherited. Longitudinal studies of children from birth to maturity and studies of twins and adopted children suggest that people inherit the inclination to be friendly and outgoing or shy and withdrawn. Sociable Sam shouts hello and makes friends with the mail carrier, the neighbors, the neighbors' dogs, and every soul who walks, jogs, or runs past his front door. Natalie rarely speaks above a whisper and smiles a shy hello even to people she knows well.

Also inherited may be tendencies to be empathic, nurturant, and altruistic, adaptable, dominant, assertive and self-confident; conforming; or depressed (Daniels and Plomin, 1985; Goldsmith, 1983; Matheny and Dolan, 1975; Rushton, 1984; Vandenberg, 1968). In her comprehensive review of studies of identical twins raised apart, Susan Farber (1981) concluded that the evidence of similarities was most convincing for the following personality traits: characteristic mood, emotional expressiveness, pattern of anxiety, and styles of talking, laughing, and moving. Intriguing case studies of the very few identical twins and triplets who have been reared apart since infancy hint that inheritance may govern even more specific traits.

This does not mean, however, that the die of personality is cast with the genes. Children with identical temperaments do not always develop in identical ways. Identical twins do not always act alike. What happens to the dispositions found in infancy and the personality styles found in childhood depends to a large extent on children's interactions with their parents and other people. For example, two of the infants in the longitudinal study by Thomas, Chess, and Birch were twins adopted into different families that were similar socially and financially. As newborns, the infant girls had similar temperaments. They slept on very irregular schedules, and they cried a lot when they weren't sleeping and were hard to console. These were temperamentally cranky, or ''difficult,'' infants. The parents in one of the adoptive families pleasantly checked their daughter for cold or wetness each night when she woke up and then, if nothing was wrong, left her alone. She soon stopped waking up at night. The parents in the other family responded to their daughter's nightly crying by picking her up, fussing over her, and feeding her. When she was 27 months old, this child had become so difficult to manage that the parents sought counseling. Two different families' styles had fostered two patterns of development.

Intelligence

There is also evidence from twin, family-tree, and adoption studies that intelligence and the abilities making up intelligence, such as comprehension of words and fluency of speech, mathematical and spatial abilities, reasoning, and memory, are to some extent genetically influenced. After reviewing many studies, conducted over the course of half a century and comprising 30,000 correlations, Loise Erlenmeyer-Kimling and Lissy Jarvik (1963) concluded that the closer the family relationship, the more similar people's IQs (see Figure 3.4). In their review, the concordance between IQ scores averaged 0.49 for brothers and sisters, 0.53 for fraternal twins raised together, and 0.87 for identical twins raised together—making a strong case for the contribution of genetics to IQ.

But, as we have already suggested, environment also affects intelligence. When identical twins were raised apart, the correlation between their IQ scores was only 0.75 (or even lower, according to an analysis of these cases by Farber [1981]). To what extent does environment affect the expression of intelligence? Overall intelligence seems to be rather deeply canalized. The difference in IQ scores that can be produced by differences in normal environment is only about 25 IQ points, by one estimate (Scarr-Salapatek, 1975).

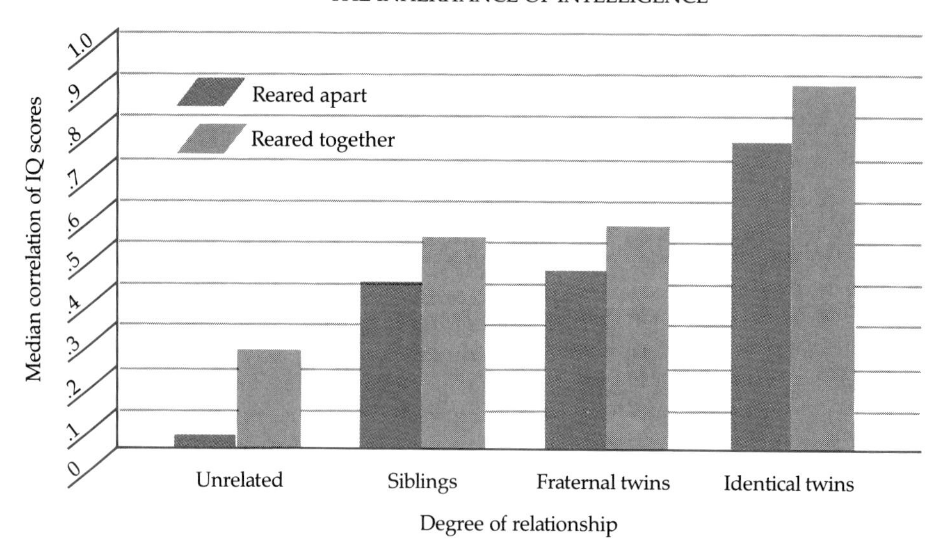

Figure 3.4 ***The contribution of heredity to intelligence is shown by comparing the correlations of IQ scores for unrelated individuals, siblings, fraternal (DZ) twins, and identical (MZ) twins, raised together (colored bars). The contribution of environment is suggested by the difference between these correlations and those of children raised in different homes (white bars) (Erlenmeyer-Kimling and Jarvik, 1963).***

focus *Twins and triplets reared apart*

In 1981, Robert Shafran went to college in New York State, and strange things began happening to him. Other students called, ''Hi, Eddy,'' a young woman whom he didn't know kissed him on the lips, and a fellow insisted that he was a dead ringer for Eddy Garland. His curiosity aroused, Robert looked at pictures of Eddy. He was astonished at how similar they looked. Robert got in touch with Eddy, and the facts of their births emerged. The monozygotic brothers were born 27 minutes apart and had been adopted when they were 6 months old. They found that their IQs were the same, that they liked the same music, food, and sports. Pictures of the reunited brothers appeared in newspapers and happened to be spotted by a family named Kellman. David Kellman, it turned out, was their third identical brother. As his adoptive mother said of the triplets, ''They talk the same. They laugh the same. They hold their cigarettes the same—it's uncanny'' (*New York Times*, September 19 and 23, 1981).

In another remarkable case, twins George and Millan were separated immediately after birth and adopted into different families. They, too, met for the first time when they were 19 years old. When they met, they looked so much alike that neighbors mistook one for the other. Both were handsome; both were athletic. Both had won boxing championships, and both were artistically inclined. Millan was musical; George worked as a commercial artist. They even had cavities in the same teeth. On personality tests, their scores were nearly identical. Within a few months of each other, George and Millan developed a rare crippling spine disease. In both twins, the disease responded to treatment. Only on social views, consistent with their different adoptive backgrounds, did the twins differ.

In yet another case, identical twin Oskar Stohr stayed in Europe with his mother while his brother, Jack Yule, went to Trinidad with their father. The boys were raised in different religions, spoke different languages, and had different kinds of schooling. But when they met each other at age 47, they

Inheritance of disorders

Abnormal numbers of chromosomes and defect-producing genes produce various physical disorders that afflict individuals across the lifespan.

Physical defects

When something goes wrong during meiosis and the zygote ends up with more or fewer than 23 pairs of chromosomes, problems may ensue. When chromosome pair 21 does not separate, and the zygote has 47 instead of 46 chromosomes, the individual suffers from **Down's syndrome**. (Chromosome 21 also has been implicated in certain cases of dementia among older adults.) These individuals have retarded mental and motor development, broad and flat faces, eyes slanted upward and a prolonged fold along the inner corner of the eye, straight and thin hair, short feet and hands, and heart defects.

As women get older, they face increasingly greater risks of bearing a child with Down's syndrome. Women under 30 bear only one Down's syndrome child in 3000, but women between 30 and 34 bear 1 in 600; women between 35 and 39, 1 in 280; women 40 to 44, 1 in 80; and women over 44 bear Down's syndrome babies once in every 40 births (Apgar and Beck, 1974). One

both spoke quickly, angered easily, and shared the same quirky habits, like flushing the toilet before they used it and fiddling with other people's rubber bands (Chen, 1979). Heredity, it seems, may extend even to the most minute of human characteristics.

These informal reports are only suggestive, and there are other possible explanations for the coincidences observed. For one thing, when many characteristics are examined, some are likely to be identical in a pair of twins just by chance. For another, the twins' environments might have been more alike than reporters realized. We need more research before we can state the extent and scope of genetic influence.

When these triplets, who had been separated in infancy, discovered one another as young men, they found that they shared more than physical appearance. They smiled and talked in the same way, had similar food preferences, and listened to the same kind of music. They had identical IQs, smoked too much, and held their cigarettes the same way.

lifespan focus

Genetic continuity of intelligence

Recent findings from the Colorado Adoption Project suggest that genetic influences on intelligence in infancy and early childhood are highly correlated with genetic influences on intelligence in adulthood. The same genetic factors that cause individual differences in infants' intelligence may cause individual differences in adults' intelligence. In other words, the individual differences that appear in infants' and adults' IQ test scores probably are influenced by many of the same genes. Here is a good example of the ties that bind development over the life span. In the Colorado Adoption Project (DeFries, Plomin, and LaBuda, 1987), researchers obtain data from adopted children, their biological and adoptive parents, and from members of nonadoptive families. One- and 2-year-old children are tested at home on the Bayley Scales of Infant Development; 3- and 4-year-olds are given the Stanford–Binet Intelligence Scale at home. Seven-year-olds and adolescents are given an intelligence test in a laboratory. Adults take tests of cognitive ability that tap their verbal, spatial, and perceptual abilities and memory.

The correlations between the test scores of adopted children and their biological and adoptive parents show a pattern of genetic continuity that increases as the children get older. This finding of genetic continuity is quite remarkable because the tests in infancy, childhood, and adulthood tap such different abilities. In infancy, tests measure abilities like the speed of putting pegs in a pegboard; in childhood, the number of words that a child understands; in adulthood, abstract reasoning. The finding does not mean that the actual physiological and psychological processes that contribute to intelligence at different ages are necessarily the same. But the networks of brain cells established in childhood may affect cognition in adulthood. Another possibility is that structural differences in the brain cells themselves may develop in infancy and contribute to performance on both infant and adult tests. Still another possibility is that genes that affect infant abilities may be biologically linked with genes that affect adult abilities. One important direction for future research is to identify the processes in childhood and adulthood through which genetic continuity occurs.

Down's syndrome results when the zygote has an extra chromosome 21. Children with this disorder have retarded mental and motor development and distinctive facial features.

possible explanation for this increasing risk is based on the fact that women are born with all the ova they will ever have (although in immature form). Because women's germ cells are exposed to environmental hazards, viruses, and even emotional stress from the time of their birth, this long period of exposure increases the risk that chromosome pairs will not separate during meiosis. But in up to one-third of the cases of Down's syndrome, the man and not the woman contributes the extra chromosome (Magenis, Overton, Chamberlin, Brady, and Louvrien, 1977).

Other physical disorders arise when harmful genes are passed on from parents to their children. **Sickle-cell anemia** is a painful and sometimes life-threatening disease found among American blacks. About 10 percent of American blacks are carriers of the sickle-cell gene. That is, they have one sickle-cell gene and one normal gene for blood cell shape. They inherited the sickle-cell trait from their African ancestors. In Africa, where malaria was a common and deadly disease, people whose blood cells formed a sickle shape were likelier than others to survive an attack of malaria. The sickle-cell trait thus was adaptive, and people who had a sickle-cell gene were likely to survive to the age of sexual maturity and thus to pass on the gene to their children. But transported to North America, where there is no malaria, the sickle-cell trait is maladaptive. In people with the sickle-cell trait, when the level of oxygen is low, the red blood cells become sickle-shaped and clump together, blocking the flow of blood through the person's small blood vessels. Depending on the location and extent of the blockage, the disease causes circulatory problems, pain, or death. People with the sickle-cell trait must be protected from conditions that reduce the oxygen in their blood, such as illness, strenuous exercise, and high altitude.

Phenylketonuria (PKU) is a metabolic disorder that is also genetically based. Infants who receive the genes for this recessive trait from both parents cannot make an enzyme necessary for digestion (phenylalanine hydroxylase). They therefore cannot digest many foods, including milk. The substance then accumulates and prevents normal brain development. The infants are irritable and hyperactive. One-third have seizures. As children, they are mentally retarded. But this disease now can be controlled. Once a newborn has consumed milk for about a week, an excess of the undigested phenylalanine can be detected in a blood test. Infants for whom the disorder is detected at this early age are given milk substitutes. Later they eat a special diet that contains only the smallest necessary amount of phenylalanine. They stay on this restricted diet until brain development is complete, at the end of adolescence.

Although this diet limits the abnormalities that PKU causes, it cannot prevent all ill effects. Infants and children with PKU are prone to emotional and learning disorders (Berman and Ford, 1970; Dodson, Kushida, Williamson, and Friedman, 1976; Steinhausen, 1974). Parents' reactions to their children with PKU may contribute to the development of these problems (Kopp and Parmelee, 1979). Anxious over children whose mental and physical development is threatened if the prescribed diet is not suitable or not followed, parents may be overprotective or rejecting.

A second kind of congenital problem involves disorders that come from defect-producing genes located on the sex chromosomes. **Sex-linked** defect-producing genes usually are recessive and carried on the X chromosome. Because males have only one X chromosome, they are more susceptible to these disorders. Whenever their single X chromosome contains the disorder, it will be expressed, because there is no corresponding normal dominant gene from the Y chromosome. For a female to exhibit a sex-linked disorder, she must inherit the defective gene from *both* parents. If a female inherits a harmful recessive gene from just one parent, she has a second X chromosome with a parallel gene to counteract the expression of the recessive gene. Hemophilia (lack of blood-clotting factors), color blindness, and baldness are sex-linked defects that appear almost exclusively in males. Girls may carry one gene for hemophilia without being bleeders, but boys who inherit the gene for hemophilia will have the disease. Chances are that half of the daughters of mothers who carry the gene for hemophilia will themselves be carriers, but half of the sons will be hemophilic.

These are just a few examples of genetically transmitted physical disorders. Virtually all diseases (cancer, arthritis, diabetes, multiple sclerosis) seem to have *some* genetic component, but the links are less clear than in the disorders we have discussed.

Behavior disorders

Certain of children's behavior problems, such as sleepwalking, car sickness, bed-wetting, constipation, and nail-biting, also seem to have a genetic component, for the concordances are higher among identical than fraternal twins (Bakwin, 1970, 1971a, 1971b, 1971c, 1971d). More seriously, **schizophrenia**, a mental illness characterized by jumbled thoughts and speech, hallucinations, and delusions, appears to have a genetic base. Children of schizophrenic parents are 15 times likelier to have schizophrenia as adults than are adults in the population at large. Even if they are adopted into normal families, these children are likely to become schizophrenic (Rosenthal, 1970). Identical twins are concordant for schizophrenia four times as often as fraternal twins. Relatively few identical twin siblings of schizophrenics are found to be normal. Most of the twins who do not have schizophrenia have symptoms that border on it. These ''schizoid'' people are suspicious and rigid of thought, suffer panic attacks, and grow afraid in the face of ordinary challenges.

Another disorder that has genetic roots is **dyslexia**, a difficulty in reading and spelling that affects people of normal intelligence who have no emotional disturbances or known neurological or sensory handicaps. Some form of dyslexia is estimated to affect as many as 10 percent of school-age children. Studies of twins and other relatives have shown that some forms of dyslexia have a strong hereditary basis (Decker and DeFries, 1981; DeFries and Baker, 1983). Among identical twins, the concordance rate for dyslexic reading disability is over 90 percent. Among fraternal twins, the concordance is about 30 percent. Siblings and parents of dyslexics are also poorer readers than relatives of unaffected children. In families of dyslexic children, over 60 percent of the stability in the children's reading scores over a period of years can be attributed to heredity, compared to less than 1 percent of the stability of normal children's reading scores.

Testing for defects

Diabetes, cystic fibrosis, clubfoot, cleft palate, Down's syndrome, spina bifida, sickle-cell anemia, hemophilia, PKU—all these are genetic abnormalities that can be detected before a baby is born. Some can be identified through tests of the mother's blood. Others can be identified through tests of the amniotic fluid surrounding the developing fetus, in a procedure called **amniocentesis**. During amniocentesis, a doctor first locates the fetus with **ultrasound**, in which high-frequency sound waves bounce off the contours of the fetus and form a detailed picture on a video screen. The doctor then draws amniotic fluid through a needle inserted into the mother's abdomen and uterus, without touching the fetus. The amniotic fluid is cultured for a month and the cells are then examined for the presence of any genetic abnormalities. The problem for prospective parents is what to do if, at this point, the tests reveal the presence of a disorder. Unfortunately, amniocentesis cannot be performed until the 14th or 15th week of pregnancy, because only then is there enough amniotic fluid to sample, and the process then takes another four weeks to complete. Parents who learn that the fetus does have a genetic defect must decide whether to end a five-month pregnancy, when the fetus is already perceptibly moving in the uterus, and the pregnancy is too far advanced for simple, safe abortion techniques.

A newer test, which can be performed as early as the 10th week of pregnancy, offers advantages to parents who want to know about—and, if necessary, do something about—their developing child's genetic condition. More and more hospitals are performing **chorionic villus biopsies**, in which a sample of chorionic villus is snipped or suctioned through the mother's cervix. Chorionic villi are tiny protrusions of the chorion, the membrane that surrounds the fetus and eventually forms part of the placenta. The tissue sample is larger and the results of analyzing it are quicker than with amniocentesis. Although chorionic villi biopsies also have risks, eventually they may allow scientists to detect any of the 3800 diseases and disorders for which genetics may be responsible (Schmeck, 1983).

In amniocentesis a syringeful of amniotic fluid is removed during the woman's fourth month of pregnancy. The chemical contents of the fluid are analyzed for evidence of certain diseases, or cells sloughed off from the fetus are cultured and later checked for chromosomal abnormalities.

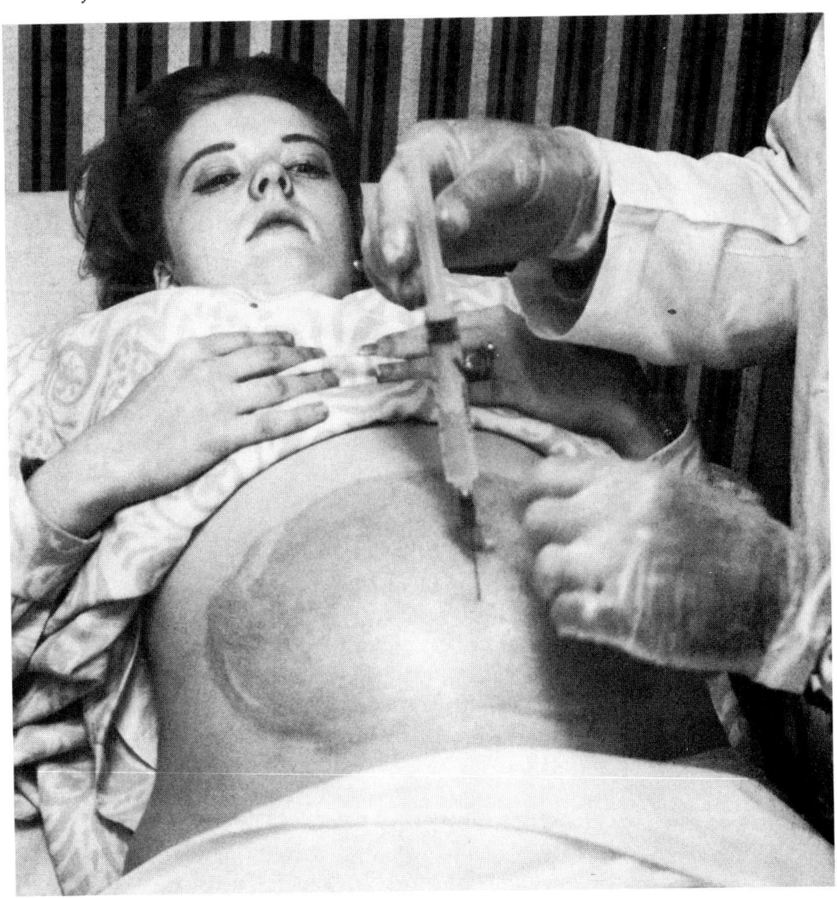

Summary

1. Both environment and heredity contribute to the development of each human being. The genotype is an individual's genetic inheritance. Its expression in visible characteristics and behavioral patterns is the individual's phenotype. Some human characteristics are the outcome of single genes; most are affected by many different genes.
2. The chemical instructions of heredity are transmitted by deoxyribonucleic acid (DNA). All an individual's thousands of genes, which lie on the 46 human chromosomes, are made of DNA. The DNA molecule is shaped like a spiral ladder. It can reproduce itself and thereby transmit genetic material from parents' germ cells to a new individual.
3. Chromosomes within the germ cells divide by the process of meiosis. Each sperm or ovum is left with 23 chromosomes. When sperm and ovum unite at conception, the number of chromosomes is doubled to 46, the normal number for human body cells.
4. The sex chromosomes determine an individual's sex. An ovum always contains an X chromosome. If it is fertilized by an X-chromosome-bearing sperm, a female results. If it is fertilized by a Y-chromosome-bearing sperm, a male results. The autosomes supply most of the genetic information for the inheritance of physical and behavioral characteristics.
5. The inheritance of behavioral characteristics is complex and polygenic. By studying the presence of certain behavioral characteristics in fraternal versus identical twins, in adopted children and their adoptive and biological parents, and in family trees, researchers try to find out whether behavioral characteristics are inherited. Intelligence, activity level, sociability, and schizophrenia are partly inherited.
6. Tests of blood and fetal cells are used to detect inherited disorders and to provide prospective parents with information upon which to decide whether to prevent or prepare for the birth of a defective child.

Key terms

genotype
expression
phenotype
reaction range
canalized
chromosomes
genes
deoxyribonucleic acid (DNA)
mitosis
ribonucleic acid (RNA)
meiosis
testes
ovaries
sperm
ova
homologous
crossover
conception
zygote
autosomes
X chromosome
Y chromosome
recessive
dominant
polygenic
monozygotic (MZ)
dizygotic (DZ)
concordance
personality
temperament
sex-linked
sickle-cell anemia
phenylketonuria (PKU)
Down's syndrome
schizophrenia
dyslexia
amniocentesis
ultrasound
chorionic villus biopsy

Suggested readings

APGAR, VIRGINIA, and BECK, JOAN. *Is My Baby All Right?* New York: Pocket Books, 1974. The major congenital problems carefully explained, with personal accounts to humanize the complex medical and psychological impact of birth defects.

FARBER, SUSAN. *Identical Twins Reared Apart: a Reanalysis*. New York: Basic Books, 1981. A comprehensive review of all existing studies of twins raised apart, in which the author tries to track down the evidence for genetic effects on intelligence and personality.

LEWONTIN, RICHARD. *Human Diversity*. New York: W. H. Freeman, 1982. A scholarly treatment of human variety and its basis in evolution, heredity, and environment, including an up-to-date overview of the nature of genetic mechanisms.

WATSON, JAMES D. *The Double Helix*. New York: New American Library, 1968. A thrilling suspense story of the competition to unlock the secret of life by discovering the structure of DNA. One of the few lively descriptions of the scientific community.

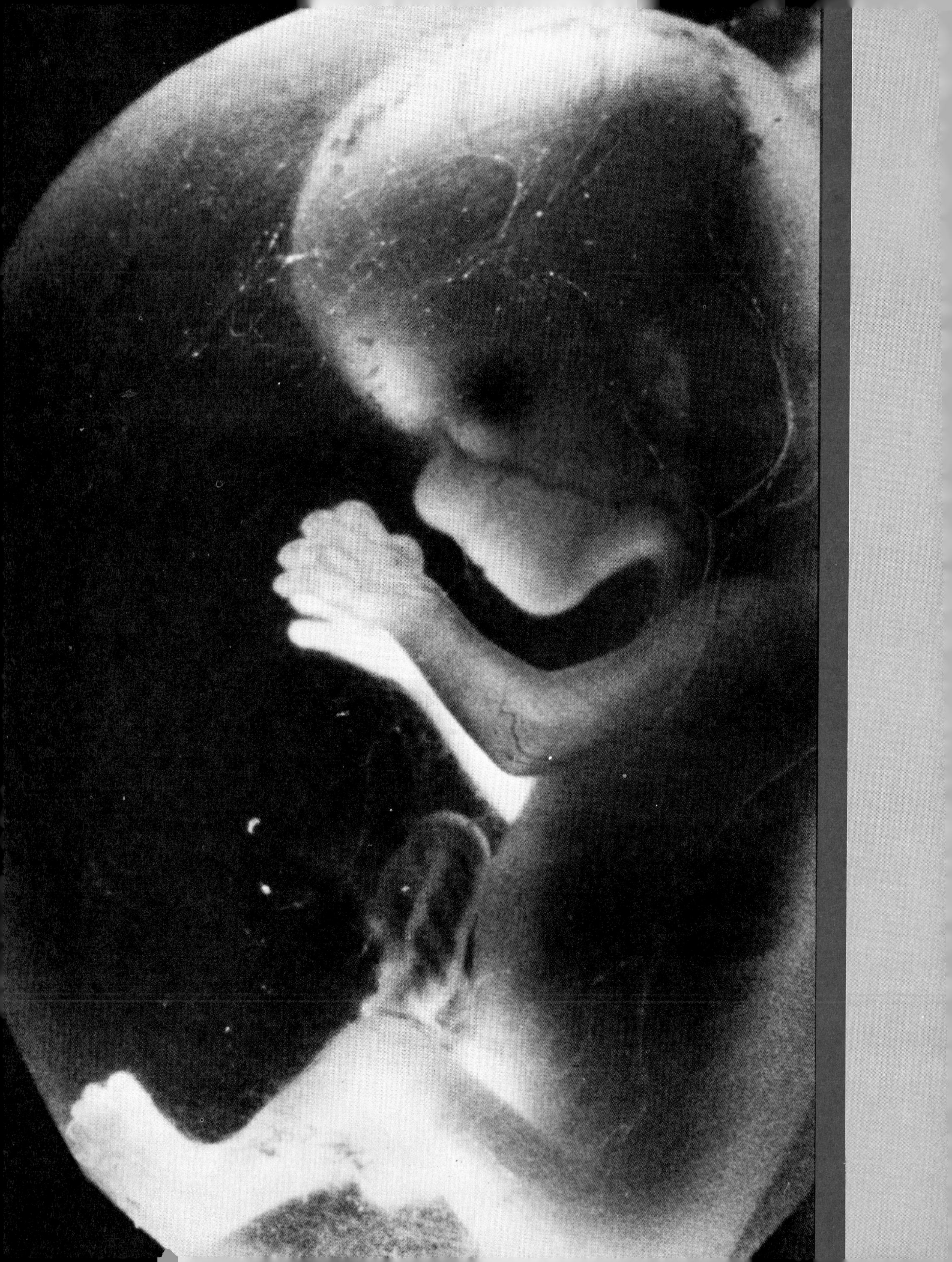

chapter four
Prenatal development and birth

Beginning new life

The conception of a child is the moment that a new life begins—the beginning of cells dividing and multiplying so that in mere months an infant is ready to greet the world. Conception also begins a new stage of life for the man and the woman whose sperm and ovum have united. The parents-to-be take on roles and embark on a relationship like no other. They cannot divorce themselves from the growing fetus, take a vacation from their 24-hour-a-day connection to it, or postpone the day of its arrival. They, too, go through developmental changes as the birth of the infant grows closer. In this chapter we will describe the development in the fetus and parents during the eventful nine months of pregnancy and prenatal growth.

The parents' choices

Not so long ago, couples were expected to have large broods. Parents seen walking with their ''only two'' children raised comments—unflattering ones. Even as recently as the late 1960s, reference books listed information about ''childlessness'' under ''sterility''—for the assumption was that people without children suffered from a medical malady. Whereas in the 1950s, 1 couple in 100 did not have children, by the early 1980s, that figure had risen to 10 childless couples in 100—some voluntarily and some involuntarily. Today, for many people in our culture, parenthood has become a matter of choices. To a large extent, couples choose whether, when, and how many children to have.

> I was in my 30s, my career was launched, and I knew that I was mature enough to welcome a new person into our lives. My husband felt pretty much the same way. He wanted a child he could do all sorts of wonderful things with—go to baseball games, see movies, cook supper, go to the beach.

The major reason for the trend toward deliberately planned children is simple: better contraception. Contraceptives have been available for centuries. Pitch, animal earwax, oxgall, and elephant dung all have been touted for their abilities to prevent conception. But more appealing and more effective contraceptives became available only in the mid-1800s, when the rubber-manufacturing process was perfected. Then, middle-class people could approach their pharmacists for rubber diaphragms and condoms. Not until the mid-1900s did the more reliable methods of contraception—birth control pills and intrauterine devices (IUDs)—become available and widely used.

In addition to better contraception, recent social pressures have made many women question whether to become parents. The majority of women now need, want, and hold jobs. They want to fulfill themselves through their work and may want to postpone or avoid the demands of an infant. Yet even now, the internal and external pressures on young adults to have children remain strong. Some religious groups forbid the use of contraception or abortion. Many employers grow suspicious of employees who do not start families. Parents push subtly (or not so subtly) for grandchildren, and other adults wonder what their childless friends are waiting for. Women themselves may feel as if a biological clock is ticking off their childbearing years. The decision to have a child is complex, often conflicted, and probably unique for every couple.

The choice of when to have a child is also complex. Many couples today postpone childbearing until they are financially and socially settled.

> Peter was born when I was in my late 30s. By then my career was established, I could support a baby, and I had the emotional reserves to spare that I hadn't had when I was trying to make a place for myself

professionally. It was also time to have a baby before the hourglass simply ran out.

Yet many other couples rush into parenthood, either unintentionally or because they let nature take its course. There are pros and cons to either choice. The couple whose first child appears when they are 20-year-old students, newcomers to marriage, and perhaps near strangers to each other, may be staggered by parenthood. But as still energetic 40-year-olds, these same parents have launched the child into the world and can face their own middle age relatively free of child care. In contrast, the couple whose first child arrives as they approach 35 or 40 faces a middle age of active child care. But they have had perhaps two decades as adults to weave a marriage and careers that will embrace that much wanted child.

Conception

All pregnancies begin when sperm and egg unite. Roughly 14 days before her menstrual period, a woman **ovulates**, and a mature egg, or ovum, is released from one of her ovaries. The ovaries contain thousands of immature ova, called **oocytes** (which have been present since the woman's birth), but usually only one mature egg is released (ovulated) each cycle into the fringed ends of one of the two **fallopian tubes**. These microscopically thin tubes are lined with thousands of waving cilia (see Figure 4.1), which move the ovum toward the **uterus**. During sexual intercourse, a man ejaculates millions of sperm, or **spermatozoa**, his reproductive cells. These sperm are produced in the testes, then mixed with seminal fluid and ejaculated. Each cubic inch of ejaculated semen may contain over 300 million swimming sperm. Out of the millions of sperm that are ejaculated into the woman's vagina and that swim through the

CONCEPTION

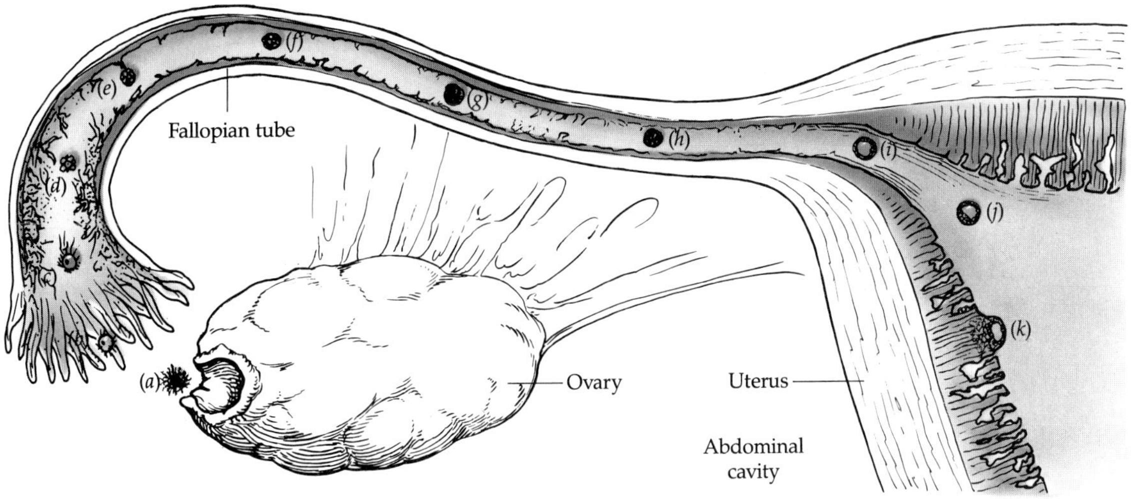

Figure 4.1 ***The story of conception is this: (a) an ovary releases an ovum into the abdominal cavity. (b) The waving movements of millions of cilia on the fringed ends of the fallopian tube draw in the ovum. (c) The ovum begins the second stage of meiosis. (d) When a sperm penetrates the ovum, it stimulates it to complete meiosis. (e) Chromosomes from sperm and ovum mingle. (f) The chromosomes go through mitosis. (g) The fertilized ovum, the zygote, divides. (h, i) The zygote divides repeatedly as it moves through the fallopian tube and (j) into the uterus. (k) On the seventh day the sphere of cells attaches to the uterine wall (Grobstein, 1979). From I. Arbel in "External Human Fertilization" by Clifford Grobstein. Copyright © 1979 by Scientific American, Inc. All rights reserved.***

cervix and into the uterus, only one will merge with the mature ovum in the fallopian tube. Why then are there so many sperm? The millions of sperm ejaculated are insurance that a few hundred will actually near the ripe ovum. Most sperm swim into the wrong fallopian tube or get lost in the vagina or uterus. Sperm live for only 48 hours or so after ejaculation, and the ovum is receptive for only 12 to 24 hours. Conception, therefore, is limited to a 72-hour period in each menstrual cycle. During this brief period, the environment in the female reproductive tract is especially "friendly." The rest of the time it is too acidic for sperm to survive. At ovulation, too, the mucous plug in the cervix, which protects the uterus from infection, is thinner and easier for sperm to swim through.

Fertility

But things do not always go smoothly. Not everyone who decides to have a baby conceives easily.

> We have been trying to conceive for six years now. We've tried everything. People say that we should go ahead and adopt, but I want *our own* baby, not someone else's.
>
> With Mikey and Erica, I think I was pregnant within an hour of our decision to put away the diaphragm.

Lucky couples, like Mikey and Erica's parents—one-third of those who are trying—do conceive in their first month. Half of the couples who are trying conceive by the third month, three-quarters by the sixth month. But some couples—one in six—cannot conceive a child. After a year of regular sexual intercourse without contraception, these couples are diagnosed as being infertile. Many then begin a sometimes slow and painful course of treatment.

The problem of **infertility** has become increasingly common in the past 20 years. There are many causes of infertility. One cause is the postponement of parenthood into a woman's middle or late 30s, because fertility drops off sharply after the age of 30. Stress is another cause of infertility; more than one woman has found herself missing menstrual periods during stressful times at home or the office. Another cause is sexually transmitted **pelvic inflammatory disease (PID)**, which leaves scars on the woman's fallopian tubes, ovaries, and uterus. IUDs also have come to light as a cause of infertility. Some women who have had prolonged exposure to toxic chemicals become infertile. Being underweight is another cause of infertility. Various physiological problems, too, can cause a woman to be infertile. Ovaries may not release mature ova, fallopian tubes may be blocked, cervical mucus may prove impenetrable to sperm, fertilized ova may not implant or remain implanted in the uterine wall.

But women account for just 60 percent of all infertility problems. The other 40 percent of infertility problems reside with men. Sperm may be too few, abnormally shaped, or weak swimmers. Blocked ducts may fail to deliver sperm into the urethra leading out of the body, and varicose veins in the testicles may interfere with sperm production. Men's fertility also declines with age.

So many factors can operate to prevent normal conception that it sometimes seems a miracle that anyone is born! Infertility can be a devastating problem. It can undermine a couple's sexual and personal relationship. It can undermine their self-esteem and confidence in their masculinity and femininity. It tears away at friendships, careers, and even at savings accounts. Clearly, infertility has high personal and social costs. But nearly all cases of infertility are diagnosed, and about half ultimately are cured. One established solution for infertility is surgery to tie off varicose veins in a man's testicles or

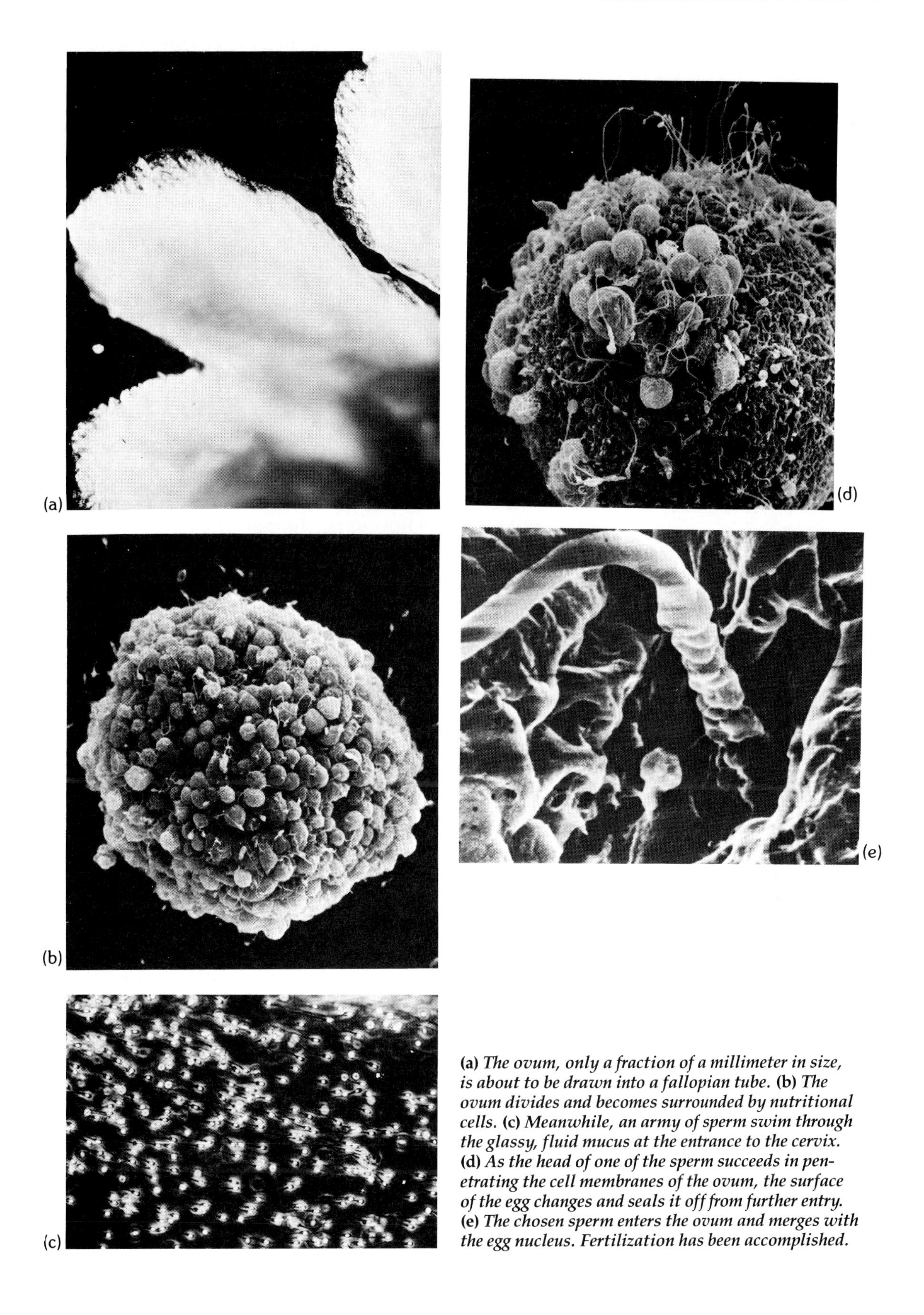

(a) *The ovum, only a fraction of a millimeter in size, is about to be drawn into a fallopian tube.* **(b)** *The ovum divides and becomes surrounded by nutritional cells.* **(c)** *Meanwhile, an army of sperm swim through the glassy, fluid mucus at the entrance to the cervix.* **(d)** *As the head of one of the sperm succeeds in penetrating the cell membranes of the ovum, the surface of the egg changes and seals it off from further entry.* **(e)** *The chosen sperm enters the ovum and merges with the egg nucleus. Fertilization has been accomplished.*

to unblock a woman's fallopian tubes. Drugs can help some men with low sperm counts or abnormally shaped sperm. Another method for dealing with infertility is **artificial insemination**. Artificial insemination has been practiced throughout this century and now accounts for some 20,000 births every year (Fleming, 1980). It is a simple, painless procedure in which a syringe is used to deposit semen at the entrance to the woman's uterus. A man's semen can be collected and pooled if the concentration of sperm is low. An anonymous donor's semen can be used if a man cannot produce normal sperm in enough quantity.

To combat infertility, many couples are turning to methods of **alternative reproduction**. These methods are designed to bypass a woman's blocked or damaged fallopian tubes. One method of alternative reproduction is **in vitro fertilization**, the creation of the so-called **test-tube baby**. Louise Joy Brown, the first test-tube baby, was delivered in England in 1978. Since then, hundreds of test tube babies have been born. Clinics that treat infertile couples with in vitro fertilization (literally "in glass," that is, in a glass laboratory dish) have been established all over the world, and with the increase in infertility, it is likely that still others will appear. In this procedure, the mother first takes fertility drugs to ripen several of her ova. (The chances of a successful pregnancy are higher if more than one ovum is recovered.) Doctors recover the ripe ova by **laparoscopy**, surgery during which general anesthesia is given, small incisions are made in the woman's navel and at the base of her abdomen, and a telescope and fiber-optic illuminator—the "laparoscope"—are introduced through the incisions. The ripe ova are removed by vacuum suction and incubated in a glass dish for three hours or so. Then a few drops of concentrated semen are added to the dish, and a number of the ova are fertilized. Within 36 hours, each fertilized egg begins to divide. When each zygote consists of eight cells, the doctor inserts them into the woman's uterus. From this point, the pregnancy can proceed normally, as the zygotes attach to the wall of the uterus and continue to grow.

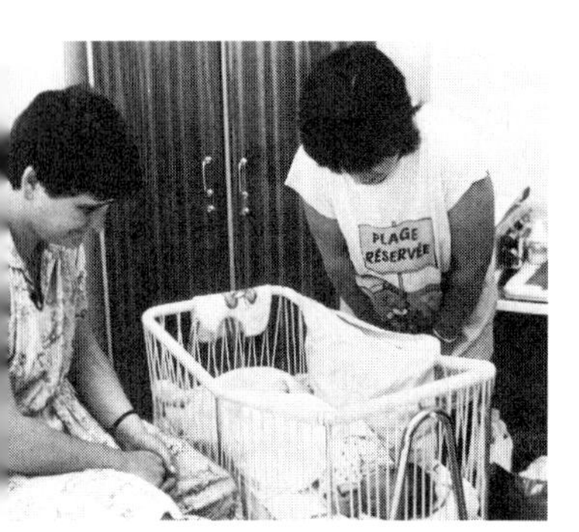

Surrogate mothering has become a controversial option. Here, surrogate mother and adopting mother meet at the cradle of their infant girl.

A second method of alternative reproduction is **surrogate mothering**. In this procedure, a fertile woman is hired by a couple in which the wife is infertile. The surrogate mother then is artificially inseminated with the husband's sperm and carries the resulting fetus to term. After she gives birth, she turns the baby over to the couple. **Ovum transfer** is a third method of alternative reproduction. In this procedure, a fertilized ovum is removed from a donor woman's uterus and implanted into the uterus of the infertile woman. Donors are given thorough physical and psychological examinations, and their genes are checked for abnormalities. After a suitable donor has been found, doctors use drugs to synchronize her menstrual cycle with that of the woman who will be the recipient. When the donor ovulates, she is artificially inseminated with sperm from the infertile woman's husband. Five days later, the fertilized egg, or zygote, reaches the uterus and can be removed to the uterus of the infertile woman, whose own body is at the same receptive point in her menstrual cycle. The zygote implants itself in her uterus, and the pregnancy continues normally. In 1984, the first American child resulting from an ovum transfer was born in Long Beach, California. In 1986 the implantation of a frozen fertilized ovum resulted in a successful birth. Compared to "old fashioned" reproduction, these alternative methods are used only rarely. But they are becoming more common—and more ingenious—as time passes.

Prenatal development

The prenatal period begins with conception and ends with birth. During these nine months, the development of one fertilized cell into a complete human infant seems wondrous. Yet for each infant, development proceeds according to a master blueprint. Body parts and functions develop at roughly the same rates and in the same order for all infants (Figure 4.2).

> When I was pregnant with what turned out to be Sam, I loved to look at pictures of fetuses at various points of development. Then I'd announce at dinnertime, "Well, the baby just got fingers and toes," or "This is the week for the chambers of the heart to develop." When the organs all were forming in the first weeks, I often felt scared that something might go wrong.

We have been given a peek at this development in the womb by such modern marvels as photo optics.

Stages of prenatal development

The months of pregnancy can be divided into three stages. The first, the **germinal stage**, begins when the ovum is fertilized and ends some two weeks later, when the zygote attaches to the wall of the uterus. The second, the **embryonic stage**, lasts from the second through the eighth week of pregnancy. During this period, the organs form. The third, the **fetal stage**, lasts until birth and is the period when the fetus grows in size and gains function in its organs and muscles.

Germinal stage

Between the first and 14th days after conception, the fertilized egg first divides into two identical cells, then doubles to four, eight, and so on. Cell division by mitosis begins within 36 hours after fertilization and continues

PRENATAL MILESTONES

Weeks	
-2	(Last menstruation)
0	Conception
2	First missed menstrual period Implantation; cell differentiation begins
4	Heart and central nervous system forming
6	Reproductive organs and ears, eyes, arms and legs forming
8	Pregnancy detectable by physical examination
10	Responds to stimulation; form recognizably human
12	Vital organs basically formed Circulatory system operating
14	Mother's abdomen visibly distended Skeleton visible in X rays
16	Movements can be felt by mother Sex organs distinct
18	
20	Lanugo and head hair forming
22	Sucking movements, vigorous body movements Eyelashes and eyebrows appear
24	Appearance: thin, wrinkled translucent skin
26	Eyes open Survival outside womb possible
28	
30	Layer of fat forming beneath skin
32	Survival outside womb probable
34	
36	
38	Normal birth
40	

Figure 4.2 ***This chart shows the timing of milestones in prenatal development for infant and mother.***

ever more rapidly. Sixty hours after fertilization, a mulberry-shaped, 12- or 16-celled **morula** is floating in the mother-to-be's fallopian tube. Each cell has until this point been **totipotent**. Separated from the others, any cell can develop into a whole human infant—one way in which identical twins get their start in life. By the time the morula has formed, the cells are no longer totipotent. Those on the inside of the "berry" are large, those on the outside, small. Specialization has begun.

Floating slowly through the fallopian tube, the morula gently descends to rest on the uterine wall about four days after fertilization. The rounded **blastula**, as it is now called, is composed of over 100 cells, with fluid at its center. Within a few days the blastula has developed into a **blastocyst**, a more thoroughly specialized and organized sphere of cells. At one side are larger cells forming the embryonic disk, which will turn into the embryo and, later, the fetus. The smaller, outer cells will form themselves into a life-support system for the fetus made up of an outer membrane, the **chorion**, the inner "bag of waters" or **amnion**, the **placenta**, and the yolk sac.

Having floated in the uterus for a day or so, the blastocyst settles down in one spot. Rather like a space capsule that touches down on the moon's surface, the embryo has "landed." The uterine lining is rich with blood, spongy, and receptive to the blastocyst. The outer cells of the blastocyst project tiny roots, or **villi**, directly into the mother's blood vessels as the blastocyst burrows ever more deeply—**implants** itself—into the uterine wall. Nearly half of all blastocysts are abnormal and do not implant themselves in the uterine wall (Roberts and Lowe, 1975). But for those blastocysts that do implant, the cell wall of the uterus then closes the opening through which the blastocyst has burrowed. Implantation is complete.

Embryonic stage

Two weeks after conception, the embryonic disk has folded and formed the distinct layers of cells called an **embryo**. From the outer layer of cells, or **ectoderm**, a bulge called the **primitive streak** forms—the primitive tissue from which brain, spinal cord, nerves, sense organs, and skin will form. From the inner layer of cells, or **endoderm**, will form the lining of the gut, salivary glands, pancreas, liver, heart, lungs, and respiratory system. From a soon-to-develop middle layer of cells, the **mesoderm**, will form cartilage and bone, muscles, blood vessels, heart, and kidneys.

The embryo's life-support system already has begun to form. Amniotic fluid is filling the amnion and making a warm cushion that protects the embryo against injury or shock. The yolk sac is generating blood cells and the germ cells that will someday let this tiny embryo itself be a mother or father. The placenta, the source of nutrients and oxygen for the fetus, is beginning to function. The placenta—like the astronaut's support system—breathes, digests, and excretes for the multiplying cells of the embryo. It is connected to the embryo through the pulsating **umbilical cord**. Along the cord's ropy length, a vein carries oxygen, sugars, fats, rudimentary proteins, and minerals to the embryo while two arteries carry waste products, carbon dioxide, and urea from the embryo, eventually to be disposed of by the mother's lungs and kidneys. Through the placenta, the mother's and embryo's blood vessels are brought into close contact. The bloodstreams of mother and embryo do not actually mix, but oxygen, nutrients, and wastes pass between them through the thin capillary walls of the embryo. The placenta screens out some harmful substances, including most bacteria, but viruses, gases, and many drugs pass through the placenta from mother to embryo.

Third and fourth weeks. In the third and fourth weeks after conception, the cells of the pea-sized embryo multiply rapidly and organize themselves into

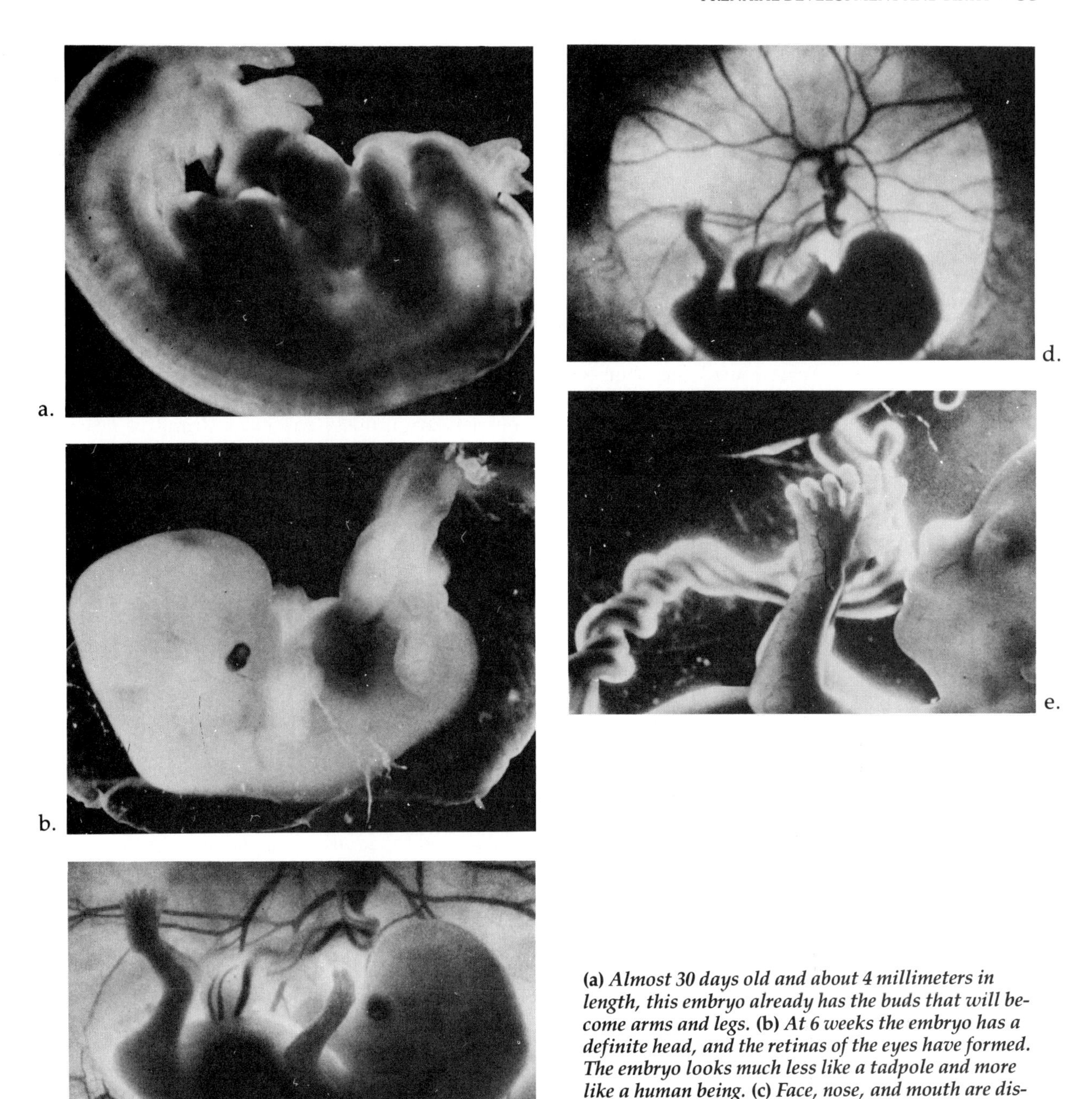

(a) *Almost 30 days old and about 4 millimeters in length, this embryo already has the buds that will become arms and legs.* **(b)** *At 6 weeks the embryo has a definite head, and the retinas of the eyes have formed. The embryo looks much less like a tadpole and more like a human being.* **(c)** *Face, nose, and mouth are distinguishable at 11 weeks, and the muscles have formed.* **(d)** *At 3 months the muscles go into action and the fetus actually moves.* **(e)** *Nourishing blood flows to this 4-month-old fetus at the rate of 0.25 liter per minute.*

functional units. The groove that runs along the primitive streak forms a tube and then develops into foundations of the brain, spinal cord, nervous system, and eyes. The heart develops first into a tube and then into a chambered pump. A system for digesting food and kidney-like structures begins to form. All these developments take place according to a master plan, by which development starts at the head and moves to the tail, in **cephalocaudal** order. First the head, then the trunk, and then the lower extremities develop. The plan also calls for development to proceed **proximodistally**, from the midline of the body—spine, heart, face—outward to the shoulders, arms and legs, hands and feet.

Second month. I was never so tired in my whole life as I was during my second month of pregnancy. The midwife said that it was because of hormones, but I think it was because my baby was developing so fast. I'd tease my husband, "Hey, I can't make dinner tonight. This baby's got me busy making his *brain*!"

During the second month of prenatal life, new physical structures develop with astonishing speed from the foundations laid in the first month. During a three-day period at the beginning of this month, for example, buds for arms, legs, and the visual system all take form. During the month, the stomach and esophagus form; the heart moves from near the mouth into the chest cavity and a valve is created that separates its upper and lower segments. Nerves grow and form connections between brain, nose, and eyes. Primitive ovaries and testes form.

A photograph of the translucent embryo at the end of the embryonic period shows a creature that looks like a human being. Its face has eyes, nose, mouth, and lips, ears and jaw in the proper places. Its hands have fingers and thumbs, its legs, knees, ankles, feet and tiny toes. With a powerful microscope, one can see whether the embryo is a boy or a girl. Flexible cartilage begins to be replaced by bone in the center of the long arm and leg bones in a process called **ossification**. This process marks the beginning of the third and final phase of prenatal development.

The embryonic stage is a critical period when organs and body structures must form if development is to proceed normally. If heart, eyes, and lungs do not appear now, they never will. If the round buds at the end of the arms and legs do not form themselves into fingers and toes, they never will. The embryonic cells are also particularly susceptibile to their environment during this stage of development, which poses both grave risk and great biological advantage. The risk is that the embryo is vulnerable to abnormal development if it is exposed during this critical period to radiation, toxins, or infection. The advantage is that the embryo is also very responsive to chemical messages from genes and other cells that organize the sequence of specialized cell development. The embryonic cells can respond readily to the message of the DNA molecules in each cell and to the blueprint for interaction between specialized systems of the body.

Fetal stage

During the fetal stage, which lasts from the end of the second month of prenatal life until birth, the structures and systems that have developed grow in size and in efficiency.

Third month. Nine weeks into its prenatal existence, the fetus is an inch long, weighs one-tenth of an ounce, and has a disproportionately large—but entirely human-looking—head. The appearance of the fetus, with its large head, mysterious eyes, and perfect innocence, has long fascinated people and captured their imaginations. In movies like *E.T.* and *Close Encounters of the Third Kind*, for instance, the wonderful alien creatures who befriend earthlings look much like fetuses. In the third month after conception, the fetus's eyelids form and are sealed shut. The roof of the mouth closes. The fetus digests amniotic fluid, excretes urine from functioning kidneys, and breathes fluid in and out of its lungs. A male fetus develops a penis. Nerves connect to muscles, and the fetus begins to make reflexive kicking, darting, and dodging movements. In a few weeks, the fetus can kick and turn his or her feet. Toes now curl, fists form, thumb opposes to fingers, head turns, mouth opens and closes and swallows, and the fetus can even suck his or her thumb. Earlier the fetus responded to stimulation with the whole body. Now the fetus can move and respond with specific parts. Stroke a hand, and the arm moves. Touch an eyelid, and the eye squints. The fetus also can move spontaneously, and he or she already has a distinct level of activity.

Fourth month. During the fourth month, the fetus grows more quickly than in any other month: from 3½ inches and 1 ounce to 6 inches and 4 ounces. Now the mother can feel the **quickening**, the first perceptible movements of the stronger, larger fetus kicking against her abdomen.

> I first felt a faint tap-tap-tapping of the baby in the 16th week. "So soon?" my mother asked. We should have known then that this was going to be one very active baby.
>
> When I felt Sam bumping around inside me, I sometimes pictured him as swimming in a pool and kicking against the sides when he changed directions.

Stronger neck muscles and a bonier skeleton help support the fetus's ever more human-looking head. Finger- and footprints have formed on the tiny touchpads of palms and soles. The eyes sense light. A female fetus develops inner and outer genitals—uterus, vagina, clitoris, and associated structures. Meanwhile, the placenta has begun to produce most of the hormones that prepare the mother's body for producing milk and the infection-fighting substances to protect the fetus.

Fifth month. During the fifth month, the fetus grows to 1 foot in length and a weight of 1 pound. The mother typically begins to "show" her pregnancy during this month as the abdominal walls expand outward to accommodate the developing baby. The fetus's sweat glands, eyelashes and brows, and hair on the head all form. **Lanugo**, a downy hair, begins to grow over most of the body. Like the astronaut freed from gravity, the fetus bounds and turns in the fluid-filled uterus. The mother feels periods of quietness—as the fetus sleeps or rests—between periods of waking activity. The fetus sheds old cells and develops new ones. These dead skin cells mix with fat from the oil glands and cover the body with a protective cream called **vernix**. Vernix keeps the fetus's skin supple as it floats in mineral-rich amniotic fluid—very "hard water" indeed.

Sixth month. During the sixth month, the fetus's eyelids first open, and the eyes can look up, down, and to the sides. The intestines descend into the abdomen. Cartilage continues to turn into bone. Development of the six layers of cells in the cerebral cortex—that part of the brain responsible in adults for complex conscious thought—is completed. The fetus can reflexively grasp, breathe, swallow, hiccough, and taste.

> One morning I was awakened from sleep by violent, rhythmic movements at the base of my abdomen. For a minute I was terrified, but then I realized that the baby had the hiccoughs!

One ingenious obstetrician treated a patient who had too much amniotic fluid by injecting a sweetener into the amnion. The fetus promptly swallowed some of the fluid, absorbed it, and passed it to the mother for her to excrete (reported in Montagu, 1962).

The 6-month fetus is still a vulnerable creature, barely equipped to survive outside the womb. The fetus has not yet gotten immunities from the mother's system. A fetus may breathe regularly for up to 24 hours at a time, but is still so immature that infants born at this stage will not survive without intensive medical care. Lacking fat under the sensitive skin, the fetus needs to be kept warm.

The relatively new medical specialty, **neonatology**, care of newborns (neonates), has greatly increased the chances of survival for fetuses born prematurely. Today, many 6-month fetuses survive and have a fair chance at functioning normally because of the intensive medical care provided by neonatologists. **Fetology** is the even newer medical specialty devoted to treating problems in fetuses before birth. Blood transfusions, the insertion of drainage tubes into kidneys, brains, and collapsed lungs, and surgical repairs on urinary tracts all have been performed on fetuses while still in their mothers' uteruses. One fetus with **hydrocephaly**, excess fluid in the brain, for example, had his brain surgically punctured and drained before he was born. When he was born, he showed no hydrocephaly and no signs of the surgery. In another case, doctors successfully drained the fluid from a fetus's collapsed lung. It is likely that fetologists will be called on to treat more and more selected cases in the future. Perhaps 1 in 400 to 500 fetuses may be candidates for surgery *in utero* (Kotulak, 1981).

Seventh month. At the beginning of the seventh month after conception, the fetus weighs about 2 pounds and has organs mature enough to offer a 50–50 chance of surviving outside the womb if birth comes prematurely and intensive care is provided. The fetus's brain now can regulate breathing, body temperature, and swallowing. It contains trillions of connected nerve cells, specialized into sections devoted to hearing, seeing, smelling, vocalizing, and moving. Many reflexes, including sucking and grasping, are established. During the seventh month, the testes of most male fetuses begin to descend into the scrotum from the abdominal cavity, where, protected from the body's internal temperature, the sperm produced after puberty will be able to survive. The ova of the female fetus already are formed in the ovaries.

Eighth month. In the eighth month of prenatal development, the fetus's body prepares for life outside of the uterus. For fetuses weighing 3 pounds or more, the chances of survival outside the uterus increase to 85 percent. The lungs are still immature, the **alveoli**, or tiny air sacs in them, not prepared to turn oxygen into carbon dioxide. The fetus's digestive and immune systems are also immature. But under the skin, a vital layer of insulating fat is laid down. This fat increases the fetus's chances of survival outside the womb.

In the eighth month, nerve cells in the brain develop branches and neurotransmitters so that messages can be passed from nerve to nerve. At this point, the nerve cells begin to function. As a result, in the eighth month, the fetus begins to lurch, roll, startle, or lift his or her head in response to a loud or sharp noise. The fetus can be soothed by the mother's heartbeat and rhythmic walk and aroused into vigorous kicking by the sounds of a piano, television, even a dishwasher.

Apparently, the fetus at this age can also learn. Studies of learning by fetuses were first carried out over 40 years ago. In one such study (Sontag and Newbery, 1940), a loud noise was made repeatedly outside the uterus. At first the fetus responded with a quickened heartbeat, but after hearing the noise a number of times, the fetus stopped reacting, apparently having learned to ignore the noise. In a second study (Spelt, 1948), a fetus was subjected to a loud and startling noise and at the same time to vibrations. After about 20 trials, the fetus startled upon feeling the vibrations without the noise. Apparently, this fetus had learned the connection between the vibrations and the noise. Unfortunately, these early studies did not separate the reactions of the mother from those of the fetus, and so the results are difficult to interpret. They have, however, gained support from recent research with both animals and humans (Kolata, 1984). In one recent study, for example, Anthony DeCasper and his associates (Spence and DeCasper, 1982) showed that newborn infants preferred hearing their mother read a nursery rhyme that she had read aloud twice a day during the last six weeks of pregnancy to one that she had never read before. This finding suggests that the infants had learned to recognize the cadences of the nursery rhymes read to them in utero.

It also suggests that fetuses can hear voices while they are in the uterus. In one study of the sounds that penetrate the uterus (Armitage, Baldwin, and Vince, 1980), hydrophones implanted in the uteri of two pregnant ewes revealed that loud sounds—like shouts and bangs—could be heard clearly. Normal conversational tones from outside the uterus were somewhat dulled. Sounds from inside the ewes, of drinking, eating, swallowing, and heavy breathing were also audible, although heartbeats were not. The researchers concluded that in the fluid environment of the uterus, fetal sheep can hear sounds from both inside and outside of the mother. When hydrophones were placed in the amniotic sac of human mothers (Querleu and Rennard, 1981), speech sounded audible but muffled. It must be heard over the ambient noise, which is as loud as the noise in a factory (Aslin et al., 1983). But even if speech is muffled, in another study, speech sounds made outside the mother's abdomen while she was in labor elicited more marked change in the infant's heart rate than did pure tones (Macfarlane, 1977). These studies demonstrate that sounds do penetrate the uterus and affect the fetus. The question of how clearly the fetus actually *hears*—with its ears full of amniotic fluid—has not been settled yet, however. A related question that also has not been settled is whether pregnant women who expose their fetuses to classical music and stimulating language routines actually give the infants a head start at learning.

Pregnancy is a time when moodiness and anxiety are common and the mother-to-be feels fear and concern for herself and for the developing fetus.

Ninth month. Because Sam was facing forward rather than backward, like most babies, he wedged his toes under my ribcage in a thoroughly uncomfortable maneuver. So I would push his toe down, and he would snap it right back.

In the home stretch, with just a few weeks before birth, the average fetus is 20 inches long and weighs 7 pounds. Now growth finally slows down. If it didn't, the infant would weigh 200 pounds at age 1! Crowded into the uterus now, the fetus folds up into a ball and squirms a bit, moving only hands, feet, and head. A heavy head makes most fetuses settle head-down, with skull wedged into the mother's pelvic girdle. As the fetus settles in, the mother feels a "lightening." The bulge in her abdomen is lower. She feels less pressure on her diaphragm and lungs. The formerly spongy placenta toughens, and immunities—against measles, mumps, whooping cough, and other illnesses the mother has been exposed to—pass from mother to fetus. As labor begins, the placenta releases the hormone **oxytocin**. It prepares the mother's body to make milk and ushers in the birth process. In 266 days, a single cell has grown into a unique human infant, ready for life outside of the womb.

How pregnancy affects parents

During the nine prenatal months when so much is happening to the fetus, the expectant parents, too, go through changes. Nearly all expectant parents go through a period of emotional flux, when mood and anticipation run high one moment and low the next. Not surprisingly, pregnancy has been termed a "maturational crisis" (Bibring, Dwyer, Huntington, and Valenstein, 1961), a time for psychological growth and adjustment.

Emotional strains

Most pregnant women, research has shown, feel vulnerable, sensitive, and in need of support (Leifer, 1977; Shereshefsky and Yarrow, 1973). They are moody at times and worry about the health and well-being of the developing fetus. Few expectant mothers develop severe emotional problems, however, especially if they are well adjusted and looking forward to the "blessed event." Well adjusted women are more likely to worry about the health of their fetus than about their own health, and they are likely to grow more confident and sure of themselves as the pregnancy progresses. Women who are emotionally unstable even before they conceive, whose pregnancies are unplanned or unwanted, and whose relationships with the father are shaky are more likely to develop strong anxieties about themselves during their pregnancy. In both expectant parents, the pregnancy may revive childhood conflicts and anxieties. Some expectant parents abstain from sexual intercourse during the last months of the pregnancy, afraid that they will harm the fetus during intercourse and orgasm. Sexual abstinence can put yet another strain on the relationship between expectant parents. Affectionate and cooperative couples can most successfully navigate the sexual difficulties and emotional strains brought on by impending parenthood.

Physical strains

I was so sick to my stomach during the first three months that I was pregnant with Sam that one night I showed up at the dinner table with a blanket over my head and managed to get down one spoonful of plain boiled rice.

lifespan focus

Adjustment to pregnancy

One way of looking at pregnancy is to see it as a period of development when a woman responds to an unfolding series of demands (Gloger-Tippelt, 1983). These demands are biological, social, and psychological. During the first trimester of pregnancy, a woman may experience sudden and dramatic disruptions in her accustomed ways of feeling and acting. The first disruptions of a woman's system are hormonal and physiological. Menstruation stops, and the woman may feel tired and nauseated, have sensitive breasts, and need to urinate often. She also may find elements of her identity threatened. She may feel psychological disruptions in her sense of maturity ("I'm a woman now, not a girl"), her sense of responsibility ("I will have to be a good mother"), her sexual identity ("How do I balance being a lover and a mother?"), her work identity ("Should I leave my job to stay home with the baby?"), and her sense of creativity and power ("I have created life"). She may also feel disruptions in her social relationships with her partner, friends, relatives, and employer. When the risk of miscarriage and many of the physical discomforts diminish, this phase ends.

In the second phase of pregnancy, many of the unpleasant symptoms like nausea and fatigue give way to relief and satisfaction. Obstetrical devices make it possible to hear the fetus's heartbeat or to see it by means of ultrasound, and these signs reassure the mother-to-be. Most women by now have committed themselves to, and feel familiar with, their pregnancy. Increasingly then, the woman becomes focused on the rapidly developing fetus. The fetus begins to signal his or her presence with movements and kicking. The fetus becomes a living presence rather than the symptom or abstraction it had been earlier in the pregnancy. Anxieties tend to be at low ebb, and the pregnant woman is likely to start making the practical arrangements in her education, work, and relationships that ready her for active motherhood. Now that she

Pregnancy brings with it physical strains as well as psychological ones. For one thing, in the early months, biochemical changes lower women's threshold to nausea and vomiting. Some psychologists have speculated that women who are anxious about their pregnancies are more vulnerable to nausea, but this speculation has never been proved. It makes sense, however, that anxiety can magnify the physical discomforts of pregnancy. For another thing, the early months of pregnancy bring fatigue.

> When I was first pregnant with Tim, I couldn't do much work because I was so tired. I'd close my office door and take an hour's nap with my face on my desk. At night when I got home, I slept some more. Was I ever relieved when I started to feel better.

Pregnant women also may crave or abhor certain foods, a tendency that is likely to be physiological in origin but aggravated by psychological reactions. Because the sense of taste is dulled during pregnancy, some women crave strong-tasting—sharp, sour, salty, or spicy—foods. Between one-third and two-thirds of pregnant women apparently suffer from cravings or aversions to certain foods.

A more extreme physical strain comes to some women: **toxemia** is a potentially life-threatening condition during pregnancy in which a woman retains fluid, vomits, gains weight rapidly, and has high blood pressure. If toxemia is unchecked, a woman may have seizures, enter a coma, and die. No one knows the cause of toxemia. As with nausea, people are divided in their opinions about whether its cause is psychological or physiological.

looks pregnant, the mother-to-be evokes certain predictable responses from others.

> People were so friendly to me when I was pregnant. They'd smile, or they'd give me their place in line, or they'd offer me a seat. Some people tapped my belly. One woman said she tapped it for good luck. Little kids stared. Several little boys informed me that they had babies growing in *their* tummies.

> When I was pregnant with Sam, the lady next door wouldn't let me wash the windows of the house, the people at the corner store wouldn't let me carry my own groceries to the car, and the guy who cut my hair told me it was bad luck to raise my arms over my shoulders.

By the eighth month and final phase of pregnancy, her large abdomen, the heavy baby settled in her pelvis, and the increased demands on her circulation and digestion can make a pregnant woman uncomfortable during sleep, eating, and moving around in general. Psychologically, the woman faces imminent birth. She may worry about pain, helplessness, losing self-control, or even about losing her partner, dying, or bearing a deformed child. Now is when she is likely to gather information about childbirth, to enroll in prepared childbirth classes, and to engage with her partner in relaxation exercises. She may also actively prepare for the new arrival by buying clothes and equipment, by learning about feeding, furniture, and the like.

> We put off buying baby furniture or clothes until my last month. I was superstitious about buying them any earlier than that, but also the baby wasn't real enough to either one of us until very late in my pregnancy.

Socially, the woman anticipates and prepares for the birth with others in her life. She may leave her job and increase her visits to the doctor.

How the prenatal environment affects the fetus

The developing fetus may be affected by any substance that can pass through the placenta and into the bloodstream. Whatever a pregnant woman eats or drinks, whatever drugs she takes, whatever she breathes in may be transmitted to the fetus. Whether she is well nourished or not, whether she is generally healthy or ill, whether she is serene or anxious, whether she is exposed to radiation or other harmful substances all may affect the development of her fetus.

The mother's emotional state

Pregnant women may be subject to stress from various sources—poverty, marriage problems, family illness, problems at work, and so on. When a pregnant woman feels continued anxiety, distress, trauma, or extreme fear or grief, her body reacts with profound, involuntary changes. Heart and respiration rates and secretions of glands all respond to emotional turmoil. If a pregnant woman is fearful, for example, her brain may signal her adrenal glands to secrete the hormone cortisone. When cortisone enters her bloodstream, it can direct blood flow toward her own internal organs, effectively diverting blood away from the fetus and reducing the amount of oxygen the fetus gets. Although the chemical effects of pregnant women's emotions can be traced by blood tests, it is less easy to assess how these changes affect the fetus's behavior. An experiment in which pregnant women are deliberately subjected to stress and their unborn infants examined cannot be justified ethically. Researchers have used indirect methods of investigating this question. They have experimented with animals and have studied large numbers of pregnant women who have (and have not) experienced stress naturally. The results of all these studies together suggest that pregnant women's extremely stressful emotions do influence their infants' prenatal development. Further research is needed, however, before we will be able to predict the likelihood, the exact nature, or the extent of these effects.

Illness

The effects of pregnant women's illnesses on their fetuses have been documented more clearly. Two chronic conditions that pose perhaps the greatest danger to the developing infant are **diabetes** and **Rh incompatibility**. Diabetes causes the mother's blood sugar level to rise, and many diabetic women take the hormone insulin to lower their blood sugar. Both high blood sugar and insulin increase the chance that the mother will miscarry or that the baby will be overweight, have physical and neurological problems, or be stillborn. The Rh factor is a protein in the red blood cells of about 85 percent of the population. When a fetus's blood contains the Rh protein and the mother's blood does not, Rh incompatibility occurs. If any Rh positive blood from the fetus enters the mother's Rh negative blood—perhaps during the process of birth itself—the mother's system forms antibodies to the Rh factor. Problems therefore are unlikely to affect a firstborn child, but the red blood cells of fetuses during later pregnancies are subject to attack by the antibodies of the now sensitized Rh-negative mother. The possible consequences to the infant include jaundice, premature birth, stillbirth, and brain damage. Some affected infants require blood transfusions immediately after birth or even before if they are to survive. In most cases, however, the Rh negative mother is given an injection of Rh immune globulin (RhoGam) right after each delivery, miscarriage, or abortion, to prevent her immune system from creating antibodies to the Rh factor.

> We had been quite lucky with our three children. I had been given RhoGam, and all of them were born healthy. But when I got pregnant—by accident—with our fourth child, we ran into problems. She needed three complete transfusions right after delivery, and we weren't sure she was going to make it. But make it she did, thank heavens.

Acute infectious diseases caught by pregnant women also affect the fetus. German measles, or **rubella**, is particularly dangerous. Following the German measles epidemic of 1964–1965 in the United States, 30,000 fetuses and newborns died, and 20,000 were born blind, deaf, mentally retarded, or with heart defects. Although having German measles does no harm in the first two weeks of pregnancy, in the following two weeks, when organs are forming in the embryo, it harms one-half of the embryos whose mothers contract it. After that the risk diminishes. During the second month of pregnancy, German measles harms only 22 percent, and during the third month, only 6 to 8 percent of the fetuses whose mothers get the disease. Women of childbearing age can be tested for immunity to German measles; almost 85 percent in the United States are immune. Those who are not immune should be vaccinated 6 months or longer before getting pregnant.

Nutrition

> "What do you eat for breakfast?" the midwife asked me when I was pregnant.
>
> "I usually have toast," I answered.
>
> "And what's on the toast?" she pressed on.
>
> "Butter," I confessed, suspecting that this was the wrong answer.
>
> "*Butter*? Butter has no protein. Use peanut butter, or melted cheese, but make sure you get protein. Protein, protein, protein! That baby needs protein!"
>
> (I got the picture.)

Whether a woman is well or poorly nourished before, during, and after pregnancy also affects the fetus's development. Because pregnancy itself places so many demands on her system, a woman should be well nourished as she begins her pregnancy. It is very difficult to overcome nutritional deficiencies during pregnancy, a time when caloric needs increase by about 20 percent, when the needs for protein and riboflavin increase by 45 percent, and the need for vitamin C increases 100 percent. A pregnant woman needs to eat about 2000 wisely chosen calories a day.

So important is nutrition to fetal development that children conceived during the cool autumn and winter weather—when their mothers eat heartier, protein-rich roasts and stews—are heavier, healthier, more likely to go to college and to appear in *Who's Who in America* than children conceived in warmer weather—when their mothers are more likely to skip heavy meals in favor of fruit and salads. Likewise, more mentally retarded children are conceived in spring and summer than in winter, and the hotter the summer, the likelier this is to be (Knobloch and Pasamanick, 1966). Infants conceived during periods of famine also may have impaired intelligence (Montagu, 1962). Newborn infants of chronically malnourished mothers lag in motor and neurological development. They suffer from malnutrition themselves, weigh less, and have lighter and less protein-dense placentas than normal infants (Bhatia, Katiyar, and Agarwal, 1979).

A generation or two ago, physicians warned pregnant women against gaining more than 10 or 15 pounds during their pregnancies. The belief was that a higher weight gain might cause toxemia.

> When I was pregnant with Sam's father, I gained 12 pounds. That was all they allowed us in those days. Some women asked for diet pills to curb their appetites. Heaven knows what that did to their babies.

But physicians today realize that women who gain too *little* weight may have stillbirths or infants that are too small. They advise women to gain at least 24 pounds during the course of a pregnancy, and women who begin their pregnancies overweight are cautioned against trying to reduce until after their babies are born.

Mother's age

Although most adolescent girls can get pregnant in their early teens and women remain fertile for 40 years or so, the years between 22 and 29 are physiologically the best time to have a baby. In this prime time, both mothers and infants are more likely to survive and to go through pregnancy and delivery free of complications. Yet social trends have increased the number of births to women both older and younger than this prime age range. Risks to older women include difficulties in conceiving and delivering and an increased probability of having a child with Down's syndrome. Another consequence is an increased likelihood that the child will have problems in developing fine-motor skills (Gillberg, Rasmussen, and Wahlstrom, 1982).

Teenage pregnancy is also risky. Physically immature, and often psychologically unprepared for pregnancy, many teenage mothers have long labors and premature deliveries; their infants are often underweight and may not survive their first year. Many factors contribute: a teenager's uterus is immature, her body is still growing, she may be poorly nourished, and she may lack prenatal medical care. This powerful combination of disadvantages makes early childbearing as dangerous to mother and child as late childbearing.

Mother's work

Today, the majority of women of childbearing age hold jobs that take them outside their homes. Many women work through their pregnancies, right up to the time of delivery. Does a pregnant woman's work affect her health or that of her fetus? It may well do so. In a survey of 7700 pregnancies, it was found that women who continued working in their third trimester bore infants who weighed from 5 to 14 ounces less than infants born to women who remained at home (Naeye and Peters, 1982). The weight deficit was greatest when the mother herself was underweight or had high blood pressure. The women whose jobs required them to stand for long periods when they were in their third trimester had fetuses who were most severely underweight, probably because the blood supply to these mothers' uteri and placentas was poor. Of course, these risks are just statistical probabilities. Many women work throughout their pregnancies without ill effects to their babies. The decision about how long to continue working has to be based on a variety of factors like the woman's health, the kind of work she does, and the leave policy where she works.

Parity

The number of children a woman already has borne, her **parity**, also affects the course of pregnancy and prenatal development. It takes a woman's endocrine system some four years to return to its previous level after a pregnancy

(Maccoby, Doering, Jacklin, and Kraemer, 1979). Any infant born sooner than that may be at a disadvantage. Beyond this period needed for recuperation, later-born infants seem to have a better prenatal environment than firstborns. Blood circulation to the placenta is richer after a first, "practice-session" pregnancy. Many later-borns are heavier at birth and suffer from fewer malformations and birth complications. Many women have easier labors and deliveries of later-born children. Said one mother,

> With my second child, my labor was about one-third as long as it had been with my first. It was as if my body had learned "the route." It somehow cooperated with the contractions rather than fighting them, as it had done the first time. I recovered faster, too. Everything about the second time was easier.

But the advantages of the first "practice session" are lost if later births follow either too quickly or too late.

External factors

Although the uterus is a protected environment for the fetus—with its warm fluid to cushion shocks and its layers of membrane and abdominal muscle—it is not impervious to the external environment. Through the placenta, the fetus is affected by whatever the pregnant mother eats, drinks, smokes, or breathes. Certain substances are now known to cross the placental barrier and to cause physical deformities and behavioral disorders in the fetus. More are being discovered all the time. **Teratology** is the study of structural and functional deformities in children. It takes its name from the Greek word *teras*, meaning "monster" or "marvel." The goal of teratologists is to trace the causes of deformities and to anticipate risks to prenatal development. Teratologists point out the critical periods in prenatal development and the external factors—**teratogens**—that may interfere with normal development. Unless they cause a **mutation** in the genes, teratogens affect only the individual fetus and are not passed on to his or her own children, should there be any.

In the sections that follow, we will describe some of the most important of these teratogens.

When this 7-year-old's mother was pregnant, she took an antinausea drug for morning sickness. Her son was born without hands or arms, and his legs are short because he has no thighs. Yet he feeds himself, colors, draws, writes, and plays several instruments with his feet.

Drugs

Until 1961, people did not know that chemicals could cross the placenta and damage the fetus (Wilson, 1977). That was the year after large numbers of infants were born without arms or legs, a condition called **phocomelia** (seal limbs). Their mothers, it turned out, had taken the drug thalidomide to quell the nausea of early pregnancy, and it had prevented the normal development of the fetuses' limbs. Before the connection between the drug and the deformity was made, 10,000 infants with phocomelia were born, of whom one half have survived to adulthood. The thalidomide disaster made scientists and laypeople alert to the previously unsuspected dangers that drugs may pose to the developing fetus.

Still, pregnant women continue to take drugs every day. One study showed that women took an average of ten drugs, not even counting vitamins, iron, caffeine, nicotine, and alcohol (Hill, 1973). Over half took aspirin or other painkillers, diuretics, or antihistamines; one-third took sedatives; one-fifth took hormones. Most hormones, sedatives, antibiotics (tetracycline, streptomycin, etc.), tranquilizers (Valium, Thorazine, etc.), and anticoagulants now are known to have potentially harmful effects on the embryo or fetus.

Street drugs also can hinder fetal development. Pregnant women who are addicted to cocaine, heroin, or methadone pass their addiction on to their fetuses, who are born addicted and must go through withdrawal after birth. Many of these infants are born prematurely, underweight, and irritable (Householder, Hatcher, Burns, and Chasnoff, 1982; Jeremy and Hans, 1985). They are tense, fussy, resist cuddling, and sleep irregularly. At the age of 1 year, their motor and cognitive development still may be lagging.

Drugs hit the fetus as hard as they do for two reasons: first, because a small amount of a drug for an adult is a huge amount for the tiny fetus; and second, because liver enzymes, necessary for breaking down drugs, do not develop until after birth. The drugs therefore stay in the body of the fetus. Women may stop taking drugs when they find out that they are pregnant, but many do not even realize that they *are* pregnant until the embryo already has been damaged. Tranquilizers, street drugs, and birth control pills all can cause deafness, cleft palate, heart and joint defects, arm and leg defects, neurological and behavior disorders that may show up at birth or later. Drugs taken in combination may have even more disastrous effects on a fetus than one drug taken alone (Wilson and Fraser, 1977). Total abstinence from all drugs that are not absolutely necessary for her own health is the best policy for any pregnant woman.

Smoking

> When I got the news that I was pregnant, I reacted by treating myself to my last cigarette. Mikey and Erica's babysitter smoked all the way through her pregnancy. I spoke to her one time about the harm it does to the baby, but she said she was too nervous to quit.

Smoking is the source of a dangerous drug (Ericson, Kallen, and Westerholm, 1979; Evans, Newcombe, and Campbell, 1979; Frazier, Davis, Goldstein, and Goldberg, 1961; Jacobson, Fein, Jacobson, Schwartz, and Dowler, 1984). With every drag on a cigarette, pregnant smokers bathe their own and their fetuses' systems in toxic substances. Pregnant smokers have 28 percent more miscarriages, stillbirths, and newborn deaths. During the last trimester of pregnancy, their fetuses gain an average of 6 fewer ounces than those of nonsmokers. Fetuses whose fathers smoke are also at risk for low birth weight.

Are there long-term harmful effects on children if their parents smoke during pregnancy? The evidence here is not so clear. Some researchers indeed have found that these children continue to be small and are likely to do poorly in school and on tests of attention and orientation as well (Butler and Goldstein, 1973; Fogelman, 1980; Streissguth et al., 1984). Other researchers have not found significant physical or intellectual effects (Hardy and Mellits, 1972; Lefkowitz, 1981). Those infants who survive the prenatal and neonatal periods may not have major problems later. To some extent what happens depends on the overall quality of their lives as children. Nevertheless, some minority of children are impaired. The risk is not worth taking. Pregnant women—and their partners—should stop smoking and stay out of smoke-filled rooms.

Alcohol

> A woman who is one of my patients had a baby right around the same time as I had Natalie. The poor baby had fetal alcohol syndrome. The mother said she drank a six-pack of beer every night after work while she

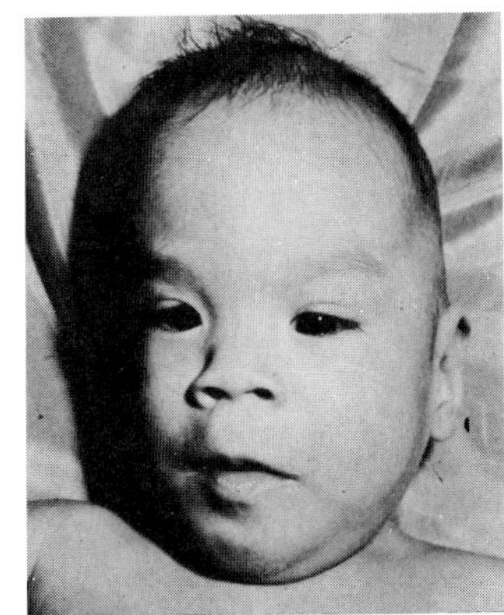

When a pregnant woman drinks 12 ounces of beer or 1.5 ounces of liquor, the fetus receives a full and debilitating adult dose of alcohol. Continued drinking during pregnancy can affect a baby's intelligence and appearance.

was pregnant. Why? She'd always drunk that much before, and no one had told her it was too much.

No one knows exactly how much alcohol during pregnancy is *too* much, but heavy drinkers run a 17 percent risk of having a stillborn baby and a 44 percent chance of having one that is deformed. The **fetal alcohol sydrome** was identified in 1973 among infants born to pregnant women who drank heavily (Jones, Smith, Ulleland, and Streissguth, 1973). A similar syndrome was later observed in infants born to pregnant women who smoked marijuana (Hingson et al., 1982). Infants with fetal alcohol syndrome are born short for their weight and do not go through a normal period of catch-up growth after birth. They are mentally retarded and slow in motor development, and they may also suffer from defects of the eye, heart, joints, arms, or legs. Many have small skulls and a distinctive look to their face: eyes spaced far apart, flat nose, and underdeveloped upper jaw. It has been estimated that as many as 20 percent of the children in mental retardation centers suffer from fetal alcohol syndrome (Rawat, 1982).

Alcohol affects the fetus directly. Alcohol crosses the placenta and, like other drugs, because the fetus's liver is immature, remains in its system for a long time. Because fetuses are so sensitive to alcohol, even small amounts may cause abnormalities in development. During the last three months of pregnancy, when the fetus's brain is developing, alcohol may be extremely dangerous. The effects of alcohol, like smoking, are most severe later in pregnancy. Mothers who drink heavily but who can cut down their alcohol intake during the last three months of pregnancy bear significantly fewer babies who are abnormally small (Rosett, Weiner, Zuckerman, McKinlay, and Edelin, 1980). When fetal alcohol syndrome was first described, doctors thought that pregnant women could drink moderately without harming the fetus. But they have since come to revise this idea. There may be no safe level of alcohol intake during pregnancy.

Radiation

I had a dental checkup very early in my pregnancy with Sam. So when the hygienist asked if they could take X rays, I said no, because I was pregnant. She wrote ''pregnant'' across the front of my chart. I guess I was being pretty careful.

The radiation from X rays and other sources can cause harmful mutations in genetic material. It can cause mutations of chromosomes in the unripe ova stored in a female's ovaries or in the sperm-producing cells in a male's testes. For this reason, unless they are absolutely necessary, X rays of the lower abdomen and pelvis should be avoided in the childbearing years and earlier. When a woman is pregnant, exposure to radiation before the zygote implants in the uterine lining is likely to end the pregnancy. If the zygote survives, however, it is likely to be normal. But exposure to radiation after implantation affects development in a variety of ways. If an embryo is exposed to radiation, it is likely to develop deformities of the central nervous system that kill it soon after birth. If a fetus is exposed to radiation, it may later develop malignant tumors or leukemia or have stunted growth. Extreme doses of radiation, from radiation therapy or atomic explosion, cause **microcephaly** (small skull size), mental retardation, Down's syndrome, hydrocephaly, defects of skull formation, or death (Joffe, 1969). Radiation fallout in the air and ground around us may also, according to some investigators, raise the rates of malformations (Joffe, 1969). Other investigators disagree, saying that natural radiation levels in most parts of the world are not high enough to harm fetuses' development (Brent, 1977). Atomic waste buried under land or water and accidents at nuclear power plants are more likely to pose hazards to fetal development.

Prenatal vulnerability

Although the teratogens we have discussed sometimes cause death or severe malformations, damage is not inevitable. Human beings have a **self-righting tendency**, a tendency to develop normally under all but the most damaging conditions. Even when exposed to possible harm, most fetuses develop normally. Despite all of the harmful substances to which fetuses may be exposed in our modern world—viruses and bacteria, junk food and alcohol, drugs and radiation—only 3 to 6 percent of the babies born today suffer actual malformations (Heinomen, Slone, and Shapiro, 1976). In addition to this self-righting tendency, a ''self-cleansing tendency'' operates to eliminate severely malformed fetuses by spontaneous abortion in the first three months of prenatal development.

When damage from a harmful substance does occur, it may range from slight to fatal. The severity of the effect rests on three factors:

1. *The constitution of the fetus.* Different fetuses react differently to the same teratogen. One might die, one might survive with severe malformations, and a third might have only a slight malformation.
2. *When exposure occurs.* The same teratogen affects a zygote, an embryo, and a fetus differently. Before cells are differentiated, exposure to a toxin is likely to kill the organism. After cells are differentiated, during the embryonic period, damage affects specific organs or systems at the particular time they are being formed (see Figure 4.3). In the fetal period, the toxin is most likely to affect behavior or intelligence.
3. *The amount of exposure.* The more extreme the exposure to the teratogen, the more severe are its effects. Evidence from pregnant women who

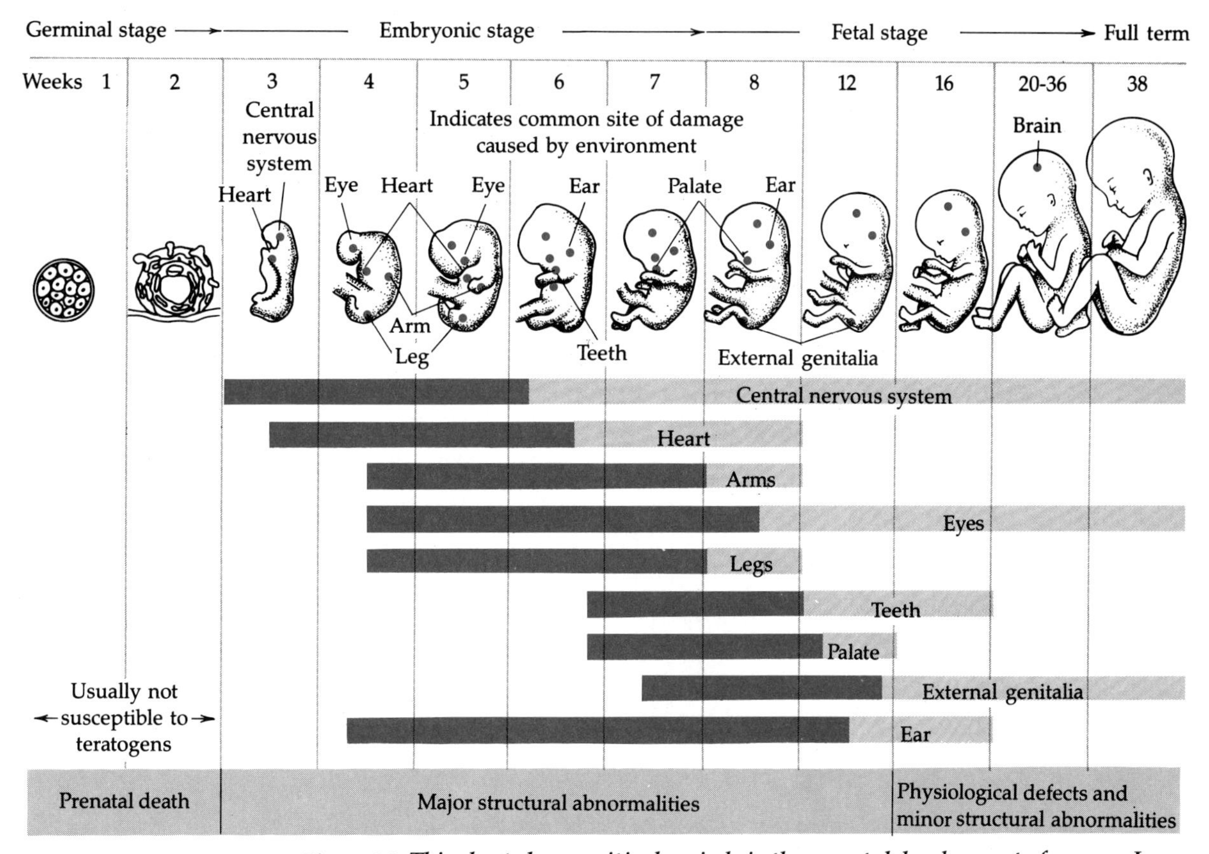

Figure 4.3 ***This chart shows critical periods in the prenatal development of organs. In the embryonic period, the likelihood of structural malformation is greatest (dark color), because organs are being formed. After organs are formed, the likelihood of structural defects declines (light color) (Moore, 1982).***

lived through the atomic bombings of Hiroshima and Nagasaki shows, for example, that no women who were within a mile of the center of the bombing bore live infants. Three-quarters of those who were 1 to 4 miles from the center miscarried, had stillbirths, or bore severely malformed infants. The infants of women farther away suffered mental and physical retardation and abnormal skull size (Joffe, 1969; Wilson, 1977).

Thus the timing and amount of exposure to a teratogen, in combination with the individual's constitution, all operate together to determine the extent of the damage.

Delivery and birth

Labor

For nine months, the mother has carried, protected, and thought about the creature developing within her. Now the big moment finally has arrived. Filled with anticipation and curiosity—and at least a little anxiety—she prepares for her moment of truth, her first glimpse of the intimate stranger. She

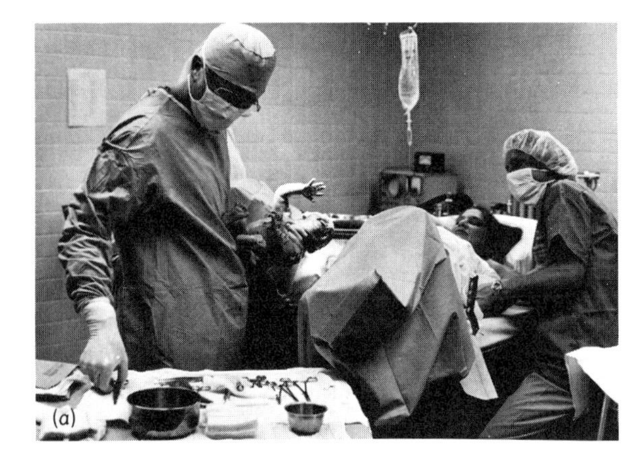

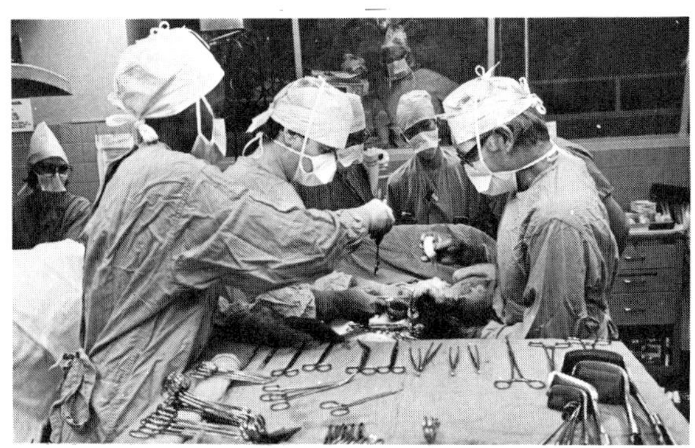

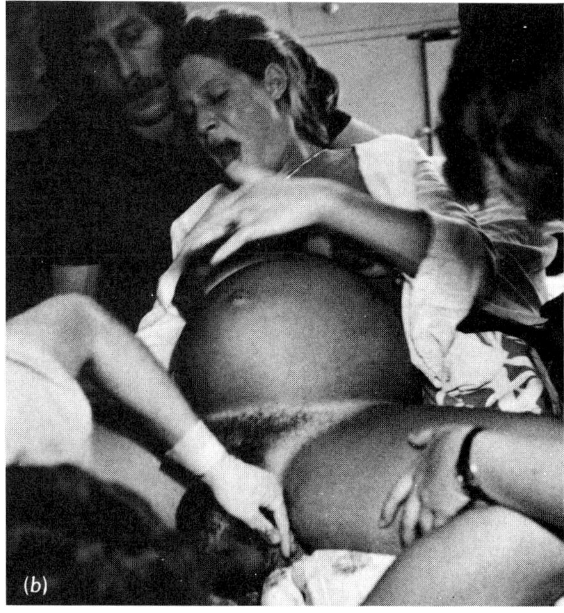

The means of delivery affects the parents' experience of birth. **(a)** *In the traditional hospital delivery room, childbirth becomes one kind of experience.* **(b)** *In a home, or a home-like birthing center, birth may seem somewhat different.* **(c)** *A cesarean delivery is another kind of experience for parents.*

begins the hard work that will end in the birth of the infant. For many mothers, **labor** begins at night, when the mother feels her uterine muscles contracting. These contractions, which are usually about 15 to 20 minutes apart and last for 15 to 60 seconds each, stretch the cervix—the opening to the uterus through which the fetus will pass into the vagina and, from there, into the outside world. The cervix gradually opens from 0.2 inch to 4 inches. As labor intensifies, the uterus contracts more intensely every two to five minutes, and these powerful contractions open the cervix the last inch. When the cervix has opened, the infant's body—in most cases, the top of the skull—begins to push through into the vagina. This stage of labor lasts an average of 14 hours for a first baby, 8 hours for later ones, although there are wide individual variations.

> I listened with envy as a friend described her labor. After the contractions started, she called her husband and went outside to bring in the cat. That done, she waited a few minutes for her husband to arrive and then drove with him to the hospital. By the time they got there, she was so fully dilated that the nurse had her do breathing exercises to try to slow things down until the doctor could get there. The baby was born half an hour later. My labor lasted 37 hours!

The second stage of labor begins once the crown of the infant's head appears at the vagina and does not recede with each contraction—a point called, naturally enough, **crowning**. Crowning is an exciting time.

> Once Sam's head crowned, I could feel a rush pass through everyone in the room. ''Brown hair! It's got brown hair,'' my husband told me almost incredulously. ''It's time to push now,'' the midwife told me. ''Push that

> baby out.'' Pushing was such a relief after the intense contractions I had been having that it was actually almost pleasant.

During the second stage of labor, which lasts between half an hour and two hours, contractions wash over the laboring woman every minute or two and last for a minute each. With each wave, she may feel a profound urge to breathe deeply, tense her muscles, and push the baby out.

After the infant's head is out of the vagina, the doctor turns it sideways and the shoulders and body slide out quickly. When the umbilical cord hits the air, a jelly-like substance inside it makes it swell and the blood vessels tighten. More contractions separate the placenta and other membranes from the uterus and push them out through the birth canal as the **afterbirth**.

Changing attitudes toward childbirth

The final outcome of any successful birth is a normal infant and intact mother. But the social conventions and attitudes with which people surround the biological events of childbirth transform its meaning from one culture and one historical era to another.

> My grandparents were all born at home between the 1890s and 1910 or so, some with a midwife and others with a doctor in attendance. My parents were born in the hospital, their mothers under general anesthesia. As one of my grandmothers said, ''I didn't want to know what happened. I wanted to wake up 24 hours later and have someone bring me a nice clean baby.'' My mother gave birth to me in a hospital, too. My father wasn't allowed in with her. She said she inhaled some kind of ''gas'' for pain and was in and out of consciousness. When I had Sam, my husband coached me all the way through labor. I had no painkillers, but I was attended by an obstetrician and a nurse-midwife. The whole routine was set up to make us feel comfortable and together.

When Fernand Lamaze, a French doctor, visited Leningrad in 1951, he saw unmedicated women going through labor with an entirely different approach from that of the French women he was used to. Pregnant women in Russia, he discovered, learned to concentrate on their breathing and muscle tension so that they worked with rather than against the natural contractions of labor. Lamaze imported the method to France. There he trained people to coach laboring women so they could regulate their breathing, relax, and concentrate on the work to be done. Soon the method was imported to the United States. Here, in the six weekly classes that begin in the mother's seventh month of pregnancy, the husband (or another relative or friend) is trained to coach and monitor the woman and to support her physically and emotionally during childbirth. Now every year in this country, over 500,000 couples take a Lamaze training course, usually taught by a trained nurse or physiotherapist. A recent survey of 400 hospitals throughout the country showed that Lamaze preparation was widespread (Wideman and Singer, 1984). Virtually all the hospitals allowed fathers into labor and delivery rooms, and almost all obstetricians recommended Lamaze training to their patients. In over 70 percent of the hospitals, more than half of the women—from all regions, cities, income levels, and ethnic backgrounds—had Lamaze preparation. The numbers increase every year. Some hospitals have converted space into homelike birthing rooms, where unmedicated women can labor, deliver, and recover in the same place, and where they and their labor partner have a degree of comfort and privacy.

How effective is Lamaze training? In one study comparing women who had taken Lamaze classes and women who had medicated deliveries without this preparation, the researchers found that although the women did not

Some research suggests that training in the Lamaze method leads to fewer birth defects and complications. In this Lamaze class, pregnant women and their partners practice for labor and delivery.

differ significantly in their levels of anxiety, those who had taken the course felt more positively about their pregnancies, needed less pain medication, remembered their labors as less painful, and felt better about themselves (Tanzer and Block, 1976). All the women who had experienced childbirth as rapturous or ecstatic were in the natural childbirth group.

Despite Lamaze preparation, of all the women who give birth in hospitals in the United States, it is estimated that 95 percent accept some form of pain medication (Brackbill, 1979). This is a mixed blessing. Although these drugs relieve the mother's pain, they also pass into the fetus's bloodstream and reduce its oxygen supply. In consequence, some infants are born too weak to breathe on their own and may suffer from a lack of oxygen to their brain. Drugs can also depress the newborn's attentiveness and vigor of sucking (Brazelton, 1961). Even a local, **epidural anesthesia**, which numbs the mother between chest and knees, slows the newborn's motor abilities somewhat, especially the ability to control the head when the infant is pulled into a sitting position (Scanlon, Brown, Weiss, and Alper, 1974). More severe effects are observed when the mother is given more and stronger drugs (Lester, Als, and Brazelton, 1982).

Not all these effects disappear right away. Before birth, the placenta can cleanse the fetus's system of some drugs. But when the umbilical cord is cut, the level of a drug in the newborn's bloodstream is 70 percent that of the mother, and as we have seen, the newborn's internal organs are inefficient at cleansing its system of drugs. For at least four weeks after birth, newborns of medicated mothers tend to see somewhat more poorly and to lag slightly behind newborns of unmedicated mothers in muscular and neural development (Brackbill, 1979).

How might these effects of medication affect the course of an infant's development? Understandably, this is a question to which parents and obstetricians would very much like the answer. But the answer is not easy to come by. Because so many different factors affect the newborn's progress, it is difficult for investigators to isolate specific causes of long lasting effects. But newborns whose mothers have been heavily medicated during childbirth generally interact differently with their mothers and other people from new-

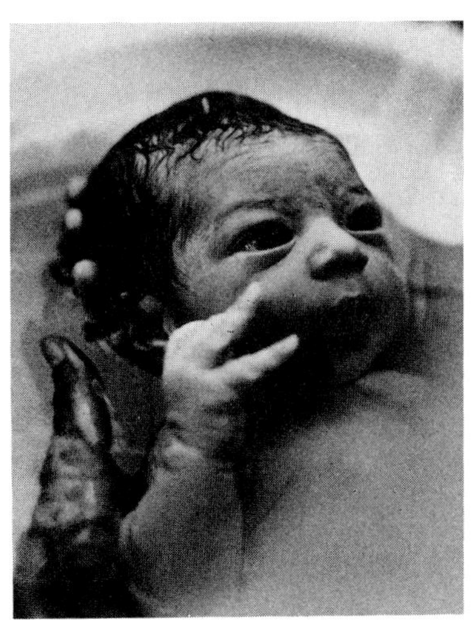

This newborn infant is being gently lowered into a warm bath similar to the amniotic waters he has just left. French obstetrician Fredrick Leboyer recommended this practice to ease the infant's transition from womb to room.

borns whose mothers have not been medicated. In one study, for example, newborns whose mothers had been more heavily medicated during childbirth opened their eyes significantly less while they were nursing and responded less to sounds than newborns of less medicated or unmedicated mothers (Brown et al., 1975). If these early behaviors harden into patterns of interaction, the relationship between infant and parents may suffer.

> When Jason was born, he did not breathe right away because of the anesthesia I had been given a few minutes before. Someone was supposed to be waiting right there with a stimulant for the baby just in case he wasn't breathing. But no one was there, and it took a minute or more until they finally got him breathing. He was sluggish and hard to get started sucking. He'd drop off to sleep and then wake up hungry in just a little while. It was hard to get in tune with his needs. He was also very irritable.

The most recent innovation in delivery procedures comes from another French obstetrician, Frederick Leboyer. Leboyer felt that the conventional methods of delivery were unnecessarily harsh and violent for the infant. He decided to do away with the bright lights, the clanking instruments, and the quick cut of the umbilical cord, to stop dangling infants upside down and slapping them on the buttocks right after birth. In Leboyer's method of **gentle birth**, the transition from soft, warm, dimly lit, muffled womb is eased with soft lights, quiet voices, a warm room, and gentle handling. Immediately after birth, the newborn is placed on the mother's abdomen and its back gently massaged. In this way, amniotic fluid is expelled from the windpipe. After a few little cries, the baby breathes naturally, and only after the umbilical cord stops pulsing is it cut. The baby is then gently raised upright and carried to a warm bath and, from there, to a warm diaper. Leboyer (1975) found that infants handled with such gentleness and respect radiated contentment.

Does the Leboyer method really make a difference? Systematic studies are scant. One French investigator (Rapoport in Salter, 1978), found that 120 8-month to 4-year-olds delivered by the Leboyer method were free of sleeping and emotional problems, had highly developed interests in people and things, and were socially well adapted. Their scores on a test of infant adaptiveness were above average. But the study lacked an all-important control group of infants delivered in the usual way. In a study in which a group of 17 infants delivered by traditional hospital methods was compared with 20 delivered by the Leboyer method, it was found that the infants differed significantly in their first 15 minutes (Oliver and Oliver, 1978). The traditionally delivered group showed more physical tension, blinking, sucking, trembling, and shuddering. The Leboyer-delivered infants were more relaxed, opened their eyes more, and made more soft sounds. In their baths, they relaxed their muscles, opened their eyes, moved around, and did not cry. But did these effects last beyond the period soon after birth? One investigator who followed newborns delivered by the Leboyer method for their first three months found that the effects were relatively short-lived (Sorrells-Jones, 1983).

Summary

1. During sexual intercourse around the time of ovulation, sperm from the man enter the woman's vagina, swim into the uterus and up the fallopian tubes, where one penetrates the ovum and merges its genetic material with the ovum's.
2. Infertility affects about 15 percent of the couples trying to have children. Many of them can be helped to have children through surgery, drugs,

artificial insemination, in vitro fertilization, ovum transfer, or surrogate mothering. About 40 percent of the cases of infertility are attributed to the man, particularly to the quantity or quality of the sperm. About 60 percent are attributed to the woman, particularly to a failure to ovulate or blocked or damaged fallopian tubes.

3. Conception marks the beginning of pregnancy and prenatal development; birth marks its end. In the earliest stage of prenatal development, the germinal stage, the fertilized egg divides into cells, which form a sphere, and implants itself in the lining of the uterus. In the embryonic stage, from 2 weeks to 2 months after conception, organ systems form. In the fetal stage, which lasts from 2 months until birth, the organs grow and physical functions develop.
4. Pregnancy brings changes and adjustments to both expectant parents. Both may feel emotional and physical strains. A pregnant woman proceeds through a succession of social, physical, and psychological adjustments in preparation for childbirth and motherhood.
5. The development of the fetus is affected by the mother's emotional state, health, diet, age, number of previous pregnancies, and exposure to drugs, radiation, and other environmental hazards.
6. Environmental substances affect prenatal development by causing genes to mutate, by interfering with normal cell division, or by delaying or distorting growth. These effects range from mild to lethal, their severity depending on the timing and dosage of the exposure and the constitution of the individual fetus.
7. Prenatal development is protected by a self-righting tendency, which means that it proceeds normally except under the most adverse conditions. It also is protected by a self-cleansing tendency, which means that severely abnormal embryos and fetuses are miscarried.
8. Around the end of the ninth month of pregnancy, labor begins. In the first stage, strong contractions of the uterine muscles thin and open the cervix. In the second stage, the baby is pushed out of the uterus, usually head first. In the third stage, uterine contractions deliver the afterbirth—placenta and membranes. Labor may be induced or speeded up if it does not begin on its own or its pace seems, to a doctor, too slow for the safety of mother or child.
9. Attitudes toward childbirth vary from one culture and one historical era to another. Today in our culture, many people favor natural births during which the mother actively participates by controlling her breathing and relaxing. In many such births, the father or another labor coach is present at the delivery. Some people also favor ''gentle birth'' into a warm, dimly lit room, with calm procedures designed to relax infant and parents. The effects of gentle births may not last beyond a few hours. But the amount and type of medication that women are given during childbirth does affect their newborns.

Key terms

ovulate
oocyte
fallopian tube
uterus
spermatozoa
cervix
infertility
pelvic inflammatory disease (PID)
artificial insemination
alternative reproduction
in vitro fertilization
test-tube baby
laparoscopy
surrogate mothering
ovum transfer
germinal stage
embryonic stage
fetal stage

morula
totipotent
blastula
blastocyst
chorion
amnion
placenta
villi
implant
embryo
ectoderm
primitive streak
endoderm
mesoderm
umbilical cord
cephalocaudal
proximodistal
ossification
quickening
lanugo
vernix
neonatology
fetology
hydrocephaly
alveoli
oxytocin
toxemia
diabetes
Rh incompatibilty
rubella
parity
teratology
teratogen
mutation
phocomelia
fetal alcohol syndrome
microcephaly
self-righting tendency
labor
crowning
afterbirth
epidural anesthesia
gentle birth

Suggested readings

ANNIS, LINDA F. *The Child before Birth*. Ithaca, N.Y.: Cornell University Press, 1978. A thorough and clearly written presentation of the prenatal physical development of the child, with explanations of Rh disease and other possible prenatal complications.

COREA, GENA. *The Mother Machine*. New York: Harper and Row, 1986. A powerful examination of the damaging and frightening implications of new reproductive technologies—artificial insemination, ovum transfer, in vitro fertilization, surrogate mothering, and sex determination.

GOLDBERG, SUSAN, and DIVITTO, BARBARA A. *Born Too Soon: Preterm Birth and Early Development*. San Francisco: W. H. Freeman, 1983. An up-to-date consideration of significant aspects of preterm infants' development, including methods of caring for them and aiding their development.

GUTTMACHER, ALAN. *Pregnancy, Birth and Family Planning*. New York: Signet, 1984. Written for expectant parents—facts and reassurance from a medical authority.

MACFARLANE, AIDAN. *The Psychology of Childbirth*. Cambridge, MA: Harvard University Press, 1977. A warm and revealing account of the feelings of expectant parents before and during childbirth.

NILSSON, LENNART. A *Child Is Born: the Drama of Life before Birth*, with text by A. Ingelman-Sundberg and C. Wirsen. New York: Dell, 1981. Remarkable photographs taken in the womb between conception and birth.

SHAPIRO, HOWARD. *The Pregnancy Book for Today's Woman*. New York: Consumers Union, 1984. Sound advice about nutrition, drugs, and hazards to the expectant mother.

PART THREE

Infancy

chapter five

Physical and perceptual development

An infant's first month is an exciting, exhausting, exhilarating time. Contrary to parents' expectations—fed on baby food advertisements—their tiny infant is not exactly the clear-eyed, rosy-cheeked, smiling moppet that they fantasized about during pregnancy. No, the real infant squawks, snores, and sneezes. When they lay the baby in the crib for a quiet nap, all they hear is a barrage of wheezes and wails. The eyes that parents want to gaze into are usually shut tight in sleep or squinting from crying, and when they do open, the infant seems to look everywhere but *at* the poor parents. Smiles come and go for no apparent reason. But when they do come, they are so thoroughly enchanting that Mother and Dad will work for *days* just to get another one.

This fascinating, puzzling, habit-forming infant is also exceptionally vulnerable in this first month of life. More people die in the first month of life than at any other time before old age. Newborns who weigh at least five pounds and have developed for at least 28 weeks in their mother's uterus have the best chances of surviving this risky first month. Newborn girls have better chances of surviving than newborn boys, and newborn blacks have better chances of survival than newborn whites of comparable weight and gestational age. They are probably less vulnerable to environmental assaults because physiologically they are more mature.

The stress of birth to the newborn is so great that doctors and nurses use a quick and simple test to judge the baby's condition. The **Apgar test**, developed by Dr. Virginia Apgar in 1953, is used in delivery rooms all over the United States and in other countries. At one minute and again at five minutes after birth, the newborn is scored on heart rate, respiration, muscle tone, reflex irritability, and color. For each category, the newborn receives a 0, 1, or 2. The healthiest babies get a total score of between 7 and 10 points. These babies have hearts beating strongly over 100 times a minute, lusty cries, firm muscle tone, and pink rather than ashen skin. Tickled on the soles of their feet, they quickly cry, cough, or sneeze. About 90 percent of the babies born in the United States pass their first test, the Apgar, with flying colors (a 7 or better).

> I had heard of the Apgar, but I was still surprised when the delivery room nurse told us that Natalie had a 9 at one minute and a 10 at five minutes. Naturally, I immediately wanted to know why she hadn't gotten a 10 the first time. Her hands and feet were bluish, as it turned out.

Those who score 4 or below need immediate medical attention if they are to survive.

> Jason didn't breathe right away after he was born, probably because I had been given some Demerol only a few minutes before that. He was limp and blue. The doctor and nurses got very quiet and intense as they tried to start him breathing. I heard the nurse murmur, "Still not breathing" what seemed hours later but was only a minute or two. They worked on him, and he sort of spluttered and fussed. Then his whole body transformed from a pasty gray to a healthy pink. His first Apgar was a 3, but his second was a 7.

Tasks for the newborn

With birth, an infant moves from one kind of environment—the dim, warm, protected liquid of the uterus, where breathing and feeding are automatic—

into a radically different environment—the bright, noisy, cool, and airy world, where for the first time the infant must breathe and feed on his or her own. Within moments of birth, each newborn must abandon one system of respiration—the placenta—and make functional another—the lungs. Until that sudden indrawing of air into the lungs occurs, the baby's life is in doubt. If the baby is to live, it must begin breathing air within minutes after emerging from the birth canal. That first breath comes only after mucus and amniotic fluid have cleared the lungs. Some of the fluid is squeezed out through the infant's mouth and nose by pressure in the birth canal during birth itself. Some escapes by gravity if the infant is held upside down. Some may be suctioned by a waiting nurse. Some evaporates or is absorbed into the infant's bloodstream. Once in the air, and even before the umbilical cord is clamped, the newborn begins to snort and sneeze, trying to clear the air passages and to inflate the thousands of air sacs in the lungs.

As the umbilical cord is exposed to the cool air, to sponging, and to handling, it stops pulsing and constricts. The doctor clamps and cuts it, and the flow of blood from the placenta stops. The infant's own circulatory system now must function on its own. Blood flow in the infant's heart now actually changes directions. As blood flows for the first time to the infant's lungs, the pressure in the chambers of the heart alters. It increases in the left atrium and closes a valve. The infant's blood begins to flow through the heart from left to right rather than from right to left, as it had before birth. This critical change may plunge the infant's blood pressure downward and race the heartbeat to 140 counts a minute. For even the healthiest of infants, it may take ten days for blood pressure to stabilize.

After nine months in the tropical heat of the uterus—98.6 degrees Fahrenheit—the newborn also must adjust to the cooler, changeable temperatures of the outside world. Most newborns have only a thin layer of fat under their delicate skin, and their temperatures may drop as much as 5 degrees right after birth.

> As soon as Sam was born, the nurse laid him on my stomach, dressed in a little cotton cap to conserve his body heat. A while later, the nurse put him under red warming lights. Finally, when he was wheeled into my room on the maternity ward, he was swaddled in several cotton blankets and sleeping peacefully inside what looked like a toaster oven! Alarmed, I asked what was wrong. The nurse calmly told me that Sam was having trouble regulating his temperature and that the warmer would help him. By the next day, he was out of the toaster and doing fine on his own.

Most infants' temperatures stabilize within eight hours or so after birth.

Not only must newborns breathe and warm themselves, but they must also begin drinking, digesting, and excreting. Healthy newborns can suck and swallow, taking in an ounce or two of liquid every two hours or so during their first days. Their bowels first excrete a tarry waste left over from fluid and cells swallowed in the uterus. Soon their stomachs and intestines begin to secrete the substances necessary for digesting nutrients.

With birth, the newborn's protection from germs in the sterile uterus comes to an end. Now the infant's vulnerable system is assaulted by all the microorganisms—harmless and harmful alike—that live in our world—in Father's sneezes, on Grandma's hands, in an unsterilized bottle of formula, or hanging around the diaper pail. Many newborns seem to lose ground for a time, as their immune systems fight to create antibodies to a flood of bacteria and viruses. Although newborns receive immunities to some diseases from their mothers before birth and after, through the mother's breast milk, their immature immune systems must nonetheless take on a host of invaders.

Risks to infants

Although newborns are most vulnerable during the first month, the whole first year poses special risks. Prematurity, slow growth before birth, disease after birth, respiratory problems, malnutrition, poor medical care, poverty, neglect, even a mother's poor health all take their toll on infants during the vulnerable first year. In the United States, the survival rate for infants of all races has been increasing. Today more than 990 out of 1000 infants survive their first year of life. But the rate is still lower than the rate in many other nations that have more widely available health services. In Japan and Finland, for example, 994 out of 1000 babies survive the first year.

Premature birth

Babies born after pregnancies that last between 35 and 40 weeks after conception are **full-term babies**. Babies born earlier than this are called **preterm babies**. Those born later are called **postterm babies**. Of all the complications of birth in this country, prematurity is the most common. It affects about 7 percent of all newborns.

Why are babies born prematurely? There are many possible reasons, and we have mentioned a number of them already. Mothers who are very young, fatigued, poorly nourished, or in poor health are likely to deliver prematurely; mothers who smoke, take drugs, have uterine problems, infections, toxemias, or who get no prenatal care are also likely to deliver prematurely. Many of the women who deliver prematurely are poor and without financial or social resources. In one-half of all cases of preterm births in this country, there is no known cause (Annis, 1978).

The number of preterm babies who survive has increased, and the severity of the problems they face has declined greatly in this century and continues to do so. Although most babies born after fewer than 26 weeks of prenatal development are too young to survive, some do make it. There are no hard and fast rules, and a preterm baby's chances depend heavily on the available medical care. At the best hospitals, almost all preterm babies weighing over 3½ pounds survive; 80 to 85 percent of preterm infants weighing between 2½ and 3½ pounds survive; and so do 50 to 60 percent of those weighing only 1½ to 2¼ pounds. In the last few years alone, the survival rate of infants weighing between 1 and 2 pounds has risen from 20 to 40 percent.

The complicated equipment that helps premature infants survive can also be intimidating to new parents. Many hospitals try to help parents overcome their fears and, most important, encourage parents to begin touching and caring for their infants while they are still in intensive care.

About 10 percent of the babies born prematurely have serious problems, like blindness, mental retardation, neurological problems, or cerebral palsy. Another 15 percent have moderately severe problems, and 75 percent—the majority—have no serious problems, if they have good hospital care (Goldberg and DiVitto, 1983; Kopp, 1983). Their development may be somewhat delayed, however. Even calculating the age of these children from conception rather than from birth, prematurely born infants compared with full-term babies are, on the average, somewhat delayed in early motor development. This delay may be the result of the relative inactivity imposed by a bed in an intensive care nursery compared with the uterine swimming pool. As for intellectual development, 85 percent of prematurely born infants are in the normal range, but on the average their IQ scores are somewhat lower, at least during the first few years of life, than those of full-term infants (Goldberg and DiVitto, 1983).

Boys seem more vulnerable to the problems associated with prematurity than girls. One study of preterm and full-term babies, all of whom were born to poor black mothers, showed that at 13 months of age, the preterm boys scored significantly lower on tests of motor development and intelligence than preterm girls of the same birth weight, although there was no comparable difference between boys and girls born at normal weight (Braine, Heimer, Wortis, and Freedman, 1966). Boys' slight lag in neurological development is apparently magnified into a noticeable problem by the stress of a premature birth.

In the hospital, preterm babies' physical functions are carefully charted. Their heart rates, breathing rates, blood pressure, blood sugar, and urine, among other indicators, are all monitored. Placed in sterile **incubators**, the tiny preterm babies automatically are provided with warmth, oxygen, and humidity. Pumps deliver nutrients and fluids to the babies through tubes to the stomach. Many preterm babies cannot breathe alone, because their lungs are immature. They are therefore placed on ventilators.

Today neonatologists are concerned not only with the physical survival of the preterm infant, but also with the quality of later development. They are concerned that preterm babies not develop problems from their treatment in the intensive care nursery. Without fat beneath the skin, they do not look like the plump, cute creatures to which parents or nurses are accustomed. Their heads may be covered with cotton stocking caps to conserve heat; their eyes may be blindfolded. They are surrounded by intimidating equipment. The intensive care unit is noisier than a typical business office. Fluorescent lights burn day and night. The babies are handled often for tests and medical care, but most of these contacts are brief and unpleasant for the baby. It is hard for parents to feel at home in such a stressful place. Many parents must travel to the nursery from far away. All these factors mean that the hospitalized preterm infant rarely gets the same kind of loving care that the term infant does.

People who work in hospitals have begun to make intensive care nurseries more human. When this happens, premature infants gain weight faster and do better on tests of motor and visual responses to stimulation (Goldberg and DiVitto, 1983). Babies benefit if the hospital nursery is made more like the womb—with tape-recordings of mothers' heartbeats, a rocking hammock lined with warm, soft sheepskin, or a waterbed—or more like home, with lots of loving handling and bright, attractive mobiles over the incubators. In one recent study, for example, massaging and moving premature babies' arms and legs for 45 minutes a day while they were in the intensive care unit resulted in a 47 percent greater weight gain, significantly more time awake and active, and more mature behavior (Scafidi et al., 1986). In fact, the infants who were given the treatment could leave the hospital six days earlier than those who were not. Another approach to making the intensive care unit more like home is to get parents more involved with their babies. When parents take part in

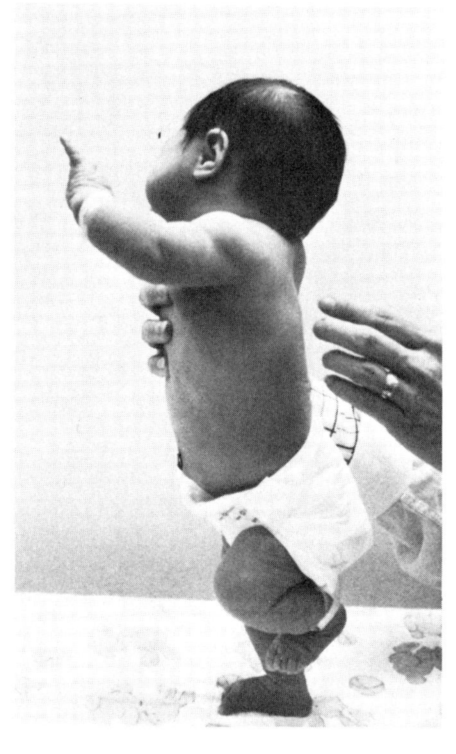

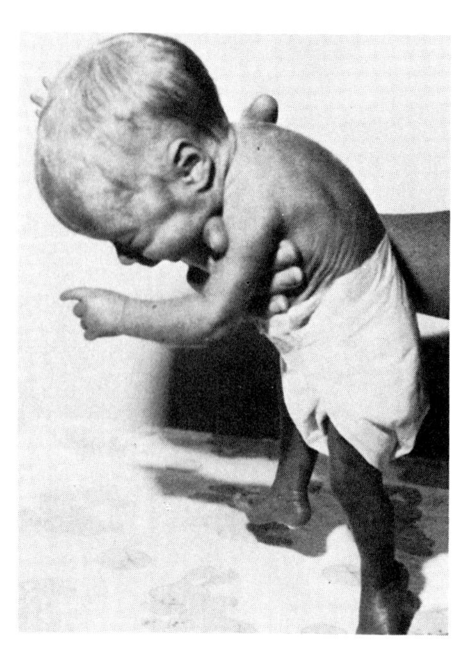

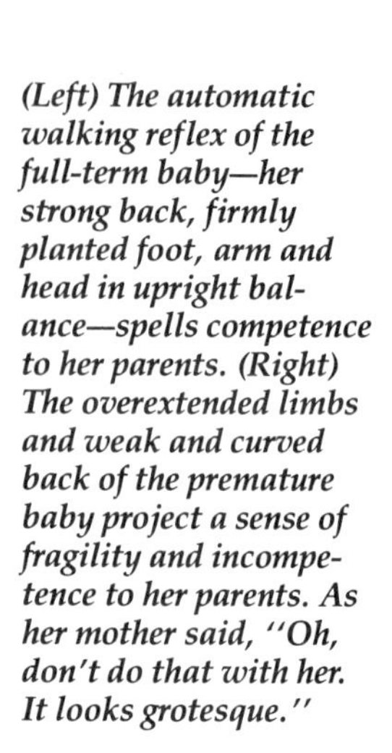

(Left) The automatic walking reflex of the full-term baby—her strong back, firmly planted foot, arm and head in upright balance—spells competence to her parents. (Right) The overextended limbs and weak and curved back of the premature baby project a sense of fragility and incompetence to her parents. As her mother said, "Oh, don't do that with her. It looks grotesque."

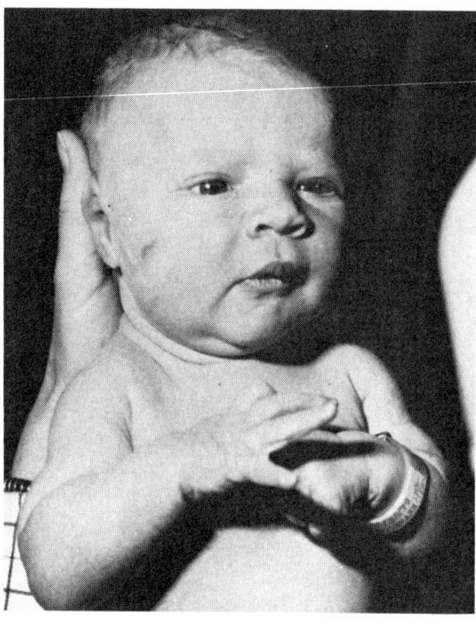

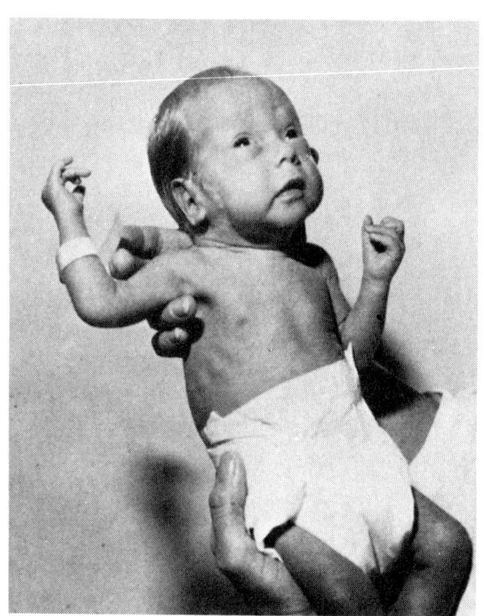

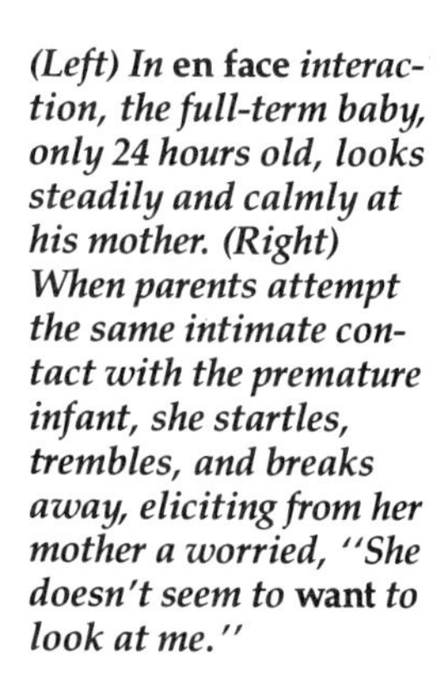

(Left) In en face *interaction, the full-term baby, only 24 hours old, looks steadily and calmly at his mother. (Right) When parents attempt the same intimate contact with the premature infant, she startles, trembles, and breaks away, eliciting from her mother a worried, "She doesn't seem to* want *to look at me."*

discussion groups with other parents of premature babies, they visit more, feel more comfortable, and better understand the problems that their babies face. Parents of premature infants themselves experience the birth prematurely and have not gone through all the "stages" of pregnancy.

Low birth weight

Full-term babies come in a varied assortment of sizes, shapes, weights, and degrees of physical maturity. One little peanut may weigh as little as 5½ pounds. The string bean in the next crib weighs 7 pounds stretched out to a long 22 inches. Next door is a hefty 10 pounder who looks more like 3 *months* old than 3 days. The average newborn weighs a few ounces over 7 pounds and extends 20 inches from head to heel. Some variation in birth weight is normal. But at the extremes—newborns who weigh under 6 pounds or over 10 pounds—there can be problems.

Full-term babies who are underweight are called **small for gestational age** or **small for dates**. The various reasons they are small—mothers' smoking, poor nutrition, poor prenatal care, and so forth—are the same as for premature babies. Like premies, these babies are vulnerable to a range of risks. At one extreme, the risks are deadly: stillbirth or death soon after birth. Serious but not deadly risks include cerebral palsy, epilepsy, and brain damage of other sorts. Many small-for-dates babies, especially if they were under 3½ pounds, have lower than average intelligence (Wiener, 1962; Wilson, 1985). Others have normal IQs but develop minor or temporary problems in school or in learning.

Low birth weight rarely acts in isolation. Its harmful effects intensify when it combines with other risks to development. Such risks may be biological, such as poor central nervous system functioning; or they may be environmental, such as oxygen starvation during birth, crowded, chaotic living conditions, unresponsive or uninvolved parents, unsafe or unstimulating play spaces. When small-for-dates babies are born into families with these problems, they are likely to perform increasingly poorly on IQ tests as they get older, but the performance of those born into families with more educational and economic resources improves with age (Wilson, 1985).

Low birth weight puts an infant at risk for serious problems, but just how serious those problems turn out to be depends on other conditions—both biological and environmental. Major efforts are now underway in this country to develop programs that will provide the environmental support and stimulation that will help low-birth-weight infants develop normally.

Malnutrition

Both before and after birth, human beings need nourishment if their bodies and minds are to grow and thrive—calories for energy, protein for building, vitamins and minerals for health and strength. A good deal of prenatal care is devoted to a pregnant woman's diet, because it has been amply documented that a good and balanced diet can make the difference between health and illness in both mother and child. After birth, infants who are severely malnourished may suffer a form of starvation called **marasmus**. These infants barely grow. Their muscles atrophy. If they live, they are unresponsive and never catch up to adequately nourished children. They learn less and, later, perform poorly on intelligence and psychological tests.

Infants with marasmus, a form of starvation, barely grow, are unresponsive, and, if they live, learn less than adequately nourished infants.

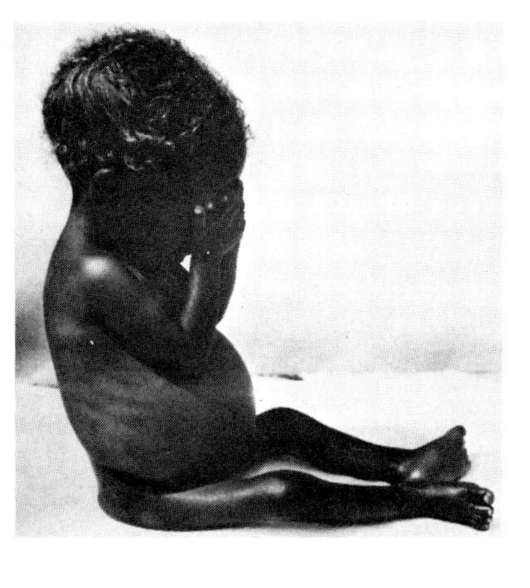

Such extreme malnourishment is uncommon in this country. But in a world where starvation threatens to wipe out whole groups of people—as it now does in the barren regions of northern Africa—it is critical that we understand all the factors that contribute to disease, death, and developmental delay by starvation. Why, for instance, in countries where malnutrition is common do some children suffer more than others? A study of a Mexican farming village showed that a mother's health was the single most important determinant of an infant's nutrition and growth (Cravioto, Birch, DeLicardie, Rosales, and Vega, 1969). Another determinant was family size: an infant with more brothers and sisters, who themselves needed food, got less to eat; an infant with lots of productive adults in the family got more.

One of the best predictors of infants' growth and health is whether they are breast-fed by their mothers or get a commercial formula from a bottle. Mother's milk is specially suited to human infants. It is clean and digestible

lifespan focus

Why mothers don't breast feed

A mother's milk protects infants against allergies, diarrhea, respiratory infections, and helps them to gain weight. Breast feeding also brings mother and infant into frequent close contact.

Despite its advantages, only about one-half of the mothers of newborns in this country decide to nurse. Why? Many working women find breast feeding incompatible with the demands of their jobs. Few employers offer women places where they can nurse. In fact, in some places women may be fired if they insist on nursing during working hours.

> Just after Sam was born, the newspapers were full of stories about a fire fighter who had been let go for nursing her baby at the fire station when she was on her breaks. I was grateful that I never ran into any problems like that.

Other women cite inconvenience, discomfort, embarrassment, worries about having enough milk, and just plain distaste as reasons for not nursing. Attitudes toward nursing in public also may deter mothers from nursing.

> When our third baby was born, my husband and I occasionally went out to a restaurant just to have some time to talk. I was nursing the baby then, so he came along too. At one restaurant, the captain came over and asked me please to nurse the baby in the ladies' room. I had covered up with a shawl, but apparently some customer had complained. That was a real nuisance.

Mothers' reluctance to breast feed is not new. At one time in Europe, for example, many mothers gave their infants to wet nurses. Wet nurses were women who were paid to breast feed infants other than their own. Some infants were sent far from home to wet nurses for up to a year or two. To combat mothers' reluctance to breast feed and convinced of its physical and psychological benefits, a strong movement arose in the 1960s and 1970s in this and other countries. The La Leche League was one such group. It educated doctors and nurses to the benefits of breast feeding so that they, in turn, would encourage new mothers to do so. It also provides mothers with information and practical support for breast feeding.

The benefits of breast feeding are real, but not every mother can or will breast feed her infant. Mothers who bottle feed should not be made to feel guilty, just as mothers who breast feed should not be made to feel embarrassed.

and confers immunities on the vulnerable infant. Sometimes called "nature's vaccine," mother's milk protects infants against allergies, diarrhea, upper respiratory infections, and other illnesses. Soon after birth and for two to four days thereafter, **colostrum** is secreted from the mother's breasts. It is a clear liquid, rich in protein, from which infants receive white blood cells that destroy bacteria and viruses and produce antibodies. It also carries a sugar that helps the infant's gut to fight disease—a nutrient that can mean the difference between life and death to infants in areas where food is scarce and disease common. While they are breast feeding, mothers have to eat properly, because not only essential nutrients reach the infant through their milk but harmful substances too. Breast milk carries chemicals, alcohol, and nicotine to the developing infant. These substances can have the same harmful effects right after birth as they did when transmitted by the mother before birth.

Mother's milk helps infants not only to gain weight faster in their first month but to avoid obesity later on (Neuman and Alpaugh, 1976). For one thing, breast-feeding mothers do not urge their infants to take more to drink. Their infants simply suck until they are sated, and therefore are not overfed. For another thing, breast-fed infants are more active than formula-fed infants.

They wake up more often at night and move about more often when they are awake (Bernal and Richards, 1969). This activity level also helps keep them thin. Of course, it is also possible that mothers who breast-feed have other attitudes that later keep them from overfeeding their children, and this, too, contributes to the children's later leanness (Weil, 1975).

Some public health experts consider breast feeding in poor families to be necessary for infants' survival (Jelliffe and Jelliffe, 1982). Although prepared formula can offer an infant adequate nutrition, many pediatricians urge new mothers to consider breast feeding for its nutritional and immunizing advantages. And although bottle feedings also can be times of close and loving contact between parents and infants, breast feeding by necessity brings mother and infant into close and frequent physical contact. Breast feeding may be physically and psychologically rewarding for both mother and infant and so perpetuate itself and lead to continued close contact.

Failure to thrive

If infants are to grow and thrive, they need loving attention no less than they need milk. Leave an infant in a disorganized home, with a caregiver who is unresponsive and who ignores or restricts the infant's activities, and you have set the stage for trouble. For under conditions like these, infants are likely to stop growing, to lose weight, and to fall behind other infants in reaching important developmental milestones like sitting up, creeping, or walking. Often when infants suffering from this condition, called **failure to thrive**, are examined, no organic cause is apparent. A sequence something like the following has been proposed as a model of how failure to thrive develops (Bradley, Casey, and Wortham, 1984). First, the pattern of interaction between parents and infant is not harmonious. The parents are unresponsive and unaccepting of the infant. The infant, too, may be difficult or unresponsive. Then, the parents fail to provide the infant with needed nurturance. They do not feed the infant enough, and the infant often is left hungry, dirty, wet, and unattended. Finally, miserable, under great stress, apathetic, and isolated, the infant begins to develop hormonal and cellular abnormalities. Normal growth and signs of responsiveness wither.

Physical growth

> When we brought Jason home from the hospital, Kristy met us at the front door and said, "But his skin doesn't fit." Later it was hard to remember how skinny he'd been.

The vast majority of infants, in this country at least, are born healthy, hungry, and energetic. They slurp their milk so noisily that it gladdens even a hovering grandma's heart. The wrinkled and wizened newborn fills out within a matter of weeks into a plump, round-cheeked, dimpled little charmer who increasingly resembles those idealized portraits on the cereal packages. Tissues and organs already formed before birth continue to develop and enlarge.

Brain growth

The infant's central nervous system continues to mature according to an internal timetable that is affected very little by the infant's environment. Brain wave records of 6-week-old infants born after 28 weeks in the womb are almost identical to those of infants born after 34 weeks and tested immediately, even though the 6-week-olds have been exposed to as many weeks of sights, sounds, human contact, and handling (Dreyfus-Brisac, 1975).

Brain activity in the first month of life occurs primarily in the sensorimotor cortex, brain stem, and cerebellum. The first brain cells to function are those in the primary motor area that control arms and trunk. By the time the infant is 3 to 4 months of age, brain activity in the cerebral cortex is common. At this point, the infant's cortex has enough mature cells to direct voluntary movements of the arms, and primary sensory areas of the cortex also develop. First touch and then vision and hearing are affected. Brain activity resembling that seen in adults, which occurs predominantly in the frontal cortex and association areas, takes place by about 7 months. Then infants can integrate sights, sounds, and voluntary movements as they reach for a ball, examine toys, and react to strangers (Conel, 1939–1967; Chugani and Phelps, 1986).

Body growth

The average American newborn, as we have seen, weighs in at 7½ pounds. What determines how much a baby weighs at birth? The size of the parents is one factor. The mother's size exerts a **restraining effect** on the size of the developing fetus. Thus even if the father is large, the size of the mother's uterus and pelvis restrains the fetus's growth to a size that her body can usually sustain and push through the birth canal.

> My sister and I look much alike, and we're exactly the same build, height, and weight. I'm married to a 6 footer, and she's married to a fellow who's 5 feet 4. Our babies were exactly the same length and weight—20 inches, 7 pounds 5½ ounces—at birth. But at 1 year old, her daughter was a short 34 inches, and mine was a tall 37 inches. Now the girls are 8 years old, and you can still see the difference. Lainie is short and broad like her father, and Nancy is tall and lanky like *her* father.

An infant's sex and order of birth also affect weight and maturity at birth. On average, boys are half an ounce heavier, have larger skulls and more muscle, and are longer than girls, whereas girls' skeletal and nervous systems are two weeks more mature than boys'. Firstborns usually weigh less than later-borns and grow faster after birth until they make up the deficit (Tanner, 1974). Most twins and all triplets are born underweight because they are born prematurely.

As the newborn begins to ingest and digest on his or her own, typically about 10 percent of body weight is lost. After about five days and ready to solo, the newborn begins to gain about an ounce a day.

> Sam weighed 7 pounds 5 ounces and was 21 inches long at birth—both 50th percentile. Within two weeks, though, he had begun to grow extremely fast. He's been off the charts ever since. At 1 year, he weighed over 30 pounds. He looked a bit like Winston Churchill. My husband swore that if we were quiet enough, we could *hear* him growing!

On average, babies double their birth weight by 4 months and triple it by 1 year. Smaller than average babies gain even faster. The faster **catch-up growth** of a baby kept small by a small mother's restraining womb takes place largely in the first five months of life. The 4-pound newborn may weigh 8 pounds at 2 or 3 months and 20 pounds at 1 year. Heredity has surprisingly little effect on prenatal size, but it has a sizable effect on catch-up growth (Tanner, 1974).

The quick weight gain of the infant's first year is mainly body fat, which is why many infants look round and dimpled. Although some of this padding is necessary insulation against cold and a food reserve during teething, infants need not be fat. Overfeeding actually may predispose them to weight problems later on (see discussion in Chapter 8).

> Natalie was 7 pounds 10 ounces at birth and 19 inches long. But she's been getting progressively rounder and shorter, roughly the 25th percentile in height and 75th in weight. We're starting to worry that her "baby fat" won't go away.

During their second year, toddlers' bodies lengthen, and they gain less weight. By 24 months, an average child in this country weighs 30 pounds and stands 34½ inches tall (Figure 5.1).

Reflexes of the newborn

The healthy newborn infant shows a remarkable range of **reflexes**, swift and finely coordinated involuntary responses to falling, to being stroked on the cheek or the sole of the foot, suspended by the fingers, or "walked" across a flat surface. More than 20 reflexes in the newborn have been identified. In the moments after birth, a doctor usually puts an infant through his or her paces to check these reflexes, for their presence is assurance that the nervous system is well developed. Abnormal or missing reflexes are warning signs of neurological problems.

Reflexes offer infants survival advantages, and investigators continue to turn up intriguing possibilities for the purposes that reflexes serve. Two reflexes that are necessary for survival are the rooting reflex and the sucking reflex. In the **rooting reflex**, the newborn automatically turns his or her mouth toward the nipple that touches cheek or lips. In the **sucking reflex**, the newborn sucks on anything that touches his or her lips. With these two reflexes, the infant may feed—an act critical to survival. Sucking is the first step in a complex pattern of behavior. Infants form a seal around the nipple with their lips and create a vacuum in their mouths by moving their jaws. This vacuum helps to draw milk from the nipple. At the same time, infants use their tongues to draw milk from the nipple and toward the back of their

GROWTH CHARTS, HEIGHT AND WEIGHT

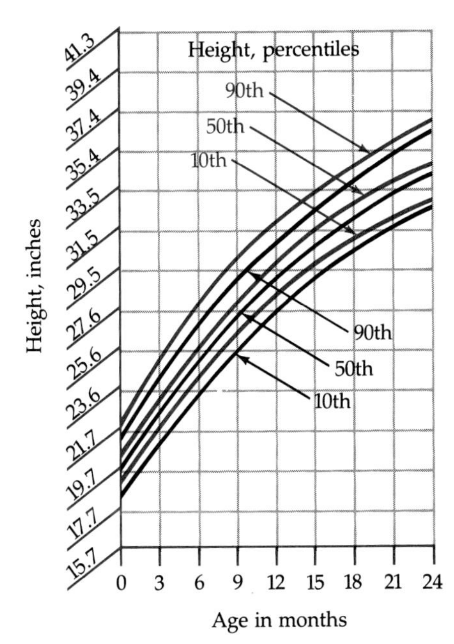

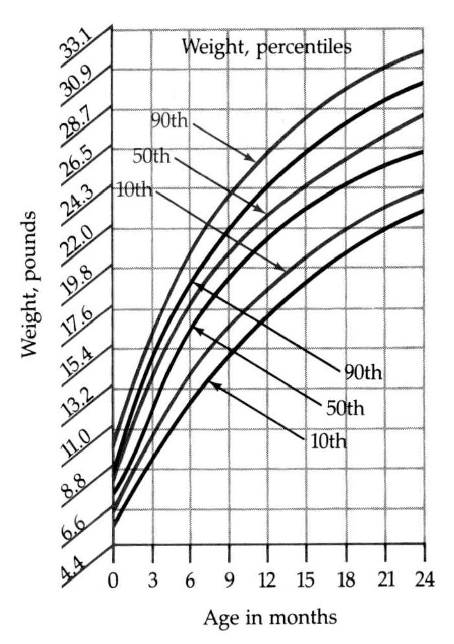

Figure 5.1 ***These charts show American infants' ranges of heights and weights over the first two years. Fiftieth percentile represents the average; 90th percentile represents infants who are larger or heavier than 90 percent of those their age. Girls (colored lines) average somewhat shorter and lighter than boys (black lines).***

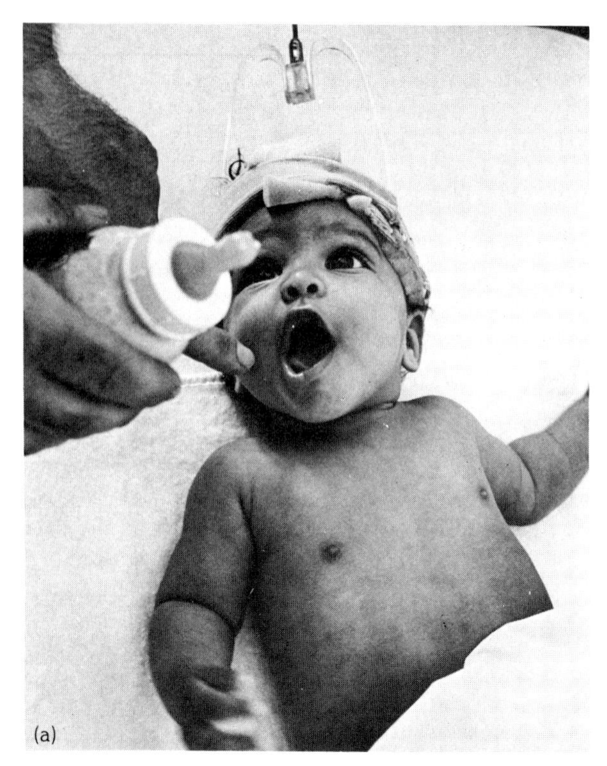

(a)

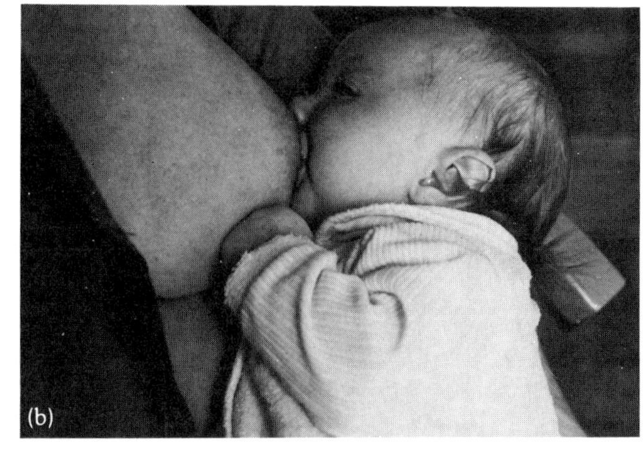

(b)

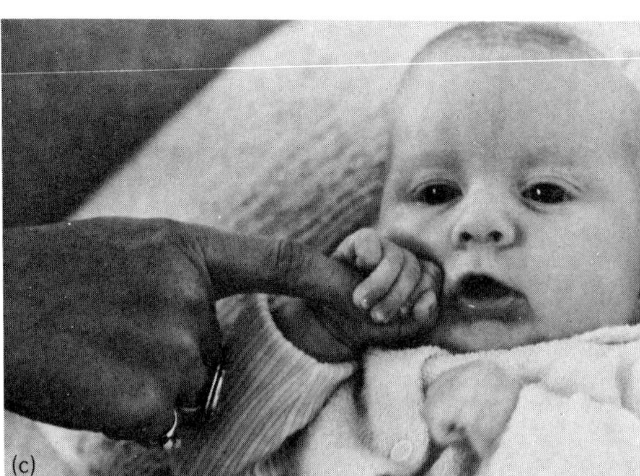

(c)

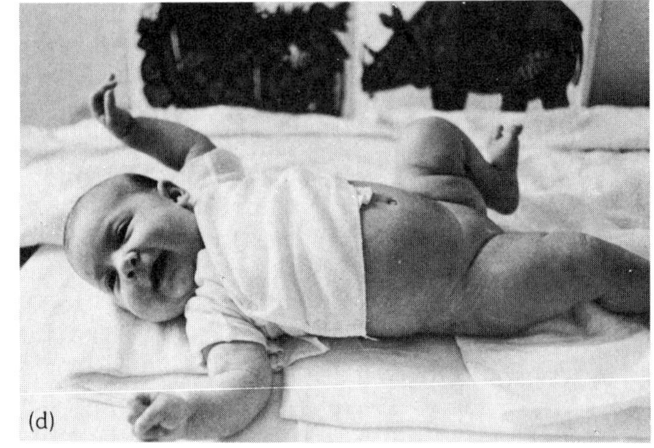

(d)

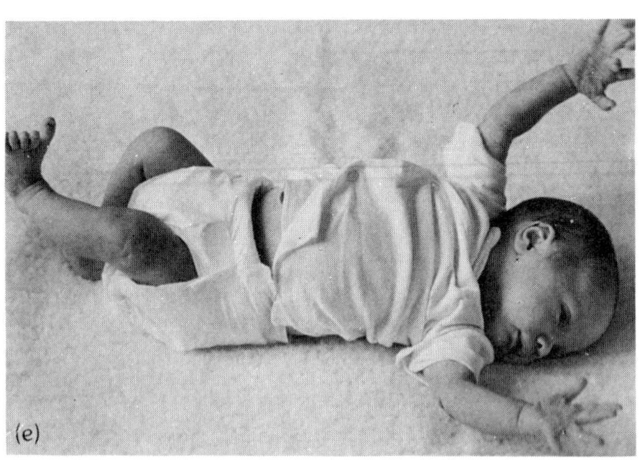

(e)

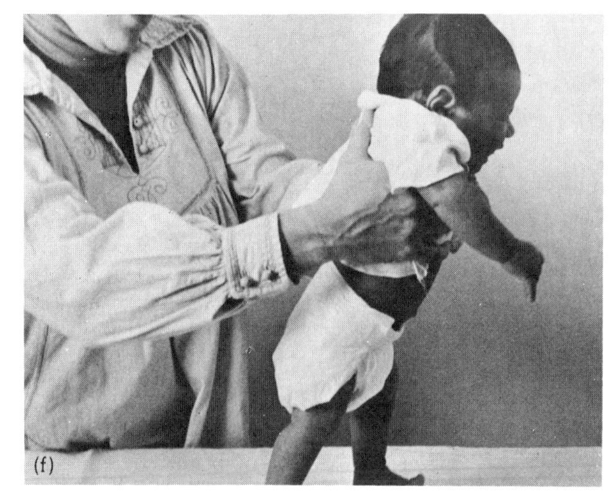

(f)

The reflexes of the newborn help the not-so-helpless infant to survive: a) rooting, b) sucking, c) palmar, d) tonic neck, e) Moro, and f) stepping reflexes.

mouths. Infants can suck and inhale simultaneously, swallowing between breaths three times faster than adults can. After a week or two, most infants have mastered the complex pattern of synchronizing sucking, swallowing, and breathing.

The **palmar grasp** is another reflex in the newborn's repertoire. Also called the ''automatic hand grasp,'' this reflex appears when someone's finger or another narrow object touches the newborn's palm. The newborn's fingers then fold so tightly around the finger or other object that they can support the baby's entire weight. The palmar grasp grows stronger for several weeks after birth, then weakens, and finally disappears altogether by the time the infant is 3 or 4 months old. Not until the age of 5 years do children have the same strength of grasp as in the newborn period.

When the newborn is startled or begins to fall, arms and legs fling outward, hands open, and fingers spread in what is called the embracing, or **Moro reflex**. As the Moro reflex continues, the infant draws arms to body, clenches his fists, arches his back, fully stretches out his legs, and opens his eyes wide. Then comes a loud wail. If the Moro and certain other reflexes do not disappear within a few months, it indicates a problem in the central nervous system. In cerebral palsy, for example, a condition that affects motor centers in the brain, early reflexes persist. Reflexes are under the control of centers in the newborn's brain stem. As the higher level cortex of the brain develops, it acts on the reflexes, and the infant's voluntary, willed actions come to replace the early reflexes.

Why do newborns possess these fleeting reflexes in the first place? Some help the newborn to feed. As we have seen, the rooting and sucking reflexes allow newborns to drink milk. The palmar grasp probably evolved among primates so that infants could cling to their mothers as they moved about. Several of the newborn's reflexes look much like actions that will appear later after much effort. Walking and swimming, for example, have antecedents in the reflexes of early infancy. Hold an infant under her arms, put her feet on a flat surface, and she ''walks.'' Hold the infant on her stomach in a tub or pool of water, and her arms and legs make swimming motions. Researchers are looking into the possible connections between these early actions and the later skills that they resemble. The **stepping reflex**, for example, in which the infant seems to walk, traditionally has been explained as one of the primitive reflexes that disappear after a few months, when the cerebral cortex is mature enough to inhibit them. But perhaps stepping disappears out of simple disuse. In one study, when a group of parents ''walked'' their infants every day, not only did the stepping reflex not disappear, but it grew more frequent, and the infants walked on their own a month earlier than infants who had not practiced walking (Zelazo, 1976, 1983; Zelazo, Zelazo, and Kolb, 1972). Stepping may not be a reflex at all, but part of a more complex innate pattern of movement that also includes stepping (Thelen, Fisher, and Ridley-Johnson, 1984).

Infant states

From the time they are born, infants spend their days in a range of states of consciousness, from deep sleep to frantic squalling. What infants are capable of doing is very much affected by their state at the time. These states are largely determined by the infant's central nervous system and other physiological systems, rather than by external stimulation. Infants can sleep—or wail—through earthquakes and concerts.

The state of consciousness most common to the young infant is sleep. Newborns sleep nearly 70 percent of the time, in seven or eight segments over

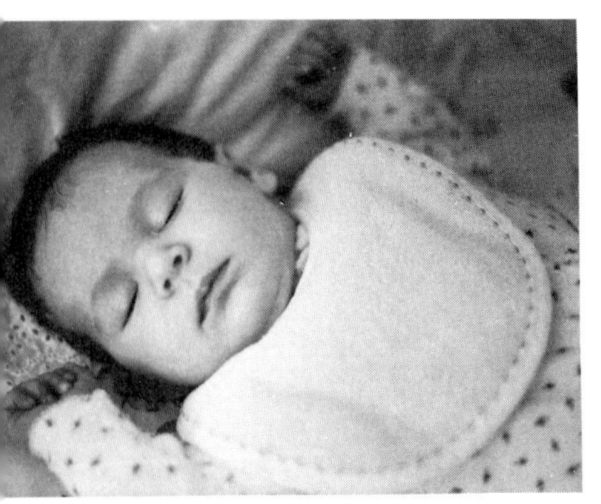

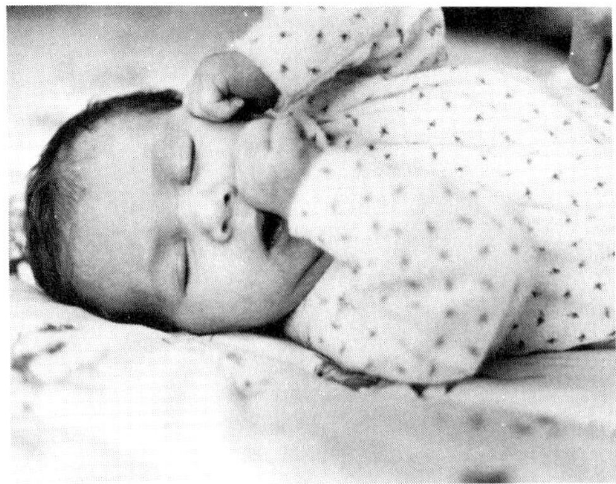

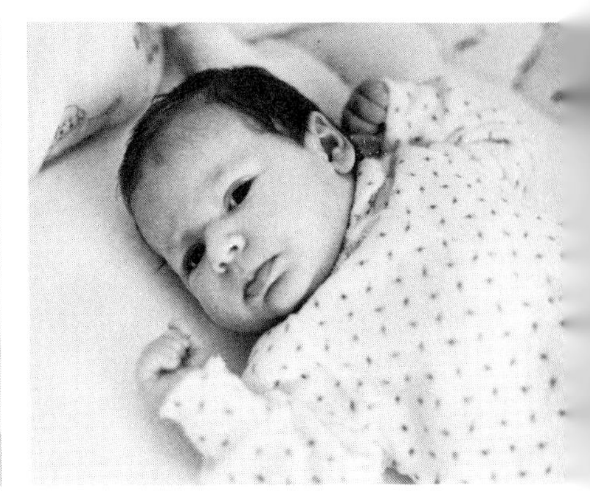

The three common states of sleep–wakefulness are (left to right) quiet, regular and restful sleep; active sleep, during which the baby moves, smiles, grimaces, and has rapid eye movements; and alert inactivity, the best state for learning.

the course of 24 hours (Figure 5.2). In fact, most parents, who are eager to play with the infant they have been awaiting for nine months, are surprised at how much the newborn sleeps. Gradually, the segments of sleep grow fewer and longer. In a matter of months, most infants need only two or three naps a day, and after they reach the age of about 1 year, their naps are even fewer. When do infants reach that blessed milestone, sleeping through the night? There is no hard and fast rule, although many weary parents find encouragement in looking forward to ''12 weeks or 12 pounds, whichever comes first.''

> Do I remember when Sam first slept through the night? I'll never forget it. I heard him cry and groaned as I hoisted myself out of bed to go and feed him. But I literally did a double take when I saw the clock on my night table. It was 5 A.M. He had slept for *seven* hours, right through his 2:30 feeding. Maybe only an exhausted new parent can understand how one could positively *rejoice* at being awakened at 5:00 in the morning. But rejoice I did.

Bottle-fed infants are likely to sleep through the night earlier than breast-fed infants. Observational studies of breast-fed and bottle-fed infants show that the 24-hour sleep patterns of the two groups begin to diverge in their second six months of life (Bernal and Richards, 1969; Elias, 1984). At 2 years, three-quarters of bottle-fed babies sleep through the night. But breast-fed babies sleep less overall and wake up more at night. Breast-fed infants probably are not as full after a feeding as bottle-fed infants are, and so they wake up hungry.

Infants experience two different kinds of sleep: active, irregular sleep, and quiet, regular sleep. In active sleep, the infants' blood pressure, heart rate, and breathing rate are high, brain waves are speeded up, and the level of arousal, indicated by the galvanic skin response, is at the same level as when infants are awake. The infants may smile, grimace, pucker their lips, and move their arms and legs during active sleep. Their eyes, though closed, make rapid movements from side to side and top to bottom. Natalie's mother is describing active sleep when she says,

> Natalie is so quiet when she's awake that it always surprises me when I hear her noisy sleeping. She flails her arms, tosses and turns. The covers fly off. Sometimes she whimpers like a kitten, and I wonder if I should waken her from a bad dream.

Newborns spend about half of their sleep time in active sleep, half in quiet sleep. By their first birthday, they spend only one-quarter of their sleeping time in active sleep.

Figure 5.2 ***This chart pictures the amounts of time that infants spend, on the average, in different physiological states (Berg, Adkinson, and Strock, 1973).***

NEWBORN STATES

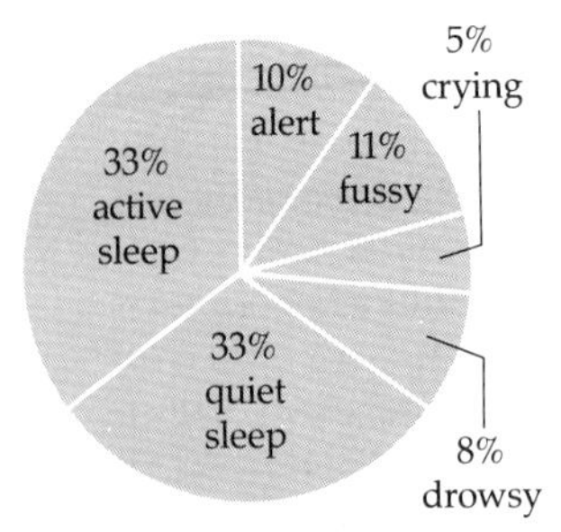

We know that in adults, rapid eye movements (REMs) occur during dreaming. It is possible that newborns spend as much time as they do in active sleep because the higher parts of their brains are working through the stimulation they have received when they are awake. As their brains grow practiced at coordinating and consolidating sensory stimulation, infants come to need less ''exercise'' time (Roffwarg, Muzio, and Dement, 1966).

If they are not sleeping, infants may be drowsy; awake and active; awake and inactive; or crying. Infants who are crying or actively moving around are less likely to be listening to or looking at what's happening around them. Infants in a state of alert inactivity can focus on and follow things with their eyes, listen, and learn (Korner, 1970, 1972).

Motor development

Infants acquire motor abilities in the same cephalocaudal-proximodistal—head to toe and center to periphery—direction in which the body grew before birth. First, infants lift their heads and necks. They control their shoulders, then their elbows, then their fingers. They control their knees before their toes. These tendencies traditionally have been ascribed to the direction in which the nervous system matures. Nerves are **myelinated**—insulated with sheaths of a protective substance called *myelin*—in the following order: head, shoulders, arms, upper chest, abdomen, legs, and feet. But this explanation of motor development recently has been challenged. When 4- to 8-month-old infants are allowed to move freely in water, they kick their feet more than they move their arms—an apparent reversal of the arm-before-leg order (Weiss and Zelazo, 1984). It just might be that in the usual course of things—on dry land—infants use their legs after their arms because their legs are heavier and harder to move than their arms are, and not because their nerves are myelinated in that order (Thelen et al., 1984).

Infants also perform gross motor acts before fine motor acts: their whole arm swipes at the toy before their hand can grasp it. They push their big toys across the floor by using their whole upper body, their shoulders, and both arms and only months later become engrossed in trying to pinch a penny from the floor between fingers and thumb.

> Natalie held up her head on her own three weeks to the day after she was born. We instantly snapped a picture of it. The day she first grasped a little toy bird with her fingers we photographed as well. One of the amazing things about parenthood is how engrossed you get in the tiniest developments in your baby.

Motor milestones

Although, as we shall see, infants vary somewhat in how fast their motor skills develop, the order in which these skills develop (see Figure 1.1) and the age *ranges* when they develop are quite consistent from infant to infant.

Birth to 3 months

From birth, many newborns can turn their heads and lift their chins as they lie on their backs. Put them on their stomachs, and they can lift their heads just enough to turn from side to side. Soon they're raising shoulders and chest, and by 2 months usually they can lift their heads up when they're lying on their backs. Sit these 2-month-olds up though, and their heads soon droop or loll to the side. At 3 months, most babies can keep their heads slightly forward but upright. Put them on their stomachs, and they rest on their knees, abdomen, chest, and cheek. Turn them onto their backs, and the head turns to one side, arm and leg on that side of the body extend forward, arm and leg on the other side are pulled in—a fencing pose, the **tonic neck reflex**. Still too young to try creeping or walking, 3-month-olds nonetheless push their feet against your hand. Newborns have no control over their arms. But within a few weeks, they can use their arm, hand, and fingers together, rather like a flipper. Put a shiny toy in front of him, and the 1-month-old does not even try to grasp it. The 2-month-old can hold his fingers against his palms and grasp objects handed to him, at least for a short while. The 3-month-old swipes at the shiny toy with his curled fist.

4 to 6 months

At 4 months, most infants can roll onto their sides from their stomachs. By this age, the infant's cortex can direct voluntary movements to the arms. The 4-month-old watches fascinated as his fingers reach. Over and over again, he glances at his hand and then at its target, bringing them into alignment. His concentration is intense. At 5 months, infants can roll onto their backs. They grasp objects by holding them to their palms with their fourth and fifth fingers. At 6 months, they can touch their hands together in front of them and their hands go up and down, side to side at will. They can reliably reach, grasp, and hold things between their palms and fingers. They can coordinate their posture and gaze with the movements of their arms and hands. The tonic neck reflex has disappeared, and most babies can move more freely on their own. Sit them up and their heads stay upright; their trunks have better muscle tone. By the end of the period, they can even sit for a short time without support, and their perspective changes from horizontal to vertical.

7 to 9 months

By 7 to 8 months of age, most infants can roll from their backs to their stomachs, pull themselves up to standing, and stand by holding onto the edge of crib or playpen.

> When Erica was 7½ months old, I heard her making noises a few minutes after my husband supposedly had put her down for a nap. Erica was standing in her crib, babbling away. I thought my husband had neglected to lay Erica down. I would certainly speak to him about *that*. So I lay her down in the crib, said "Night night," and left the room. Three minutes later I heard more babbling. There was Erica, standing up again in her crib—proud as a peacock at her new trick.
>
> From the moment he was born, Peter was happier in an upright position than lying down. None of those reclining infant seats for him. He struggled to raise himself upright. Even to get him to stop crying, the best thing was to hold him up under his arms and let him swing freely. He finally managed to pull himself up at about 7 months, and the smile on his face went from ear to ear.

Seven-month-olds can get around by wriggling forward on their stomachs with a little help from arms and feet and can hold objects in their hands. They use their opposable thumbs against several fingers; 8-month-olds can shift the toy from one hand to another. These new physical abilities allow babies to play alone for a few minutes for the first time. Because their hands now can grasp a bottle or a crust of bread, some of these independent 8-month-olds may try to feed themselves. By 9 months, most babies can rise up on tiptoes or play with their feet as they lie on their backs. They can sit unsupported for ten minutes or more and right themselves if they begin to fall over. Some even manage to push themselves up into a sitting position from a prone position.

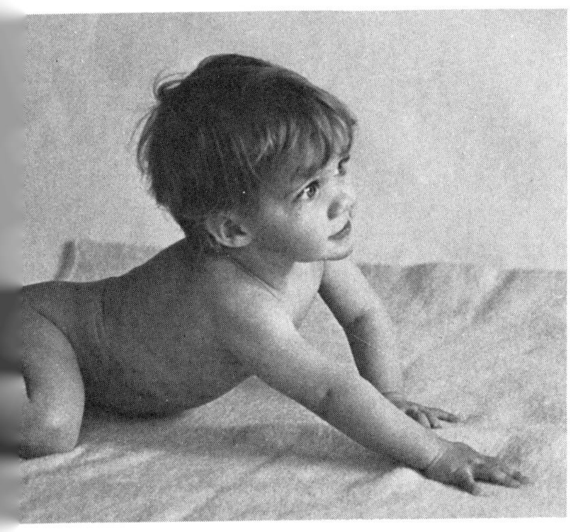

a.

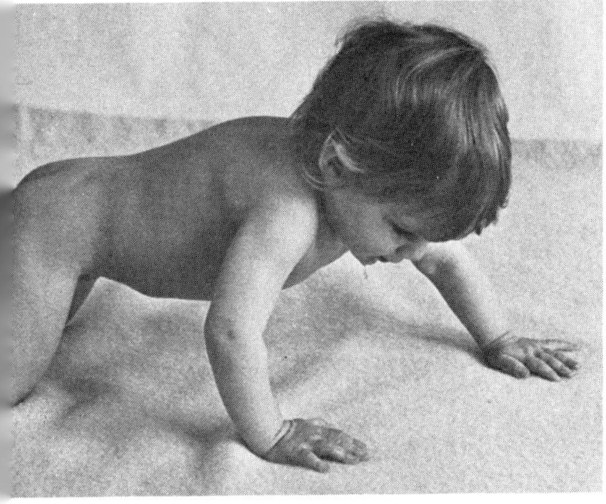

b.

c.

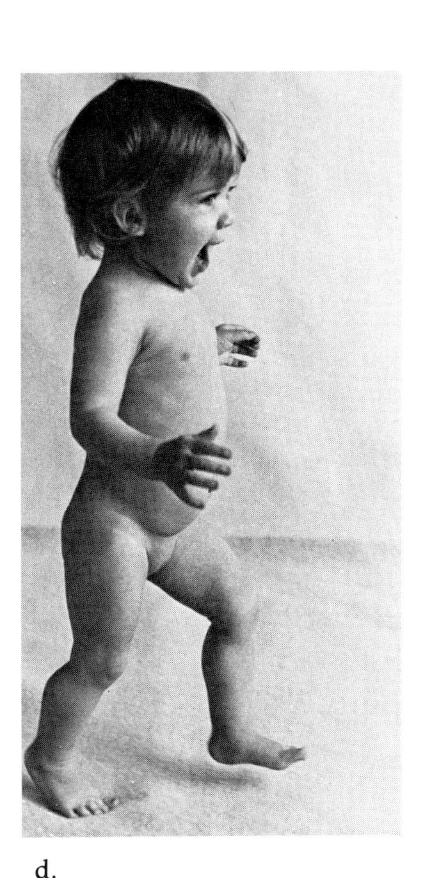

d.

Motor milestones. **(a)** *By 7 or 8 months, most babies can creep forward on their stomachs.* **(b)** *By 9 months, they can move on all fours and, soon afterwards,* **(c)** *pull themselves up to stand alone.* **(d)** *By 12 months of age, most babies can walk alone—unsteadily but with great pride.*

Now babies begin to move around on their own. Some crawl, some creep, some scoot, some walk like bears on all fours, some edge sideways like crabs. They are endlessly inventive as they embark on this wonderful new game of getting around. They can also pick up an object between thumb and forefinger, and this gives them an enormous advantage over younger infants. Infants at this age can get about, explore, and do things for themselves. Their combined motor skills let them explore and manipulate objects as they learn about three-dimensional space, actions and reactions, and the connections between events. Now is the time for them to learn about the noise a pot on the stove makes when it hits the kitchen floor, the interesting swinging action of opening and closing a cabinet door, the joy of playing peekaboo with Mommy, and seventeen different things you can do with an egg whisk.

10 to 11 months

Lay a 10-month-old on her back, and she is likely to squawk in protest. She'd much rather be sitting or standing up. She can sit long and comfortably now, pivot from side to side, move from sitting to lying and back. Hold her hands, and she'll walk with you. On her own, she'll walk along by holding the edges of cabinets and furniture. Baby handprints begin appearing along the bottoms of the windows. There is so much to see! By 11 months, she can stand alone. She takes her first step—and quickly plops to the floor.

12 months onward

By 12 months, most babies in this country begin to walk alone, without holding on. They're not too steady at it yet, and they topple often, but they keep at it with enormous pride and excitement.

> Sam began walking at 11½ months. He was so thrilled to be able to walk that he almost never sat down for nearly six months. I even had to put his food at the edge of his table so that he could peck at it in his travels. If we tried to contain him in his high chair, he would stiffen his legs and throw a tantrum. Boy, did he start to shed fat in those months!

Soon babies at this age can walk well enough to try to get themselves to certain places. Whether on their own or in a walker, babies who can walk are likely to approach adults (Gustafson, 1984). They look often at the adults, smile, and vocalize, and the adults respond to them. Walking also helps babies to explore new kinds of things—the cracks between the kitchen floor tiles, the dog's tail, the underside of chairs—and to engage other people in their explorations.

> Natalie was a bit late in walking—nearly 15 months. She always preferred to sit quietly and use her fingers, sometimes to hold a picture book, sometimes to explore the texture of things like a doll's hair, a piece of fabric, a puddle of milk. But when she did start to walk, it opened up vast vistas for her to explore. She went from one room to another in her very deliberate way.

After 12 months, walking and balance get smoother and smoother. By the end of the second year, the toddling toddler is able to run and climb stairs.

Variations in development

As Sam's and Natalie's ages of first walking show, infants reach milestones of motor development at somewhat different ages. Knowing the age range within which most children reach certain milestones of motor development—sitting alone, standing, and walking—helps parents and doctors to spot a child who is lagging behind and to inquire into the cause. Some lags in development are normal variations in a healthy child; others signal problems.

> Tim didn't hold up his head on his own until he was 6 months old, and he sat up late as well. But the pediatrician thought, because Tim could bring his hands together at the midline and because his skull was growing, that he was just developing at his own pace. We've had no other problems with Tim since then. He just needed some extra time.

> A woman I know has a 2-year-old who is just beginning to walk. He also shows some repetitive, self-stimulating behaviors. I think he has some problems that a specialist should check out, even though the mother hasn't indicated that *she* thinks there's a problem with her son.

From one generation to another, from one culture to another, the age at which normal infants achieve milestones in motor development varies by as much as three months. In two studies done in the 1930s, for example, babies in a California sample walked on the average at 13 months, and babies in a Minnesota sample walked at 15 months (Bayley, 1971a). In a study done in the 1960s, babies in a sample from Colorado walked on the average at 12 months (Frankenberg and Dodds, 1967).

When we try to figure out the reasons that some infants—whether individual infants like Sam or groups of infants like those in California—develop faster than others, diet, practice, parents' encouragement, and the chance to move freely all suggest themselves. A study of two twin boys conducted half a century ago (McGraw, 1935) offers support for the importance of practice and encouragement. One of the twins, Jimmy, was allowed the usual opportunities to walk, run, and climb and was moderately encouraged in motor skills. But his fraternal twin brother, Johnny, was given more freedom, daily practice in physical skills, and lots of encouragement. Johnny grew into a more athletic and better coordinated child than Jimmy, and the advantage lasted into adulthood. Clearly, practice makes for precocity. But even with exercise, infants achieve motor milestones like walking on a relatively fixed maturational timetable. The size of the reaction range (see Figure 3.1) for basic physical abilities is quite small because these abilities are under strong genetic control.

Heredity clearly plays a role in an individual infant's rate of motor development. Identical twins reach motor milestones at the same age more often than fraternal twins do (Wilson and Harpring, 1972). American babies of African ancestry, who inherit one body type, are likely to walk a couple of months earlier than American babies of European ancestry, who inherit another body type, even though they are reared similarly (Tanner, 1970). An infant's activity level and adventurousness, also inherited, contribute to the pace at which he or she reaches motor milestones. A slower-moving, more sedentary, and cautious infant may not walk until 15 months while an active, energetic, adventurous one will be walking at 11 months, climbing ladders at 16 months, and jumping off fences at 5 years.

Sensory and perceptual abilities

By watching infants at their daily activities—reaching, walking, climbing—one can easily see the development of their motor abilities. But what about the development of sensory and perceptual abilities? These abilities are much harder to observe. Infants cannot speak, and so everything about their sensory abilities must be inferred indirectly. Researchers face formidable problems in finding out what infants can see, hear, smell, and taste. Not only must they come up with clever tests to probe the infants' sensations, but they also must deal with subjects who don't know the meaning of cooperating with experimenters.

Researchers persevere even so, devising ingenious procedures to use with their tiny subjects. One procedure they follow is to record infants' eye movements. Like anyone else, infants look longer at some things than others. To monitor an infant's visual responses, researchers film the light reflected from the infant's eye. This reflection tells them precisely where the infant is

During the first year, infants acquire fine hand skills. They progress from an interest in their own tiny fingers, to grasping a toy, to manipulating and making a toy work.

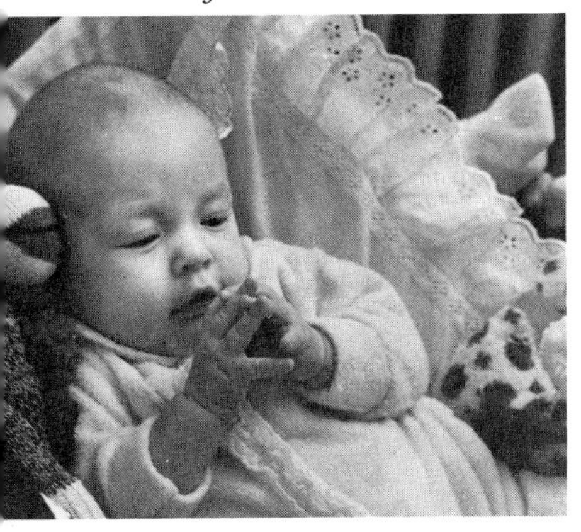

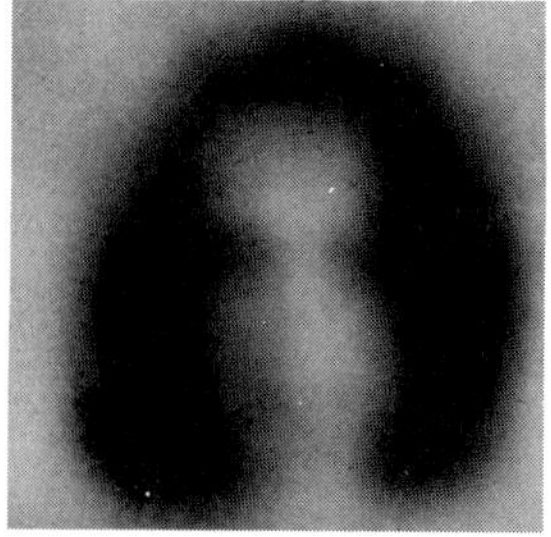

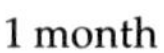

1 month

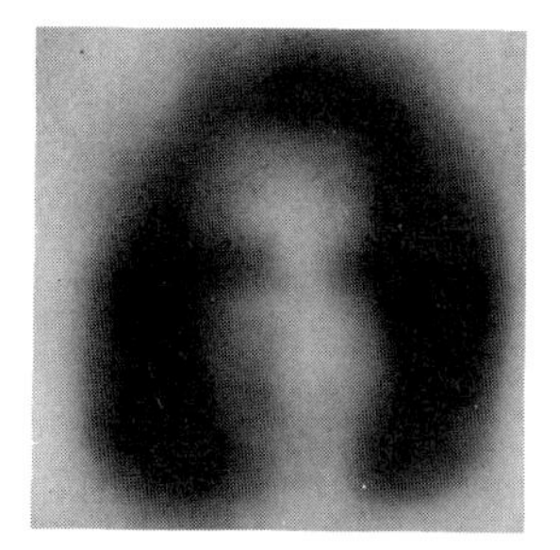

2 months

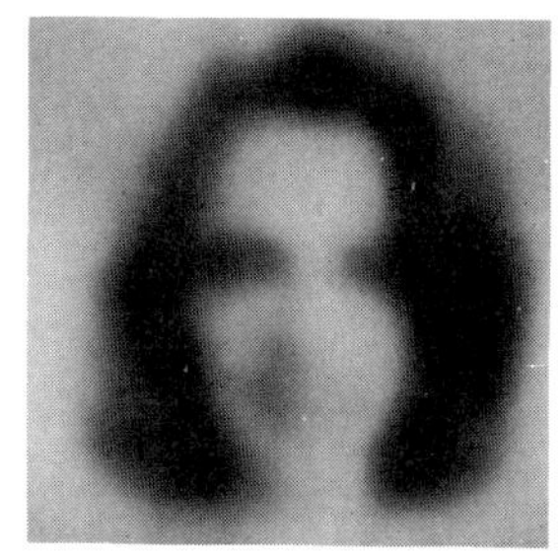

3 months

Adult

These photographs simulate what mother looks like to her infant at one month, two months, and three months. By three months the child has sufficient spatial resolution and experience to recognize the mother's face.

looking. In another procedure, researchers give an infant a pacifier wired to record the infant's sucking. By first recording a baseline rate of sucking, the researcher can compare the infant's faster or slower sucking in response to particular sights and sounds. When the sucking rate returns to the baseline, the infant has gotten used to or bored by—**habituated** to—the stimulus. If the researcher then shows the infant another stimulus and the sucking rate stays the same, the researcher can assume that the infant probably does not notice any difference between the two stimuli. If the sucking rate changes, the researcher can assume that the infant does notice the difference between the two stimuli. In a different procedure to assess infants' hearing and vision, pacifiers are wired to devices that let the infant turn on a recorded sound or bring a picture into focus by sucking. In yet another procedure, stabilimeters are used to measure the infants' movements, pneumographs to measure their breathing, or electrocardiographs to measure changes in heart rate when they see or hear a stimulus. Reaching and head turning may be used to indicate that the infant sees or hears as well. By using all these procedures, researchers have learned important things about how infants see, hear, taste, and smell.

Vision

Until about 20 years ago, people did not really know just what or how very young infants could see. Some doctors claimed that infants couldn't really see until they were 3, 4, or 6 weeks old, although they might flinch at a bright light shone in their eyes. Parents, in contrast, argued that their infants were quite obviously gazing at their faces during nursing or other periods of cuddling. In the last 20 years, scientists have learned much about the functioning, the anatomy, and the physiology of infants' vision (see Figure 5.3).

Visual capacities

The infant's **retina**, the part of the eye that turns light into nerve signals to the brain, has been found to be like an adult's in its components—rods, cones, and synapses—but it has no distinct **fovea**, the area in the adult retina on which central visual images are formed. Without a fovea, the central retina's handling of spatial resolution is poor. Therefore, researchers infer, the newborn's sight is not sharp. The retina matures fairly quickly, though, and by 11 months or earlier, its major structures are adult-like. During this period of maturation, the part of the nervous system that relays visual impressions between the retina and the cortex of the brain also gets more efficient until it is like an adult's. In the visual cortex, **dendrites** at the ends of the nerve cells form more branches, and the myelin sheaths that insulate the nerve cells continue to form. At about the age of 2 months, the infant's visual cortex begins to process visual stimuli (Hoffman, 1975).

Eye movements change, too. In young infants, the **saccadic eye movements**, which are the rapid movements between one point of visual fixation

ADULT AND NEWBORN EYES

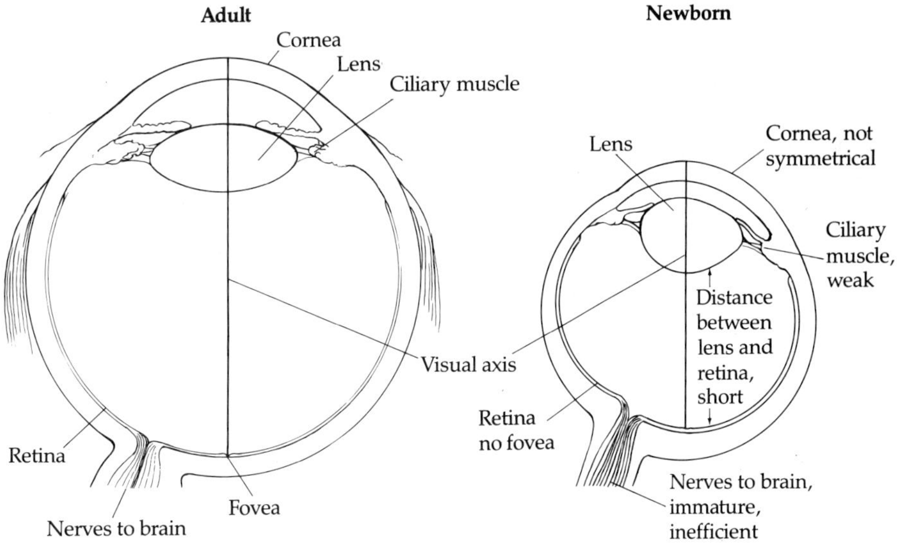

Figure 5.3 ***This diagram compares the eyes of an adult and a newborn infant. It shows some of the ways in which the infant's visual system is limited. The diagram does not show the infant's jerky, inefficient eye movements and the immaturity of nerves in the visual cortex of the brain, which also limit the newborn's vision.***

and another, are slower and more numerous than those of adults. In an adult, a saccade moves the eye 90 percent of the way to the visual target, but infants need several saccades to reach their visual target. The infant's eyes also do not move as smoothly as an adult's in tracking a moving target. Although they actually seem to prefer to look at moving objects over still ones (Haith, 1966), infants trying to track a moving stimulus refixate their eyes often, first gazing at one spot and then another. This jerkiness may be necessary because convergence, the focusing of both eyes on the same point so that only one stimulus is seen, is difficult for newborns. By 48 hours after birth, most infants can track a slowly moving object (Haith, 1966). But only at 6 weeks of age do infants track a moving object smoothly (Dayton et al., 1964). By 2 to 3 months, infants' perception of movement apparently is as good as adults'.

The shape and structure of the eye also are important in determining how well someone can see. The newborn's eyeball is short and the distance between the retina and lens reduced. Therefore, most infants are quite farsighted. Many also have an **astigmatism** during their first year, a difficulty in focusing that arises because their cornea is not symmetrical. Because of weak ciliary muscles, moreover, newborns cannot change the shape of the lens of the eye to **accommodate** to the shifting plane of focus for visual targets. Until they are 1 month old, babies can adjust their focus only for targets that are between 5 and 10 inches away. Although it seems logical to expect that infants at this age could see objects more clearly at a distance of 5 to 10 inches, they cannot. Apparently the infant's **depth of focus**, the distance that an object can be moved without a perceptible change in sharpness, is so large because other inadequacies in the visual system keep even substantial focusing errors from causing a noticeable increase in blurring. By 4 months of age, infants can accommodate to shifting planes of focus as well as adults can.

As a result of all these limitations in the visual system, but particularly because of the lack of a fovea and the immaturity of neurons, **visual acuity** is poor early in infancy and improves greatly in the first year. In one investigation of infants' visual acuity (Allen, 1978), infants between 2 weeks and 6 months old were shown two slides at the same time that were equal in size, brightness, and color. One had no pattern on it, and the other had high-contrast stripes. The experimenter progressively increased the width of these stripes from narrow to wide. By recording the point at which infants first discriminated between the unpatterned and the striped slides, the experimenter measured the infants' visual acuity. At 2 weeks, infants could make out stripes that were one-eighth inch wide; at 6 months they could make out stripes one-fortieth of an inch wide. In another study, newborns' visual acuity was shown to range from about 20/150 to 20/800 (Fantz, 1961). (A person with 20/150 visual acuity sees an object 20 feet away as if it were 150 feet away.) By the second half of their first year, infants have 20/20 vision.

But even in the first six months, infants are far from blind. Martin Banks and Philip Salapatek (1981) suggested a useful way of thinking about the visual limitations of infants. They referred to the amount of information in a pattern that an infant can see as his or her visual "window." Early in life, an infant's visual window is quite limited, but the contours of many common objects are well within this "window." Young infants can see the contrast between hair and skin on a parent's face, for example. Although infants cannot see small things on the other side of a room, they can see large objects close up—the distance at which much of the interaction between young infants and their parents takes place.

Figure 5.4 ***Newborn infants prefer to look at a moderately complex stimulus—in this set of stimuli the drawing of a face (Fantz, 1963).***

NORMAL AND SCRAMBLED FACES

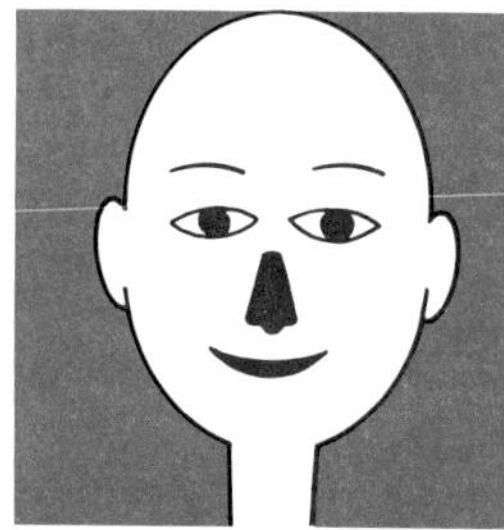

(*a*)

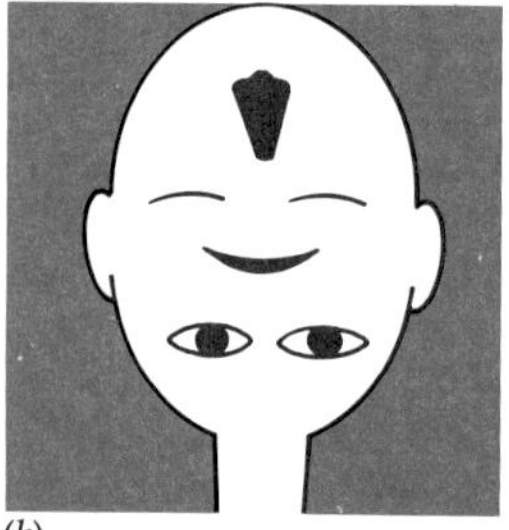

(*b*)

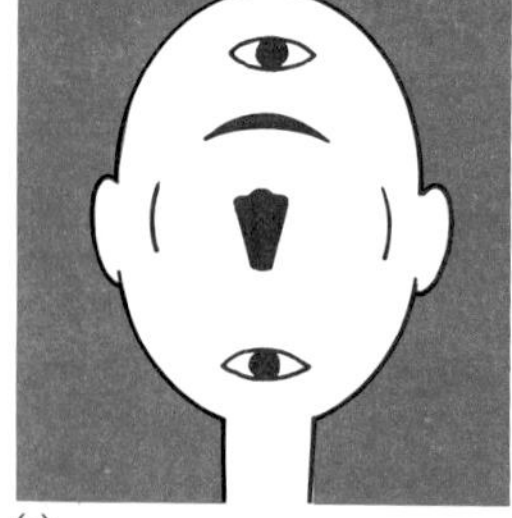

(*c*)

Visual preferences

Infants show clear preferences for looking at certain *kinds* of objects up close. They look longer at a pattern than at a plain surface (Fantz, 1965). But too much pattern can make them tune out. In the first month, infants look longer at a pattern made up of a four-square checkerboard than at a more complex checkerboard (Brennan, Ames, and Moore; 1965; Hershenson, 1964). Infants also like to look at pictures of faces (Fantz, 1963; see Figure 5.4).

It is now generally thought that faces have no innate attractiveness for babies but instead contain attractive elements—like curved contours, a moderate amount of complexity, and movement (Olson, 1983). When researchers showed newborns pictures of faces with normal or scrambled features, they found no difference in how far the infants turned their heads or their eyes to follow the pictures (Maurer and Young, 1983). Infants younger than 2 months do not look at normal-looking faces in preference to scrambled faces (Fantz, 1961; Fantz and Nevis, 1967; Maurer and Barrera, 1981; see Figure 5.5).

Infants prefer patterns that are the most salient and visible and have the largest discernible elements that fit their visual "window" (Banks and Salapatek, 1983; see Figure 5.6). This appears to be true for infants up to 3 months old (Gayl, Roberts, and Werner, 1983). Because infants have only a limited ability to perceive complex patterns, they tend to look at the most striking aspect of a pattern, the part that is moving, or largest, or of greatest contrast—the part they can see best. For such young infants, this is usually the outside contour of a figure or face (Figure 5.7). At 2 to 4 months, when infants see a face, they focus on the now visible eyes and mouth as well. Infants focus on contrast, it has been suggested, to keep the neurons of the visual cortex firing at a high level. This firing seems to prod the development of the visual cortex. Looking at contrast also allows infants to get the most information a stimulus affords. Later, at 5 to 7 months, infants also gaze at the nose when they see a face. By then they seem to know something about what faces are and distinguish between scrambled and normal arrangements of features (Caron, Caron, Caldwell, and Weiss, 1973; Haith, Bergman, and Moore, 1977; Salapatek, 1975).

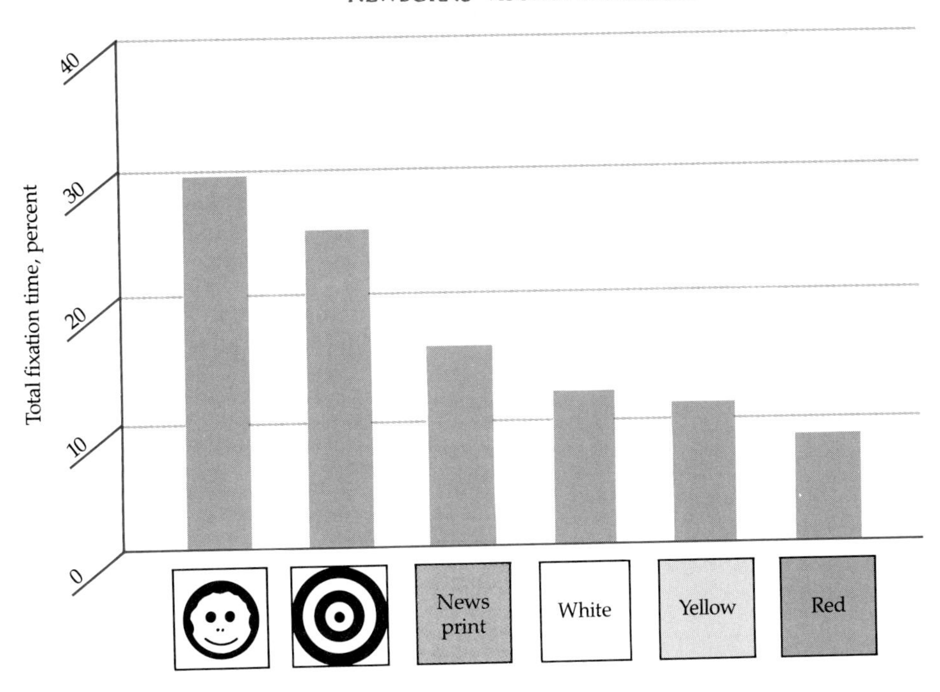

Figure 5.5 ***Newborn infants show no preference for* (a) *normal versus* (b), (c) *scrambled faces.***

Perceiving distance

Psychologists long have been interested in whether babies can tell when things are near or far away. Because newborns have had no experience with objects and have lived in the dimly lit uterus, how can they know about visual cues to distance? Some researchers (Bower, Broughton, and Moore, 1970) have suggested that newborns do perceive distance. In their studies, infants were

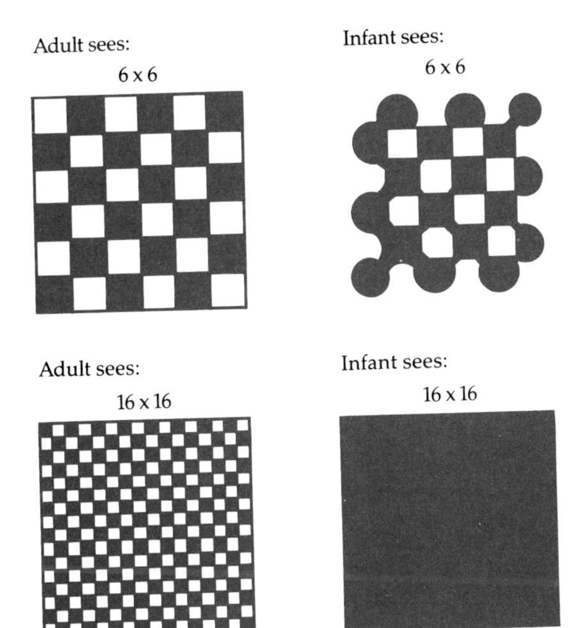

Figure 5.6 ***No wonder 1-month-old infants prefer to look at a 6×6 checkerboard instead of one with 256 squares. In the infant's limited visual system, the complex patterning of a 16×16 checkerboard is lost (Banks and Salapatek, 1983).***

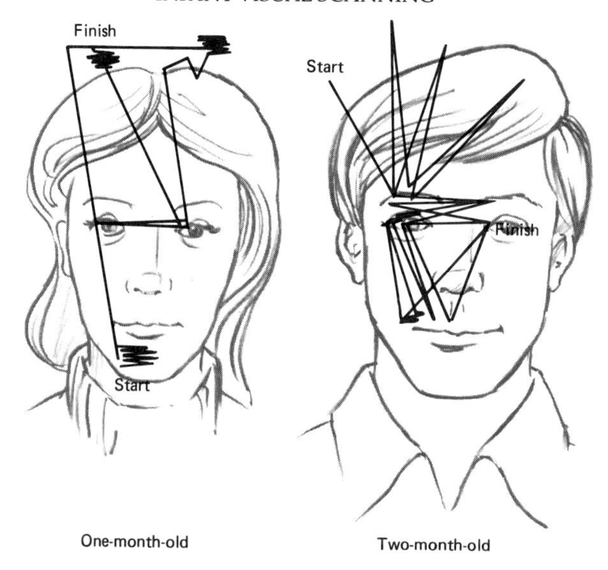

Figure 5.7 Infrared marker lights placed behind a stimulus reflect off babies' eyes, permitting the photographing of their pupils' movements and the tracking of their visual scanning. One-month-old infants concentrate on areas of greatest contrast in faces, particularly the edges of the face and head. Two-month-old infants scan features, especially the eyes and mouth (Maurer and Salapatek, 1975).

put at the end of a table, and a large box was moved toward them. As the box approached, the infants pulled in their heads, raised their arms protectively, put their hands in front of their faces, and widened their eyes or grew distressed. The researchers inferred that when babies saw the box getting larger, they understood that it was getting closer. This interpretation is speculative, however. Another interpretation is that infants were not reacting to an impending collision but were simply moving their arms or heads to parallel the rising or falling contours of the visual stimulus as it got closer (Yonas and Pettersen, 1979).

Other researchers have tested infants' reactions to objects that seem to be coming at them and that then veer away. They use blinking to indicate a defensive reaction to an impending collision. In one study, for example, researchers tested 6-week-old infants for their reactions to a triangle that loomed toward them (Pettersen, Yonas, and Fisch, 1980). The infants blinked on only 16 percent of the approaches. Apparently distance perception is not present at birth.

Another way of testing whether infants can tell that things are near or far away relies on the **visual cliff**. Infants old enough to crawl are put on a slightly raised runway in the center of a large glass table several feet high. On one side of the runway, a checkerboard pattern is fixed to the underside of the glass. On the other side, the checkerboard extends from the runway down and across the floor, creating the illusion of a steep cliff. Infants are beckoned by their mothers to cross the glass over the "cliff." In one of the early studies with the visual cliff, 6- to 14-month-old infants would not cross the cliff. They probably could recognize that the deep side was farther away (Gibson and Walk, 1960).

But what about younger infants? In a study of 2- to 4-month-olds (Campos, Langer, and Krowitz, 1970), infants were placed on their stomachs on the cliff side. Their heart rates slowed noticeably, they cried less, and they were more attentive to what lay below them—all signs that they may have perceived the drop-off beneath them. Some researchers remain skeptical, however, that these findings prove that infants so young can perceive distance. An alternative explanation for the slowed heart rates and other signs is that the infants were responding to a perceived difference in the patterns' visibility rather than to distance.

This young fellow is clearly aware of depth and does not want to venture from the "safe" runway of the visual cliff.

On the basis of such studies, researchers have concluded that infants develop sensitivity to distance cues in their first six months of life. In sum, it seems unlikely that infants are born with the same appreciation of distance that adults have. More likely, they develop the abilities to perceive distance during their first six months of life, as their visual and motor systems mature and as they learn to interpret what they see.

Hearing

Newborns clearly are physiologically equipped to hear. As we saw in Chapter 4, in fact, even fetuses can hear. Newborns' ears have a nearly adult-sized **tympanum** and a well formed **cochlea**, and their auditory system is functioning (Northern and Downs, 1974). But, just as clearly, they do not hear as well as adults do. Although data from various investigations are not completely consistent, it seems that the softest sound that a newborn can hear is 10 to 20 decibels louder than the softest sound an adult can hear. There are several reasons for this deficiency in infants' hearing, which is approximately the amount of hearing an adult loses during a head cold.

At first, the amniotic fluid that remains in the outer ear plus the fluid and residual tissue left in the middle ear may impair the newborn's hearing. In the first few days after birth, the outer ear drains and the middle ear absorbs nearly all the fluid and tissue. At that point, researchers can investigate infants' hearing by recording changes in their heart rates, sucking rates, or brain-wave patterns when a sound is presented. Infants 1 to 2 weeks old can discriminate between loud and soft sounds and high and low ones. They can detect the difference between two notes on a musical scale (Weir, 1979). The softest sounds they respond to are, on the average, about 20 to 30 decibels—the sound of a soft voice—although there are significant differences among individuals, and the exact level of sound responded to depends on the measurement technique used (Acredolo and Hake, 1982). Past the newborn period, hearing gradually improves. But even in their second six months of life, infants are about four times less sensitive to differences in sound frequencies than adults (Aslin and Sinnott, 1984). Hearing improves to adult levels over the first two years of life.

Speech perception

Of special interest to developmental psychologists is infants' ability to hear the sounds of speech. Newborns have better hearing for sounds in the range of speech than for higher or lower sounds (Eisenberg, 1965, 1979).

Although they show large body movements and a faster heartbeat when listening to single tones and noises, when they hear someone talking, newborns show small movements, such as grimaces, crying (or stopping crying), dilation of the pupils, and movement of their eyes toward the source of the sound (Eisenberg, 1976, 1979). By 2 weeks of age, infants notice the difference between a human voice and other sounds, such as the sound of a bell. At 4 weeks, they can discriminate between small differences in spoken sounds. In one study, babies sucked on a nipple that turned on a recording of /*b*/ sounds (Eimas, Siqueland, Jusczyk, and Vigorito, 1971). After they had heard the sound repeatedly, they grew used to it and sucked less. Gradually the recording changed to a /*p*/ sound. When the sound was clearly /*p*/ and not something in between /*b*/ and /*p*/, the babies sucked harder again. The study showed that infants perceive speech sounds in the same categories as adults do (such as /*b*/ and /*p*/). Infants can discriminate between even subtly different sounds, like the /*a*/ in "c*o*t" versus the /ɔ/ in "c*au*ght," other researchers have shown (Kuhl, 1983). They can discriminate between these sounds despite changes in speaker—man, woman, or child—and, in some cases, despite changes in rising or falling pitch.

The ability to discriminate between sounds like these probably is innate rather than learned, for it precedes infants' exposure to much language. Cross-language studies (for example, Lasky, Syrdal-Lasky, and Klein, 1975; Trehub, 1973) show that infants from many different linguistic backgrounds can discriminate between similar sounds. Young infants can even hear contrasts in a foreign language that adults who do not speak the language cannot hear. Japanese babies, for instance, have no trouble with the /*l*/–/*r*/ distinction that their parents find difficult. Adults have had long practice at learning *not* to hear many sounds that are not used in their own language. This effect sets in early. Even 1-year-olds have more trouble than younger infants distinguishing sounds that are not used in their own language (Werker and Tees, in Friedrich, 1983).

Infants discriminate between speech sounds because some acoustical attributes are more striking than others, rather than that humans have some special "speech processor." For one thing, infants discriminate between nonspeech sounds as well as speech sounds (Jusczyk, Rosner, Cutting, Foard, and Smith, 1977). For another, lower species of animals, like chinchillas, as well as humans, discriminate between speech sounds (Kuhl, 1981).

Even though infants *can* discriminate between similar spoken sounds in an experiment, psychologists still do not have the answer to how babies actually *do* segment the speech they hear. Do they segment it into **phonemes**—the smallest units of language, such as /*b*/*ed* versus /*r*/*ed*—into syllables, into words, into all of these, or into none? Although researchers have shown that infants *can* discriminate between phonemes like /*b*/ and /*p*/, syllables like /*bah*/ and /*pah*/, and words like *ball* and *Paul*, they have not yet demonstrated that infants *do* divide the speech stream they hear in these ways. At some point, infants must begin to segment speech in some meaningful way, or they would never learn to talk. But they may not begin to do so until they have communication as their goal and have received feedback from adults about the meaning of their utterances. It seems plausible that syllables or one-syllable words are the first divisions that infants make (Bertoncini and Mehler, 1981). Finer distinctions may not happen until children are learning to read (Jusczyk, 1977; Liberman, Shankweiler, Fischer, and Carter, 1974).

Baby talk

No matter how they segment sounds, babies often hear speech that is especially adapted to their hearing. Adults the world over talk to babies in a special way—in "baby talk," of course. **Baby talk** is a whole range of changes

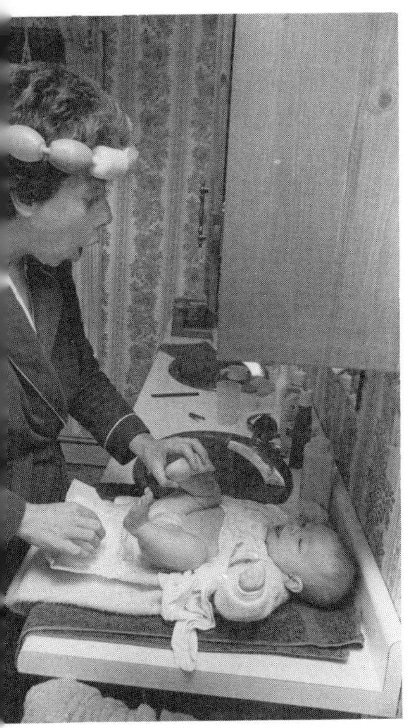

Adults around the world talk to babies in a special way—in baby talk. Infants prefer the high pitch and exaggerated intonations of this special speech.

worked on the usual patterns of adult speech. People speak to babies at a high pitch, exaggerate their tones from bass to squeaks, and swoop exaggeratedly from whispers to shouts. ''Hi baby-baby. Izzata baby-baby?'' They speak quickly or slowly, repeat words and phrases, and draw out their vowels. They make nonsense sounds and speak in a singsong rhythm. ''''Im luva Mommy? *At's* a goo' boy.'' In languages from Arabic to Latvian, people from grandfathers to older sisters pitch their voices high when they speak to infants (Ferguson, 1964; Ruke-Dravina, 1977; Fernald and Simon, 1984; Fourcin, 1978). This higher pitch is not universal. It is not found among Quiche Mayan mothers, for example (Ratner and Pye, 1984), but it is very widespread.

Infants love this baby talk. Infants from 4 to 18 months old who could choose between hearing adult conversation and baby talk chose to listen to the baby talk (Fernald, 1981, 1985; Glenn and Cunningham, 1983). Judging by their listening choices, infants like its rising tones—''This little piggy went to market'' rather than ''No, no, don't touch'' (Sullivan and Horowitz, 1983). They like its expressive and exaggerated intonation—''And *this* little piggy ran *all* the way home'' (Lieberman, 1967; Turnure, 1971; Kearsley, 1973). They particularly like its high pitch (Fernald, 1982). A mother's voice is probably especially appealing to her infant because it has all these qualities. When 4- to 6-week-old infants heard brief speeches by their mothers and by another woman, some of which were highly intonated and some of which were monotonous, they preferred their mothers' voices, but only when they were intonated (Mehler, Bertoncini, Barriere, and Jassik-Gerschenfeld, 1978).

> Even before he could focus his eyes clearly, Sam would stop, listen, and perk up whenever he heard my voice as I entered the room he was in.

> After Jason took a bottle, he usually was responsive and not irritable. I'd entertain him with an old high-school cheer that I knew. It was full of nonsense syllables and heavily rhythmical.
>
> *Ah-ba-la-ba, goo-ba-la-ba, goo-ba-la-ba, vee-stay.*
> *Ah-ba-la-ba, goo-ba-la-ba, goo-ba-la-ba, vee-stay.*
> *Oh, no, no, no, no, vee-stay, vee-stay.*
> *Eenie-meanie dixie-deenie ooh-la thumbeleenie,*
> *Aatchie, patchie Liberace, I mean* you!
>
> Jason would stare at me, transfixed, and I had a wonderful time sing-songing to him. Of course, anyone who heard us would have thought that we had lost our minds.

Infants also like to listen to baby talk because baby talk is salient, tuned to infants' sensory capacitites, and infants may be biologically predisposed to hear and pay attention to it (Fernald, 1984; Sachs, 1977). For one thing, the pitch of baby talk makes it subjectively louder than speech at lower pitch, so it stands out against background noise. For another, the contours of baby talk—long, smooth, gradual—are simple, separated by pauses, and easy to track. Baby talk is like a simple melody, and because infants' auditory systems are immature, simple patterns are easier for them to perceive. Infants react physiologically to the pitch contours of baby talk. They are aroused by a high, rising pitch and soothed by a low, falling one—another reason that baby talk stands out for them. Baby talk is tuned not only to babies' hearing but also to their vocal capacities. Baby talk contains many syllables that begin with the very initial-stop consonants—*baba, papa, dada*—that infants burble themselves (Pierce, 1974). The high pitch, special tones, patterns, and rhythms that characterize baby talk are used in infants' own speech. When infants hear a high-pitched voice, they immediately raise their own pitch, for example

(Webster, Steinhardt, and Senter, 1972). Baby talk is clearly an appropriate way to talk to infants.

Looking and listening

In real life, infants see and hear at the same time, integrating the information from the two domains to make sense of their world. One way they do this is by looking to locate the source of a sound. Even newborns do this in a rudimentary way. In one study of 2- to 4-day-old infants (Muir and Field, 1979), a tape-recorded rattle sounded from one of two loudspeakers on either side of the infant. Three-quarters of the time, infants turned toward the sound, although it took them about 2½ seconds to begin responding and nearly 6 seconds to turn their heads all the way. By 4 months of age, infants can look toward the right person when their mother or father talks to them (Spelke and Owsley, 1979). They can demonstrate their coordination of sight and sound in other ways as well. When shown a film of a bouncing toy donkey on one screen and a bouncing toy kangaroo on another, infants preferred to watch the film of the animal that bounced in time to the sound track (Spelke, 1976, 1978, 1979a, 1979b).

Taste and smell

Newborns can tell water, sugar, and salt from milk, and they respond differently to different concentrations of sweet, salty, and bitter solutions (Ganchrow, Steiner, and Daher, 1983; Jensen, 1932). Newborns suck longer and pause for shorter periods when they are given a sweet solution (Lipsitt, Reilly, Butcher, and Greenwood, 1976). They also suck more slowly, as if they are savoring the sweetness. In addition to sucking, newborns smile and lick their upper lip when they taste a sweet solution (Ganchrow et al., 1983).

Newborns also have a good sense of smell, although they cannot discriminate smells as acutely as they will later on. Infants turn away from a strong smell, such as ammonia. They breathe faster and move around more when they smell asatfetida, which smells like garlic (Lipsitt, Engen, and Kaye, 1963). They turn their faces toward a sweet smell, and their heart rates and breathing slow down (Brazelton, 1969). They smile at the smell of vanilla and banana but protest the smell of rotten eggs and shrimp. Within days after birth, breast-fed newborns prefer the odor of their mother's milk to that of another mother (Russell, 1976).

Perceptual processes

The traditional view in philosophy and psychology is that we begin life with minimal perceptual capacities and build them up only with years of experience. We now know that infants have some perceptual capacities even at birth. Much of what we know about infants' perceptual development is the result of the theory and research conducted and inspired by James and Eleanor Gibson (Gibson, 1966, 1969; Gibson and Spelke, 1983). In their view, infants are endowed with powerful perceptual capacities from birth. With experience, infants become increasingly aware of information in the world, and their perception becomes more exact and differentiated. They learn to extract more and more of this information from objects and to make more and more subtle distinctions among objects. Unlike Piaget, who held that infants construct their perception of objects by acting on them (see Chapter 6), the Gibsons theorized that the information is in the object and that all infants have to do is seek, and they will find.

Exploring

The first step in this seeking and finding is attending to the environment, exploring its features, and focusing on some action or object out of the vast array available. Even newborns pay attention to the sights and sounds in their world, although often the only sign that they have noticed something is a slowed heartbeat. They notice especially the *onsets* of sights and sounds and, as we have discussed, sights and sounds that are contrasting and relatively intense.

> As I came up to 2-month-old Natalie's stroller and said hello, her eyes flickered. Then she seemed to stare hard at my face for almost a minute. But she looked past my shoulder when a noisy truck passed by.

Very young infants can attend to the environment for only short periods of time, but as they mature, their periods of attentiveness lengthen. In these attentive periods, infants actively explore, with eyes and ears, hands and mouths, the sights, sounds, objects, and events that have caught their attention. With age their exploration becomes more swift, efficient, systematic, and focused in direction. But right from the beginning, it is purposeful and directed. Even newborns actively scan visual objects, moving their eyes back and forth systematically when they come across contrast, contours, or corners. At this age, though, they move their eyes back and forth in one place rather than searching whole scenes. As they get older, they look at whole objects. In the early months, infants also explore by putting things into their mouths, and for the first six months or so, they prefer mouthing to exploring with their hands. By 8 or 9 months of age, with improved motor control and coordination they explore more with their hands, bang and squeeze things. They also coordinate their touch—fingering, turning, poking—with their visual examination (Ruff, 1986). The fact that similar developmental changes

Infants explore the features of their surroundings. This 8-month-old looks, touches, and tastes.

occur in scanning by touch and by sight suggests that central processes direct exploration in all the senses.

Infants also become increasingly selective about which features of the environment they explore and focus on. As we have already seen, they prefer at first to look at edges and angles, contours and corners because of their cortical and visual capacities rather than the properties of the stimuli. Beginning at the age of 3 or 4 months and thereafter, however, the properties of the stimuli themselves affect infants' attention and exploration. Some infants attend to color more than size; others attend to shape more than color (Odom, 1978). An infant may attend first to color, then to shape, then to size, and so on. But whatever the differences among individuals, all infants grow increasingly selective and focused as they get older. They learn to tailor their perception to the particular object and situation.

Figure 5.8 ***Adults see a square only in stimulus (c). Researchers using habituation techniques have found that 5-month-olds respond as though all three arrangements are alike, whereas 7-month-olds pay more attention to stimulus c, suggesting that they perceive a square or part of it (Bertenthal et al., 1980).***

PERCEIVING A SQUARE

Habituation stimulus

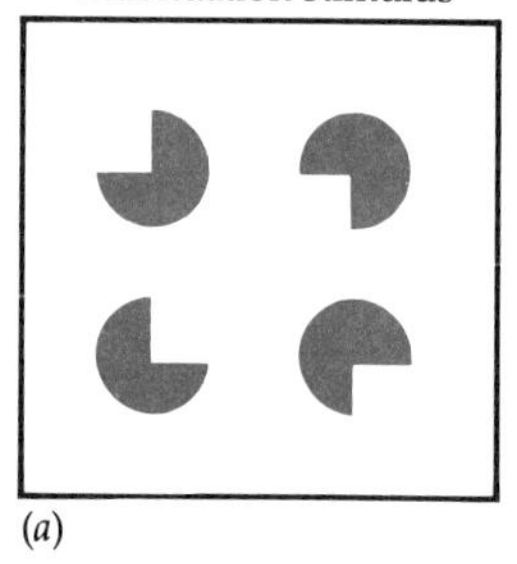

(*a*)

Novel stimulus

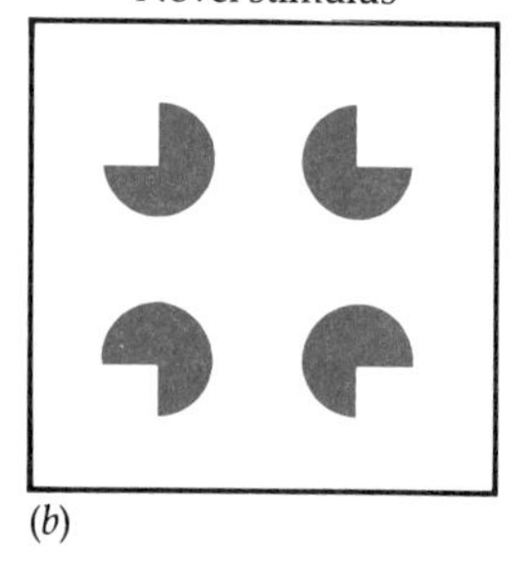

(*b*)

Novel stimulus

(*c*)

Detecting invariances

According to the Gibsons' theory of perception, infants perceive events, objects, and places by detecting **invariants**—that is, actions or things that look or sound the same every time they appear. The ability to perceive these invariants is based on innate perceptual abilities and, like the ability to explore the environment, they suggest, improves with physical maturation and with experience.

The process by which infants come to know that things are objects typifies the process. To study infants' perception of objects, a researcher might show the infants an object in which part is hidden, then show them the complete object, and then see whether they discriminate between the two. Results of experiments like this suggest that, from the beginning, infants know something about the nature of objects. They seem to know that if two points move together or touch each other, they are part of the same object (Spelke, 1982). They perceive an object when they see connected surfaces that remain connected while in motion, and they perceive the difference between object and background as the object moves and the texture of the background is added to and subtracted from along the moving object's edges. By perceiving the connected movements of a caterpillar crawling along a leaf, a ball rolling along a sidewalk, or Dad entering a room, infants come to perceive caterpillar, ball, and Dad as objects. Infants fail to track objects moving in tandem with their backgrounds, perhaps because they perceive object and background as a single unit moving together. Only after experience with many objects do infants learn that objects are separate from their backgrounds.

Part of perceiving invariants is recognizing that an object's size and shape remain constant even if the object is moved closer or farther away or tilted to the left or right. By the time infants are 3 or 4 months old, they have this knowledge (Day and MacKenzie, 1973, 1981). In fact, even at birth infants appear to have **shape constancy** for simple shapes (Slater, 1986). That is, they respond to identical simple shapes in the same way even if the shapes are tilted at different angles. The perception of more complex shapes seems to take time to develop (Gibson and Spelke, 1983; see Figure 5.8). Experts still disagree to some extent about what is inborn, what matures, and what is learned in the first months of life. But there is agreement that by the end of infancy, the basic perceptual competencies are in place.

Recognizing affordances

In the Gibsons' theory, infants learn not only about invariances but about the properties of objects. They learn what they can *do* with objects and what they can expect from them. Water affords wetness, and floors afford support. Fires afford warming and light to read by. A wall is an obstacle that affords collision, but a doorway in the wall affords passage. **Affordances** also are related to the observer: a floor affords support to people but not to elephants, and a leaf affords shade to a person but support to a caterpillar. Affordances may derive from an object's texture, shape, or substance and may involve more than one sense. To judge affordances, we use information that we take in with our eyes—a floor must be level, solid, not too slippery to be walkable upon—plus information from our ears—the creak of the floor—and from touch—the give of the floor under our weight.

When do infants first begin to perceive the properties of objects, from which they can judge the objects' affordances? Even very young infants have rudimentary abilities to perceive the properties of objects. At 1 month, infants can differentiate between rigid and flexible substances in their mouths, like teething rings and nipples. At 3½ months, they can see the difference between rigid and flexible motion—the difference between a ball being rolled and a ball being squeezed, for instance. They can see the difference between flat and solid objects and between objects of different simple shapes like spheres and cubes. At 6 months, infants can feel the difference between rough and smooth textures. One-year-olds distinguish between rigid and elastic substances by banging the hard objects and squeezing, pressing, and wiping with the spongy ones. They recognize by sight or by touch simple shapes they have manipulated (Gottfried, Rose, and Bridger, 1977). As they get older and develop strategies of exploring and manipulating, infants' perceptions of objects become increasingly differentiated. In addition to perceiving the properties of flexibility, shape, and texture, infants perceive the color, temperature, scent, size, and animacy of the objects around them (Gibson and Spelke, 1983). From their perceptions, they predict objects' affordances.

Sam never had any trouble knowing which things were good to eat. Later on he got good at knowing just what to do with a new toy—bang with the hard ones and squeeze the soft ones.

Natalie's babysitter arrived one morning with a cake of soap that looked exactly like a piece of layer cake, complete with "frosting" and a "cherry" on top. Poor Natalie bit off a big mouthful, then let out a startled wail, and made the most horrible grimace of disgust I have ever seen.

Summary

1. An infant's first month after birth is a risky time, when untested biological systems must function independently in a germ-filled environment. The newborn's immediate tasks include breathing, eating, stabilizing body temperature, and circulating blood. The Apgar test is used to assess the condition of the newborn immediately after birth.
2. The most common birth complication is prematurity. Premature birth is associated with conditions that are especially common among poor women. Most preterm babies can be saved with good hospital care, and the majority suffer no serious permanent handicaps, although their development may be somewhat delayed. In efforts to reduce these delays, hospital staff have tried to humanize the stressful, noisy atmosphere of the intensive care nurseries where preterm babies are cared for. Other risks to infants' survival in their first year of life are low birth weight, disease, malnutrition, and neglect. An infant's sex and birth order influence weight and maturity at birth.

3. The average newborn weighs 7½ pounds and is 20 inches long. Newborns whose size was restrained by their mother's small uterus go through catch-up growth after birth. Heredity has a smaller effect on prenatal size than on catch-up growth.
4. Healthy newborns have more than 20 reflexes. Some reflexes help the newborn to feed, some to maintain contact with the mother, and others anticipate actions—like swimming and walking—that are learned later only by deliberate effort. Tests of reflexes help doctors to detect central nervous system disorders. Most reflexes disappear within a matter of months after birth.
5. Infants go through a range of states of consciousness from deep to active sleep, drowsiness to quiet alertness, calm activity to restlessness, soft crying to loud squalling. Newborns sleep up to 70 percent of the time; half of that sleep time is active sleep, half quiet sleep.
6. Infants acquire motor abilities in the same order that body parts grew before birth. In the first months, infants are able to lift and support their heads. Later, they learn to roll from stomach to side or back and to sit unsupported. At 6 months, they can touch their hands together in front of them and move them up and down and side to side. At 7 months, they may pull themselves into a standing position. Soon comes creeping, crawling, and sometime around the first birthday, the first independent steps. Individual variations in reaching motor milestones are common, and both heredity and environmental conditions affect infants' rates of motor development.
7. The infant's visual system matures gradually over the first two years. In early infancy, the abilities to focus, to track objects, to accommodate to the shifting planes of visual targets, and to discriminate details all are worse than adults'. Infants can see clearly only objects that are large, close to them, and high in contrast—and a parent's face during feeding, for instance, precisely fits this description. Infants also prefer to look at things with high contrast, contours, a moderate degree of complexity, and movement.
8. Researchers have used several different techniques to determine that the perception of distance develops in infants over the first six months of life.
9. Newborns can hear the difference between loud and soft, high and low tones. But their hearing is not as acute as adults' throughout their first year of life. Infants prefer human speech to other sounds. They especially like baby talk, which typically has a high pitch, exaggerated tone, slow and repetitious wording, and drawn-out vowel sounds.
10. Newborns coordinate what they hear with what they see. They can taste and smell at birth.
11. Newborns pay attention to their environments and notice the onset of sights and sounds. As they mature, they can pay attention for longer periods. They actively explore with their senses and select things to focus on. They detect the invariance of objects in the environment and perceive what they can do with these things.

Key terms

Apgar test
full-term babies
preterm babies
postterm babies
incubators
restraining effect
small for gestational age (small for dates)
marasmus
colostrum
placebo
failure to thrive
restraining effect
catch-up growth

reflex
rooting reflex
sucking reflex
palmar grasp
Moro (embracing) reflex
stepping reflex
myelination
tonic neck reflex
habituation
retina
fovea
dendrites
saccadic eye movements

astigmatism
accommodate
depth of focus
visual acuity
visual cliff
tympanum
cochlea
phoneme
baby talk
invariants
shape constancy
affordances

Suggested readings

BOWER, T. G. R. *The Perceptual World of the Child.* Cambridge, MA: Harvard University Press, 1977. An easy to read discussion of the perceptual capacities and development of infants and young children.

GESELL, ARNOLD, ILG, FRANCES, and AMES, LOUISE. *Infant and Child in the Culture of Today.* New York: Harper and Row, 1974. Milestones of motor and perceptual development in infancy and childhood.

KLAUS, MARSHALL and KLAUS, PHYLLIS. *The Amazing Newborn.* Reading, MA: Addison-Wesley, 1985. Discussion of the awareness, sensory capacities, and abilities of the newborn, amply illustrated with photos.

chapter six

Cognitive development

What goes on inside that fuzzy head with the angelic smile? What do babies remember from yesterday? What do they think about their toys? their parents? their first steps? How do they learn that something delicious is in the bottle? How do they learn to play with spoons, blocks, cars? How do they learn that they are unique and separate beings? Developmental psychologists have asked these questions—and have designed experiments to try to find the answers. In the last chapter, we discussed how infants take in the sights and sounds and smells around them. In this chapter, we will discuss how infants learn, remember, think, and know about the world.

Learning in infancy

For his first few months, Sam slept in a bassinet with a string of bright red, yellow, and blue plastic horses draped over one side. The horses rattled when they were moved. One night at about midnight, I was awakened by the sound of the horses rattling. Sam's hand must have brushed against them by accident in the dark. The sound stopped. An hour later, I heard them rattling again: once, twice, three times. The sound stopped. I fell asleep again. Soon the horses were rattling again—and again and again. Sam rattled the horses in bursts all night long. He had learned his first trick. The next night, I made sure that the horses were draped *outside* of the bassinet!

Ingenious experiments have offered developmental psychologists a baby's-eye view of learning. In these experiments researchers have probed the ways in which babies learn. One way infants might learn is by classical conditioning, associating events that repeatedly happen together. Ivan Pavlov, you will recall, demonstrated that animals can learn in this way. In his classic experiment, Pavlov rang a bell (the **conditioned stimulus**—CS) just before food (the **unconditioned stimulus**—the UCS) was put into a dog's mouth.The dog responded to the food by salivating (the **unconditioned response**—UCR). After the food and the bell had been paired repeatedly, the dog began to salivate (the **conditioned response**—CR) as soon as it heard the bell. Later, John Watson demonstrated that infants, too, could learn by classical conditioning (Watson and Rayner, 1920). In his most famous demonstration, Watson chose a 9-month-old named Albert as his subject and set out to make Albert afraid of a white rat. First, he demonstrated that Albert was not afraid of the white rat, a white rabbit, cotton batting, or other white things. He then demonstrated that when Albert heard a steel bar rapped loudly behind his head (UCS), he jumped (UCR). Two months later, Watson began to condition Albert. Whenever Albert saw the white rat (CS) and reached out to play with it, Watson rapped the steel bar behind the baby's head. After seven repetitions, Albert drew back at the sight of the rat, cried, and tried to crawl away (CR). Five days later and at two other sessions, Albert reacted fearfully to the white rat and to some extent to the other white things, too. He now was conditioned to fear previously positive stimuli, and his fear generalized to other, similar objects. Classical conditioning can take place in everyday situations, too, not just in the laboratory. The infant whose doctor comes to be associated with painful injections or the aunt who comes to be associated with painful, cheek-pinching greetings make the infant recoil in fear, just as Albert recoiled from the white rat.

But how old do infants have to be before they begin to associate things in this way? Do even newborn infants learn through classical conditioning?

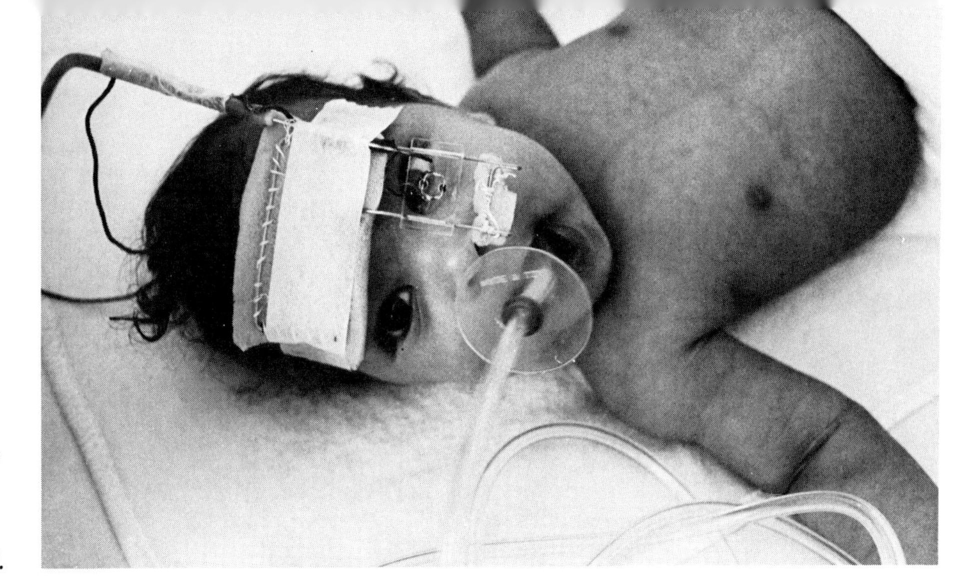

In this classical conditioning experiment, researchers attempted to condition the infant to blink, a reflex (UCR), when a tone (CS) was played. They originally elicited the blink by directing a puff of air (UCS) into the infant's eye as the tone sounded.

Psychologists have tried to condition very young infants under controlled conditions to provide answers to these questions. There is little evidence that infants under 2 or 3 months old can be classically conditioned. Most attempts to classically condition very young infants have failed (Sameroff and Cavanaugh, 1979). Psychologists now tend to be skeptical that infants in their first few months learn through classical conditioning.

Another way that infants might learn is through operant, or instrumental conditioning. In this kind of learning, a person or an animal does something, finds the consequences of the action rewarding, and therefore repeats the action. A laboratory pigeon, for example, pecks at a bar on the side of its cage (the operant response). The peck releases a pellet of food (the reinforcer) into the cage. The pigeon pecks again at the bar and receives another pellet of food. The rate at which the pigeon pecks the bar is increased by the presentation of food pellets. As long as food is presented at least some of the time after the bar is pressed, the pigeon will press the bar at a high rate. If reinforcement is discontinued, bar pressing will decline. Pigeons are not the only creatures to learn to do things for rewards. People do, too, of course. But do infants learn through instrumental conditioning?

With young infants, researchers' attempts to demonstrate operant conditioning have been more successful than their attempts to demonstrate classical conditioning. In one study, for example, newborn infants' sucking on a nipple became stronger or faster if only strong or fast sucking was rewarded with milk (Sameroff, 1968). To keep the milk flowing, infants also learned to change the way they sucked. In another successful demonstration of instrumental conditioning, newborn infants learned to turn their heads to the side if the experimenter brushed their cheeks and gave them sweetened water every time they did so (Siqueland and Lipsitt, 1966). These successful demonstrations of instrumental conditioning in young infants relied on the infants' sucking and rooting reflexes. Researchers now suspect that the reflexes of newborns can be conditioned because newborns are already physiologically prepared to take in stimuli and to learn from the environment by using these reflexes (Sameroff and Cavanaugh, 1979). If this is true, the instrumental learning of newborns seen in these experiments may differ in kind from that seen in older infants and adults, reflecting simply a basic adaptation of reflexes rather than the modification of a voluntary or operant action.

Beginning when infants are 3 months old, however, a broader range of behavior—not just reflexes—can be modified through instrumental conditioning. At 3 to 6 months, infants coo, smile, look, reach, press, and touch if they are rewarded for doing so. But even at this age, infants seem to have been prepared either by past experience or by biological predisposition to respond in certain ways. In one experiment, for example, 6-month-olds were seated in front of a clear panel on which they could press (Cavanaugh and Davidson, 1977). Infants in one group were rewarded by seeing lights behind the panel and hearing a bell each time they pressed the panel. Infants in a second group saw lights behind the panel and heard a bell, but the sound of the bell was not timed to their pressing. A third group of infants saw lights go on every 20 seconds, but the lights were not behind the panel. Infants in *each* group pressed the panel more often as time went by—even though only those in the first group were systematically rewarded for doing so. Why? Presumably, the infants had had lots of practice pressing things like buzzers and busy boxes, and so when they saw something to press, they reacted naturally by reaching out and pressing, and the pressing itself was rewarding. Infants in the first group did show some effects of their conditioning, however. During the *extinction* phase of the study, when the infants saw no lights and heard no bell, their rate of pressing fell below what it had been before the experiment began. Infants do learn through instrumental conditioning, but their learning at this age is still closely tied to a limited array of familiar behaviors.

Memory

Parents know from their everyday experiences what experiments demonstrate—that babies in their first year remember things: the 9-month-old baby smiles when Mother or Father, but not a stranger, appears by the crib; the 8-month-old baby who once tensed up now splashes happily in the familiar bathtub.

> Sam and I had a raucous routine that we both loved. It was a rollicking rhyme that we bounced and swung around in time to. When Sam was looking at me and I knew I had his attention, I'd begin with a loud, ''Ah *one*, ah *two*, I'm talking to *you*.'' After a couple of times, all I had to say was ''Ah *one*,'' and he would chortle with delight.

The problem for developmental psychologists is to demonstrate systematically and scientifically what infants can remember.

Methods of studying infants' memory

Researchers can use any of several different methods to find out how much and for how long infants remember. One method they have used is to show infants a pair of patterns. Then the infants are shown one of the patterns they have just seen plus a new one. If the infants look at the new pattern in the pair longer than they look at the familiar one, it can be inferred that they remember the familiar pattern. Infants can be shown pairs of patterns after hours, days, or weeks have passed to test the length of their memories.

Researchers capitalize on the human tendency to grow bored with the old in a second method of studying infants' memory. Researchers show infants a pattern and see how quickly they habituate. When they habituate, the assumption is that they remember having seen the pattern. Researchers rely on infants' retention of learned responses as a third measure of their memories. In one ingenious study, for instance, 3-month-olds were put in their own cribs at home, with a ribbon tying one of their feet to a lever that turned a mobile (Rovee-Collier, Sullivan, Enright, Lucas, and Fagen, 1980). As the infants kicked, the mobile turned, and this interesting motion reinforced their kicking. They had two learning sessions one day apart; during each of them, the infants moved the mobile for about ten minutes. Several weeks later, the infants got a three-minute "refresher" session with the mobile. A day after the refresher session, the infants all could turn the mobile like pros. Infants who had not had the learning experience could not. In short, 3-month-old infants retained what they had learned for several weeks, if their memories were jogged by reminders of the learning experience.

A fourth method of studying infants' memories relies on observation of the infants' spontaneous behavior. Common sense tells us that infants do not behave in the same way at home as they do in a laboratory—where much of the research on their memory abilities has taken place. The laboratory is unfamiliar, devoid of informal social stimulation, and as carefully controlled for outside influences as the investigator can arrange. Some researchers therefore have looked for signs of infants' memories at home, in their natural surroundings. Home is rich in stimulation and support, and the kinds of learning and remembering that children show there can be quite different from what they show in more sterile and objective assessments. When the test materials are familiar and meaningful, and when the surroundings are comfortable and relaxed, infants are more likely to show that they retain complex memories.

After putting together evidence from all these different methods of research—comparing infants' reactions to pairs of familiar and unfamiliar patterns, seeing whether infants habituate to a pattern, finding out how long infants retain a learned response, and observing what infants spontaneously recall—we begin to get a picture of what infants can and do remember.

Phases and functions of infant memory

Although some kind of memory is possible right from the start, there is a vast difference between the memory of the newborn, who can recognize a simple

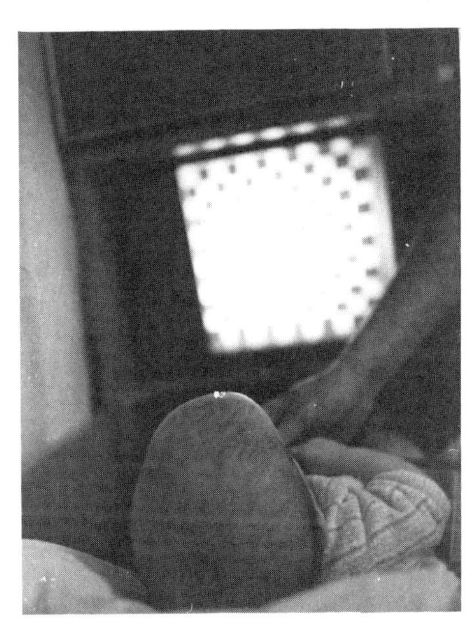

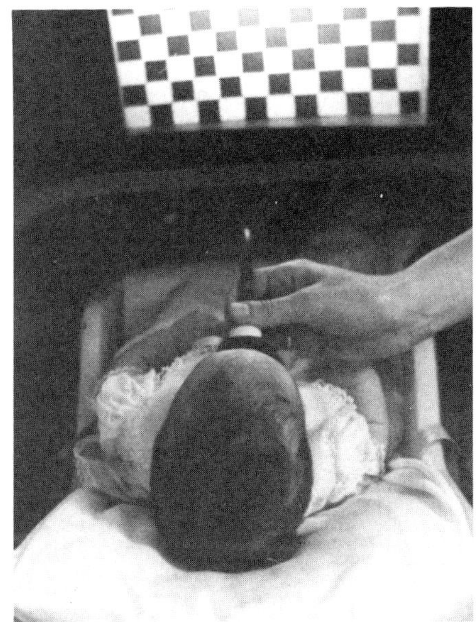

In this study, the infant sees a checkerboard pattern on a television-like screen. By sucking on a pacifier connected to the projector, he can keep the pattern in sharp focus. For a while the infant sucks with great energy and interest, but like most babies, he eventually becomes bored with what he is attending to. This is called habituation. As he slows his sucking, the checkerboard fades. When the screen offers another picture, he will suck energetically again.

pattern, and the 1-year-old, whose face lights up as Father's approaching footsteps sound on the stairs and who greets the babysitter by diving into Mother's lap. In the first year of life, memory develops into increasingly elaborate and sophisticated patterns—carrying along the infant from the largely helpless, speechless being of the early weeks to the remembering, speaking, active reasoner of a year later.

The development of memory during this crucial first year can be divided into several phases (Mandler, 1981; Olson and Strauss, 1981; Schacter and Moscovitch, 1981). In the first phase, the infant's memories apparently depend on neurological "wiring." Show the newborn an object—Mother's face, a bright toy, a nipple, a pattern—and neurons fire in the brain's visual cortex. Infants fix their eyes on new and interesting patterns, and their neurons fire. But as the infant continues to stare, the firing decreases, and the infant becomes habituated to the pattern. Once habituated to this particular pattern, the infant may turn away. "Oh, that," the infant may seem to be saying, "I've seen that before. How about something new?" In fact, the infant probably isn't thinking at all, and the seeming recognition of the pattern is simply the effect of tired neurons.

At about 3 months of age, infants begin a second phase in the development of memory. Their memories seem to become cognitive rather than sensory. Now they can both control their attention and actively explore with it. They can learn and take in information readily and rapidly—sometimes after only a few seconds of experience—remember for relatively long periods, and build on what they have learned. Able now to reach, touch, grasp, turn, look, and pull, they actively explore their world. They scan the patterns that they see and **encode** them in their brains. By the end of this phase, 6- to 7-month-old infants recognize not only bold patterns and sharp contrasts but subtleties and details. They remember them for weeks.

Another major change in infants' abilities to remember takes place between the ages of 8 and 10 months. Infants start to organize the vast array of information that they absorb into systematic categories. They can generalize across time and place. Their memories now include general categories like ducks and geese and clowns and puppies, not just specific objects like Daffy Duck, Mother Goose, Bozo, and Fido. In the laboratory, infants' memory for categories can be tested. First they are shown several objects from one category—perhaps women's faces—and then an object from a new category—a man's face. If they look longer at the new object, the infants are assumed to have noticed the abstract properties common to the first category, women's faces. When just such a test was run, 8-month-old infants did remember the women's faces; 4-month-olds did not (Cohen and Strauss, 1979).

By 8 to 10 months, infants have been found to categorize shapes regardless of their size, color, or orientation in space and faces regardless of their poses; to differentiate between faces with familiar features and faces with unfamiliar features; to categorize kinds of movement and toys representing men, letters, animals, food, furniture, and moving vehicles (Olson and Strauss, 1981). Their ability to form categories of things that are wet versus things that are dry, things that are good to eat versus things that are bad to eat, things that are rough versus smooth, and so on provides the basis for an organizational framework for memory. It allows infants to represent information in a summary fashion, thereby significantly reducing their memory load.

These changes in classification and memory abilities have important implications for how infants act. Beginning between 8 and 10 months, infants approach familiar people and things but are wary of strangers. They remember previous departures, and so now the sight of their parents leaving is crushing to them. They begin to recall things that are not immediately available to their senses, and so now they can look for their toys in familiar places.

The simplest kind of remembering is recognizing an object or an image as one seen before. Infants do this from an early age.

Although they are probably still without any conscious notion of the past, they have begun to build on their backlog of remembered experiences. By the end of the first year, the walking, seeing, hearing infant is also a thinking and remembering infant. The memory systems that will last a lifetime are in operation. Although the abilities to recognize, encode, recall, and retrieve information from memory will grow more elaborate, efficient, complex, and better integrated, no new memory systems will appear. The neural equipment that underlies the ability to remember the past is in place.

Sensorimotor intelligence

One person who was especially interested in infants' abilities to think and remember was Jean Piaget. Piaget carefully investigated, charted, and tried to explain the fascinating journey infants take over their first two years, from early reflexive (unthinking) behavior to later (thinking) mental images. Although he was mistaken in some particulars, it is important that we know what Piaget said, because his was the first truly careful, sustained, empirical and theoretical exploration of infant cognitive development.

Schemes

Piaget called the period from birth to about 2 years the **sensorimotor period**, because, he claimed, infants' mental activity then depends solely on sensory and motor abilities and experiences. During the sensorimotor period, according to Piaget, babies do not yet reflect on or consciously think about their world. But they do have notions about the world that Piaget called sensorimotor **schemes**. Reflexes like sucking and grasping and simple sensory acts like looking and hearing are basic building blocks for the development of sensorimotor schemes. Schemes are primitive mental structures, the most basic units of knowledge. As babies experience regularities—the constant properties of objects, the consistent reactions of people—they form schemes that incorporate the information they have gleaned. Schemes are generalizations built up from specific, repeated, single events and experiences. They are built out of sensorimotor actions like sucking on a nipple, looking at Mother's face, or nibbling one's toes (a favorite activity of the 6-month-old set).

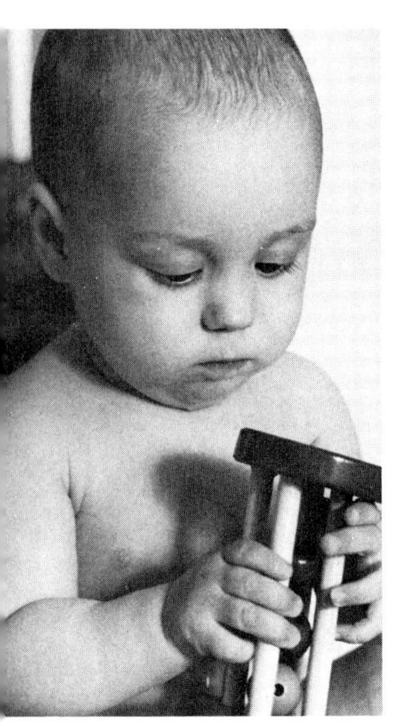

During the sensorimotor period, the infant's schemes revolve around actions such as grasping, shaking, patting, and looking at objects.

Schemes are many, complex, and varied. As infants' senses, muscles, and memories mature, their schemes grow and change. Primitive action schemes built upon reflexive sucking and grasping grow into voluntary mouthing, hitting, shaking, waving, banging, sliding, swinging, dropping, and tearing schemes. Give the 7- to 9-month-old a block and a cup, and he hits, pats, and shakes them, puts them in his mouth, bangs them on the table. Later, as a 10- or 12-month-old, he examines them intently and turns them around and around. He slides them along the highchair tray, rubs them together, and places the block in and out of the cup. He puts the cup on the saucer and the spoon to his lips. Soon he graduates to letting go with his fingers, dropping the block onto the floor. The 14-month-old engages in building schemes—turning over the cup and putting the block on top of it; drinking schemes—putting the cup to his own and to his parents' lips; and grooming schemes—applying the hairbrush (none too expertly) to his hair, as he uses objects for their intended purposes. He seems to know what things are *for*. He looks first. Then he rolls the truck along the tabletop, or, in perfect sequence, puts the cup on the saucer and stirs with the spoon before he "drinks." He can combine actions into longer and more coherent sequences. He stirs the "food" in a pot, then eats it from a plate, and then "washes" the pot with water.

In addition to advances in perceptual, motor, and memory development, in Piaget's theory, there are three mental processes that guide the infant's development of schemes. **Adaptation** is the process by which infants modify their schemes. Adaptation has two complementary aspects: **accommodation** and **assimilation**. To assimilate is to take in. Infants take in new objects to fit their existing schemes. Give the 9-month-old a cup with a spout on it, and he will be sucking on it in a flash, for spouts are not much different from nipples. Nipples are not much different from pacifiers, or drinking straws, or garden hoses, or pipestems, or (to Mother's horror) pens—all the common household objects that infants can and do apply their sucking scheme to. Luckily, there is more to development than assimilation, however, or you might be sucking on the corner of this textbook right now. Infants also accommodate, or modify, their existing schemes when they encounter new objects that do not fit. Breast-fed babies given a bottle for the first time change their sucking scheme slightly, moving their heads a bit, adjusting their mouths to the new nipple, and modifying the movements of their jaws and tongues. The 9-month-old learns that pipes taste yucky and eliminates them from his sucking scheme. The 1-year-old discovers that books are not suitable for sucking but have pages that turn and tear in an especially interesting way. Assimilation and accommodation are complementary processes of adaptation. The second process that guides the infant's development of schemes, according to Piaget, is **organization**. This is the process by which the infant combines and integrates separate schemes. The 3-month-old stares at and then reaches for his rattle, combining the staring scheme with the reaching scheme. The 5-month-old can combine three schemes—mouthing, staring, and reaching—as he puts his rattle to his lips. The third process, **equilibration**, is the way that the infant moves from one stage to another. When the infant reaches the end of one stage, he feels the inadequacies of his present stage. This produces cognitive disequilibrium, or uncertainty and conflict. To resolve this uncertainty and to restore cognitive equilibrium, he moves to the next higher stage of thought. With these processes, adaptation, organization, and equilibration, the infant makes rapid progress to higher levels of sensorimotor intelligence.

Stages

Infants move along in their developmental journey in an unvarying sequence of stages, according to Piaget. Each advance emerges from earlier experiences,

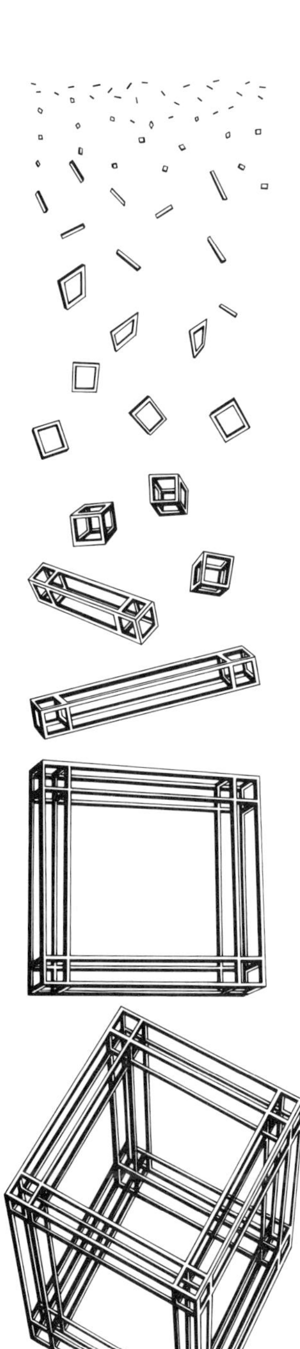

Figure 6.1 ***In this illustration, each structure is formed from the simpler structures above it. According to Piaget, sensorimotor intelligence develops in this way. Abilities and knowledge from previous stages do not disappear; they are integrated into a new, more complex structure as the infant develops.***

and each stage incorporates the mental structures that developed in the previous stage (see Figure 6.1). For the sensorimotor period, Piaget outlined six successive stages of development. The age ranges for these stages are approximate. More important is the order in which the stages occur.

Stage 1: practice and repetition of reflexes (birth to approximately 1 month)

In the first stage, the newborn practices reflexive actions like sucking, grasping, looking, and listening. A toy touches her lips, and she begins sucking. A nipple is put to her mouth, and she will suck. Her own hand touches her lips, and she sucks. She spends her time fitting the world into her limited range of reflexes.

Stage 2: the first acquired adaptations (approximately 1 month to 4 months)

At about 1 month, the infant begins exercising simple actions for pleasure. Her hand moves, and the movement is pleasant, and so she repeats it. She kicks her legs gleefully over and over again. Piaget called these repetitions of what begin as random or reflexive actions **primary circular reactions**. Circular reactions are the first kind of accommodation, for they are the willful modification of behavior through repetition.

At this stage also, infants begin to develop the notion that objects are separate from themselves. When they drop their rattle, they stare at their fingers, a bit perplexed. But within moments, they have gone on to something else. They do not search for objects they cannot see; out of sight is literally out of mind. They do not realize that objects are permanent, a realization Piaget called **object permanence**.

Stage 3: procedures for making interesting sights last (approximately 4 to 8 months)

At the age of 3 months, Piaget's daughter, Lucienne, noticed that shaking her leg, which she'd been doing just for fun, also made some dolls attached to her bassinet swing back and forth.

> At 0;3 (5) Lucienne shakes her bassinet by moving her legs violently (bending and unbending them, etc.) which makes the cloth dolls swing from the hood. Lucienne looks at them, smiling, and recommences at once. These movements appear simply to be the concomitants of joy. . . . The next day. . .I present the dolls: Lucienne immediately moves, shakes her legs, but this time without smiling. Her interest is intense and sustained. . . .
>
> 0;3 (8). . .a chance movement disturbs the dolls: Lucienne. . .looks at them. . .and shakes herself with regularity. She stares at the dolls, barely smiles and moves her leg vigorously and thoroughly. . . .
>
> At 0;3 (16) as soon as I suspended the dolls she immediately shakes them, without smiling, with precise and rhythmical movements with quite an interval between shakes, as though she were studying the phenomenon. Success gradually causes her to smile (Piaget, 1952, pp. 157–158).

When the baby began repeating her action, shaking her legs to see its effect on the movement of the dolls, she was engaged in what Piaget called a **secondary circular reaction**. Secondary circular reactions extend beyond the baby's own body, to an object or a person who responds to the baby's acts. In this stage, babies try to prolong interesting sights, like dolls bouncing, although it takes them some experimentation to learn which of their actions cause which reactions in the world outside of them. Piaget's son, for example, pulled a rope hanging from the top of his cradle, and the rope made an interesting rattle. But he also pulled the rope in a vain attempt to make his mother stay in the room with him (Piaget, 1952). His mother, of course, was not attached to the rope, a mistake in inferring causality that Piaget called **magico-phenomenistic thinking**. By this term, Piaget referred to the way that infants, with their imperfect understanding of cause and effect, often act as though things happened by magic. To someone so new to this complex world, magic must seem to be everywhere: *presto!* and the room is lit up; *presto!* and Mommy appears by the crib; *presto!* and flames appear under the shiny pots; *presto!* and the telephone bell rings.

At this stage, babies begin to understand something about the permanence of physical objects. If they see their ball on the floor, peeking out from beneath a blanket, they may move the blanket aside. With this action, they show that they can reconstruct the whole object from seeing just part of it, a sign that they have developed a primitive mental image of the ball. But if the ball is invisible beneath the blanket, they will ignore it completely. It will seem like magic when someone pulls the blanket off the hidden ball. The magic with which people and things disappear and reappear makes games like peekaboo and jack-in-the-box lots of fun for babies at this stage.

Stage 4: coordination of means and ends (approximately 8 to 12 months)

The budding scientist of 8 to 12 months old is quickly learning more and more about cause and effect. Play peekaboo with her now, and she'll pull your fingers from your face (a month or so ago, she waited until *you* took your hands down). She's no fool; she knows you're in there! Now she watches as Daddy picks up her ball, tosses it into the toy chest, and closes the cover. She goes to the toy chest, finds the cover too heavy to lift, and bangs on it. Daddy is oblivious. So she goes to Daddy, pulls him to the toy chest, and presses his hand on the cover. The light dawns; Daddy opens the cover; and she reaches

As the infant's understanding of cause and effect increases, she is able to use objects as means to an end.

in for her ball. Her understanding of cause and effect is much more sophisticated now, although she still apparently believes that it is she who initiates and causes all actions.

She knows lots about the properties of different kinds of objects now, and her schemes are more varied and appropriately used. No longer is every object sucked, shaken, or dropped to the floor. Now her stuffed bear is carried, hugged, and propped on the living room couch. Her toy telephone is babbled into. Pots and pans are nested, drummed upon, and their lids—after Daddy is led by the hand to perform—are spun like tops. Cabinets doors are swung, opened, and closed. She has begun to learn about the positions of objects in three-dimensional space.

> When Sam was just over a year old, he loved to put things inside one another. I always found it very funny to discover things like the dog's brush, a toy xylophone, a plastic ladle, and one mitten inside of the soup kettle, all piled in front of the back door and covered with a towel.

But move a ball from one hiding place to another, and the baby cannot follow your sleight of hand. She may watch your every move as you hide the ball under the pillow, take it out, and then hide it under the blanket. Ask her where it is, and she goes to the pillow. Even so, this mistake shows that she has some grasp of object permanence. For now she searches for the ball even when it is completely hidden; even when she cannot see any part of it, she knows that the object still exists. Until now, she had to be able to see at least some part of the object to credit its existence. Although she now has a mental image of the ball, it is bound up with the *act* of searching for it and with the specific place where she first saw it hidden. According to Piaget, at this stage the infant still does not have a fully mature image of the object as *separate* from all other objects and from her own actions.

The year-old infant is an enquiring scientist who performs endless experiments—drop the cup, open the door, smear the dirt—just to see what will happen.

Stage 5: experiments in order to see (approximately 12 to 18 months)

At around her first birthday, the baby is a confirmed scientist. Now that walking has freed her hands to explore the world at will, she performs endless experiments in order to see what will happen. She pulls the cabinet doors open, pulls the dog's fur, pulls the highchair across the floor, pulls her own hair, pulls Mommy's and Daddy's hair. Through **tertiary circular reactions**, she repeats her pulling action, modifying it slightly each time to see the effects of the modification. Pull Mommy's hair, she learns, and Mommy squawks; pull the cabinet doors and they swing open; pull her own hair, and she can't see it but it hurts.

> When Sam was 12 months old, he deliberately spilled his food, bit by bit, onto the floor. We knew that he was experimenting, and we knew that he wasn't being naughty on purpose. But it was hard to live with the mess our little scientist was creating at every meal.

Babies now can infer the cause from seeing just its effect. Bedroom door opening? Mommy coming in! Kettle whistling? Hot! Stroller rolling? Someone pushing it! Now infants can follow the path of the hidden ball being moved from under the pillow to beneath the blanket and can retrieve it from its hiding place—as long as they saw it being hidden. Hide things when they are not watching, and they will not be able to find them. Their concept of objects is still tied to their immediate perceptions.

Stage 6: invention of new means through mental combinations: the beginnings of thought (approximately 18 to 24 months)

In the sixth and final stage of the sensorimotor period, children form mental representations of cause and effect independent of their immediate perceptions. A few months ago, they would open a door even though their chair was in the way—and knock over the chair. Now they move the chair out of the way before they open the door, for now they understand how the door will affect the chair. They can perform mental combinations as they invent solutions to their problems. If they want out of the playpen, they may cry, call, or pretend they want a drink. If they want to go outside, they may grab their shoes and coat and present them to the nearest available parent.

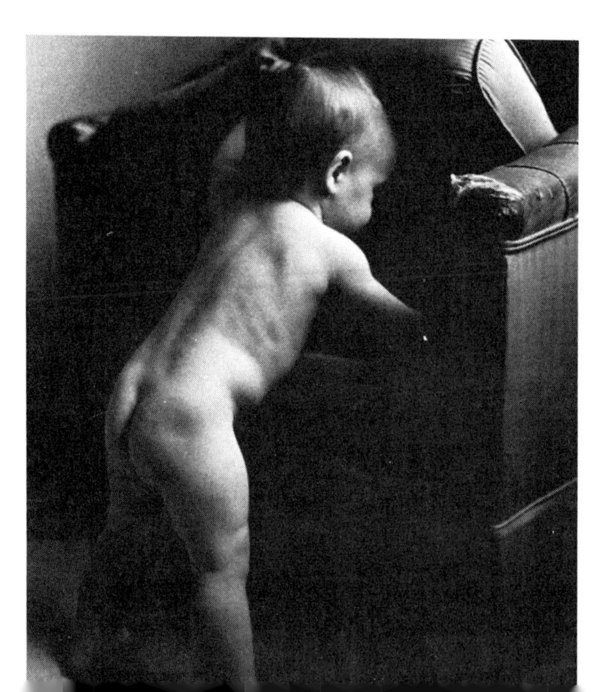

When infants achieve object permanence, they can search for hidden objects even when they have not seen them being hidden. This ability means that the infants have mental representations of objects that transcend their immediate perceptions.

The concept of object permanence is mature enough now that toddlers can search for a ball hidden under the couch even when they haven't watched it being hidden there. If the ball rolls under the couch, the child can understand that it may have rolled out the other side. She can walk around the couch, taking a path different from the ball's, and find the ball a few inches behind the couch. Her ability to follow such an invisible path is evidence that her mental representations of objects have transcended her immediate perceptions. Now when her parents forget to take her favorite blanket on a car ride, she protests. When her mother disappears into the bathroom and closes the door, she finds the baby impatiently waiting outside—or pounding on the door to be let in. The ability to create mental representations is the culminating achievement of the sensorimotor period. It ushers in a new world for children and their parents, who will always remember fondly the happy days of distraction when, for their infant, out of sight was out of mind.

Problems in Piaget's theory of sensorimotor development

Piaget's descriptions of changes in babies' behavior during the sensorimotor period are vivid and appealing. But not everyone is persuaded by Piaget's explanations of what is going on in babies' heads. Some developmental psychologists have questioned Piaget's explanation of how infants develop object permanence, for instance. They have questioned both the way Piaget tested object permanence and the inferences he made about his results.

Piaget's test of infants' object permanence gave infants a choice of two locations in which they could search for a hidden object. One location was where the object first was hidden, and the other was where the object was hidden second and remained. In this test, Piaget found that Stage 4 (8- to 12-month-old) infants were likely to continue to search where they had found the object earlier—at point *A*—rather than at the location where the object actually was hidden—at point *B*. Recently researchers have suggested that such *A not B* errors may be artifacts of the particular test, rather than a milestone in cognitive development, as Piaget suggested. In one series of studies, researchers lined up five locations for hiding objects, with the points *A* and *B* at either end (Bjork and Cummings, 1979; Cummings and Bjork, 1981). Infants 9 to 12 months old first had five tries to look for the object when it was hidden at point *A*. Most of them looked at point *A* or very near it. Then, in plain view, the experimenter moved the object and hid it at point *B*. But the infants did not make *A not B* errors, as Piaget would have predicted. Instead, they searched at or near point *B*. The researchers suggested that the reasons infants succeeded on this test and not on Piaget's were that they had more experience with the hiding-and-seeking procedure and that the *A* and *B* locations were clearly separated. They failed on Piaget's test, these researchers suggested, not because they did not realize that the object was permanent, but because they simply had forgotten where they had seen the object last hidden. Evidence from the laboratory supports this idea that babies get confused when they have to remember in which of two locations something disappeared. In another study, for example, 8-month-olds—who clearly knew something about the permanence of Mother—watched their mothers leave a room through door *A*. After the mother had disappeared, the infants looked first at that door—a sign that they had registered the spot where their mothers had disappeared—but then—apparently forgetting—they looked randomly at doors *A* and *B* (Zucker, 1982).

Another source of evidence that throws doubt on Piaget's interpretation of the *A not B* error is research demonstrating that infants tend not to make *A not B* errors if they can begin searching for an object immediately after it has been hidden rather than having to wait for several seconds (Fox et al., 1979; Gratch, Appel, Evans, Le Compte, and Wright, 1974). This evidence suggests that Piaget's infants may have failed his test not because they had no idea of object permanence but because their memory span for where the object was hidden had been exceeded. In support of this suggestion, research also shows that infants do better at finding hidden objects when there is a marker—a reminder—at the place the object is hidden and no marker at the place it is not hidden (Freeman, Lloyd, and Sinha, 1980; Lucas and Uzgiris, 1977).

As Paul Harris (1983) points out, finding a hidden object requires two things: (1) mentally representing the hidden object and (2) figuring out where it might be. Piaget did not allow for the possibility that an infant might be capable of the first but not the second, that an infant might know that an object exists without being able to find it. Developmental changes on Piaget's hidden-object test may reveal infants' ever improving search strategies rather than their dawning realization of the permanence of objects.

Evidence that infants have some notion that objects are permanent has been uncovered even before the 8- to 12-month period. In one experiment, 5-month-olds first were shown briefly an attractive object. Then the lights in the room suddenly went off, and the object disappeared completely from their view. Even so, the babies reached right toward the place where they had just seen the object, something that they would not be likely to do if they had no mental image of the vanished object (Bower and Wishart, 1972; Hood and Willatts, 1986). In other experiments, babies as young as 3½ months showed that they had some notion of the permanence of objects because they stared longer at physically impossible actions (like a toy car rolling along a ramp and apparently *through* a solid box) than at physically possible actions (a car rolling *in front of* the box) (Baillargeon, 1984, 1986).

Altogether, then, a substantial amount of evidence makes us cautious about accepting as fact Piaget's interpretation of what infants think and know in the first year of life. Many American researchers today doubt that infant intelligence is limited to sensory and motor procedures and routines, doubt that thought plays so little part in the first two years. Research evidence from the last fifteen years suggests that infants have acquired some semblance of object permanence by 9 months or even earlier and that thereafter their ability consolidates, becomes more widely and consistently applied and becomes better connected to other response systems such as search routines. Despite the disagreement between these researchers and Piaget about the age when infants first can mentally represent objects to themselves, there is no doubt that infants are not born with this ability, that they do develop it sometime during the first two years, and that when this ability *does* develop, a profound change in thinking and understanding has taken place. The infant has moved to a higher level of cognition.

Testing intelligence

All infants do not develop at the same rate. Even children within the same family show wide differences in their rates of cognitive development. To chart individual children's rates of mental development, psychologists need tests that reflect the relative abilities of different infants at the same age. By observing how infants at different ages behave when they are presented with particular objects under controlled and standardized conditions, psychologists can measure the infants' development against established norms for all infants of that age. This is the goal of infant intelligence tests.

Arnold Gesell was the pioneer of infant intelligence testing, and his work has influenced the content of all existing infant intelligence tests. Gesell tested infants' hand–eye coordination by giving them objects like a red ring on a string and seeing whether they could grasp them. He tested infants' fine motor skills by giving them objects like red one-inch cubes and seeing whether they could pick them up and build with them. He tested gross motor skills by seeing whether infants could sit up, walk, and climb stairs. He tested language abilities by seeing if infants could label or point to objects, and he tested personal–social behavior by seeing how infants responded to the tester. How a particular infant compared to the norms for these items collected on hundreds of infants of the same age was used as an indication of the child's developmental maturity. Other tests have been devised that follow and expand on Gesell's. They, too, are useful for evaluating infants' overall levels of physical and mental development.

Language

By the time he was 2½, Jason was still saying only a few words—''baba'' for bottle, ''guck'' for duck, ''caw'' for car, and some others. He seemed to understand well, but he couldn't express himself. He preferred running around to sitting still for a story, and he couldn't have cared less about looking at books.

Natalie started saying recognizable words when she was 9 or 10 months old. She could use simple sentences by the time she was 1½—''Go bye-bye car,'' and ''She talkin' me.''

One of the most impressive developments in infancy and one that clearly reveals individual differences among children is the beginning of language. How and when infants begin to talk and to understand speech is a story that has fascinated linguists, developmental psychologists, and parents. But it is a story that is still being written, as we shall see in this chapter and in Chapter 9. Learning language poses awesome challenges for infants. How do infants learn that the sounds of language are different from the surrounding din of humming, coughing, chuckling, and harrumphing? Even more difficult, how do they learn what the language sounds mean? How does infant Sam, for example, learn that ''dog,'' means dog, when he hears the word ''dog'' applied to a Great Dane, a dachshund, a poodle, and a stuffed toy, and the family pet referred to as ''Airedale,'' ''Rosie,'' ''Good girl!'' and ''Bad dog!''? How does he figure out whether ''dog'' means this dog only, all dogs, all animals, all four-legged or brown or long-tailed creatures, or this dog's hind leg? The fact is, though, that infants do meet these challenges.

Before words: basic abilities

Sometime around 10 to 13 months, most infants utter their first word. But the infant's first word does not appear out of the blue. To reach this milestone, infants have been practicing sounds for months and listening intently to those around them. Both nature and nurture contribute to the achievement of these first words. From the very beginning, as we discussed in Chapter 5, infants are tuned in to the sounds of speech, and the speech they hear is tuned to their capacities. They are innately prepared to perceive the phonemes in spoken language and to listen to the high pitch of baby talk. They also apparently have innate cognitive abilities to categorize the world and to map ideas onto language. They seem to assume, for example, that each word they hear represents a single concept (Gleitman and Wanner, 1982; Mervis, 1985; Slobin, 1973). Infants' developing knowledge of language parallels and de-

lifespan focus *Infant tests and later intelligence*

Infant intelligence tests have been used not only to evaluate infants' level of development but to predict their later level of intelligence. Unfortunately, these predictions have not proved especially accurate. Correlations between intelligence test scores in infancy and later are usually not significant for boys and only marginally significant for girls (McCall, Hogarty, and Hurlburt, 1972; Bornstein and Sigman, 1986). When significant correlations between infant and later test scores have been found, it is only because subjects with extremely low scores have been included in the study. People who score at the extreme low end in infancy continue to get low scores, probably because they have a serious and permanent retardation that can be observed even in the infant test. But within the normal range of intelligence, there is little predictability from infant to later tests.

> When Jason was 1½, we had his intelligence tested. When the psychologist who tested him told me that he had gotten a score of 85, I was scared. But she said not to worry, that some boys are slower to develop and that how they do at 1½ doesn't have much to do with how well they'll do in school.

Intelligence test scores from the early years rarely agree with scores later on for several reasons. For one thing, development in infancy occurs in fits and starts. Sometimes infants develop rapidly, sometimes slowly, and so it can be difficult to get a fix on their abilities. For another thing, performance on infant tests may be affected by individual differences in temperament. Researchers who conducted one study (Brucefors et al., 1974) suggested that less active 3-month-olds responded less to test objects and tasks and therefore scored lower than their more active counterparts but were learning in ways that allowed them to consolidate their gains when they were older. Yet another reason for differences between infant and later test scores is the difficulty of giving tests to infants. Accurate measurement of infants' rapidly developing mental abilities depends not only on the adequacy of the test, but also on the skill of the tester and, perhaps most important, on the state of the infant. The ideal time to test an infant, of course, is when the infant is quietly alert, interested, and responsive. The ideal time is hard to catch, however. But perhaps the most powerful explanation is that it has been difficult to identify the precursors in infancy of the abilities we consider intelligence later in

pends on the development of the ability to mentally represent objects and actions that we discussed previously. The understanding that is reflected in object permanence, for example, enables the infant to know that words are symbols and to try to express meaning in words.

Long before they produce intelligible words on their own, many infants give signs that they understand other people's words. As many a parent has remarked, the baby "doesn't talk yet, but she understands everything." "Everything" is probably an exaggeration, but year-old infants do understand lots of things that people say to them, especially when they can rely on other cues. When Daddy says, "Roll the ball to me," the baby does so. She uses the ball itself as the cue, and there's not much else she could do with the ball. Daddy might also say, "Where's the ball?" and look or point at it with large, plain gestures. By removing prompts and cues like these, researchers have shown that infants first understand the meanings of some proper names—*Mommy, Baby,*—and object names—*ball, egg, bottle*—at around 10 months (Hut-

The Bayley Scale of Infant Mental Development is often used to test infants' power of perception, sensation and cognitive skills. Unfortunately, predictability from scores on this test to later IQ scores is limited.

childhood and adulthood. Infant intelligence tests focus on perceptual and sensorimotor abilities—because that is what infants have. Intelligence tests for older children and adults focus on quite different skills, particularly on verbal and abstract skills—because that is where intelligent adults excell.

Recently, researchers have hypothesized that an infant's quickness to recognize a stimulus as familiar or new reflects the same ability that later intelligence tests tap, because both involve the ability to process and remember information. They have tested this hypothesis with a number of samples of children and found that, indeed, infants' abilities to recognize stimuli are significantly correlated with how they do on later tests of verbal intelligence (Fagan and Singer, 1983; Bornstein, 1984). The relation holds for samples of children with normal intelligence and those with Down's syndrome, and it holds for all socioeconomic groups and for both boys and girls. It has been replicated in other studies (Bornstein and Sigman, 1986). These results suggest that psychologists soon may be able to predict with greater confidence how a child will develop intellectually from the time he or she is an infant.

tenlocher, 1974). They recognize words that their caregivers use all the time—''light,'' ''bottle''—as early as 5 to 8 months.

Besides basic abilities for understanding speech, infants also have basic abilities for producing speech. At first, when they are very young, infants coo little relaxed sounds. Then, when they are about 4 months old, they begin to produce more and different sounds, vowels especially. Over the next few months, they discover and invent many sounds, alternating vowels and consonants—the ''Mamamamama'' and ''Babababa'' that so delight parents. This **babbling** is the earliest of an infant's speech-like sounds. Repeating syllables gives infants practice in combining sounds that eventually they will use to form meaningful words. Included in the stream of sounds, however, are those that the infant has not heard and will not use later on. The babbling of hearing and deaf infants, the world over, is the same. Only the accident of birth turns an infant into an English-speaking, Swahili-speaking, or Japanese-speaking specialist. At about 9 months, the babbling of English-speakers-in-the-making begins to lose its German gutturals and French nasals. Babbling grows increasingly complex, taking on the intonation and stress patterns of adult speech, and infants seem to speak in long "sentences" that mimic the sounds of the language they are beginning to learn. By listening to these wordless "sentences," adults can identify the language that the infants soon will speak (DeBoysson-Bardies, Sagart, and Durand, 1984).

But the continuity between babbling and talking is not entirely straightforward. Although infants babble many sounds, they do not babble *all* the sounds later used in words. Conversely, they cannot later say in words *all* the sounds they once babbled. The relation between babbling and speech seems to be indirect. To a great extent, babbling is simply exercise for the mouth and vocal tract and practice in producing sequences of sounds and tones (Clark and Clark, 1977).

At around 9 months, infants begin to use sounds like "duh," "da," and "ma." These new sounds soon replace babbling. Although they bear little or no relation to any adult words, they *seem* very much like words. Their combination of short utterance, communicative intention, and appropriateness to context makes them seem as though they should be understood. In this period between babbling and the first real words, infants use these shortened speech sounds repeatedly, on purpose, and in specific contexts (Dore, 1978a). They express emotions: "ooh wow," to convey joy, for example, and "uh-uh-uh" to convey anger. They express desires: short, urgent little cries to get a parent to reach something on a high shelf that the infant wants. They point things out: "duh" or "dah" plus a pointed finger to catch an adult's attention and to indicate something like "this" or "look" or "I'm seeing this." They designate similar objects: a single sound applied to any one of a group of objects, for example, "oo" referring to boots, shoes, slippers, and socks. These expressions are not words, but unlike babbling, they act like words. These expressions also show that in addition to their basic abilities to hear phonemes and words, to understand meaning, and to produce speech sounds, infants have basic communication skills. As we will discuss further in Chapter 7, through interaction with ready partners like Mom and Dad, infants learn to vocalize when others talk to them, to use vocalization to express feelings and wishes, to refer to interesting objects, and to take turns in "conversation."

First words

Finally, infants begin to use real words. These words are accompanied by lots of cues to make the message clear—pointing, reaching, grunting, staring. With the combination of word, gesture, facial expression, and action, the year-old infant requests and demands, labels people and things, describes actions, and expresses joy, displeasure, and surprise. These one-word utterances plus associated gestures, called **holophrases**, can do the work of whole paragraphs.

> Natalie would place herself squarely in front of me, reach her arms over her head, assume a supplicating face, and command, "Up." Once perched in my arms, she showed me by looking and pointing, emitting guiding grunts, and saying something that sounded like "mih" what she was after.

Pronunciation

Infants' first words sound imperfect because their speech sounds remain limited. Their first words use relatively easy sounds. They are likely to include

This 18-month-old makes her wishes clear as she reaches for her sister's pen and says ''peh.'' This is a typical early use of language.

the consonants *p, b, t, d, m,* and *n,* which are formed with the tongue at the front of the mouth, and the vowels *a* and *e,* which come from the back of a relaxed mouth. Perhaps because of these limitations in infants' articulation, in most languages, the names for the people dearest to an infant's heart are made of these easy sounds: ''nana,'' ''baba,'' ''mama,'' ''papa.''

For ease of pronunciation, infants often drop final consonants—duck is ''duh,'' and bed is ''beh,'' milk is ''mih,'' and ball is ''bah,'' or they drop initial consonants—cup is ''up'' (de Villiers and de Villiers, 1979). Only later do infants manage to produce whole sound clusters beginning with a vowel—''appuh'' for apple, ''itty'' for kitty. Clusters of consonants are difficult for infants to say and may be simplified—spoon to ''poon'' and stop to ''top'' (Greenfield and Smith, 1976; Waterson, 1978). Infants also tend to use voiced consonants (*b, d*) at the beginnings of words and unvoiced (*k, p*) at the ends. Pie becomes ''bie,'' and dog becomes ''dok.'' In trying to say two-syllable words, infants often repeat one of the syllables—''Zeezee'' for Rosie and the familiar ''baba'' for bottle. They can't produce two different syllables in one word, but hearing that the word has two syllables, they do their best. Simplifying the word by using the same vowel or consonant twice is another solution. Thus doggy becomes ''goggy'' or ''doddy.'' In words with two or more syllables, they may simply drop the unexpressed one and say the stressed syllable: ''raff'' for giraffe '''mote 'trol'' for remote control. Later, they may substitute an ''uh'' sound (ə)—''tape uhcorder''—for the unstressed syllable.

Content

The particular words that infants first say usually are social or functional expressions like *hi, bye-bye, thanks, yes, ouch, this, up,* and *on*; names for objects that are important, familiar, permanent, and usually movable, like *ball,* ''Dada,'' and ''b'anket''; simple adjectives like *hot* and *big*; and action words like *push* and *give*. They do not use less salient, filler words like *and, be,* or *the*. They use words to comment when the object has just changed in some way. Someone or something may just have appeared or disappeared, opened or closed, moved or brightened. Infants chirp ''Daddy'' as he walks in the door and ''ight'' as the lamp clicks on. Early vocabularies of object words are quite similar for different children, covering the categories of foods, animals, and toys. Infants usually acquire nouns for basic categories—flower or dog—

before they acquire more general words—plant or animal—or more specific ones—rose or collie (Brown, 1958; Rosch, Mervis, Gray, Johnson, and Boyes-Braem, 1976).

It is often difficult for others to know what infants understand and mean by certain of the words they use. Does the infant's "goggy" mean dog, animal, furry thing with a tail, or dog's hind leg? The problem for the adult in interpreting the infant's language is the same as for the infant in interpreting the adult's language.

Process

Infants learn the meanings of words by induction. They hear others speak words or react to their own utterances, and they draw inferences about the words' meanings (Carey, 1985; Figure 6.2). Like young interrogators, infants learn word meanings by pointing to possible referents, saying the word in question, and observing the reactions of others. Infants formulate hypotheses—*dog* means dog? *dog* means pet? *dog* means hind leg?—that are constrained to converge on the correct meaning relatively quickly, sometimes after just one trial. Constraints on an infant's induction are imposed by the infant's understanding of objects, causes, people, and intentions. Although the infant may make the category of dog slightly too narrow—only this dog is *dog*—or too wide—*dog* refers to all four-legged animals—the infant is extremely unlikely to mistake bone or floor for *dog*, even though he or she often sees them together.

Figure 6.2 What does Mother mean when she says "bowwow!"? This is the challenge that confronts the infant who first hears the sound. He attempts to find out by saying a few "bowwows" himself.

At first, infants often **overextend**, or overgeneralize, their words, using the words they do know to cover extra meanings. These overextensions are based on common properties or functions of objects—shapes, movements, sizes, sounds, textures. ''Fly'' may mean all small, dark objects, from insects to raisins. ''Bowwow'' may mean all small and furry animals, ''dance'' any kind of turning around (Brown, 1958; Clark and Clark, 1977; Rosch, 1975). Infants also may **underextend**, or undergeneralize, some words (Anglin, 1975). ''Dog'' may mean only *their* dog, not any other dogs.

Are overextensions the result of problems in making categories or in knowing words? Does the infant use the overextended term to find out what the right word is, if he knows that when he says ''dog,'' his father will say something else, perhaps ''cat''? Does he overextend his words because it's the best he can do to communicate?

The main reason for overextensions seems to be that infants have limited vocabularies. They use overextensions even when they understand more specific terms. In one study, for example, even infants who could point correctly to pictures of a dog and cat called them both ''dog'' (Fremgen and Fay, 1980). In another study, researchers found that like adults, infants categorized objects by applying the same word to all members of the category. But because their vocabularies were so limited, they overextended the name of one of the category members to the other members of the category rather than using the category name (Rescorla, 1981). In the study, six infants were followed from the time they were about 1 year old for six months. Their mothers kept diaries of the children's new words, and the researcher visited with the infants at home. When she visited, she brought a variety of small toys, and she recorded the names the children gave the toys and the names they understood, to see how they categorized these objects. All the infants organized vehicles, at first, around the word *car*. For about a month, cars were just cars, but thereafter the children overextended the word *car* to denote trucks, buses, planes, trains, and strollers. After the period of overextension, the children's concepts of car narrowed, and they acquired a new word—*truck*—and a new category—large commercial vehicles. Then they overextended the word *truck* to label buses, trains, bulldozers, cement mixers, and fire engines. They also discovered, before the end of the study, the word *plane* and used it for all aircraft, including helicopters, blimps, rockets, and gliders.

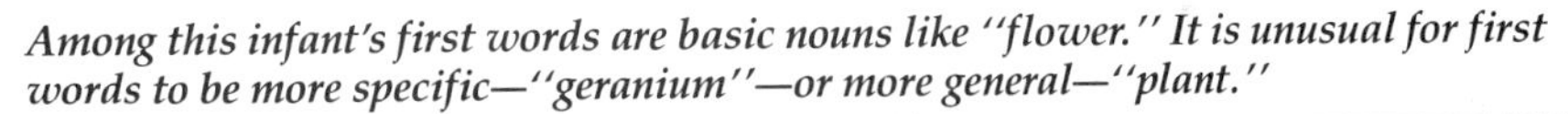

Among this infant's first words are basic nouns like ''flower.'' It is unusual for first words to be more specific—''geranium''—or more general—''plant.''

When the researcher used the specific vehicle words in talking to the children, however, she found that they could *understand* finer distinctions among vehicle words than they could actually say.

Eve Clark (1983) has described the processes that young children use to increase their vocabularies. When children notice a gap in their ability to talk about something, they actively search for words to fill it. They look for words to label objects that are salient to them. A child interested in cars and trucks will attend more to words in this category than to words for animals, for example. Later, this child may set up several categories for trucks—moving trucks, delivery vans, mail trucks, big trucks, noisy trucks, dump trucks, and so on—and listen for new words to describe them. Hearing new words applied to trucks, he also starts looking for the kinds of trucks they name. In the meantime, he fills gaps in his vocabulary by overextending the words he has, by relying on general-purpose words like "that," and by coining new words whose meanings contrast with those of words he knows already. Children often coin words for actions—"Mommy *nippled* Anna," "I'm going to *gun* you"—and for objects—"chop" for an axe or "knock-thing" for something to knock with. They also coin words for people—"a smile-person" for someone friendly who smiles or a "hitter-man" for someone who hits things.

> Peter's expressions were wonderfully inventive. "It's not pepping," he complained about an empty pepper shaker. "Can I sprink the lawn?" "The fan breezed me." I still call sirens "ers," from Peter's baby word for the sound they make.

What part do parents play in children's acquisition of early words? For one thing, as we discussed in Chapter 5, parents' use of baby talk simplifies the language the baby hears. For another, parents' labeling of objects draws the baby's attention to the connection between words and things. Also, parents usually use nouns from the basic categories when they name things for their toddlers, and these are the words that the toddlers themselves use (Mervis and Mervis, 1982). Asked to label pictures for their 2-year-olds, one group of mothers labeled a collie a "dog," a sandal a "shoe," a pigeon a "bird," and a dime "money," although they used the words "collie," "sandal," "pigeon," and "dime" when speaking to another adult (Anglin, 1977). But mothers in this study also labeled some pictures with specific words: "pineapple" and not "fruit," "typewriter" and not "machine," "butterfly" and not "insect." What determines the kind of label that mothers use and children learn? Common sense, perhaps. Mothers use common labels for objects that are similar or that people use or respond to similarly. Thus mothers give both a collie and a miniature poodle the label "dog," because both are pets, both gnaw bones, and both walk on a leash. But mothers call a cockroach an "insect" and butterfly a "butterfly," because cockroaches, like most insects, are disgusting and butterflies, unlike other insects, are beautiful.

Parents' responsiveness to their infants' attempts to talk also affects the infants' language development. Responsive parents accept and reinforce their children's attempts at pronouncing words. They treat their children's speech as meaningful, they echo their words, and they give them examples of more words. Here is a dialogue between a responsive mother and her 14-month-old daughter.

MOTHER: Is that a car?
JANE: Bah.
MOTHER: Yes, car. Here's another car.
JANE: Gah.

MOTHER: Car, yes.
JANE: Bah. Daddy.
MOTHER: Daddy. Daddy's car is all gone (Nelson, 1973, p. 105).

Unresponsive parents, in contrast, are critical of their children's language, offer no reinforcements or appropriate models, and do not accept their children's language as meaningful. Here is a conversation between Paul, age 17 months, and his mother as they look at a picture book.

MOTHER: What's that? A dog. What does the dog say? One page at a time. Oh, that one over there. What's that one there?
PAUL: Baoh.
MOTHER: What? You know that.
PAUL: Bah.
MOTHER: What?
PAUL: Ah wah.
MOTHER: What?
PAUL: Caw.
MOTHER: Car?
PAUL: Caw, awh.
MOTHER: Little kitty, you know that (Nelson, 1973, pp. 105–106).

Children of unresponsive, critical mothers tend to lag behind in their language development.

Stimulating cognitive development

Developmental psychologists have not been content simply to describe infants' language and cognitive development. They have wanted to learn how factors in the environment stimulate them. What helps and what hinders cognitive development? Given the constraints of an infant's genetic potential, what are the limits on enhancing the rate or level of cognitive development?

The stimulation of toys

We can infer from what is known about sensorimotor development that normal cognitive development requires infants to have practice in manipulating different objects. These little scientists undoubtedly need the chance to collect information as they experiment with objects and interpret their growing cache of information according to their ever-changing working knowledge of their surroundings. Be they bright or dull, colorful or drab, natural or synthetic, bought or homemade, designed for adults or for children, objects are likely to stimulate children's vision, exploration, schemes, and play. All the available research evidence suggests that children's early cognitive growth depends to some extent on the availability of a variety of stimulating objects.

In one study, for instance, when 5-month-old black infants from various socioeconomic groups were observed at home, the variety of objects available to the infants was related to their scores on tests of intelligence, problem solving, object permanence, and exploration (Yarrow, Rubenstein, and Pedersen, 1975). A number of other studies, too, suggest that young children's mental abilities are stimulated when they can play with a variety of interesting things, when they are allowed to explore their surroundings, and when their

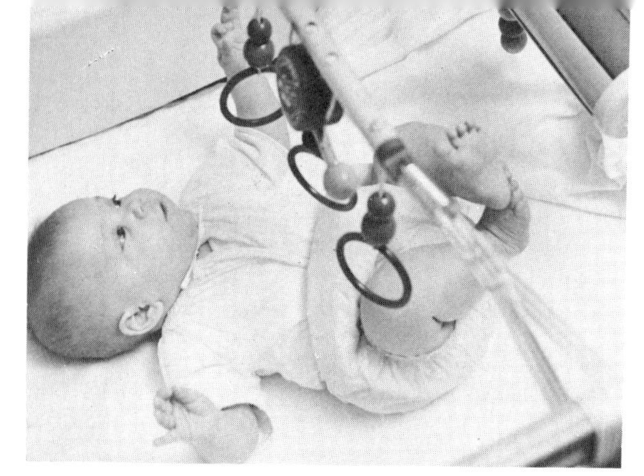

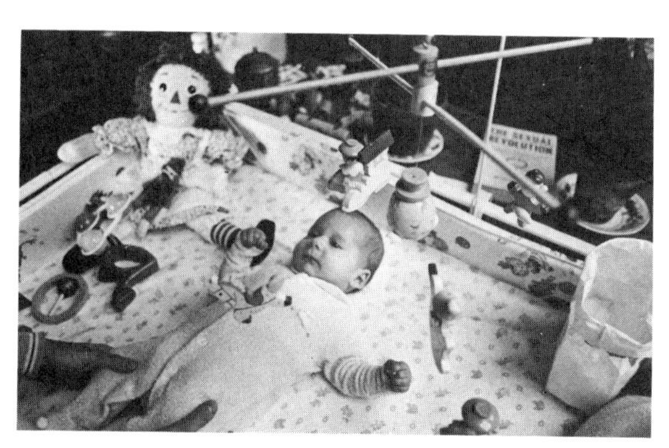

Moderate stimulation is best for infants' cognitive development; it arouses interest. Overstimulation is confusing, and understimulation is boring.

parents encourage them to do both these things (Wachs and Gruen, 1982). It also may be beneficial if the objects respond to infants' actions—if jack-in-the-boxes pop open, squeeze toys squeak, and bath toys float and bob (McCall, 1977). Of course, it is impossible to say, from correlational studies like these, that toys *alone* influence infants' development. There is likely to be a strong genetic component linking advanced infant development with greater availability of toys. But the results do strongly hint that objects that are varied, interesting, and responsive, objects that force infants to reach a bit but are not overwhelming, are best for fostering infants' cognitive development.

Parents

For most infants, toys and other play objects are part of the stimulation provided by parents. These objects may be especially important if parents play with them with the infant (Smith, Adamson, and Bakeman, 1986). Many psychologists have studied the relations between infants' cognitive development and parents' stimulating play—with and without toys. In one observational study, 25 pairs of mothers and infants were observed for 3½ hours when the infants were 3 months old (Crockenberg, 1983). What the mothers did with their babies—smiling, making eye contact, responding to the baby's crying, and the like—was recorded. Then, when the babies were 21 months old, they were tested on a standarized test of infant mental development. The test results were clear-cut: infants who scored high on the test had mothers who were more educated and more responsive to the infants' needs, who smiled and made more frequent eye contact with the infants, and who spent less of their time with the infant in such routine caretaking as diapering and feeding. In short, there was a clear connection between the quality and quantity of the mothers' stimulation of their infants and the infants' cognitive development.

This study is fairly typical of the kind of research that has been done in this area in the past 15 years. Many other studies (reviewed by Appleton,

A mother stimulates her 3-month-old baby with a toy. There is a clear connection between the parents' stimulation and their infants' cognitive development. But at this age it is unlikely that the parent is having an immediate or irreversible effect on the baby's development.

Clifton, and Goldberg, 1975; Clarke-Stewart, 1977; Yarrow et al., 1975) have shown that mothers' education, responsiveness, and stimulation of their infants are related to how the infants develop. Mothers with more education talk more to their babies. They respond to their babies' babbling by talking. They provide more interesting toys for them, and they are more effective teachers. (Better educated mothers also are more likely than less educated mothers to have a husband living at home, more money, fewer children, and quieter households—factors that make it easier for them to provide this kind of attention, responsiveness, and stimulation to their infants.) When parents act in these stimulating and responsive ways with their infants, the infants do better on tests of cognitive development. These parents are providing their infants with interesting information about the world and, presumably, teaching them that the world is a place of predictable actions and reactions, a place worth exploring and learning about.

When psychologists observe how infants' cognitive development is related to their parents' behavior, they do not really know whether more stimulating and responsive parents are *causing* the infants to develop quickly or more intelligent infants are *eliciting* more stimulation and more appropriate responsiveness from their parents. (This is a problem with correlational research in general, as we discussed in Chapter 2.) To try to separate these two possibilities, researchers at the University of Colorado (Hardy-Brown, Plomin, and DeFries, 1981) studied 50 adopted 1-year-olds and their biological and adopted parents. The *adoptive* parents' socioeconomic status and educational level, cognitive ability and vocabulary level, overall frequency of speaking or reading to their 1-year-olds, they found, were not related to the infants' language development. In contrast, the *biological* mothers' cognitive ability, especially memory ability, was significantly related to their 1-year-olds' language competence. This finding suggested that language development in the first year of life is influenced by genetic factors more than by parents' behavior. This finding is consistent with other studies (some of them discussed in Chapter 3) that suggest that the link between parents' behavior and individual differences in infants' intelligence in the first two years of life is largely genetic. Infants are by no means empty vessels into which parents pour intelligence merely by their stimulation and responsiveness. But parents do encourage their infants' natural intelligence to come out, and their influence increases as infants get older. Moreover, although parents' behavior may not always be correlated with individual differences in infants' intelligence, some stimulation and responsiveness from parents are still necessary for infants' development. It just may be that most parents provide *enough* stimulation and responsiveness to foster their infant's development so that *more* stimulation does not advance development faster.

To add to the complexity of the nature–nurture issue, other innate characteristics, such as the baby's sex and temperament, also are likely to affect both parents' activities with their infants and the consequences of these activities. Apparently not all ''positive'' early experiences enhance development, and not all ''negative'' early experiences interfere with it. In one study to consider this possibility, easygoing infants whose parents allowed them to explore freely, in homes where the toys and decorations were plentiful and where mothers smiled and responded to the infants' vocalizations, did well on tests of cognitive development (Wachs and Gandour, 1983). In contrast, more difficult and irritable infants reacted negatively to interactions with people and were more sensitive than easygoing infants to noise, confusion, and other stressful aspects of their physical environments; they did worse on tests of cognitive development. What is moderate stimulation for one infant may be overwhelming to another infant. Parents must consider the unique qualities of their own baby as they try to offer that baby experiences that will stimulate cognitive development. An infant's innate resilience or vulnerability to assaults from the environment also affects cognitive development. For infants who are relatively invulnerable, cognitive development is likely to proceed well even in a difficult environment. But for those who are vulnerable, cognitive development may proceed well only if the environment is favorable. The process of stimulating infants' cognitive development is clearly a reciprocal one, to which both infants and parents contribute.

Summary

1. Investigations of how early and in what ways infants learn show that classical conditioning generally is unsuccessful with infants under 2 or 3 months old. Operant conditioning with very young infants involves only reflexes; after 3 months of age or so, a broader range of behavior is affected by reinforcement.
2. By using various methods—paired comparison, habituation, retention of learned responses, and observing spontaneous behavior—researchers have learned what and for how long infants remember. At first, apparently, memory consists of primitive, sensory, wired-in recognition, without thought. At about 3 months, memory becomes more cognitive, as infants begin to control what they look at, listen to, and pay attention to. By 8 to 10 months, infants form mental categories, which allow them to summarize information, thus reducing their memory load.
3. According to Jean Piaget, the early mental activity of infants, called sensorimotor intelligence, is limited to what can be sensed and acted on. Out of their sensorimotor experiences with objects, infants develop action schemes. They organize—combine and integrate—their schemes and adapt them as their experience increases. Infants assimilate new objects and experiences into their existing schemes; they accommodate, or modify, schemes when they encounter new objects and experiences that do not fit their schemes.
4. According to Piaget, infants progress through an unvarying sequence of developmental stages, each of which incorporates the knowledge gained in the previous stage. During the sensorimotor period, Piaget claimed, infants progress from a stage of practicing and repeating reflexes, to acquiring adaptations, to making interesting sights last. During this stage, they engage in magico-phenomenistic thinking. They also begin to develop object permanence, the knowledge that even when objects are out of sight they continue to exist. Later stages bring fuller understanding of object permanence, cause and effect, the physical properties of objects, and of the self as a separate being.
5. Other developmental psychologists have found problems in Piaget's theory of sensorimotor development and his suggestion that infants do not begin to form mental representations of objects until the middle of their second year. They have found evidence that infants form mental representations of objects in the first year and suggest that Piaget's hidden object task is difficult for infants not because they don't realize that objects are permanent but because they are not good at searching for objects or remembering where they are hidden.
6. Not all infants develop cognitively at the same rate. Psychologists test individuals' cognitive levels with infant intelligence tests. These tests are useful for assessing whether infants are progressing normally, but they generally do not predict individuals' later IQ scores. One reason is that

intelligence tests in infancy center on perceptual and sensorimotor abilities, whereas later intelligence tests center on verbal and abstract abilities. Infants' recognition of new and familiar stimuli is more predictive of later IQ scores than these infant IQ tests.

7. Language develops in infancy, too, out of basic capacities like speech perception and categorization, understanding of the meaning of words, the articulation aspects of babbling, and a conducive learning environment. At first, because of their limited vocabularies infants may overextend words (using ''dog'' for cats, dogs, and rabbits) or underextend them (using ''dog'' just for Spot).
8. Unresponsive and unstimulating care in infancy retards cognitive development. Although parents cannot create intelligence in their children, they can foster its development.

Key terms

conditioned stimulus
unconditioned stimulus
unconditioned response
conditioned response
encode
sensorimotor period
schemes
adaptation
accommodation
assimilation
organization
equilibration
circular reactions (primary, secondary, and tertiary)
object permanence
magico-phenomenistic thinking
A not B phenomenon
babbling
holophrase
overextend
underextend

Suggested readings

Clarke, Ann M., and Clarke, A. D. B. (Eds.) *Early Experience: Myth and Evidence.* New York: The Free Press, 1976. A collection of articles demonstrating the remarkable ability of children to recover from extremely depriving and depressing early experiences. This book takes the position that early childhood is not the only important period for intellectual and social development.

McCall, Robert C. *Infants.* Cambridge, Mass.: Harvard University Press, 1979. Developmental milestones of the infant's cognitive and social progress, clearly explained for parents.

Piaget, Jean. *The Origins of Intelligence in Children.* New York: International University Press, 1966. Piaget's own account of the evidence for and theory behind sensorimotor development in infancy.

Weir, Ruth H. *Language in the Crib.* The Hague: Mouton, 1970. An in-depth analysis of a child's very early speech.

chapter seven

Social and emotional development

A new baby *seems* helpless. But look more closely, and you will see this same baby's formidable capacity to attract people to herself. The baby cries for food—and her parents come, meal at the ready. She whimpers for comfort—and arms scoop her into an embrace. She opens her eyes—and her parents fix their gaze on her tiny face, waiting for eye contact. She hiccoughs after her milk—and someone props her up, taps her back, and waits for the air bubbles to come up. She fusses—and hands come to check the sogginess of her diaper. Over and over again, hundreds of times a day, the new baby draws people to herself with gazes, cries, snuggles, and sniffles. She participates actively in a world prepared to support and encourage her social inclinations. This tiny creature has begun in her own way to build the social ties that will shape her lifelong experience. In this chapter, we will follow the infant's progress in the social world from the first moments after her birth to her first love affair.

The first meeting

> Look at his little face. His little nails. Oh. His little squashed up nose like your nose. He has red hair.
>
> Little baby got big feet—he has got big feet, hasn't he?
>
> He's blowing bubbles. His little hands are all wrinkled—looks like he's done the washing up, doesn't he?
>
> Yes (laughs). Oh dear!
>
> Oh, he's opened his eyes, there—look.
>
> Hello (as baby opens his eyes for the first time)! (Macfarlane, 1977, pp. 91, 94, 95).

When mother and infant first look at each other, they are not really strangers. They have been intimately connected for the infant's entire existence in the womb. Now, when they meet face to face, the mother gets to know this new arrival by looking, touching, and talking (Klaus, Kennell, Plumb, and Zuehlke, 1970). Most mothers follow a similar pattern in getting to know their infants in these first minutes after birth. They reach out and tentatively touch the infant's fingertips, then the arms and legs. Within a few minutes, they place their palms on the infant's trunk and massage and stroke the tiny body. They look, the infant looks, and their eyes meet. Even right after birth, infants will follow the mother's face. Mothers of blind infants say that they feel ''lost,'' because their infants cannot return their gaze (Fraiberg, 1974). For as long as an hour after birth (unless the mother has been heavily sedated during the delivery), the infant is quiet, alert, and open-eyed.

> Right after she was born, Natalie was wide awake and nursed hungrily at my breast. I kept gazing at those beautiful blue eyes and stroking her all over. It was hard to believe that we had created anything so wonderful as that baby in my arms.

The first hour after birth can be an important time, because the infant is alert and gazing, and the mother's receptivity is intensified by high hormone levels. The two ''converse'': Mother talks to the infant, and, some observers have suggested, the infant moves in time to her words (Condon and Sander, 1974). Seeing her infant move and look into her face encourages the mother to keep on speaking, touching, and looking. It pleases her deeply to hold the infant. When the infant is attentive and the mother welcoming their relationship is off to a good start.

The bond between mother and infant

The special feeling that mothers have for their new infants even in the first moments after birth usually deepens over the first few months into a strong emotional bond. Most mothers feel nurturant and caring toward their infants. They are disturbed by separation or the thought of separation, and they are willing to make the sacrifices necessary for their infant's survival—like getting up several times every night to feed and comfort the infant. But this bond does not always form. In the 1970s, Marshall Klaus and John Kennell (1976), two pediatricians, observed that mothers who were separated from their newborns immediately after the birth treated their infants differently from mothers who remained in contact with their newborns. Mothers who were separated from their infants, usually either because mother or infant was ill or because the infant was premature, were more likely to neglect or abuse their infant. According to Klaus and Kennell, these mothers seemed to have missed an early chance to form deep emotional bonds with their infants. When they handled them, the mothers seemed fearful and unsure. Some felt uncertain that the infants really were theirs. ''Are you mine? Are you really mine? Are you alive? Are you really alive?'' asked one mother reunited with her baby (Klaus and Kennell, 1976, p. 10). Perhaps the moments immediately after birth, Klaus and Kennell suggested, are a **sensitive period** when mothers easily form emotional bonds with their infants. Preventing the pair from **bonding** in these early moments, they argued, hampers the development of this important tie.

Researchers began to test this suggestion. At the time the researchers began their investigations, newborns in American hospitals were routinely removed from their mothers immediately after birth and kept in infant nurseries, except for feeding visits to their mothers. By increasing the contact between mothers and infants beyond this usual level, investigators could gauge how early contact affected the bond between mother and infant. Researchers compared mothers whose infants went to the newborn nursery with those who kept their infants with them, a procedure called **rooming in**. They experimented with keeping infants and mothers together for several hours right after birth and with keeping infants and mothers together for extra hours every day. Several of the studies showed that mothers who had more early contact with their infants reported feeling more competent than mothers who were separated from their infants by ordinary hospital routine. They were less willing to turn their infants over to anyone else. They kept their infants closer to them, touched and soothed, kissed and caressed them more often—not only when their infants were newborns but a year later as well (Greenberg, Rosenberg, and Lind, 1973; Hales, Lozoff, Sosa, and Kennell, 1977; Kennell et al., 1974).

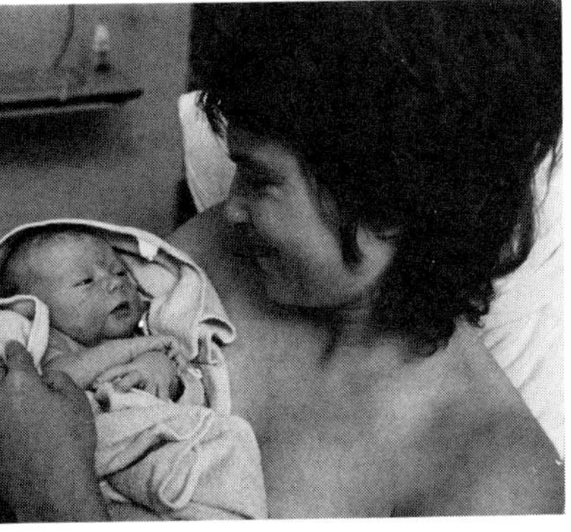

The mother–infant bond has its earliest expression in mutual gazing, which begins in the first hour after birth as the mother holds, touches, and studies her newborn.

These effects of early contact were not apparent in all studies, however. Middle-class mothers in particular did not always benefit so clearly from extra early contact with their infants (Svejda, Pannabecker, and Emde, 1982). Effects of early contact that last beyond the period of infancy have been especially difficult to demonstrate (Grossmann, Thane, and Grossmann, 1981; Sostek, Scanlon, and Abramson, 1982). Early contact, then, *sometimes* affects mothers' feelings toward their infants. These effects are not universal. Human mothers differ from mothers of many other animal species in that they form bonds with their infants even if the early sensitive period is missed or disrupted. Many human mothers miss the early moments after an infant's birth because they have been under anesthesia for some obstetrical procedure or because the infant is whisked away to an incubator or intensive care. Most of these mothers do form strong and abiding emotional bonds to their infants. So, too, adoptive parents, fathers who missed their infant's birth, and many other people form emotional bonds with the infant. Early contact may start the bonds forming earlier, but later contact is likely to lead to an attachment that is just as deep and enduring.

The main value in Klaus and Kennell's original ideas about the danger of separating mothers and babies immediately after birth may turn out to have been practical: the humanizing of hospital procedure.

Postpartum blues

No matter whether they have been separated from their infants in the hospital or not, mothers face a difficult transition when they arrive home and take on the enormous task of caring for a newborn infant. Most new mothers find their moods swinging and swaying. From half to three-quarters of new mothers have crying jags during the first ten days after delivery (Davidson, 1972; Pitt, 1973; Yalom, Lunde, Moos, and Hamburg, 1968).

> I was so happy after Sam was born that I found it amazing when I started to cry at the oddest things. When we brought him home from the hospital, I cried. I chalked that up to joy. A few days later, I cried miserably because I'd gotten a drop of ketchup on my bathrobe.

New mothers feel vulnerable and sensitive, a consequence of both physical and psychological changes following childbirth. They are tired from the hard work of labor and delivery and from meeting the round-the-clock demands of their newborn infant. In the days and weeks following childbirth, mothers need time to rest. They also need the support of family and friends. In one study it was found that new mothers who lacked emotional and practical support from relatives and friends were more likely to need psychiatric care (Gordon and Gordon, 1960). Expectant mothers who had been told to set up a support network—of husband, grandparents or other relatives, friends, and babysitters—and who had been taught how to minimize stress while they were pregnant suffered fewer emotional problems in the adjustment period after childbirth than mothers who did not (Gordon, Kapostins, and Gordon, 1965).

> Our Lamaze teacher told our class of expectant parents that the best gifts for new mothers were phone calls that said, ''I'm going grocery shopping now. Tell me what you need,'' or, ''I'll be at your house at 4 o'clock to do your laundry and make supper. You rest.''

Not only social support, but the mother's personality and expectations also affect her adjustment to the new baby. Mothers adapt most readily if they are nurturant, self-confident, interested in child rearing, see themselves as mothers, and remember their own mothers as close, warm, and happy (Shereshefsy and Yarrow, 1973). They also adapt well if they really wanted the baby (Field, Sandberg, Vega-Lahr, Goldstein, and Guy, 1985).

How the infant acts also has a lot to do with the new mother's postpartum adjustment. One pair of investigators found that three-quarters of the mothers who reported strong positive feelings for their infants in the first three months explained that the reason was their infant's responsiveness (Robson and Moss, 1970). Mothers want healthy infants—infants who nurse well and gain weight, infants who sleep well and are quiet. But they also want infants who are responsive and communicative—smiling, laughing, making eye contact. A responsive, communicative infant makes the postpartum adjustment easier and deepens the mother's bond to her infant.

Infant communication

Infants communicate in a number of ways, active and passive, noisy and quiet, by vocalization, facial expression, and gesture.

"Babyness"

Parents and grandparents, brothers and sisters, even strangers who peek into the baby carriage nearly all smile and coo, open their eyes wide, exaggerate the expressions on their faces, and talk baby talk in high voices. Why do otherwise sober people act so dotty? The cute "babyness" of that tiny infant does it, with a big round head on a little body, a rounded forehead, big eyes, plump cheeks, small face and mouth. The baby's helplessness is touching, her grins and bright eyes are charming, her sweet pink mouth is alluring, and her squeaks and gurgles are endearing. Human babies, like puppies, kittens, calves, and colts, have a special appeal.

This cute "babyness" equips infants with a powerful means of attracting the nurturance they need. Ethologists maintain that "babyness" helps to ensure the infant's survival (Bowlby, 1969; Lorenz, 1971). It keeps adults nearby and interested in feeding, sheltering, and stimulating their offspring.

Gazing

Babies also communicate with the people around them by gazing. When infants are held close, they can see their parent's face and hair or a proffered hand or toy. During feeding and play, parents hold their infants at a distance that is well within the infant's visual "window." That interesting face, as Mom or Dad talks to and looks at the infant, turns feeding and playtime into excellent opportunities for forming social and emotional ties.

When infants are about 6 weeks old, their visual and motor systems have matured enough so that they can focus on their mother's eyes, with their own bright eyes wide open (Wolff, 1963). For the first time, the mother feels that the baby is looking *at her*. Even mothers who are not consciously aware of this wonderful new development step up the pace of their interactions noticeably—doing more playing, talking, and bouncing.

The enrapt mother and her infant are clearly communicating, but without a single syllable being exchanged. Their mutual gazing tells it all.

> Jason was such a handful in the first month and a half. He cried and cried; we paced the floors with him for hours on end. He stopped crying only long enough to take a bottle. Then he'd spit it all up. But one day, as I was changing his diaper, my hand on his stomach, Jason stopped squirming, opened his eyes wide, and looked *at me*. I was so thrilled! It had been weeks of torture, but now it was okay. There was someone inside that baby head.

When they are about 3 months old, infants become able to maintain eye contact with their mothers, to gaze into their eyes for several seconds, and the mother's feelings of attachment to the infant deepen. Now mothers are likely to feel so strongly attached that being away from their infant feels uncomfortable, and imagining its loss is unbearable (Robson and Moss, 1970). Mothers of blind infants, who do not maintain eye contact, have difficulty forming an attachment (Fraiberg, 1974). They find their infants perplexing and unresponsive.

The infant's intent gaze not only affects how the mother feels, but it also communicates a useful message about how the infant is feeling. Videotapes of middle-class mothers show that on the average these mothers spend three-quarters of the time that they are feeding or playing with their infants gazing at the infant's face, and the infant's gaze signals the mother about how to proceed (Peery and Stern, 1976; Schaffer, Collis, and Parsons, 1977). So long as the infant is looking at her, the mother continues to play. But when the infant turns away, the mother stops playing. Mothers interpret the turning away as a signal that the infant wants a change. Thus, gazing is an effective means by which the infant communicates with parents.

Vocalizing

Vocalizing is another way that the infant communicates. For the first few months, the infant's vocal sounds are coos and gurgles, uttered after meals and naps, when the infant is relaxed and being held. By about 4 months, infants utter relaxed consonant and vowel sounds when they hear their parents' voices or see their faces approaching. They string these sounds together—''babababababa,'' ''dahdahdah''—in sentence-like patterns. This babbling is vocal play, not a deliberate attempt to say something. Infants vocalize in this way even when there is no one around to listen. But although infants do not intend these sounds to convey meaning, parents react to them—as they do to yawns, burps, smiles, and hand movements—as though they were *full* of meaning. While the infant is vocalizing or right after, the mother is likely to speak (Jones and Moss, 1971; Lewis and Freedle, 1973). ''Oh, is dat so,'' she may say, ''Is oo trying to tell me somethin'?''

Soon, infants are imitating the sounds they hear and synchronizing their babbling with their parents' speech. They grow quiet when someone nearby speaks, then babble excitedly, and then listen again. Parents treat this ''conversational'' babbling, too, as though it were meaningful and respond with talk, delight, and affection. Although vocalization at this age is not language, it can communicate an infant's intention. One 6-month-old vocalized when he was playing with objects in a higher pitch than he used when he wanted his mother. When his pitch dropped, his mother checked to see what he wanted. Another baby's vocalizations grew sharper when he wanted something beyond his reach, and his mother responded to this change in sound by fetching the thing he wanted (Bruner, 1978). The infant's vocalization is a means of communicating with parents who are willing to listen.

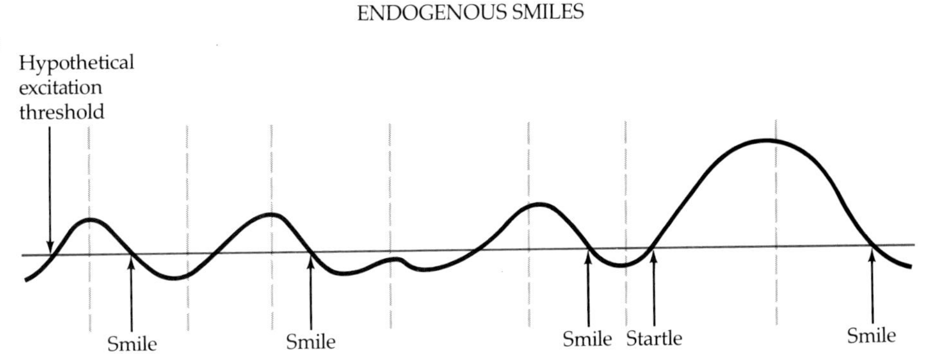

Figure 7.1 This schematic drawing of excitation and arousal shows an infant's behavior in relation to a hypothetical threshold of arousal. When infants are aroused and then relax just below the threshold, they give a tiny spontaneous (endogenous) smile (Sroufe, 1979).

Smiling and laughing

Smiles and laughter are arguably the infant's most irresistable means of communicating. During the first two weeks of life, infants ordinarily smile when they are drowsy or sleeping lightly rather than when they are awake. These smiles are not triggered by social stimulation. Instead, they come from completely internal events within the brain: the infant's rising and falling levels of arousal. They are, therefore, called **endogenous smiles**. As infants relax after being aroused, their muscles relax, and a smile appears on their face (Figure 7.1). As infants sleep lightly, their level of physiological arousal stays near this ''smile threshold,'' and little endogenous smiles play across their faces. Startle them with a loud noise, and they grow too highly aroused to smile. But touch them softly or whisper, and in six to eight seconds an endogenous smile may appear.

In the second two weeks of life, smiles can be coaxed from infants by gentle stimulation while they are awake. At the end of the first month, bouncing and other forms of moderate stimulation coax those smiles from the infant. In each case, the infant is stimulated to a point above threshold, and then, six to eight seconds after the stimulation ends, a smile appears. Although parents get great pleasure from seeing these endogenous smiles, they do not really signify pleasure or sociability in the infant. These early smiles *look* like later smiles, and like later smiles they arise from arousal and relaxation. But it will be another few weeks until the infant is smiling with real pleasure.

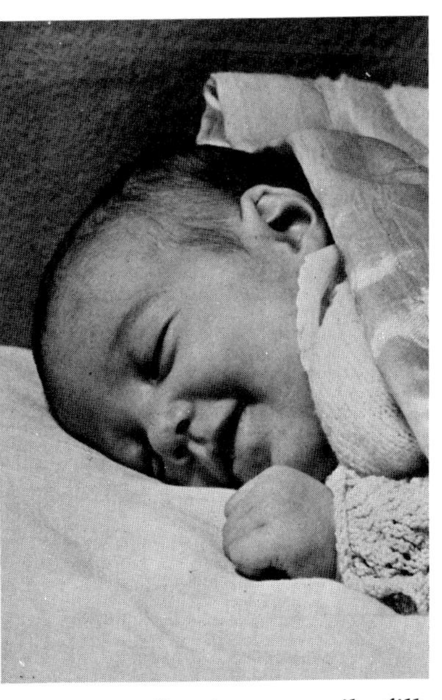

Spontaneous smiles fill the infant's face in sleep.

When infants are about 1 month old, they begin to show **exogenous smiles** in response to events in the outside world. They smile when heads nod at them silently or speak in a high voice, when lights blink, and when they perceive rhythm or repetition (Sroufe, 1979). By 3 months infants smile when they see a human face, a doll, or other familiar object. Now their smiles are truly social. Three-month-olds smile when they see Mom or Dad. Their smiles may reflect pleasure, or they may signal that the infants have recognized someone or something (Kagan, 1971). When babies are 4 months old, the motionless face that just a few weeks ago made them smile now does not, perhaps because it's too easy to recognize. But a moving face is more grist for their mill, and so they smile. The 5-month-old smiles after he has done something like move a ball or a mobile and then shown his mastery by moving it again. His smiles show that in his earliest months of life, he already takes pleasure from learning and doing.

Young infants' smiles draw others to them because they are so rewarding. The more infants smile, the more their parents are eager to watch, interact, and smile back. The upward spiral is quite lovely: the more the infant smiles, the more his parents smile back, and so the more he smiles, and so the more they smile. . . .

> Sam grinned when I came into his room to pick him up after his nap, and he grinned when he first saw me in the morning. He grinned when he saw his father, or his rice cereal, or the brightly colored horses on the side of his bassinet. He grinned when he was changed, and he grinned when he was sung to. We were so in love with those grins of his that his father at one point said, "That kid's got us trained like seals."

Crying

Parents may love to hear their infants laughing—but not crying, another powerful means of communication. The sound or sight of an infant crying galvanizes most people into action. A nursing mother has only to hear her own or another infant cry, and her breasts fill with milk. A mother has only to watch a *silent* videotape of an infant the same age as her own crying, and her heart races (Donovan, Leavitt, and Balling, 1978). If she watches a videotape of her own infant crying, or if a soundtrack is added to the tape, her heart races faster still (Wiesenfeld and Klorman, 1978).

Why infants cry

New babies cry when they are hungry or cold. Infants whose rooms are heated to a toasty 88° Fahrenheit have been found to cry less and sleep more soundly than those in a merely tropical 78° (Wolff, 1969). Warmth and sleep both may insulate infants from crying. No matter how warm their room, infants fuss when they are undressed. Quick, sharp noises or movements that startle the infants often make them cry as well. Babies also cry if they are overstimulated. Even gazing too long at a bright toy or mobile may overstimulate them, and they cry to break away from it (Sroufe, 1979; Tennes, Kisley, and Metcalf, 1972).

Babies 2 to 3 months old cry for different reasons—if someone interrupts their feeding, takes something out of their hand, approaches suddenly, makes a loud noise, or stares at them unresponsively (Bernal, 1972; Stern, 1974;

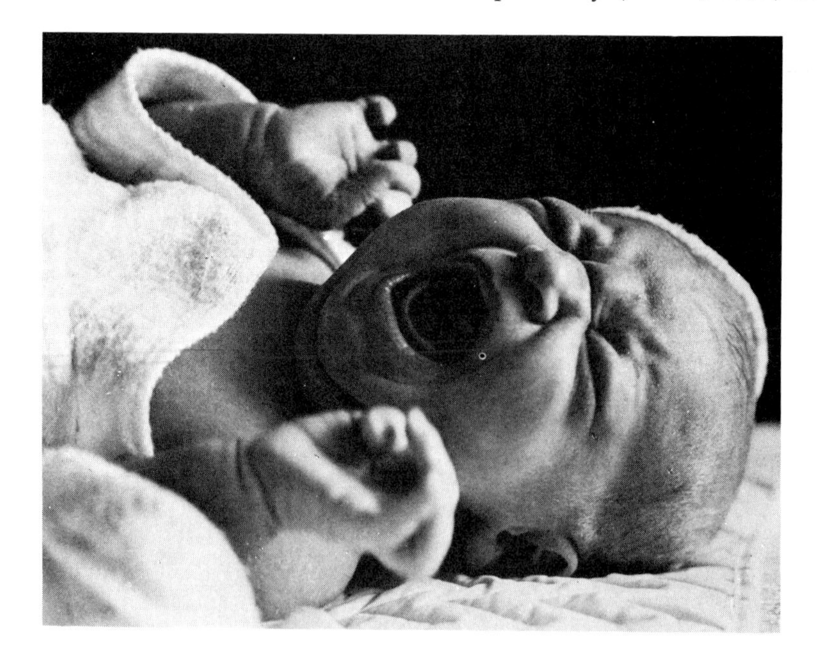

The infant's cry: a frowning face; eyes closed and mouth open; a lusty wail.

Wolff, 1969). Babies at this age also sometimes cry for a moment when someone leaves them.

By 4 to 6 months, infants develop intentions, the motivation to complete an action, and cry if they are frustrated. Infants begin to distinguish between familiar and unfamiliar people. When an unfamiliar person appears, the baby studies the person's face intently, frowns, takes a deep breath, and usually looks or turns away; he then may cry (Bronson, 1972). This *wariness* of strangers is an important new development that demonstrates the infant's growing awareness of the social environment. The infant's interest is caught by the interesting human face of the stranger. His mind engaged and his body aroused, the infant works at recognizing, or assimilating, the stranger's face. If he cannot assimilate the person, he acts wary. Between 6 and 12 months, as cognitive awareness increases, the likelihood that the infant will be wary increases (see Figure 7.2). The baby also acts wary if someone familiar acts strangely or looks different—perhaps wearing new glasses or a mask (Brazelton, Tronick, Adamson, Als, and Wise, 1975). The *incongruity* confuses and upsets the infant (Bronson, 1972).

Sometimes during this period infants act not only wary but actually fearful. They cling to their parent, burst into tears, scream, or scoot away. Earlier, crying is brought about by physical conditions like hunger or cold, or it is an immediate reaction to an event like a face not responding. But this **stranger anxiety** or **fear of strangers** involves cognitive processing and therefore may be qualitatively different from the crying caused by pain or overstimulation. Developmental psychologists now think that infants' fearful reaction to strangers depend on their interpretation of the stranger and the situation—who else is there, how the stranger acts, and so on. Safe in his mother's arms when a stranger appears, an infant acts wary (frowns, looks away, has a racing heart) but usually does not cry fearfully. But if the infant is alone, upset, or in an unfamiliar place, or if the stranger comes too close, too quickly, looks too serious or scary, or does not back off when the infant looks away, the infant is likely to react fearfully (Bronson, 1978; Brooks and Lewis, 1976; Emde, Gaensbauer, and Harmon, 1976; Skarin, 1977; Sroufe, 1977).

If the stranger reminds the infant of some past experience with a stranger, the infant also may act fearful. Beginning at 8 to 10 months, as we

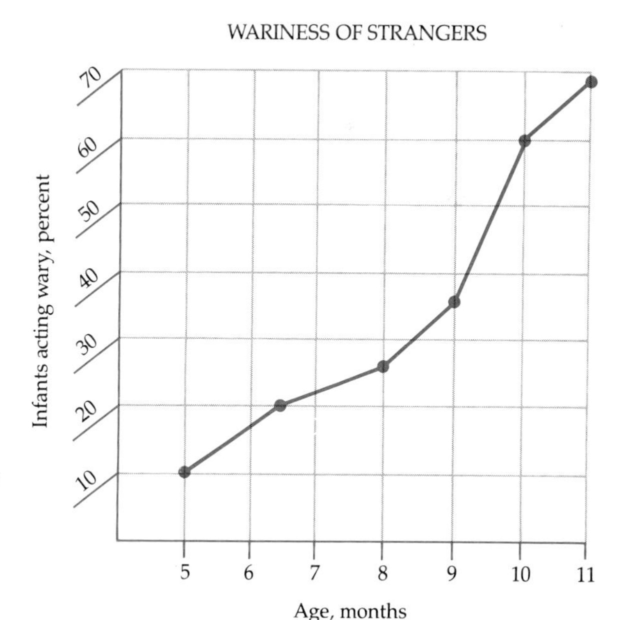

Figure 7.2 ***The likelihood that an infant will act wary of a stranger increases noticeably from 5 to 11 months of age (Decarie, 1974; Skarin, 1977; Waters, Matas, and Sroufe, 1975).***

discussed in Chapter 6, infants can form basic categories—like ''scary stranger'' or ''nice new playmate.'' They can make guesses about the new stranger based on their past experiences. Frightened by one stranger, they are more likely to be frightened by a second.

> When Sam was 9 months old, he stayed with his grandparents one afternoon. They couldn't understand why he cried so hard when the teenaged girl from next door stopped by for a few minutes. He probably thought she was a babysitter, and he got scared that his grandparents were going to leave him alone with her. After all, most of the other teenaged girls he'd seen *had* been babysitters.

Not all infants show fear of strangers, and not all strangers are feared. But most infants can be intimidated by an insensitive newcomer. It takes a certain sensitivity to make friends with a baby.

> **Parents' response to crying**
> My mother was from the school of child rearing that believed in letting babies ''cry it out.'' But I don't think it's good for them to cry it out. So pretty much every time Natalie or the other kids cried, I tried to comfort them.

Whether caused by pain, hunger, cold, fatigue, frustration, overstimulation, incongruity, or fear, crying is not good for babies. It saps their energy, increases the pressure on their brains, removes oxygen and adds white cells to their blood. In newborns, crying may even return blood circulation to its fetal direction (Anderson, 1984). Even so, the question of whether it is best to respond every time a baby cries remains controversial. Experts in child care give conflicting advice. Most experts agree that cries of pain and hunger should be met promptly. But what about all those other cries? Under controlled circumstances in the short run, babies can be conditioned to cry less and smile more if their crying is ignored and their smiling is rewarded or, conversely, to cry more if their crying is rewarded (Etzel and Gewirtz, 1967; Gewirtz, in press). But at home, in the long run, if the infant is ignored, crying spells gradually grow longer and more frequent (Bell and Ainsworth, 1972; Clarke-Stewart, 1973; Belsky, Rovine, and Taylor, 1984). Perhaps if the parents respond right away, they reinforce a short cry, and so the infant's crying spells over time grow shorter. It has been observed that infants develop a ''request

This baby is showing her ''stranger anxiety'' as she turns away from the unfamiliar woman to the arms of her mother.

cry,'' a signal that begins with a short cry and a pause for the parent's response (Bruner, 1978). If the parent responds to the request cry, the crying ends. If the parent does not, the infant moves into a longer, full-blown ''demand cry.'' In fact, a very alert parent may nip a crying spell in the bud by responding to the infant's intensifying signs of distress *before* he begins to cry. During the half hour before he starts crying, the baby may put his hand to his mouth, suck on his hand, stick his tongue in and out, or whimper quietly (Gill, White, and Anderson, 1984). A parent who responds consistently and promptly to these early warning signals teaches the infant that he does not have to cry to get what he wants.

How then to stop infants' crying? Providing food and warmth and eliminating pain are the first steps. Beyond this, during the first two weeks, a baby may be comforted by rhythmic motion—rocking or gentle patting (Ambrose, 1961; Korner and Thoman, 1972). Swaddling the new baby in a blanket also stops crying, by preventing overstimulation from kicking and arm waving. Picking up a swaddled baby and carrying it upright often can stop crying (Korner and Thoman, 1972). So can white noise, recordings of sounds heard in the womb, the noise of an air conditioner, vacuum cleaner, or blender, or Beethoven played loudly (Brackbill, 1958; Rosner and Doherty, 1979).

> One night Sam was really tough to soothe. His father put earphones on him and played a recording of Mozart. No go. But when the music was changed to the Rolling Stones, Sam calmed down right away.

Some babies quiet if they can suck on a pacifier. The rhythm of their sucking seems to soothe them. Nursing also soothes because it has the qualities of sucking and movement, provides food, warmth, faces, and voices. It is a biological pattern well adapted to soothing the crying infant.

Developing emotions

Infants express their emotions in smiles, cries, frowns, and grimaces from the very beginning of life. These facial expressions are part of the infant's inborn repertoire. Even blind infants have these facial expressions (Eibl-Eibesfeldt, 1973). But the emotions of infants differ from those of adults. They are elicited by different events, experienced in different feelings, and expressed in different ways. Adults, for example, usually do not cry when they are angry; infants do. Adults usually mask their smiles of delight when a competitor stumbles; youngsters do not. Over the course of childhood, the rules for expressing and responding to emotions must be learned. This learning begins in infancy.

Children must learn how to modulate and control their expressions of emotions. They must learn when to express, to exaggerate, to mask, to pretend, and to neutralize emotions, consistent with the rules of their culture. Learning about emotions begins as infants interact with adults (Lewis and Michalson, 1982).

> One morning when Sam and Natalie had been playing together, Natalie's mother said that it was time for Natalie to think about getting her coat on to go home. ''I don't want to,'' Natalie complained.
>
> ''You go home *now*,'' Sam said, tired of sharing his toys.
>
> ''Sam, we don't say things like that to people,'' his mother interjected. ''It hurts their feelings.''

Children learn when and where they may express feelings. Learning these display rules, it seems, is a slow and gradual process. The first rule about feelings that infants learn is to smile often (Demos, 1982). Later, they learn to modify or suppress socially unacceptable feelings like displays of temper, dislike, and distress. Even infants just under 1 year old may do some of this when they interact with people they know well (Feinman and Lewis, 1981). In their second year, they try to control expressions of distress by pressing, sucking, or funneling their lips (Demos, 1982). As they learn language, infants learn the labels for emotions.

Children probably learn most about feelings from their parents. Parents are the ones who model expressions of emotion. It has been estimated that infants between the ages of 3 and 6 months are exposed to 32,000 facial expressions of emotion (Malatesta and Haviland, 1982)—not a trivial learning experience. Parents also talk about emotions to their infants, saying things like "Don't be afraid, baby," "I know you love Mommy," and "Oh, you're embarrassed" (Pannabecker, Emde, Johnson, Stenberg, and Davis, 1980). A parent can turn any emotional incident into a learning experience. For example, a 1-year-old starts to cry after a tower of blocks she has built falls down on her leg. Her mother may say, "You're frustrated" and help to rebuild the tower. The mother has labeled the feeling of frustration. Instead, the mother may bark, "Stop crying! If you can't build it that high, don't try. You're driving me crazy." The mother has labeled her own emotions. Maybe the mother says, "Oh, you hurt yourself" and cuddles the baby. She has labeled the baby's response to pain. Three quite different learning experiences have followed from the infant's crying, a single emotional expression.

From an early age, infants also respond to differences in other people's emotional expressions if these are reflected in direct interaction with the infant. In one study, for example, researchers tested the responses of 3-month-old babies to the emotional expressions of their mothers while interacting with the infants (Cohn and Tronick, 1983). Some mothers were asked to act as they normally would, some to act depressed—to slow their speech, keep their face still, and minimize touching their infant or moving their own body. The babies responded quite differently to normal and "depressed" mothers. Babies of "depressed" mothers spent half the time protesting, reacting warily, or giving but fleeting smiles. Babies of normal acting mothers showed much more variety in their play, protested or acted wary only rarely, and when they smiled, did so for significant periods of time.

Young infants also respond to others' emotional expressions if the expressions are clear and salient—like crying. Investigators have observed, for example, that 6-month-olds are responsive to the distress of other babies (Hay, Nash, and Pedersen, 1981). When infants were placed in a room with another infant who began to cry, nearly all of them (84 percent) looked at the distressed infant for nearly the whole time he or she cried. Some infants also

Infants pay attention and respond to the emotional expressions that they see on the faces of other people.

leaned, gestured, or touched the crying baby. Moreover, if there were no toys present and the infant went on crying, the other infant was likely to grow distressed too. Clearly, infants respond to crying. But what about more subtle emotional expressions, such as smiles and frowns?

Researchers have looked into the abilities of babies to read these emotional expressions (Caron and Myers, 1984; Caron, Caron, and Myers, 1985). In some studies, the researchers used a habituation design in which the infants were first habituated to a picture of a woman posing one facial expression and then shown a picture of another woman posing either the familiar expression or a new one. At all ages, the babies could distinguish faces with toothy smiles from those with closed mouths (either angry or smiling). But they could not distinguish toothy smiles from toothy angry expressions. Clearly, infants pick up some information about other people's emotions from what they see on their faces, but that information is incomplete. To know whether Mother is happy or angry, the infant must depend on movement and sound as well. Tone of voice is an emotional expression that infants respond to. In one study, when researchers looked angry and uttered a nonsense word in an angry tone of voice, 10-month-old infants inhibited their touching of an attractive object. Infants who heard the same word uttered without emotion did not inhibit their touching (Bradshaw, Campos, and Klinnert, 1986).

Infants use information about other people's emotions to guide their own behavior. In an unfamiliar situation, infants tend to orient themselves so that they can see their mother's face (Carr, Dabbs, and Carr, 1975). Perhaps they do so because they can get important cues about how to act in the unfamiliar situation from their mother's face and gestures. If the mother is out of sight, infants will rely on cues from her voice (Campos and Stenberg, 1981). In one study of this **social referencing** (Sorce, Emde, Campos, and Klinnert, 1981, 1985), one group of infants, who had to cross a visual cliff with an apparent drop-off of four feet, refused to cross the glass. A second group, who faced no apparent drop-off, crossed immediately without checking their mothers' faces. But a third group, confronting a drop-off of two feet—an ambiguous situation—looked at their mothers for cues before they did anything. When the mother looked afraid, not a single infant crossed the glass. When she looked happy, 80 percent of the infants crossed, but when she looked angry, only 10 percent crossed. When mothers looked sad—an inappropriate response—one-third of the infants crossed but vacillated before doing so. In short, it was shown that infants not only referred to their mothers' expressions for cues in an unfamiliar situation, but also behaved differently depending on the sort of emotional messages they got from their mothers' faces. As they continue to develop cognitively and socially, children come to depend less on social referencing and more on their own, internal strategies for evaluating events. But it is likely that people never completely outgrow referring to others when they need to evaluate ambiguous situations.

A social partner

It is clear that infants communicate with the people around them. They gaze and babble, smile and cry, frown and reach. They react to the behavior of others. But when do infants first turn into real social "partners"? When do they first hold up their end of a dialogue or a game?

The parent–infant "system"

From birth, as we mentioned in Chapter 5, the infant's physiological systems are "prewired" in such a way that the infant pays attention to parents' faces.

The infant's eyes rest on the contrasts at the edges of the face, on the moving, blinking eyes, on the animated red mouth. But the infant is still too young to be a true social partner. Parents therefore work at enfolding the infant within an interactive system. They project their own feelings and intentions onto the infant's built-in rhythms and actions, and in doing so, they begin to create shared routines with their infant (Kaye, 1982).

Even during pregnancy, parents speak for their infant and draw the still unborn participant into the family. As they feel the fetus moving, parents say things like

> "She says, 'It's getting crowded in here!'"

Once the infant is born, the parents continue to speak for it:

> "He says, 'I'm sleepy, Daddy; don't bounce me so much.'"

They interpret the baby's hunger cries or restless sleepiness as though the infant were trying to scare up a meal or find a resting place. As infants begin to show their intentions—by reaching for the toy that Daddy holds up or turning their head to search for a voice, parents continue talking for them.

> "He says, 'Give it to me, Dad. I want to put that in my mouth.'"
>
> "She says, "'Where'd it go?'"

Parents interpret their infant's expressions and behavior as meaningful and communicative. They act *as though* their infant were a little person and an intelligent partner. As they treat the infant as a contributing member of the family, they gradually draw the infant into the family system.

Taking turns

The evolving social system of parents and infant shows in the way parent and infant take turns. Since investigators have been able to use stop-frame and slow-motion **microanalysis** of films and videotapes, they have been able to pick out the fine details of these turns. They have seen that mothers and infants begin these turns from the moment the infant is born. From birth, for example, babies suck in a regular pattern of bursts—suck, suck, suck—and pauses. Mothers pick up on this pattern and act in time to it. When the baby sucks, Mother is quiet. When the baby pauses, Mother bounces, touches, and talks. The baby sucks and pauses whether the mother responds or not, but her response makes feeding a turn-taking *interaction*. She treats the baby's bursts of sucking as a turn, and she treats the baby's pause as though he were saying, "There, my turn is over. It's your turn, Mom." During her turn, she bounces the baby and says, "Hi there, cutie" (Kaye, 1977).

Mothers also may act as if the baby were taking turns in "conversations." To give their babies turns, mothers generally wait the length of a pause in adult conversation and listen for an imagined response. They then wait the length of another pause in adult conversation before they talk. Mothers have been observed in these "conversations" with babies as young as 3 months (Stern, 1977).

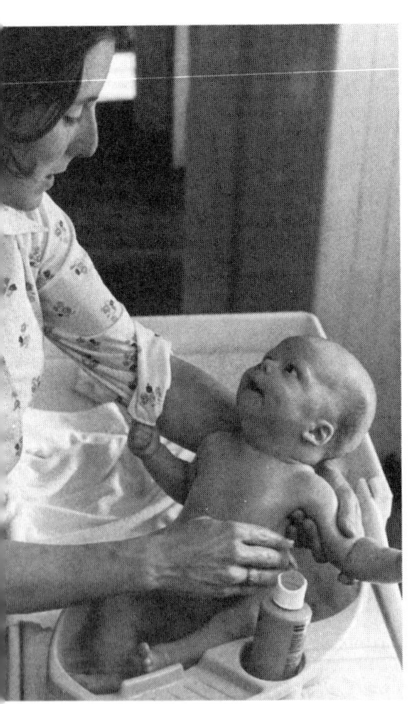

Mother–infant play makes an enjoyable break from the bath for the young infant. The behavior of mother and baby is finely meshed and in tune.

MOTHER: "Aren't you my cutie?"
Pause: (0.60 second)
Imagined response from baby: "Yes"
Pause: (0.60 second)
MOTHER: "You sure are!"

These "conversations" are even clearer once the baby is old enough to babble (Trevarthan, 1977).

MOTHER: "Are you talking to me?"
BABY: (Babbles)
MOTHER: "Are you telling me a story?"
BABY: (Babbles)
MOTHER: "Are you telling me a story?"
BABY: (Babbles)

When mothers do respond to their infants' babbling with this turn taking, the infants' babbling increases (Bloom, Russell, and Davis, 1986). Turn taking in these first **pseudodialogues**, of course, is up to the parent. Infants' rhythmic behavior—suck, suck, suck, pause, suck, suck, suck; babble, pause, babble, pause—allow the appearance of turn taking. But parents still have to speak babies' parts for them—to fill in the blanks.

By the time they are about a year old, babies can speak their own parts. They can vocalize when their parents pause expectantly. Pseudodialogues have become true dialogues and infants, true social partners (Newson, 1977).

BABY: (Looks at a toy top.)
MOTHER: "Do you like that?"
BABY: "Da!"
MOTHER: "Yes, it's a nice toy, isn't it?"
BABY: "Da! Da!"

Individual differences

Most infants are born with regular cycles of behavior that offer their parents cues for social interaction. Infants who feed and behave predictably are easier for parents to engage in social exchanges than those whose inborn cycles of behavior are disorganized and hard to predict. In the developing social system of infant and parent, it takes two to tango, and not all parents and infants dance together smoothly. Sometimes it is the infant who is clumsy, sometimes the parent.

Skilled and unskilled mothers

A mother may be depressed and understimulating; she may be hyperactive and overstimulating; or she just may not know how to play with her baby. She may not be able to muster the energy for play because she is so preoccupied with her other responsibilities at work and at home. She may be under psychological strain, or she may be inhibited or fearful around the baby. She may resent the baby and all the care required. She may be personally insecure and interpret the baby's cries or averted eyes as a personal rejection. There are many reasons that a mother may be a less than perfect partner to her baby.

In the dance between parent and infant, some parents continually misread the cues that their infants send out. If during spirited play, the infant frowns or looks away, parents should read those reactions as cues for them to back off and ease the stimulation. But an intrusive parent instead acts *more* intensely, and the infant has lost a chance to regulate his environment, to learn that his wishes matter and that he can exert control over what happens to him. After enough such frustrating social exchanges, the infant may stop trying. Parents who are insensitive to their infants' signals may end up teaching the infants that the world is beyond their control. The smoothness of the exchanges between parents and infants depends on parents' abilities to interpret their infants' signals and intentions, complete their actions, and anticipate their reactions.

Mothers who are young, poor, and have problems feeding their infant are classified as "at risk" for neglecting or abusing their infants. These mothers may not be as good at interpreting infants' emotional expression as other mothers. At-risk mothers, in one recent study, were more likely to see their infants as expressing sadness, anger, shame, and joy and less likely to see them as expressing fear and interest (Butterfield, 1986).

In some cases, mothers who have trouble interpreting their infants' signals can be helped. All parents need support and encouragement in their attempts to care for their infants, but programs specifically for troubled mothers are particularly important. Professionals can teach these mothers how to stimulate difficult or listless infants and how to read the signals of quiet infants. Troubled, disturbed mothers may not know how to nurture their infants because they themselves were nurtured poorly. They may misunderstand things that their infants do because they do not know how normal infants develop. For example, one mother of a 4-month-old interpreted his arching his back when she tried to hold him as rejection (Peterson, 1982). In fact, the infant's muscles simply were too immature for him to support his back. He wasn't rejecting his mother at all.

Easy and difficult infants

All problems in the social interactions of mothers and infants cannot be laid at the mothers' feet, of course. Some infants are more difficult than others. Even the most responsive and sensitive of parents may have a hard time getting along with these infants.

> Jason was brought from the hospital nursery into my room. As he lay in his crib, he seemed fussy and fretful. His mouth grimaced, his face and scalp went red, and he grumbled as his eyes flickered open. We thought that he might be hungry. Charlie picked him up tenderly, and I started to arrange myself to try and nurse him. Within a few seconds, Jason was flailing his arms, kicking his legs, and screaming so hard that we both got scared. His little body was rigid with tension, his eyes were clenched shut, and I had no idea how to get him to suck. I called for a nurse, and she showed us how to quiet him by swaddling him and propping him on her shoulder. She helped me put Jason to my breast. But he spluttered and started to cry again—fiercely. The nurse said she'd take him back to the nursery and we could try again later. I looked at Charlie and burst into tears.

Alexander Thomas and Stella Chess (whose study we mentioned in Chapter 3) were among the first researchers to study individual differences in infants' behavior during the first weeks of their lives. They knew about the prevailing inclination to blame mothers for children's psychological problems. But how, they asked, could behavior that appeared so early be attributed to mothers? To explore this issue and to trace the development of temperament from infancy onward, they began the New York Longitudinal Study. Parents in the study were interviewed when the infants were 2 months, 6 months, and 1 year old. The questions were specific. For instance, in the past several weeks, how had the infant acted in the bath, when a diaper was wet, when tasting a new food? The researchers asked for descriptions of infants' behavior in

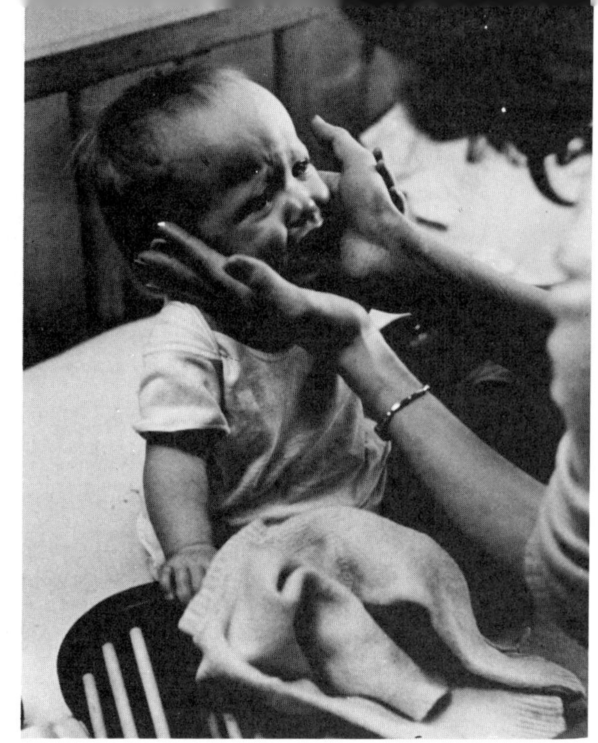

The infant's easy or difficult temperament makes a profound difference in the parents' "quality of life."

specific situations in an attempt to steer as clear of parents' biases as possible. When the children were older, the researchers observed them directly and questioned their teachers. When the children were 16 years old, they and their parents were interviewed separately. The data collected from all these sources strongly suggested that individual differences in temperament were apparent in infancy and left a trail right through childhood and adolescence (Thomas and Chess, 1981).

Three broad patterns of temperament emerged.

1. *Easy babies* usually were happy and received people and things enthusiastically. They did not react to small discomforts or frustrations. They got hungry and sleepy at pretty much the same times every day. About 40 percent of the babies in the New York study were easy.
2. *Difficult babies* took most things hard. They cried often, ran on irregular schedules, and adapted to new people and places with difficulty. Their mood tended to be down. About 10 percent of the babies in the study were difficult.
3. *Slow-to-warm-up babies* took a while to adjust to new people, places, and events. They reacted mildly to things and ran on moderately irregular schedules. But once they got used to something or someone, they liked and enjoyed them. About 15 percent of the babies in the study were slow to warm up.

Babies with these three types of temperament were found in all kinds of families, regardless of social class or child-rearing style and regardless of the babies' sex.

Evidence from other studies, in which unbiased observers (and not parents) watched infants, supports these findings (for example, Korner, 1973; Schaffer and Emerson, 1964b). From the first week of life, babies show individual styles of responding. Some babies, for example, are more cuddly than others. Cuddly babies are more placid and sleep longer and are more likely than other babies to take comfort from people, thumb sucking, or a soft blanket. In contrast, uncuddly babies are restless and active. They are quickly aroused and dislike physical restraints—and to a restless ''noncuddler,'' even a mother's hug may feel like a restraint. When it comes to soothing, some babies can soothe themselves by sucking their thumb or fingers. But others need to be soothed by someone else. Some babies can be calmed easily and for long periods; others are difficult to calm. More active babies express their feelings in cries, grimaces, and noises. They like new things and strong excitement. Quiet babies spend more time looking around and making small movements. They need gentler stimulation that they can sift through, reduce, or turn away from.

Despite this seemingly unbiased evidence for individual differences in infants' behavior, some psychologists have suggested that temperament is not really a characteristic of infants. Instead, they say, temperament is something that parents project onto their infants. Parents are not unbiased observers trained to collect objective data about infants' behavior. Their assessments of their infant's temperament are related to their own characteristics, social class, race, attitudes, and mental health (Bates, Freeland, and Lounsbury, 1979; Crockenberg and Acredolo, 1983; Sameroff, Seifer, and Elias, 1982). Although temperamental patterns may be exaggerated in parents' eyes and reports, infants do exhibit individual characteristics that are relatively stable over time. This stability shows up in the assessments of objective observers as well as parents. One-year-olds, for example, who approach new situations cautiously also tend to act that way as 3-year-olds (Gibbons, Johnson, McDonough, and Reznick, 1986). Difficult babies continue to protest, fuss, and cry as toddlers (Riese, 1986; Worobey, 1984). Boys who are difficult as infants are usually still difficult at 3 years (Bates and Bayles, 1984), have more behavior problems at 4 (Guerin and Gottfried, 1986), and often still are difficult as young adults (Korn, 1984).

> Jason was a fussy and stubborn infant, and he turned into a fussy and stubborn toddler. He hit and fought with the other kids so often that even his nursery school teachers, who were extra patient, agreed that he was ''high strung.'' He's always been alert, but he's touchy and aggressive, and so he gets into lots of fights.

Becoming attached

Out of these bumpy or smooth interactions between infants and their parents develop relationships. In the first year of life, infants form their first relationships—lasting, loving ties, or **attachments**, to the most important people in their world. These attachments are the infant's first love affairs, and like any love affairs they have their pleasures and heartbreaks—the delight of being together, the pain of being apart. Love this deep does not blossom overnight. It develops gradually (Ainsworth, 1973; Bowlby, 1969):

1. During the *preattachment* phase, new babies look and smile at and can be comforted by everyone. This phase ends when babies can tell people apart and recognize the difference between familiar folks and strangers. Although babies can tell the difference between familiar and new people by smell, touch, and hearing even earlier, the preattachment phase is generally considered to be over when the 2- or 3-month-old baby consistently responds differently when he or she *sees* Mom and Dad or a stranger.

lifespan focus

Temperament over Time

Infants may start out with individual temperaments, but temperamental traits are not cast in stone. The way temperament is expressed changes as children get older, and parents' feelings about it may change as well. The extent to which a temperamental trait is transformed into either problem or productive behavior is likely to depend on how it fits with the parents' goals, values, expectations, and styles of behaving. Infants' traits may mesh well or poorly with their surroundings. If the fit is good between the behavior of infants and their parents' expectations and behavior, the infants are likely to develop well. If the fit is poor, development may be compromised (Thomas and Chess, 1977; Lerner, 1984).

We need more research on goodness of fit to confirm or disconfirm its importance for infants' development (Windle and Lerner, 1986). We also need more research to find out exactly how infants' temperaments affect the behavior and attitudes of those around them (Crockenberg, 1986). In one study, new parents who perceived their infant as difficult felt less in control than they had before the infant was born (Sirignano and Lachman, 1985). In another study, simply being told beforehand that an infant was difficult made women less effective at stopping the infant's crying (Donovan and Leavitt, 1984). The belief that they would be unable to soothe the infant apparently made these women feel helpless, and their feeling of helplessness in turn made them less likely to console the infant. In yet another study, the effort that mothers put into teaching their children varied according to how easy or difficult the children had been as infants (Maccoby, Snow, and Jacklin, 1984). Mothers of boys who had been difficult infants later put less effort into teaching their sons. Thus temperament—real or imagined—may feed into a vicious circle. Either a difficult infant or an unskilled parent may start things off on the wrong foot, and the interactions between infant and parent that follow are likely to be bumpy. It undoubtedly takes especially sensitive, patient parents to avoid falling into difficult interactions with difficult infants, and especially easy, responsive infants to smooth the interactions with troubled parents.

2. The second phase of a baby's developing social attachment, when the baby responds differently to familiar and unfamiliar people, is called the *attachment-in-the-making* phase. The baby smiles and "talks" more to the familiar people and later greets them and cries when they walk away. Familiar people can console the baby better than unfamiliar ones can. When the baby is 6 or 7 months old, he or she begins to stay close to one of these familiar people, and an attachment is said to be formed.
3. During the third, *clear-cut attachment* phase, the baby is truly attached to that specific person. She tries to stay near the person by crawling, calling, pulling, and hugging. She protests when the person leaves. In families in which the mother is the infant's primary caregiver, this person is usually the mother (Cohen and Campos, 1974; Kotelchuck, 1976; Lamb, 1976a, b). But although most babies grow attached to their mothers first, soon they also are likely to grow attached to their fathers, grandparents, older brothers and sisters, or even a neighbor or other relative they see often. By the time they are a year old, babies usually have grown attached to one, two, or possibly three people. The phase of clear-cut attachment lasts until the baby is 2 to 3 years old.

Theories of attachment

Many developmental theorists have noted the importance of the infant's attachment to the mother. Freud called it "unique, without parallel, established unalterably for a whole lifetime as the first and strongest love object" (1938, p. 85) and proposed that it develops out of the infant's oral pleasure at the mother's breast. Erikson (1968) made attachment the cornerstone of his theory of psychosocial development when he claimed that the first and most basic human crisis is the development of a sense of trust or mistrust, based on the infant's expectation that the mother will or will not meet his or her needs. For social learning theorists, the infant's attachment to the mother was important because it illustrated learning principles. They explained the infant's desire to be close to the mother as a secondary drive, learned through association of the mother with satisfaction of hunger, a primary drive. Each of these theories, however, has been found wanting in some way or another. The theories have been criticized on the grounds that the details of how the attachment develops are sketched incompletely and on the grounds that the proposed course of development does not fit with the facts (for example, that sucking and feeding do not form the bases for attachment).

The most persuasive description of attachment comes from John Bowlby (1969). Bowlby suggests that attachment has its origins in inherited behavior that is characteristic of the human species. Through crying, sucking, clinging, smiling, vocalizing, and following, infants draw their mother or other caretaker to them and keep the adult nearby. Nearness to a caring adult is essential to an infant's survival. Although all human infants inherit the tendency to seek this kind of physical closeness, how much closeness they need depends on their experiences and their circumstances. Through repeated interactions with the attachment figure, infants construct mental models of the adult and of themselves. These models help infants to appraise new situations and guide their behavior. If their experiences have led them to construct a model of the attachment figure as someone who gives necessary support, infants do not have to stay nearby constantly to monitor the person's whereabouts. But if infants cannot count on this responsiveness, they may seek more closeness. The need for closeness, Bowlby suggested, is a **set goal** that each infant strives for. If the amount of closeness infants get falls below their set goal, they strive for more. Infants whose parents have been away may seek more closeness when they return than infants whose parents have been available. Infants who are sick or tired may seek more closeness with their attachment figure than when they are well and bursting with energy. Infants who feel threatened by, say, the appearance of a stranger or sudden darkness also seek closeness. Infants may crawl or call to their parents when the parents move away from the infants or even when they make the first signs of moving away—for example, putting on their coat or taking keys in hand. If crawling, calling, and clinging do not bring the parents close enough to reassure them—and meet their set goals—infants may start to cry.

Others have added to Bowlby's theory of attachment. Mary Ainsworth (1973) suggested that the infant's set goal of seeking closeness to the mother must be balanced against a second set goal, the infant's wish to explore the surroundings. Alan Sroufe (Sroufe and Waters, 1977) suggested that the infant's set goal is not simply physical closeness, but the feeling of security that closeness brings.

Assessing attachment

Babies *feel* attached to their mothers. But how can psychologists measure these feelings? One procedure has become the accepted standard for assess-

In the standard procedure for assessing a 1-year-old's attachment behavior, the "strange situation," (1) the infant first is allowed to explore the unfamiliar room while his mother is present. (2) He briefly approaches her. (3) Left alone in the room, he cries and rocks himself. (4) When the mother returns to the room after a brief absence, this securely attached infant hugs her.

ing the feelings of attachment in 1- to 1½-year-old babies. In the so-called **strange situation,** the baby is brought by the mother into an unfamiliar but unthreatening playroom with a one-way window in it. The mother has been coached ahead of time about how she is to act. She puts the baby down a little way from some toys and sits in a chair. A stranger comes into the room, remains quiet for a while, then talks with the mother, and eventually tries to play with the baby. The mother hears a tap on the one-way window, her signal to leave the room for three minutes. Then she goes back into the room, the stranger leaves, and the mother stays with her baby until she hears another tap on the window. The mother then leaves the room for a second time, and the baby is alone. Next, the stranger goes back into the room and tries again to play with or console the baby. Finally, the mother goes back in, speaks with, and picks up the baby.

Most babies studied in this country (for example, Ainsworth, Blehar, Waters, and Wall, 1978) do not protest when their mothers put them on the floor, and over three-quarters make for the toys in less than a minute. Few babies are distressed at this point. Most begin exploring their surroundings, keeping their mother in sight, smiling and calling to her, and showing her the toys. But when the mothers leave the room for the first time, about one-fifth of the babies start crying, and about half cry at some time during the separation. Some babies try to find the mother at the door, some at the chair she was sitting in. When the mothers return, over three-quarters of the babies smile, go toward, reach for, touch, or talk to them. One-third of the babies cry.

The second time the mother leaves the room, babies are more likely to get upset. Over three-quarters cry; half cry so hard that their mother has to return before the allotted three minutes are up. Almost none of the babies pays attention to the toys, and the stranger can stop only about one-quarter of the babies from crying. When the mothers come back a second time, half of the babies cry, and three-quarters hold onto her tightly (compared to one-third who held on tightly after the first separation).

Patterns of attachment

The reactions of 1-year-olds to the strange situation have been grouped into three patterns of attachment (Ainsworth et al., 1978). Babies with a **secure attachment** try to stay near their mother and pay more attention to her than to the stranger. The mother's presence in the room offers them a secure base from which to explore the room and the toys. When their mother returns to the room after the brief separation, the babies act happy, greet her, and play nearby. Most babies in this country have secure attachments to their parents (two-thirds of the infants in the Ainsworth study, for example). Babies with an **insecure-avoidant attachment** ignore and turn or look away from their mother when she returns to the room after the separation. These babies, about one-fifth of Ainsworth's sample, do not try to stay near their mother, cry little or not at all when their mother leaves, and can be comforted by the stranger. The babies with an **insecure-ambivalent attachment** are quite distressed when their mother leaves the room. When the mother returns, some of the babies approach her. But they are not comforted when she picks them up, and they soon wriggle to be put down. These ambivalently or resistantly attached babies made up just over one-tenth of Ainsworth's sample.

Forming a secure attachment

All infants form some kind of attachment to the primary person in their lives. But what qualities in that person are most likely to foster a *secure* attachment? What kind and how much contact work best? Bowlby (1951) suggested that to develop a secure attachment, an infant needs a loving relationship with the same person, which continues unbroken from birth, through infancy, and childhood. What is the evidence for this claim?

Rx: loving care

Secure attachments do not grow automatically within infants with ever-present caregivers who provide adequate food and physical care (Ainsworth, 1973; Rutter, 1974). It is the *quality* of the interactions with the caregiver that seems to be crucial. Infants with secure attachments are more likely to have caregivers who treat them warmly and affectionately, smile at them and play happily with them, hold and cuddle them often and tenderly, and enjoy the everyday details of taking care of the infant. Mothers of securely attached infants give their infants orders more pleasantly and are more supportive, helpful, affectionate, and playful with their infants than mothers of insecurely attached infants (Arend, Gove, and Sroufe, 1979; Clarke-Stewart, 1973; Pastor, 1981).

Loving caregivers are not always Pollyannas, however. Love intensifies *all* emotions, both the pleasant ones and the unpleasant ones (Schaffer and Crook, 1978). The most loving of mothers can and does lose her temper. Anger and upset also reflect a mother's emotional involvement with her infant. For infants, the important thing is the relative proportion of warmth and affection to coolness or anger. Infants whose mothers are *often* angry, critical, rejecting, and interfering are likely to be insecure and avoidant with them (Ainsworth,

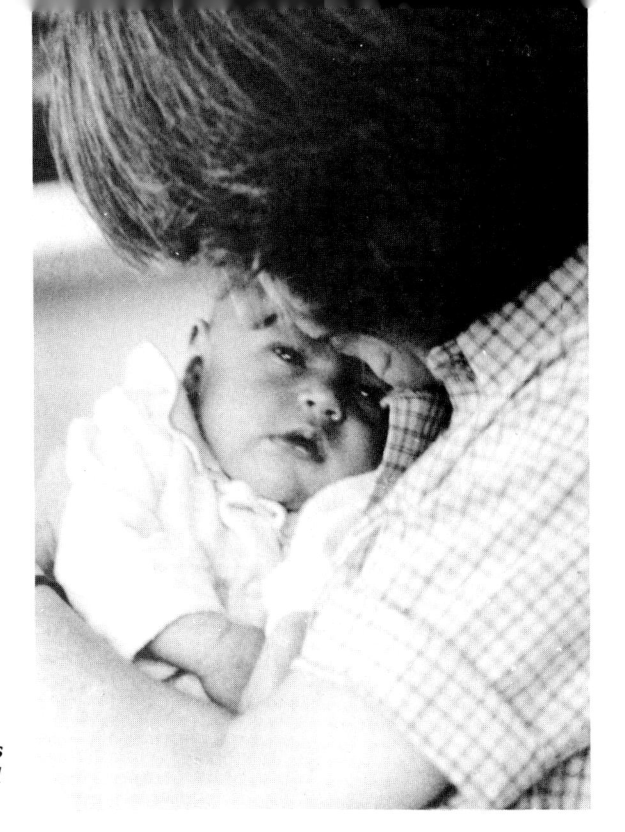

A loving relationship with mother or another caregiver is essential for the baby's social and emotional development.

1973; Beckwith, 1972; Egeland and Farber, 1984; Radke-Yarrow, Cummings, Kuczyncki, and Chapman, 1985). Infants whose mothers are actually abusive or neglectful are especially likely to be insecure and anxious (Egeland and Sroufe, 1981; Schneider-Rosen and Cicchetti, 1984). The loving care that babies need to form secure relationships is likelier to flourish in a family setting than in an institution. It is the *love* that is important, though, not where or by whom it is offered. Babies thrive on love from adoptive parents just as well as on love from biological parents and are just as likely to form secure attachments to them (Singer, Brodzinsky, Ramsay, Steir, and Walters, 1985; Tizard and Rees, 1974).

Not only affection but also empathy is important in the loving relationship of mother and infant. The empathic mother sees things through the infant's eyes, picks up and understands the infant's cues, and responds quickly, reliably, and appropriately. Sensitive, empathic mothers are likely to have children who are securely attached, happy, and sociable (Ainsworth, 1973; Clarke-Stewart, 1973; Egeland and Farber, 1984; Schaffer and Emerson, 1964a). Mothers of securely attached infants also are moderately stimulating and playful with their babies (Belsky, Rovine, and Taylor, 1984). They are not too stimulating—overstimulation makes infants avoid their mothers—nor are they understimulating—understimulation makes babies angry. It is not a simple matter of "more is better," but a matter of a happy medium. As we have pointed out with respect to cognitive development, stimulation has to be geared sensitively to the infant's needs.

Mothers who are psychologically healthier—strong, self-confident, affectionate—in situations not involving their infants may also behave more sensitively with their infants and have securely attached infants (Belsky and Isabella, in press; Benn, 1985). In contrast, mothers of avoidant infants have reported themselves to be relatively intense, and mothers of resistant infants have reported themselves to be less adaptable to new situations (Weber, Levitt, and Clark, 1986). Mothers who recall their childhoods as painful also

focus Attachment and the working mother—the baby's view

Mothers who work full-time and put their infants in some kind of day care test some of the assumptions about the formation of attachments. After reviewing the available studies of infants in day care, Alison Clarke-Stewart and Greta Fein (1983) drew several conclusions about the effects of these programs. First, infants in day care, like those reared exclusively at home, do become attached to their mothers. Their relationships with the caregivers in day care do not replace their attachments to their mothers. But are children in day care as *securely* attached as children cared for at home? Some studies show that day care causes no ill effects on the security of infants' attachment. For example, in one study researchers inquired into how mothers' working affected the attachment of firstborn children in intact, middle-class families (Chase-Lansdale, 1981). In all these families, the mothers had returned to work before their infant was 3 months old—therefore, before a true attachment had formed. No relation between the mother's working and the security of her child's attachment was found. But some studies have shown that infants, boys especially, in full-time day care keep more distance between themselves and their mothers in the strange situation than home-reared infants do. They are more likely to show an avoidant attachment pattern with their mothers than home-reared children (Barglow, Vaughn and Molitor, in press; Schwartz, 1983; Vaughn, Joffe, Egeland, Dienard, and Waters, 1979). This effect is especially likely if day care begins in the first half of the first year, before a strong and stable attachment to the mother has formed. Day-care infants are not more likely than home-care infants to be anxious or angry or to protest their mother's leaving, however.

What can explain the observed tendency for day-care infants to distance themselves more from their mothers? Is it because their mothers spend less time with them than home-care mothers? This explanation is unlikely. Working mothers may have less time at home, but they spend just about as much time as homebound mothers do in actual interaction with their children. A likelier explanation is that working mothers are less emotionally and psychologically accessible to their children than home-care mothers are. Working mothers tend to feel overworked and overtired. Although working often improves their self-image and even their health, they may feel guilty about taking time from their children and feel that their mothering is rushed and harried. Could this account for the physical distance between day-care children and their mothers? *Home-reared* children who avoid their mothers, we have seen, have mothers who are rejecting and angry or who do not reciprocate their children's initiation of physical contact. Yet the few studies that bear on this issue suggest that day-care mothers do not, as a group, differ from home-care mothers on measures of rejection or reciprocity with their children.

Infants in day care probably are too young to interpret their daily separations from their mothers as rejections. Day-care children probably get used to being with many different people outside of their families and become more independent of their mothers. Their greater distance from their mother may reflect generally advanced development. For *most* children, greater distance from their mother probably reflects an adaptive, realistic reaction to the situation rather than an emotional disturbance. Even so, because most available studies have been focused on high-quality day care, and the measures of attachment they have included are not necessarily the most sensitive, the question merits further investigation. As more and more working mothers rely on day care for their infants, effects on attachment should be monitored closely.

may be rejecting and have avoidant infants. Mothers in one study who reported being rejected by their own mothers, who could not remember their own childhoods, or who idealized their own rejecting mothers were more likely to reject their own infants. Mothers who did not feel rejected by their own mothers or who could express their anger and resentment toward their mothers were less likely to reject their own infants (Main and Goldwyn, 1984).

These studies provide suggestions about how parents' personalities and experiences affect infants' attachments. But we still do not understand the whole picture. For example, we still do not know whether sensitive parents foster secure attachments by reinforcing their infants' social skills, by creating positive expectations, by encouraging a sense of personal efficacy, by all (or none) of these (Lamb, Thompson, Gardner, and Charnov, 1985).

An easy baby

One factor that Bowlby did not consider in predicting the quality of infants' attachments was the baby's own temperament. We now have evidence that infants' attachments to their caretakers depend on the infants' characteristics as well as on their caretakers'. In one study, for example, children's attachment to their mother was related to how difficult they were as newborns (Holmes, Ruble, Kowalski, and Lauesen, 1984). Securely attached 1-year-olds as newborns cried relatively little; insecurely attached 1-year-olds were more likely to have been difficult newborns. Other researchers, similarly, have reported finding a link between infants' difficult temperament at 6 months and insecure attachment at 1 year (Egeland and Farber, 1984; Maslin and Bates, 1983).

In a recent review of all the studies in which researchers have looked for links between temperament and attachment (Goldsmith, Bradshaw, and Riesser-Danner, 1986), it was concluded that, although the link is not strong, infants who are temperamentally prone to distress are more likely to develop resistant, ambivalent attachments, and infants who are temperamentally more interested, persistent, and happy are more likely to develop avoidant attachments. But only further research will explain how temperament influences the development of attachment. For example, we do not know whether temperament affects infants' reactions to the strange situation, their reactions to their caretakers, or their caretakers' reactions to them. In one study, after infants developed secure attachments, their mothers regarded them as having become less difficult, whereas if infants developed insecure attachments, their mothers regarded them as more difficult (Belsky and Isabella, in press). What we do know is that the links between attachment and temperament are many and complex.

An easy life

Another potentially important influence on infants' development of attachments is the circumstances surrounding the caretaker–infant pair. When families are poor, when mothers are chronically stressed, when they are without social support, when neighbors are felt to be unsupportive, when a marriage is unsatisfying, or when the quality of a marriage declines after an infant is born, the chances increase that infants will develop insecure attachments (Crockenberg, 1981; Goldberg and Easterbrooks, 1984; Belsky and Isabella, in press). Conversely, when circumstances are more favorable, infants and caretakers have better chances of developing secure relationships. In general, good marriages and sensitive parents produce healthy, secure infants, and poor marriages are associated with insensitive parents and insecure infants. In sum, when all three systems—parents' personalities, infant's temperament, and surrounding circumstances—are positive, most infants develop secure attachments. When all three systems are negative, infants are likely to develop insecure attachments.

Significance of a secure attachment

Do secure attachments have far-reaching consequences? Possibly so. In a number of studies, it has been found that infants who are securely attached to their mothers generally are more socially and emotionally competent than infants who are insecurely attached, and these advantages persist well into childhood (Cassidy and Main, 1984; Schneider-Rosen and Cicchetti, 1984; Waters, Wippman, and Sroufe, 1979). Securely attached toddlers have been observed to be more sociable than insecure ones when playing with others their age (Pastor, 1981) or with adult strangers (Main and Weston, 1981; Thompson and Lamb, 1983). They are more cooperative, enthusiastic, persistent, and competent at solving problems than insecurely attached toddlers, more compliant and likelier to have internalized controls—they obey past commands even when no one is watching (Frankel, 1984; Londerville and Main, 1981; Matas, Arend, and Sroufe, 1978). At age 3, children who were securely attached to their mothers as toddlers have been found to attract more positive responses from unfamiliar peers, children with avoidant attachments to elicit less positive responses, and children with ambivalent attachments to provoke more antagonistic and resistant responses (Jacobson and Wille, 1986). In preschool, children who were ambivalent and resistant as infants have been described by their teachers as impulsive and tense or as helpless and fearful, whereas those who were avoidant have been described as either hostile or restrained and socially isolated from other children (Arend et al., 1979; Sroufe, 1983). In kindergarten and first grade, children who were securely attached to both parents are more trusting and open with an unfamiliar woman (Weston and Richardson, 1985) and have fewer psychological problems (Lewis, Feiring, McGuffog, and Jaskir, 1984).

But attachment is just one of a number of factors—including stressful events and family characteristics—that affect the course of children's social and psychological development (Lewis et al., 1984). The link between attachment and later competence probably appears consistently not only because secure attachment gives a good emotional start to infants' development but also because it reflects infants' competence and because the infants observed in most studies were growing up in relatively stable circumstances, which foster the development of secure attachments and other forms of competence as well (Lamb et al., 1985). A secure attachment to one's parents does not guarantee later social success in the world; an insecure attachment does not condemn the child to later psychopathology. Attachment is a significant part of a whole constellation of social competencies, which can be modified over time.

Fathers' behavior

Once upon a time, developmental psychologists focused exclusively on infants' relationships with their mothers. Mothers were central in the theories of development and responsible for virtually all the daily infant care. But recently researchers have trained their focus on fathers, too. Their work tells something about how fathers take care of, enjoy, and think about their infants, and it suggests that a father plays a special role in his infant's development. Mothers and fathers, it seems, typically offer their children somewhat different kinds of experiences.

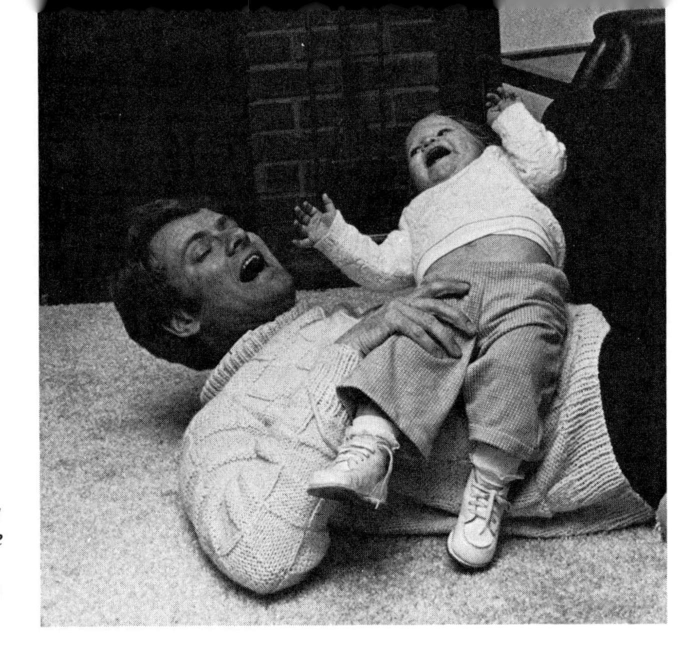

Fathers are more likely than mothers to engage in rough-and-tumble play with their infants and young children.

> In our family, when Ron comes home from work, the kids squeal with delight. He gets down on the floor with them and roughhouses; he tosses the baby in the air; he gives them horsey rides on his back. Then he says, "Okay, kids, that's it," and he hands the baby back to me and shoos Natalie and her sister off to their toys. When *I* come home from work, the kids say, "Feed me, feed me."

Most new fathers are thrilled, absorbed, and preoccupied with their new babies (Greenberg and Morris, 1974). In their fond eyes, their baby is beautiful and special. They like to hold and touch, gaze and smile, talk and stroke their new baby just as much as new mothers do (Parke and O'Leary, 1976). Fathers are no less sensitive to their baby's cries, sneezes, and coughs than mothers are; they may stop a feeding to soothe, and they are likely to feed a baby just as much milk as new mothers do (Frodi, Lamb, Leavitt, and Donovan, 1978a; Parke and Sawin, 1975). But on the average, fathers do not spend as much time actively "on duty" with their infants as mothers do. American mothers are still the ones who spend the most time and do the lion(ess)'s share of the feeding, bathing, and dressing. Observations of both parents at home with their infants have shown that fathers are more likely to watch television and read, while mothers are actively engaged with their infants (Belsky, Gilstrap, and Rovine, 1984). Over the first year, these differences decrease somewhat. Fathers grow more comfortable with their infants and give them, especially firstborns, more stimulation, and mothers grow less involved in breast feeding and other aspects of physical caretaking.

When fathers do interact with their infants, they play more roughly and physically, doing more tapping and poking, bouncing and tossing, than mothers (Clarke-Stewart, 1978a, 1980; Kotelchuck, 1976; Lamb, 1977; Parke and Sawin, 1975; Yogman, 1977). Mothers are more likely than fathers to use a soft, repetitive kind of talking in their play with the baby. Mothers sing and rhyme; fathers hoist and toss. Few differences have been found in the length of time mothers and fathers play with their infants when the infants are playing happily with their toys. But when the baby's attention and interest wane, researchers have observed, mothers and fathers react differently (Power and Parke, 1983). Mothers tend to take cues from the baby, following the baby's gaze at a particular toy and then playing with that toy. They spend more time watching or holding toys when the baby's interest flags. Fathers do not so consistently follow the baby's gaze, and they often try unsuccessfully to get the baby to play with a toy in which the baby is not interested. When the

lifespan focus *Family development*

Just as each child grows from one tiny cell into a complex being, the family itself grows in complexity and richness as, over time, individual members join it and leave it. When an infant is born, or an adolescent leaves the nest, or the mother gets a job, or the couple gets divorced, or the father remarries, or the grandparents move in, or a grown child returns—the family shrinks or expands. In families, the behavior of each member ripples through the entire structure. The actions of one member are felt by all the others. Few events make such oceanic waves in a family as the arrival of a new baby. Somehow all routines stop, and all relationships change in the wake of this powerful new creature.

> Our marriage was very strong before we had Sam. It's different now, because we're different people. When Sam is going through a difficult phase, we all suffer. We bicker and slam doors. But when he's peaceful, we're a pretty happy group.

> Looking back on it, Jason's birth put strains on what was already a crumbling marriage. I didn't want to think that it was crumbling, so I ignored the signs for a long time. But when Jason was just a tiny baby—and difficult—Charlie began working longer and longer hours. He began missing not just week-night suppers, but important family times—weddings, holidays, birthday parties. He was never there to walk the floor to get Jason to sleep, or to watch him doing the nice, normal kid things like learning to walk and talk, or, later, to talk to the teachers about things that came up at school. I ended up taking care of everything to do with Jason, and finally it was just as easy to ignore Charlie altogether.

For the first six months after a baby is born, a family undergoes many changes—some for better and some for worse (Collins, 1985). In most cases, the husband and wife act less affectionately toward each other and have less free time together; they lose some of the romance from their marriage. Yet despite the waning of romance, they feel a greater sense of partnership and shared caretaking. Couples suffer most in making the transition to parenthood if their expectations of parenthood are unrealistic, if one spouse has job-related problems, or if the division of household chores falls much more heavily on the wife. When husband and wife feel supported and encouraged by their coworkers who are parents, positive feelings about parenthood and marriage increase. Although the new parents go through a sea change after the baby arrives, there are important continuities in the quality of a marriage. "The best marriages seem to stay the best after the arrival of the baby, and the worst marriages stay the worst" (Belsky, in Collins, 1985, p. 44).

The families who seem to adjust best to the birth of an infant are those in which the quality of the marriage stays high before and after the birth and in which the father actively participates in daily care of the infant (Grossman Eichler, and Winikoff, 1980; Goldberg, Michaels, and Lamb, 1985; Heinicke, Diskin, Ramsey-Klee, and Oates, 1983).

baby's interest in toys flags, fathers begin to play physically with the baby. Not surprisingly, babies react to their mothers and fathers somewhat differently—frowning or giggling at fathers' poking and tossing, gazing and smiling at mothers' soft sounds (Clarke-Stewart, 1978a; Lamb, 1977; Yogman, 1977).

Are the differences between mothers' and fathers' behavior with their babies and the parallel differences in the babies' behavior the results of learning or biology? To examine this question, one researcher (Field, 1978) compared the behavior of fathers who were the primary caretakers for their babies with the behavior of fathers who were secondary caretakers. Primary caretakers presumably would have had more chances to learn how to interact with their infants. No matter which their role, fathers poked and played physically more often than mothers. Fathers probably are *both* biologically and culturally primed to play physically with their babies. Not only human fathers, but monkey fathers, too, like to roughhouse with their offspring (Suomi, 1977). In Sweden, where it is more common for fathers to be primary caretakers for their babies than it is here, researchers have observed that mothers and fathers play similarly with their babies, but the mothers hold, pick up, and tend to the babies' needs more than the fathers do, no matter whether the fathers are primary caregivers or not (Lamb, Frodi, Hwang, Frodi, and Steinberg, 1982).

Today, in our society, the range of fathers' behavior toward their infants is wide. Fathers can be caretakers or teachers, playmates or strangers. Attitudes toward fatherhood are changing, as the shape of the traditional family—working father, housewife, and their children—changes to encompass working mothers and single fathers. These changes have given some fathers an added incentive to be involved with their children. But fathers lack a single clear model of how they are to act and have no set biological programs to guide them in their caretaking activities. What is more, fathers have not been uniformly socialized, as most mothers have been since they were young, to play with dolls and look after babies. As a consequence, fathers vary widely in the extent and content of their activities with their infants. Although, on the average, mothers are more involved in caring for their infants than fathers are, many fathers today are deeply involved in infant care.

This kind of involved fathering, not surprisingly, improves the family relationships. The more the father is involved with the infant, and the more sensitive he is to the infant's needs, the more likely the relationship between father and child is to be joyful and secure, and, it has been found, the more likely the *mother* is to talk and play with the infant (Clarke-Stewart, 1980; Easterbrooks and Goldberg, 1984).

Summary

1. In the first hour after birth, when an infant is alert and gazing and the mother's receptivity is heightened by high hormone levels, the mother begins to form emotional ties to the infant. Mothers who miss this sensitive period for bonding have many opportunities later on to grow emotionally attached to their infants.
2. Infants communicate with those around them with their appealing babyness, gazes, vocalizations, smiles, laughter, and cries. Mothers interpret babies' expressions and gestures as communication, even when the baby is not intentionally sending messages. Eventually, the infant does learn to communicate and becomes a truly interactive partner in these social exchanges. Mothers and infants take turns and play "duets" in feeding, vocalizing, and playing games.
3. In their interactions with adults, infants begin to learn their culture's rules about when and where they may display emotions. They learn to smile often and to control expressions of distress. Infants also respond to other people's expressions of emotions and use them to guide their own behavior.

4. The newborn is physiologically primed to pay attention to the kinds of stimulation provided by parents' faces—the contrasts between skin and hair, the moving eyes and red mouth. In turn, parents project their own feelings and rhythms onto the infant and act *as though* those actions were social. So begins the interactive ''system'' of parents and child.
5. Some mothers and infants mesh well, others not so well. Mothers may be overstimulating, anxious, insecure, depressed, even mentally ill. Some babies are temperamentally unresponsive, difficult, irritable, and rejecting of mothers' attempts at social communication.
6. Temperament is generally stable over time, although the course of any infant's development depends to some extent on the fit of his or her temperament with parents' expectations and behavior.
7. The most important social and emotional development of an infant's first year is the formation of a lasting, loving tie with the person he or she interacts with the most. This attachment develops in the second six months of life. By observing infants and mothers in the ''strange situation,'' psychologists have found that infants are either securely or insecurely attached to their mothers. Insecure infants avoid contact with their mothers or act ambivalently toward them. Mothers who are loving, responsive, and sensitive are likely to offer the infant appropriate levels of stimulation and to foster secure attachments in their infants. Mothers who are tense and angry are likely to foster avoidance in their infants.
8. Secure attachments are likely to flourish when a baby has one regular caretaker who gives loving and sensitive attention and who keeps long separations to a minimum.
9. Securely attached infants typically become children who are more socially and emotionally competent than those who are insecurely attached.
10. Mothers do most of the physical care of infants. Fathers are more likely to play and roughhouse. There are, of course, marked individual differences in how fathers behave with their infants.

Key terms

sensitive period
bonding
rooming in
endogenous smiles
exogenous smiles
stranger anxiety (fear of strangers)
social referencing
microanalysis
pseudodialogues
attachment
set goal
strange situation
secure attachment
insecure-avoidant attachment
insecure-ambivalent attachment

Suggested readings

AINSWORTH, MARY D. S., BLEHAR, MARY C., WATERS, EVERETT, and WALL, SALLY. *Patterns of Attachment: a Psychological Study of the Strange Situation.* Hillsdale, N.J.: Lawrence Erlbaum Associates, 1978. Details Ainsworth's procedure for studying qualitative differences in infant–mother attachment and reviews the results of research using the strange situation assessment procedure.

BOWLBY, JOHN. *Attachment and Loss*, Vol. 1, *Attachment.* New York: Basic Books, 1969. The first of Bowlby's influential trilogy on the theory of attachment; a seminal book that inspired a whole line of research on children's early social relations with their mothers.

CHESS, STELLA, and THOMAS, ALEXANDER. *Your Child Is a Person.* New York: Viking Press, 1965. An excellent book integrating the notion of temperamental differences in children with practical advice on child rearing.

KAYE, KENNETH. *The Mental and Social Life of Babies: How Parents Create Persons.* Chicago: University of Chicago Press, 1982. An original theory of the roles of mothers and infants in their early face-to-face interactions.

KLAUS, MARSHALL H., and KENNELL, JOHN H. *Parent-Infant Bonding*. St. Louis: C. V. Mosby, 1982. Support for the critical period explanation of maternal bonding, gathered by its chief proponents from animal and human studies.

LAMB, MICHAEL, THOMPSON, ROSS, GARDNER, WILLIAM, and CHARNOV, ERIC. *Infant-Mother Attachment*. Hillsdale, N.J.: Lawrence Erlbaum Associates, 1985. A comprehensive review of the research on the development of infants' relationships with their mothers.

PARKE, ROSS D. *Fathers*. Cambridge, Mass.: Harvard University Press, 1977. A comprehensive, readable presentation that reviews the research on the roles of the father, how they differ from those of the mother, and what effects they have on children's development.

SCHAFFER, H. RUDOLPH. *Mothering*. Cambridge, Mass.: Harvard University Press, 1977. A lucid review of psychological research on the different roles that an infant's mother fills and how they relate to the infant's development.

SLUCKIN, WLADYSLAW, HERBERT, MARTIN, and SLUCKIN, ALICE. *Maternal Bonding*. Oxford: Basil Blackwell, 1983. A brief but balanced discussion of the issues and research on the formation of emotional bonds between mothers and their infants.

PART FOUR

Early Childhood

PHYSICAL GROWTH
Growth Patterns
Changing Size and Proportions
FACTORS AFFECTING GROWTH
Genetic and Inborn Factors
Malnutrition and Neglect
BRAIN GROWTH
Brain Structure and Organization
Brain and Behavior
Brain Lateralization
PHYSICAL ABILITIES AND CHANGES
Child Abuse
Lifespan Focus: Child Abuse—a Vicious Cycle

chapter eight
Physical development

Physical growth

In Chapter 5, we saw the marvelous changes in size and motor abilities that take place during infancy. At the end of that discussion, we left the infant of about 2 toddling around, getting into things, having stretched to some 34 inches and 27 pounds, and sleeping about 12 hours a day. In this chapter, we look at the physical growth and development that take place in early childhood, the ages from 2 to 6 years.

It is with mixed feelings that parents watch their infants and children put on inches and pounds. Parents take pride in these outward signs of their children's robust good health. They take satisfaction in knowing that all those proddings—"Finish your milk before you go out to play, dear," "But broccoli's *good* for you," "I don't *care* if George can have cookies for breakfast. In this house . . ."—have paid off. But parents also sigh as they set off once again to buy shoes for toes that seem always to be poking through the front of the sneakers and jeans to cover ankles that seem always to be showing. And they sigh again when their armful of an infant has so quickly grown into more than a comfortable lapful of young girl or boy. Between the ages of 2 and 6, the average American child grows 3 inches and gains 4½ pounds a year. The average 5-year-old weighs 40 pounds and stands 43 inches tall. That's a long way from the newborn of 7 pounds and 20 inches. But the visible does-a-parent's-heart-good growth of a child in inches and pounds masks complex and varied patterns of cell and tissue growth.

Growth patterns

Within the child's body, growth creates a state of constant change. Nutrients are continually ingested, broken down into their components, and used for fuel. Some cells divide, others stop dividing and change in size, and still others stay relatively stable. At any time during infancy and early childhood, one part or organ of the body is likely to be growing faster than another. This **asynchrony** is typical of how human beings grow both before and after birth. Vital organs grow at different rates because their cells divide and grow at different rates. The physical growth that we notice—the longer arms sticking out of the sleeves that were just the right length a few months ago, the broader rib cage stretching taut the sweater that was loose just last season—is the result of microscopically small changes in body cells.

Cells grow in one of three ways. Some cells, such as those that line the gut and make up skin and blood, continuously die and are replaced. Other cells, such as those in glands and parts of the liver and kidneys, live for long periods. The third kind of cells, such as those in nerves, muscles, and fat, form only during certain periods of growth. Most of the nerve cells in the central nervous system have formed by the 18th week of prenatal life. The newborn has all of the muscle cells it will ever have. Fat cells generally multiply only in the weeks before birth, during the first two years of life, and in early adolescence (Hirsch, 1975).

All this activity, with cells asynchronously dividing and multiplying, goes on *inside*. Outside, to the pediatrician who weighs and measures infant and child, the process of growth proceeds gradually and, for the most part, regularly. A **growth curve** is a record of a child's pattern of physical change over time. If the curve of a child's growth from birth to maturity is drawn on a graph, it shows how far and fast the child has grown. The two curves in Figure 8.1 are the earliest known growth curves recorded. They plot the growth in height of an 18th-century French nobleman's son from the time he was 6 months old until he was 18. The upper curve, which shows his *height* at different ages, rises gradually, as de Montbeillard's son grew taller. The lower

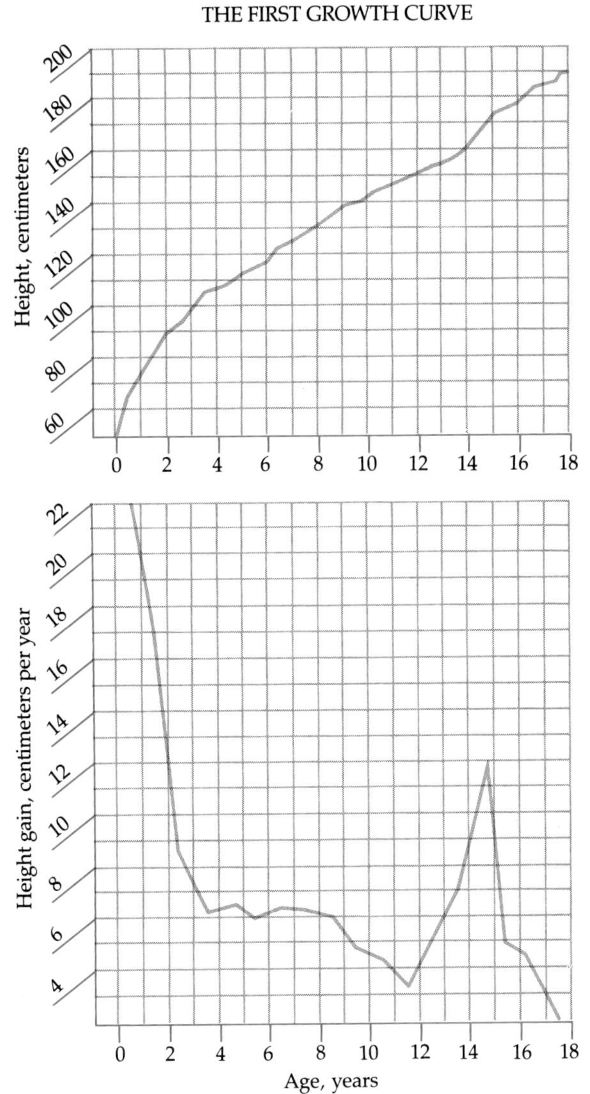

Figure 8.1 ***The growth curve at the top traces the increasing height of de Montbeillard's son, who turned 18 in the year 1777. The bottom growth curve shows that his rate of growth decreased, except for a spurt in adolescence (Tanner, 1978). (10 centimeters = 3.9 inches)***

curve, which shows *changes* in height from one age to the next, declines over time, as the boy's *rate* of growth slowed down after infancy. Then it rises sharply, as the boy reached the growth spurt of early adolescence. Had the boy's rate of growth been plotted from conception, the slowdown would have begun in the fourth month of prenatal development (Tanner, 1978).

Growth curves can be plotted for groups of children as well as for individuals like de Montbeillard's son. Although individual children grow taller or heavier at slightly different rates, plotting these average curves for large groups of children can help in identifying children whose growth is markedly abnormal (Figure 8.2). It is not uncommon for parents to talk about their children in terms of their places on growth curves for height and weight.

> "Tim is 25th percentile in weight and 90th in height—tall and skinny like his father."
>
> "Denny is a pip-squeak—10th percentile height and weight."

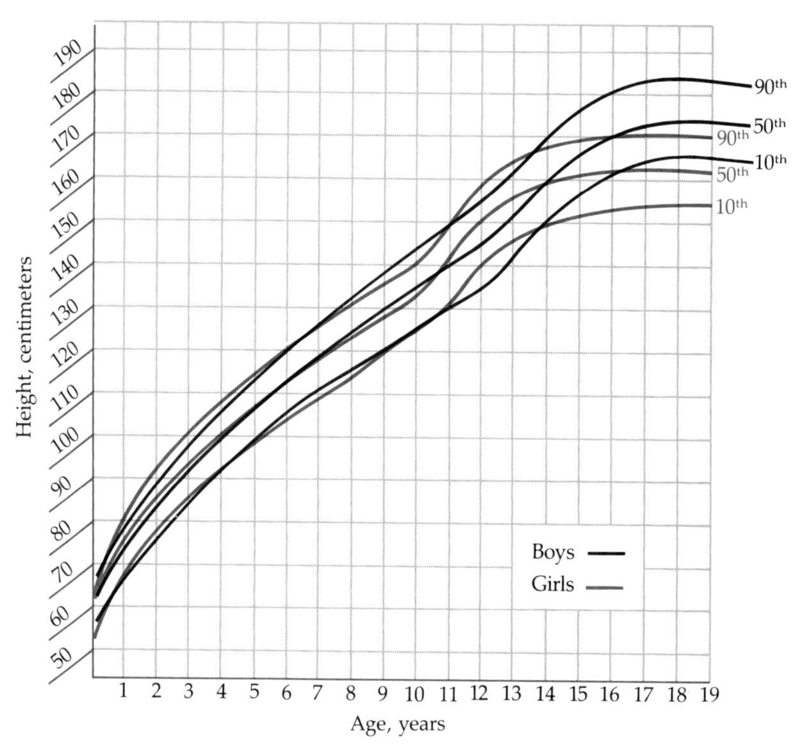

Figure 8.2 ***The darker lines trace the norms of height from birth to 19 years of age for tall American boys, those in the 90th percentile (taller than 90 percent of their contemporaries); average American boys (50th percentile); and short American boys (10th percentile). The lighter lines show the norms of height for tall, average, and short American girls from birth to 19 years of age. (10 centimeters = 3.9 inches)***

A child whose growth departs radically from these growth norms may have a genetic problem or may be getting poor care. A child who is below the third percentile in height, for example, is so small that his or her condition needs investigating. Of course, if the assessment of abnormality is to be meaningful, the norms themselves must be based on a broad and up-to-date cross section of the population. In countries like the United States, which are made up of many different ethnic and racial groups, separate growth norms for various groups are needed.

If children from genetically small populations—such as the Vietnamese and Cambodian—are measured against norms set for children from genetically large populations—such as Scandinavian and Afro-American—they may be singled out as abnormally short when they are actually growing normally. Similarly, if children from groups of large people are measured against norms set for children from groups of small people, they may be judged normal when they are actually growing abnormally slowly because of illness or malnutrition. It is also possible that a whole subgroup of the population is not growing at a normal rate because of a dietary deficiency. Only if the growth of this whole subgroup is measured against a national or international sample will the growth problem of an entire subgroup show up.

Changing size and proportions

Children grow faster during their first few years than they will at any other time in their lives, except during the growth spurt of adolescence (see Table 8.1). They gain 15 pounds and 10 inches in the first year, 5 pounds and 5 inches in the second year, and 4½ pounds and 4 inches in the third year. They grow faster both in absolute terms—inches—and in relative terms—proportion of baby height.

Table 8.1
Expected increase in height per year

Age, years	Growth, inches	
0–1	9.8	
1–2	4.8	
2–3	3.1	
3–4	2.9	
4–5	2.7	
5–6	2.7	
6–7	2.4	
7–8	2.2	
8–9	2.2	
9–10	2.0	
10–11	1.9 (boys)	2.3 (girls)
11–12	1.9 (boys)	2.5 (girls)
12–13	2.2 (boys)	2.6 (girls)
13–14	2.8 (boys)	1.9 (girls)

SOURCE: Lowrey, 1978.

> Sam grew so fast in his first few years that I used to joke that I had to keep the car running so that I could dash out fast enough to buy him bigger clothes.

The 3-year-old, who grows 3 inches, is gaining a greater proportion—one-twelfth compared to one-nineteenth—of her height than the 10-year-old who grows 3 inches. But her growth is slowing down. If it continued at the same rate as in the first year, the toddling 2-year-old would be a mammoth 63 pounds—much too heavy for mother to lift into the shopping cart.

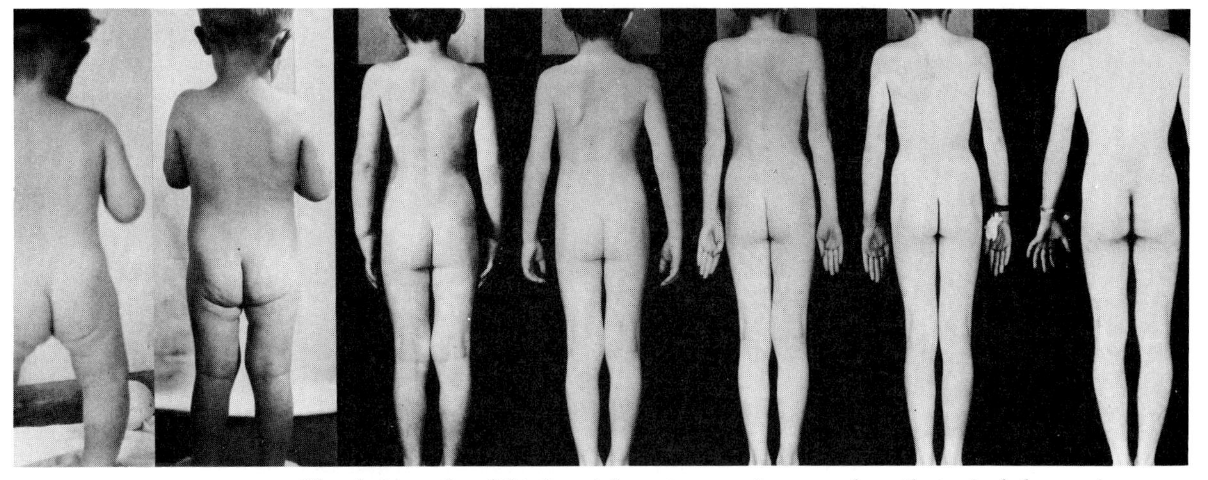

The photographs of this boy, taken at successive ages, show the typical changes in proportions from infancy to maturity. His head, which was half his body at 2 months after conception, makes up less and less of his body size. The ages, left to right, are 15 months, 30 months and 6, 7, 11, 14, and 18 years.

Because growth of different body parts is uneven, as children grow they change in physical proportions as well as size. The newborn's head is fully one-fourth of its body size, but its legs are short—just a bit longer than that big head—and its feet are proportionally the tiniest part of its body. After birth, the baby's head continues to grow most rapidly. Toward the end of the first year, muscle and bone growth quickens, and the baby's trunk grows fastest of all. After the first year, the baby's arms and legs grow more quickly than the body. With their longer limbs and increased mobility, most 2- and 3-year-olds lose much of their baby fat (and pot bellies). The chubby infant usually thins down to the slimmer, more active toddler, who may have a much smaller appetite and very definite ideas about what is good to eat. By age 2, a boy is about half his adult height, a girl slightly more than half her adult height (Lowrey, 1973). By age 6, children are about 70 percent of their adult height.

Many children, especially those whose parents are very different in size, do not fall neatly into predictable slots, however. There are marked individual differences in growth patterns, height, and weight.

> Sam was so regularly off the growth charts that we worried he'd grow up gigantic. But then he slowed down, thank goodness. Now the doctor tells us that he'll probably top out at about 6 feet 3 inches.

> Erica has stayed proportionally pretty much the same since she was a few months old—30th percentile for height and 80th percentile for weight. We figure she'll keep on being short and solid.

Factors affecting growth

Genetic and inborn factors

Normal physical growth, as we mentioned in Chapter 5, is largely under genetic control. The brain's pituitary gland follows genetic instructions that cause it to produce growth hormone and other hormones that affect growth. We know from longitudinal studies like the Berkeley Growth Study that parents and their children are likely to follow similar patterns in their rates of growth, although they may differ slightly in their final sizes (Tanner, 1970).

In rare cases—about 1 child in 1000—the instructions that control growth go awry, and the children grow extremely short or extremely tall. **Dwarfism** and **giantism** may be genetic in origin or the result of abnormalities in prenatal development.

A particular kind of genetic influence over growth shows up in the differences between males and females. From early on in prenatal development, there are physical differences between boys and girls. Newborn girls' skeletal systems are several weeks more mature than newborn boys'. Their permanent teeth come in sooner than boys' do. By adolescence, girls' skeletal systems are, on average, three years more mature than boys'.

As a result of this difference, girls tend to excel at different sorts of motor activities from boys. Girls are likely to do better at fine motor skills like writing, drawing, and buttoning, and they do better than boys when the fine motor tasks require quickness.

> Natalie is very dextrous. At 2, she could button her sweater and buckle her belt. At 3, she could tie her shoelaces, comb her hair, and zip her jacket. She's always liked toys with little pieces and parts—toy phones with little holes in the dial, spelling games with lots of keys to press, beads and buttons to string for necklaces.

At age 4 or 5, girls may also skip and run with greater coordination than boys. But not all differences favor girls. For their first seven years, boys are slightly taller and more muscular than girls. Their baby fat disappears sooner than girls', and they are a bit leaner throughout childhood. Boys throw, bat, and hit better than girls. Some boys do better at gross motor skills, especially those that use the muscles of the upper body. In the schoolyard, while the girls are jumping rope and playing kickball, the boys are playing hockey and baseball.

> Sam spent lots of time at nursery school building towers and castles with sturdy wooden blocks, clambering around a jungle gym, running toy ''hepticopters'' and ''hairplanes''—and asking when he'd be old enough to ice skate, play hockey, and be as strong as He-Man.

Whatever the inborn differences between them, boys and girls differ in the amounts of experience they have with particular physical activities. These experiences may enhance inborn differences—with boys climbing, pedaling toy cars and tricycles, throwing and batting balls; and girls turning the pages of books, writing, and painting—but not always. Differences in group averages do not tell about the skills of individual girls and individual boys. Many boys excel at painting and writing, and many girls are very adept at pitching balls and riding bicycles. Even the group differences in abilities between boys and girls tend to be quite small. In a study comparing kindergarten boys and girls in running and jumping, for example, 50 percent of the boys and 45 percent of the girls could do a standing jump of 35 inches and dash 400 feet in 50 seconds (Milne, Seefeldt, and Reuschlein, 1976). Although the averages for the boys were slightly ahead of those for the girls, in each class there were some girls who could run faster and jump farther than most of the boys in the class.

Malnutrition and neglect

Without nutrients, children's bodies cannot grow. The earlier and the more severe their malnutrition, the worse the prospects for their normal development. One kind of severe malnutrition that children between 2 and 4 years may suffer from is a protein deficiency called **kwashiorkor**. With bloated bellies, thin and colorless hair, skin lesions, these children are apathetic, withdrawn, and irritable. Their motor skills are poor, and their intellects are impaired (Thomson and Pollitt, 1977). Malnutrition also lowers children's

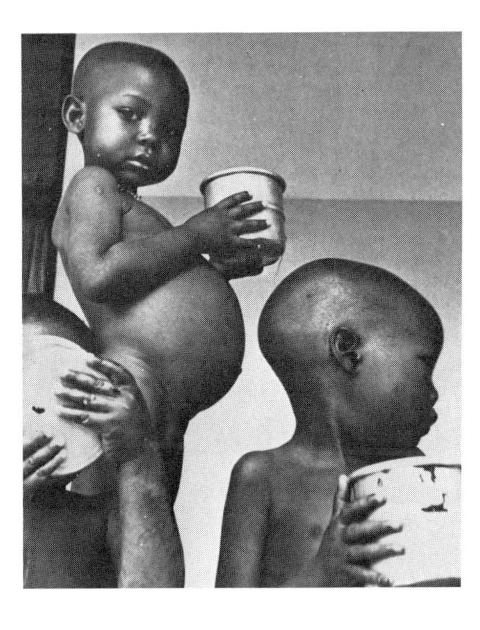

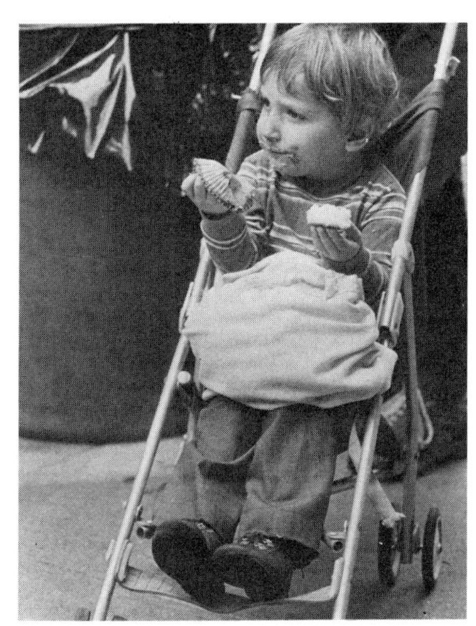

Malnutrition is a worldwide problem. In Africa, many children live on too few calories and too little protein. In the United States, many children live on junk food and ''empty'' calories.

resistance to disease. They are less able to resist viruses, bacteria, and fungi than other children (Edelman, 1977; Edelman, Suskind, Sirisinha, and Olson, 1973). Because they are susceptible to diarrhea, malnourished children may not be able to use what little food they take in. The poor sanitation of a poverty-stricken environment means that malnourished children are exposed to many infectious diseases.

A survey of Asian, African, and South American children found 80 percent of them to be suffering from moderate to severe malnutrition. Estimates are that one hundred million children under 5 years of age suffer from malnutrition (Suskind, 1977). In the United States, surveys have found chronically poor nutrition in 20 percent of the children under 6 years and in 30 percent of the children from poor families. Many preschool-aged children do not get enough iron, and nearly half of them get less vitamin A and vitamin C than is recommended (National Center for Health Statistics, 1975). In the United States, a study of children between 6 and 11 showed that children from wealthy families were 1.2 inches taller than children from poor families (Goldstein, 1971). In contrast, a study of children in Swedish cities showed no connection between children's height and family income, presumably because in Sweden poor and rich alike have access to food and social services, and the link between poverty and malnutrition is broken (Lindgren, 1976).

In this country, where shelves are stocked with prepared and packaged foods, even parents who have money to buy their children enough food face the problem of checking food additives. Too much sugar, salt, or chemical preservatives? Too few minerals and vitamins in processed foods? Too many junk foods—lurid orange puffed cheese, greasy chips, sugary soda and cookies—competing for the preschool-aged child's already limited appetite? As we've seen, once they are past the extremely rapid growth of infancy, children need fewer calories per pound, and so their appetites decrease. Feeding preschoolers can be a problem even for affluent parents.

Severe emotional stress is another factor that affects children's growth. Unloving care or extreme neglect can cause a child to stop growing, even if the child is fed adequately. The child's pituitary gland stops producing enough growth hormone for normal physical development. The child is short and immature in bone development. A child of 6 might have the bone age of a 3½-year-old, for example. In one strife-torn family, the mother of infant twins unhappily found herself pregnant again when the twins were 4 months old (Gardner, 1972). Their father lost his job, fought with the mother, and finally left the family. The mother took out her rage on one of the twins—the boy. Although his twin sister continued to grow normally, the boy did not. When he was examined at 13 months, he was the size of an infant six months younger. He was placed in the hospital for treatment and began to grow normally. After his father returned to the family, he was discharged from the hospital, and when the twins were examined 2½ years later, both were of nearly normal height. Children in such dire circumstances are often treated by being placed in foster care, itself a dire step. But in their new surroundings, without any hormone therapy, and eating an ordinary diet, the children usually begin to develop and grow more lively and sociable. Children who have been growing only about 1½ inches a year suddenly gain, on the average, 6.3 inches a year (*How Children Grow*, 1972; Powell, Brasel, Raiti, and Blizzard, 1967).

Brain growth

For psychologists, perhaps the most important aspect of physical growth is the growth of the brain. Recent technical improvements in recording the activity and mapping the structure of the brain have made it possible for the first time to peek at how this most important organ develops. Brain development and function depend on several different properties and processes:

- The number, size, and structure of neurons, or nerve cells, in the brain.
- The branching dendrites at the ends of the nerve cells.
- The arcs and connections among the nerve cells.
- The covering of myelin on the nerve cells.
- The location and arrangement of the nerve cells.

Each is important to the development of the brain and behavior.

Brain structure and organization

Neurons first appear in the brain during the second prenatal month and, in the eighth prenatal month, develop branching dendrites, effective **neurotransmitters**, and the ability to metabolize glucose. Neurons continue to develop, grow, and change in all these ways from that point into childhood. First, dendrites increase in length and number of branches. There are individual variations in the rate at which this occurs, and some evidence suggests that stimulation from the environment can increase the growth of dendrites. In one study, for example, kittens were trained to pull back one of their front paws to avoid a mild electric shock (Spinelli, Jensen, and Viana Di Prisco, 1980). Examination under the microscope showed that in the corresponding area of the brain, the dendrites had developed more branches. This kind of branching of dendrites is the most important clue to the functional capacity of the brain. The more branches, the wider the range of behavior possible. Second, during the course of development, the types and strengths of the chemical transmitters between neurons at various sites in the brain change (Axelrod, 1974; Iverson, 1979). These changes make the brain more efficient. Third, glucose metabolism increases until children are 11 years old, which indicates an increase in the brain's functioning (Farkas-Bargeton and Diebler, 1978).

Neurons also grow in physical size. At birth, the infant's brain contains all the neurons it will ever have, but it weighs only about one-fourth of its adult weight. The infant's brain must be small to allow the head to pass through the birth canal. After birth, nerve cells increase in size, and other, supporting cells called **neuroglia** continue to multiply until one or two years after birth (Tanner, 1978). The brain doubles in volume during the first year and reaches about 90 percent of its adult size by age 3 (Trevarthan, 1983). During childhood and into adulthood, myelin, the white fatty substance that improves the transmission of signals along the neurons, is deposited around the axons of these cells. Myelination also contributes to the increased size and weight of the brain and to changes in its appearance. In the first six years, the brain's surface area increases, and **fissures** develop as the neurons enlarge and are myelinated. The substances and structures *within* the neurons also develop gradually during childhood. Neurons include a cell nucleus and, surrounding the nucleus, **Nissl substance**. This substance often appears at the same time as a new behavioral function appears. Finally, neurons in the brain congregate into functional groups.

In sum, brain maturation is a system of closely coordinated developments at many different levels: the growth and formation of brain cells, myelination, complexity of branching dendrites, and the connections between cells. But brain development is not all growth. Brain regions originally develop an excess of cells. In the course of normal development, the extra cells die off and the extra connections between cells are reduced. After the first two years of life, unnecessary branches are discarded and replaced by more efficient ones (Huttenlocher, 1979; Mark, 1974). Thus, brain development also includes the elimination of redundant or inefficient pathways.

Brain and behavior

Increases in brain size, complexity, and myelination are related to advances in children's thinking, acting, and talking. As myelin forms in the brainstem before and soon after birth, the infant develops the ability to babble, for example (Lecours, 1975). During a second cycle of myelination, which lasts from soon after birth until a child is between 3½ and 4½, the child's earliest actual speech develops. Areas in the cortex, which control the muscles for producing speech and the neurons involved in hearing and understanding speech, mature and are myelinated. And from infancy until the age of 15 or so there is a third cycle of myelination, in the association areas of the cortex; cognitive functions develop during this cycle (Bay, 1975; Yacovlev and Lecours, 1967). Unfortunately, although it has been suggested—and seems sensible—that these changes in brain maturation are *causing* the advances in language and cognitive development that occur at the same time, it has not been proved. Research shows only that the changes in the brain's structure and the development of abilities occur at the same time (Pappas, 1983). It may be that the behavioral changes are causing changes in the brain. We have not yet been able to trace the links from physiology to psychology, from brain to behavior, from nerve cells to thoughts.

Brain lateralization

The cortex is divided into two halves, or hemispheres, connected by a tough band of myelinated tissue called the **corpus callosum**. Each hemisphere has its own specialized functions, a characteristic called **lateralization**. For example, muscles on one side of the body are controlled by areas of the brain in the opposite hemisphere. Although the hemispheres seem physically symmetrical, closer analysis shows that they are not completely so. The area governing language is larger in one hemisphere than in the other, for example. In most people, electrical activity is greater in the left than the right hemisphere when they listen to spoken sounds, and it is greater in the right cortex when they listen to music. This division is related to processing and producing language and music. In right-handed people, language is usually governed by the left hemisphere, and music and nonspeech sounds, emotions, and spatial abilities like map reading and figure drawing by the right hemisphere. Studies have shown, for example, that people whose right hemisphere has been damaged show inappropriate emotional responses, misperceive other people's emotions, and cannot draw or build a model from a plan (Geschwind, 1979; Kimura, 1975). In most adults, the left hemisphere is more efficient at logic, sequential tasks, and the processing of rapidly changing stimuli, and the right is more efficient at spatial processing, emotion, and intuition (Ornstein, 1978).

But what about children? In one study, researchers simultaneously played different sentences into children's two ears to see which ear they favored for processing sentences with emotional content and sentences with just verbal content (Saxby and Bryden, 1984). Even 5-year-old children favored the right hemisphere for processing emotional material and the left hemisphere for processing verbal material. In another study, researchers found that infants showed greater electrical activity in the left hemisphere when they heard words that they knew (Molfese and Molfese, 1985). Some scientists now suggest that lateralization develops before birth (Levine, 1983). However, research on children who suffer brain injuries in the first two years of life suggests that there is considerable plasticity in the development of lateralization over this period.

Physical abilities and changes

Between the ages of 2 and 6 years, children shed much of their babyishness. Physically skilled and coordinated children emerge—good jumpers, runners, pitchers, catchers. Curious, energetic, resourceful, and eager, they dance, prance, climb, run, push, dart, throw, stack, and fall down in a heap—only to be back up within seconds dancing and prancing. All this play exercises and develops muscles, senses, and coordination. Large motor skills improve in the preschool years. In general, the 2-year-old who falls and bumps things emerges as the 6-year-old who can carry dish and cup without mishap. The 6-year-old is thinner, stronger, taller, and better coordinated than the 2-year-old. The 6-year-old not only may be able to pedal a tricycle, catch a ball, and climb a ladder, but may do so gracefully. These skills develop according to a maturational timetable, helped along by experience. The balance of the 6-year-old allows for riding bicycles, skiing, skating, and skateboarding.

Preschoolers also develop their fine motor and perceptual skills. The average 5- or 6-year-old can both hold a pencil and focus on a line of print. (Before this age, children tend to focus more easily on objects at a distance than up close.) To most children, the worlds of reading and writing now open for the first time. Of course, children vary widely in their fine motor skills. One 4-year-old can tie her shoelaces and cut her food with the side of her fork, but another cannot even manipulate the large pieces of a puzzle or string chunky beads. In general, preschool children—with their chubby fingers and farsighted eyes—are more adept at large than fine motor skills. Much of their play—stacking blocks, holding pencil and paintbrush, playing with the dials on telephone, television, radio, and stove—is an exercise in developing these important fine motor skills.

After the age of 2, children's heart and breathing rates slow. Average body temperature falls, and even fevers get lower. The average body tempera-

Preschoolers' play and exercise develop their muscles, senses, and coordination. Both large and fine motor skills refine noticeably.

ture of the 1-year-old is 99.7 degrees Fahrenheit. That of the average 5-year-old is, like adults', 98.6 (Lowrey, 1978). Kindergartners play for longer periods in one spot, sit still longer without fidgeting, need fewer, shorter naps, if any, and even sleep less restlessly (Routh, Schroeder, and O'Tuama, 1974). Their physical systems "tune up" and function more smoothly and efficiently. A number of the small physical nuisances of the early years begin to disappear. Most children who wet the bed at 2, 3, and 4 years are dry at 5 or 6, their bladder capacity having enlarged or their sleep patterns now allowing them to awaken in response to a full bladder. During the day, too, the children need to make fewer trips to the bathroom, and there are fewer "accidents." Sensitivities to foods, dust, and other allergens seem to die down in many children. As middle and inner ears grow larger, ear infections wane. As windpipes grow larger, too, upper respiratory illnesses are less frequent. Children who as toddlers always seemed to have flu, colds, throat and ear infections as preschoolers stay healthy for longer spells. Those who as infants were colicky or subject to stomachaches now digest more smoothly.

One winter, Kristy stayed home from nursery school every day except three between February 3 and March 22 because she had ear or strep infections. But in kindergarten, she missed only six days all winter. What a *relief*.

Child abuse

Unfortunately, sometimes the physical well-being of children is directly threatened by their parents. Each day in this country abusive parents are responsible for the deaths of one or two children. The number of children under 5 killed each year by their own parents may be greater than the number of those who die from disease (Kempe and Helfer, 1972). An estimated 500,000 children are beaten, burned, thrown, kicked, and battered without losing their lives. Exact figures are not known. Since 1967 all 50 states have had laws requiring that suspected cases of **child abuse** be reported; most require doctors, clinic and hospital personnel, teachers, social workers, psychologists, lawyers, police officers, and coroners to do so. In 1974, the U.S. Congress passed the Child Abuse Prevention and Treatment Act. The numbers of cases reported jumped by tens of thousands after these laws were enacted.

Most abusive parents say that they love their children. Few have any specific psychiatric illness. Nevertheless, research suggests that abusive parents share certain characteristics (Bee, Disbrow, Johnson-Crowley, and Barnard, 1981; deLissovoy, 1973; Dibble and Straus, 1980; Kempe and Helfer, 1972). Abusive parents may be impatient, immature, and know little of child development. They may feel personally inadequate and have little self-esteem. They may fear their children, especially when the children act fearful or sexual. They may consider severe physical punishment necessary for their children to behave "properly." They respond angrily to infants' crying and may be violent toward their spouse as well. These parents are plagued with stress yet lack the psychological or social resources for dealing effectively with that stress.

In an estimated one-quarter of the cases of child abuse, qualities in the child trigger the abuse (Gil, 1971). All children put stress on their parents, but overburdened parents are more likely to abuse a particularly troublesome child. We can also compile a profile of the abused child (deLissovoy, 1973; Hoffman-Plotkin and Twentyman, 1984; Main and George, 1985; McCabe, 1984).

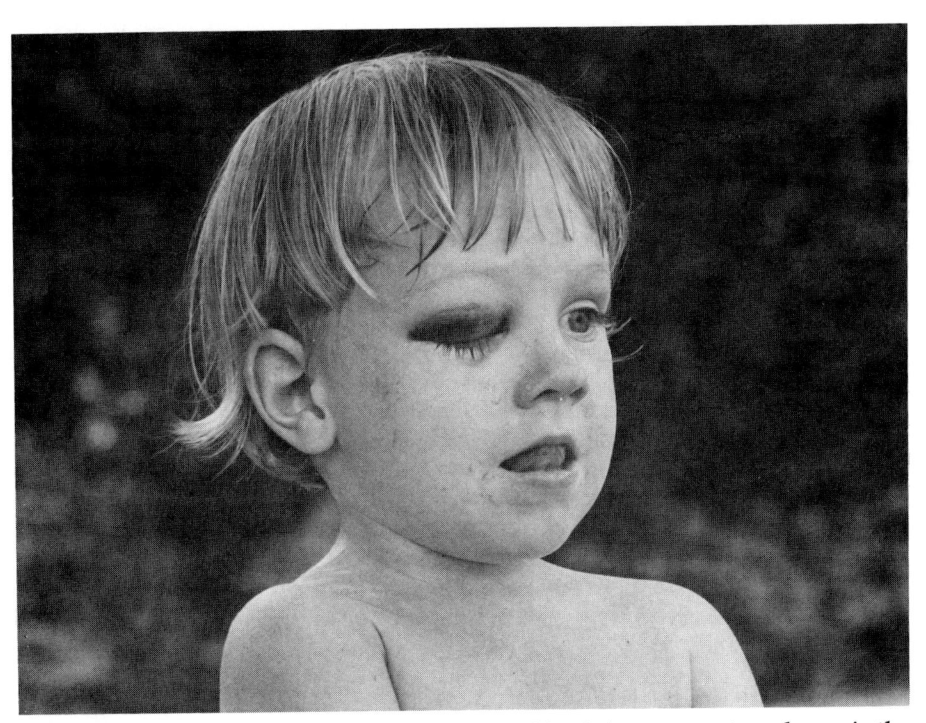

This child is one of 500,000 children who are abused by their own parents each year in the United States.

Abused children often were underweight at birth and therefore as infants cried often, matured slowly, needed special feeding and care, and looked thin and pinched. As toddlers, they often are aggressive, annoying, hyperactive, mentally retarded, learning-disabled, sleep poorly, cry excessively, and vomit frequently. They may be unresponsive, difficult to manage, give unclear social cues, and react poorly to discipline.

Infants and children who demand lots of attention and unusual sensitivity from parents may be abused by parents who lack even the ordinary social and psychological resources. The important question, though, is whether the abused children's physical, psychological, and social shortcomings *result* from the abuse they receive or whether they *invite* that abuse. As with so many of the problems that children have, separating cause from effect is difficult. The answer, however, is probably that these shortcomings are both causes *and* effects of the abuse that children receive. Abuse is part of a vicious cycle set in motion by problems in both parents and children, a vicious cycle that is perpetuated across generations.

To break the vicious cycle of abusive parents raising children who later abuse their own children, parents must receive therapy to raise their low self-esteem and to learn—for the first time—how to take competent care of children. With training, most parents can learn gentler means of discipline and can break out of their self-imposed isolation from family and friends. Parents Anonymous, a self-help organization with chapters in major cities, allows child abusers to admit what they have done, to get social support, and to learn how to stop the cycle of abuse. Luckily, most parents *can* handle the demands of their children even under stress, especially when they have the support of caring family and community members. When they themselves have the necessary physical, psychological, and social resources, parents can raise healthy, thriving children.

lifestyle focus

Child abuse—a vicious cycle

Abusive parents describe their own childhoods as filled with physical punishment and abuse. In a cycle that repeats itself in one generation after another, the abused children often grow up and abuse in turn.

In order to understand and, they hope, ultimately to break the vicious cycle, psychologists have tried to uncover the origin of child abuse. In one recent study, Mary Main and Carol George (George and Main, 1979; Main and Goldwyn, 1984) compared a group of abused children and their mothers with a matched group of nonabused children and their mothers. The abused children, only 1 and 2 years old, already had begun to imitate the destructive ways of their parents. They more often hit, slapped, kicked, and assaulted their playmates than nonabused children. They also assaulted and threatened their caregivers more often than the other children did. They seemed to *harass* their caregivers, to act aggressive without provocation, apparently just to hurt the other person. An abused toddler might suddenly spit at a caregiver, threaten another child with a shovel, or slap another child after having been scolded. Seven out of ten abused children harassed their caregivers, compared to two of the ten nonabused children. The abused children also avoided other people's friendly overtures, by moving away or turning away as another child or a caregiver approached them, or ambivalently approached others—crawling toward someone with head averted and then suddenly turning away. None of the children in the comparison group of nonabused children showed this ambivalent behavior.

Although several of the nonabused toddlers responded to another child's distress with sadness, concern, or empathy, none of the abused toddlers did so. What is more, all but one of the abused toddlers responded more than half of the time to another child's crying with fear, anger, or physical abuse—slapping, hitting, or kicking.

> Martin, a 2-year-old boy who'd been abused, slapped a crying child on the arm. Then he turned away from her, looked at the ground, and said, "Cut it out! Cut it out!" with more and more agitation and vehemence. He then patted the girl on the back, but she didn't like that, and so he backed away, hissing and baring his teeth. He patted her back again, but the patting turned into beating, and he beat the little girl as she screamed (Main and Goldwyn, 1984, p. 207).

What can we learn from these observations of abusive toddlers? We learn that hostility and abuse are passed from one generation to the next, beginning with the aggression evident in early childhood and continuing as the abused toddlers grow up and abuse their own children.

Summary

1. Growth creates a state of constant, asynchronous change in children's bodies. Cells die and are replaced, live for long periods, or form only during certain periods of growth.
2. Growth curves chart the changes in children's height, weight, and other physical signs of maturity. Growth curves plotted for subgroups in the population help in identifying children whose growth is abnormal.
3. Children grow faster in the first three years than at any other time until the adolescent growth spurt. As they change in size, they change in physical proportions, too—from the infant with large head and tiny feet to the 5-year-old with long legs and arms.

4. Normal physical growth is largely under genetic control. Abnormalities in growth, such as dwarfism and giantism, usually are genetic in origin or result from abnormalities in prenatal development. Occasionally, children do not grow because they are severely neglected by their parents. Inborn sex differences account for some differences in physical growth, and these differences are enhanced by differences in boys' and girls' experiences.
5. Nutrients are necessary for children's physical growth. The earlier and more severe a child's malnutrition, the worse the prospects for normal physical growth or full development of intellectual abilities.
6. Brain development and function depend on the number, size, structure, location, covering, arrangement, and branching of nerve cells and the connections among nerve cells.
7. As various areas of the child's brain develop, behavior changes. Although it seems sensible that brain maturation *causes* these concurrent changes—for example, advances in language and cognition—it has not yet been proved.
8. Each half, or hemisphere, of the brain has its own specialized functions, which appear from infancy onward.
9. During early childhood, children grow more competent in a wide range of physical skills, and many of their earlier physical sensitivities and physiological vulnerabilities diminish.
10. Child abuse threatens the physical well-being of many children. Child abuse is part of a vicious cycle set in motion by both child problems (irritability, unattractiveness, aggressiveness) and parent problems (ignorance, impatience, inadequacy). It passes from one generation to the next.

Key terms

asynchrony
growth curve
dwarfism
giantism
kwashiorkor
neurotransmitters
neuroglia
fissures
Nissl substance
corpus callosum
lateralization
child abuse

Suggested readings

FEATHERSTONE, HELEN. *A Difference in the Family: Living with a Disabled Child.* New York: Penguin, 1981. A discussion of the difficulties confronted by the rest of family when there is a handicapped child.

KEMPE, RUTH, and KEMPE, C. HENRY. *The Common Secret: Sexual Abuse of Children and Adolescents.* New York: W. H. Freeman, 1984. A book that helps explain the increase in sexual abuse in this society and sheds new light on the family's role. It is vividly illustrated by case studies.

KEMPE, RUTH, and KEMPE, C. HENRY. *Child Abuse.* Cambridge, Mass.: Harvard University Press, 1985. A discussion of the reasons for all kinds of child abuse and neglect and suggestions for their prevention and treatment.

chapter nine

Cognitive development

Two-year-olds are on the threshold of a whole new phase of thinking and knowing. Soon they will be able to count, to make believe, and to imagine things in ways that take them far beyond the here-and-nowness of infancy. No longer will their imaginations be bound by their immediate perceptions of people and things. Soon they will be talking about dreams, playing superheroes, and pretending that dolls are real. In this chapter, we will look at the cognitive development of young children: at the symbolic thought they show in fantasy play, drawings, and words; at the intuitive thought they show in dreams and animistic beliefs. We will discuss the stages of cognitive development and language development that young children go through and describe their attention and memory capacities. Finally, we will discuss the context within which cognitive development unfolds—the contexts of toys and television, parents and peers.

Symbolic thought

During infancy, as we saw in Chapter 6, children develop the ability to mentally represent objects and actions. In early childhood, they begin to use **symbols** to represent them. This ability to symbolize opens up new worlds for children. They begin to draw, to pretend, and most important, to use language meaningfully. They use symbols to remember things that have happened and to imagine those that never will.

According to Piaget (Piaget and Inhelder, 1969), symbolic thought is the major new cognitive activity in the years from 1 to 4.

Fantasy play

> Kristy at 2½ was always underfoot and as bossy as could be—a difficult combination to take sometimes. When I needed a few minutes off, I wheeled her miniature plastic shopping cart over to her and commanded, "Okay, Kristy. Go shopping." She usually responded by taking the cart all over the kitchen and living room, tossing things in, and naming all her groceries.

Toddlers' pretend play is the earliest evidence of symbolic thinking. Pretending to be a bird, or pretending that a postcard is a car or that a piece of cloth is a pillow are examples of the simplest kind of fantasy play, **symbolic play**. A 2½-year-old is looking at a book showing a picture of a child licking an

In their symbolic play—here, games of "office" and "school"—young children show their understanding of objects and roles.

ice cream cone. She puts her lips to the picture of the ice-cream and says, "Yummy." Young children pretend that they are Mommy or Daddy or Big Bird. They may have an imaginary playmate or make up a new brother or sister.

Pretending, at this age, can be serious business. Children often cannot separate fantasy from reality. Piaget's daughter stood beside him making the sound of bells at the top of her lungs. When asked to move and be quiet, she refused angrily and stated, "Don't. I'm a church." And woe be to anyone who tries to sit where an imaginary playmate has already settled himself.

> Three-and-a-half-year-old Sam told his mother about his doll, Betsy, "She was just talkin' to me." "She was? That's nice. When I was a little girl, sometimes I had trouble knowing whether my dolls could talk and whether they really couldn't. Do you have that trouble?" "No," Sam announced firmly, "Betsy can talk."

According to Erik Erikson (1968), children use their symbolic play to work through fears and frustrations, relive unpleasant events, or carry out in play acts forbidden in reality. One child pretends to cook at a make-believe stove after being scolded for going near the real stove. Another "shaves" with an empty razor. Another guides her doll through a reenactment of a scene from the previous day when she had cut her lip, telling the doll not to be upset, because her lip would be all right.

Fantasy play lets children feel in control of situations that threaten to overwhelm them. In play they can push the mother down the stairs, drown the baby in the bathtub, and make the bedroom unsafe for monsters.

Between the ages of 1 and 4, the nature of children's fantasy play changes. They progress from symbolic play with objects, to pretend play in which they act like a grown-up, and finally to dramatic play with roles announced ahead of time—"Let's play. . ." (Field, DeStefano, and Koewler, 1982). The 2½-year-old sticks a miniature plastic man in the driver's seat and loads the truck with miniature pieces of furniture. The 3½-year-old puts on a hat that turns him into a "moving man" who uses his toy hammer and screwdriver to move the "family" into their cardboard box "new house." The 4½-year-old says, "Let's play 'moving man.' I'm the driver. Peter, you're my helper. Sarah, you're the mother. George, you're the baby. This box is the suitcase we have to move into the new house. This cup is your baby bottle, George. . . ."

Drawing

Children's drawings, which festoon the walls of nursery schools, bedrooms, and kitchens, also reflect the development of symbolic thought. The earliest drawings are scribblings that reveal children's pleasure in putting crayon to paper and watching marks appear. Scribblers cover the paper with marks that are random and repetitive, first with straight lines, then curves and spirals (Kellogg, 1970). But soon young children are explaining what their pictures show. "This is a house and a mommy and a daddy going for a walk," said one 3-year-old of his page of large and small lines.

Children's drawings of figures, at first, leave out parts of the body—the torso, for instance—but include others—head, eyes, legs, arms, hair, and hands. The "tadpole" drawing (see Figure 9.1) is an example. It shows how a 4-year-old commonly draws a person: a round body—which looks like a big face—out of which stick two vertical lines for legs and two horizontal lines for arms. According to Piaget, these incomplete drawings reflect young children's incomplete and global mental images. Other developmental psychologists have suggested that children omit details because their memories or their

"TADPOLE" DRAWINGS

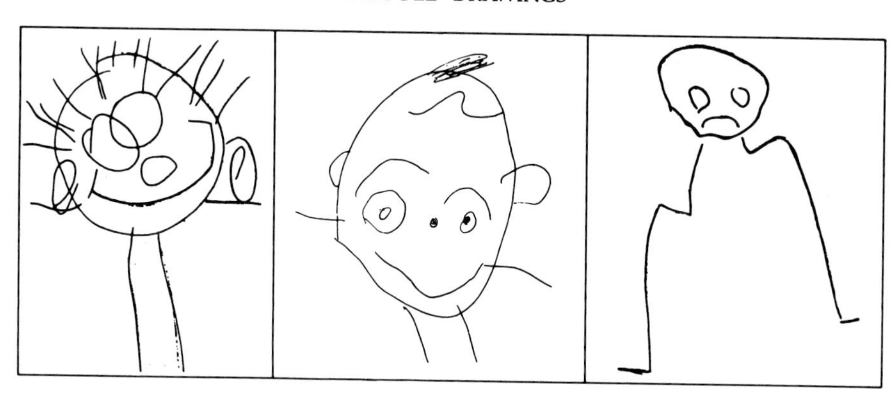

Figure 9.1 ***Four-year-old children's drawings of people usually consist of a large head or face, to which they attach sticks or other thin lines for arms or legs (Golomb, 1974). Why do they leave out the body?***

physical abilities are limited or because they have trouble turning what they imagine into a drawing (Freeman, 1980; Golomb, 1974; Goodnow, 1977). Perhaps children have accurate mental images of what to draw but inadequate strategies for drawing them. In support of this view, researchers have shown that children can be taught to draw complete figures. In one study (Brown, 1981), kindergartners first were asked what they would include in drawings of themselves to grace their work folders, and they and their teacher went through the details of the body to be represented. With questions and answers, beginning with the top of the head and moving to hair, eyebrows, eyes, ears, neck, and so forth, the children suggested details. They clearly knew parts of the body and what people look like. Then they drew pictures of themselves—with a wealth of detail. Their teacher was astonished; all the children's earlier drawings of people had been global and barren.

Words

The most important form of symbolic representation is language—the use of words to stand for things and actions.

As we described in Chapter 6, children usually have acquired their first recognizable words by their first birthday. By the time they are 1½ to 2, they are adding an average of five to seven new words each day (Carey, 1982). Increasingly, their words are restricted to conventional signs and accurately match accepted meanings. By the time they are 5, children have vocabularies of 5000 to 10,000 words. The words that children utter and understand clearly reflect their symbolic thought. Only when they understand the connection between language and the objects, events, and relationships it represents can they use words appropriately.

Intuitive thought

The thought of 4-, 5-, and 6-year-old children, in contrast to that of 2- and 3-year-olds, according to Piaget (1970) is dominated not by fantasy but by **intuition**—by guesses about reality rather than rational inferences. Preschool children know many things about people, animals, toys, and food. They work at tasks and at increasingly complex forms of reasoning. But the solutions they propose often are wrong when judged by adult standards. Although recent research suggests that Piaget underestimated the knowledge and reasoning abilities of preschoolers, it still is true that preschoolers do not think like older children.

Dreams

The intuitive way that young children think shows in the way they describe their dreams. Piaget found three developmental stages in children's understanding of dreams. First, at around 4, young children think that dreams are real events that happen outside of themselves.

> When Natalie was 4, she told us with complete conviction that she had driven the car to her cousin Jerry's house. She described the trip and the route she had taken in perfect detail, right down to having put on the parking brake at Cousin Jerry's house. When we told her that she had dreamed it, she did not believe us.

They sometimes wake up frightened because their dreams are so real to them.

> At 3;7, when [the child] was trying to overcome a tendency to bite her nails, she said when she woke, but was still half asleep: "When I was little, a dog bit my fingers," and showed the finger she most often put in her mouth, as she had probably been doing in her sleep (Piaget, 1962, p. 178).

One characteristic of intuitive thought is this blending of the real and the imagined. Young children cannot separate dreamed from real events: "We had a circus in my room last night," "I get pictures on my windows at night when it's dark." When Piaget asked 4-year-olds about their dreams, these were typical replies:

> When do you dream? *At night*. Where is the dream when you are dreaming? *In the sky*. Can you touch the dream? *No, you can't see and besides you're asleep*. When you are asleep, could another person see your dream? *No, because you're asleep*. Why can't one see it? *Because it is night*. Where do dreams come from? *From the sky* (Piaget, 1975, pp. 93–94).

Children believe that inanimate objects are alive and have feelings and motives. Thus the moon seems to follow the children across the sky, and the sun sets so that they can go to sleep.

In the second stage of understanding dreams, 5- and 6-year-olds understand that dreams come from inside a person's head but, even so, think that they happen outside of it:

> What is a dream? *You dream at night. You are thinking of something*. Where does it come from? *I don't know*. What do you think? *That we make them ourselves*. Where is the dream while you are dreaming? *Outside*. Where? *There (pointing to the street, through the window)* (Piaget, 1975, p. 107).

Only when they are about 7 years old do children clearly understand that dreams are thoughts and imaginings that come from within a person.

Animism

Another characteristic of intuitive thought, according to Piaget, is that children think that inanimate things are living and have feelings, intentions, and thoughts just as they do. Piaget called this belief **animism**. Balls roll and waves break because they want to. Snow-covered trees feel cold, and empty cars feel lonely.

> The moon moves; it moves because it's alive.
>
> The clouds go very slowly because they haven't any paws or legs; they stretch out like worms and caterpillars; that's why they go slowly.
>
> The moon's hiding in the clouds again. It's cold (Piaget, 1962, p. 251).

Because an object is alive, its motions are intended. The sun sets to go to sleep; the rain falls to bring us flowers; the highway stretches to reach Grandma's house; and the oven makes our supper. Someone—divine or human—has created all things in the universe. Thus, stones were planted and grew into mountains. People built cities, then scooped out lake beds and filled them with water so that the people in the cities could go swimming and have pretty lakes to look at.

When children do not have factual information about things and the way they work, they use their intuition to come up with explanations like those that Piaget heard. But when children do have information, researchers have found, they do not appear so naive. Preschool children may well say that inanimate objects they know little about or that they have heard about in fairy tales—clouds, wind, moon—are alive. But apparently they do not think that *all* inanimate objects are alive. Even in this early period, children's intuitions are readily modified by observable, concrete facts. When researchers have interviewed preschool children about familiar objects—not remote things like clouds and wind—they have found that the children's responses do not show the same belief in animism. In one study, researchers asked preschoolers about the differences between a person, a doll, and a rock (Gelman, Spelke, and Meck, in Pines, 1983).

> Can a person make a wish? *Yes*. Can a rock make a wish? *No*. How is it that a person can, but a doll and a rock can't? *A rock is just chemicals, made of little stones. Dolls are made out of plastic sometimes*. What about a person? *A person is made out of a seed* (Pines, 1983, p. 50).

Children answered in ways that showed that they believed that objects made to move from outside are inanimate and objects that can move from within are animate. None of the children questioned said that rocks had thoughts or feelings, although some said, perhaps in make-believe, that dolls had feelings.

In another study, 3- to 5-year-olds were shown videotapes of moving objects—animate objects—a girl, a rabbit—and inanimate objects—blocks, a windup worm—and were asked questions about what was happening (Bullock, 1985). The questions were designed to elicit the children's understanding of animacy. They were asked whether the objects had a brain, could grow, would get hungry, and were alive, whether they could be fixed with glue, thrown, or bought in another color. The findings suggested that preschoolers do not have an animistic view of all objects. They were just as likely to say that animate objects had inanimate properties as the reverse. Their problem seemed to be a lack of knowledge about the properties of many objects, especially those they had not had the chance to explore.

Perspective taking

Another characteristic of intuitive thought, according to Piaget, is its **egocentrism**. Preschool-age children rarely stop to think that others do not see things through their eyes. "What dis?" the 3-year-old asks from two rooms away, as if her mother can see through walls. She holds a picture book to face herself and asks her mother across the room, "Why is dis doggy running?"

Piaget called this apparent inability to take the perspective of another person *egocentrism* and documented young children's egocentric thought with his "three mountains" task. In this task, a three-dimensional model of three mountains of different heights is set on a table that is surrounded by four chairs (Figure 9.2). A child looks at the mountains from all sides and then sits in one chair, and a wooden doll is placed on another chair. The child then looks at ten photographs to find the one that shows the mountains as the doll would see them. Four-year-olds, Piaget found, chose any picture at random. Five-year-olds chose the picture that showed their own view. Six- and 7-year-olds understood that point of view affects what one sees, but they could not

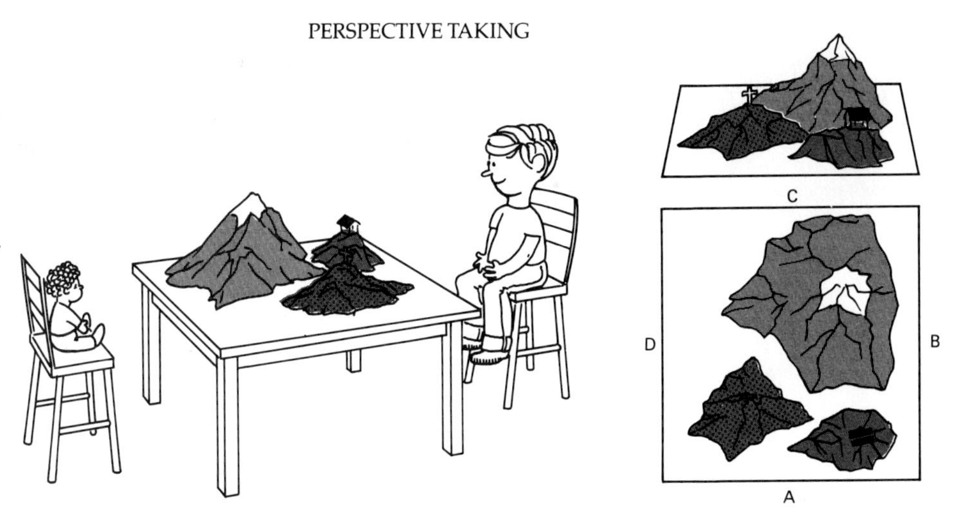

Figure 9.2 In Piaget's three-mountain task, the child is asked how the scene appears to the doll, which the experimenter seats in different positions (B, C, D) around the table. The experimenter may ask the child to describe the scene that the doll sees, to construct it from cardboard models, or to pick the correct picture from a collection of ten (Piaget and Inhelder, 1967).

always choose the picture showing the doll's point of view. They guessed, changed their minds, tried another picture, and then went back to their first choice. Only after age 7 or 8 did children reliably choose the right picture. Other researchers have shown, similarly, that preschoolers have only a limited appreciation of another person's point of view in this kind of task (Flavell, Botkin, Fry, Wright, and Jarvis, 1968; Flavell, Everett, Croft, and Flavell, 1981; Flavell, Shipstead, and Croft, 1978).

But young children are not always egocentric. When the task is simple, preschool children show that they realize that other people have different points of view from theirs. In one study, when 1- to 3-year-olds were given a cube with a picture pasted to the bottom and asked to show it to a person across the room, for example, all the children over 2 could do it (Lempers, Flavell, and Flavell, 1977). In another study, many 2- and 3-year-olds could hide an object from someone else behind a screen, even though they themselves could still see it (Flavell et al., 1978).

The understanding of *how* things look to another person develops later than the realization that they *do* look different (Flavell et al., 1981). The age at which children can figure out how things look to another person largely depends on the difficulty of the task and the kind of response the children must give. For example, in one study (Borke, 1975), when the display that the child was shown contained familiar objects, such as a toy sailboat on a lake, children as young as 3 said correctly what the doll could see. In other studies, too, it has been shown that preschool children can take another's perspective if they can actually turn the display instead of picking a picture or if they can say what the other's point of view is without having to look at the display at the same time (Hardwick, McIntye, and Pick, 1976; Huttenlocher and Presson, 1973). Children often do *act* egocentric—in real life as well as in experiments. But the findings from all these studies suggest that young children are limited by their ability to perform mental operations (like rotating a scene in their minds) rather than that they simply do not realize that other people have perspectives that are different from their own.

Preoperational logic

Unlike older children, young children do not rely on logical rules. Young children show in numberless ways that they have yet to develop mature

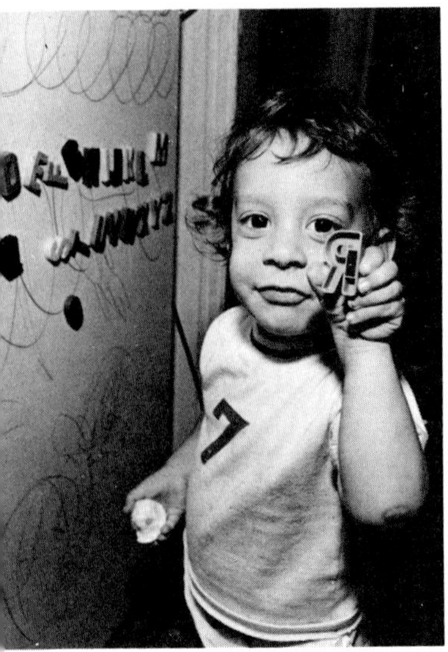

This 2-year-old gives away his egocentric thinking by showing the photographer his "R."

abilities to reason logically. While they are in this period of early childhood, they cannot perform logical **operations**, like addition, multiplication, subtraction, which transform information and form an organized network of knowledge. For this reason, Piaget called the logic of 2- to 6-year-old children *preoperational* and the developmental stage they are in the **preoperational period**. Although Piaget to some extent underestimated young children's knowledge and capacities in this area (as in some others we have discussed) his influence on thinking about child development has been so strong that it is useful to present his views and the evidence on which they are based. According to Piaget, preoperational logic shows clearly in children's abilities to reason and to understand conservation.

Reasoning

As Piaget listened to children's attempts to reason things out, one kind of preoperational reasoning he heard was **transductive reasoning**—reasoning from one particular to another rather than from the general to the particular.

> J. had a temperature and wanted oranges. It was too early in the season for oranges to be in the shops, and we tried to explain to her that they were not yet ripe. "They're still green. We can't eat them. They haven't yet got their lovely yellow color." J. seemed to accept this, but a moment later, as she was drinking her camomile tea, she said, "Camomile isn't green; it's yellow already. Give me some oranges" (Piaget, 1962, p. 23).

Confusing oranges, which ripen as they turn yellow, with tea leaves, which turn the tea water yellow, Jacqueline failed to understand that not all things turn from green to yellow for the same reasons. Children are trying hard to put things together, and so they often string together thoughts that apparently have no logical relation.

> The sun does not fall down. *Why*? Because it is hot. The sun stops there. *How*? Because it is yellow. *And the moon, how does it stop there*? The same as the sun, because it is lying down on the sky. Because it is very high up, because there is [no more] sun, because it is very high up (Piaget, 1962, p. 229).

From Piaget's observations, one gets the impression that preschoolers' reasoning is a muddle of illogic. More recent research, however, demonstrates that children as young as 4 sometimes can reason correctly and deductively. In the exchange that follows, a 5-year-old girl correctly deduces that there must be two Mr. Campbells. She has heard of Donald Campbell, who had died not long before in an attempt to break the world's water speed record. She also has heard of another "Mr. Campbell," a research worker named Robin Campbell, who had come to her school.

> CHILD: Is that Mr. Campbell who came here. . .*dead*? (Dramatic stress on the word "dead.")
>
> ADULT (SURPRISED):
> No, I'm quite sure he isn't dead.
>
> CHILD: Well, there must be two Mr. Campbells then, because Mr. Campbell's dead, under the water (Donaldson, 1983, p. 233).

When children listen to stories or look at picture books, their comments reveal a rich store of valid deductions:

> But how can it be [that they are getting married]? You have to have a man too." (The book contains an illustration of a wedding in which the man looks rather like a woman. The child thinks it is a picture of two women.)

First premise: You need a man for a wedding.
Second premise: There is no man in the picture.
Conclusion: It can't be a wedding (Donaldson, 1983, p. 234).

In fact, children even younger than 4 have been heard to reason deductively—especially when it's in their best interests. A 3-year-old child tries to keep her younger sibling from jumping on and off the couch with her. First she tried physical means, but then she gets out the big guns:

> People with shoes on can't jump. I haven't got them on, so I can jump. but you can't; you've got no shoes. So there (Mills and Funnell, 1983, p. 167).

From exchanges like these, we can deduce that young children can reason deductively, even though they may not be able to do so when asked about things they know nothing about—like sun and moon or tea and oranges.

Even in the laboratory, they can reason. In a recent study of 4-year-olds' ability to reason, children were presented with different kinds of logical problems (Hawkins, Pea, Glick, and Scribner, 1984). They could reason correctly when the problem was purely fictional.

> Merds laugh when they're happy.
> All animals that laugh like mushrooms.
> Do merds like mushrooms? [Yes.]

They also could reason correctly when their real-life knowledge did not conflict with the right answer. But they could not correctly solve problems like the following:

> Glasses bounce when they fall.
> Everything that bounces is made of rubber.
> Are glasses made of rubber? [Yes.]

Preschool children have trouble ignoring the evidence that they see in front of them or facts that they know to be true; they have trouble concentrating on just the logic of a proposition.

Piaget also claimed that young children's thinking was illogical because they did not understand the difference between cause and effect. Children, he reported, sometimes think that the cause is the effect.

> The man fell from his bicycle because he broke his arm.
>
> I had a bath because afterwards I was clean.
>
> I've lost my pen because I'm not writing (Piaget, 1926, pp. 17–18).

The boy who said these things clearly sensed that the events were connected, but he either did not understand or could not express their causal relation.

Researchers since Piaget have investigated children's understanding of cause and effect by means other than interviews. In one study by Rochel Gelman, for example, 3- and 4-year-olds were shown how to tell a story from a series of three picture cards from left to right (Gelman et al., 1980). Then they were shown two picture cards—perhaps one of a dog and one of water pouring from a pail—and were asked to pick a third card that would complete the story about the first two cards. (The choices might be pictures of a cat, a wet apple, and a wet dog.) The children were very good at choosing a third card that told a logical story—in this case, the picture of the wet dog. "First you have a dry doggie, and then the water puts water on that doggie, and you end up with a wet doggie," said a 3-year-old to explain her quite logical choice. As children got older, they made fewer mistakes and gave fuller explanations, but even the youngest children could infer cause and effect in these stories.

Preschoolers can also understand quite a lot about causal relations between objects and events. In another experiment (Gelman and Bullock, in Pines, 1983), children saw a contraption that looked as though two runways led inside a box where there was a pop-up Snoopy. Actually, the runways came to dead ends, and the Snoopy appeared when the researcher touched a hidden lever. First the researchers inserted a ball into a hole at the top of one runway that led to Snoopy's box, then they made Snoopy appear, and then they dropped a second ball into a hole at the top of the second runway. When they asked the children, "Which ball made Snoopy come up," 75 percent of the 3-year-olds, 87 percent of the 4-year-olds, and 100 percent of the 5-year-olds said that the first ball had made Snoopy come up. They understood that a physical cause precedes its effect. Next, the researchers removed the runway that led to Snoopy's box and then inserted a ball into a hole at the top of the unconnected runway. When the children saw Snoopy pop up, they were amazed. Apparently they also understood that physical causes do not act at a distance. When the researchers dropped a second ball into the hole at the top of Snoopy's runway, though, the children insisted that the first ball had made Snoopy come up (even though the first ball had rolled down the unconnected runway). What the children did not understand was that for something to be the cause of an event, it must *always* be followed by the event. When the runways were connected to Snoopy's box again, the ball dropped down the first hole, and Snoopy popped up, the children once again claimed that the ball down the hole caused Snoopy to pop up—even though they had seen Snoopy pop up when no ball was put down the hole. Preschool children know more than Piaget gave them credit for, but they do not completely understand cause and effect.

Conservation

Conservation is the name given to the understanding that even though an object's outward appearance may change, its length, quantity, mass, area, weight, and volume do not change as well. When children know that milk in a tall, narrow glass is the same amount after it is poured into a short, wide glass, they understand conservation of liquid quantity. When they know that four cookies are four cookies, no matter whether they are spread out or squeezed together on a plate, they understand conservation of number.

Piaget's experiments on conservation suggest that there are three stages in the development of children's understanding of conservation (Piaget and Inhelder, 1969). In his classic experiment on conservation of the amount of substance, Piaget showed children two balls of clay and asked them to add more clay to the balls until both were the same. Once the children said that the balls had the same amount of clay, Piaget rolled one of the balls into a long snake and asked the children, "Now, do both these pieces have the same amount of clay?"

Four- and 5-year-old children, in the first stage, thought that the long snake had more clay and could not be convinced otherwise. They could only focus, or *center*, on the one dimension—the salient dimension of length; they could not take thinness into account at the same time. They did not appreciate that the act of making the snake could be reversed and the clay turned back into a ball. Five- and 6-year olds, in the second, transition stage, could not decide whether the snake or the ball had more clay. They might say that the amounts were the same when the snake was short, but when the snake was long, they would say that the snake had more clay. They might guess correctly before Piaget rolled out the clay, but when they saw the snake, they would say that it had more clay. They were still strongly influenced by the way things *look*. Six- and 7-year-olds, in the third stage, realized that the ball and the

A psychologist tests a boy's understanding of the conservation of liquid quantity. When the liquid is poured into the tall, thin beaker, it looks to the boy as though it is more than the liquid in the short jar. The psychologist then encourages the boy to pour the liquid into the other jar. But until he understands conservation, the boy continues to think that the liquid has changed in amounts.

snake had equal amounts of clay, no matter how long the snake looked. They understood that the process of rolling the clay was reversible and the snake could be rolled back into a ball. They also understood that the dimensions of length and thickness were reciprocal—that the snake was long—but also skinny. They could explain why the amounts of clay were the same. They understood conservation of substance.

In similar experiments on conservation of number, length, weight, and so on (Figure 9.3), Piaget found that children went through the same stages of mastering conservation for all these quantities. He described children's thinking in the preconserving stage as preoperational because children at this stage could not perform mental operations like **reversibility**, understanding that some transformations are reversible, and **reciprocity**, understanding that dimensions like height and width or length and thickness are reciprocal or complementary.

Since Piaget carried out his experiments on conservation, many researchers in this country have tried to duplicate, understand, and explain his results. In one study (O'Bryan and Boersma, 1971), for example, researchers recorded the eye movements of children who were taking part in experiments like Piaget's on conservation of length, area, and liquid quantity. Nonconservers, they found, spent most of their viewing time looking at only one dimension, the ''dominant'' one. Children in the second, transitional stage shifted focus from one dimension to another now and then. But conservers shifted their focus often and quickly, supporting Piaget's claim that **decentra-**

Type of Conservation	*Child sees*	*Experimenter then transforms display*	*Child is asked conservation question*
Length	 Two sticks of equal length and agrees that they are of equal length	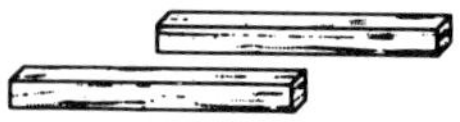 Moves stick over.	***Which stick is longer?*** *Preconserving* child will say that one of the sticks is longer. *Conserving* child will say that they are both the same length.
Liquid quantity	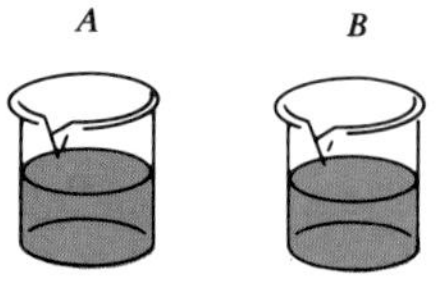 Two beakers filled with water and says that they both contain the same amount of water.	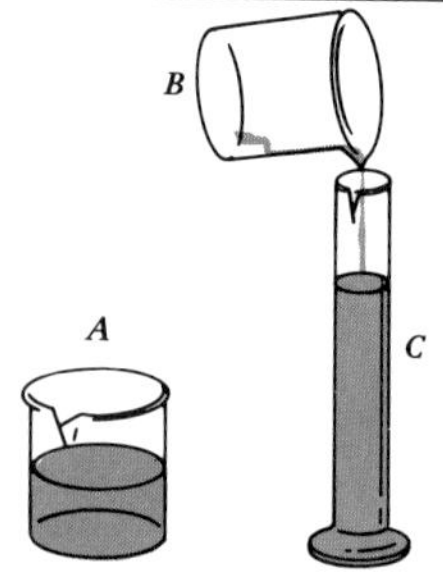 Pours water from *B* into a tall, thin beaker *C*, so that water level in *C* is higher than in *A*.	***Which beaker has more water?*** *Preconserving* child will say that *C* has more water: "See, it's higher" *Conserving* child will say that they have the same amount of water: "You only poured it!"
Substance amount	 Two identical clay balls and acknowledges that the two have equal amounts of clay.	 Rolls out one of the balls.	***Do the two pieces have the same amount of clay?*** *Preconserving* child will say that the long piece has more clay. *Conserving* child will say that the two pieces have the same amount of clay.
Area	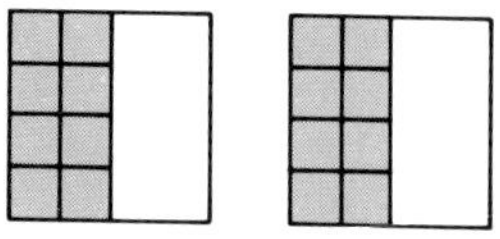 Two identical sheets of cardboard with wooden blocks placed on them in identical positions. The child acknowledges that the same amount of space is left open on each piece of cardboard.	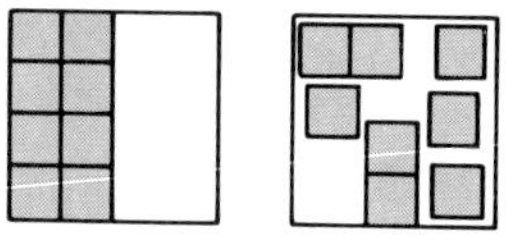 Scatters the blocks on one piece of cardboard.	***Do the two pieces of cardboard have the same amount of open space?*** *Preconserving* child will say that the cardboard with scattered blocks has less open space. *Conserving* child will say that both pieces have the same amount of open space.
Volume	 Two balls of clay in two identical glasses with an equal amount of water. The child acknowledges that they displace equal amounts of water.	 Changes the shape of one of the balls.	***Do the two pieces of clay displace the same amount of water?*** *Preconserving* child will say that the longer piece displaces more water. *Conserving* child will say that both pieces displace the same amount of water.

Figure 9.3 ***These procedures are traditional methods of testing children's abilities to conserve length, liquid quantity, substance amount, area, and volume.***

tion—being able to focus on more than one dimension at a time—is necessary if children are to conserve quantity.

Other investigators have been more critical of Piaget's work (see for example, Scholnick, 1983). They have suggested that sometimes the wording of Piaget's questions actually hid children's conservation abilities rather than revealing them. For example, when children see two plates, each of which has the same number of cookies, they may say "yes" to the question, "Does one have more?" To the young child, "more" may describe the amount of space the cookies take up, not the number of cookies. "Bigger" may mean "longer" or "taller" to young children, not "having greater mass or quantity." In one study of third graders' conservation abilities, researchers first asked Piaget's standard questions and then tried to reword them to make sure that the wording was not obscuring the children's conservation abilities (Pennington,

Wallach, and Wallach, 1980). When the children were asked, ''Are there still 13 objects?'' and not ''Are there the same as before?'' the number of children who answered correctly increased. Even questioning a child right before and right after the experimenter transforms the objects may suggest to the child that the first answer was wrong. Children more often answer questions about conservation correctly when they are asked only once how many objects there are (Rose and Blank, 1969; Samuel and Bryant, 1984).

Piaget's standard tasks for assessing conservation abilities may also make children seem less cognitively advanced than they, in fact, are because they distract the children unnecessarily. When a screen was put in front of the containers in a conservation of liquid task so that the children would not be distracted by seeing the different liquid levels, half of the 4-year-olds, 90 percent of the 5-year-olds, and all the 6- and 7-year-olds who had previously failed on the task answered correctly, saying something like ''It's still the same water'' or ''The water was only poured'' (Bruner, 1964). When the screen was taken away, and the children were asked again which container had more water, the 4-year-olds said that the tall container had more, but most of the older children stuck by their correct answers.

To some extent, children can be tricked into answering as though they can or cannot conserve by the way an experimenter presents a task, asks questions, and encourages a child to agree. Therefore it is difficult to say just when children develop the ability to conserve or exactly what conservation means. Piaget's age ranges are no longer accepted as invariable. But it does seem that in the preschool years, most children do *not* figure out the implications of reciprocity and reversibility and number and size, use mental operations consistently, back up their judgments with sound, logical reasoning, or understand conservation problems so that they can answer correctly, regardless of the wording of the questions or the appearance of the objects.

Stages of cognitive development

According to Piaget, the developmental stages observed for conservation and seriation—preoperational, transitional, and operational—occur across all domains. They appear in a fixed and irreversible sequence, such that no child who achieves the stage of operational thought ever backslides to preoperational thought. They are discontinuous; that is, the shift from one stage to the next is relatively sudden. And they are universal; all children everywhere go through these stages.

Researchers have tested the accuracy of these claims that cognitive development occurs in stages. One thing they have tested is whether children reach the same stage in different areas at the same time. For example, does a child acquire conservation of weight, length, and volume, seriation, and transitive inference at the same time? Most of the researchers who have looked for consistency in stage level across areas have found inconsistencies (Gelman and Baillargeon, 1983). Children develop particular concrete operations at different times, they have found, and not necessarily even in the order that Piaget proposed. Piaget himself observed that children acquire conservation of different quantities at different ages. Piaget named this phenomenon **horizontal décalage**—after the French *décaler*, meaning to unwedge or displace—to suggest that children learn conservation of different quantities at different times. But giving it a name does not reconcile this phenomenon with a theory of general, cross-domain cognitive stages.

Another claim about stages that researchers have tested is whether children in Europe, Africa, China, and Australia all go through the same developmental stages in the same order. In general, these tests have been positive. Cross-cultural studies show that although environment affects the *rate* at which children develop, it rarely and minimally affects the *sequence* of stages. Children everywhere develop the same mental operations, although children may grow more skillful in using the operations that are valued and practical in their particular culture (Cole, Gay, Glick, and Sharp, 1971; Dasen, 1977).

A third test of Piaget's theory of stages has been to see whether children can be taught the operations from a cognitive stage more advanced than their own. According to Piaget, children must work out these operations at their own pace and within the context of their own everyday activities. Deliberate training as a way to speed children's development is not necessary or helpful. Piaget did allow that the child in the transitional stage of mastering an operation might find training useful but only if the training were geared to the child's present level of understanding and if it offered the child information about concepts at the next stage of development. Short-term training, according to Piaget, could not create new cognitive abilities or structures.

American researchers have conducted many studies to test this aspect of Piaget's theory. In one study (Gelman, 1969), the researcher tried to train 5-year-olds to conserve number and length. She showed children rows of chips or sticks with different numbers and spacing (Figure 9.4). The children were asked to point to the rows that had either the same or different numbers of chips or sticks. When children in one group got it right, they got a prize. Children in another group were not rewarded or told when their answers were correct. The children whose right answers were rewarded quickly learned to tell the difference between number and length; those who were not rewarded learned little. When the children were tested the next day and again two to three weeks later, those who had been rewarded could conserve number and length 95 percent of the time and mass and liquid two-thirds of the time. Their reasoning showed clearly that they understood conservation. The children

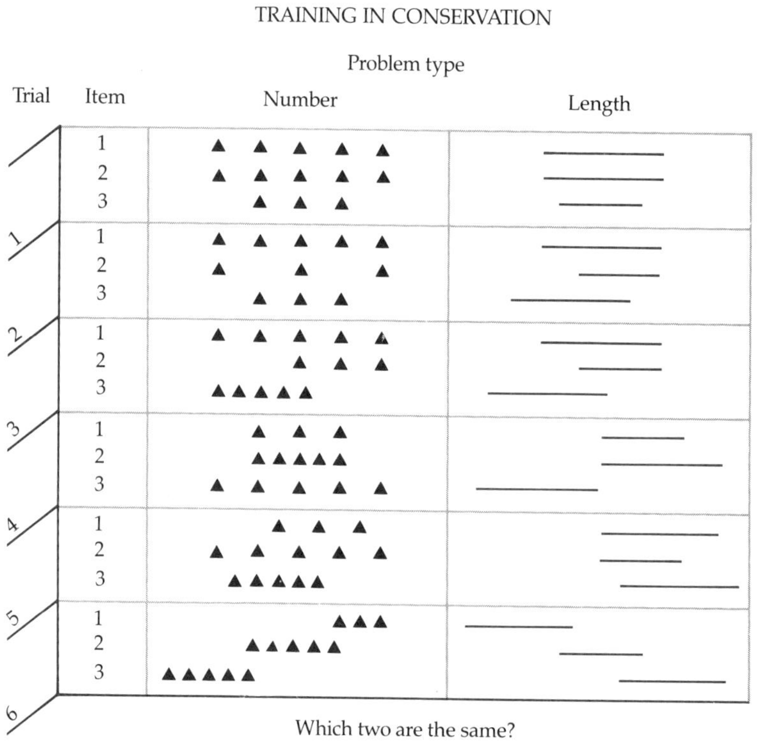

TRAINING IN CONSERVATION

Problem type

Trial
Item
Number
Length
Which two are the same?

Figure 9.4 These sets of chips and sticks are examples of the groups of items that Gelman (1969) used in training children to discriminate number and length.

who had not been given feedback did not acquire these abilities. Training apparently did help preoperational children to learn conservation. Mere exposure to the materials on number and length did not allow the children to discover conservation on their own.

In another training study, third graders who understood conservation were paired with third graders who did not (Silverman and Stone, 1972). After the pairs had talked about conservation of area, 11 of the 14 original nonconservers could understand conservation of area. Retested a month later, the children still understood the concept.

Researchers have tested whether children can be trained in virtually every ability that Piaget described as developing in stages (Brainerd, 1983). Although children in middle childhood are easier to train than preschoolers, even the younger children can be trained to apply rules and strategies that help them to acquire these abilities. In general, training studies have not supported Piaget's proposal that only children in transition between stages can be trained. Children at lower stages have benefited from training just as much as those in transitional stages (Brainerd, 1978). Although most normal children, who do not have the advantages of training by a researcher, likely do proceed through the cognitive stages as Piaget proposed, training studies demonstrate that cognition can be affected by many factors besides a child's everyday interactions with the physical world. In fact, for teaching concrete operations, training seems to be a more effective teacher than a child's own discoveries.

Piaget provided a useful framework for thinking about children's cognitive development. But it now seems clear that the ages at which children can succeed at particular tasks can be modified when the tasks are presented in a simpler way or when children are given training. Of course age was never the important thing to Piaget. The fact remains that younger children do not spontaneously succeed on Piaget's tests, whereas older children do, and this fact reflects real differences in the thinking of younger and older children.

Much of the difference between the thinking of children in early and middle childhood is a result of increased knowledge. But three qualitative changes seen on Piaget's tasks do seem to differentiate the thinking of preschoolers and school-age children.

1. Preschool children depend on the appearance of things rather than inferred reality. School-age children are more sensitive to distinctions between what *seems* to be and what really *is*.
2. Preschool children tend to concentrate their attention on one particularly interesting or salient feature and neglect others. School-age children pay attention to more features.
3. Preschool children do not understand reversibility; older children do.

Most developmental psychologists today do not subscribe to a strong, monolithic stage theory as was laid out by Piaget. But because of developmental differences such as these three, many psychologists still are convinced that there are stages in cognitive development (for example, Fischer, 1980; Case, 1985). Whether children can perform at their highest stage, these psychologists suggest, depends on the difficulty or complexity of tasks and the amount of help the children get in solving them (Fischer and Canfield, 1986). In most situations, children do not function at their highest level. When children are given help, and when the task is simple, development does appear to take place in discrete stages, as Piaget suggested. Neo-Piagetian psychologists support a more liberal view of stages in which children's progress in different domains is somewhat independent but constrained by certain age-related factors (Case, 1985).

One such constraint is working memory. One explanation of why training succeeds is that it acts to improve children's working memory (Brainerd, 1983). Children need to encode and remember the facts of the situation—like the height and width of two glasses—while they figure out the problem. Training that improves children's abilities to encode or to remember facts thus may help them to solve problems in cognition. The issue of what preschool children can encode and remember is taken up in the next section.

Information processing

Piaget has provided one framework for describing preschool children's thinking. Another framework is provided by the information-processing approach (see Chapter 1). This approach emphasizes quantitative changes in children's abilities, not stages. The preschool child's thought is not seen as qualitatively different from the school-age child's but as constrained by context, complexity, the verbal demands of tasks, by general abilities such as memory, and by specific limits like lack of knowledge. According to the information-processing approach, as children get older, their skills and their ability to apply them to more tasks increase. Researchers who have followed this approach have focused on how children attend to information in the environment and store that information in memory. They have studied the amount of information children can take in, how quickly they process it, and how well they remember it. In this section, we discuss the results of research on children's processing of information.

Attention

Preschool children attend to sights and sounds by extensively scanning the environment, but their exploration is not systematic.

> For a long time, I was worried at how poor Jason was at finding things. He would open the toy cabinet but consistently miss the toy he was after. He would rummage through the paper on the top of his father's desk but miss the toy car. I thought there might be something wrong with his eyesight. But as he got older, he got much better at finding things. The other day he even found a ring I had lost.

In one study, 4- to 8-year-olds were shown pictures of houses in pairs (Vurpillot, 1968; Figure 9.5). They were asked whether the houses were the same or different. The most efficient way to decide this is to compare the houses systematically, looking from one house to the other and back, window by window, row by row. The 8-year-olds did follow this strategy, and when they had found a difference between two windows, they efficiently stopped searching and said that the houses were different. The 5-year-olds, however, did not scan systematically. They did not compare the two houses window by window; they did not start at one corner and work toward another. They did not even look at every window. In another study (Elkind, 1977), children were asked to name the pictures that they saw pasted onto a card—hat, parrot, chair, and so on. Eight-year-olds named all the pictures systematically from left to right and top to bottom. Five-year-olds left out some pictures and named others twice because they jumped from one picture to another. Only when a system was imposed on the pictures by the researcher and they were pasted in a triangular shape did the 5-year-olds, like the 8-year-olds, name the pictures in order.

Clearly, preschool children are not as systematic as older children in paying attention to the environment. They are also not as selective as older

Figure 9.5 ***A pair of different houses (top) and a pair of identical houses (bottom) (Vurpillot, 1968). When children are asked to say which pair of houses is the same, preschool children do not scan as systematically as older children.***

children in what they pay attention to. In studies of children watching television, for example, 2-year-olds have been found to be more easily distracted, to talk more to other people, to play more with toys, and to look around the room more than 4-year-olds (Anderson and Levin, 1976). As they get older, children get better at picking out one object or picture from a clutter and listening to one voice while ignoring another (Geffen and Sexton, 1978; Sexton and Geffen, 1979). They become increasingly skilled at selecting stimuli to attend to and focusing their attention at will. In one study, children were shown a row of cards, each containing a picture of a common household object and a picture of an animal (Hagen and Hale, 1973). Then the cards were turned face down, and the children were asked to locate the animals. Later, after the game was over, the children were asked which objects had been on the same cards as the animals. Younger children did better than older ones on this task, because the task-oriented older children had focused their attention only on the animals.

Why are preschoolers less selective about the information they attend to than older children? For one thing, young children may be less *efficient* at processing information and so, because they use more of their limited ability to pay attention to process irrelevant stimuli, they cannot pay as much attention to important stimuli. In support of this explanation is the finding that young children are more easily distracted by irrelevant stimuli (Smith, Kemler, and Aronfreed, 1975). For another thing, perhaps young children do not understand what experimenters want from them. Perhaps they do not select as important the elements that an experimenter intends as important and instead focus on the elements that are more salient for them. Research has shown that if the stimuli the experimenter intends to be important are also salient to the child, young children do not differ from older children in what they pay attention to or remember (Odom, 1982). Finally, perhaps young children do not have the *strategy* of consciously controlling or directing their attention so that they focus only on relevant aspects of tasks. Older children know that they should be selective and have conscious strategies for paying attention.

In brief, 3-year-olds are explorers. They approach the environment curiously and playfully, move quickly from one thing to another, are easily distracted, and respond to whatever attracts them. Six-year-olds are searchers. They investigate the environment systematically, with plan and purpose.

Memory

After children have attended to information and taken it into their information-processing system, they must remember it. Attention determines which information enters the ''computer,'' and memory determines which information stays in it and how it is filed.

Recognition of an object or image one has seen before is the simplest kind of remembering. Even preschoolers are quite good at this kind of remembering—if what they are asked to remember is simple and familiar. In one study, researchers showed 2- and 4-year-olds 18 small, familiar, attractive objects (Myers and Perlmutter, 1978). Later, they showed them 36 objects, the original 18 plus another 18. The 2-year-olds recognized over 80 percent of the original objects; the 4-year-olds recognized over 90 percent. Both groups were extremely accurate at realizing which objects they had never seen. Preschoolers are less accurate at recognizing complex, abstract forms, however (Nelson and Kosslyn, 1976). Presumably, this is because they are not as good at encoding information into their information-processing system as older children. It is clearly not that they cannot retain information in memory once it is there.

Recall is a more difficult process than recognition, because information must be retrieved from memory with no prompts from the environment. No person or object is present to trigger memory, as in recognition.

> Whenever I asked Peter about something he couldn't remember—''What was the name of the boy you were playing with when I picked you up?'' ''What was that snack you wanted me to buy for you?'' ''Where does Sam live?''—he'd answer, ''Do you know? Tell me all the names [or snacks or streets] you know.'' Then, when I hit the missing word, his face lit up, and he crowed, ''You got it!''

How well people recall information depends, for one thing, on how much practice they have had in recalling. Unlike recognition memory, which is similar in all cultures (Cole and Scribner, 1977), recall varies widely from one culture to another. In our society, we rely heavily on written material to help us remember.

How much can preschoolers in this society recall in tests of memory? In one study, researchers showed preschool children nine small objects one at a time from nine different conceptual categories, such as animals, utensils, vehicles, and the like, labeled them, and put them in a box (Perlmutter and Myers, 1979). Even though they were told that they could keep the objects they could name, the 3-year-olds recalled only two of the objects, the 4-year-olds three or four objects. Most of the children recalled the last object they had seen. When the nine objects were from just three categories, the children remembered more of them, perhaps because by this age, children's recall is organized into conceptual categories. When the experimenter prompted the children with a clue—"Do you remember any other animals?"—they recalled even more. Marion Perlmutter (1980) suggests that preschoolers recall poorly because they lack language for encoding information and strategies for memorizing and retrieving it. Preschool children lack the organization of memory that older children have. They do not yet have a filing system, so their memories may be jumbled and inaccurate. Preschoolers also have a more limited memory capacity (Brown, Bransford, Ferrera, and Campione, 1983; Case, 1985). They can remember fewer units or bits of information than older children, and they use up more of their mental "space" encoding information, so less is left over for actually storing information.

In familiar surroundings, preschoolers can recall better. They recall social interactions and cartoon characters (though only rarely objects—the usual target of laboratory research), and their recall extends back for several months (Todd and Perlmutter, 1980). They also recall "scripts" for familiar events and routines like going shopping, taking a bath, going to school, or having a party. Katherine Nelson and Janice Gruendel (1979) interviewed children between 3 and 8 years old about "what happens" when you have a birthday party, plant a garden, and other familiar situations. The youngest children's scripts were sketchy and general—"You cook a cake and eat it." The older children included social aspects in their scripts, such as children arriving at a birthday party and playing games, and they described them with more complex details. Scripts are representations of what *usually* happens rather than descriptions of what happened last week.

Preschoolers can tell you about having dinner at a restaurant or having birthday parties, but they may be incapable of describing last night's dinner or cousin Lainie's third birthday party. In a study of one group of 5-year-olds' school scripts (Fivush, 1984), the children had lots to say about what happens at school but little to say about what happened "yesterday" at school.

> Why was it that every time I asked Kristy what she'd done at nursery school today, she answered with utter silence or, at most, "Nuffin'"? But if I asked her, "What do kids do at school?" she could tell me they line up and go inside, play, eat snacks, and go back outside to the swings.

These "scripts" are one way that preschoolers have for remembering things. They offer them some organization, as one study in which 3- and 4-year-olds were asked to recall lists of words showed (Lucariello and Nelson, 1985). Children recalled more words when the lists had been made up from children's scripts (for getting dressed in the morning, having lunch, going to the zoo) rather than from the adult categories of clothes, food, and animals. They remembered the most words when they were given the script list and were prompted with questions such as, "Tell me which of those things you could put on in the morning (eat for lunch, see at the zoo)." Thus, although, as Perlmutter showed, preschool children find adult category organization helpful, because script organization is *more* helpful, children's knowledge of general events may be more accessible than their knowledge of categories.

Language

We have already described something of children's acquisition of words. But language, of course, is more than a grab bag of words. The remarkable speed

focus Children as witnesses

In court cases concerning both the rights and the abuse of children, many children have taken the stand and told their tales to judge and jury. Just how accurate are children as witnesses? Should their testimony be allowed? What about 3-year-old Lori?

> One summer day a man in an orange Datsun pulled up in front of the neighbor's yard where Lori was playing and told her to get in. She did. Three days later the police found Lori, crying and bruised, in the pit of a deserted outhouse. Lori told the police that the ''bad man'' had hit her and left her there. The next day, when the police showed her a set of twelve photographs, she gasped and identified one man as her abductor. The man Lori identified was arrested, and a week before the case was scheduled to come to trial, he confessed (Goleman, 1984, p. 19).

But when a case of sexual abuse of children in Minnesota was dropped because of insufficient evidence, people began to ask whether the complaining children had lied.

> A tense and deeply embarrassed 11-year-old boy testified that the Bentzes had sexually abused him over the previous two summers. But the next day, the youth recanted part of his testimony, saying he had lied the day before when he described the sex acts with Robert Bentz. When defense attorney Earl Gray asked, ''Was that a big lie or a little lie?'' the child answered, ''A big one'' (Siegel, 1984, p. 28).

Lawyers on one side held that adults had threatened the children, put words into their mouths, and educated them about sex by using anatomically correct dolls as models. The other side countered that no one could have coached the children to make up such sexually explicit stories.

Research suggests that in some ways, children make better witnesses than adults. Because young children do not attend to information as selectively as older children or adults, they may notice and recall details that older people ignore. For example, in one study, when children and adults watched a basketball game on videotape, 75 percent of the first graders, 22 percent of the fourth graders, and no adults recalled seeing a woman with an umbrella walk through the gym (Neisser, 1979). Although their stories tend to come out in bits and pieces, children's recall of simple events may be no less accurate than adults'. Children are also good at recognizing faces. Children under 12 do, however, remember fewer details than adults, especially those they do not understand or follow well (Goleman, 1984). They also sometimes mix reality and imagination. In one study, children 6 years old distinguished an imaginary from a real, remembered event as well as adults, but they confused what they had really done in the situation with what they had only thought about doing (Foley and Johnson, 1985). They also may mix a memory of a single event with their general knowledge of what usually happens (their ''script'') (Myles-Worskley, Cromer, and Dodd, 1986).

Four-year-olds may not know how to count or name the days of the week. They may give literal answers that sound contradictory. Asked if he had gone to the defendant's *house,* one child witness said ''no.'' It turned out that he had gone to the defendant's *apartment.* Judges need to be alert to lawyers' questions that exceed children's cognitive capacities. Then, as long as they are asked simple, straightforward, nonsuggestive, and unthreatening questions, children can make reliable and useful witnesses.

with which preschoolers learn words partly depends on their having a framework on which to hang those words. **Syntax** is the name for this framework. Syntax is how words are combined into sentences to express who did what to whom.

Before children can manage complex sentences with subjects, predicates, and objects, though, they use strings of related one-word utterances (Scollon, 1976), or they say two words and then pause (Branigan, 1979). They do not take a flying leap from uttering their first words to putting them together into many-word utterances. They do not develop an entire system of syntax in one swoop. Instead, they make their way in small steps along a winding path, combining words in ever longer and more complex sentences. In this section, we will first describe the child's progress toward complex sentences and then discuss the processes underlying the child's acquisition of syntax.

Two- and three-word sentences

Children's first "sentences" are two words long and very basic. They include only the most important words in the sentence—like a telegram sent by a miser or a newspaper headline: "Doggy bark," "More car." If she wants her mother to give her a picture book, the child may say, "Give book" or, "Mommy give" or even, "Mommy book," but she does not say, "Mommy, please give me the book that is on the shelf." She is limited to her two word **telegraphic sentence** because of her limited memory and information-processing skills. She expresses the core of the sentence and omits "frills" like the copula verb *(is)*, the article *(the)*, the preposition *(on)*, and the conjunction *(that)*. Why are these words omitted? For one thing, they are difficult to learn because there is no one-to-one correspondence between them and their meanings. For another, they are omitted because they are unstressed sounds, and children at this stage often omit unstressed syllables—"e-phant" for elephant, "'mato" for tomato, and "'puter" for computer (Gleitman and Wanner, 1982).

With their two-word sentences, children comment on actions and objects, possession and location. "Milk fell," "Go car," "Molly room," "Daddy home." What was once expressed with one word and a gesture is now handily expressed in two little words, "See doggy." "Bottle here." With rare exceptions, the word order in these two-word sentences is the same as that an adult would use in a longer sentence: "Doggy bark" and not "Bark doggy" (de Villiers and de Villiers, 1973). The child apparently knows something about syntax as well as meaning even at this early age. She knows that in English sentences, "doers" come before actions; that actions come before objects. She shows this in the word order of her two-word sentences. (In fact, even in the one-word stage children apparently understand this. In an ingenious study of 16- to 19-month-olds who could say only two or three words themselves, researchers found that when children heard "Cookie Monster is tickling Big Bird," they looked more at a videotape of Cookie Monster tickling Big Bird than at a videotape of Big Bird tickling Cookie Monster shown at the same time [Hirsh-Pasek, Golinkoff, Fletcher, Beaubien, and Cawley, 1985]. If they had understood only the meaning of "Cookie Monster" and "Big Bird" and had not been sensitive to the significance of word order, they would not have shown a preference). Two-word speakers appreciate well-formed, complete sentences—with all the frills—even though they cannot yet speak them. In another study, two-word speakers were found to obey more often if their mother said, "Throw me the ball" than if she said, "Throw ball" (Shipley, Smith, and Gleitman, 1969). Children clearly are aware of syntax even at this early stage of language development.

Children's three-word sentences are still telegraphic and still leave out the nonessential words. But as before, they follow conventional English subject–verb–object form. The child now asks for her picture book, "Mommy give book."

Complex sentences

As sentences become more complex, syntactic "frills" begin to be added, generally in this order (Brown, 1973; de Villiers and de Villiers, 1973):

ing ("Cat drink*ing*.")
in, on ("Cat *in* basket.")
s ("Cat*s* playing.")
's ("Cat*'s* milk in bowl.")
is ("Cat *is* fat.")
a, the ("*The* cat is drinking.")
ed ("The cat play*ed*.")

Thus in trying to communicate, "The cat is drinking," the child who once said, "Cat drink," is more likely to say, "Cat drinking," than to say "Cat is drink" or "The cat drink."

Elements are added in this order not because of semantic complexity but the frequency with which the child hears these elements in the parents' speech and their acoustical distinctiveness (Moerk, 1983). The child's own need for language may also contribute. Thus the child may learn to say "my" before "his," because the child needs to talk about herself before she talks about someone else.

At first, in their early complex sentences, although children know how to make a sentence, they do not always do so. They may know how to say "Daddy drive car," a complete sentence, but still use a sentence fragment like "Daddy car." They also may know how to add *-ed* or *-ing*, for example, but not do so. Michael Maratsos (1983) has suggested that at first children may not assume that a grammatical form that is expressed *sometimes* has to be expressed *every* time.

By the time they are correctly using the *-ed* ending to indicate the past tense about half of the time, though, young children do figure this out. They begin to **overregularize** their verbs and tack *-ed* onto irregular verbs that they once said correctly—"I breaked my crayon," "We goed to the store." Similarly, they overregularize words with irregular plural endings—"My foots hurted," "Daddy and me are mans." The child seems to be searching for general rules of language that operate consistently.

Most children relearn irregular past tenses and plurals by the time they start school—through tedious memorization. But even then for unfamiliar words, they are likely first to try out regular rules ("wring, wringed" "bleed, bleeded").

Along with pronouns and suffixes, children add adjectives and adverbs to their sentences. They learn the most general ones first: *big, nice, little,* before *tall, naughty, short*. The former are the adjectives the child probably hears most often, and they are useful in a variety of situations and sentences.

Children also learn to ask questions. In their two-word sentences, children signal questions simply with a rising intonation: "Me go?" "Doggy eat?" When they begin to use auxiliary verbs, they first say: "Doggy can eat?" "Where I am going?" They keep the verb and the auxiliary verb together. Only later do they learn to reverse the order of auxiliary and subject: "Can doggy eat?" "Where am I going?" This is a syntactically more complex form. Children also ask the simpler "wh-" questions: "where," "what," ''who,'' and ''whose'' first. Locations (where), objects (what), and agents (who), you will recall, were the first concepts that children put into their two-word sentences. Causation (why), manner (how), and time (when) are more difficult concepts and so are learned later and appear in sentences later (de Villiers and de Villiers, 1979; Ervin-Tripp, 1970). Children continue to think that ''when'' means ''where'' as late as 3 years of age:

MOTHER: When are you having lunch?
CHILD: In the kitchen.

It takes several more years of listening to and using language before children have figured out how to ask indirect questions. When preschool children in one study were asked to relay a question to a third party, for example, the children were likely to make mistakes (Tanz, 1980). When the experimenter said, ''Ask Tom where I should put this,'' the children were likely to say, ''Where should you put this?'' or ''Where should I put this?'' rather than ''Where should she put this?'' They did much better with, ''Tell Tom where I should put this'' (C. Chomsky, 1969).

As their language increases in complexity, children learn how to put more than one idea in a single sentence. They join ideas with *and, but,* and *or.* These conjunctions can make a sentence go on forever. ''Daddy took me to the grocery store, *and* we bought carrots, *and* then we bought peanut butter, *and* Daddy said, 'Where's the bread?' *and* we found the bread, *and* we put it in our cart, *and*. . . .'' Three-year-olds also use the conjunctions *if, when,* and *because*: ''If I brush my teeth, can I have two stories?'' ''When it gets dark, I go to sleep.'' ''I like you because you are my friend.'' They use *before, after,* and *until* less often, because these words express temporal relations that are quite sophisticated.

Until they are about 5 years old, children describe events in their actual order of occurrence: ''I brushed my teeth, and then I got in bed.'' ''The mailman rang the bell, and I got a letter.'' Not until later will they say, ''I got in bed after I brushed my teeth'' or ''Before I got a letter, the mailman rang the bell.'' Of practical importance, children *understand* compound sentences more easily if they mirror the real order of events and if the order is logical or irreversible. It is easier for them to understand—''Feed the baby before you put her to bed'' than ''Feed the baby before you pick up the phone'' (E. Clark, 1971; Kavanaugh, 1979). Knowing this can help adults when they give children instructions. Giving directions in the actual order they want them followed—''Wash your hands, and then come to the table''—is likely to be more effective than saying, ''Before you come to the table, wash your hands.'' During the preschool period, children gradually learn to use even more complex sentence structures like relative clauses—''Here's the doll that I need''—and embedded relative clauses—''The doll that I need is up there.''

By first grade, most children have at their disposal all the basic sentence forms in their native language.

How do children acquire syntax?

In just a few short years, most children have become fluent sentence makers. They learn language with an ease and eagerness that older people, struggling over their foreign-language texts, can only envy. Just how *do* young children acquire syntax? On one level it is obvious that both innate abilities and environmental input are involved in the acquisition of language. Innate abilities clearly are involved because only children, not their pets, learn to speak in sentences, despite the fact that they both hear sentences. Environmental input clearly is involved because American children learn to speak English whereas Mexican children learn to speak Spanish, despite the fact that they both have the same innate language abilities. The *emphasis* given to innate versus environmental factors in explaining how children acquire language, however, has been a matter of some debate.

Behaviorist B.F. Skinner (1957) suggested that children learn language as adults systematically reinforce their efforts. When parents hear their baby babble "nananana," Skinner proposed, they smile with delight and talk back to the baby. Pleased, the baby babbles again. Mothers give big hugs and lots of praise when babies first say "Mama." They fill the bottle with milk when babies say "baba," and later, as their children are learning to put words together, parents are responding, mirroring, modeling, and correcting. Their children then imitate the correct forms they hear from their parents.

Noam Chomsky (1957, 1965, 1968), a linguist, had a different explanation of how children learn language. He admitted that reinforcement and imitation were important but claimed that they were not sufficient to account for children's acquisition of language. Children, he pointed out, utter sentences for which they have never been reinforced. In fact, they constantly utter expressions they have never even heard before—original creations, "We go store now buy present Jerry baby," and mistaken constructions found nowhere in adult speech—"foots," "runned," and "Allgone wet." Children also vary considerably in the degree to which they imitate the speech they hear from others. Some children imitate hardly at all. Others, much to an embarrassed parent's dismay, are amazing in their ability to echo something someone has said, even days later. Chomsky called attention to the remarkable capacity of the child to understand and create sentences, to generalize, and to process language in a variety of very special and complex ways that must be largely innate.

Imitation may be how children say words for the first time. But they then apply these words to new objects and use them in new sentences. In learning syntax, children first must understand the form of the sentences they hear; then they may imitate the form; finally, they use the form in their own sentences (Whitehurst and Vasta, 1975). Children seldom imitate entire sentences verbatim. Hearing "I love you, Peter," the child does not say, "I love you, Peter." If he imitates it at all, he is more likely to say, "I love you, Mommy," imitating the *form* of the sentence. Even with encouragement, though, children do not imitate syntactic forms that they do not use spontaneously.

ADULT: Adam, say what I say: "Where can I put them?"
ADAM: "Where I can put them?" (Slobin, 1971, p. 52)

ADULT: This one is the giant, but this one is little.
CHILD: Dis one little, annat one big (Slobin and Welsh, 1973, p. 490).

Children seem to imitate constructions that are just entering their repertoire, not those that are totally new (Bloom, Hood, and Lightbown, 1974).

Just as children seldom imitate their parents' sentences, their parents seldom correct their children's sentences—at least not their syntax. If they try, usually they are unsuccessful.

CHILD: Nobody don't like me.
MOTHER: No, say "Nobody likes me."
CHILD: Nobody don't like me.
MOTHER: No, say "Nobody likes me."
CHILD: Nobody don't like me.
MOTHER: No, now listen carefully: "Nobody likes me."
CHILD: Oh! Nobody don't likes me (McNeill, 1966, p. 69).

Motherese is an attempt to talk so that the young child can understand. Sentences are short, simplified, and repetitive.

Inspired by Chomsky, developmental psychologists now ask *how* the child's innate language abilities interact with environmental factors to create a speaking, comprehending child.

How parents talk to children

Day in and day out, as parents and children do things together, parents talk. They talk about people, about things, and about events. Parents comment on what is obvious and salient to their children. We have already discussed (in Chapter 6) the process of early *word* learning. Here we are concerned with how children learn *syntax* from their parents' speech. As parents talk to their youngsters, the children may learn something about the order of words. Parents format conversations with their young children: "What's that? You know. A kitty." They also mend their children's failed messages: "What's 'itty'? Do you mean 'kitty'? No? Oh, you mean, 'Sit here'" (Bruner, 1983).

But parents do not usually give specific lessons in syntax to their children. Instead, parents attend to the messages their children are trying to convey (Brown and Hanlon, 1970; Hirsh-Pasek, Treiman, and Schneiderman, 1984; Slobin, 1975). As in adult conversation, the goal is communication. When a girl with sopping overalls says, "I not wet my pants," her mother is more likely to respond, "Yes, you did," than to ask the child to say, "I did not wet my pants."

Adults typically speak to young children in a special, simplified way called **motherese** (Newport, 1976). Motherese consists of short, simple sentences that may be repeated in different forms for emphasis—"Don't fall, Mikey. Don't fall. Watch out. You'll fall." "Megan, come here. Put the ball down. Good girl." In contrast to speech to adults, motherese has shorter sentences separated by longer pauses, more commands and questions, more present tenses and references to the here and now, fewer pronouns, modifiers, and conjunctions. Pronouns, which are difficult for young children, may be dropped in favor of names:

Did *Natalie* make that?

Tell *Mommy* a story.

As children learn more language, their parents' speech to them changes (Kavanaugh and Jirkovsky, 1982; Slobin, 1975). The parents' sentences get longer; declarative forms replace questions and commands (Maratsos, 1983; Newport, 1976). But there is no *simple* relation between parents' and children's syntax (Chesnick, Menyuk, Liebergott, Ferrier, and Strand, 1983; Kavanaugh and Jirkovsky, 1982; Nelson, Denninger, Bonvillian, Kaplan, and Baker, 1983). Motherese is not a deliberate attempt by adults to teach children language. In fact, adults talk to their dogs with the same kind of language as motherese, and clearly they are not trying to teach their dogs syntax (Hirsh-Pasek and Trieman, 1982). Motherese is mainly an adult's attempt to speak so that a child can *understand*. In one study, for example, adult subjects told stories to a 2-year-old boy. Whenever the boy said ''What?'' or ''Huh?'' to indicate he hadn't understood, the adults shortened their next sentence—in good motherese fashion (Bohannon and Marquis, 1977).

The shorter utterances of motherese very likely ease children's comprehension, because young children's memories are limited. Pauses between sentences likely help children to pick out syntactic units (Hirsh-Pasek, Nelson, Jusczyk, and Wright, 1986). Utterances repeated in slightly different forms illustrate different syntactic arrangements and are related to increases in children's use of longer and more complex sentences (Hoff-Ginsberg, 1986). Simpler sentences probably help children to pick out important words and sentence forms. But although some simplification probably does help children to learn language, it does not follow that the simpler the language spoken to children, the better. Children are not encouraged to advance linguistically if the language they hear is reduced to their level; it should be somewhat more complex (Clarke-Stewart, VanderStoep, and Killian, 1979; Gleitman, Newport, and Gleitman, 1984).

In helping children to learn language, extending or expanding their speech is more effective than speaking simply, it seems.

CHILD: Doggy out?
MOTHER: Doggy wants to go out?

Expanding children's utterances does seem to be related to their language competence—as long as the expansions are simple and to the point. Keith Nelson and his colleagues (Nelson, 1980; Nelson, Denninger, Bonvillian, Kaplan, and Baker, 1983) found that some mothers of 2-year-olds expanded just one part of the children's utterances:

CHILD: Broke.
MOTHER: The truck broke.

Other mothers expanded on two or three parts of the children's utterances:

CHILD: Broke.
MOTHER: The big truck broke its wheel.

Two-year-olds whose mothers made simple expansions of what they said spoke in longer sentences and used more auxiliary verbs at younger ages. The 2-year-olds who heard complex expansions progressed more slowly. Simple expansions help children to notice and analyze a new syntactical structure. When mothers expand on their children's speech simply and reasonably, they hold the children's attention and apparently help them to acquire language.

Parents' speech to children provides the children with information about language and prods them to analyze it. Parents' complex utterances illustrate regularities in language, and parents' questions to their children prod the children to produce new utterances. In a recent study, it was found that when mothers' language to their children contained more complex utterances, more repetitions, and more questions that required more than yes/no answers, children's language developed more quickly (Hoff-Ginsburg, 1986).

Although parents' expansions, questions, and repetitions speed up their children's early language growth, it is worth noting that even the children of parents who do not expand their children's sentences turn into fluent speakers by 4½ or 5 years. What is more, in some other cultures, parents do not modify their speech in the ways that American parents do. Among the Pacific Kaluli tribe, for example, there is no motherese (Schieffelin and Ochs, 1983), because parents there believe that if adults spoke baby talk, children would sound babyish—not a desirable state of affairs to the Kaluli. But Kaluli children still learn to talk. Given the wide variety of conditions under which children learn to speak, from occasional comments to endless expansions, from casual conversations to deliberate instruction, it is clear that children have many chances to learn this crucial skill.

Innate abilities and constraints

Ultimately the task of learning language is up to the child. No matter how much or how little adult speech is directed to children's ears, no matter how much or how little it is modified, children must still make sense of the underlying patterns within the surrounding babble. How much of language learning is the result of children's innate language abilities? Cases in which children have never been exposed to language tell us something about innate predispositions to learn language.

Deaf children, for example, do not hear language. Yet when the spontaneous gestures of deaf children who have not been taught sign language are analyzed, they are found to have a number of properties of spoken language (Goldin-Meadow and Feldman, 1977; Goldin-Meadow and Mylander, 1985). Like hearing children, those who are deaf begin with single gestures and use them to point to familiar objects. Deaf children acquire a vocabulary of gestures that, like hearing children's words, refer to actions, objects, and people and are used in a consistent order. The abilities to use and order signs may be innate. Unlike hearing children, though, deaf children never spontaneously develop "frills," like articles. These aspects of language depend more on environmental input.

Occasionally, even hearing children are deprived of normal language. One child, Genie, for example, was discovered and rescued at the age of 13, unable to speak or understand language, unable even to stand up (Curtiss, 1977). Genie had been isolated, abused, and neglected by her father. After she was discovered, Genie was given intensive therapy and language lessons. By the age of 19, she spoke in sentences and expressed normal feelings. But she had not mastered all aspects of language. She could not use words like *what, that,* and *which.* She could not use the passive voice ("The child was found") or questions with an inverted auxiliary verb ("Is she talking?"). She used few auxiliary verbs at all and said, "She gone home" rather than "She *has* gone home." She could not put more than one simple idea into a sentence. Despite the great progress that Genie had made in language and in social behavior, she had the telegraphic speech of a typical 2- or 3-year-old.

It has been suggested that children must learn language during a critical period between infancy and puberty, if they are to speak normally. The preschool years may well be a period when children are especially primed to acquire language (Lenneberg, 1967). Once beyond this critical period, the very physiology of the brain may prevent full and easy language acquisition. Genie did learn to speak after she was 13. But she did not master an adult's level of complexity in her language.

Research on both deaf children and deprived children thus suggests that children have innate abilities to make up sentences from the syntactic units of subject, action, and object. More complex sentence forms, however, such as those with auxiliary verbs and relative clauses, are apparently more fragile. They appear later or not at all in a child's speech if the environment does not provide models as the child is learning to put sentences together. Other abilities that seem to be innate and that affect children's acquisition of language are children's predisposition to look for regularities in language, to generalize about syntax from the sentences they hear, and to hypothesize about the rules of language—"*-ed* means it happened in the past." According to Chomsky, innate constraints limit these hypotheses. Young children's limited memory capacity also innately constrains the length of the sentences that they can utter. In their short sentences, children therefore say the words that they hear best—those that are stressed in adults' speech—and those that are the most informative and communicative. Mature, complex speech clearly requires both innate abilities and environmental input.

The context of cognitive development

Just as language development depends on both innate abilities and environmental input, so does cognitive development in general. Even children who inherit genes for brilliance will stagnate mentally if they have no chances to learn first-hand about people and things. In this section, we discuss how several factors in the environment affect children's cognitive development in the preschool years.

Toys

Sam stayed in bed unusually late one morning. When I peeked in, he was sitting up in bed "reading" a story to his animals and dolls, fireman's helmet on his head, slippers on his feet, and a necklace made of dry noodles around his neck.

Natalie took a coffee tin from the kitchen counter and banged on the lid with two spoons. "This is the drum," she told her younger sister. "You follow me and be a horn."

The best toy Erica ever got was the crate from a washing machine. She turned it into a house, a fort, a hospital, a school—an endless array of spaces for playing in.

Toys need not be elaborate, complex, and expensive to satisfy children's needs to explore, experiment, and learn. Two- and 3-year-olds rifle through drawers and boxes, pull books off shelves, push brooms and vacuum cleaners, crawl into closets and under beds, scribble with crayons, and dress dolls. They try to ride the family dog, wear Daddy's glasses to "read" the *TV Guide*, swathe themselves in Mommy's scarves and high heels, and bathe their dolls in the sink. They slide down sliding boards, climb jungle gyms, pump swings, tunnel through sand, and run after ants and butterflies. As they do, they satisfy their curiosity, add to their knowledge about the world, and adapt to it.

Curiosity bubbles from most children like water from a garden hose, especially when they are exposed to stimulating surroundings and intriguing new objects (Berlyne, 1950; Sussman, 1979; Switsky, Haywood, and Isett, 1974). New objects should be complex enough to catch and keep children's interest but not so complex as to be overwhelming. The 3-year-old faced with a 10-piece mouse puzzle is likely to be fascinated but faced with a 50-piece abstract puzzle is likely to be overwhelmed and frustrated. When their toys and play spaces are stimulating and right for their age, they help children in both their make-believe and exploratory play (Bradley, Caldwell, and Elardo, 1977).

Toys also can teach children about colors, shapes, sizes, numbers, and words. Puzzles teach that things can be broken down into pieces—and rebuilt.

Stimulating toys satisfy children's needs to explore, experiment, and learn. Two-year-olds may favor playthings that allow simple manipulation and movement. Four-year-olds favor objects that allow for more advanced symbolic play.

Crayons, paint, and clay allow children to create symbols. Researchers who have observed children's play with different kinds of toys have found that building materials like blocks and boards, pretend materials like dolls and dress-up clothes, and academic materials like books and puzzles are most likely to promote rich and complex play in preschoolers (Sylva, Roy, and Painter, 1980, and other studies reviewed in Clarke-Stewart, 1982; Minuchin and Shapiro, 1983). Sand, clay, buttons, and other materials for "messing around" encourage creative, experimental, but somewhat less complex play. Small toys like guns or checkers, microscopes or gyroscopes, encourage still less complex play, for children simply do what the materials suggest. When materials are scarce or inflexible, children do less playing, and their play is neither complex nor intellectually challenging.

In their play with toys, young children learn about the way things work, and they grow ever more skillful and adept. Children may first explore the properties of unfamiliar objects—looking, touching, moving, opening, closing—and wondering, "Hmm, what does this thing *do*?" Later, as the object becomes familiar, they engage in play that answers, "Now what can *I* do with this?" Exploration and play with objects also can help children to solve simple problems. In one study, for example, 3- to 5-year-olds were asked to remove a piece of chalk from a box they could not reach (Sylva, Bruner, and Genova, 1976). To get the chalk, the children had to clamp together two sticks and extend them toward the box. Children who had been allowed to play with the sticks first were as successful in reaching the box with the sticks as were children who first saw an adult demonstrate how to solve the problem. By exploring and playing at home, children learn how to get the mail, reach cookies on top of the refrigerator, open doors, turn on the oven, stack pots, nest spoons, and dial the telephone.

Television

> Mikey used to go to a babysitter who *seemed* to take excellent care of him. But I found out that she parked him in front of the television for most of every afternoon, especially on days when they couldn't go outside, and so I found other day care for him. We just felt that watching television wouldn't stimulate him the way we thought he needed.

In many households, the television plays as long as anyone is at home, and it is often a child who perches intently before the screen, waiting for the next scene. Nearly every household in the United States—97 percent—contains a television set. No wonder that many children watch television more than they do anything else except sleep (Keye, 1974). Although children average close to four hours a day in front of the television, they get little intellectual stimulation from the programs they watch most often (Liebert and Poulos, 1975). The young children who spend the most time watching television, in fact, develop intellectually more slowly than other children (Carew, 1980; Nelson, 1973). In one study, a group of children randomly assigned to restricted television viewing were found to spend more time reading, to be more careful and reflective on tests, and to improve in their performance scores on IQ tests (Gadberry, 1980).

Spending time watching television generally does not speed children's language development. Children of deaf parents, for example, do not learn language if all they have to listen to is television (Sachs and Johnson, 1976), and probably for many of the same reasons, children do not learn to speak a foreign language simply by watching foreign television programs (Friedlander, Jacobs, David, and Wetstone, 1972). But although television *in general* does not teach children cognitive or language skills, certain children's television programs actually may (Rice, 1983). Some educational shows ("Mr. Rogers' Neighborhood," "Sesame Street," "Electric Company") highlight words and phrases, repeat, and refer to things that are present, concrete, and explicitly clear. Children who watch "Sesame Street" learn the alphabet, parts of the body, and properties of objects such as size, amount, and position earlier than children who do not. After a year of watching "Sesame Street," they do better on alphabet and vocabulary tests than other children (Bogatz and Ball, 1971; Minton, 1972). Television may be good or bad for children's intellectual development. It all depends on what the children watch.

Parents

It is, of course, parents who buy the toys and turn on (and off) the television set. They are their children's first teachers, and so they have the enormous responsibility for making their home stimulating enough to keep their preschool children's minds alert and active. It must be safe yet challenging, organized but not too restrictive, intriguing but not overwhelming. Parents also have the responsibility for offering their children the activities that foster exploration and learning, and they are children's first models for intellectual interests. Several researchers have shown that parents who provide appropriate and varied play materials, who are responsive, who participate in their children's games and activities, and who help them with exploratory play encourage their children's development (for example, Bradley, Caldwell, and Elardo, 1977; Bradley and Caldwell, 1984)

Parents can deliberately guide and educate older preschool children as well as provide them with a stimulating environment. Parents who show their children how to complete puzzles, to give the doll a ride in the carriage, to build a skyscraper out of blocks, and who read stories to their children are helping them develop intellectually. Mothers who rehearse specific content with their children, as opposed to just chatting or playing with them, foster their children's learning of specific content—for example, knowledge of letters and numbers—research with 4-year-olds shows (Price, Hess, and Dickson, 1981). When mothers ask their preschoolers to remember things and help them to do so, the children remember better. In one study of 3-year-olds, for example, the children whose mothers asked questions that made demands on their memories remembered better than other children (Ratner, 1984). When parents link their children's current experiences to a broader family and social context, children are more likely to reason deductively. When a child points to

Parents foster the cognitive development of children by providing a stimulating environment, by being responsive, and by deliberately guiding and educating them. Studies of transracial adoptions can help tease apart the relative influences of heredity and environment on intelligence.

a picture and says, ''That a cow,'' all mothers are likely to respond by saying, ''Yes, and it says 'moo' '' or ''Yes, and there are two of them. Can you see the other cow?'' But only some mothers create ''world links'' and say, ''Yes, and do you remember when we all went on a picnic and the cow nibbled your hat?'' or ''Yes, and it's the same color as the cow who jumped over the moon. You used to love that rhyme when you were a baby.'' In one study, only mothers who made these world links were found to have preschoolers who reasoned deductively (Mills and Funnell, 1983).

In Chapter 6, we said that it was not so clear that the environment had an effect on infants' cognitive development. The primary link between infants' and parents' intelligence seemed to be genetic. In the preschool years, the effect of home environment and parents' behavior is more marked. In fact it is in this age period that parents may have their greatest effect on their children's cognitive development. In middle childhood, instruction and experiences at school are likely to contribute more than parents to children's intellectual progress.

Programs

In the past 20 years, many attempts have been made to improve the learning opportunities of preschool children. These attempts have been focused either on changing the children's home environments or on enriching their education. One such program designed to enrich preschool children's education was the Abecedarian Project carried out at the Frank Parker Graham Center in Chapel Hill, North Carolina. This program provided full-time day care in a high quality center for children from poor black families. Children were assigned at random to attend the program or to remain at home with their mothers from infancy to school age. The program was a significant success. Children who attended the program scored higher than the control group on IQ tests at ages 2 and 3 (by 10 and 15 points) (Ramey and Campbell, 1979).

A model preschool program can modify preschoolers' intelligence. Many investigators have reported IQ increases of from 5 to 15 points in children who attended educational programs.

When they were 4 and 5 years old, children in the program still scored higher than the control children (by 12 and 7 IQ points) (Ramey and Haskins, 1981).

The **Head Start** program is perhaps the best known and certainly the most comprehensive enrichment program for preschoolers. It began in the 1960s as an eight-week summer program for children who were about to enter school but soon turned into a year-round, center-based program for disadvantaged preschool children. Currently, about 450,000 poor and minority children attend Head Start every year. Head Start programs provide medical and dental examinations and immunizations for children, a hot meal and a snack, as well as educational activities for the children and classes for parents in home economics, buying and preparing food, and child care. The educational activities for children, in some Head Start programs, consist of traditional middle-class nursery school routines such as coloring and dress-up, being read to, learning rhymes, drawing, coloring, cutting, playing with blocks and puzzles, riding tricycles, and learning about animals. In other programs, children's education comes from a strictly scheduled and highly structured academic curriculum. Still other Head Start programs have adopted a low-key ''discovery'' approach to education. Children are free to explore and learn from a smorgasbord of materials—wooden blocks, sand, water, weights, science materials, geometric forms, books—laid out for them by the teacher. The teacher responds with a ''mini-lesson'' only when specifically asked by a child. There also are programs that stress language and communication skills, others that emphasize emotional expressiveness and creativity, and still others that focus on social interaction and adjustment.

A number of evaluative studies have been conducted to measure the effect of Head Start on preschool children. One of the first, the Westinghouse Learning Corporation and Ohio University study (1973), indicated that Head Start had a significant effect on children's IQ scores, raising them 10 points, but that the gains made did not last once the children entered school. Other studies revealed that Head Start children, compared to other disadvantaged children not in the program, were more attentive and less impulsive, were more receptive to language, and had greater curiosity and motivation (Brink, Ellis, and Sarason, 1968; Lesser and Fox, 1968).

Many other smaller and more carefully controlled programs have also produced IQ gains of 5 to 15 points for the children attending them. These IQ gains last a year or two and then begin to diminish. If no further intervention occurs, the gains have disappeared by the time children are 10 to 17 years old (Gray, Ramsey, and Klaus, 1982; Lazar, Darlington, Murray, Royce, and Snipper, 1982; see Figure 9.6). Indirect benefits like improved self-esteem and

PROGRAM EFFECT FADE-OUT

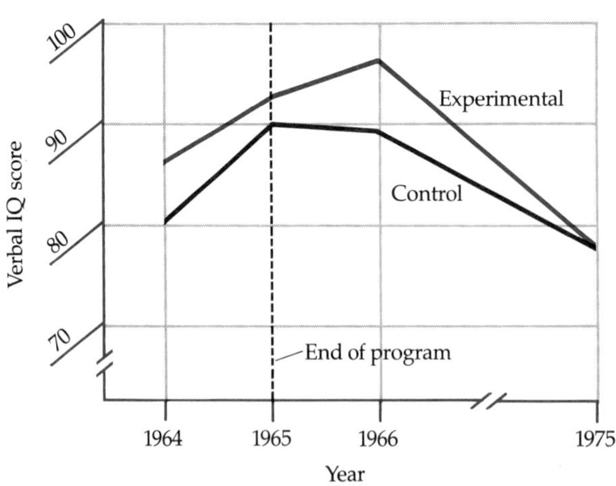

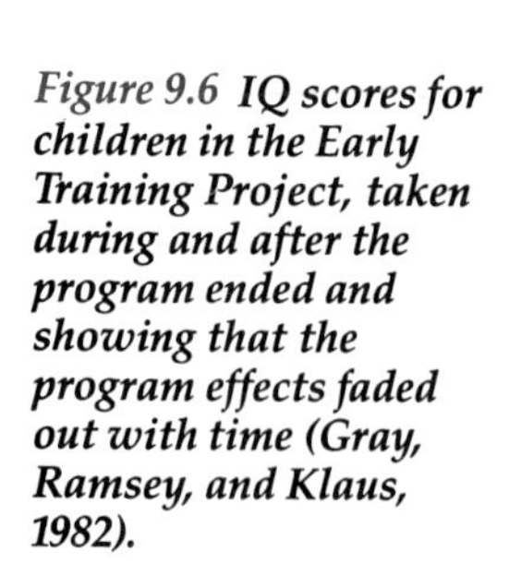
Figure 9.6 ***IQ scores for children in the Early Training Project, taken during and after the program ended and showing that the program effects faded out with time (Gray, Ramsey, and Klaus, 1982).***

lifespan focus

Is early childhood a "critical period"?

Many people, both psychologists and the general public alike, think that children's experiences in their first several years of life forever after affect the children's minds. But *is* early experience critical, and are its effects really permanent? Some studies have shown that the effects of early experience are at least partly reversible. Most developmental psychologists now believe that psychological development in children is not characterized by the strict *critical periods* that characterize physical development before birth. There may be *sensitive periods* when specific forms of development occur most quickly and easily. But most forms of development, they believe, can be acquired later, if with more difficulty, once the sensitive period has passed.

Children from deprived early environments who were later adopted into middle-class families provide one source of evidence about the reversibility of early experience. Studies (reviewed by Clarke and Clarke, 1976; Clarke-Stewart and Apfel, 1979) show, for example, that institutionalized infants who are adopted in infancy recover from their early deprivation completely by age 4. But children who are over 2 when they are adopted recover greatly but not completely. Another source of evidence about the reversibility of early experience comes from children who go in the opposite direction. In enriching early childhood programs, children from poor families flourish. But once out of the programs and in a regular grade school, their intellectual gains drop off (Lazar et al., 1982). Reversibility clearly goes both ways.

Finally, evidence about reversibility comes from studies in which the environments of young children are substantially altered in the normal course of child rearing. Studies of the development of children in Guatemala by Jerome Kagan and his associates (Kagan and Klein, 1973; Kagan, Klein, Finley, Rogoff, and Nolan, 1979) provide one example. Infants in the Guatemalan village of San Marcos spend most of their first year inside a dark hut. No one talks or plays with them much, and they look depressed and withdrawn. As 1- to 2-year-olds, they are retarded in the development of speech, concepts of objects and people, and other forms of behavior. But once they venture outside the hut, in their second year, they begin to meet the variety that the world offers, and their lives become more stimulating. Compared to children from a nearby village, in which infants are not so severely restricted, the San Marcos children under age 10 or 11 do poorly on tests of memory and reasoning. But by age 15 to 18, the San Marcos children have more or less caught up to the children in the other Guatemalan village and to children raised in the United States.

The results of these three kinds of research support the view that both early and later experiences are important. It is not the case that early experience is critical and its effects irreversible. There are continuities in development, but what happens at age 1 does not totally determine the way a person will be at 17. Lack of early stimulation can be compensated for in later opportunities. The benefits of early enrichment can fade away—almost completely—if the environment becomes barren or threatening. Later experience is not the same as early experience, but it may be no less important. In lifelong human development, it is the whole that is important.

more positive feelings about school are likely to endure, however. Disadvantaged children who attended preschool enrichment programs are less likely to stay back a grade in school or to need special education than those who did not (Lazar et al., 1982). Even their brothers and sisters benefit indirectly from the effects of their participation in an early intervention program (Klaus and Gray, 1968). Later, Head Start children are more likely to graduate from high school and get full-time jobs than disadvantaged students who did not attend Head Start (Deutsch, 1985).

Middle-class children as well as poor children, of course, can benefit from educational enrichment. By the time they are 4 years old, most children in the United States are attending some form of **nursery school** or day-care program, and with each passing year, so are more and more 3-, 2-, and even 1-year-olds. Nursery schools traditionally have prepared children for kindergarten and first grade. Over the years, many evaluations have shown that when children start school, those whose parents have sent them to nursery school are advanced over children without this experience (Clarke-Stewart and Fein, 1983). Day-care programs with an educational component, research shows, also temporarily accelerate children's intellectual development. Day-care-center children are more advanced than home-care children on both verbal and nonverbal skills: language comprehension and fluency, eye–hand coordination, and drawing. In the Chicago Study of Child Care and Development (Clarke-Stewart, 1984), for example, 2- to 4-year-old children in nursery school or day-care-center programs were 6 to 9 months ahead of home-care children on tests of intellectual development. This intellectual advancement may stem from the stimulation offered in these programs, the availability of a variety of materials and equipment, and the more systematically educational environment of the early childhood program compared to a home. Like the advantages that children get from Head Start and other intervention programs, however, the intellectual acceleration of children who were in day care relative to that of children who stayed at home diminishes by the time first grade rolls around.

Summary

1. Children use symbols that are verbal—a word, a made-up name—and physical—a drawing, a doll, a stick used as a gun. Toddlers' pretend play is the first evidence of their symbolic thinking. During the preschool period, pretend play changes from symbolic play with objects, to play in which children act out roles, to dramatic play with roles announced ahead of time.
2. Children between 4 and 6 years old think intuitively and therefore sometimes confuse the real with the imagined, the animate with the inanimate, and their own perspective with others'.
3. Young children cannot perform logical operations such as addition, subtraction, or multiplication, and so their thought, according to Piaget, is "preoperational." They sometimes reason transductively—from one particular to another rather than from the general to the particular—although they are capable of simple deductive reasoning. They cannot solve conservation problems.
4. Preschool children are neither as systematic nor as selective in paying attention as older children are, perhaps because they are less efficient at encoding information, because they do not understand what others think it is important for them to pay attention to, or because they cannot consciously direct their attention so that they focus only on relevant aspects of tasks. Memory improves with age, as children get better at encoding information.
5. Children begin speaking in one-word and then two-word telegraphic utterances. Overregularization is a common kind of mistake children

make as they are learning to speak in sentences and shows that they have learned the rules for making past tenses and plurals. Language is the result of both innate abilities and environmental input.

6. Although adults do not give children deliberate language lessons, they speak to them in motherese, a special and simplified language adapted to young children's limited language abilities, and they expand their simple utterences into more complex sentences.
7. The number and kind of toys they play with, television shows they watch, and the stimulation they receive from their parents and peers all affect the cognitive development of children.
8. Programs to improve the learning environments of young children can improve children's performance on intelligence tests. Young children's cognitive development is also affected by their attendance at nursery school and day-care programs with an educational component.

Key terms

symbol
symbolic play
intuition
animism
egocentrism
operations
preoperational period
transductive reasoning
conservation
reversibility
reciprocity
decentration
horizontal décalage
recognition
recall
syntax
telegraphic sentence
overregularize
motherese
Head Start
nursery school

Suggested readings

CLARKE, ANN M., and CLARKE, A. D. B. (Eds.). *Early Experience: Myth and Evidence.* New York: The Free Press, 1976. A collection of chapters demonstrating the remarkable ability of children to recover from extremely depriving and depressing early experiences. This book takes the position that early childhood is not a critical period for intellectual and social development.

DE VILLIERS, PETER A., and DE VILLIERS, JILL G. *Early Language.* Cambridge, Mass.: Harvard University Press, 1979. A concise guide to the major advances between birth and six years of age as children learn language. The book includes recent findings in psycholinguistics and engaging examples of children's speech.

FLAVELL, JOHN. *Cognitive Development* (2nd ed.) Englewood Cliffs, N.J.: Prentice-Hall, 1985. Well written textbook on cognitive development by an eminent researcher in the field, who is sympathetic to Piaget's views but at the same time responsive to the findings of contemporary researchers in cognitive development.

GARDNER, H. *Artful Scribbles: the Significance of Children's Drawings.* New York: Basic Books, 1980. This book takes children's artwork seriously and ties the development of drawing, from scribbles to abstract forms, to other aspects of child development.

PERLMUTTER, MARION (Ed.). *Children's Memory.* San Francisco: Jossey-Bass, 1980. The studies reported in this volume of the New Directions for Child Development series focus on infants' and young children's cognition and memory as demonstrated in everyday life.

PIAGET, JEAN. *Play, Dreams and Imitation in Childhood.* New York: Norton, 1962. A classic that weaves together Piaget's theory with wonderful examples of children's play, dreams, and imitation.

chapter ten

Social and emotional development

Between the first and sixth birthdays, the totally dependent infant is transformed into a proudly independent child ready to venture forth and meet the rest of the world. The bravery of the young child who dares, for the first time, to stay alone at nursery school, or at a friend's house, or in the doctor's examining room is touching evidence of this progress. During the preschool years, young children develop autonomy and extend relationships beyond the family. They learn social skills and social rules, how to interpret social situations, and how to express and control their emotions. They find out how boys play and girls play. And they find out more about themselves, as their self-concept and understanding of gender roles deepen.

Relations with parents

During the preschool years, the close, constant physical contact between parents and infant, so important to the infant's physical and emotional well-being, gradually gives way to a more autonomous relationship. Toward the end of the first year, the infant crawls away from the parents for brief periods of exploration. The toddler ventures out of the parent's sight, to find a toy or to follow someone into another room. The preschool child becomes less distressed by a separation from the mother or father, and the distance that the child can tolerate away from their comforting presence gradually increases.

Between the ages of 2 and 4, children comfortably play and explore farther away from their mothers and need less physical contact than they did in their earliest years.

Developing autonomy

To study the gradual growth of autonomy, researchers have asked mothers to leave their children briefly in a university playroom either with a stranger or alone (Feldman and Ingham, 1975; Maccoby and Feldman, 1972) or have observed naturally occurring separations between parents and children (Clarke-Stewart, 1973; Clarke-Stewart and Hevey, 1981). Children between 1 and 4 years old, they have found, show less and less distress at these brief separations and seek less physical contact and closeness to their mother. As long as they can still see and hear their mother, 2-year-olds are willing to leave her side (Carr, Dabbs, and Carr, 1975; Mahler, Pine, and Bergman, 1975). Even the sound of their mother's voice or the televised image of her face helps 3-year-olds to enter and play more comfortably in an unfamiliar playroom than if they are alone. But let a stranger be present or the situation be stressful in some other way, and 3-year-olds tend to be upset at separations from their mothers (Adams and Passman, 1979). One-third of the 3-year-olds in one study (Murphy, 1962) had trouble separating from their mothers to leave home or to go upstairs with an examiner. Only among 4-year-olds is distress at brief separations from their mothers really rare (Marvin, 1977).

> Robbie was 2½ when he started day care in the home of a woman who took care of three other children. For several weeks, he stood at the door, crying "Mommy, Mommy" as I left. As he got to know the babysitter and the other children, he stopped crying. But he acted subdued when he was dropped off every day until he was about 4.

When they are playing outdoors—in a park or playground, for example—children establish physical boundaries and stay within them without having to be retrieved. They move around in brief bouts, move, stop, look, and return, as long as their mother is in view and facing them. In these unfamiliar outdoor settings, children venture from their mothers a distance that increases with impressive mathematical regularity: for every month of increasing age, the children will venture about one foot farther from their mothers in 15 minutes of play (Rheingold and Eckerman, 1971).

Although they venture farther away from their parents and tolerate longer separations from them, older children are no less emotionally attached. Physical contact may diminish, but the attachment to parents remains strong. Older children show their attachment in new ways. Now they show their mothers a toy, smile, and talk to her from a distance rather than hanging on to her hand or sitting in her lap (Clarke-Stewart and Hevey, 1981; Maccoby and Feldman, 1972).

As young children are venturing ever farther from their parents, their parents are gradually letting go of the children. In small ways, parents encourage their children to be independent. First they lessen their physical contact with the child, then they relax their constant attention and surveillance, and finally they start fewer interactions (Clarke-Stewart and Hevey, 1981).

> One evening when Sam was 2½, I realized that for the very first time in his life, 20 waking minutes had passed in which he hadn't needed me or his father. Sam was intent upon sitting under a chair, which he pretended was a hotel. From then on, those brief periods of independence—and relaxation for us—grew more frequent and longer.

The development of autonomy has its rocky moments at first. When 1½- or 2-year-olds get a whiff of independence and strike out farther on their own, they turn their backs on their parents not only literally but figuratively, too. They want to do everything by themselves and for themselves. They want to see, feel, *test* everything for themselves. "*Me* do it!" is the rallying cry of the 2-year-old. If they meet resistance, as they often do, they may explode in anger, throw their food on the floor, kick and bite, yell and wail, hold their breath until they turn blue. Tantrums may rain down like hailstones, and the shrieks emanating from the home of a normal 2-year-old may make the neighbors wonder just what is *going on* in that family. These moments—which led some psychologists to refer to this period as the **terrible twos**—reflect the child's bid for autonomy raised to fever pitch. For 2-year-olds, the world is full of temptation and frustration.

During the so-called terrible twos, the young child may occasionally become a tiny tyrant, all in the quest for autonomy.

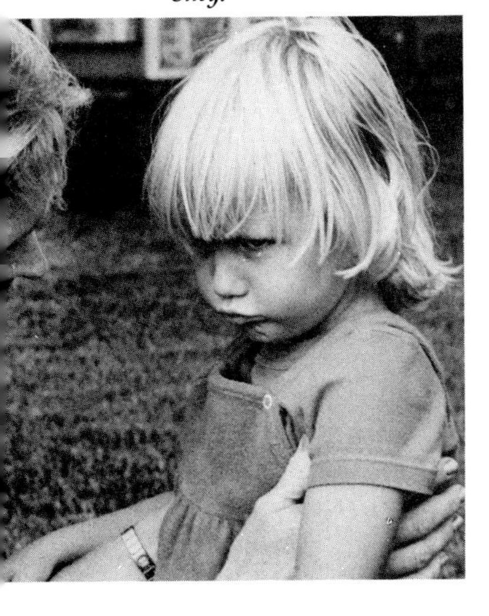

> When Kristy was 2, in the space of one morning she jammed something in the dishwasher so that it wouldn't work, spilled the contents of the refrigerator into the trash can, peeled all the labels off the cans of food in the pantry, spilled the dog's water on the floor, and threw a tantrum when I wouldn't let her "skate" on it. She threw another tantrum when she couldn't mop it up.

Luckily for children and their parents, this phase passes. In secure parent–child relationships, it passes within a year or so, and children move toward a point of balance between their wishes and those of their parents. By the time children are 3, mothers and children are approximately equal partners in the initiation of social exchanges (Clarke-Stewart and Hevey, 1981). Once children can communicate in words, their parents try to mesh their plans with those of their children—"It's going to be bath time in five minutes, Tim," or "Would you like to go in the car with me?" or "Please get your sneakers, Robbie. They're in your bedroom." The pattern of child as master and parent as slave, which prevails during infancy, gradually breaks down. Young children can wait until their parents finish what they are doing to have their demands satisfied. When Mom says that she has to leave her child alone so that she can go and make a phone call—and the child agrees—then the child is not disturbed at the brief separation (Marvin, 1977).

Continuing patterns of attachment

Not only do young children remain emotionally attached to their parents during the preschool years, but their individual patterns of attachment usually remain stable over time as well. In general, unless there are major changes in the family, children's patterns of attachment—secure, avoidant, ambivalent—remain stable between 1 and 6 years of age (Cassidy and Main, 1984). Infants who were securely attached to their mothers continue to be securely attached as young children, and the same tendency holds true for infants who were ambivalently attached or avoidant.

What is more, as we mentioned in Chapter 7, the pattern of the child's attachment to parents is related to his or her interactions with other people in the preschool period. Securely attached preschoolers, in contrast to those less securely attached, have been observed to enter a relationship with an unfamiliar man more quickly and easily, to be more relaxed and productive in competing with him to build a tower and to be more openly sad when they failed the competition (Lutkenhaus, 1984). They elicit more positive reactions and play in more interesting and innovative ways with an unfamiliar child (Jacobson and Wille, 1984), and they score significantly higher on measures of initiative and skill in getting along with other children at nursery school than insecurely attached children (LaFreniere and Sroufe, 1985; Waters, Wippman, and Sroufe, 1979). There are clear continuities in children's social and emotional behavior—over time and across partners.

Socialization and discipline

In the preschool years, the relationship between parents and children changes in another way: parents begin more actively to socialize and discipline the children, trying to shape and mold them into mature, responsible, productive, functioning members of society. As children near their second birthday, parents tend to perceive them as increasingly competent and assertive. They refuse more of the children's demands and expect from them at least some compliance and cooperation. As children grow more independent and venturesome, parents become less forgiving and more demanding. As children extend the boundaries of their experience, parents begin the process of setting limits and training their children to exercise self-restraint and respect for other people's boundaries. As children become capable of learning the habits, skills, and standards that adults consider important, parents teach them which behavior is acceptable and which unacceptable and impart to them the accumulated knowledge of their culture. There is little correspondence between the controlling behavior parents use with 1- and with 2-year olds (Dunn, Plomin, and Daniels, 1986).

Patterns of authority

How do parents try to influence their preschool children? In extensive research on parents' disciplinary styles by Diana Baumrind (1971, 1979), three distinct patterns of discipline emerged. **Authoritarian parents**, it was found, are firm, punitive, unaffectionate, unsympathetic, detached, and sparing in their praise. They give edicts and expect their children to obey them. Rules are not discussed ahead of time and are not subject to bargaining. The parents try to curb their children's will rather than encouraging their independence. Children who disobey may be punished severely and, in many cases, physically. Children in these families have few rights. Their parents make demands on them but do not reciprocally accept their children's demands. Children are

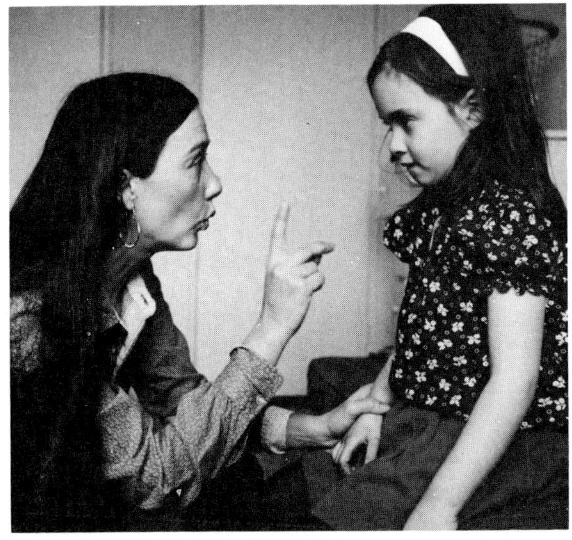

Authoritative parents set firm limits on their children's behavior but treat the youngsters with respect, explaining carefully and emphatically the reasons for restrictions. Authoritarian parents simply apply force to shape their children's behavior.

expected to curb their demanding and begging. In extreme cases, children may not speak until spoken to.

> When we were bad as kids, my mother would promise, "Just wait until your father gets home. He'll give you such a licking." She couldn't do anything with us. But my father would walk in, ask how things were, and when he heard that we'd done such and such, he got the strap and walloped our backsides.

Permissive parents do not exert control over their children. The children are given considerable freedom, few responsibilities, and rights similar to adults'. Permissive parents avoid laying down rules, asserting authority, or imposing restrictions. They tolerate and accept their children's impulses, including their aggressive and sexual impulses. Children are expected to regulate their own behavior and to make decisions on their own—when and what to eat, what to watch on television, what to read, when to sleep.

> The two girls who live across the street from us are spoiled rotten. Their parents cater to them and give them anything they want. All one of the girls has to do is pout, and the mother caves in. I've heard them yell at their parents, "Get over here right now!"—and the parents do it!

Authoritative parents see the rights and responsibilities of parents and children as complementary. As children mature, they are allowed greater responsibility for their own actions. Authoritative parents reason with their children, encourage give and take, and listen to objections. They keep the level of conflict with their children low. They are firm, loving, and understanding. They try to make their demands reasonable, rational, and consistent. They set limits, but they also encourage independence. They let their children know that they are loved, and they give them a clear idea of what is expected of them.

> If I told my parents that I *really* objected to some law they'd laid down, they always listened to my side of things. Sometimes they would even change their minds, but not often.

For their part, children actually *prefer* parents who explain, reason, and occasionally back up their reasons with physical punishment. When researchers have read children little stories about how parents might discipline their children, the children generally have said that they do not like permissive discipline and that they want parents to set limits (Siegal and Cowen, 1984; Siegal and Barclay, 1985). Children themselves, in other words, favor an authoritative style of disciplining, based on parents' sensitivity *and* firmness.

In the real world of parents and children, the distinctions in disciplinary patterns blur. Although it is true that some parents never hit and some hit a lot, that some parents often reason and some never do, even the most dogmatic parent can be lenient at times, and even the most permissive one can occasionally clamp down. In moments of calm reflection—reading a textbook or talking with another adult about how parents *ought to* enforce discipline—most people favor the path of setting firm limits for children and reasoning with them. But in moments of heat and passion, of which there are *many* in the life of even the best adjusted, most secure and loving, perfectly normal children, reasoning may fly out the window, limits may crumble, and usually measured voices may cause the dishes to rattle.

> Natalie had refused to nap, refused to get dressed, refused her lunch, sulked, hit the baby when I wasn't looking, and was generally a pain all day. When she took a big kettle and spilled water all over the kitchen floor, I slapped her bottom so hard that she was stunned. I stunned myself, too.

Despite these occasional lapses, parents hold strong—and divergent—views on whether children should be physically punished. In a recent survey of 500 readers of *Psychology Today* (Stark, 1985), 49 percent thought that children should never be physically punished, and 51 percent thought that they should be, although most were against anything more than a simple spanking on the behind. The respondents argued against physical punishment for children on the grounds that "violence begets violence" and that it gives children the message that bigger, more powerful people can act abusively. Many said that physical punishment should be reserved for occasions when children run into the street, play with matches, or reach on top of a lighted stove. Others said that spankings are warranted when children talk back, throw tantrums, bite, or lie.

Social and emotional consequences

Although in real families, disciplinary styles are not always crisply defined, the three general styles Baumrind found—authoritarian, permissive, and authoritative—have been found to be related to consistent patterns in preschool children's behavior at home, in nursery school, and in the laboratory (Baumrind, 1971, 1979; Becker, 1964; Crockenberg, 1984; Hoffman, 1970; Maccoby and Martin, 1983). Children reared by authoritarian parents tend to be suspicious and withdrawn, unfriendly and discontented. Although they may be well controlled, they are also likely to be fearful, dependent, and submissive, slow to explore and less likely to strive intellectually. Authoritarian discipline, with its threat of severe punishment, may arouse so much emotion in children that they cannot concentrate on moral reasoning, such as how wrongdoing affects other people. For this reason, these children are more likely to cheat and less likely than other children to feel guilt or to accept blame. Children heavily disciplined in their first two or three years are likely to be conforming, dependent, and inhibited; children given heavy discipline only later are likely to be hostile and aggressive.

Like the children with authoritarian parents, children with permissive parents tend to be dependent and unhappy. They are more likely than children with authoritarian parents to be outgoing and sociable and to strive intellectually, but like these children they tend to be immature and aggressive and to lack persistence and self-reliance.

Children whose parents are authoritative are most likely to be friendly, cooperative, competent, intellectually assertive, self-reliant, independent, happy, and socially responsible—an ideal combination of qualities. The best discipline for parents who wish to foster their children's social, cognitive, and moral development seems to be to create an atmosphere of warm approval, praise, and acceptance, and then to explain why the child *should* act in particular ways and to intervene actively when the child acts otherwise. Children are most likely to behave as their parents wish if the parents point out how the child's misbehavior will affect other people—''If you throw snow on their walk, they will have to clean it up again''; if they stop the child from continuing the misbehavior—''Stop throwing that snow!''; and if they enlist the child's help in repairing the damage—''Now help me shovel the snow off again.'' It also helps if parents explain other people's motives or needs to the child—''Don't hit him. He didn't mean to push you. He tripped.'' ''She's afraid of dogs, so don't let Rusty jump on her.'' ''No loud playing while Daddy is sleeping. He needs his sleep.'' This kind of discipline appeals to children's pride, competence, and concern for others. It shows them how their own behavior affects other people, and it helps them to integrate this knowledge with their capacity for empathy.

Within these associations between parents' disciplinary styles and children's behavior, though, what is cause and what is effect? It has long been assumed that discipline causes the differences observed in children's behavior. But it is entirely possible that parents of more mature and competent children follow their regimen of moderate, authoritative discipline *because* their children act compliantly and cooperatively, rather than the reverse (Lewis, 1981). Discipline is undoubtedly a two-way street. On one side, it is affected by parents' attitudes, values, and circumstances and, on the other side, it is affected by characteristics of the child like temperament and sex. Researchers have found, for example, that authoritarian discipline is related to aggressiveness and social incompetence in boys but not in girls (Baumrind, 1979; Maccoby, Snow, and Jacklin, 1984; Martin, 1975). Girls are more likely to lack social competence if their parents are permissive and do not challenge them or make demands on them to act maturely. Many gaps remain in psychologists' attempts to chart the course of disciplinary styles and their long-term social consequences, but there are clear suggestions that the effects of parents' discipline are neither simple nor one-sided.

Looking and learning

In attempting to socialize their children, parents often use less direct methods than telling children what to do and punishing them if they disobey. One way in which children learn appropriate social behavior is through being rewarded for social acts. The power of instrumental conditioning has been amply demonstrated in laboratory experiments. Parents as well as experimeters often use tangible rewards to shape their children's behavior: ''Behave in the store, and I'll buy you a balloon.'' ''Good girl; here's a piece of gum.'' But parents have to be careful about how they dole out the rewards, for the effect is not always to increase the frequency of the rewarded behavior. Children come to dislike foods and activities they once liked, researchers have shown, if they are rewarded for them (Birch, Marlin, and Rotter, 1984; Lepper and Greene, 1975, 1978). It's as though the child thinks, ''If she has to give me something for eating this stuff, I must not like it,'' or ''If I'm going to get a prize for doing this, it must be work.'' In contrast, children who are praised are likely to keep doing the things they are praised for. Whenever possible, parents should stick to compliments—''What a helpful boy you are,'' ''That's a grown-up thing to do''—rather than chocolate chip cookies or quarters for walking the dog or playing quietly.

Children also learn vast quantities of information incidentally, casually, as they observe everyday events, with no cookies, quarters, or compliments to motivate them. Children learn many things simply from watching their parents. Adults' influence on children's behavior has been shown by exposing children to an adult in the laboratory who acts in a certain way and then observing the children's behavior in a similar situation. In one experiment, over 90 percent of the 4-year-olds studied imitated the adult experimenter knocking over a doll, and 45 percent marched around the room in the same unusual way as the adult had (Bandura and Huston, 1961).

> When Jason was 2½, he turned around to watch me just as I was popping an aspirin into my mouth and started to drink some water. All he saw was my hand cover my mouth; he didn't see the aspirin. For the rest of the day, he drank by first tapping his mouth and then drinking.

In a typical experiment, the adult seen by some children is selfish; she does not donate her pennies, candy, or toys to (unseen) children who are ''in need''; the adult seen by other children generously donates her winnings. It has been shown consistently in such experiments that watching an adult demonstrate generosity increases the willingness of many children to help and share—especially if the model combines prosocial behavior with warmth, interest, and the exercise of power—as parents do at home (Mussen and Eisenberg-Berg, 1977). Do these effects last? Children in one study watched an adult donate generously and then were given the chance to donate themselves. A week later the children were more likely to donate, and three weeks later they behaved generously in a different situation. Even two to four months later, the effects of watching the model were still evident (Rushton, 1975).

In one study of observational learning in the real world, researchers watched mothers with their children during half an hour of free play with a set

Children learn many things just by watching their parents and other people. They do not need any kind of tangible reward for imitating what they see.

of toys (Waxler and Radke-Yarrow, 1975). *All* children imitated the way their mother played with the toys at least once. They were especially likely to imitate her if she was enthusiastic and demonstrated only a few ways of playing with the toys and if the child had a good relationship with her. Children do not automatically copy their parents' behavior. As children observe other people's actions, they encode their observations in words or images; they form mental representations of what they see. Then, in *some* instances, depending on how they interpret what is wanted or needed or will be rewarded, and depending on their relationship with the person, children imitate what they have observed (Bandura, 1977).

Family stressors

When we imagine "the typical American family," most of us probably conjure up an image of a father who kisses his wife and two children—Sis and Junior—goodbye every morning before he drives to work. Mom, still in her bathrobe, clears the breakfast dishes and then gets Junior and Sis ready for school. The problem with this image is that it no longer represents the typical family. Today fewer than one in five American families fit this image. The rest of us live in families where Dad may well go off to work—but Mom does too. The children—or *child,* for many families stop at one child—goes to day care or to preschool, and preschool may well have a special group for children of divorced parents. After their parents divorce, children may live with one parent and visit the other. They also may have to cope with stepparents, stepsiblings, and stepgrandparents. Divorce and full-time jobs for both parents are stressful circumstances for adults and children. They affect parents' well-being, children's development, and family dynamics.

Divorce

> When the kids' father and I got divorced, Kristy and Jason had a few rocky years. But I knew that Kristy at least had turned the corner when she saw a *Newsweek* cover story on "children of divorce" and joked, "Well, we may be miserable, but at least we're typical." It's been harder for Jason not to have his father around. I'm still not sure how it's all going to turn out for him.
>
> As I pushed Sam and some of the other 4-year-olds on the swings, I heard one kid ask, "How many dads do you have?"

Over one million couples in this country divorce every year, and 70 percent of them have children. Among parents of preschoolers, the divorce rate climbed from 8 percent in 1950 to 20 percent in 1980. At the present divorce rate, half of the children born in this decade will live at least temporarily with one parent, typically their mother. Not only do children suffer when they are separated from a parent, but the parent does too. Even the parent with custody of the children usually suffers, by losing the economic, emotional, and practical support of the other parent.

In one of the most comprehensive studies yet done of the psychological effects of divorce on children and parents, researchers collected data on 96 families with 4-year-old children through interviews, personal records, psychological tests, and observations at 2 months, 1 year, and 2 years after the divorce (Hetherington, Cox, and Cox, 1976, 1978, 1982). In all families, the mother retained custody of the child or children. In the year after the divorce, it was found, mothers became more authoritarian and less affectionate with their youngsters. Family routine became more chaotic. Children became more unruly. Boys especially were aggressive and ornery at home and in nursery school. For fathers, the first year after the divorce marked the beginning of decreasing emotional attachment to their children; the emotional bond deteriorated and the number of contacts diminished. Two years after the divorce, mothers were more patient than they had been the year before, children were more cooperative, and family routines were more stable. But there were still problems. Fewer than one-half of the fathers saw their children as often as once a week—although all lived nearby—and the boys were suffering. Most of them were more feminine and less mature than boys whose parents were not divorced. Their aggressiveness, which had been physical the first year, had been transformed into a more feminine form, verbal aggression. Many preschool sons had become ''sissies.'' Only the boys who continued to see their fathers were unlikely to have developed this problem.

One reason that preschool children may find their parents' divorce so difficult is that they lack a firm sense of the continuity of family relationships. They tend to think that a family consists of people living in the same house. If the father moves out, it seems to the child that he is no longer part of the family. Preschool children also may blame themselves for the divorce. In one study, researchers interviewed children in 60 middle-class families at the time of the divorce and one year later (Wallerstein and Kelly, 1980). Between ages 3½ and 6, children's main feeling about the divorce was self-blame. They saw the difficulty in their parents' marriage not as an issue between mother and father but as something that had gone wrong between themselves and their parents.

After a marriage breaks up, parents often become absorbed in their own difficulties, and the quality of the attention and care they offer their children deteriorates, which is another reason that children suffer from their parents' divorce. Divorce itself creates stress, and stress preoccupies parents. Parents who are preoccupied play less with their children and stimulate, support, and help them less. Their interactions are more curt, critical, and interfering (Zussman, 1980). Stress causes parents to withdraw from interaction or to act irritably and aversively with their children (Forgatch and Wieder, cited in Patterson, 1982). Continued conflict between divorced parents is still another reason that children suffer. When parents do not get along, when they fight in front of their children, no matter whether they are divorced or married, their children suffer and are likely to act out and act up. They are more likely to develop behavior problems (Richman, Stevenson, and Graham, 1983), and they act more disruptively, if their parents fight openly (Hetherington, Cox, and Cox, 1982). Exactly what makes children act up is not clear. Perhaps they imitate their parents' aggressive actions toward each other. Perhaps they absorb and reflect their parents' anger. The finding that children who lose a parent to death do not usually act disruptively points to discord between the parents as the decisive factor. Whether children develop long-term problems depends in large part on how their parents handle the divorce—whether they part amicably or bitterly, whether they fight openly or not—and on the relationships that they maintain with their children (Wallerstein and Kelly, 1980).

Mother's work

Mother's work, too, can be a source of stress for the child and for the family. We all cherish the image of a mother who greets her child's return from nursery school with an enveloping hug and a batch of warm cookies. But nowadays, about half of the mothers of preschool children work away from home. Working mothers spend less time taking care of their children than do mothers who are housewives (Hill and Stafford, 1978), and they spend less time within calling distance or in the same room but doing somethin their children, like ironing, cooking, reading, or watching televisio berg, 1977). But there is generally no difference in the amount of time working mothers and housewives spend *alone* with their children (Goldberg, 1977). Several investigators have suggested that the time working mothers spend playing and talking with their children is especially intense and that they set aside uninterrupted chunks of time to spend with their children (Hoffman, 1980; Pedersen, Cain, Zaslow, and Anderson, 1983). Others have found that when differences in the nature of mother–child interaction are observed, working mothers do pay more attention and talk more to their young children (Hoffman, 1984).

But work is not the same for all women. Some researchers have hypothesized that a job that improves a mother's morale will improve her relationship with her child and that a job that strains her will cut into her relationship with her child, and the data bear out this hypothesis (Hock, 1980; Schubert, Bradley-Johnson, and Nuttal, 1980; Stuckey, McGhee, and Bell, 1982). The cause, however, is not yet clear. It may be that happy work makes happy mothers, or it just may be that some women are more competent and happier wherever they are, at home or at work.

What are the effects on their young children of mothers' working? For one thing a mother's work is likely to enlarge the opportunities that her children can imagine for themselves. Children of working mothers, for example, can think of more possibilities in answer to questions like "What would you like to be when you grow up?" than children of housewives (Zuckerman, 1985).

> I'm an eye doctor, and Natalie's father is a surgeon, so our kids just take it for granted that both parents work. It *is* hard on them sometimes, like when we have a new babysitter or when I have to miss some event at school that's important to them. But the kids know that mommies and daddies go to work, and they've never questioned it. I'm sure Natalie will have a career when she grows up.

For another thing, many daughters of working mothers are more independent and outgoing; they are socially and personally better adjusted; they achieve more and have higher career goals; and they admire their mothers and women in general more than do daughters of housewives. Sons of working mothers, however, do not fare so well. They tend to be less well adjusted socially and to achieve less in school (Hoffman, 1984).

Some psychologists (for example, Chase-Lansdale, 1981) have suggested that the increased training in independence that working mothers give their children is too much or too soon for sons but not for daughters. But they have not yet tested the hypothesis empirically. Working mothers may favor their daughters. In one study of 2- to 6-year-olds, it was found that in families in which mothers worked, daughters got more attention than sons. But in families in which mothers did not work, the sons got more attention (Stuckey, McGhee, and Bell, 1982). Similarly, in a study of 3-year-olds and their parents, fathers and mothers who worked full time described their daughters in more positive terms and their sons in less positive terms than parents who did not both work (Bronfenbrenner, Alvarez, and Henderson, 1983; Bronfenbrenner and Crouter, 1982). Why would parents describe their sons and daughters so differently? Perhaps families in which the wife has a more egalitarian role simply do not subscribe to the traditional pattern of favoring sons and paying more attention to them than to daughters (Hoffman, 1979). Perhaps preschool boys actually are more ornery and difficult than girls and present more of a strain to their working, overburdened mothers. There is some evidence to support this (Bates, 1980b). Perhaps boys are affected more negatively than

lifespan focus

Stress and the single parent

Marsha Weinraub and Barbara Wolf (1983), of Temple University, wanted to know what it is like for children to grow up in a family with only one parent, their mother. They wondered whether single mothers face more life changes and stresses and fewer social supports or community ties than married mothers. They wondered whether single mothers have more difficulty coping with their stresses and responsibilities than married mothers. They wondered also whether interaction between parents and children is influenced by social supports and whether the relationship among stresses, support, and mother-child interactions differs in single- and two-parent families. Their subjects were 14 single mothers and their children and 14 married mothers and their children, matched for the child's age (2 to 4½ years) and the mother's education and income. Half of the single women were divorced; half had chosen to raise a child alone from the beginning. Only women who had raised their child alone from birth or shortly thereafter were included. The women were interviewed in depth and observed with their children.

For both parent and child, life in a single-parent family can bring special stresses and pain—and pleasure.

girls when their mothers work because, as we described earlier, they are more vulnerable to environmental stresses. Perhaps girls benefit more from stress because it makes their parents tougher on them, and as we have already mentioned, that toughness fosters social competence. There are many possibilities.

One researcher in this area, Lois Hoffman (1979, 1984), suggests that the reason that daughters of working mothers do better than the daughters of nonworking mothers is that they have a female model of social competence and high status and they get more training in independence within their family than daughters of nonworking mothers. Both these factors are likely to benefit girls' development. But for the sons of working mothers, a mother's work may carry negative connotations: it can mean that the father has failed to

Single mothers, the researchers found, worked substantially more hours a week than the married mothers. They had experienced significantly more stressful life changes in the previous year than the married mothers had, and the nature of the changes differed as well. Single women faced changes in jobs, living arrangements, and personal goals; married women faced pregnancy, birth, and home mortgages. Single mothers had fewer social contacts and were less likely to confide in people they saw often than married mothers were. Single mothers also considered their friends, relatives, and people in social or church groups less supportive than married mothers did and were less satisfied with the support that they received. Their social networks changed more often than those of the married mothers. Single mothers used twice the amount of day care that married mothers did. Although single mothers had to cope on their own with finances, child care, and household responsibilities, they reported no more difficulty coping than married mothers. Only in the area of household responsibilities did single mothers have more trouble.

Still, despite these differences in stressful circumstances, the *relations* between mothers and children did not differ appreciably in the one- and two-parent families. Single mothers did no better or worse at controlling, nurturing, or communicating with their children, and the children seemed equally compliant and mature. In both kinds of families, mothers who were more stressed were less nurturant. Unexpectedly, Weinraub and Wolf found that single mothers who had the most social contacts tended to be less nurturant and to have less control over their children. These single mothers generally were of two types: young divorced women who had moved in with their parents and whose social contacts included critical relatives, and mothers who lived alone with their children and who sought social contacts through work or dating. The former were placed in a position of immaturity themselves. The latter faced the dilemma of having to choose between spending time with their children and satisfying their own needs for intimacy and support. Social contacts do not assure social support.

In short, Weinraub and Wolf found substantial differences in the lives of single and married mothers. Single mothers are under more stress from life changes and from the longer hours they work. Their social networks offer them less support for their role as parent. Married mothers can more easily integrate their roles as mother, worker, and adult woman. Despite the increased pressures they operate under, however, single mothers are much like married mothers in their ability to handle their children.

provide financially for the family or that the mother has failed to conform to the traditional feminine role. It may mean that when mothers work, they do not so intensely socialize their sons to do well in school as stay-at-home mothers do. It may also be that a mother's work creates more stress in families, and boys, being particularly vulnerable to all kinds of environmental stresses, react negatively. The effect of a mother's working is especially damaging to sons if the work itself is stressful. Sons, then, are likely to feel that their mothers are uninvolved or unsupportive to them, and they feel depressed and inadequate (Piotrkowski and Stark, 1985).

Not all children let their mothers go easily off to work, but knowing that their mothers like to work, knowing that most mothers do work, and knowing that they have somewhere to go after school seems to help children cope with the situation (Etaugh, 1974; Hoffman, 1979). School-age children can understand why their mothers want and need to work better than they could when they were younger, and they can feel both pride and worry on the mother's behalf. Some children worry that their mothers have too much to do, and they long for the times—usually on weekends—when they can have some undivided attention.

> My Mom drops me off at school on her way to work every morning at 8:00. After school I go home on the bus and do my homework or watch TV. Mom gets home at 6:00 on nights when she doesn't have a meeting or some dinner to go to. Then she gets home much later. Dad gets home at 7:00 if the train's on time. We have supper together, and I have to go to bed at 8:00. But we spend Saturday mornings "mousing around," just me and Mom doing errands or hanging around the house while Dad sleeps late.—Sarah, age 10

When their mothers work, schoolchildren are more likely to take responsibility for themselves and for doing things around the house. They are also more likely to value women and their abilities. When mothers are pleased with their lives—whether they work outside the home or not—their contentment is reflected in their children's contentment and stable adjustment.

Relationships with brothers and sisters

Most American families have two children. An important part of many young children's social development, therefore, is learning to get along with another child. Four- to 6-year-old siblings have been observed to spend twice as much time alone together as they spend with their parents (Bank and Kahn, 1975).

Siblings show their interest in each other and learn from each other in many ways. One way is by imitating each other. In one study, nearly one-third of the interactions between 2-year-olds and their older siblings consisted of imitation (Pepler, Corter, and Abramovitch, 1982). Younger children imitated their older siblings far more than the reverse, but older siblings imitated their younger siblings as well. Siblings also show their interest in each other by doing things together. They do things in tandem with great excitement and pleasure (Buhler, 1939; Dunn and Kendrick, 1982a; Greenwood, 1983). Interactions of brothers and sisters are marked by strongly positive feelings.

> He loves being with her and her friends—he's very fond of one of her friends. He trails after Laura. . . . They play in the sand a lot . . . making pies. She organizes it, and swipes away things that are dangerous and

When a new brother or sister is born, the older sibling gains a new playmate and new responsibilities but to some extent also loses a parent.

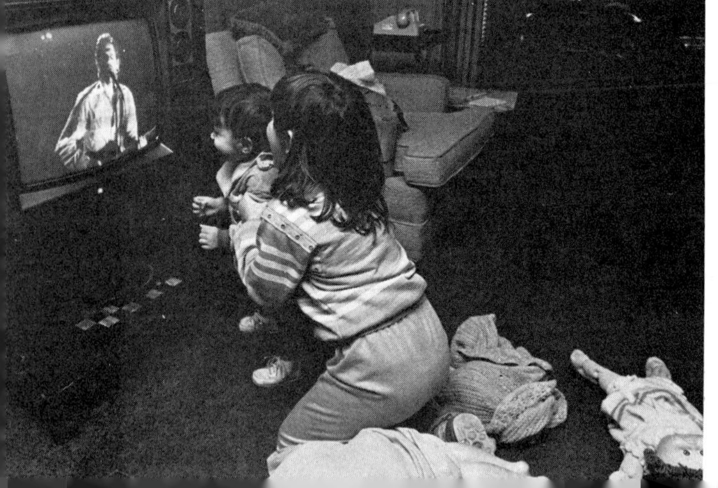

> gives him something else. They go upstairs and bounce on the bed. Then he'll lie there while she sings to him, and reads books to him. And he'll go off in a trance with his hanky [comfort object]. The important thing is there are becoming games that they'll play together. He'll start something by laughing and running towards some toy, turning round to see if she's following. He'll go upstairs and race into one bedroom and shriek, and she joins him (Dunn and Kendrick, 1982a, p. 39).

By the time the younger child is 3 years old, siblings often cooperate in games and play, show physical affection and concern, and try actively to help and comfort each other (Dunn, 1983; Dunn and Kendrick, 1982a).

Most siblings are friendly and comforting with each other, but is their relationship like the attachment between children and their parents? Two early studies (Ainsworth, 1967; Schaffer and Emerson, 1964a) showed that many babies acted distressed in the absence of their older siblings and greeted them joyfully on their return at the same age as they showed this sort of attachment behavior to their mother. Similarly, half of the 14-month-olds in another sample missed an absent older sibling, and one-third went to the older sibling when they felt unhappy (Dunn and Kendrick, 1982b). In a strange situation or outdoors, older siblings can comfort younger ones and serve as a secure base when a stranger approaches or when the mother is out of reach (Clark and Krige, 1979; Samuels, 1980). Thus, for the younger child at least, the relationship with an older sibling may be characterized as an attachment.

Not only do older siblings act like parents in providing security to younger ones, but they also speak in the same ways as parents do to young children. Speaking motherese, older siblings repeat, explain, and help a younger child to practice speech. Here a 31-month-old tries to dissuade his younger brother from eating a piece of candy that fell on the floor. Telling him that the family dog, Scottie, will eat it, he urges the baby into the kitchen with ever shorter sentences:

> No don't you eat it. Scottie will eat it. No not you. Scottie will eat it. Not you. Scottie. Not you. Shall we go in door? Right. Come on. Come on. In door Robin. In door (Dunn and Kendrick, 1982a, p. 50).

It has been suggested that the ability to take the sibling's perspective, which these children's actions imply, comes from the children's familiarity with each other and from their reciprocal relationship. Because siblings share so much and understand each other so well, they are well placed to understand each other. Older children often interpret their younger siblings' wishes or feelings; "Kenny wants cakey, Mom" or "Jo-Jo likes monkeys." They clearly understand the younger child's point of view. As one older brother said as he watched his younger brother playing with a balloon, "He going pop in a minute. And he going cry. And he going be frightened of me too. I *like* the pop" (Dunn and Kendrick, 1982a, p. 46).

Most older siblings show interest in younger brothers and sisters, worry when they cry, and try to entertain and take care of them (Brody et al., 1985; Dunn, Kendrick, and MacNamee, 1981; Legg, Sherick, and Wadland, 1974; Trause et al., 1981). Nevertheless, siblings' relationships are very different from parent–child relationships. No matter how positive sibling interactions are, it is unlikely that they are as positive as interactions with parents. Researchers have shown that children interact more positively with parents than with siblings (Baskett and Johnson, 1982). They also have more negative encounters with siblings.

Brothers and sisters understand not only how to comfort their distressed siblings but also how to provoke them, and their interactions are characterized by strongly negative feelings as well as positive ones.

> It's worse now he's on the go. He annoys her. They fight a lot—more than four or five big fights a day, and every day. They're very bad tempered with each other. He makes her cry such a lot (Dunn and Kendrick, 1982a, p. 39).

Siblings can annoy and compete with painful accuracy.

> When our family went on car trips, I sat in the back seat next to my younger sister. Under my breath and too quietly for my parents to hear, I would make a buzzing sound to pester her. It drove my sister crazy. She'd yell to our parents, "She's *bothering* me!" But they would say, "Cut it out now, Lynn. She isn't doing anything at all."

Nearly one-third of the interactions between siblings have been observed to be antagonistic, with older siblings more likely to antagonize younger (Dale, 1983; Dunn and Kendrick, 1982a; Pepler, Corter, and Abramovitch, 1982). With age, younger siblings themselves increase not only their positive actions toward their siblings, but their antagonistic and aggressive actions as well. By the time they are 2 years old, younger siblings can tease their older siblings quite cleverly.

Older siblings also may act as teachers to their younger siblings. Children of 6 and 7 have shown that they can teach younger siblings—more effectively, in fact, than unrelated teachers (Cicirelli, 1972, 1977, 1978). Even preschool children can teach their younger siblings quite well, teaching them physical skills, games, and how to use toys, labels, words, and numbers. They draw the younger siblings' attention and adapt their instructions to the younger ones' actions—at least when their mothers are supervising the proceedings (Pepler, 1981; Stewart, 1983). Older sisters are especially good teachers (Dunn and Kendrick, 1982b; Minnett, Vandell, and Santrock, 1983).

In sum, sibling relationships are both *reciprocal*—full of mutual imitation, play, talk, comforting, and fighting—and *complementary*—as older siblings take care of, help, teach, and manage their younger brothers and sisters. They are both *positive*—as siblings cooperate and share—and *negative*—as they compete, hit, and fight. The relationships between siblings may be described as ambivalent, but they are seldom lukewarm (Dunn, 1983).

Interactions with other children

Brothers and sisters are not the only playmates that preschool children have. Most preschoolers have chances to play with other boys and girls their own age—cousins and neighbors, nursery school classmates, and family friends. Together, they do many of the same things that siblings do: play, teach, cooperate, and help; punch, kick, hit, and wrangle over toys and turns.

Patterns of peer play

Toddler play

With or without siblings, from the time they are infants, children are interested in others their own age. Their small size and interesting antics make them attractive to each other (Lewis and Brooks, 1974). During their second year, toddlers begin to act less neutral. Their expressions of positive and negative feelings increase. In one study of peer play, the meetings of unacquainted 1- and 2-year-olds were videotaped (Ross and Goldman, 1977a; Goldman and Ross, 1978). Even though the children had not played together before, in just 13 hours of tape, researchers identified over 2000 clearly positive social overtures—positive vocalizations, giving toys, and so on. Even at this early age, the children started games of tickling, touching, give and take, and laughing at antics. The play of the 2-year-olds was more intentional, lasted longer, and was more varied, intense, coordinated, and complex than the 1-year-olds'. Their play also consisted of more games with a greater number of turns. Positive overtures appeared more often in the interactions of these 1- and 2-year-olds than withdrawing, hitting, fighting, crying, or swiping. But conflicts do occur. Most were struggles for possession of a toy. But even these conflicts had a social flavor. Even when identical toys were available, children still quarreled over a single object; one child's possession seemed to whet the other child's appetite.

In these studies, researchers examined the play of children who were strangers. In another study, researchers followed the play of five boys who met two mornings a week for three months, starting when they were 13 to 18 months old (Mueller and Lucas, 1975). They identified three stages of peer contact as the boys became older and better acquainted. In the first stage, one child's curious examination of an object attracted the attention of the other boys, but there was no interaction. If one child acted, the others watched; if one boy gestured or vocalized to another, he was ignored. In the second stage, these social bids were responded to. Often, in the beginning, a second boy imitated the first; later there were longer and longer interaction chains. In the third stage, the boys tried to elicit not imitation but the appropriate complementary response—a take to a give, a catch to a throw, a run to a chase.

Some social interactions, like give and take, catch and throw, require the presence of an object. But even beyond this, toys were important for these toddlers. The boys spent most of their time playing with toys, and 90 percent of their activities involved toys. Toys are the main avenue for toddlers' social exchanges, and they allow them to show, give, fetch, and grab. At first, the toy itself is the focus of the interaction. The toy acts either as a carrot, attracting a peer, or a stick, forcing a child to notice the child who takes away his toy (Mueller and Vandell, 1979). Later, the toy is used to mediate social interaction, and the peer becomes the focus. All the while, though, the toy is a necessary prop.

Preschool play

In the preschool period, children's social contacts grow more frequent, and their social skills grow more sophisticated. Observations of groups of children between 2 and 5 years show that as children get older, they do less

Toys are central in the play of preschool children. A jack-in-the-box brings two little girls together, in admiration and in a struggle for possession.

staring, crying, pointing, sucking, fleeing, and playing alone and more talking, smiling, laughing, and playing together (Mueller, 1972; Blurton Jones, 1972; Smith and Connolly, 1972; Parten, 1932). They make more eye contact and talk more, and their communication becomes more effective (Savitsky and Watson, 1975). They can collaborate more and play together with others more successfully and cooperatively—dividing roles and sharing goals—than they could when they were younger (Cooper, 1977; Parten, 1932; Tieszen, 1979).

In nursery school, Sam and Natalie played beautifully together. One morning, Sam helped to trace Natalie's outline on a long piece of brown paper. Then she got up and helped to trace Sam's outline. Then the two of them drew in their own eyes and mouths.

Over the preschool period, children's play becomes progressively more positive than negative. One group of researchers found the ratio of positive to negative interchanges among preschoolers to be 8:1 (Walters, Pearce, and Dahms, 1957). Although toys are still often used to mediate social interaction, over the preschool period children become increasingly likely to interact without toys. One kind of purely social play that 3- and 4-year-olds engage in is **ritual play**—repetitive, rhythmic exchanges or turns, with exaggerated intonations, distorted rhythms, and broad gestures. "I can see Amy." "I can't see Ben." "I can see Amy." "I can't see Ben." "I can see Amy . . ." Five-year-olds also engage in ritual play, but their rituals are shorter than the younger children's. Older preschoolers are more likely than younger preschoolers to engage in **language play**—rhyming, saying nonsense words, and playing with words. "Mother dear mother near mother tear mother hear," they singsong. They also play out complex roles together in pretend fantasies—house, school, doctor, and other play dramas (Garvey, 1984). Of course much of preschoolers' play does still include toys. **Parallel play**, in which children play independently with toys that are like the toys of a nearby child but in which they do not actually play together, predominates with 2- and younger 3-year-olds, but older 3- and 4-year-olds still play this way occasionally (Parten, 1932, 1933; Rubin, Watson, and Jambor, 1978; Smith, 1978).

Sharing and caring

Young children love to share things that they find interesting. As they wander through an unfamiliar playroom, for example, toddlers will leave their mother's side to explore and to show or give her the interesting toys they discover. In one series of studies (Rheingold, Hay, and West, 1976), this early form of sharing cropped up in every single 1- and 2-year-old observed. The children showed and gave food and toys to parents and peers, whether they were asked to or not, and whether the other person responded positively or not.

I took a sip of water from Kristy's glass once when she was about 2. She looked at me, and I expected a scolding. But instead she smiled broadly and said, "We sharing," and offered me some more.

The development of children's caring actions is more gradual. Only slowly do children develop social skills for responding to other people's needs. In the second year of life, researchers have found, many children begin to perform caring acts (Hay and Rheingold, 1983). In one laboratory study designed to bring out children's caring acts (Rheingold and Emery, 1986), all the 1½- and 2½-year-old boys and girls acted caring, and most of them acted out a wide range of caring acts—putting a doll to bed, feeding, disciplining, grooming, carrying, giving affection. As they wheeled their dolls in carriages, brushed their hair, diapered, bathed, and carefully seated the toy animals and dolls, the children were affectionate and loving, spoke appropriately—"Night-night, bear." "Oh, poor baby. Too cold. Too cold. I'll wrap you around."

By this age, children respond empathetically to another person's distress at least part of the time (Zahn-Waxler, Radke-Yarrow, and King, 1983). **Empathy** is a vicarious emotional response in which the emotions of the person observing, to some extent, match those of the person being observed. An empathic response is one in which a person responds *as though* he or she were feeling what another person is feeling, like feeling sad when someone else is crying. Researchers Carolyn Zahn-Waxler and Marian Radke-Yarrow (1979; Zahn-Waxler, Radke-Yarrow, and Brady-Smith, 1977) have found that the earliest emotional reaction most children show to others' distress is empathic—crying or sadness. These researchers had mothers keep diaries in which they recorded their children's responses to others' distress, and they also tested children with simulated incidents in the laboratory that might elicit empathic responses—such as spilling papers or having a mother "accidentally" bump her elbow. They found that at both 2 and 7 years of age, children had about the same intensity of response to distress. But individual children behaved differently. One child at 2 blocked her ears and ran away when she heard crying or saw someone angry. At 7, the same girl complained, "I can't take much more of this crying." A more compassionate child at 2 ran to her mother and buried her head in her lap when her friend got hurt. Later she comforted a crying baby, her own lips quivering. Still later, she gave her sandals to her younger friend to keep her friend's feet from burning as they walked on hot pavement. A third child was detached and rational. At 2 she remarked as she looked at a friend's face and wiped away her tears, "Annie's crying. She's sad. See her tears." At 7, she asked, "Where does it hurt? How much does it hurt? Why does it hurt?"

What distinguished the 2-year-olds from the 7-year-olds was how they expressed their concern in **prosocial** actions. Although empathic, 2-year-olds lack the skills to be really helpful. They may try to offer comfort to people who are crying or hurt by snuggling, patting, hugging, or offering to feed them. But they offer what they themselves find comforting—a bottle, a doll, a

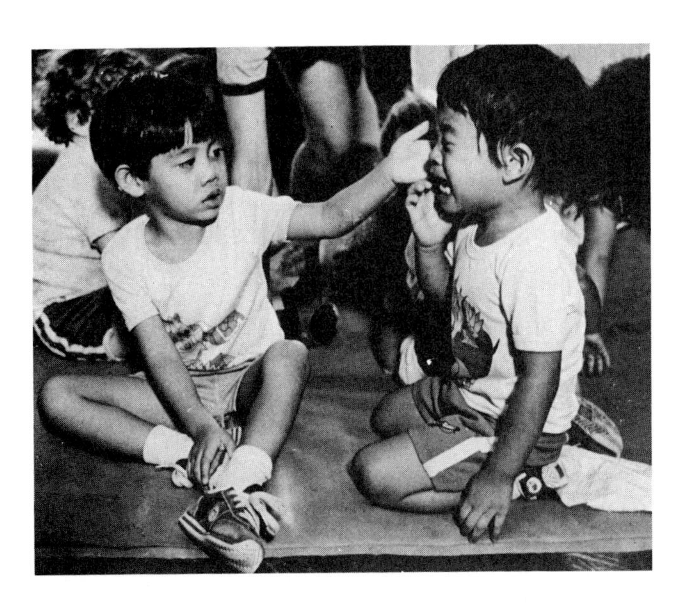

Even preschool children express sympathy and try to comfort someone who is hurt. This young boy tries to pat away the pain felt by his twin brother.

cracker, their own mother's hand. Clearly sympathetic and wishing to be helpful, these very young children do not yet know the practical steps that people take to help others. Later, children's offers of comfort grow more elaborate and appropriate. They help directly, offer suggestions, attempt rescues, say sympathetic things, referee fights, and protect victims (Zahn-Waxler and Radke-Yarrow, 1979; Zahn-Waxler, Radke-Yarrow, and Brady Smith, 1977).

Hitting and hurting

> I worry about Natalie. If someone takes a toy away from her, she just looks lost and never even tries to get it back. Maybe nursery school will toughen her up.
>
> I wish some of Natalie would rub off on Sam. When he wants something, he bulldozes it away from whoever is holding it.

At one time or another, virtually every parent worries about a young child's aggressive behavior—or lack of it. Where do children's antisocial actions come from, when are they expressed, and how do they change over time?

Between the ages of 2 and 4, children become first more and then less aggressive (Blurton Jones, 1972; Hartup, 1974; Walters, Pearce, and Dahms, 1957). After the third year, physical actions like hitting and stamping decrease, and verbal insults and attacks increase (Goodenough, 1931; Jersild and Markey, 1935). Most quarrels are over playthings and possesssions that have to be shared with others. Frustration long has been regarded as a major cause of aggression (Dollard, Doob, Miller, Mowrer, and Sears, 1939), and preschoolers are likely to feel frustrated and aggressive when they are left out of a game at nursery school or on the playground or when they cannot play right away with the toys they want because someone else already has them. Preschoolers also are sensitive to anger in others around them. In one recent laboratory study (Cummings, Iannotti, and Zahn-Waxler, 1985), actors simulated an angry interaction while pairs of 2-year-old children played with toys nearby. During the angry exchange the children were clearly distressed; immediately afterward they were more likely to hit, kick, push, or physically

Aggression between 2- and 3-year-olds, most often over sharing toys, is direct and physical. Five-year-olds hit and pull less and engage in more verbal attacks and insults.

attempt to take things from each other. One month later, after witnessing another heated argument, the children were even more upset and aggressive. Anger itself may be contagious, and its effects may be cumulative.

Children learn aggressive acts by directly imitating adults, other children, and even television characters who act aggressively. At home, children may learn to act aggressively through being selectively reinforced for doing so. Parents may pay attention only to the child's naughtiness, for example, and in this way selectively reinforce the aggressive behavior. Naturalistic observations of parents and preschoolers reveal that parents of nonaggressive children communicate often with them—''What do you want to do?'' ''Do you want a piggyback ride?'' ''Do you want to play with the dolls?'' They listen to what the children answer and pay attention to how they behave (Montagner, in Pines, 1984). They are not aggressive themselves, nor are they overprotective, and they threaten the child only in possibly dangerous situations. Their interactions with children are sensitive, friendly, accepting, and happy (Stevenson-Hinde et al., 1986). Parents of aggressive children, in contrast, are likely to be either aggressive and punitive themselves or overly permissive. A mother may call for her child at day care but turn away to ask the teacher, as the child approaches, ''Was he naughty today? He's always hitting people at home.'' As the child veers away, the mother calls again. She calls, yells, and finally chases him down, grabbing him roughly, and hauling him from the room. Parents who punish their children physically stimulate the children's anger and provide them with a model of aggressive behavior. The long-term effect on children is likely to be increased aggressiveness. Parents who do not attempt to control their children's aggression also encourage children to have aggressive outbursts. Families with one permissive parent and one punitive parent are especially likely to have children who act aggressively.

Children are most likely to learn to control anger and aggression when discipline is consistent and mixed with love and reasoning and when parents are not often angry with them or each other. It is also helpful if parents foster acceptable outlets for the anger children inevitably feel. For example, they might suggest that the children express the anger verbally. Parents of aggressive children often fail to provide positive encouragement for replacing aggression with other behavior, and so their children know what they *cannot* do but not what they *should* do.

Children also pick up aggressive behavior from watching television.

> I was amazed the first time Peter, age 3, saw an episode of ''Spiderman'' on TV. When the bad guys started beating up Spiderman, Peter walked right over to the set and began hitting it.

It has been demonstrated that exposure to television violence increases children's subsequent aggression in the laboratory. But it is difficult to generalize from the laboratory situation to real life; children in these studies may just have been trying to do what they thought was expected of them. In a more naturalistic study (Friedrich and Stein, 1973), groups of preschoolers watched one of three types of television program daily for four weeks at their own nursery school: violent cartoons like ''Batman'' and ''Superman''; episodes of the prosocial ''Mr. Rogers' Neighborhood''; or neutral programs, such as children visiting a farm. Two weeks after their television treatment, children in the first group were less willing to exert self-control, less able to tolerate delays in obtaining their desires, and less likely to obey the rules of the nursery school. If they had been aggressive before the four weeks of television viewing, they were more so afterward; but children who were not aggressive did not become so. The differences were not very large, and we do not know how long they lasted beyond the two weeks. Given the extensiveness of most children's television viewing, however, we should not dismiss the possibility that the effects of actual television viewing are profound and persistent.

A child's hormones, temperament, physical appearance, muscularity, and social skills also contribute to aggressiveness. Hormones play an important role during prenatal and later development and may modify children's behavior. In one study, males and females who had been exposed prenatally to synthetic progestins were found to act more aggressively than their same-sex siblings who had not been exposed to progestins (Reinisch, 1981). In another study, researchers found a positive relation between blood levels of testosterone and males' aggressive and antisocial behavior (Kreuz and Rose, 1972). As far as physical appearance goes, a kind of self-fulfilling prophecy may work to make unattractive children more aggressive than their attractive peers. Although the aggressiveness of attractive and unattractive 3-year-olds is not different, by the time they are 5 years old unattractive children tend to be more aggressive and boisterous than attractive ones (Langlois and Downs, 1980). As for temperament, children who develop aggressive behavior problems are more likely to have been difficult infants—more active, irregular, sensitive, nonadaptive and intense (Thomas, Chess, and Birch, 1968). The influences on aggressive behavior in young children are many and complex.

Friendships

After they have played together for a time, preschool-age children begin to form friendships with a few other children. Friends usually are of the same sex, age, energy and activity levels as the children themselves (Gamer, Thomas, and Kendall, 1975). Preschoolers like to spend time with these friends, to be close to them, and may feel sad when they are separated.

> When Sam's best friend, another 3-year-old boy, left the day-care group, Sam was weepy for several days every time he thought about it. He had to be reassured often that he could still play with his friend—"my Robbie"—on weekends.

If they do not have real friends at this age, preschoolers may create imaginary friends (Manosevitz, Prentice, and Wilson, 1973). The imaginary friend is a companion in play and a scapegoat for the child's naughtiness: "No! I not do it! Burdette doed it."

Compared to the ways they act with unfamiliar children, preschoolers engage in more pretend play with their friends, have more connected dialogues, and are more likely to talk about what they have in common than how they are different (Gottman, 1983). With friends, preschoolers are more agreeable and compliant. They offer friends sympathy and help and ask about their feelings. By what social processes do children become friends? In one study, a number of 3- to 6-year-old children were observed at home as they visited several times with another child their age whom they did not know (Gottman, 1983). The children who hit it off and progressed toward friendship could tell each other things, play at the same activities, talk about themselves and how they were alike or different, resolve their few conflicts, joke, gossip, fantasize, and play dramatic games of house, doctor, and superheroes. Children who did not hit it off did not have these kinds of interactions.

During preschool years, though, friendships are fleeting things. Robert Selman (1981; Selman and Selman, 1979) called this the period of *momentary playmateship*. Friends are valued for their material possessions and for living close by. As one little boy said in an interview with Selman, "He's my friend because he has a giant Superman doll and a real swing set." A little girl explained, "She's my friend—she lives on my street." By the end of the preschool period, the child enters a stage Selman called *one-way assistance*. At this point the friendship may be based on more than Superman dolls; chil dren begin to understand that feelings and intentions, not just things, keep friends together. But they take into account only their *own* needs and satisfaction. Said one child: "She's not my friend anymore because she wouldn't go with me when I wanted her to." And another: "You trust a friend if he does what you want." This is the highest level reached in the preschool period; only later do friendships become reciprocal and intimate.

Social understanding

Understanding the people who make up one's social world, interpreting their emotions and intentions, and communicating with them effectively are important aspects of social understanding. During early childhood, children develop skills that help them to accomplish these tasks. Preschool children begin to learn to read people's emotional expressions and to infer which emotions are likely to be provoked in particular situations.

> Peter has always been sharp as a tack about figuring out what people are feeling. As a 2-year-old, he read the facial expressions on the characters in his storybooks. If he didn't recognize an expression, he asked about it. The people on cereal boxes, butter and syrup containers, you name it, and Peter had to talk about whether "Him happy" or "Dem scared" or "Her laughing." Even now, at 4, if the slightest frown crosses my face, he asks anxiously, "You mad, Mommy?"

By the time they are 3, children can recognize and label facial expressions that signal happiness, sadness, anger, and fear (Camras, 1977). They also can give examples of being happy, sad, angry, and scared (Harter, 1979). As they get older, they can name and recognize a wider range of emotions and can predict how a person will feel in emotional situations. In one study, children between 3 and 8 years old were read short stories about a child who ate a favorite snack, lost a toy, got lost in the woods, and was made to go to bed (Borke, 1971). Then the children were asked to choose the picture of a face that best expressed how the child in each story felt—happy, sad, afraid, or angry. The older the children, the more accurately they identified the character's feelings. The 3-year-olds could identify happy emotions. Of the 4-year-olds, 60 percent recognized stories about fear, as did all the 6-year-olds. At 5½ children were substantially correct about the sad stories, but only later did children correctly identify the emotion in stories about anger.

In another study, children were interviewed about emotions and emotional situations (Harter, 1979). Preschoolers could not imagine how someone might feel *two* emotions at the same time or in close succession. "Can you feel good about going to visit grandma but grouchy about having to pack?" They answered, "No way!" or "It's hard to think of this feeling and that feeling because you have one mind," or "I've never done that, you know; I've only lived six years." Similarly, when they were shown conflicting cues, like a smiling face on a child at the dentist's office, preschoolers had only one interpretation. They took things at face value, and their interpretations were consistent with the facial expression rather than the situation. They had trouble reconciling conflicting cues (Gnepp, 1983). As we saw in Chapter 9, preschool children usually do not concentrate on more than one cue or dimension at a time. By preschool age, children clearly are more skillful than toddlers at interpreting emotional expressions and social situations. But their interpretations are still relatively unsophisticated, superficial, and unsubtle.

Self-concept

By the age of 2, children realize that they are distinct from other people. A little boy now may grab a toy and insist, "Mine!" He may see his photograph and pipe, "Me!" He begins to tell people, "I a boy," or "I a baby," or "I Mikey." Very young children usually know themselves by their informal "everyday" names—"Mikey," "Kristy," "Sam"—and later learn the difference between given names, nicknames, and polite forms of address such as "Mr." "Miss," "Aunt," or "Uncle" (Garvey, 1984). Older preschool children avidly learn their full names and recite them with pleasure.

> "What's a nickname?" Sam asked when he was 3.
> "It's usually a short form of someone's name." 'Sam' is the nickname for 'Samuel.' 'Tim' is the nickname for 'Timothy.' 'Beth' is your babysitter's nickname.
> "Oh," Sam replied, light dawning in his eyes, "so Beff's real name is Beffnick."

Not only do children learn their names, but they begin to learn their place in the social universe. Preschool children know family members in terms of their relationships to themselves. They call them "Mommy" and "Daddy," not "husband" or "wife," Mary or Jim (Garvey, 1984). Children ask lots of questions about family relationships. Over the course of several months when he was 3, Sam asked his mother, "Who's your mommy?" ("Grandma.") "Who's Grandma's mommy?" ("Grandma's mommy was my 'Nana,' but she died.") "Who's my sister?" ("You don't have a sister.") "Who's your sister?" ("Aunt Laura.") "My aunt's your sister?" ("That's right.")

Although preschoolers know their names and nicknames, they still do not see themselves as others see them. Their **self-concept** is based on fleeting, sometimes faulty, ideas. When asked to describe themselves, preschoolers usually mention something they like to do: "is a helper who does the dishes," "can work hard," and "sits and reads stories." They do not mention more enduring characteristics like "friendly" or "bright." In a study of 3- to 5-year-olds, for example, researchers found that the children thought of themselves in terms of things they did (Keller, Ford, and Meacham, 1978). Asked to tell the researcher about themselves and to complete the sentences, "I am a ________" and "I am a boy/girl who ________," more than half of the children's responses described actions such as "I play ball" and "I walk to school." They mentioned their likes and dislikes only 5 percent of the time. Preschool children define themselves in terms of physical actions and possessions rather than psychological attributes.

Sex differences and gender roles

One important aspect of the preschool child's self-concept is his or her sex. Researchers have long been interested in how children develop concepts of themselves as boys or girls. They have studied children's actions and words in interviews, experiments, and observations, looking for differences between boys and girls. They have tried to figure out how much is biological, how much learned. From their research, we have learned quite a bit about the development of sex differences and gender roles in the preschool period.

Play styles

At 2, boys and girls look much the same, play much the same, and are interested in many of the same things. But by the time they are 3, boys and

Differences between boys' and girls' play styles become increasingly noticeable during the preschool period. Girls like to paint, draw, play with dolls, dress up, and hear stories. Boys like to hammer, ride vehicles, stack blocks, and play tough guy.

girls often act differently, and these differences grow more marked during the preschool period (Pitcher and Schultz, 1984). In nursery school, most girls like to paint, draw, help the teacher, play with dolls, dress up, look at books, and listen to stories. Most boys like to hammer, ride tricyles, and play with cars and trucks. Most girls like stuffed animals, dolls, cooking utensils, and boys like puzzles, blocks, and tools (Fagot and Leinbech, 1983; Sprafkin et al., 1983). Both boys and girls engage in fantasy play, but girls play house, and boys play superheroes. Girls' fantasies often are tied to everyday, domestic realities. Boys' fantasies often are magical, bizarre, supernatural. A banana becomes a little girl's "telephone" and a little boy's "magic wand" (Haney, 1984).

In their play, boys act out masculine roles—warriors, mailmen, fire chiefs, plumbers, and fathers—and they love grandiose themes. They are Masters of the Universe, GI Joe, Superman, monsters, and dinosaurs. The few realistic masculine roles that appear in preschool children's play are global and imprecise. Fathers are always off at some vague "work" or at home eating, reading the paper, watching television, or sick in bed, tended by mothers. So boys make up their own, more exciting masculine world. They do not give in to girls' romantic overtures. They back away from playing marriage and curl their lips—"Yuck!"—at the thought of a kiss. Boys explore friendship by being "brothers" and "blood brothers," by exchanging roles—bad guys and good guys, Batman and Robin, father and little boy—and by protecting one another from danger, fighting wars together, conquering death and destruction on the same side. In their play, preschool girls are more likely to act out masculine roles than boys are to act out feminine ones, but most girls express only a brief interest in the power fantasies that fascinate boys. Most girls like to play at domestic routines, getting married, cooking meals, shopping. Girls do not like to play killers and destroyers, just as boys cannot long tolerate their roles in girls' domestic fantasies. Boy "babies" cannot act babyish for long, and a boy who "irons" soon is chasing robbers from the house (Pitcher and Schultz, 1984).

Boys engage in more rough-and-tumble play than girls do. Teasing, rough-and-tumble play peaks in 5-year-old boys, but few girls play roughly together. Boys tease and say silly words, wrestle, bump into and fall on top of each other, push and pull, shout, make noises like machine guns, chase each other with space guns, fall down laughing at themselves when they make toy horses "sneeze," smear clay in one another's hair, tickle, fall dead, slide from piles of blocks, fall over chairs, and eat fire. Boys love to talk dirty.

Girls enjoy telling boys the "rules," and loudly announcing when a boy has broken a rule: "You marked on the table!" "Sit up straight in your chair when you're coloring." "Well, a man's gotta marry a woman and a woman's gotta marry a man." Boys' disagreements and reprimands center on intrusion, territorial and property rights, and details of work procedures: "You stepped on my car!" "Hey, you're wrecking my house." "You're stacking those chairs the wrong way." When they disagree with girls, they forthrightly refuse—"Nope, I won't"—or make a general judgment—"You're not right."

As early as 2 years of age and increasingly thereafter, boys and girls segregate themselves into same-sex groups. During free play in nursery school, an invisible curtain divides the girls in the crafts areas from the boys in the block corner (LaFreniere, Strayer, and Gauthier, 1984). At storytime, boys and girls spontaneously arrange themselves on opposite sides of a circle (Paley, 1984). Girls are the in-group, and boys are the intruders. Girls are intimate confidantes, and to them boys are "crazy" and "mean." They ignore boys' aggressive overtures. Meanwhile, boys egg on other boys to be He-Man to their Skeletor, to play good guys and shoot bad guys, to be rowdy and loud. They strongly discourage other boys from playing with girls. By age 4, boys direct their aggression only to other boys and get exciting retaliation in return (Pitcher and Schultz, 1984). They may deliberately break the rules of docile nursery school behavior just to get other boys' approval (Dweck and Bush, 1976).

Preschool children create their own male and female worlds—places where they can unambiguously define their own gender, places where they can be unforgiving of children who stray from accepted practices, and places where they can almost rigidly pursue gender-typed play. No matter how many times teachers praise them for playing with children of the other sex, no matter how much consciousness raising their earnest parents attempt to involve them in, most boys still insist on being cowboys and kings of the universe, and most girls still insist on dressing their baby dolls and playing at marriage.

Of course not all girls play girl games all the time, nor boys boys' games. But there are clear and pervasive differences in the ways boys and girls play. There are also, as we shall see, other differences in their behavior.

Nurturant girls

Put a baby in a room with a group of 3- and 4-year-olds, and both girls and boys will pay attention to the baby. But put the baby in a room with 5- and 6-year-olds, and the girls pay much more attention to the baby than the boys do (Berman and Goodman, 1984). Over the preschool period, girls become increasingly nurturant toward other children. Boys become less so. And so begins a major difference between males and females that will last throughout their lives.

From the first months of life, girls have certain sensitivities that may turn them toward social cues and, ultimately, make them more nurturant than boys (see Lever, 1976; Maccoby and Jacklin, 1980). On the average compared to infant boys, girls' skin is more sensitive, and they pay more attention to faces, patterns of speech, and subtle changes of voice. They are more likely to cry when they hear a baby cry. They can recognize their mother's face earlier. They speak earlier, more often, and more fluently and so are likely to be more sociable and readier to start, join, and respond to social exchanges.

By the end of the preschool period, girls play in small groups and emphasize intimacy, mutual support, and sharing secrets, whereas boys in their peer groups tend to emphasize solidarity, loyalty, and doing things together. Girls stay physically closer and make more eye contact with each other than boys do (Hoffman, 1977; Restak, 1979). Both boys and girls are equally able to understand the emotional reactions of others. When assessing an emotional situation, such as seeing pictures of a child who has lost a pet dog, boys and girls are equally adept at inferring how the child feels. But girls are more likely to respond empathically. Their faces, posture, and words of sympathy are more likely to reflect the distress of the person they watch (Cummings et al., 1985; Hoffman, 1977). Boys may try to solve another person's problem when they see that person in a difficult situation, but girls tend to imagine themselves in the other's place.

Girls' nurturance shows up in nursery school play, too. Although both boys and girls play at cooking food, it is mainly the girls who play at serving it. Boys are more taken by the stirring and the mechanics of preparing food. Girls also protect, bathe, diaper, and doctor their dolls and each other. When one girl is the "baby," another girl is likely to kiss, hug, pat, stroke, and speak to her endearingly—"Come, honey, let's go to the house to eat your cake," or "How are you doing, sister?" (Pitcher and Schultz, 1984, p. 63). They help boys put on painters' smocks, comb their hair, hang up their coats and put their pictures in a safe place so they "won't get lost." Girls adjust each other's aprons, veils, hats, and dresses. They wrap bandages, give "medicine," and tend "the wounded" (Pitcher and Schultz, 1984).

Aggressive boys

In all cultures, boys tend to be more aggressive toward their peers than girls are. They are more likely to intentionally hurt or harm each other. As we have described, they also are more likely to play rough, mock-fight, and have aggressive fantasies than girls. These differences appear as early as age 2, when boys are likelier to get angry and strike an obstacle or another child to get what they want. In a review of 32 observational studies of preschool children, Eleanor Maccoby and Carol Jacklin (1980) found that in 24 of the studies boys were more aggressive, and in 8 studies neither sex was more aggressive. But in no study were girls more aggressive than boys. Although these findings on aggression are reliable, the average differences are small and getting smaller over time as society places more emphasis on gender equality (Hyde, 1984). Not every boy is more aggressive than every girl, and not every girl is more nurturant than every boy. The differences apply to averages only. The sex differences also are not in *feelings*, research suggests, but in aggressive actions. Girls have just as many hostile feelings as boys do, but they are more likely to express them in words or indirect actions than with physical force (Mallick and McCandless, 1966).

The bases of gender roles

Why do boys and girls differ in playing with dolls and guns, in nurturance and aggression? Are the differences biological or learned? The answer to this question is: probably both.

> When Natalie's mother learned that her third child was going to be a boy—her first—she asked Sam's mother how she had learned about cars and trucks. "Sam taught me, and your boy will teach you, too. Before Sam was born, I really believed that the differences between boys and girls were almost totally socially determined. But now I *know* that they aren't. Sam loves his dolls and stuffed animals, and he plays at cooking the supper—'Me the cooker tonight.' But from the start, he's been fascinated with things that move and roll. He likes to put his face close to the wheels of his toy cars and watch them turn. My husband jokes that there's a 'truck gene' on the Y chromosome. You'll see."

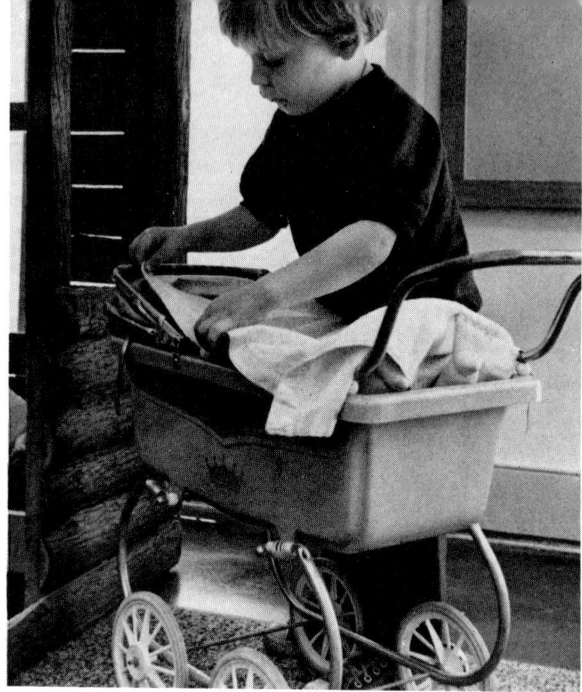

Child rearing takes some special effort if parents do not want their children to develop stereotyped gender roles. Providing toys and materials that are not traditionally gender-typed and encouraging their use can help.

There are certainly biological differences between males and females in muscle structure, build, hormones, and abilities. But contrary to what one might expect were biology the full answer, in the animal kingdom, female lions hunt and kill, and male marmosets nurture their young, and in some non-western societies, females hunt and are not nurturant, and males garden and are not aggressive. Studies exploring the possibility that hormonal differences underlie sex differences in behavior also are not consistent, and behavioral differences between boys and girls appear earlier than physical differences like muscles and build. These facts suggest that learning is an important part of the distinct patterns of masculine and feminine behavior found in American children and adults.

Gender roles are the ways of behaving that are socially prescribed for males and females in a particular culture. They are encouraged, or socialized, in children by parents, peers, teachers, and the community at large. Prescribed gender roles are so intertwined with possible biological tendencies toward physical aggression in males and nurturance in females that the effects are difficult to unravel. Learning experiences may be mediated from within; boys on their own may seek out chances to sharpen their combative skills and girls ways to practice caretaking skills. But in a culture prepared to offer guns to boys and dolls to girls, even the slightest inherited predisposition could be built into a major psychological difference. How do parents, teachers, and peers socialize gender roles in preschool children?

Parents

From birth, boys and girls are treated differently by their parents. Girls are seen as smaller, weaker, prettier; boys as firmer, better coordinated, stronger, and more alert (Rubin, Provenzano, and Luria, 1974). Parents have these biased perceptions, even though newborn boys and girls are not measurably different. Even neutral physical activity of infant boys tends to be interpreted by their parents as aggression.

Later, parents generally encourage their young sons and daughters to act differently in many areas. Dancing, playing with dolls, and dress-up are for girls; climbing, building, and exploring are right for boys. Parents buy vehicles, educational materials, and military toys for their sons, and they decorate their rooms with pictures of animals. They provide dolls, dollhouses, purses, and housekeeping toys such as rolling pins, dishes, and stoves for their daughters, and they decorate their rooms with lace and flowers. They give sons toys that encourage inventiveness, manipulation, and feedback about the physical world. They give daughters toys that encourage imitation, are used near a caretaker, and give less opportunity for innovation or variation (Fagot, 1978; Fein, Johnson, Kosson, Stork, and Wasserman, 1975; Rheingold and Cook, 1975; Rosenfield, 1975). They expect children to play with toys considered fitting for their gender (O'Brien and Huston, 1985; Schau, Kahn, Diepold, and Cherry, 1980). Children are encouraged to choose gender-typed activities—"Let's arrange piano lessons for you, Natalie," "Let's go get you some soccer shoes, Tim"—and, once engaged in these activities, children's gender-typed behavior increases. The structure of these activities themselves further influences children's development. In highly structured activities—which girls engage in—children interact more with adults, imitate, obey, help, and negotiate with them; in unstructured activities—which boys engage in—children are more likely to initiate the interactions, give orders, and act aggressively (Carpenter, 1983). Over time, then, as girls participate in more structured activities and boys in less structured, children and their caregivers, in effect, create distinctive small environments for boys and girls within which gender-typed behavior flourishes.

Children's gender-typed behavior also is fostered by their parents' child rearing attitudes and practices. Reviewers of research done in the 1950s, 1960s, and 1970s (for example, Block, 1983; Shepherd-Look, 1982) note that in general parents describe themselves as encouraging sons more than daughters to achieve, compete, control their feelings, act independently, and assume personal responsibility. Parents claim to punish sons more often than they punish daughters. Fathers especially are more authoritarian with their sons; they are stricter, firmer, less tolerant of behavior that diverges from the traditional masculine stereotype. Fathers encourage instrumentality, mastery, and orientation to tasks with their sons; with their daughters, they encourage expressiveness and dependence. In raising daughters, parents are less punitive, warmer, physically closer, more confident of the child's trustworthiness and truthfulness, expect more "ladylike" behavior, and expect more reflectiveness than they expect from their sons. Mothers supervise their daughters more strictly than they do their sons. Girls are pressured to be nurturant, obedient, responsible, unselfish, kind, well mannered; boys to be self-reliant and successful, ambitious and strong willed (Hoffman, 1977; Tudiver, 1979).

Boys generally are given more room to explore than girls are (Lewis and Weinraub, 1974). The chores they are assigned more often take them outside the house and away from the direct supervision of a caretaker. Girls often are assigned chores as "helpers" (Duncan, Schuman, and Duncan, 1973; Whiting and Edwards, 1975). Parents respond more often to boys' large movements than to girls'; boys are handled and played with more roughly than girls; and girls are treated as if they were more fragile.

Even the subtle ways that parents communicate with their children can reinforce gender roles. When they read a story together, for instance, parents—fathers particularly—are more likely to interrupt daughters than sons (Greif, 1979). The extent to which fathers act in gender-typed ways at home is correlated with what their children think about gender roles. The behavior of mothers at home, in contrast, does not appear to have the same effect. Only when mothers work outside the home do their children seem more aware of gender labels and gender-role differences (Weinraub et al., 1984).

In the area of aggression, the findings are not entirely consistent, but there is certainly no evidence that the greater aggressiveness that has been observed for boys is the result of parents' deliberate encouragement. Some researchers have found no differences in the socialization of boys and girls as far as displaying or returning aggression or settling fights goes (Newson and Newson, 1968; Sears, Rau, and Alpert, 1965). Other researchers have found that boys are punished more severely than girls for aggression (Serbin, O'Leary, Kent, and Tonick, 1973). Although parents do not want their children to be taken advantage of, they do not consider aggression an acceptable way for girls *or* boys to solve problems (Shepherd-Look, 1982). Although boys apparently receive no more direct encouragement of aggression from their parents than girls do, subtle differences in the treatment of boys and girls may affect their aggressiveness. Researchers have found that boys become markedly aggressive when they begin with aggressive temperaments and are raised by punitive fathers (Eron, Walder, and Lefkowitz, 1971; Olweus, 1980). Boys who are severely punished for aggression are more aggressive, and the more they are punished, the more aggressive they become. Children probably learn gender roles not only through their parents' deliberate efforts, but through their indirect influence as well.

Fundamental changes in values could reduce the divergence in parents' socialization of boys and girls. So far researchers studying children's development in families in which mother and father have nontraditional gender-role values and practices have examined only a few, atypical families. But even in traditional families, parents can socialize their children in less gender-stereotyped ways. In a study by Baumrind (1979), some boys were observed to act in the typically feminine way—to be nurturant and friendly. These boys' mothers were particularly responsive to their needs and wishes. Some girls were observed to act in the typically masculine way—to be independent and assertive. Their parents pushed them by being less warm and nurturing and by making strong demands for independent behavior. A recent study of children in the Stanford, California, area suggests that in the last decade, rigid gender socialization by parents may have diminished. After observing and questioning parents and preschool children at home, at school, and in the laboratory, Jacklin and Maccoby (reported in Turkington, 1984) found that parents did not stereotype their children as much as parents in previous studies were reported to do. They treated their sons and daughters similarly and gave them equal amounts of warmth and nurturance, acceptance and restriction. The one exception was that fathers still tended to play more roughly with their sons than with their daughters.

Parents who do not want their children to develop traditional, stereotyped gender roles will still have to make special efforts in the 1980s and beyond. Despite the growing opportunities for women to work at jobs not open to them in the past, traditional gender roles remain firmly entrenched in our society. The forces combining to perpetuate them are formidable: biological predispositions, a society that continues to place most child-rearing responsibility with women, obvious and subtle cultural reinforcements, and the traditional upbringing of most of the people who are rearing children. For all these reasons, the socialization of boys is likely to remain different from that of girls. Even in the 1980s, mothers in single-parent families, who said that they held nontraditional attitudes toward gender roles and who hoped that their sons would develop into ''thoughtful'' men and their daughters into ''self-confident'' women, gave their children gender-typed toys and asked the children to do gender-typed chores (Richmond-Abott, 1984).

Knowledge of sex and gender

In the preschool period, children not only *act* in the gender-stereotyped ways we have described, but they begin to *recognize* and understand the differences between boys and girls, masculine and feminine. Although a boy of 2 or 3 might be able to tell you that he is a boy, he believes that he could be a girl or a mommy if he wanted to be. All he would have to do is play girls' games or wear dresses and grow his hair long. "Boy" may mean no more to him than the name "Tim" or "Jack." Children this young do not understand that sex is anatomical and cannot be changed.

> "When I grow up," Peter told me, "I'm going to be a girl." He also expressed an interest in growing up to be a rabbit and a professor.

The development of **sex constancy** has been tested when researchers have asked children to choose pictures, drawings, or dolls of a boy, a girl, a man, a woman (Emmerich, Goldman, Kirsh, and Sharabany, 1977; Thompson, 1975; Thompson and Bentler, 1971). In these studies, researchers have found that 2-year-olds can select which of the pictures is the same sex as they are, but they are not sure whether it is a boy or a girl. Three-year-olds usually select correctly and know that the picture is of a boy or a girl. Children between 4 and 6 can choose appropriate clothes for a man and a woman, but they make their choice according to hair length, not body build or genitals. Few children think that a girl remains a girl if she changes her haircut, clothing, and activities. The immutability of sex is not an easy notion for preschool children to grasp, for it must be distinguished from other personal attributes that *do* change, such as age and size. Children may see themselves as having more in common with a cross-sex age-mate than with a same-sex adult, if they do not yet understand the significance of sex. Learning this takes considerable experience and a degree of cognitive maturity not achieved by most children until age 6 or even older.

Even at age 2 or 3, however, although children do not know that sex is constant, they do have some sense of **gender identity**. In the ways that people talk to young children—"Aren't you a lovely girl!" "What a big boy you are!"—children constantly hear gender labels and soon can label their own gender accurately (Kuhn, Nash, and Brucken, 1978). They are likely to get upset if someone mistakes whether they are a boy or a girl. Research with very young children whose genitals have both male and female characteristics (and for whom corrective surgery often is considered) suggests that by the age of 3, a child has a sense of gender identity that is difficult to alter later (Money, Hampson, and Hampson, 1957).

Young children also have some rudimentary knowledge of gender roles, of what is appropriate behavior for boys and girls, men and women. When 3- and 4-year-olds in one study were questioned about which toys they thought they themselves, another boy, or another girl would like to play with and why, they often justified their choices on the basis of gender roles and stereotypes (although when it came to choosing toys for real play, they rarely resorted to such justifications) (Eisenberg, Murray, and Hite, 1982). Preschoolers in another study believed that girls are more likely to play with dolls, help their mothers, be talkative, "never hit," ask for help, clean house, and grow up to be nurses, whereas boys are more likely to help their fathers, hit, mow the lawn, and grow up to be bosses, doctors, and governors. According to the boys, girls cry, cook dinner, and act slowly, whereas boys act naughty and make noise. According to the girls, girls kiss, care for babies, look pretty, and say, "I can do it best"; boys are mean and fight (Kuhn, Nash, and Brucken, 1978). Preschoolers know that men wear suits and shave and that women wear dresses, carry purses, and wear makeup; that men drive trucks, fix cars, and put out fires, and women cook, wash, iron, and clean. Not surprisingly, for preschoolers whose lives are dominated by women—mother, babysitters, nursery school teachers—the masculine gender role stands out. Preschool children's knowledge of male stereotypes develops earlier, and they know

focus Gender identity

Identical twin boys were born in 1963. When they were 7 months old, the boys were circumcised. But when the doctor was operating on one boy, he cut an artery, and part of the penis was lost. When the twins were 17 months old, on the advice of doctors, the parents decided to reassign the twin as a girl. They changed the child's name, clothing, and hairstyle, and the first surgery to make the child's genitals look female was done. Psychologists offered the parents guidance, and it was decided not to tell the twins what had happened. If the affected twin knew of the gender reassignment, it was reasoned, her gender identity would be in danger (Money, 1975; Money and Ehrhardt, 1972).

When she was 4, the little girl preferred dresses to jeans and loved her long hair. Her mother reported that she was very feminine—neat and tidy, proud of herself in a new dress, happy to have her mother set her hair. The parents' teachings were taking hold, for the girl was being raised in a traditional gender-typed fashion. Her mother was stricter about sexual modesty in the girl than the boy. When the girl took off her panties in the front yard, her mother spanked her and told her that "nice little girls didn't do that." The mother discouraged the girl's tomboyishness and "bossiness." Her brother became protective of his sister. The mother also encouraged her daughter, but not her son, to help her, gave the girl dolls and feminine toys and the boy, cars, tools, and garages.

Until a few years ago, psychologists had concluded these two boys, perfectly matched genetically and hormonally during their first 1½ years, had been shaped into one child with a thoroughly masculine gender identity and one with a thoroughly feminine gender identity. But then new news appeared. At age 13 and the onset of puberty, the girl was in psychological trouble. She walked and looked masculine. At school she was taunted as "Cavewoman." She had few friends and said that boys have a better life than girls. She planned to become a mechanic. The shaping had not been so successful as people once had hoped.

Because this true story of the identical twins is still unfolding, no one yet knows how it will turn out. This tantalizing case shows how complex is the question of gender identity in childhood and beyond, and it shows once again how development is a matter of both nature and nurture.

more masculine than feminine traits (Best et al., 1977). Boys are more aware of gender roles than girls are (O'Keefe and Hyde, 1983; Weinraub et al., 1984).

> Tim made the mistake of telling his pal, Donny, that he wanted to be an artist when he grows up. Donny ridiculed him, "An *artist*! Only sissies are artists."

After they have gleaned this knowledge about gender roles, children at first are likely to subscribe to gender stereotypes with considerable rigidity and oversimplification. For example, 5-year-old Andy may think that only men can be doctors, only women can cook, and only boys can play ball, despite the fact that his mother is a surgeon, his father is a chef, and his sister is captain of the softball team. Gradually, as experience in the school years modifies their views, children allow for more variations. It may be that during this modifying phase parents, siblings, peers, teachers, television, and textbooks can reduce stereotypes significantly. Researchers who have examined the effects of mothers' work on their children's gender-role stereotyping have found stereotypes to be most rigid at 5 to 6 years; influenced most by the mother's work at

7 to 8; and the least stereotyped, regardless of the mother's kind of work, at 9 to 11 (Marantz and Mansfield, 1977). Apparently, 5-year-olds feel that they must conform to gender-role expectations to maintain their gender identity (Ullian, 1976). Only later, with sex constancy firmly established, do children understand that playing a boyish game will not change a girl, that putting on an apron will not change a boy.

Summary

1. As they get older, children become more autonomous, get less upset at brief separations from their parents and seek less physical contact with them. Yet young children remain strongly emotionally attached to their parents. Patterns of attachment—secure, avoidant, and ambivalent—usually remain stable over time.
2. In the preschool period, parents begin to exert more control and discipline over their children. Parents with an authoritarian style of discipline are firm, punitive, unsympathetic, and do not value independence in their children. Their children tend to be suspicious, withdrawn, unfriendly, and unhappy. Permissive parents do not lay down rules or assert their authority over their children. Their children tend to be immature, dependent, and unhappy. Authoritative parents reason with their children, set few but firm limits, and keep conflict to a minimum. Their children are most likely to be friendly, competent, socially responsible, and happy.
3. Parents' interpretations of their children's behavior and children's characteristics affect the parents' disciplinary style and effectiveness. Mothers of difficult young children do more controlling, warning, and forbidding than mothers of easy children, and difficult children ignore, protest, and talk back more to their mothers than other children.
4. Many families in this country experience the stress of divorce. Whether children have long-term problems as a result of divorce depends in large part on whether their parents part amicably or fight openly and on the relationships they maintain with the children.
5. Mothers' work also is a source of stress—and strength—to their children's development. Daughters are more likely than sons to benefit from their mother's working.
6. In families with more than one child, siblings interact from infancy onwards. They imitate each other, compete for toys and their parents' attention and affection, and have many positive interchanges. Older siblings may act like parents to their younger siblings by providing them with security and even speaking motherese to them. Sibling interactions are both reciprocal and complementary.
7. Over the preschool period, children have more chances to play with other children. As 2- and 3-year-olds, their play is mainly parallel and ritualistic—repetitive, rhythmic exchanges or turns. Older preschoolers are more likely to engage in language play and pretend play, to play interactively and cooperatively.
8. Preschool children act in both prosocial and antisocial ways. They help, hit, hurt, disrupt, and fight with other children. Determinants of young children's aggressive behavior are many and complex. They include frustration; selective reinforcement; imitation of aggressive parents, teachers, other children, and television characters; certain hormones; and physical unattractiveness.

9. Preschoolers make friends with children of the same sex, age, energy and activity levels as they, and with their friends they talk, agree and comply, offer sympathy and help, and inquire about one another's feelings.
10. As young children's social understanding improves, they better understand other people's emotions.
11. The young child's self-concept is based on fleeting, sometimes inaccurate, perceptions. Young children usually define themselves in terms of physical actions, not psychological traits.
12. By the time they are 3 years old, boys and girls begin to act in different ways. Girls play house, pay attention to people's feelings and relationships; boys have grandiose power fantasies and games. Girls are more likely to be nurturant; boys are more likely to be aggressive. These differences stem, in part, from parents', teachers', and other children's implicit and explicit teachings about gender roles. Children also acquire knowledge about sex constancy and gender roles and develop their gender identity.

Key terms

terrible twos
authoritarian parents
permissive parents
authoritative parents
ritual play
language play
parallel play
empathy
prosocial
self-concept
gender roles
sex constancy
gender identity

Suggested readings

CLARKE-STEWART, ALISON. *Daycare*. Cambridge, Mass.: Harvard University Press, 1982. All about day care—its history, politics, ecology, effects—with some practical suggestions for parents on how to find and select a good day-care arrangement.

DUNN, JUDY. *Sisters and Brothers*. Cambridge, Mass.: Harvard University Press, 1985. A fresh new discussion of how siblings influence one another's personalities, ways of thinking and talking, and perceptions of themselves, their families, and their friends.

GARVEY, CATHERINE. *Play*. Cambridge, Mass.: Harvard University Press, 1977. An excellent discussion of the many forms of children's play. Quotations from children illustrate why children play, what play is, and how it fosters learning and social development.

MURRAY, JOHN. *Television and Youth: 25 Years of Research and Controversy*. Boys Town, Nebr.: Boys Town Center for the Study of Youth Development, 1980. A compilation of the results of thousands of studies done over the past 25 years on the controversial subject of the effects of television on children.

MUSSEN, PAUL H., and EISENBERG-BERG, NANCY. *Roots of Caring, Sharing, and Helping*. San Francisco: W. H. Freeman, 1977. An excellent review of research on prosocial behavior in young children. Suggests how children's generosity, cooperation, and altruism are encouraged by factors in the family, the media, the immediate situation, and the child's own nature.

RUBIN, ZICK. *Children's Friendships*. Cambridge, Mass.: Harvard University Press, 1980. A lucid and lively description of friendships between children in the preschool and early school years. Suggests how their friendship patterns are affected by children's educational and social programs.

TAVRIS, CAROL, and WADE, CAROLE. *The Longest War: Sex Differences in Perspective*. (2nd ed.). San Diego, Calif.: Harcourt Brace Jovanovich, 1984. A witty but serious scientific review of the battle of the sexes and the nature and nurture of sex differences in behavior.

WALLERSTEIN, JUDITH S., and KELLY, JOAN B. *Surviving the Breakup: How Children and Parents Cope with Divorce*. New York: Basic Books, 1980. A detailed account of the results of the California Children of Divorce Project, which shows how children of different ages cope with this stressful event.

PART FIVE

Middle Childhood

chapter eleven
Physical development

Physical growth and skills

One day Natalie was a short, chunky preschooler and the next day, it seemed, she had turned into a lovely, well-coordinated child who loved to pedal her bike around the neighborhood, jump rope, and draw with chalk on the sidewalks.

As they enter middle childhood, boys and girls begin to grow more slowly than they did as younger children. Each year, they grow an average of 2½ inches and gain about 5 pounds. The typical 10-year-old weighs 70 pounds and stands 4½ feet tall. From 7 to 10 years, girls are taller than boys, but by age 10, boys have caught up and become taller than girls, and this difference in height continues into adulthood. Boys and girls do not differ much in their physical shape and abilities during middle childhood, although boys' forearms are stronger and girls' bodies are more flexible (Tanner, 1970)—an edge to boys on the pitcher's mound and girls on the balance beam. School-age children are less clumsy and more physically coordinated than younger children. They can throw a ball twice as far as a preschooler, manage a screwdriver, hammer, and saw, knit, sew, and crochet, draw, write, and print legibly, cut their own nails—no more fights with Mom, thank heavens—button buttons, zip zippers, and tie laces, ride a bicycle, climb ladders and trees, swim and dive, roller-skate, ice-skate, and skateboard, jump rope, play hopscotch, baseball, football, board games, cards, and jacks.

All this physical activity helps children to develop. As physically active children master skills, they raise their self-esteem and increase their competence in the eyes of others. Children who can run fast, pitch well, do handstands, or jump rope without being ''out'' are popular with other children, and they like themselves. Physically active children do their bodies a favor by developing muscle, keeping weight down, and preventing the chronic diseases of adulthood, such as heart disease, obesity, and high blood pressure, from gaining an early hold.

Physically active children also do something for their minds. More than 30 years ago, French doctors and educators became concerned that children's health was being harmed by too much schoolwork and too little physical exercise and recreation. Schoolchildren between 6 and 11 years old were divided into two groups (Bailey, 1973). The experimental group got schoolwork all morning, and physical exercise, recreation, art, and music all afternoon. They were not assigned written homework overnight. The control group did schoolwork all day and homework at night. The children were given achievement tests and physical examinations; their absence rates from school were recorded; and their teachers rated the children's behavior problems and observed the children's interactions with other children and with adults. Even though the children in the experimental group had four fewer hours of schoolwork every day, these tests showed, their academic progress surpassed that of the children in the control group. Their health, fitness, behavior, and enthusiasm were also superior. Clearly, physical activity is important for development in middle childhood.

Handicaps

Not all children develop through middle childhood with physical grace and ease. Some are afflicted with disease, disorders, or disabling conditions that affect both their physical and psychological development.

Diseases that strike adults often begin in childhood. Risk factors for cardiovascular disease, cancer, and stroke—the cause of two-thirds of deaths

Of the children in this country, 10 to 20 percent have physical handicaps. Adjusting to these handicaps requires accommodations on the parts of the children themselves, their parents, and their peers.

among adults in the United States—are common in children. These risk factors include overweight, high levels of cholesterol in the blood, high blood pressure, poor physical fitness, and diabetes. In one sample of elementary schoolboys from California, for example, nearly half had one risk factor associated with heart disease, and over 10 percent had two (Wilmore and McNamara, 1974). In a New York sample, 43 percent of the schoolchildren surveyed had abnormally high cholesterol levels; 10 percent smoked cigarettes; 15 percent were overweight; and 2 percent had high blood pressure (Williams, Arnold, and Wynder, 1977).

Many researchers believe that prevention of heart disease, cancer, and stroke should begin in childhood. One preventive screening and education program, the "Know Your Body" program, has been instituted in New York to determine whether certain risk factors for chronic disease can be reduced in schoolchildren (Williams et al., 1977). Results over the long term will tell the research team whether the program has motivated the students to change their behavior enough to reduce their risk and, ultimately, their incidence of fatal disease.

In addition to these risks of later health problems, schoolchildren may suffer from handicaps during childhood. At any given time, 10 to 20 percent of the children in this country have a physical handicap (Pless and Satterwhite, 1975). If sensory impairment, mental retardation, and behavior disorders are included, then 30 to 40 percent of children suffer from one or more chronic disorders. The most common of these are asthma, heart problems, cerebral palsy, orthopedic problems, and diabetes (Mattson, 1972). About 25 percent of elementary school children have vision problems, 10 to 20 percent have reading problems, 3 percent have hearing problems, 3 percent are mentally retarded, 1 percent have epilepsy, and 1 percent have a major speech disorder.

Behavior disorders

> Our younger son was a beautiful baby. But by the time he was about a year old, we were worried about the way he wasn't responding to us and wasn't making any progress in talking. By the time he was 2, he not only wasn't talking, but spent hours hitting his head against the side of his crib or rocking back and forth. It was impossible to catch his eye. He seemed to be going backward, into some dark isolated place where we couldn't reach him. We took him to specialists, and the word came back: "autism." He's 8 now and in a residential school for autistic children near our home. We take him home on weekends, and he has made some progress. He can feed himself with a spoon and say a few words. The head banging and rocking are less. But he isn't normal, and God only knows whether he ever will be.

It is not only physical handicaps and illness that present problems for children and parents. Psychological handicaps and behavior disorders can be just as devastating. These behavior disorders range in severity from problems in sleeping and concentrating and outbursts of temper, to extreme hyperactivity and excessive withdrawal or autism.

Hyperactivity

Thomas Edison's teachers complained so about his unrestrained behavior that his mother had to take him out of school and tutor him at home. Winston Churchill's teachers considered his overactivity a problem (Aries, 1962). Today **hyperactivity** is considered a behavior disorder to be treated by psychotherapy, psychoactive drugs, and special education (Ross and Ross, 1982).

At what point does a child's level of activity deserve the label "hyperactive"? Hyperactive children are restless and have difficulty concentrating. They cannot quiet themselves when someone asks them to. They are distract-

Hyperactive children are distractible and restless; they are often aggressive, destructive, and intolerant of frustration.

ible, impulsive, easily angered and frustrated, often destructive and aggressive. Unpredictable and disruptive, hyperactive children often are referred by a teacher to a physician for diagnosis and treatment. In the United States, hyperactivity is the most common behavior disorder that child psychiatrists see, affecting some 1 to 6 percent of all schoolchildren (Ross and Ross, 1982). It is a syndrome that has attracted the attention of many researchers (Abikoff, Gittelman-Klein, and Klein, 1980; Campbell and Paulaukas, 1979; Ceci and Tishman, 1984; Douglas, 1980; Kinsbourne, 1973; Loney, Langhorne, and Paternite, 1978; Routh, Schroeder, and O'Tuama, 1974; Whalen and Henker, 1984).

Specific situations seem to affect hyperactive children in different ways. Some children are hyperactive when they are in groups but not when they are with just one other person. For others, the opposite is true. Some children are hyperactive only in unfamiliar situations; others, only in familiar ones. But almost all hyperactive children have trouble at school. When they can set their own pace, hyperactive children often seem quite competent, but when they must follow someone else's pace—as often happens at school—they seem out of step. In the classroom, hyperactive children are more physically active than other children and act up at inappropriate times. Compared to normal children, hyperactive children make more mistakes on work that requires focused attention, and they do not control their impulses well. As hyperactive children try to learn, their attention is scattered and focused on peripheral attributes—a condition called **attention deficit disorder**. Some hyperactive children also show poor judgment.

Many of the symptoms seen in hyperactive children are associated with delays in maturation. The behavior of hyperactive children is normal for children three to four years younger. As a result of their impulsiveness, poor judgment, and scattered attention, hyperactive children often are in conflict with others, and so many hyperactive children develop a poor self-image and think of themselves as "bad." They have few friends, and others tend to think badly of them. When hyperactive children are also aggressive and hostile, they tend to develop more and more problems as they get older. Hyperactive children often become poorly adjusted adults—regularly changing jobs, homes, and relationships.

What causes hyperactivity? It is possible that hyperactivity stems at least in part from genetic factors. Parents of hyperactive children tend to have more psychiatric problems, including hyperactivity as children, than parents of normal children (Morrison and Stewart, 1971, 1973). Hyperactivity also occurs almost exclusively in boys. For these reasons, and because hyperactivity is strongly related to a deficit in the ability to attend to stimuli, researchers now believe that hyperactivity has a physiological basis. But environmental variables also may contribute. Some people have suggested that hyperactivity is associated with a child's blood sugar level. They recommend eliminating refined sugars from the hyperactive child's diet. This suggestion is based on personal testimony rather than objective data, however. Another school of thought is that food additives set off hyperactivity (Feingold, 1975). Many investigators have not found evidence to support this hypothesis (for example, Conners, 1980). But others have found that food dyes and preservatives have an effect on *some* hyperactive children (Swanson and Kinsbourne, 1980). Still other researchers have suggested that food allergies relate to hyperactivity. The consensus among researchers at present seems to be that dietary factors are likely to be the cause of hyperactivity in at most only a small fraction of children (Henker and Whalen, 1980; Kerasotes and Walker, 1983). Other environmental factors that have been blamed for hyperactivity include exposure to radiation, heavy drinking by the mother during pregnancy, stressful child rearing, stressful experiences in school, chronic low levels of lead in

lifespan focus

We all recognize them. They are the people who read while they are driving, who chomp their food so quickly they seem to inhale it, who fly off the handle, who rush through the day. They are **Type A** people. Compared to their slower-paced **Type B** fellows, Type A people act, think, and feel more quickly and intensely. Compared to a Type B person with the same intelligence score, a Type A person gets higher grades and undertakes more activities. Type A behavior seems to enhance professional achievement and productivity—a great advantage in a competitive, industrialized society like our own. But Type A behavior has its dangers as well. Type A people often feel hostile, angry, urgently rushed, and anxious (Brody, 1984). These qualities, uncomfortable enough in their own right, have been linked to heart disease in adult men. In fact, it was two San Francisco cardiologists who first described Type A behavior (Friedman and Rosenman, 1974). Although Type A and Type B adults differ on certain neurological, hormonal, and physiological measures, the specific mechanisms by which Type A behavior is translated into heart disease have not yet been uncovered. The connection between Type A behavior and factors associated with heart disease has shown up in studies of young people as well as of adults, however. In an investigation of children between 10 and 17 years old, it was found that the Type A personality traits of eagerness and energy were correlated with blood levels of cholesterol and triglycerides (Hunter, Wolf, Sklov, Webber, and Berenson, 1981). Type A behavior seems to be a lifelong pattern.

To explore the differences between Type A and Type B children, researchers have investigated how elementary school children react to a variety of tasks. In one study, for example, the researchers hypothesized that Type A children would perform tasks more hurriedly and intensely than Type B children and that Type B children would perform hurriedly and intensely only if they were so instructed (Wolf, Sklov, Wenz, Hunter, and Berenson, 1982). First, researchers asked the children to rate themselves on scales describing their own behavior. From the children's ratings on scales like "I am easygoing—I am hard driving," "It does matter if I am late—It doesn't matter if I am late," "I walk fast—I walk slowly," the investigators classified them as Type A or Type B. Then, the children were given a number of tasks, which included reading an emotionally charged passage, eating two graham crackers, delivering an envelope to a box, playing marbles, estimating when one minute had passed, and crossing out numbers on a page. The Type A children read more loudly than Type B children, ate their graham crackers and delivered the envelope faster, were more competitive in playing marbles,

the body, and fluorescent lighting (Henker and Whalen, 1980; Ross and Ross, 1982). Research to establish these links is continuing. But like so many other kinds of behavior, hyperactivity is doubtless a result of both biological and environmental factors.

What can be done to control hyperactivity? Stimulants like **Ritalin** calm hyperactive children and help them to focus their attention. But the side effects of the stimulants—at first they interfere with sleep and appetite, and for a while, they may stop a child from growing—have led people to question their value as a treatment. Because of this concern about treating children with psychoactive drugs, psychologists have tried to find alternative treatments for hyperactivity. To date, their research suggests that psychoactive drugs are most effective in diminishing the intensity of children's hyperactive behavior, crossed out more numbers, and estimated that one minute had passed more quickly than Type B children did. All the findings supported the hypothesis that Type A children would act more hurriedly, intensely, aggressively, and competitively than Type B children. The research team concluded that Type A people try harder to exert and maintain control over environmental demands and challenges.

Because it is widely accepted that Type A behavior can contribute to the development of heart disease, researchers have tried to uncover the psychological and social factors that foster Type A behavior. Some have observed how parents encourage their children's competitiveness, aggression, and impatience. For example, one mother encouraged her blindfolded son as he tried to pick up a pile of blocks. When time was up, she said, "Next time try for six blocks." Said another mother to her son, after the same task, "Next time go a little faster." Both sons had been classified as Type A (Brody, 1984).

Signs of the influence of both nature and nurture on Type A behavior emerged clearly in another study of children, some of whom were rated high in Type A characteristics and others, low in Type A characteristics (Thoresen, Eagleston, Kirmil-Gray, and Bracke, 1985). Many of the high Type A children had parents who also were high in Type A behavior. These Type A children reported often feeling stressed and tense and suffering from symptoms of stress such as cardiovascular and sleep problems. They also were more often angry and hostile than the low Type A children. When the children attempted to build a tower of blocks, parents of both high Type As and low Type As praised their children. (Praise for success, research has shown, is likely to instill in children a need to strive for achievement.) But the parents of the high Type A children also criticized their children's failures, gave them more instructions, and made more decisions for them on a ring toss problem. Chronic interference and criticism for failure are likely to impair children's self-confidence.

Other researchers have investigated the contribution of constitutional factors. In one study using subjects from the New York Longitudinal Study, it was shown that Type A behavior in young adult subjects had links to assessments of temperament made when they were children (Steinberg, 1985). Young adults who showed the striving for achievement typical of Type As had been adaptable but irritable and negative in mood as young children. Young adults who were impatient and angry, another characteristic of Type As, were unadaptable, sensitive to stimulation (had low sensory thresholds), and not persistent as young children. It is likely that Type A behavior, like many other behavior patterns that we have discussed, has roots in both nature and nurture.

but cognitive and behavioral therapies are effective in improving social behavior and in helping children to manage their anger. In therapeutic programs, children are taught specific strategies for solving problems and for keeping their actions under control when they meet a new child or play a competitive game (Hinshaw, Henker, and Whalen, 1984). They are trained to recognize the external triggers that anger them and to identify the thoughts and feelings that signal their building anger. They learn how to handle teasing and provocation by ignoring it, staring out the window, or talking calmly. The best treatment for hyperactive children at present combines educating children and their parents about the disorder, structuring the children's environments at home and at school—setting firm limits, rules, and providing supervision—and giving the children drug therapy.

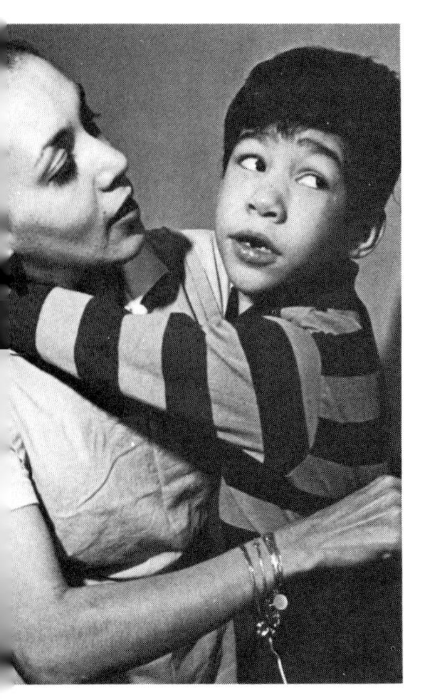

Autistic children avoid social interaction and eye contact. They may also have other sensory handicaps, as this child does.

Autism

One rare but devastating childhood disorder is **autism**. Autism is a form of psychosis that strikes 4 or 5 out of every 10,000 children, usually before they are 2½ (Brask, 1967; Rutter, 1978). In about 70 percent of cases, the autistic child is also seriously mentally retarded (DeMyer et al., 1974). Although the diagnostic criteria have been debated, generally autism is characterized by a child's profound aloneness and inability to communicate or to form social relationships. Unlike normal children their age, autistic children do not imitate others' social behavior, use objects appropriately, or play simple games. They do not form emotional attachments to their parents or other people, express empathy, or exhibit cooperation. They do not make friends, and they tend to stare at other people's eyes with an unusual gaze. Their behavior is rigid, ritualistic, and compulsive (Rutter, 1978).

Parents of autistic children long were thought to be cold and obsessive (Eisenberg, 1957; Kanner, 1949), but most psychologists now reject this explanation of autism's cause. Instead they look for an organic explanation for autism. The role of several neurotransmitters, particularly serotonin, has been investigated, as has the role of zinc deficiency (Coleman, 1978). But to date autism has not been related conclusively to any one biochemical agent. In one study of 21 same-sex pairs of twins, in which at least one twin was autistic, researchers found that in 4 of 11 monozygotic pairs but none of the 10 dizygotic pairs were both twins autistic (Folstein and Rutter, 1977). In 12 of 17 pairs of twins in which only one twin was autistic, the autistic twin had had a brain injury. But among the twins concordant for autism, there were no histories of brain injury. The researchers concluded that autism probably derives from several causes, including both brain damage and a genetic abnormality. Several investigators have pointed out that the language and cognitive problems typical of autism are problems in left hemisphere functions. Many autistic children show signs of early damage to the left hemisphere (Dawson, 1982). The right hemisphere functions—such as musical, visual, and spatial skills—of many autistic children are normal or even superior (Blackstock, 1978; Lockyer and Rutter, 1970).

Learning disabilities

In practice, people use the term **learning disabilities** as a catchall to take in everything from hyperactivity to reading disorders (Farnham-Diggory, 1978). Because it sounds better to say that Jason has "a learning disability" than to say that he is "slow" or "brain-damaged" or "retarded" or "emotionally disturbed," more and more problems have been included under the umbrella "learning disability." But properly speaking, "learning disabilities" is a label for the problems that some children have in one or more of the basic processes necessary for using and understanding language and numbers. These children can process some information perfectly well. They have normal eyesight, hearing, intelligence, and physical coordination. But on certain tasks, they cannot function. Children may have problems in reading or spelling, **dyslexia**; in arithmetic, **dyscalcula**; or in writing, **dysgraphia**.

Compared to other children of their overall IQ and age, children with learning disabilities lag behind by several grades in reading and spelling (Farnham-Diggory, 1978). To a child with dyslexia, letters on a page may seem backward. One dyslexic boy named Fred called himself "Derf" because of the way his name looked to him. Dyslexic children can remember sounds, but they cannot remember images of words. One child, for example, spells "brother" as "birth" and "helicopter" as "thracatei" (Figure 11.1).

A DYSLEXIC CHILD'S STORY

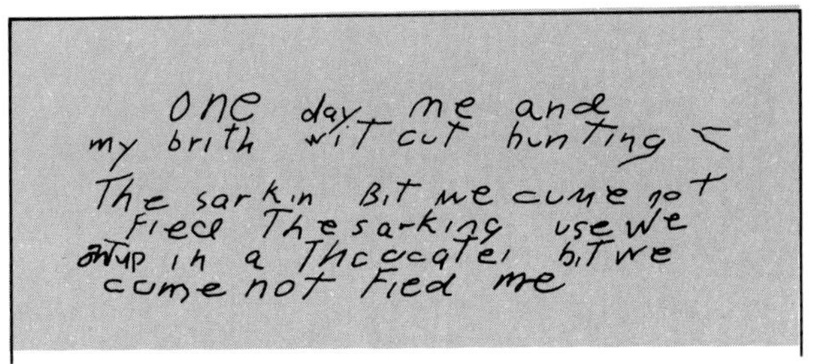

Figure 11.1 ***This story was written by a dyslexic child. It is meant to say, "One day me and my brother went out hunting the sark. But we could not find the sark. So we went up in a helicopter, but we could not find him" (Farnham-Diggory, 1978, p. 61).***

What is behind these surprising mistakes? Dyslexic children process visual information unusually slowly. In one study, children first were shown a compound figure—a cross inside a square—and then were shown the two parts separately. For normal children to see the two parts separately, the interval between the two images had to be at least 100 microseconds (Stanley and Hall, 1973). For the dyslexic children to see them separately, the interval had to be at least 140 microseconds. When they were asked to draw the two parts of the picture, the normal children needed 180 microseconds to identify the separate parts, but the dyslexic children needed 230 microseconds. In other words, the image "stayed on" in the minds of the dyslexic children longer.

Another problem for dyslexic children as they read is **masking** (Figure 11.2). Masking happens when the contours of letters are similar, the letters overlap in the visual field, and the time between the visual pickup of the first letter overlaps with the pickup of the second letter. It stands to reason that if dyslexic children process letters slowly, then when they read, masking may make them see a kind of visual jumble.

But dyslexia involves other problems as well. Dyslexic children may perseverate: a letter that they have read may echo in their mind's eye even when they are reading another letter. When 10-year-old Laura, for example,

PERCEPTUAL MASKING IN READING

t o s
nte o hsx

o
bom
sbomk
asbomku
easbomkut
geasbomkutc
wgeasbomkutcz
dwgeasbomkutczh
idwgeasbomkutczhv
xidwgeasbomkutczhvp
fxidwgeasbomkutczhvpn
rfxidwgeasbomkutczhvpnj
yrfxidwgeasbomkutczhvpnjl

Figure 11.2 ***In this display, if you look at the* o *in the top row, you can see it and the letters* s *and* t *clearly. But in the second row,* s *and* t *are masked by the letters next to them. As you continue down the pyramid of letters, looking at the central* o*, the end letters continue to be recognizable, but the letters in between are masked.***

tried to read the word "reverence," she still had not got it right after 51 seconds of trying to sound it out. First she said "rever," but then she said "renay," and then "never." The *n* apparently was perseverating in her visual memory (Farnham-Diggory, 1984). Many dyslexics have poor visual memories. Compared to normal schoolchildren, for example, dyslexic children were slower to name pictures of common objects (duck, leaf, cat, etc.), colors, and numbers (Spring and Capps, 1974). Dyslexic children also show signs of memory fatigue sooner than other children and do not recover from it when they switch from one kind of mental task to another—from visual to auditory memory, for example—as other children do.

Learning disabilities like dyslexia probably arise from minimal brain dysfunction. In some cases, the connection between the two hemispheres of the brain may be awry, and in other cases, areas controlling vision, hearing, or other senses are defective. Injuries to the brain that prevent a person from connecting the words seen via the right hemisphere with the association area in the left hemisphere may produce a reading disability. If the association area is also damaged, a writing disability may occur as well. When children have trouble putting words or letters in a temporal sequence, the problem may arise because the left hemisphere, which controls order and sequence, is inadequate in overall control (Farnham-Diggory, 1978). Some learning disabilities result from brain injuries, but most are inherited (Decker and DeFries, 1981).

Parents do not produce learning disabilities in their children, but their reactions to a child who has them can make things worse. Like all handicaps, learning disabilities can disorganize, upset, and unsettle family members.

Summary

1. Middle childhood is a period of slower growth and greater physical skills. It is also a period when skills in different areas are consolidated. Physical activity sharpens children's minds and helps keep obesity and other problems from developing.
2. Chronic physical disorders, such as cerebral palsy, asthma, orthopedic problems, and diabetes, handicap 10 to 20 percent of all children under 18. Another 10 to 20 percent suffer from sensory impairments, mental retardation, and behavior problems. Boys are more vulnerable than girls to these developmental disorders.
3. Hyperactivity is a behavior disorder of children who are so active, distractible, and excitable that they cannot attend to schoolwork or other tasks that other children their age can manage. Hyperactive children's attention seems scattered and peripherally focused. Hyperactivity is sometimes managed successfully with psychoactive drugs, sometimes with cognitive strategies that teach a child to manage actions and outbursts, and sometimes with interventions to structure the child's behavior.
4. Children with learning disabilities may have perceptual problems that interfere with their perception and processing of information. Learning-disabled children have trouble reading, writing, spelling, speaking, or calculating, even though their motor abilities, eyesight, hearing, memory capacity, motivation, and intelligence are normal. Learning disabilities may be the product of brain injury or dysfunction. They seem to run in some families and, like many childhood disorders, afflict more boys than girls.

Key terms

hyperactivity
attention deficit disorder
Ritalin
autism
learning disability
dyslexia
dyscalcula
dysgraphia
masking

Suggested readings

FARNHAM-DIGGORY, SYLVIA. *Learning Disabilities.* Cambridge, Mass.: Harvard University Press, 1978. An intriguing discussion of learning disabilities, particularly dyslexia. The discussion of the definitions, causes, and treatments of learning disabilities makes clear how much we have yet to learn about them.

FEATHERSTONE, HELEN. *A Difference in the Family: Living with a Disabled Child.* New York: Penguin, 1981. Writing from her own experience and that of other parents, Featherstone honestly describes the emotional and practical strains of having a seriously disabled child.

ROSS, DOROTHEA M., and ROSS, SHEILA A. *Hyperactivity: Current Issues, Research, and Theory* (2nd ed.). New York: Wiley, 1982. A description of the course of hyperactivity from infancy through early adulthood and of the drug therapies and special school programs by which it is treated. The hyperactive child's view and those of his or her associates are given.

WHALEN, CAROL K., and HENKER, BARBARA (Eds.). *Hyperactive Children: The Social Ecology of Identification and Treatment.* New York: Academic Press, 1980. A collection of theoretical and empirical articles on hyperactivity, including an examination of factors likely to influence the diagnosis and treatment of hyperactive children.

chapter twelve

Cognitive development

Just as middle childhood is a time when children grow more physically skillful, it is a time when children mature cognitively. Children who begin school have shed their intuitive beliefs that balls roll because they want to, that mothers can see through walls, that dreams really happen. They can fathom cause and effect and are beginning to understand the working of numbers. During the elementary school years, children will learn to plan and memorize, to read and calculate. By the end of first grade, they will know how to read simple words and to add pairs of numbers. By the end of sixth grade, they will know rules of spelling, multiplication tables and fractions, and will be able to read and understand complex stories and take tests. In this chapter, we discuss children's progress during the elementary school years and some of the factors that influence how fast and how far individual children go.

Concrete operational thought

The thinking typical of early childhood, as we saw in Chapter 9, is a charming blend of impressions, intuitions, and partial logic. But as children enter school and middle childhood, the flavor of their thinking changes. Armed now with greater understanding of people and things—after all, haven't they dropped things, built towers, aligned sticks, and asked ''why'' at least a million times?—children begin to grasp the logical relations of things, the orderly rules and constant properties that govern the ways in which things happen. Whereas in early childhood, thought was based on the child's immediate perceptions, now thought becomes more integrated and bound by rules of logic.

> When I was 5 years old, my 8-year-old cousin asked me this riddle: ''Which is heavier, a ton of feathers or a ton of bricks?'' I answered ''a ton of bricks.'' She burst into guffaws, and I was mortified.

Reasoning logically about objects was called by Piaget **concrete operational thought**: it is ''concrete'' because children can reason only about concrete, tangible things like milk and cookies, sticks and stones, bricks and feathers. It is ''operational'' because children can perform mental manipulations, or *operations*, on the things in an organized and systematic way.

Children who can use concrete operations know that multiplication is related to division, that subtraction is the opposite of addition, and that ''equals,'' ''greater than,'' and ''less than'' are all interrelated.

> When Erica was 6 we were on a long highway drive, and Erica was getting impatient. As we stopped to pay a toll, she asked, ''Are we off the highway yet?'' ''No,'' her father replied, ''we have four more tolls to pay before we get off the highway.'' At the next tollbooth, Erica said, ''Now we have three more tolls?'' Somewhere along the way she had figured out how to subtract.

They realize that certain transformations of objects are *reversible*—that milk can be poured from one glass to another and then back again, for example. They *decenter*, that is, focus on and coordinate more than a single dimension—height *and* width, for example—at the same time. (In early childhood, you recall, children focused, or centered, on just one dimension at a time.) They also recognize that one dimension may make up for another dimension; that two dimensions may be *reciprocal*, or complementary. The short, wide glass may hold just as much milk as the tall, narrow glass.

For most children, concrete operational thinking first appears between the ages of 5 and 7. It continues to develop during middle childhood, as children apply the operations of addition and subtraction, multiplication and division, reversibility and reciprocity to objects. Concrete operational thinking allows children to solve many problems that elude preschoolers. They can solve problems in conservation, seriation, and velocity. Concrete operational thinking also allows children to appreciate jokes and to understand concepts that they could not earlier.

Conservation

With the ability to perform concrete operations, children realize that the milk poured from a tall, thin glass into a short, fat glass *looks* different but remains the same in quantity. This understanding (discussed in Chapter 9) was called by Piaget *conservation*. During the school years, children master conservation of different quantities in the following order: number, length, liquid quantity, mass, weight, and volume. The 7-year-old knows that a ball of clay rolled out to make a snake retains the same amount of substance, but he may not understand that they weigh the same. The 9-year-old understands that they weigh the same but perhaps not that their volume is the same. Only at age 11 or 12 do most children master conservation of volume. Children seem to learn about conservation first with the simplest tasks, those with the most salient and visible qualities. Only later do they move on to those that are not readily apparent to the eye. Although, as we discussed in Chapter 9, preschool children can be trained to solve simple conservation problems, middle childhood is when most children figure out for themselves the implications of reciprocity and reversibility, number and size, so that they can understand conservation problems regardless of how questions are worded or how objects appear, and can back up their judgments with sound, logical reasoning.

Children's understanding of concrete operations shows up outside the laboratory as well as in researchers' tests. Children with concrete operational thought laugh at jokes that earlier would have gone over their heads (and later will be greeted with groans). Preschoolers think that riddles are questions with arbitrary answers—''What has four wheels and flies?'' ''A horse!''—or even factual answers—''Why did the boy tiptoe by the medicine cabinet?'' ''Because everyone was asleep!''

> When he was 4, Peter was always asking, ''Is that funny?'' before he would commit himself to a forced laugh. When he was 5, he thought that a joke was, ''Why did the chicken cross the road?'' ''To eat snakes,'' or ''Knock, knock.'' ''Who's there?'' ''Mickey Mouse's underwear.''

But during middle childhood, with their greater cognitive maturity, children realize that words—like glasses of water—are not always as they appear. They love puns:

> Hey, call me a taxi.
> Okay, you're a taxi.
>
> I saw a man-eating shark in the aquarium.
> So what? I saw a man eating tuna in the restaurant.

With their developing understanding of conservation, children giggle at a joke like this one:

> WAITER: Do you want me to cut the pizza into eight pieces for you?
> FAT WOMAN: No, no. Cut it into six pieces. I could never eat eight pieces.

The funniest jokes, researchers have found, pose a moderate challenge to a person's cognitive abilities and are understandable but surprising—they violate the person's expectations (Prentice and Fathman, 1975; Zigler, Levine, and Gould, 1967). School-age children who are just working through the nuances of the logical relations involved in reciprocity, reversibility, classification, and transitivity find jokes funny that depend on these operations.

Moral reasoning

In middle childhood, children's ideas about right and wrong are also concrete. Piaget was the first researcher to study children's moral reasoning. He developed his theory of children's moral reasoning by watching and questioning children as they played marbles, a game that required children to deal with issues of justice, fairness, and turn taking. He saw that children's rules for playing marbles and their respect for these rules grew more sophisticated as they grew older. At the beginning of middle childhood, children first began to play by strict rules and to play to win. They wanted to settle which players controlled the game, and they wanted all the players to play by the same rules. These children, Piaget suggested, played marbles according to an **external morality** in which rules are seen as cast in stone, handed down by authority figures. Asked who makes up the rules to a game of marbles, children at this age were likely to answer ''God'' or ''Daddy.'' Although the children sometimes bent the rules as they played, they denied that rules can be changed. Here Ben discusses rule making with Piaget:

> Invent a rule. *I couldn't invent one straight away like that.* Yes you could. I can see that you are cleverer than you make yourself out to be. *Well, let's say that you're not caught when you are in the square.* Good. Would that come off with the others? *Oh, yes, they'd like to do that.* Then people could play that way? *Oh, no, because it would be cheating.* But all your pals would like to, wouldn't they? *Yes, they all would.* Then why would it be cheating? *Because I invented it: it isn't a rule! It's a wrong rule because it's outside of the rules. A fair rule is one that is in the game* (Piaget, 1932, pp. 54–55).

Another child told Piaget that if he changed the rules, God would punish him by making him miss during his turn at marbles. Children at this age believed that obeying rules is good and disobeying rules is bad. Their very definition of goodness, in fact, was obedience and conformity to rules.

Piaget used children's understanding of the rules of games as an index of their level of moral development. Between the ages of 5 and 10, children regard rules as sacrosanct and unalterable.

lifespan focus

Stages of moral development

By starting from the external and internal moral reasoning described by Piaget, Lawrence Kohlberg (1969) developed a more complete theory of stages in moral development (Table 12.1). As Piaget had done, Kohlberg based his descriptions on people's responses to hypothetical moral dilemmas such as this one:

> In Europe, a woman was near death from a special kind of cancer. There was one drug that the doctors thought might save her. It was a form of radium that a druggist in the same town had recently discovered. The drug was expensive to make, but the druggist was charging ten times what the drug cost him to make. He paid $200 for the radium and charged $2000 for a small dose of the drug. The sick woman's husband, Heinz, went to everyone he knew to borrow the money, but could only get together about $1000, which was half of what it cost. He told the druggist that his wife was dying and asked him to sell it cheaper or let him pay later. But the druggist said, "No, I discovered the drug and I'm going to make money from it." So Heinz got desperate and considered breaking into the man's store to steal the drug for his wife. Should Heinz steal the radium? (Kohlberg and Gilligan, 1971, pp. 1072–1073)

In the first stage of moral development, Kohlberg found, people are concerned with getting rewards and avoiding punishments. What they consider moral is determined by authority figures. People at the first stage of moral development argue on the principle of "might makes right." Heinz should not steal the drug, they might argue, because he will go to jail for it. Or, they might argue, Heinz should steal the drug because he loves his wife very much. Kohlberg was interested in the *reasons* people gave for their moral judgments, not whether they said that Heinz should or should not steal.

In the second stage of moral reasoning, people think that moral action is making fair deals and trades. A person at this second stage might reason that Heinz should not steal because it would not be worth having to go to jail for, or the person might reason that Heinz should steal the drug because that way he would still have his wife. At both these stages, people are basically concerned with looking out for themselves and protecting their own interests. Kohlberg called this **preconventional moral reasoning**, because it was determined by personal interests, not social conventions.

The next level of moral development is **conventional moral reasoning**. People at this stage think that moral behavior is following social rules and conventions, and they focus on conforming to social order, family obligations, and caring for other people. In Kohlberg's Stage 3, the first stage of conventional moral reasoning, people do good things so that others will approve of them. This stage has been called the "good boy–nice girl" stage. People at this stage may reason that Heinz should steal the drug because he loves his wife and she will approve of him for stealing it for her. Family is more important to people at this stage than social institutions or outside individuals.

People in Stage 4, the "law-and-order" stage, believe strongly in rules and social order and reason that moral behavior serves the interests of society. Heinz, they may reason, should steal the drug for his wife because he has vowed to protect her until death do them part. His marriage vow and his duty as a husband are of overriding importance. Stage 4 reasoners may argue that Heinz should not steal the drug because stealing is against the law.

Kohlberg called the highest level of moral reasoning **postconventional**. In the first stage of postconventional reasoning—Stage 5—people are oriented toward social contracts. They understand that laws are written by consensus

Table 12.1
Kohlberg's stages of moral development

Moral reasoning	What is right?	How People Answer the Heinz Dilemma	
		Pro	**Con**
Preconventional Level			
Stage 1: Punishment-obedience orientation	To obey the rules of others in order to avoid punishment. Obedience for its own sake, and avoiding physical damage to persons and property.	He should steal the drug. It is not really bad to take it. It is not like he did not ask to pay for it first. The drug he would take is only worth $200, not $2000.	Heinz should buy the drug. If he steals the drug, he might get put in jail and have to put the drug back anyway.
Stage 2: Instrumental-exchange orientation	Following rules only when it is to your advantage; acting to meet your own interests and letting others do the same. Right is also what is an equal exchange.	Heinz should steal the drug to save his wife's life. He might get sent to jail, but he'd still have his wife.	He should not steal it. The druggist is not wrong or bad; he just wants to make a profit.
Conventional Level			
Stage 3: Good-boy–nice-girl orientation	Living up to what is expected by people close to you or what people generally expect of a son, brother, friend. Being "good" is important and means having good motives, showing concern for others.	If I were Heinz, I would have stolen the drug for my wife. You can't put a price on love. You can't put a price on life either.	He should not steal. If his wife dies, he cannot be blamed. The druggist is the selfish or heartless one.
Stage 4: System-maintaining orientation	Carrying out the duties that are your obligation. Laws are to be upheld except when they conflict with other fixed social duties. Right is contributing to society, the group, or institution.	When you get married, you take a vow to love and cherish your wife. Marriage is an obligation like a legal contract.	It is a natural thing for Heinz to want to save his wife, but it is still always wrong to steal. He still knows he is stealing and taking a valuable drug from the man who made it.
Postconventional Level			
Stage 5: Social-contract orientation	Being aware that most values and rules are relative. They should usually be upheld because they are a social contract. Some rights, such as life and liberty, however, must be upheld in any society, regardless of majority opinion.	The law was not set up for these circumstances. Taking the drug in this situation is not really right, but it is justified.	You cannot have people stealing whenever they get desperate. The end may be good, but the ends do not justify the means.
Stage 6: Universal-ethical-principles orientation	Following self-chosen ethical principles. When laws violate these principles, you must act in accordance with the universal principles: equal rights for all and respect for the dignity of individuals.	The situation forces him to choose between stealing and letting his wife die. Where the choice must be made, it is morally right to steal. He has to act on the principle of preserving and respecting life.	Heinz must decide whether to consider the other people who need the drug just as badly as his wife. Heinz ought to consider the value of all the lives involved.

SOURCE: Kohlberg, 1969, pp. 379–380.

and can be changed by consensus as well. Even though rules may be arbitrary, they guard human rights, and the privilege of living in a society incurs in each individual the obligation to obey its laws and to abide by its social contracts. If laws and rules threaten individual human rights, they can in some instances be superseded, however. As one person reasoned when he was asked about the Heinz dilemma,

> It is the husband's duty to save his wife. The fact that her life is in danger transcends every other standard you might use to judge his action. Life is more important than property.
> *Suppose it were a friend, not his wife?*
> I don't think that would be much different from a moral point of view. It's still a human being in danger.
> *Suppose it were a stranger?*
> To be consistent, yes, from a moral standpoint.
> *What is this moral standpoint?*
> I think every individual has a right to live, and if there is a way of saving an individual, he should be saved (Kohlberg, 1976, p. 38).

According to this postconventional, Stage 5 argument, the value of human life overrides the value of any convention. A drug to save a life can, therefore, be stolen for a wife, a friend, or a stranger.

In Stage 6, the highest stage of moral reasoning in Kohlberg's theory, people base their judgments on universal principles and unimpeachable ethics. Their ethical values rest on justice, reciprocity, equality, human life, and human rights. Stage 6 reasoners arrive at their moral decisions on their own, sometimes *in spite of* social conventions. They listen more intently to their conscience than to public opinion or social conventions. The actions of Martin Luther King, Jr., and Mahatma Gandhi exemplify this kind of moral reasoning. These men were guided by strong, individual, ethical principles, and when laws conflicted with their principles, they disobeyed the law: Martin Luther King went to jail rather than obey the laws that supported segregation, just as Gandhi went to jail rather than capitulate to the inequities of India's system of government. A Stage 6 reasoner might argue that Heinz must steal the drug to preserve life rather than obey a law that would effectively condemn his wife to death.

Kohlberg and his associates completed a 20-year longitudinal study of the moral reasoning of a sample of Chicago boys (Colby, Kohlberg, Gibbs, and Lieberman, 1983). Beginning when the boys were 10 years old, the researchers asked them about Kohlberg's moral dilemmas every three to four years, and their responses were recorded. The boys proceeded through Kohlberg's developmental stages in the order Kohlberg had observed in cross-sectional comparisons of people of different ages. No subject skipped a stage; only rarely did a subject seem to move back a stage. The subjects' moral reasoning continued to advance for the duration of the study, by the end of which the subjects were men in their 30s (see Figure 12.1).

The results of the study showed a general upward trend for all the subjects as they got older. Even so, there were clear individual differences in the rates at which the boys moved through the stages and in the levels that they ultimately reached. The most common level of reasoning for 6- to 9-year-olds was Stage 1; for 10- to 12-year-olds, Stage 2; for 14- to 24-year-olds, Stage 3; and for 24- to 36-year-olds, Stage 4. But some individuals reasoned at Stage 4 when they were only 12 years old and at Stage 5 when they were 18. A few others never rose above Stage 2. Moral reasoning is linked to increasing age and experience, but it is clearly not guaranteed by them.

Kohlberg's theory of moral development has come in for its share of criticism. It has been criticized, for example, for being a politically liberal hierarchy in which political conservatism is represented as Stage 4, law-and-

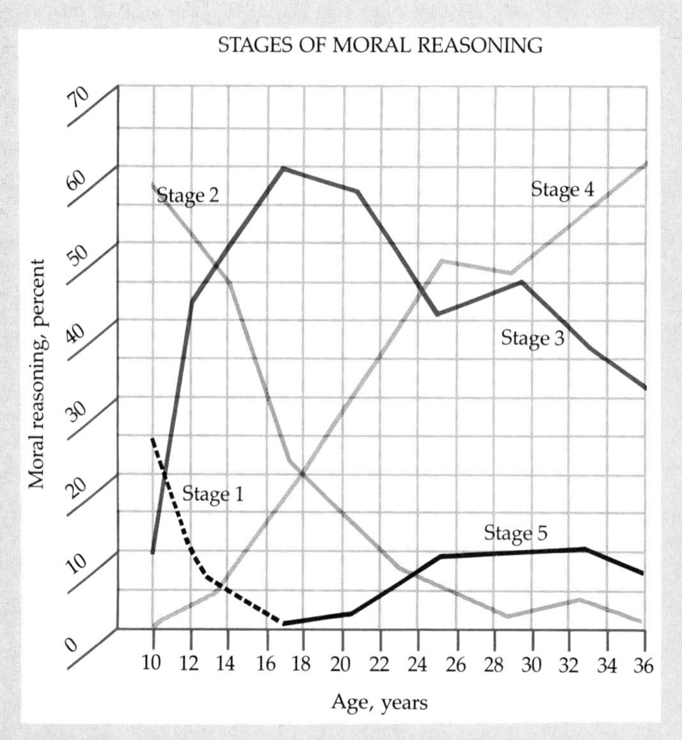

Figure 12.1 ***This graph shows the percentage of boys and men at different ages who reasoned about moral dilemmas at each of Kohlberg's six stages (Colby, Kohlberg, Gibbs, and Lieberman, 1983).***

order reasoning. It has been criticized for reflecting the values of Western culture and not universal values. It has been criticized for being based on data that may have been affected by particular historical events and conditions (Rest, Davison, and Robbins, 1978). The subjects in Kohlberg's research had lived through the civil rights struggle, the Vietnam War, Watergate, and the wave of feminism, each of which focused public scrutiny on moral issues, and this may have honed their ideas about fairness. Historical change probably accounts for at least part of the increases in the subjects' scores on tests of moral reasoning. Most recently, Carol Gilligan (1982) has aroused wide interest with her criticism of Kohlberg's theory as sexist. Gilligan opposes Kohlberg's analysis because it is based on and biased toward male responses. She suggests that males and females differ psychologically in ways that directly affect their moral reasoning. Women and girls are more attuned to the relationships and connections between people; men and boys are more attuned to individual achievement, to the separateness and distinctiveness of each person. Women take much of their personal identities from their personal relationships; men take theirs from their work. Women are likely to be morally concerned with how people are connected, obligated to one another, and generally interdependent. Men are likely to be morally concerned with justice, individual rights and obligations, and equitable solutions to each separate person's conflicting and competing claims. Says Gilligan, women's is a morality of caring, and men's is a morality of justice. These criticisms aside, Kohlberg's theory and data have been useful for showing how people's reasoning about moral issues changes as they get older.

Children's respect for rules remained rigid, Piaget found, until they reached a level of **internal morality**, in which they understood that people may agree to reason out, discuss, change, and remake rules. They did not reach this level until about age 11, at the end of middle childhood.

Piaget elaborated his theory of moral development by listening to children's reasoning when he asked them moral questions in little stories.

> A little boy who is called John is in his room. He is called to dinner. He goes into the dining room. But behind the door there is a chair, and on the chair there is a tray with fifteen cups on it. John couldn't have known that there was all this behind the door. He goes in, the door knocks against the tray, bang go the fifteen cups and they all get broken!

> Once there was a little boy whose name was Henry. One day when his mother was out he tried to get some jam out of the cupboard. He climbed up on to a chair and stretched out his arm. But the jam was too high up and he couldn't reach it and have any. But while he was trying to get it he knocked over a cup. The cup fell down and broke.

After telling children stories like these, Piaget asked them whether John or Henry was naughtier and who should be punished more severely. Children in middle childhood told Piaget that John was naughtier, because he broke more cups. They were basing their judgment on the severity of the external consequences of the act—15 cups versus one. Eleven- and 12-year-olds, like adults, told Piaget that Henry was naughtier, because he was trying to sneak some jam. They based their judgment on the internal intentions behind the act—innocent versus sneaky.

Transitions in moral reasoning

What makes children advance from one stage or level of moral judgment to the next? It has been suggested that when children meet with reasoning that is one stage higher than their own, they try to resolve the conflict between what they believe and what they are hearing by reasoning at the higher level. In one study to test this suggestion, researchers measured children's reactions to moral reasoning a stage or two below or a stage above their own (Rest, Turiel, and Kohlberg, 1969). First the children read several moral dilemmas and wrote down their responses to the dilemmas. Then they read other children's advice about what the character in the dilemma ought to do. Finally, the children were asked to decide which advice was the best, the worst, the smartest, and the most reasonable and to put the advice into their own words. The children were considered to understand the advice if their restatement of it was at an equal level of moral reasoning. The researchers found that children were most likely to understand reasoning one stage below their own, although they considered it the ''worst'' advice. The children had more trouble understanding reasoning a level or two above their own, but they preferred this more advanced reasoning.

Does this preference for reasoning one stage higher advance children's later moral reasoning? In a study to test this possibility, children were briefly trained in reasoning one stage below, one stage above, or two stages above their own (Turiel, 1966). The children who were most likely to advance were those exposed to reasoning one stage above their own. The evidence supported the hypothesis.

In yet another study, children in fifth through seventh grades heard a man and a woman reasoning about six moral dilemmas at a level that was the same as their own, one stage below it, one stage above it, or two stages above it (Walker, 1982). When the children were tested one week later, the children who had heard reasoning one or two levels above their own had advanced one stage in moral reasoning. They were still reasoning about one-half a stage above their original levels on a follow-up test seven weeks later. Children who were exposed to reasoning at their own or a lower level or to no reasoning at all did not advance. Transitions in moral reasoning, these studies suggest, can be encouraged by exposing children to higher reasoning.

Another approach to training in moral reasoning is to have children discuss moral issues with other children. Piaget himself believed that this kind of discussion contributes importantly to moral development. In one study (Damon and Killen, 1982), children between 5 and 9 years old, in discussion groups of three, talked about moral problems. Their debates, even those that were brief, did modestly—but significantly—advance some of the children's moral reasoning. The children who gained, though, were not those who disagreed, argued, contradicted, or offered contrary solutions; the children who advanced tended to agree with other children's statements and to accept, extend, and clarify them, to work with others, to reciprocate, and, if necessary, to compromise. Overt conflict or arguments apparently were not necessary for advances in moral reasoning.

Yet another approach to moral training is to capitalize on real moral problems that confront children—like fighting in class, taking someone's lunch money, or making someone feel bad. While children are reflecting on these problems, the experimenter suggests to them higher-level solutions and encourages them to try out these solutions in real situations. This approach, too, has been used successfully with school-age children to advance both social and moral reasoning (Enright, Lapsley, and Levy, 1983).

There is still some debate among researchers about the *best* way to advance children's moral reasoning, however. Is it better to present them with a cognitive conflict in which they hear a view different from their own in a hypothetical moral dilemma, or is it better to involve them in a social conflict in which they have to discuss their views with others? These two methods were compared in a recent study (Haan, 1985). In the cognitive-conflict condition, groups of friends discussed hypothetical moral dilemmas for five sessions. In the social-conflict condition, groups of friends met for five sessions and played ''games'' designed to simulate actual moral problems—for example, the friends were representatives from three nations that were in competition for food to feed their citizens. All the students were interviewed on moral issues before and after the experiment. The students who played the social-conflict games were significantly more likely to gain in moral reasoning (60 percent gained) than the students who discussed the hypothetical dilemmas (30 percent gained). Personal confrontation with moral problems may be more effective in advancing moral reasoning than mere exposure to cognitive conflict. But some researchers have questioned the value of any brief training (Rest, 1983). They have argued that training may be too short or too narrowly focused, that assessments of children's stages of moral reasoning are crude, and that real moral growth takes time. Longitudinal studies of normal children show that moving through a full stage of moral development typically takes at least four years.

Memory

Another aspect of cognitive development in middle childhood is the growth of memory. School-age children can remember lots of new information and integrate it with what they already know. They can remember more than preschoolers for several reasons, which we discuss in this section.

Strategies

One reason that school-age children remember more than preschoolers is that they use deliberate strategies for remembering. What are the strategies school-age children use?

Rehearsal

One simple strategy for remembering information is **rehearsal**, the process of repeating something until it is memorized. Actors rehearse their lines, young children rehearse their names and addresses, and college students rehearse their class notes before exams. Strictly speaking, very young children do not use this strategy for memorizing information (Perlmutter and Myers, 1979), although even they use rudimentary rehearsal strategies—talking about a hidden toy or its hiding place, staring or pointing at the hiding place, resting their hand on it and nodding yes (DeLoache, Cassidy, and Brown, 1985; Wellman, Ritter, and Flavell, 1975).

Most 5-year-olds rarely rehearse spontaneously, but they can be trained to rehearse, and rehearsal improves their recall. In one series of studies (Flavell, Beach, and Chinsky, 1966; Keeney, Canizzo, and Flavell, 1967), 5- and 10-year-olds were asked to memorize the order in which an experimenter pointed to pictures of objects. After they were shown the pictures, the children had to wait 15 seconds. During this interval, few of the 5-year-olds rehearsed the names of the objects—assessed by reading their lip movements while their eyes were hidden by a space helmet! Nearly all the 10-year-olds did. When the 5-year-olds were instructed by the experimenter to whisper the names of the objects while they were waiting, their recall improved. But when the children were later given another memorization task, the 5-year-olds again did not rehearse the items, and their recall declined. Training can make young children perform as well as older children and, conversely, if older children are prevented from rehearsing, their memories are as spotty as those of young nonrehearsers. But these effects of training are not permanent. Rehearsal is clearly one aspect of cognitive processing that needs time, and rehearsal, to develop.

By the time they are 8 years old, children asked to memorize a list of words do rehearse—but not very effectively. They repeat each word by itself (Ornstein, Naus, and Liberty, 1975). If they are given more time on each word

School-age children can remember more than preschoolers because they use deliberate strategies like rehearsal for memorizing words and actions. These children have memorized a dance routine about Andrew Jackson.

and a chance to look at the words already presented, they are more likely to rehearse appropriately and to recall more words (Ornstein, Medlin, Stone, and Naus, 1985). It seems that they are aware of the importance of rehearsal but are less skilled than older children at retrieving the words to be rehearsed together. By age 12, children can retrieve words and rehearse the whole list cumulatively.

Organization

Another strategy that people can use to increase their recall is grouping items into meaningful clusters or chunks. Most people can remember only about seven separate chunks of information. But by including several related items in each chunk, they can remember substantial amounts of information.

Even 2-year-olds remember related items better than those that are unrelated. ''Big and tall'' is easier for them to remember than ''big and sad,'' for example (Goldberg, Perlmutter, and Myers, 1974). But they do not form chunks or categories the way school-age children do. Preschool children may organize words by rhyming or by similar sounds—''big'' and ''pig,'' ''bat'' and ''pat''—but school-age children are more likely to organize words by their meanings (Bach and Underwood, 1970; Hasher and Clifton, 1974). Preschool children also organize more according to function—things that go together (bat and ball); school-age children organize more according to taxonomy—things that belong to the same category or class (bat and racquet) (Flavell, 1970). It is difficult, though, to tell whether preschool children *cannot* form taxonomic categories, or just *do not*. In one study (Smiley and Brown, 1979), researchers asked kindergartners, grade school children, and adults to say which two pictures out of three were alike. Subjects could pair the pictures either functionally (horse with saddle or needle with thread) or taxonomically (horse with cow or needle with pin). Kindergartners paired the pictures functionally, older children and adults taxonomically. But when the kindergartners were asked for an alternative choice, and to justify it, they provided and justified it easily. They had chosen according to personal preference, not ability, a preference that affected both their learning and their recall (Overcast, Murphy, Smiley, and Brown, 1975; Smiley and Brown, 1979). Whether taxonomic classification is more accessible or simply preferred, it helps older children remember more items.

Social and spatial information also seems to help children organize their memories. Children may remember things like their classmates' names according to seating plans and reading groups, groups of friends and cliques (Bjorkland and Zeman, 1982; Chi, 1981).

> Natalie reeled off the names of the kids at school each day by thinking of who sat around the table at lunch time.

> Sam remembers people by the kinds of cars they drive. People with trucks or motorcycles will live in his memory forever.

Another kind of organization that can improve memory is a story line. Children remember things better when they are organized into meaningful, logical, and coherent stories. Children remember the locations of things better, for example, if they have a story around which to organize their memories of the locations (Herman and Roth, 1984). In one study (Buss, Yussen, Mathews, Miller, and Rembold, 1983), children from second and sixth grades heard a tape of a story about a fish named Albert. Some got a straightforward version of the story, and others got a version in which the sentence order was scrambled. Children recalled less of the scrambled stories than of the straightforward ones. Then the researchers gave children training in how to order sentences into a proper story sequence—first the name of the character, then what made the story begin, then what the character did, and then how the story ended—and asked them to retell the stories in the right order. The instruction proved effective, for the children could accurately retrieve and reorder information according to a typical story form.

Preschool children can be trained to use organizational strategies, as they can be trained to rehearse, but the effects are weaker and less durable than the effects of training school-age children (Moely, Olson, Halwes, and Flavell, 1969; Williams and Goulet, 1975). The ability to organize items into meaningful chunks continues to improve over the course of middle childhood and adolescence. In one study of 8- to 21-year-olds, for example, the older subjects remembered the same numbers of chunks as the younger subjects, judging by the bursts and pauses in their recital of the names of animals and furniture. More information was included in each chunk, however, and the older subjects' total memory scores were significantly higher (Kail and Nippold, 1984). Organizing strategies for *retrieving* items from memory also improve over the years of childhood (Morrison and Lord, 1982).

Imagery

Using imagery is the strategy of imagining pictures of items to be remembered. Imagery can help people to remember names (for example, a sand-covered cat for "Sandra Katz"). Children who do not discover this mnemonic technique on their own can be trained to use it. One researcher (Levin, 1980) used imagery to help fourth and fifth graders learn the 50 capital cities of the United States. After providing the students with concrete "keywords" for each place name—"apple" for "Annapolis," "marry" for "Maryland," and showing them illustrations in which the keywords for each state and its capital were combined pictorially—for example, a judge "marrying" two apples—the experimenter found that the children could recall the names of the states and capitals better than a group of children who had been allowed to study the names on their own.

Why does imagery strengthen recall? Perhaps it underscores associations between items to be remembered. Perhaps it provides two forms—words and pictures—in which information can be recalled. Perhaps it makes learning pleasant, personal, and vivid. Perhaps it subjects the information to deeper and thus better remembered levels of cognitive processing. Perhaps it does all these things.

Elaboration

When people link together two or more unrelated items in order to remember them, they are using another memory strategy called elaboration. Elaboration is at work in the mnemonic, "In fourteen hundred ninety-two, Columbus sailed the ocean blue." Children recall better when their elaborations are active—"The lady flew on a broom"—rather than static—"The lady had a broom" (Buckhalt, Mahoney, and Paris, 1976). Again, older children use this strategy for remembering more than younger children do (Paris and Lindauer, 1976). They also use more active elaborations than younger children do (Reese, 1977). Older children remember better when they generate the elaborations themselves than when an experimenter does; younger children benefit more from an experimenter's elaborations, perhaps because the adult's are of better quality (Turnure, Buium, and Thurlow, 1976).

Metamemory

Metamemory is what people understand and know of their own memory processes. One reason young children do not use the memorization strategies we have just discussed may be that they lack this kind of awareness. Metamemory includes knowing which situations call for conscious efforts at memorization and which factors affect memory. It includes, for example, knowing that faces are easier to remember than names, that everybody forgets things sometimes, that poorly organized material is difficult to recall, and that repeating a fact will help in memorizing it.

Preschoolers do understand some things about memory. They know that noise interferes with memory; and they know that it is harder to remember many items than a few (Wellman, 1977b). They know that it is hard to remember long-ago events and easier to relearn information than to learn it fresh. But beyond knowing these things, preschool children are limited in their metamemories.

Schoolchildren know that time affects memory, that pairs of antonyms are easier to remember than unrelated words, and that they would be likelier to remember a short list studied for a short time than a long list studied for a long time. They also know about ways to improve memory—tying strings around fingers, reading notes, listening to tape recordings, and asking for information from other people (Kreutzer et al., 1975). When researchers in one study (Kreutzer et al., 1975) asked children how they might remember a telephone number, nearly all third and fifth graders, but only 40 percent of the kindergartners, said that they should phone right away. Most of the older children, but only 60 percent of the kindergartners, said that they would write down the number and rehearse it or use some other mnemonic strategy. In another study (Yussen and Levy, 1977), when children between third and ninth grades were quizzed about how they might try to recall a forgotten idea or a lost item, the youngest children gave one or two suggestions and then stopped. But older children offered many plausible suggestions. In yet another study, researchers found that second graders knew that rehearsal and categorization are useful strategies for memorizing, but only sixth graders consistently realized that categorization is more effective than rehearsal, and used it (Justice, 1985).

Just knowing about memory is not the same thing as remembering, however. Metamemory does not directly predict a person's performance on a memory task (Siegler, 1983). The accuracy of children's reports about how memory works and their actual memorization abilities often bear only a moderate relation (Brown et al., 1983). Children sometimes know *how to* remember something but still are not able to remember it (Chi, 1985). Such vagaries of children's behavior make psychologists realize that they have a way to go before they understand metamemory and its relation to memory. Some psychologists (for example, Cavanaugh and Perlmutter, 1982) have commented that the concept of metamemory has not yet contributed much to the understanding of how memory works. Others have questioned how accurately children can report on their own thinking. Further research is necessary to specify how thinking about memory affects memory itself.

Knowledge

Another way in which school-age children differ from younger ones is in the amount of knowledge they have accumulated. One kind of knowledge that children have, as we mentioned in Chapter 9, is of familiar routines or scripts—eating in a restaurant, having birthday parties, and so on (Hudson and Nelson, 1983). Another kind of knowledge is conceptual. A third kind is factual. School-age children have more of *all* these kinds of knowledge than younger children do, and this knowledge also increases their ability to remember things.

The powerful effect of knowledge on memory was demonstrated in a study of children who were experts and adults who were novices at chess (Chi, 1978). For ten seconds, the subjects were shown a chessboard with

School-age children have more knowledge than younger children, which helps them remember more. Children who are experts at chess, researchers have found, are better at remembering the arrangement of chess pieces than adults who are novice players. This difference can only be ascribed to the children's superior knowledge of chess.

pieces set up on it; then the board was covered and they were asked to reproduce the arrangement of chess pieces they had seen. The young experts reproduced the arrangement much more accurately than the adult novices, a finding that could not be ascribed to their superior intelligence or memory abilities, but only to their superior knowledge of chess. When children know as much about a subject as adults do, the usual advantage that adults have all but disappears (Roth, 1983).

How is it that knowing more about a subject helps one remember more new information? For one thing, being familiar with the terms used simplifies and speeds up the encoding of the new information. For another, more knowledge allows a person to draw more inferences and to integrate incoming information into a more complete network of facts. Remembering is not just a parroting back of random facts; it is a process of constructing, of adding within an existing framework, of checking for plausibility against existing knowledge, and of making inferences about new information (Siegler, 1983). Clearly, knowledge is an essential part of the complex process of remembering.

Basic processes

Finally, some evidence shows that the capacity of **short-term memory**—what children remember for a few seconds to a minute—increases with age. Up to a point, older children can hold more chunks of information in short-term memory than younger children. In one study, for example, 3-year-olds could recall only three things, 5-year-olds four, and 7-year-olds five (Morrison, Holmes, and Haith, 1974). Not all researchers agree that children's short-term memory capacity increases with age, however (Siegler, 1986). Some psychologists believe that more important in explaining why older children remember more is the fact that children's ability to encode information and get it into short-term memory improves with age. In one recent study that supports this idea (Howard and Polich, 1985), researchers found that 5- to 14-year-olds who could remember more items on a test of memory also encoded new information (tones) faster, as indicated by their brain waves. As processing grows speedier and more efficient, then, operating space in memory, once needed for encoding, is freed for remembering (Case, Kurland, and Goldberg, 1982).

In sum, then, there are many reasons that children in middle childhood remember more and better than younger children, and that memory continues to improve with age. Older children know numerous and more sophisticated memorization and retrieval strategies, use them in more diverse situations, and are more flexible in tailoring strategies to situations. They use memory strategies that are more appropriate to the task, pay more attention to detail, and care more about remembering. They know more about the subject matter and more about how to remember. They have a larger and more efficient memory storage system.

Individual differences in cognition

Intelligence

Not all children develop cognitive skills at the same age or rate or reach the same ultimate levels of ability. Some children are clearly more advanced at any given age than others. They talk and read earlier and better, understand mathematical and logical operations more quickly, and remember more things more clearly. This advanced rate of cognitive development in childhood and the ultimate differences in cognitive levels in adulthood are important ingredients in what we call intelligence.

Bobby and Gary are brothers. Bobby is bright, but Gary is super bright. When he was 28 months old, his grandmother asked him to count out loud for her. When he was still going strong at 129, she decided to have his IQ tested.

Sarah is one of several gifted children in her class at school. She talked early; at 16 months her mother remembers her saying, ''Daddy go bye-bye car.'' She had taught herself to read by the time she was 3, and her vocabulary was larger than most adults' by the time she was 8.

Testing IQ

In 1905, Alfred Binet and Théophile Simon published the first psychometric test of children's intelligence, a test of ''good sense'' and knowledge. (Recall the discussion in Chapter 1 of the significance of this event in the history of developmental psychology.) By testing many children, Binet and Simon found the normative age at which most could answer each question correctly. Children's mental ages (MA) were determined through comparisons of their answers to these norms. In 1912, William Stern hit upon the concept of the **intelligence quotient (IQ)**. To find someone's IQ, he divided their mental age by their chronological age and multiplied by 100. ''Normal'' children thus had an IQ of 100. Today, IQs are computed so that the mean IQ score for children at each age is adjusted to 100 with a **standard deviation** of 15 points. Thus two-thirds of all people come within the ''normal'' IQ range between 85 and 115 points, and 95 percent fall within the IQ range of 70 and 130 (Figure 12.2).

In 1916, more items were added to the Binet-Simon test to reflect children's ability to think abstractly and to use verbal symbols. Today, the Stanford-Binet scale can be used to test the intelligence of children as young as 2. It extends through middle childhood and taps children's abilities to define words, recognize verbal absurdities, identify similarities and differences between objects, propose solutions to everyday problems, and name the days of the week. Since 1938, another intelligence test has been in use, devised by David Wechsler. Wechsler considered intelligence ''the overall capacity of an individual to cope with the world'' (1974, p. 5). His test, the Wechsler Intel-

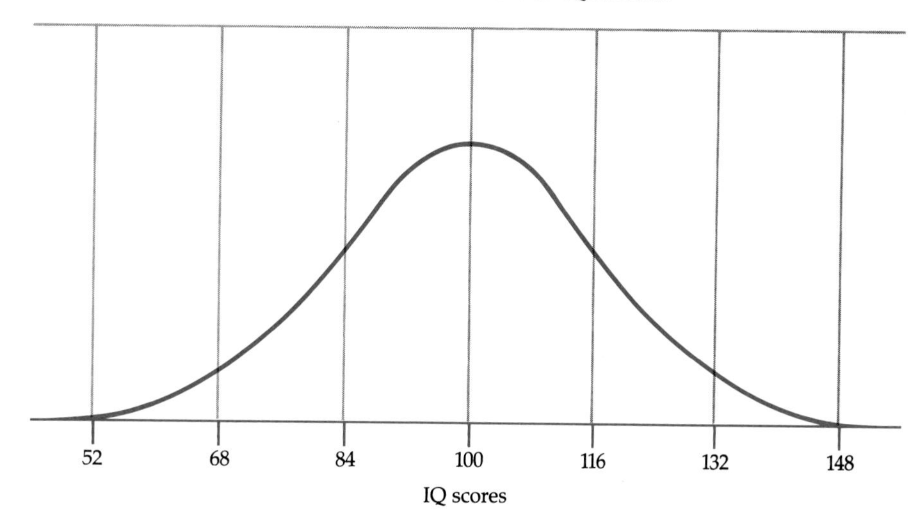

Figure 12.2 ***The distribution of IQ scores of the general population follows the normal curve.***

ligence Scale for Children (WISC), is divided into *verbal* and *performance* subtests. The former taps the child's knowledge and information, vocabulary, comprehension of everyday skills, mathematical ability, recall, and interpretation. The latter taps intelligence through a child's abilities to code numbers into symbols, copy designs made with blocks, put pictures into logical order, complete a picture, and assemble a cut-up picture of an object (see Figure 12.3). The tester begins with problems the child can answer easily and ends with those the child consistently fails.

Both the Wechsler and the Stanford-Binet tests are administered by an experienced tester to an individual child. The tester has to be trained to administer, score, and interpret the results of the test and to issue instructions neutrally and in a form that does not vary from one child to the next. A tester also must be able to establish rapport with the child and to detect when the child is confused, sick, or overly nervous.

For children in middle childhood, intelligence testing may be conducted individually, with the Stanford-Binet or WISC. More often, however, children's intelligence in the school years is assessed by tests administered to a whole class. These group tests do not allow examiners to identify children who are ill or nervous, nor to account for any child's especially good or poor performance. They make no allowances for mood or temperament. The tests also generally penalize children's creativity and even their perception of two right answers to a single question. But these tests require little special training of the tester, can be given economically to many children at once, produce objective scores, and have been standardized on large samples.

IQ tests: how fair? how useful?

Intelligence tests are relative. They give one child's score in relation to the scores of others who have taken the test. The norms to which a child's scores are compared are based on his or her age. But age does not guarantee that children have had the same experiences. Many intelligence tests are biased towards knowledge that reflects white, middle-class culture. In one extensive sampling of the ways that children speak naturally at home, middle-class children were found to use the language contained in intelligence tests significantly more than children from working-class families (Hall, Nagy, and

Linn, 1984). IQ tests also contain questions about such esoteric things as xylophones, tubas, and marimbas, which penalize children who have never seen a band.

STANDARD IQ TEST ITEMS

Verbal Subtests

1. *Information and Knowledge*
   How many wings does a bird have?
   Who wrote *Tom Sawyer*?
   What is steam made of?
2. *Comprehension*
   What should you do if you see someone forget his book when he leaves a restaurant?
   What is the advantage of keeping money in a bank?
3. *Arithmetic and Numerical Reasoning*
   Sam had three pieces of candy and Joe gave him four more. How many pieces of candy did Sam have altogether?
4. *Verbal Similarities*
   In what way are a lion and a tiger alike?
   In what way are an hour and a week alike?
5. *Digit Span and Memory*
   Repeat these numbers: 9 3 4 8 7 1
6. *Vocabulary*
   What does __________ mean or what is a __________?
   (The words given cover a wide range of familiarity and difficulty.)

Performance Subtests

1. *Digit Symbol*

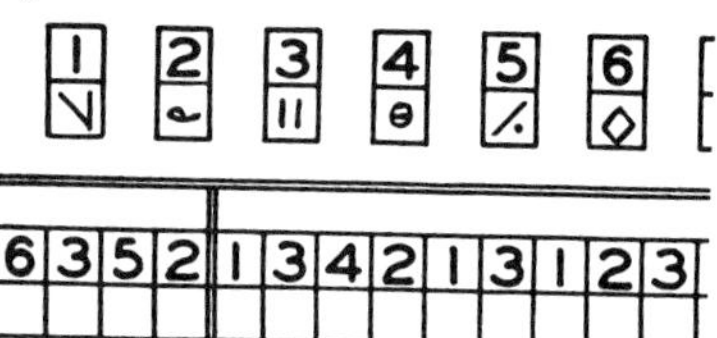

2. *Block Design*

3. *Picture Arrangement*

4. *Picture Completion*

Figure 12.3 ***These items are like those on the WISC.***

5. *Object Assembly*

To avoid the bias against children from lower-class families, some researchers have designed intelligence tests around spatial perception and reasoning. One such test is the Raven Progressive Matrices Test. The Raven test relies on 60 different designs, each with a missing piece (see Figure 12.4). The child tries to supply the missing piece from several possible choices. Other "culture-fair" tests have been designed to tap only a carefully limited vocabulary and set of ideas. Unfortunately, even these tests usually reveal differences in IQ scores for children from different cultural and economic backgrounds. What is more, they are less predictive than traditional IQ tests of how well children will do in school. For better or worse, traditional IQ scores are highly predictive of school grades (the correlation is approximately .70 in most studies [McClelland, 1973]). They are also predictive of economic success in adulthood (at about the same level [Jencks et al., 1972]). Although intelligence tests may be culturally biased, they do reflect certain realities of life today in the United States. In our society, success on intelligence tests, in schools, and, to some extent, in life is related to middle-class knowledge, values, and norms.

Middle-class norms apply to the way that IQ tests are presented as well as to their content. Intelligence tests, it should be remembered, are not perfect measures of children's basic *competence* to think, to understand, and to draw logical conclusions. They are measures of children's *performance* in one particular situation—the IQ test. They are influenced by children's skills in test taking and working fast under pressure and in their motivation and desire to cooperate and apply themselves to the test. When special efforts are made to

CULTURE-FAIR IQ TEST ITEMS

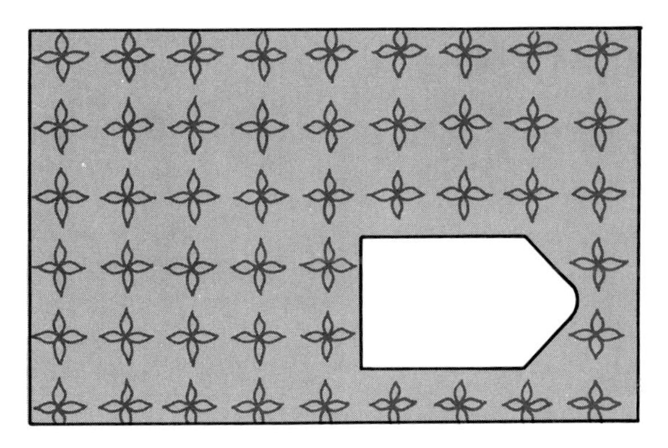

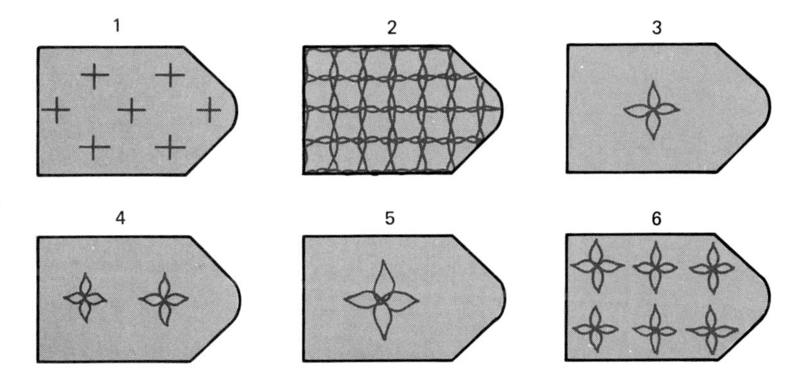

Figure 12.4 Sample item from the Raven Progressive Matrices Test. The child must pick out the piece below that is missing from the matrix above.

motivate children, their scores go up. In one experiment, for example, inner-city children were given tokens for each right answer on an intelligence test. The tokens could be exchanged later for toys. Children who had gotten tokens scored 13 points higher than children in a control group who had gotten no tokens (Bradley-Johnson et al., 1984).

Dissatisfied with traditional IQ tests, some psychologists have proposed alternative methods of evaluating children's intelligence. One such alternative they have suggested is to use individual differences in information-processing speed, choice of memory strategies, and in content of long-term memory as measures of intelligence (Keating and Bobbitt, 1978). This approach to measuring intelligence will no doubt continue to be explored and refined in the future. Another approach that will be explored is a search for different *kinds* of intelligence. Howard Gardner (1983) has proposed that in their richly various thoughts and deeds, people display many kinds of intelligence—linguistic, musical, logical-mathematical, spatial, bodily, personal.

One characteristic not tapped by standard intelligence tests is creativity. Creative children make up new theories, use new ways of seeing, and produce new creations. Their thoughts lead them in fresh directions and into uncharted territory.

> When he was 6, Sam put pieces of shrimp shells over his thumbs and called them his Band-aids. When he was 9, he wrote a story that began, "Once upon a time, there was a boy who lived inside a light bulb."

IQ tests measure what children know of the usual and expected right answers—so-called **convergent thinking**. Creative thinking is based on the unusual and unexpected—so-called **divergent thinking**. It is better measured when the creativity of children's drawing and writing and the diversity of their answers to questions are taken into account (see Figure 12.5). A question like "Tell me all the things that you can do with a brick," might elicit from creative children answers as diverse as "you can build with it, throw it, use it for an anchor, grind it for paint, use it as a step for young children by a water fountain." A theory of intelligence that incorporates creativity as well as more standard aspects of intelligence has recently been proposed by Robert Sternberg (1985). In his book *Beyond IQ*, Sternberg suggests that intelligence includes not only the abilities usually measured on IQ tests but, in addition, the abilities to allocate time, to plan, monitor, and evaluate one's activities, to have insights through selective encoding, combining, and comparing, and to adapt to the real world.

The idea of a single IQ score that sums up a child's intelligence may eventually be a thing of the past. Meanwhile, even with its problems, the IQ score has some usefulness for telling how adequately a child is developing and for predicting how well he or she is likely to do in school and later in life. Scores at the extremes of the IQ continuum are the most useful for predicting success or failure. For scores closer to 100, it is more difficult to predict future learning or performance.

Stability and change in IQ scores

When we look at what happens to a child's knowledge and skills, as measured by intelligence tests, over time, we see a general onward and upward trend. When the *raw scores* on intelligence tests of subjects in the longitudinal Berkeley Growth Study were plotted, they were found to increase rapidly until early adolescence, increase more slowly until midadolescence, and then taper off by late adolescence (Bayley, 1966, 1968). But do individual children retain the same *IQ scores* throughout these years? Not often. IQ scores, which reflect the child's position relative to other children of the same age, vary considerably over time. Only one-fifth of the children from the Berkeley Growth Study retained the same IQ scores over their first nine years. For some children, IQ scores shifted enormously. One girl's went from 133 to 77; one boy's went from the middle 90s to 160, then to 135. Over time, thirteen children varied by 30 or more points, one-third of the group varied by 20 points, and nearly two-thirds varied by 15 or more points. Some of these changes might have resulted from temporary fluctuations in children's

CREATIVE DRAWINGS

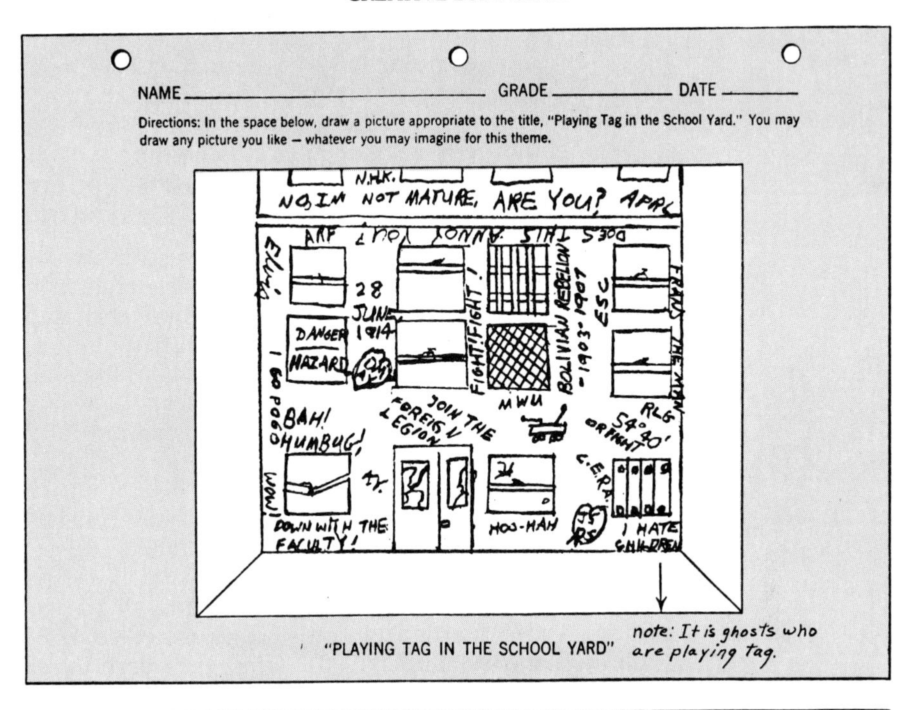

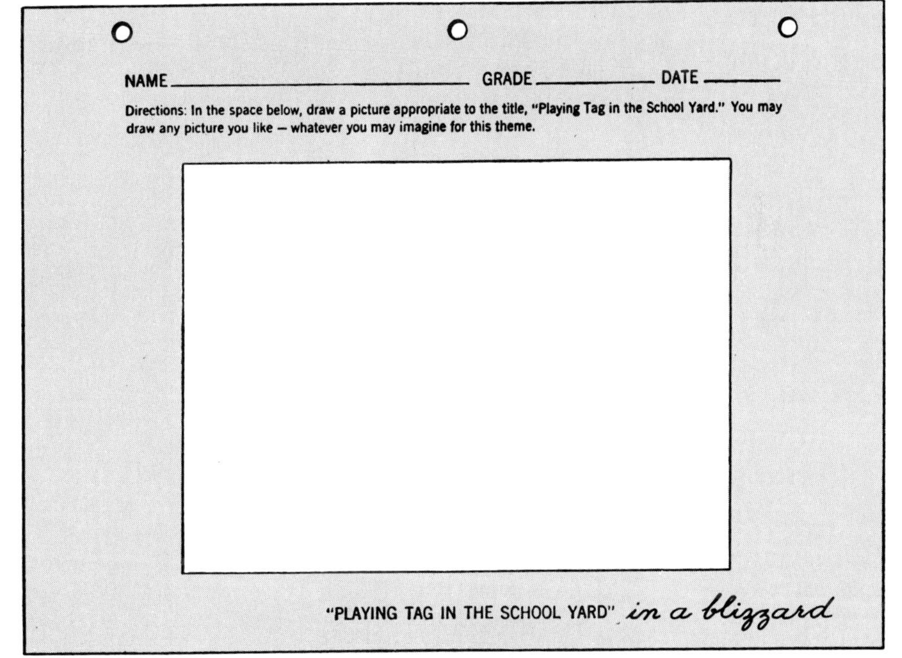

Figure 12.5 ***Creative children, when asked to draw a picture of an incident, are not literal about it. They may mock convention, depict violence, fashion new creatures, and often use humor (Getzels and Jackson, 1962).***

moods, health, feelings of stress, and so forth. Some might have resulted from changes in the content of the tests at different ages. Yet others are likely to have reflected changing circumstances in the children's families and schools. In less stable families than the white, middle-class ones in the Berkeley study, the shifts in children's IQ scores could be even greater. IQ scores tend to be more stable the shorter the intervals between tests and the older the child taking them. By age 12, IQ test scores are stable and useful for predicting current and future success.

Sex differences in cognition

> Erica is beginning to clutch in math class. She's not really sure of herself when it comes to things like working with fractions, even though she's doing great in English. I don't really understand what makes girls fall behind boys in math. If it's something biological, then why are there women who make it as stockbrokers, engineers, and physicists, and why are there men who make it as historians, actors, and writers?

In earlier chapters, we have discussed differences between boys and girls in physical size, strength, and vulnerability, in play styles and preferences, in nurturance and aggression. There also may be differences in cognitive abilities, although these are more difficult to pin down. Eleanor Maccoby and Carol Jacklin (1974) combed through 1400 studies to try to find these differences. Unfortunately, many of these studies contained subjective self-reports and potentially biased reports by parents and teachers. Some had not controlled for the sex of the researchers or testers. Some did not take into account children's ages.

In all this research, however, girls quite consistently were found to have greater verbal ability than boys. In infancy and early childhood, girls, on average, are more responsive to tones, patterns of speech, and subtle changes of voice; they usually begin to speak and sing in tune earlier; and they are more readily startled by loud noises (see also Friedman and Jacobs, 1981; Gunnar and Donahue, 1980). Later, girls learn to read sooner. By the end of middle childhood and thereafter, girls understand and use language more fluently than boys.

In contrast, boys quite consistently were shown in the research to have greater spatial ability than girls. Boy babies, on average, are more likely to ignore their mother and pay attention to a blinking light or a geometric form

In general, boys excel in visual and spatial skills; girls excel in verbal skills.

than girl babies are. They like to manipulate objects with their hands, pushing, pulling, taking apart, and putting them back together. Preschool-age boys show their greater spatial ability in the way they fold paper and twist and turn things. By school age, boys can make these spatial twists and turns in their heads (see Figure 12.6), and they do well on tests of finding embedded geometric figures. They can remember three-dimensional objects and imagine rotating them, abilities useful in mathematics. By age 12, boys do better than girls on mathematical problems, although their advantage is not so great as in spatial reasoning. When problems can be solved equally well by verbal or spatial reasoning, boys and girls do equally well. But when problems require verbal solutions, boys are at a disadvantage; when problems require spatial solutions, girls are at a disadvantage.

These differences in boys' and girls' verbal and visual skills (also documented in more recent research reviewed by Linn and Peterson, 1985), though consistent, should not be overestimated. Such differences are statistically significant, but they apply on the average, not to *every* individual, and they may be small in real terms. Among spatial abilities, for example, mental rotation is most strongly linked to sex; for other spatial abilities, sex accounts for not more than 5 percent of the variation in individuals. In addition, boys and girls do equally well when it comes to learning by rote, learning to discriminate between stimuli, shifting to a new solution to a problem, determining probabilities, and analyzing the elements they will need to perform a task.

Figure 12.6 ***Boys do better than girls on tests of spatial perception (top) and mental rotation (bottom).***

VISUAL-SPATIAL TESTS

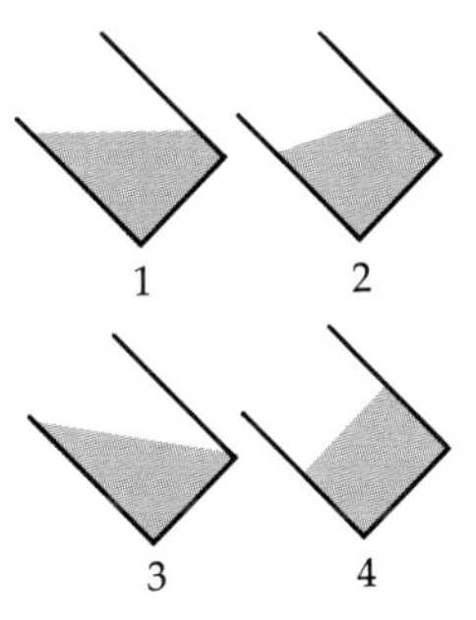

Water level: In which glass is the water horizontal?

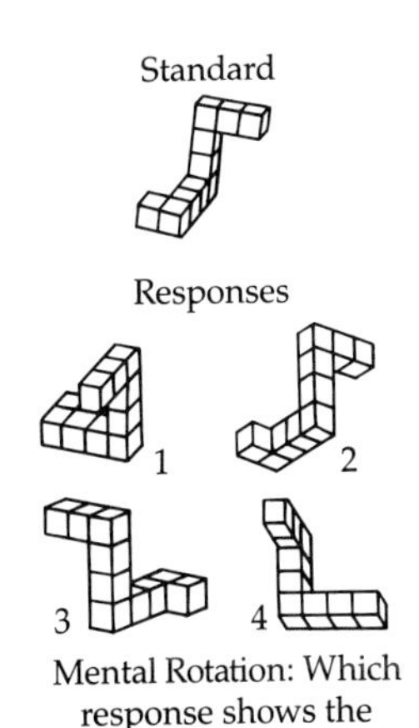

Mental Rotation: Which response shows the standard form in a different orientation?

The origins of sex differences

What are the origins of sex differences in verbal and visual abilities? Scientists have argued that question back and forth for some time. Some have suggested that the differences originate in the X chromosome. They have speculated that spatial ability, like baldness and hemophilia, is sex-linked, or X-linked. Boys, you remember, inherit only one X chromosome along with their one Y chromosome; girls inherit two X chromosomes. If visual–spatial ability came from recessive genes on the X chromosome, it would show up in boys more often than girls because girls would need to inherit the ability from both parents, boys from only one. But the results of studies of verbal and spatial abilities in girls with chromosome abnormalities (such as having only one X chromosome or having three X chromosomes) have not been consistent with the sex-linked hypothesis (Polani, Lessof, and Bishop, 1956; Rovet and Netley, 1983; Vandenberg and Kuse, 1979).

Other scientists have suggested that sex differences in verbal and visual abilities are related to brain structure. As we discussed in Chapter 8, the two hemispheres of the brain are not completely alike in their functions. The left hemisphere apparently controls verbal skills. The right hemisphere processes global configurations. The superiority of girls in verbal tasks and boys in visual and spatial tasks has led some researchers to investigate possible sex differences in brain lateralization. They have found that males generally have a sharper division of function in brain hemispheres. By age 6, a boy's right hemisphere dominates in processing visual and spatial information (Witelson, 1976). When boys work on a spatial task, such as trying to imagine which of three folded shapes can be made from a flat, irregularly shaped piece of paper, their right hemispheres are active. When girls work on a spatial task, both of their hemispheres are active (Restak, 1979). Processing with both hemispheres, as girls do, may be less efficient for solving spatial problems (Levy, 1969), but it also may be one reason that girls suffer less than boys from language disorders.

But the findings on brain lateralization are puzzling. It has been suggested that the timing of maturation, and not sex, determines brain lateralization. Because girls mature earlier than boys, their brains may be less lateralized and their verbal abilities relatively more developed. Early-maturing adolescents of *both* sexes have been found to do better on tests of verbal ability, and late-maturing adolescents of *both* sexes do better on tests of spatial ability (Diamond, Carey, and Back, 1983; Newcombe and Bandura, 1983; Sanders and Soares, 1986; Waber, 1976). But evidence to the contrary also exists. These differences do not show up in all samples or on all measures (Rovet, 1983; Sanders and Soares, 1986; Waber et al., 1985). Maturation, then, is not the whole explanation for sex differences in visual and spatial abilities.

A third possibility is that hormones, those powerful chemicals secreted by the ductless glands of the endocrine system, cause sex differences in visual and spatial abilities. Females with Turner's syndrome (born missing one X chromosome) and males with androgen insensitivity syndrome do not produce normal levels of sex hormones and are poor at tasks requiring spatial ability (Khan and Cataio, 1984). But there are inconsistencies. In one recent study (Hines and Shipley, 1984), researchers investigated the effects of prenatal exposure to hormones on children's cognitive abilities. They compared girls whose mothers had taken a synthetic estrogen during pregnancy with their sisters who had not been exposed to the estrogen. They found that the girls exposed to estrogen had a more masculine pattern of brain lateralization, but they found no differences between the two groups in verbal or spatial abilities. In another study, researchers investigated the relation between intellectual abilities and body type, which would reflect hormonal influence (Berenbaum and Resnick, 1982). They found no significant relation.

In brief, evidence that sex differences in visual and spatial abilities are biological (based on the X chromosome, brain lateralization, rate of maturation, or hormones) is not especially convincing. Nevertheless, the fact that identical twins are quite similar in verbal IQ and spatial and mathematical abilities suggests that these abilities do have some still unidentified genetic component. There may be a biological foundation for visual and verbal differences, but so far we do not completely understand what it is.

What about the influence of social and psychological factors? If boys are endowed with a slight edge in spatial and mathematical skills, it may be reinforced by the kinds of play they engage in and by the courses they take in school. Boys like to build with blocks, put models together, and practice aiming and throwing things. This kind of play is likely to sharpen boys' spatial and visual perception and coordination. In fact, researchers have found that the more both boys and girls play with "masculine" toys like blocks, tricycles, jungle gyms, Lincoln Logs, Tinkertoys, and ring toss in preschool, the better are their visual and spatial abilities in grade school (Fagot and Littman, 1976; Serbin and Connor, 1979). In an experiment to show the influence of toys on children's spatial abilities (Denier and Serbin, 1978), children were assigned to a group that met with a teacher who showed them masculine toys, suggested interesting ways in which they could play with them, and encouraged and supported their efforts. Before the experiment, the boys had outscored the girls on a test of spatial ability (geometry). After the experiment, both boys and girls who had been trained to use the masculine toys outscored children who had not been trained to use them. Play with "feminine" toys like dolls and doll furniture, crayons and paints, felt boards and kitchen toys does not seem related to children's spatial abilities, but is related to higher verbal abilities (Fagot and Littman, 1976; Serbin and Connor, 1979). Once they are in school, boys take many more courses that rely on spatial skills—math, science, computers—than girls do. Evidence from a number of different sources shows that spatial skills—in both boys and girls—can be sharpened with training in just such subjects (Sprafkin et al., 1983).

The expression of verbal and visual abilities also is likely to depend on parents' and children's expectations and attitudes. Parents usually think that math is more important for sons than for daughters, and this affects children's attitudes more than their own past performance in math class (Parsons, Adler, and Kaczala, 1982). It doesn't seem to make any difference that Mom is a mathematical wizard and Dad a poet. The web of family expectations and attitudes—"He's always been good at math." "Her? She couldn't count her way out of a paper bag"—is what counts in children's approach to math.

At school, too, teachers may reinforce sex differences. Evidence from a number of studies suggests that teachers pay more attention to boys than to girls and respond to them with more ideas about solving problems and with more specific information, which could help them in science and math (Meyer and Thompson, 1956; Sadker and Sadker, 1985; Serbin, O'Leary, Kent, and Tonick, 1973). But girls tend to live up to teachers' expectations that they will be good readers. In one study, first-grade boys whose teachers believed that boys do not learn to read as well as girls, in fact, read less well than first-grade boys whose teachers did not hold those beliefs (Palardy, 1969). When teachers are evenhanded about praising and criticizing boys and girls, both sexes have the same expectations in elementary school about how well they will do (Parson, Kaczala, and Meece, 1982).

Textbooks and tests reinforce teachers' biases and contribute to the differences between boys and girls. Many show boys and men doing math and science as more important doctors, scientists, and professors, whereas girls and women are less important school teachers and mothers (Saario, Jacklin, and Tittle, 1973; Taylor, 1973). But although texts may more often be about men and boys, they do not encourage boys to read because they are not about subjects that boys find more interesting. When boys and girls are asked to read material that *both* consider interesting, boys read as well as girls (Asher, 1980).

The context of cognitive development

Teachers and texts not only magnify or minimize the differences between boys and girls in verbal and spatial abilities, they also affect children's cognitive development more generally. In this section, we discuss some of the important environmental factors that influence children's cognitive development in middle childhood. In the preschool years, as we saw in Chapter 9, the primary context of children's cognitive development is home. Much of children's cognitive development during their school years takes place, not surprisingly, in school, where children are helped or hindered by teachers' expectations, instructional methods, classroom arrangements, courses, and curricula.

Teachers' expectations

Teachers' beliefs and expectations about their students' abilities contribute directly to children's academic self-concepts. These images influence how hard children work and their feelings about school and learning in general.

> When we got the note from Mikey's first-grade teacher that she wanted to talk about the possibility that Mikey might be having trouble with reading, I was really upset. First grade is when kids start forming ideas about how smart they are. They start getting pegged—the fast-reading group and the slow-reading group, the smart ones in arithmetic and the dumb ones, the good spellers and the poor spellers—and if you're pegged as slow in first grade, you may never get over it.

> On the first day of fourth grade, the teacher called the roll. When she got to me, she said, "Oh, I see we have another Dutton with us this year. I hope you do better than your brothers did."

What happens to children tagged by their teachers as "slow learner," "class clown," "a brain," or "another Dutton"?

Robert Rosenthal and Lenore Jacobson (1966, 1968) conducted a landmark study of the effects of teacher expectations on children's achievement. Intelligence tests were administered to students in the first to the sixth grades of a school in California. The names of 20 percent of the students were then randomly chosen, and teachers of these students were told that the tests had identified them as "bloomers" who could be expected to show unusual intellectual gains in the coming year. The IQ test was given again eight months later. The randomly selected "bloomers," particularly those in the first and second grades, had made unusual gains—the so-called Pygmalion effect. The younger children improved more than the older children for several possible reasons. First, having sketchier academic self-images, the younger children may have been more susceptible to differential treatment from the teachers. Second, less background information had accumulated about them, and their teachers may have found their promised blooming more credible. The Rosenthal and Jacobson study subsequently was criticized on several grounds. Nevertheless, after a rigorous and comprehensive reanalysis by other researchers, it was concluded that teachers' expectations had probably affected at least the first- and second-grade "bloomers" (Elashoff and Snow, 1970).

A number of related studies since have indicated the wide-ranging influence of teachers' expectations. Students with little ability who were taught French by teachers with a positive attitude toward them learned more of the language than did poor students whose teachers had a negative attitude (Durstall, 1975). The reading achievement of students whose teachers overestimated their IQ scores was significantly higher than that of students whose teachers underestimated their IQ scores (Doyle, Hancock, and Kifer, 1971). Even students' physical performance in gym class was influenced by their teachers' expectations (Rosenthal, 1984). Teachers' expectations clearly make a difference in how much children learn in school.

When teachers have high expectations of students, the environment they create is warmer and happier. The teachers smile often, act friendly, nod their head, and look the students in the eye. They talk more, teach more, and teach more challenging material. They give more clues, repeat and rephrase questions more, pay closer attention, and wait longer for students to answer. They praise the students more. Thus on several different levels at once—emotional, body language, spoken language, teaching materials, praise and criticism—teachers express their expectations (Rosenthal, Baratz, and Hall, 1974).

Teachers' expectations and their interactions with students are influenced by many factors: the teacher's own attitudes; the children's appearance, race, social class, abilities, and interest; and teachers' and children's personal, academic, and family histories (Brophy and Good, 1984). The degree of control they have over the students also influences teachers' expectations. Students who respond more predictably, answer teachers' questions, hand in their papers, take their tests, and read their assignments make a good impression and raise teachers' expectations. Students who threaten teachers' sense of control are ignored and criticized.

Open and structured classes, open and closed classrooms

> Tim's in a mixed grade class for first, second, and third graders. The room has lots of activity centers where kids can do different kinds of

> things like reading, or science projects, or computers, or writing. The classes also do things in large groups. Recently, they all worked on building a model town from the early 1900s. Tim apparently went off to a table and worked by himself all day. When he came back, he had made telephone poles for the whole town. He's really thriving in the open setting. Willy isn't old enough yet, but he's very distractible, very active, and kind of immature. I think he needs a structured environment. Otherwise, he's going to fool around and be the class troublemaker. We're going to have to find a class with walls and desks, for Willy.

In the last few decades, many school classrooms have become more open in their physical arrangements—open rooms, movable furniture, room dividers, and activity stations instead of traditional rows of desks. Many school classes have become more open in their routines and curricula—students work at their own pace, on their own projects, instead of listening to the teacher lecture.

Do **open classrooms** help or hinder children's learning? In one study, researchers compared third graders in open and self-contained **closed classrooms** on tests of vocabulary, reading comprehension, mathematical concepts, and problem solving (Lukasevich and Gray, 1978). The students had been in either open or closed classrooms for three years. Children in the closed classrooms did better on tests of mathematical concepts. Several investigators have reported problems with open classrooms, including noisiness, overstimulation, and lack of privacy. In some open classrooms, students tend to cluster around the teacher, and this tendency may contribute to students and teachers feeling crowded (Rivlin and Rothenberg, 1976). One experimental study in which children were randomly assigned to crowded or open environments showed that students in crowded spaces felt more tense, annoyed, uncomfortable, and competitive (Aiello, Nicosia, and Thompson, 1979). When classrooms have open spaces around their edges and when ceilings are high, teachers say they feel less crowded (Ahrentzen, 1981). Yet even these classrooms may be so noisy that teachers can be heard no more than 7 feet away. This noisiness is unlikely to benefit children. Children in noisy elementary schools, in the flight paths of the Los Angeles Airport, where planes flew overhead every few minutes, for instance, were found to do more poorly on tests of cognitive ability than children attending quieter schools (Cohen, Evans, Krantz, and Stokols, 1980). Their blood pressure was higher, they were more easily distracted, they felt more helpless, and they were more likely to give up trying to solve a problem before time was up. Living in a quiet home seemed to make no difference in the performance of the children from noisy schools, and even when noise at school diminished, the children did not score

Elementary schools and classrooms may be "open" or "closed." Each arrangement has its pros and cons.

much better than before. Various studies show that the effects of noise on performance are usually cumulative and long lasting (Cohen and Weinstein, 1982).

Because there are few barriers to sight or sound in open classrooms, distraction poses another problem for some students. Teachers must tailor class activities to avoid distracting students who are working on other activities: no movies, slide shows, musical instruments, outside speakers, or cooking demonstrations unless everyone can participate. It takes time for new activities to be set up in open classrooms, interruptions are frequent, and children may spend lots of time looking around doing nothing. Some students, of course, would thrive in any classroom, and some students thrive on the way that open classrooms invite them to respond actively—to move around, to manipulate things, to communicate actively (Gump, 1987). But open classrooms are worse for distractible, immature, and disadvantaged students than traditional classrooms are.

In the **open class**, scheduling is flexible, and a changing buffet of materials is available for children to explore and manipulate, at their own pace. Class visitors are likely to see children, often of different ages, scattered around the room and working either individually or in small groups. Older children may help to teach the younger ones, and the teacher studies with the children, discussing subject matter rather than lecturing. The friendly atmosphere and the wealth of educational materials are meant to make learning exciting. Children are more likely to be judged by what they make and by their individual progress than by group tests. In a **structured class**, children spend most of their time on lessons, usually on the three Rs. The teacher gives facts and keeps order; subject matter and tests are taken from a curriculum drawn up for the class as a whole. Promotion depends on the children's grades.

What difference does it make if children are in open or structured classes? In a comparison of 100 6-year-olds whose first-grade classes differed in the openness of their curricula, no significant differences in the children's year-end achievement were found to be related to the openness of the class (Day and Brice, 1977). In studies of children who have been in structured or open classes for a longer time, however, more direct, structured teaching is associated with greater student learning and achievement in math and reading (Brophy and Evertson, 1974; Lukasevich and Gray, 1978). In a major study of 871 primary schools in England (Bennett, 1976), researchers found that the more structured style was superior for teaching math, reading, and English, especially to anxious children who need to know what is expected of them. Children in open classes accomplished less academic work than children in more structured classes, but their creative activity was more advanced. Even for creative work, though, a happy medium of moderate structure seemed the most beneficial in this and other studies (Soar and Soar, 1976). An upper limit to the amount of freedom allowed children is apparently better for all kinds of achievement and development. Leaving students entirely on their own is likely to hinder their creativity as well as their academic achievement.

Learning from computers

A revolution is sweeping this country. By the end of this century, most children will be educated by computers (Kleiman, 1984). A computer that 30 years ago cost $10 million and filled a room today costs less than $1000 and fits in a briefcase. If the automobile industry had made the same strides in price and efficiency as the computer industry has done, we could all buy a Rolls-Royce for $2.75, and it would get 3 million miles to the gallon (Lepper, 1985). As computers continue to grow more affordable, more versatile, and more powerful, parents and educators will use them more to teach children many of the skills once taught only in books, lectures, and written drills.

Children are drawn to computers. Computers respond instantly, a boon to impatient children. They are also impersonal and objective. Computers do not blame children for making mistakes. Said one 7-year-old, "The computer doesn't yell" (Greenfield, 1984, p. 131). Computers do not play favorites. Children appreciate qualities like these. One of the biggest contributions that computers may make to education is simply keeping students in class and learning. In one Los Angeles school in the middle of the barrio, in which the overall absentee rate is 20 percent, the absentee rate for computer classes is 5 percent (Greenfield, 1984).

Computers have the potential for being powerful teachers—and for creating powerful problems as well. They may deepen children's understanding and powers of reasoning and introduce them to facts and situations as no other learning tools can do—or they may make children passive, reward superficiality and easy answers. Whether computers work for good or for ill rests in how *people* manage them. Computers are touted as being "fun" tools for learning, but how do they affect children's interest in learning? One set of computer games, for instance, gives children the chance to run a bicycle store, manage a lemonade stand, or rule a small kingdom. The objective is to give the children practice in arithmetic and to expose them to simple economic principles such as the law of supply and demand, the value of advertising, profit and loss, and the like. Games like these have advantages, to be sure. They help children to grasp concrete, action-oriented concepts about how economic variables operate. They may lay a foundation which later will help children to understand the concepts at a more abstract level. But the games allow children to earn endless wealth—without thinking or solving problems. Students who figure out just one simple way to earn can mindlessly repeat it as long as their fingers can punch the buttons (Greenfield, 1984). Computers, it seems, are no different from any other learning tool. Designed well, they can inspire children to learn. Designed poorly, they can rob children of the motivation to learn. When the material to be learned relates naturally and not arbitrarily to its context, students are likely to be drawn in and to learn.

Computers are also "interactive" in a way that books and films are not. Just as children like to pat the squirrels and lambs at the zoo more than they like watching caged animals from afar, they like the give and take of computers. Some people maintain therefore that computers offer children the chance

Children like to learn from computers because computers respond instantly, play no favorites, and strengthen powers of reasoning. As one child said, "A computer doesn't yell."

to engage in "discovery-based" learning, which may be more valuable than more traditional, didactic instruction. Children can learn about food chains and ecological niches in the lakes of North America from books, teachers, and class discussion. They can also "discover" the same information by interacting with a computer screen that shows a rainbow trout faced with another fish and a set of decisions about whether to flee, chase, or stop for a meal. When children can experiment and explore, they learn actively, they reason inductively, and they discover principles and facts. Whether such discovery-based learning is more effective in the long run than traditional classroom learning remains an open question.

One area of schoolwork in which computers are definitely helpful is composition. With word-processing programs, children write at the computer keyboard. They type their stories, reports, poems, or dialogues, and the work appears on the screen in front of them. They can change and substitute words, move things around, erase, check their spelling right then and there, with no blots or eraser holes on the paper. When they are satisfied with what they have written, they print a copy. Tomorrow, if they want to make any new changes, all they have to do is call their original up on the screen and edit away. When children are eager to revise, their writing improves. One study of the writing skills of classes of third and fourth graders found that those who had been exposed to computerized word processing used 64 percent more words than they had before, and their adherence to topic and organization improved as well (Levin, Boruta, and Vasconcellos, 1983).

Programming the computer offers children other kinds of learning, too. To program, children must learn to plan, to reason logically, to form models, to break problems into smaller, workable chunks, to use flowcharts, and to "debug" or isolate and correct conceptual errors (Lepper, 1985). They must learn the relation between cause and effect, concentrate, and focus their attention. In a study in which 6-year-old children were randomly assigned to a computer programming class for three months and then tested, the children proved to be more reflective and creative, to know more about thinking, and to give better directions than children in a control group (Clements and Gullo, 1984).

The power of computers for teaching children school skills and logical ways of thinking, for motivating them to learn and achieve, has just begun to be tapped by researchers and educators. More efforts to study and exploit these modern machines will undoubtedly be made in the future, as the computer revolution continues.

Summary

1. During middle childhood, children begin to grasp the logical relations among things, the orderly rules and constant properties that govern how things happen. Their logical reasoning is concrete—because they can reason only about concrete things—and operational because they can operate mentally on the things in an organized, integrated, and systematic way.
2. With these new abilities children come to understand the conservation of matter: that even though objects' outward appearances may change, their length, quantity, mass, area, weight, and volume do not necessarily change.
3. According to Piaget, children between 5 and 10 years old have an external morality, in which rules are seen as inviolable instructions from figures of authority. They do not understand that people may agree to discuss and remake rules. In Kohlberg's stages of moral reasoning, children are first concerned with getting rewards and avoiding punish-

ment. Next they are concerned with making fair deals and trades. At the level of conventional moral reasoning, children think that moral behavior means following social rules and conventions.

4. School-age children remember more than younger children in part because they can use deliberate strategies for remembering, such as imagery, rehearsal, and organizing information into systematic categories, and in part because their broader knowledge of the world gives them organizing frameworks. Short-term memory capacity also increases somewhat with age. School-age children understand their own memory processes better than younger children do, but it is not clear that this understanding helps them to remember.
5. Individual children develop at different rates and to different levels, and this is reflected in their performance on IQ tests. Children's IQ scores are affected by the cultural content of the test itself and the testing situation as well as by their cognitive competence.
6. Girls consistently average higher scores than boys on tests of verbal abilities, and boys, higher scores on tests of spatial abilities. Biological, social, and psychological factors contribute to these observed differences in intellectual abilities.
7. The openness of classes and classrooms and teachers' expectations of students' academic performance all affect how well and how much children learn.
8. Computers have the potential for being powerful teachers and motivators. They are especially valuable in teaching children how to write compositions and how to think logically and systematically as they write computer programs.

Key terms

concrete operational thought
decenter
external morality
internal morality
preconventional moral reasoning
conventional moral reasoning
postconventional moral reasoning
rehearsal
metamemory
short-term memory
intelligence quotient (IQ)
standard deviation
convergent thinking
divergent thinking
open classroom
closed classroom
open class
structured class

Suggested readings

BARTH, ROLAND. *Open Education and the American School.* New York: Agathon Press, 1972. One of many books explaining and criticizing open education. This one was written by an elementary school principal who instituted open instruction in his school.

BRODY, ERNESS BRIGHT, and BRODY, NATHAN (Eds.). *Intelligence: Nature, Determinants, and Consequences.* New York: Academic Press, 1976. Excellent historical background on intelligence testing from the earliest IQ tests to the present, plus a balanced presentation of both sides of the Jensen controversy.

DAMON, WILLIAM. (Ed.). *Moral Development.* San Francisco: Jossey-Bass, 1978. A collection of provocative essays on how the culture transmits its standards and values to the young child.

GARDNER, HOWARD. *Frames of Mind.* New York: Basic Books, 1983. A new view of human intelligence in which IQ, or verbal intelligence, is just one of several "intelligences."

GETZELS, JACOB, and CSIKSZENTMIHALYI, MIHALYI. *The Creative Vision: A Longitudinal Study of Problem Finding in Art.* New York: Wiley, 1976. An examination of the personalities, values, and special aptitudes of a group of young artists. They are followed for several years after graduation to determine how the ability to find problems is related to the creative process.

ROSENTHAL, ROBERT, and JACOBSON, LENORE. *Pygmalion in the Classroom: Teacher Expectation and Pupils' Intellectual Development.* New York: Holt, Rinehart and Winston, 1968. The original research that alerted the educational community to the possible effects of teachers' positive and negative expectations about their students' achievement.

LEMONADE
LARGE 25
MED. 20
SMALL 10
HINCKLEY
HAVEN

chapter thirteen

Social and emotional development

The world of the school-age child is peopled with family members, teachers and classmates, and "the other kids." Although children are still deeply involved in family life in middle childhood, they are independent enough to ride the school bus and play in their neighborhood. After school, they travel on foot or bicycle to local parks, movies, libraries, playing fields, YMCAs and YWCAs, Boys' and Girls' Clubs and Scout troops. Girls may babysit, paint, draw, and do their homework. Boys may make and fix things, skateboard, play team sports, and do their homework. Through all these activities, children's social skills develop. In this chapter, we will discuss changes in children's relationships with parents and peers and improvements in their abilities to cooperate and compete, play and teach, understand rules and other people, as these unfold over the course of middle childhood.

Family relationships

Finding a balance

> Robbie may be independent now and sleep at Sam's house for overnights, ride his bike all over the neighborhood, and be out of the house from 8:00 in the morning till 5:30 in the afternoon, but he still needs to be cuddled and kissed at bedtime. And he still needs a spanking when he's bratty.

When over 2000 school-age children were asked about themselves, their parents, friends, school, and neighbors, 90 percent of them said that they liked their families (Foundation for Child Development, 1976). But 80 percent expressed some worry about their families as well. Many children worried, they said, when they heard their parents argue. Many got angry "when no one pays attention to you at home." Children's feelings and thoughts in middle childhood are still focused largely on their families. Their parents are of central importance, and whether the family is happy or not profoundly influences how children feel about the world and their place within it. True, children of 8 or 10 may be more guarded with their parents than they were as chatty preschoolers who bubbled forth with all the news of the day. Older children may whisper secrets to their friends—"Promise you won't tell anyone *ever*?"—and hide some of their feelings behind an impassive face. But children in middle childhood feel deeply about their families.

> My mamma is the best. I'd like to tell her that, but I never can! All the trying comes to nothing, and then I do something that makes her sad and she knows it. I lie in bed at night and think about it. Suddenly the door opens and my mommy comes in to see if I'm asleep. I quickly close my eyes. She gives me a kiss and then I think that maybe she knows anyway how much I love her (Suzanne, age 6, in Lepman, 1971, p. 6).

Family rules are important throughout the course of children's development. In the survey of 2000 children, nearly all said that their families imposed rules. Only 3 percent were allowed to swear or curse; only 6 percent had no set bedtime; only 25 percent could snack when and on whatever they wished. But parents adapt rules to their children's changing behavior and needs. The simple, absolute "no crossing the street alone" kind of rule for young children does not work with older children, who have become more independent and who have entered into social relationships with classmates, teammates, teachers, friends, and neighbors. With preschoolers, the basic issues for parents were establishing routines, teaching children to care for themselves, and controlling temper tantrums and sibling fights. But in middle childhood, new issues arise: Which chores should children do? How should parents supervise children's social lives? Should parents help with homework? What should parents do if children have trouble in school (Newson and Newson, 1976)? Ideally, family rules in middle childhood, as earlier, take a middle course between permissiveness and restrictiveness.

School-age children need less monitoring than younger children do, of course, and involved parents have to be sensitive to the changing needs of their growing children. When parents and children trust one another, they constantly telegraph their needs, and the more controlling relationship between involved parents and immature children is gradually reshaped during middle childhood into the less controlling relationship between involved parents and their trustworthy, trusting, mature children. Parents turn over to their children many daily decisions. There is a shift from parents' regulation of the child's activities to *coregulation* by both parent and child. But some rules are still necessary—and children know this. Even when they do not like particular rules and restrictions, children know that some limits are necessary. They think that parents should step in when their children misbehave—for example, when a child hurts a puppy, refuses to return a toy, or acts rude to the grandmother (Appel, 1977; Siegal and Rablin, 1982). When family rules are too restrictive, though, children may have a hard time achieving independence (Trautner, 1972).

Rules are just one path by which parents influence their children. The emotional quality of parents' interaction with their children also continues to be related to children's behavior—with the parents and with other people. When parents are warm and loving, their children tend to be friendly and sociable, too; when parents are cold or hostile, their children are likely to act sulky, withdrawn, or hostile (Armentrout, 1972; Rohner, 1975; Schaefer and Bayley, 1963). Parents who are loving but set no limits, however, may well turn out children who lack a moral conscience and who are dependent and selfish. As with younger children, parents need to balance love and limits. Research on school-age children (as on preschoolers) supports the value of an authoritative disciplinary style rather than an extremely permissive or excessively authoritarian style (Baumrind, 1985). Authoritative parents accept and support their children's expressions of feeling and opinion but also encourage their autonomy. They care deeply about their children, as their children do about them, and they continue to be involved in their children's lives, as they were in early childhood.

A society of children

As children move through middle childhood, they come increasingly to inhabit two different worlds, the inside world of family and home and the outside world of friends and school. Researchers have studied how children make the transition to the new world of school and peers.

Peer groups

> When Ted went to first grade, he didn't have any problems, but *I* did. All of a sudden his friends were much more important to him. He started making his own plans, deciding which kids he wanted to see, hanging around with Robbie and Brandon. They were inseparable.

During middle childhood, groups, clans, and cliques of pals begin to take shape. By the end of middle childhood, children hang out in these groups to go to the movies, play games, go to the beach, or just talk. How do these groups form, and what keeps them together? A series of experiments with boys at a summer camp provides some answers (Sherif and Sherif, 1964; Sherif, Harvey, White, Hood, and Sherif, 1961). Two groups of twelve 11- and 12-year-old boys arrived at camp on separate buses and were kept apart by the camp leaders. The boys within each group were strangers to each other and spent the first few days getting to know each other. Together, group members camped and cooked, took boat trips, and worked on making a swimming hole and a playing field. In just a week, norms, leaders, and followers had emerged in each of the two groups. One group called themselves the Rattlers; the other, the Eagles. When the camp director then set up the two groups to compete against each other in contests and sports, the group members stuck together, competing vigorously against the other group. The rivalry between the groups intensified into scuffles and name calling. Finally, the groups would have nothing to do with each other. Going to movies together and eating together in the same dining hall only made matters worse; the boys had more chances to fight. Suddenly (as part of the experimenters' plan) problems at camp developed that threatened everyone's supplies of food and water. To solve the problems, members of the two groups had to cooperate. Gradually, hostility between the two groups diminished. Friendships began to form across group boundaries. One thing that makes children form cohesive groups thus is their desire to achieve a shared goal, whether that goal is beating the other team or beating a problem.

Children also form groups on the basis of sex. The sex segregation that started in the preschool years (see Chapter 10) becomes more marked during the school years (Figure 13.1). The character of girls' groups and boys' groups also differs (Hartup, 1983; Pitcher and Schultz, 1984; Waldrop and Halverson, 1975). Girls form smaller groups and play in smaller spaces than boys. In these groups, girls play at refining social rules and roles. They learn about personal relationships, subtle social cues, and unspoken rules of social con-

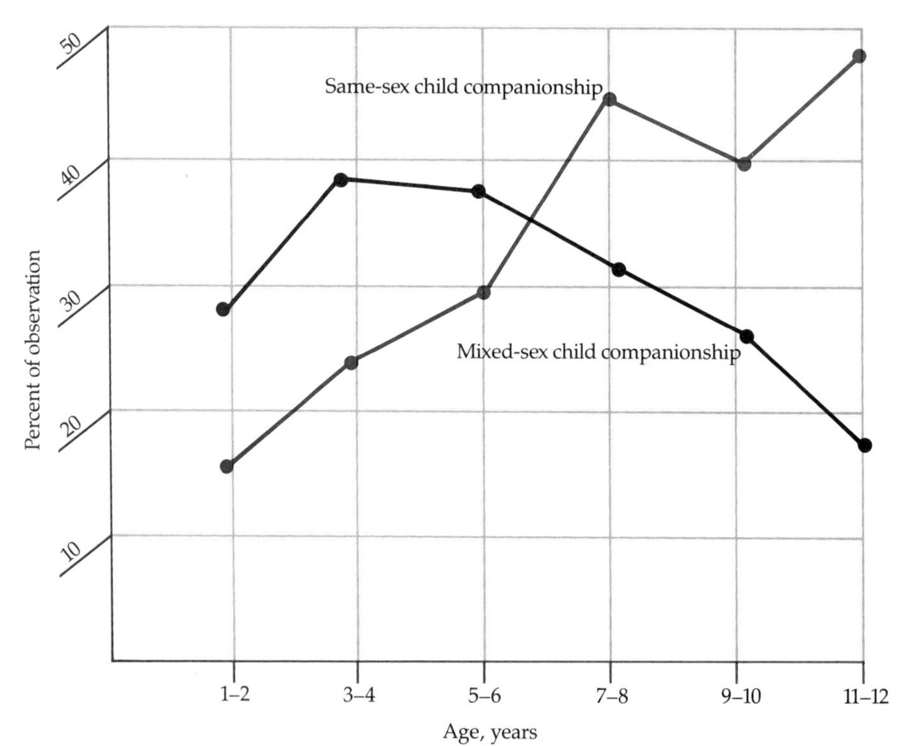

Figure 13.1 Children's segregation by sex increases during middle childhood. Data are from observations of over 400 children made during the summer vacation in a middle-income neighborhood in Salt Lake City (Ellis, Rogoff, and Cromer, 1981).

tact. They practice noticing and responding to others' needs. Their interactions are intimate and **intensive** as they play with only one or two other girls, express intense feelings about them, and share experiences and fantasies. Boys form larger groups and play more physical games in larger spaces. Their games tend to be competitive and to contain explicit rules for achieving a stated goal. In these activities, boys learn about long-range goals, demand attention and give orders, engage in conflict, rowdiness, and rough play, and experience a feeling of group solidarity. Their peer relations, as they share sports and games with three or more other boys, are **extensive**.

During the elementary school years, boys' and girls' groups generally have nothing to do with one another. Boys' groups devalue girls, and their rough-and-tumble enactment of power relationships excludes girls from the central roles. If the girls find the boys' play exciting and attractive, the boys are likely to insult, humiliate, and reject them. "Girls have cooties!" Girls may try to enter boys' groups, but they are usually unsuccessful (Pitcher and Schultz, 1984). During the early school years, the gender-stereotyped roles picked up in the preschool years harden into habits.

Not all children's groups form on the basis of sex, however. The strong segregation of boys' and girls' groups that we see in the United States seems to arise to some extent from cultural demands. The most important factors in finding friends are likely to be compatibility, similarity, and shared goals. In our culture, with its age-segregated elementary school classes, sex is the most likely basis of similarity between children. But in one study, sex was superseded by age as a basis for segregation when children had the chance to form groups from a wider age range (Roopnarine and Johnson, 1984). "I'm not gonna play with *her*. She's only in second grade." More research on children's groups—under naturalistic and under constrained conditions—would show how strong the same-sex ties are. In one study, after a year-long program of mixed-sex work groups in school, children had increased the frequency of cross-sex interactions outside the work groups, but they still *preferred* work and play partners of their own sex (Lockheed, in Maccoby, 1986).

Popularity

Mikey is very popular with the other kids his age. He's always getting calls from kids who want him to come over and play and invitations to their birthday parties. When he was in first grade, the teachers said that he was so popular that the kids would try to pull him in several different directions at once, and sometimes he ended up doing things he didn't like very much.

Nancy has one friend in her class who comes to our house to play and sometimes invites Nancy over there. But she really seems to prefer playing by herself, or reading, or just hanging around the house. Sometimes I worry that she isn't more popular.

What makes some children more popular with "the other kids" than others? For one thing, good-looking children are likely to be popular (Kleck, Richardson, and Ronald, 1974). Perhaps this is because children think that good looks go along with positive qualities. In one study, for example, when children were shown pictures of attractive and unattractive children, they all said that the attractive children were smarter, friendlier, more generous, more likable, and less mean than the unattractive children (Langlois and Stephan, 1977). In fact, physically attractive children do tend to be more socially competent (Vaughn and Langlois, 1983). Power also makes a difference in children's popularity. Children who are leaders in their own groups are more popular with other children in the class, and so are children who do well at their schoolwork. But being a top student, like being a near flunk-out, wins no popularity contests (Hallinan, 1981; Hartup, 1970).

Even more important in predicting popularity than these qualities are the skills that children bring to social interactions with their peers. When they are with others their age, some children withdraw, mumble, and act shy. Others are disruptive and aggressive. Others are confident and outgoing, active and friendly. They get along with anyone, always know just what to say and when to say it. These are the popular children.

> Sam cannot let pass a chance to greet and chat with a friend or neighbor. His face lights up when he sees someone he knows, and he instantly gives a big cheery "Hi!" Our next door neighbor's daughter, who is in Sam's grade at school, is just the opposite. She walks around glumly, her eyes on the ground, and she never calls out a hello. She doesn't seem to have many friends.

Many studies have been done in the last few years to find out just what sets these children apart (Asher and Renshaw, 1981; Berndt, 1983; Coie and Kupersmidt, 1983; Dodge, 1983; Gottman, Gonso, and Rasmussen, 1975; Putallaz, 1983; Rubin and Daniels-Beirness, 1983; Sroufe, Schork, Motti, Lawroski, and LaFreniere, 1984). Popular children, the researchers have found, know how to act friendly to a newcomer—asking questions, giving information, and inviting the newcomer to visit. With their classmates, popular children are rarely aggressive. They make good suggestions, give praise, and play constructively and cooperatively. They understand and adapt to group norms, use prosocial strategies like taking turns and compromising rather than hitting or yelling for solving problems, act positively, and know how to have fun.

Rejected children, these studies show, are likely to be socially inept. They may act inappropriately (like standing on the table in the lunchroom), or they may be distractible, hyperactive, or aggressive. They may have trouble managing conflict. Asked, for example, how they would treat a child who took away their game, they are more likely to say that they would punch or beat him. They may disrupt other children's play, play uncooperatively, and not stick to a task with other children. They may talk a lot but not chat sociably, or they may be withdrawn, working and playing quietly by themselves.

But there is hope even for rejected children. A number of experiments have been designed to change the behavior of unpopular children so that they, too, might win the respect and friendship of their peers. Sherri Oden and Steven Asher (1977) coached socially isolated children in third and fourth grades about how to play with others, pay attention, cooperate, take turns, share, communicate, and give support and encouragement to their peers. Several days after the six coaching sessions had been completed, all third- and fourth-grade children were asked how they enjoyed playing with each of their classmates. The popularity rating for the coached children had increased significantly, and their improvement was still evident one year later. In another successful coaching program (Ladd, 1981), children were taught to use positive and supportive statements, to make useful suggestions to playmates, and also to evaluate their own behavior. Some children do not realize that what they do affects whether classmates like them. Recognizing the effect one has on others is an important step toward peer acceptance, and, possibly, popularity.

Yet another approach has been to pair socially isolated children with a younger playmate for free play (Furman, Rahe, and Hartup, 1979). After ten play sessions, the withdrawn children have been observed to be more sociable when they are back in their classrooms, presumably because they have had the chance to assert themselves and to play dominantly and successfully with the younger children. This is an important area of research and practice, yet it often is overlooked.

Friendship

Even though they win no popularity contests, most children have at least one friend. Friendships are important to children's development for a number of reasons. They offer opportunities to learn social skills like communication, cooperation, and self-control; they offer emotional and cognitive resources; they provide a context for developing intimacy; and they serve as precursors of later relationships. Throughout the elementary school years, friends become ever more important to children. Children's joy in learning and doing is heightened when their friends are present.

> Sarah has been my best friend since first grade. Whenever one of our families takes a day trip on the weekend—to an amusement park or the beach or a museum or somewhere like that—we always try to get the other one invited. My parents like Sarah, and her parents like me, so usually they say okay.—Erica, age 10

Interactions with friends are more emotionally expressive and vigorous in exploring materials than interactions with nonfriends (Newcomb and Brady, 1982). With a close or best friend, schoolchildren share, talk, do things together, and form strong bonds. If a close friend moves away, they may mourn. The one-sided quality of preschoolers' friendships, when a friend was "someone who does what I want when I want it," changes into a more cooperative, sympathetic, mutual exchange in middle childhood. Friendships become two-way relationships (Selman, 1981). Friendships last longer in fourth grade than in first (Berndt and Hoyle, 1985). But although friends at this age can give and take somewhat, their friendships are still likely to crumble under the weight of serious conflict.

As children grow older, their understanding of friendship and their depth of feeling increase. Six- and 7-year-olds are likely to say that friendship consists of sharing objects and toys or having fun together (Furman and Bierman, 1984; Selman, 1981; Selman and Selman, 1979; Youniss and Volpe, 1978). Nine- and 10-year-olds say that friendship consists of sharing thoughts and feelings. They understand that friendship is different from mere proximity and that friends feel affection and respect for each other. Friends, they know, can relieve loneliness and unhappiness. When a friend chooses someone else to go somewhere or to do something with, children of this age feel

Schoolgirls have "intensive" relationships with one or two close friends; boys hang out in "extensive" groups or gangs.

left out and rejected (Selman, 1981; Selman and Selman, 1979). They expect friends to share more and argue less than "kids who simply go to the same school but don't know each other well." But when asked, "How do you know someone is your best friend?" only 2 percent of them mention trust and intimacy as qualities of friendship (Berndt, 1978a).

Eleven- and 12-year-olds think that important qualities in friendship are sharing problems, emotional support, trust, and loyalty (Berndt, 1978a; Selman, 1981). "I tell her everything that happens to me," "I can tell him secrets, and he won't tell anyone else," they say. Friends of this age feel committed to each other; they are loyal to the point of possessiveness. Said one 12-year-old,

> Trust is everything in a friendship. You tell each other things that you don't tell anyone else, how you really feel about personal things. It takes a long time to make a close friend, so you really feel bad if you find out that he is trying to make other close friends, too (Selman and Selman, 1979, p. 74).

Friends at this age have the cognitive capacity to see things from their own *and* a friend's point of view, and they can also appreciate the inherent pleasure and value of friendship. They know their friend's personal characteristics and preferences, what their friend worries about and likes to play (Diaz and Berndt, 1982).

As we have seen already, in middle childhood, boys and girls play in different ways with their friends. They also talk about their friends in different ways. Girls are likely to dwell on how their friends look and on their personalities:

> "I like Hazel because she is a nice girl who always wears a school uniform if possible. She takes great care how she is dressed."
>
> "My best friend is Vera. She always dresses nicely and walks with her shoulders back. Sometimes she wears brown slip-on shoes, a green cardigan, and a dress. She wears white ankle socks" (Opie and Opie, in Pitcher and Schultz, 1984, p. 126).

But boys talk less about their friends' looks. A friend is a boy who likes playing the same games as you do, who always plays with you on the playground, who calls for you to walk with to school in the morning and walks home with you in the afternoon, who belongs to your "gang," and whom you call your "pal," "partner," or "buddy."

> My best friend is John Corbett, and the reason I like him is that he is so nice to me and we both draw space ships, and what's more he plays with me nearly every time in the playground. Another thing about John is that he is sensible and nice. Whenever we are playing rocket ships he never starts laughing when we get to an awkward point (Opie and Opie, in Pitcher and Schultz, 1984, p. 127).

Activities with age-mates

What do children do together with their friends? In school, they work together, collaborate on projects, tutor, help, teach one another, and compete as well. Out of school, they also work together, play games and sports, form clubs, go exploring, sit and talk, and share toys, arguments, and fantasies.

Playing games and sports

The play of children in the early years of elementary school often involves sports and games.

> At recess every day in third grade a group of six girls would play "Doggy and Master," in which one partner was the doggy who pranced around and took orders and the other held a leash made of rope and barked the orders.
>
> When he was 9, Tim played Little League baseball and went to soccer camp for two weeks in the summer. We had taken down the swing set in the backyard, because the kids were too old for it. Tim and his friends liked to play volleyball and softball in our yard or race their bikes around the neighborhood.

With their improved motor skills and their burst of interest in other children, children find games lots of fun. Some games are spontaneous and informal, some elaborate and formal; some require teams, some a partner. Some are ancient, some brand new. Schoolchildren play tag, king of the mountain, Simon says, hopscotch, jump rope, jacks, ring-a-lievo, hide-and-seek, and many other games. Over the last 60 years, children's games have become less formal and, because of increased supervision from adults, less rough (Sutton-Smith and Rosenberg, 1971). More significant, school-age girls now choose to play more active games and sports—more in the style of traditional boys' games and sports—than they did 60 years ago. Girls now swim and play marbles, play on softball and volleyball teams, bounce on trampolines and over sawhorses, and take part in active bouts of fox and hounds. They like to play games that have leaders—statues, Simon says, follow the leader—and still enjoy some traditional girls' games, too, like hopscotch, dress-up, and dolls. Boys meanwhile have taken up even more vigorous games like football, wrestling, soccer, and martial arts, and they enthusiastically play war-like fantasy games. Girls seldom join in these games (Medrich, Roizen, Rubin, and Buckley, 1982).

Games help children make the transition into the larger social world. Games offer practice in following rules, in cooperating and competing. Games offer children challenges to rise to and succeed at. Even kissing games serve a purpose, bridging the gap between boys and girls and setting the stage for the eroticism of adolescence.

Helping and sharing, cooperating and competing

In the elementary school years, children find new possibilities for cooperation and generosity, for helping, sharing, and caring, for benefiting others. Yet some social values conflict with these prosocial actions. In our society the values of competition, individualism, acquiring property, and winning are also stressed, and these values may interfere with positive social actions.

Perhaps as a result, helping and showing concern in some situations decline toward the end of middle childhood (Staub, 1975). In one experiment, for example, sixth graders alone in a room when they heard a bookcase crash and a child cry for help from another room were less likely than younger children to run to the rescue (Staub, 1975). Fourth graders did try to help when they heard these distressing sounds, if they were alone at the time, but not if they were with another child. They may have been inhibited by thinking about how their peers would judge them. ("It's not cool to show concern.") They may have thought that it was not their responsibility to do anything. ("Why doesn't Sam help?") By contrast, children in first and second grade were *more* likely to act helpful when they were with another child. For them, it's still a good thing to be helpful, and having another child there makes it easier.

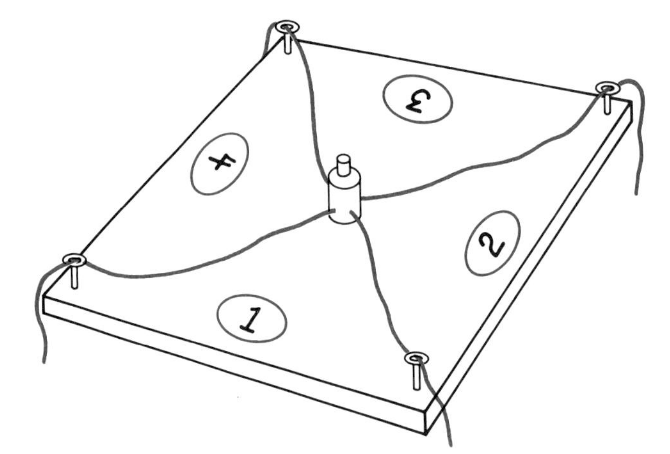

Figure 13.2 ***The object of the cooperation board is for children to draw lines through the target circles. To do so, they must cooperate in pulling their strings.***

At the same time that prosocial actions decline, competitivenes increases. In school and in play, children learn that it's good to be best. Winning is *fun*. Competitive games, from baseball to badminton, give them arenas in which to assert and confirm their developing sense of self, and getting good marks in school gives them status and approval. But competing is not always adaptive. Even under circumstances in which cooperation would benefit children more, researchers have found, elementary school children may compete. This has been demonstrated in studies using specially designed board games. In one such game, four children sit around a square, paper-covered board (see Figure 13.2). In each corner of the board is an eyelet threaded with a string. The strings are attached to a metal cylinder holding a pen in the center of the board. Each child holds a string and pulls it toward himself, trying to draw a line through the circle on the paper in front of him. But the children have to cooperate to draw the lines; any child can prevent the others from drawing lines through their circles by holding the string tight or pulling it toward himself. When the experimenter gives the instructions, "When the pen draws a line across one of the circles, only the child whose name is in the circle gets a prize," the most effective strategy for getting rewards is for children to take turns letting one another win. Preschool children in the United States cooperate in this way—"Let's help each other. C'mon." But 7- to 9-year-old school children typically choose to compete, trying to get the pen through their circle first, even though it means no one gets a prize (Kagan and Madsen, 1971; Madsen, Kagan and Madsen, 1971). School-age children from cultures in which competition is not stressed as much as it is in this one (like Mexico or the kibbutzim in Israel) are not as likely to act competitively.

To some extent, the competitiveness of children playing these board games is a result of the situation itself. When researchers have children do other things that are not so traditionally competitive—building something together, for instance—they find that school-age American children are not more likely to compete than to cooperate (Ayman-Nolley, 1985; Brady, Newcomb, and Hartup, 1983). Even so, cultural pressures clearly influence American children to be competitive in many situations.

A family of Russian emigrants whose son is in Natalie's class have told us that they feel uncomfortable with the stress American schools place on individual excellence at the expense of the group. They ask why it's so important that some kids be in the top reading group, or on the advanced

Competition, individualism, and the importance of winning are stressed in our culture and become increasingly salient over the elementary school years.

> soccer team, or in the accelerated math class. What does this all do to the children in the bottom reading group or the slow math class?

In American society, boys are especially likely to be rewarded for competing and are especially likely to pick up competitive attitudes. Girls are likely to be rewarded for cooperation, and they take a dimmer view of competition (Ahlgren, 1983; Mussen and Eisenberg-Berg, 1977; Stockdale, Galejs, and Wolins, 1983).

The pressure to compete does not, however, seem to destroy either girls' or boys' generosity. Through the school years, both boys and girls increase in generosity and become more likely to share their possessions like pencils, candy, or allowance with other children—at least when an adult demands it (Staub, 1975; Zarbatany, Hartmann, and Gelfand, 1985). More generous children are more self-confident, competent, and self-controlled, more empathic, expressive, active, outgoing, and assertive (Barrett and Radke-Yarrow, 1977; Mussen and Eisenberg-Berg, 1977). The *most* generous children are not always the *best* adjusted psychologically, however. In a study of fourth-grade boys given the opportunity to donate to other children the prizes they had earned, the most generous boys tended to be naive and prone to guilty feelings (Mussen and Eisenberg-Berg, 1977). They were willing to give away everything they had earned. The least generous boys were weak and lacked self-control and perseverence. The best adjusted boys were moderately—perhaps appropriately—generous. They could think of their own interests *and* other children's.

Like competition, though, children's generosity depends not only on these personal qualities but also on the immediate situation—how the child feels at the moment and what he thinks about the other person. Young schoolchildren, in first through fourth grades, researchers have found, are less likely to act generously when they are in bad moods than when they are in neutral or good moods. For example, children who thought that they had performed badly in a game gave less to charity than those who thought they had done well (Isen, Horn, and Rosenhan, 1973). Similarly, children who were asked to think about sad memories gave less to charity than children who were not asked to think sad thoughts (Cialdini and Kendrick, 1976; Moore, Underwood, and Rosenhan, 1973). By the end of middle childhood, however, children are like adults; out of empathy or guilty conscience, they are more inclined to be generous when their mood is negative (Barnett, Howard, Melton, and Dino, 1982; Cialdini and Rosenhan, 1973).

Video games provide a new source of influence on children's generosity. In a recent study (Ascione and Chambers, 1985), researchers had elementary school children play one of two commercial home video games: either an aggressive boxing match or a prosocial adventure in which players try to rescue a fantasy creature from danger. Then the researchers gave each child a dollar in change and told the children they could donate some or all of the money toward a fund for needy local children. Despite the fact that each session lasted only ten minutes, there were clear differences in donations depending on which game the children had played. Those who came to the fantasy character's rescue donated 41 cents, on average; those who punched it donated only an average of 28 cents each. The researchers speculate that the overtly aggressive and competitive nature of the boxing game fostered a self-centered attitude in the children who played it so that they became less inclined to help others when the time came to donate.

Lying and cheating

> Erica had a rabbit's foot, for good luck I suppose, on top of her bureau when she was in first grade. Mikey really wanted that rabbit's foot. He'd stand around and gaze at it or pat it with his fingertips, and Erica would warn him away. I made it clear to him that it was Erica's and that he was not to touch it without her express permission. Well, one day the rabbit's foot was missing. Erica was frantic. Mikey swore up and down that he hadn't seen it. It turned out that he had taken it and then, afraid to put it back, had sold it to a kid at school for a quarter. Not only had Mikey disobeyed us; he stole, lied, and got rid of the incriminating evidence. We were afraid we were harboring a criminal in the making.

On occasion, almost all children cheat or lie. This is a source of concern to their parents—and it has been to psychologists, too. In a major study of children's dishonesty, done in the 1920s, the researchers tempted some 11,000 children between the ages of 8 and 16 to do something for their own gain that they would not want other people to know about—to copy answers on a test, to cheat on a party game or at a sport, or to lie about doing their chores (Burton, 1963; Hartshorne and May, 1928–1930). Without knowing that they were being observed, the children were tempted at home, at school, at church, on playgrounds, and in clubs. Nearly all the children cheated at least some of the time. But when there was risk of discovery or when cheating took some effort, they were more likely to be cautious—and honest. Resisting temptation depended less on conscience than on circumstance. In some school classes, many children cheated, and in others few did, a finding that the researchers chalked up to differing group pressures to conform. Some children were nearly always honest and some the opposite, but everyone's honesty varied as the circumstances changed (Burton, 1963).

Whether children give in to temptation depends to some extent on what they see other people doing. In one study, fourth-grade boys were asked to sit in a chair and perform a boring job while an exciting movie was shown just out of their sight (Stein, 1967). If they stayed in their chair, they were considered to have resisted temptation. Boys in one group saw a man get out of his chair and peek at the movie. Boys in another group saw a man resist temptation. Boys in the control group saw no model. The boys who saw the man peek at the movie gave in to temptation more readily than the boys in the other two groups. But the boys who saw the unyielding adult were no more resistant to temptation than the boys in the control group. The moral of the story? Honest behavior may not be catching, but dishonest behavior is, if children learn from watching someone else that they are not likely to get caught.

How, then, can one get children to be honest and upright? Simply modeling good behavior apparently is not enough, and neither is simply telling children that cheating is bad (Burton, 1976). Using scare tactics and punishment is not always the most effective strategy either.

> Natalie was half an hour late coming home from school one day in second grade. Why? The teacher discovered that a felt animal was missing from the feltboard, and she kept the entire class after school until she found the culprit. Natalie said that the teacher really scared them. First, the teacher made them sit at their desks, hands folded, and demanded to know who did it. When no one confessed, she made them all close their eyes and put their heads down on their desks. The culprit was supposed to return the animal to the feltboard, and none of the other kids would have to know who did it. That didn't work either. Then the teacher brought in the other second-grade teacher. She lectured them about stealing and lying and shamed them with, "I can't believe that any child in this beautiful new school would stoop so low as to take something that didn't belong to him." Eventually, one of the boys in the class pulled the felt duck out of his desk, and the kids got to go home. I just hope the teachers were as effective in scaring that boy as they were in scaring Natalie.

Punishment tends to be most effective when it follows the wrongdoing right away (Parke, 1977). But this condition gets harder to meet as children grow older and spend more and more time on their own. Parents and teachers are unlikely to find out that the child skipped school, or cheated, or lied until well after the fact, and by then punishment may have lost much of its deterrent effect. If adults can recreate the situation and reconnect the wrongdoing with the punishment, they may be more successful. In one study, children saw themselves misbehaving on a videotape and heard their misbehavior described, or they were told to misbehave again and then were punished (Andres and Walters, 1970). Later on, these children were less likely to misbehave than were children who were simply punished long after the misbehavior.

Sometimes attributing moral qualities to children helps them to resist temptation. In one study, a child was left alone to watch some equipment, while a distraction was staged nearby (Dienstbier, Hillman, Lehnhoff, Hillman, and Valkenaar, 1975). When the child looked away, the equipment was made to malfunction. With some children, the experimenter then returned and said, "You feel bad because I caught you." With others, the experimenter said, "You feel bad because you did something you knew was wrong." It was children from the second group who later resisted temptation more resolutely in later tests.

> One night when Sam was about 9 he was pretending to make trick shots with a cake platter that had been in the family for several generations and acted snippy when I asked him to be careful. The platter smashed on the kitchen floor. For a minute, I was just too angry to speak. Then, instead of snarling at him, I had the presence of mind to say, "I know you didn't mean to break that, Sam. You're too careful a person to have intended that." He really was more careful after that, too.

The issue of how to foster morality in children has been important for many psychologists. According to Freud, morality develops early in childhood out of fear and anxiety at the thought of losing a parent's love and from guilt over incestuous fantasies about the parent of the other sex (Freud, 1940). When the child has developed a sense of guilt, moral sanctions emanate from within. In psychoanalytic theory, the child's ability to identify with the parent is necessary for the formation of conscience. It is *the* essential relationship for fostering moral growth, one that cannot be replaced by other identifications with teachers, peers, or siblings. Taking the role of the parent leads the child to internalize parents' standards wholly and unquestioningly.

Empirical research does not support the contention that conscience is formed only through identification with parents, however. In studies comparing the moral development of kibbutz-reared and family-reared Israeli children (reviewed by Kohlberg, 1964), for example, no differences in the guilt felt by the two groups of children and in their morality have been found, even though kibbutz-reared children received their moral socialization primarily from caregivers and teachers in the Children's House where they lived. Nevertheless, identification with some adult authority and acceptance of that authority's standards probably are necessary for children to feel guilt and behave morally.

How likely children are to behave morally, as we suggested in Chapter 10, may be related to their parents' methods of discipline. When parents simply assert their power with physical punishment, by taking away the child's toys or privileges, by applying force, or by threatening any of these, their children are less likely to behave morally (Hoffman, 1970). They are more likely to cheat, less likely to feel guilt or to accept blame after a moral transgression. The most effective disciplinary method for advancing moral development is to create an atmosphere of warm approval and acceptance for the child, to explain why the child should act in particular ways, and to stress the effects of the child's behaviors.

Acting aggressively

Perhaps the worst thing that children do together is to hurt each other. Preschool children may hit or push other children, but their actions usually are for some instrumental purpose like getting back a toy. In the school years, children begin to direct aggression toward others just to hurt them (Hartup, 1974; Lefkowitz, Eron, Walder, and Huesmann, 1977).

Children are likely to respond with aggression to incursions into their space, possessions, or movements. Older children are likelier than younger ones to react aggressively to people who insult, criticize, or tattle on them.

> I'll never forget walking home from school on my eighth birthday, followed by a bunch of girls. As we passed my neighbor on the sidewalk, I taunted her, "We're going to my birthday party, and you can't come." She got even with me later and stole all my Winnie the Pooh books.

> The sixth grader up the street is harassing Sam. Yesterday, when Sam told him he wouldn't carry his books home for him, he ground a baseball bat into Sam's toe with all his might.

This kind of hostile aggression increases steadily during middle childhood. Older children hit, punch, kick, make faces, and hurl insults and snide comments if another child gets in their way, takes their things, insults, criticizes, or tattles on them. They break into fights, disrupt class, and occasionally hurt animals or the teacher. Their aggressiveness tends to be expressed openly now, on the playground, in the classroom, on the street, where it is likely to meet with public attention and censure.

What makes schoolchildren act aggressively? Perhaps the most comprehensive attempt to answer this question is a longitudinal study of aggression carried out in New York State's semirural Columbia County (Eron, Walder, and Lefkowitz, 1971; Huesmann, Eron, Lefkowitz, and Walder, 1984; Lefkowitz, Eron, Walder, and Huesmann, 1977). After testing all the third graders in the county and interviewing many of their parents, the researchers chose a smaller sample of boys and girls for intensive study. They followed these children from then until they were 30 years old. From their observations, they found that certain qualities of children's experiences at home, and particularly their parents' behavior, were associated with aggressiveness in the children. If parents were rejecting, unloving, and in conflict with each other, their children tended to be aggressive. If parents punished their children for aggression but did not reward them for cooperation or sharing, their children also tended to be aggressive. If children identified strongly with their parents, usually they were too afraid of punishment to act aggressively, but children who did not identify with their parents grew more aggressive the more their parents punished them. Even though they preached nonviolence, parents who were themselves aggressive produced aggressive children. Children clearly learned aggression from their parents. They also learned it from their siblings. It has been suggested that severe aggressive behavior is learned most easily in a family that includes both victims—younger siblings—and models—older siblings (Patterson, 1982).

Family members figure in the socialization of aggression, but so do peers. Peers act to reinforce, model, absorb, and provoke hostility. In one experiment (Hall, 1974), for example, pairs of unacquainted 6- and 7-year-old boys played together. One child in each pair had been trained by the researchers to act either aggressively or passively. When that child acted aggressively, the probability of the other child's acting aggressively was .75. Children who were attacked by aggressive peers fought back. In another study (Hicks, 1965), children imitated their peers' aggression—even though it was not directed toward them—when they saw a film of an aggressive child.

Finally, aggression is encouraged by the structure and activities of schoolchildren's play groups. Large groups, playing outdoors unsupervised by adults in rowdy, rough-and-tumble, or competitive games that are concerned with establishing pecking orders, are most likely to bring out aggression.

But children's aggressiveness is not simply provoked by aggressiveness in the environment. Children's perceptions and cognition—perhaps built up over repeated encounters in the family or at school—also contribute. Kenneth Dodge (1980) has suggested that children who respond aggressively may be reading the worst into the other person's intentions. This researcher exposed schoolboys, rated by their teachers and classmates as either aggressive or nonaggressive, to frustration meted out by an unfamiliar peer. The boy heard the peer in the next room breaking the puzzle that the first boy had been working on. The peer's intent was hostile ("I don't want him to win." *Crash*.); benign ("I think I'll help him with the puzzle." *Crash*.); or ambiguous ("It looks like he's got a lot done." *Crash*.). All boys reacted more aggressively to the peer whose intent was hostile than to the peer whose intent was benign. But when the peer's intent was ambiguous, the aggressive boys reacted as though the peer had acted hostilely, and the nonaggressive boys reacted as though he had acted benignly.

One further cause of aggressiveness in school-age children, as with younger ones, is the violence they see on television. In research done in the 1970s (Gerbner, 1970; Gerbner et al., 1977), it was found that nearly one-third of the prime-time and Saturday programs were saturated with violence, and nearly two-thirds featured a good deal of violence, with violence defined as an actual act of physical force or one causing pain. Despite efforts by the PTA and other groups, children's programs in the 1980s continue to portray a great deal of violence. For every hour of "Sesame Street," there are many hours of violent cartoons in which good guys vanquish bad. It is estimated that by graduation from high school the average American child has seen 18,000 televised murders (Brody, 1975).

In studies of the relation between watching violence on television and aggressive behavior in inner-city and suburban Chicago children and in children in Finland, researchers have found that television violence has a particularly strong influence on children's aggressiveness beginning when they are about 8 or 9 years old (Eron, Huesmann, Brice, Fischer, and Mermelstein, 1983; Huesmann, Lagerspetz, and Eron, 1984). Beginning at this age, the effects of television seem to intensify. The more often children watch violent shows and the more violent the shows are, the more aggressive are the children. It is likely that children learn to act aggressively from what they observe on television and that televised violence justifies aggression and makes it more palatable to children, repeatedly showing people who solve their problems violently. Not only did watching violence on television beget aggressive behavior but aggressive behavior also tended to increase the amount of violence children watched on television—a two-way street.

Surprisingly, in the Chicago sample, unlike the New York sample studied earlier (Eron et al., 1971; Huesmann et al., 1984) or the Finnish sample, even though girls acted less aggressively than boys and watched less violent television, their behavior grew increasingly aggressive over time, just as the boys' did. In the New York sample, only the girls who liked masculine games had been affected as much as boys were by televised violence. The shift, the researchers suggest, may derive from social changes in this country and on television over the last two decades. Today, girls are allowed to be more active, assertive, and aggressive than they were 20 years ago, and today there are more aggressive female characters on television.

In brief, it appears that children are most likely to act aggressively if they

- Watch a lot of violent television shows.
- Believe that these shows are real, not fiction.
- Identify strongly with the aggressive characters on the shows.
- Have aggressive fantasies.
- Prefer activities that are physically rough and active.
- Have parents, especially mothers, who act aggressively towards them.
- Observe aggression of their parents, friends, and heroes.
- Are reinforced for their own aggressive behavior.
- Are objects of aggression.

Aggression has many sources, and when these sources come together in one child's life, the result inevitably is antisocial behavior.

How can parents or teachers reduce children's aggressiveness? First, they can reduce their own aggressiveness toward the children. Then they can try to change children's perceptions of those who frustrate them. In one study (Mallick and McCandless, 1966), third graders were prevented from finishing a task and winning a cash prize by a frustrator. If the children were told that the frustrator was sleepy or upset, they attacked less, both verbally and physically. Parents and teachers also can encourage children to excuse their classmates' slights or shoves and promote their children's cooperative ventures. In the boys' camp experiment that we discussed earlier (Sherif and Sherif, 1964; Sherif et al., 1961), shifting to a shared goal changed the way the campers perceived the "others." Instead of frustrators and rivals, they became friends. Much more research needs to be done to find alternatives to aggression and ways to replace it with a healthy assertiveness that helps children to cope in the face of pressure and to persist in the face of frustration.

Social understanding

Understanding other people

As children get older, they become better at understanding other people's actions and more adept at reading cues about their feelings, intentions, and motives. Preschool children, when asked to comment freely about people's behavior, comment on obvious actions and expressions. But over the school years, children learn to infer more about people's *internal* states and thoughts. In one study of children's social understanding (Flapan, 1968), for example, 6-, 9-, and 12-year-olds were shown a film of a young girl who went through various emotional situations: punishment from her father for not sharing her roller skates; a visit to the circus; a gift of a calf from her father; accidentally killing a squirrel. Children of all these ages could tell when the girl or her father was angry, hurt, or happy. But the 6-year-olds often did not understand the father's intentions. The 9- and 12-year-olds did understand the father's intentions and also understood how the girl deliberately shaped her actions to meet certain goals. The 12-year-olds could deduce what the father thought about the girl's feelings and what the girl thought about the father's.

In another study, Robert Selman (1980) used fictional stories to find out how well children understand how other people think and feel. For example,

> Eight-year-old Tom is trying to decide what to buy his friend, Mike, for a birthday party. By chance, he meets Mike on the street and learns that Mike is extremely upset because his dog, Pepper, has been lost for two weeks. In fact, Mike is so upset that he tells Tom, "I miss Pepper so much that I never want to look at another dog again." Tom goes off, only to pass a store with a sale on puppies. Only two are left, and these will soon be gone (Selman, 1980, p. 94).

The interviewer then asked the children whether Tom should buy a puppy for Mike's birthday, and why. Children's responses to these dilemmas, Selman found, form three levels of social awareness. At the first level, preschool children do not distinguish between inner, mental events and outer, physical events and observable actions. They typically replied to the dilemmas as though a person's feelings and statements were the same thing. "If he says that he doesn't want to see a puppy ever again, then he really won't ever want to." At the next level of understanding, at around age 6, children distinguished between what people felt and what they said, but they still thought that the two always were in agreement. By age 8, children understood that inner thoughts and outward expressions may be inconsistent or contradictory. In response to the dilemma, they might say that although Mike says he does not want another puppy, inside he really does want one. With the knowledge that people's facades may differ from their inner experiences, children reach a level of reflective, psychological understanding.

The shift toward greater psychological understanding that occurs during this age period was documented in another study as well (Rholes and Ruble, 1984). Children between 5 and 10 years old were read vignettes about children who had either positive or negative characteristics. For example, some children were portrayed as generously sharing a lunch, some as ungenerously refusing to share lunch with a hungry child. The subjects then rated the children in the vignettes as "nice and kind," "brave," "mean," and so on, and were asked how the children might act in other, comparable circumstances. Although children at all these ages rated the story characters as equally generous, stingy, and so on, the 10-year-olds far oftener than the 5-year-olds predicted that the fictional characters would behave consistently from one situation to another. They believed, as most adults do, that a person's behavior in different situations can be predicted by an abiding personal disposition.

At around the age of 10, children also made a rapid and pronounced shift from describing other people in terms of their behavior to describing them in terms of these inferred, stable, psychological attributes. In one study, children were asked to describe their classmates (Barenboim, 1981). Children younger than 10 would say something like, ''Billy runs a lot faster than Jason'' or ''Penny draws the best in our whole class.'' But beginning at 10, children would describe a classmate with ''He's really conceited; he thinks he's great'' or ''Linda is real sensitive, a lot more than most people.'' The findings from these studies are consistent with the findings from a sample done in England (Livesley and Bromley, 1983). When researchers in that study recorded how children between 7 and 16 years old described their classmates, they found the same progression from physical and behavioral to psychological descriptions. Younger children described their acquaintances in terms of appearance, name, age, sex, routine habits, possessions, and social roles.

> Max sits next to me; his eyes are hazel and he is tall. He hasn't got a very big head; he's got a big pointed nose [age 7 years, 6 months] (Livesley and Bromley, 1973, p. 213).

Older children's descriptions focused on more psychological aspects like personality traits, general habits, motives, values, and attitudes.

> Andy is very modest. He is even shyer than I am when near strangers and yet is very talkative with people he knows and likes. He always seems good tempered, and I have never seen him in a bad temper. He tends to degrade other people's achievements and yet never praises his own. He does not seem to voice his opinions to anyone. He easily gets nervous [age 15 years, 8 months] (Livesley and Bromley, 1973, p. 221).

Learning social rules

Another way that children grow in social understanding over the school years is by learning social rules. Part of the task of growing up and getting along with other people involves learning the many rules that govern everyday behavior. There are trivial rules and important rules, implicit rules and explicit rules, rules for school and rules for home. By school age, children are quite familiar with the culturally defined and arbitrary rules establishing dress codes, forms of address, and table manners—the so-called **social conventions**. Researchers have tried to determine just when children acquire these social rules by interviewing them intensively. In one study the children were told a story about Melissa:

> Melissa is a new girl at school. She comes from a country in the Far East. In her country there are different eating habits, and the people there eat many kinds of food with their fingers. One day a teacher saw her eating some spaghetti with her fingers and got upset (Damon, 1977, p. 243).

Preschool children seemed to think that Melissa's behavior was fine. They did not feel bound by the rules of American table etiquette and said that whatever people wanted to do, they could.

INTERVIEWER: Do you think it is okay for Melissa to eat with her fingers?

KAREN (4 YEARS, 2 MONTHS): Yes.

INTERVIEWER: It is? How come that is okay?

KAREN: Because different people eat different ways (Damon, 1977, p. 250).

Kindergartners no longer confused rules with personal habits, but they thought rules should be respected to avoid the displeasure of authority figures.

INTERVIEWER: Was it wrong for Melissa to eat with her fingers in school?

ANDREA (5 YEARS, 1 MONTH): Yes. . .Because she can't use her fingers.

INTERVIEWER: Why not?

ANDREA: The mother will hit her (Damon, 1977, p. 255).

At the beginning of middle childhood, children recognized that rules could have exceptions.

INTERVIEWER: Do you think it was wrong for Melissa to eat with her fingers?

TOM (6 YEARS, 2 MONTHS): No, she didn't learn how to eat with other stuff.

INTERVIEWER: Do you think it was wrong?

TOM: No, because she was new to this country anyway.

INTERVIEWER: Why do you think the teacher got really upset when she saw her doing that?

TOM: She may not know that she was new to the country (Damon, 1977, p. 261).

Finally, by the end of middle childhood, children come to respect rules because they understand their social function. They realize that social conventions are necessary for maintaining order.

INTERVIEWER: What if there was a rule about eating with a fork at Melissa's school?

HEATHER (8 YEARS, 8 MONTHS): If there is a rule then she would have to eat that way.

INTERVIEWER: Why is that?

HEATHER: Because then everyone would have to do the same thing no matter how they were taught at home or it wouldn't be fair.

INTERVIEWER: Couldn't Melissa just ignore the rule?

HEATHER: No. Because then everybody would start looking at her and start doing it. . .What a mess! Then if only one person didn't do it they would make fun of him, and they might not think it's fair (Damon, 1977, p. 271).

One kind of social rule that school-age children follow is about gender roles.

''Why do you think people tell George not to play with dolls?''

''Well, he should only play with things that boys play with. The things that he's playing with now is girls' stuff.''

''What makes it girls' stuff?''

''Because it's pictures of girls. But the boys' things are boys'.''

''Can George play with Barbie dolls if he wants to?''

''No, sir.''

''How come?''

''If he doesn't want to play with dolls, then he's right, but if he does want to play with dolls, he's double wrong.''

''Why is he double wrong?''

''All the time he's playing with girls' stuff.''

''Do you think people are right when they tell George not to play with girls' dolls?''

''Yes.''. . .

> "What should George do?"
>
> "He should stop playing with the girls' dolls and start playing with the G.I. Joe" (Pitcher and Schultz, 1984, p. 114).

In one recent study (Stoddart and Turiel, 1985), school-age children were shown pictures in which children were doing things appropriate for the other gender (a boy wearing a barette or putting on nail polish, a girl with a crew cut or wearing a boy's suit). The 5- and 6-year-olds thought that breaking gender-role rules was worse than the 8- to 10-year-olds did.

Children respond differently to social rules—like playing the right way—and moral rules—like not stealing and lying (Nucci and Nucci, 1982). But in the early school years children do not have quite the same appreciation as in the later elementary school years of the distinction between social conventions, which are arbitrary, and moral rules, which are absolute. In one study, first and second graders were interviewed about rules and heard stories about social transgressions—a boy who gets up in the morning and washes his face but does not comb his hair—and moral transgressions—children refusing to share or hitting (Shantz and Shantz, 1977). The children were then asked to point to the face that best represented what they thought about the naughty child—from a broadly smiling "very good" to a frowning "very bad" face. Although, on the average, children ranked the moral transgressions as worse than the social transgressions, the children did not rank all moral transgressions as more serious than all social transgressions. They thought that some of the social transgressions, like not combing your hair, were as serious as the moral transgressions, whereas others were not as bad. Development brings with it increasing awareness of subtle distinctions between types of rules and greater flexibility in complying with arbitrary rules.

Development also brings an appreciation for more subtle social rules—like rules of deference and politeness that are implicit in the ways we speak. In many European languages, the pronouns express deference. For example, a French "you" can be expressed by the familiar form, *tu*, or the more formal *vous. Tu* not only means "the person to whom I am speaking" but also indicates familiarity and closeness between the speaker and listener. Children, intimates, and people of low status are addressed with *tu*, and *vous* is used to convey respect and formal distance. European children develop an awareness for these social differences in pronouns at about the same age that American children begin to make requests politely—6 years (Hollos, 1977). Children of this age are capable of learning the social conventions of polite speech because they know that other people have perspectives different from their own. They also have had some practice with polite forms. Young children try out the polite phrases that they hear around them. One 4-year-old, on meeting "George" for the first time, said, "My, you look nice today, George." Said 4-year-old Peter to departing guests, as he roused himself from a nap on the living room couch, "Come back again." But not until the elementary school years are children able to *understand* linguistic rules of politeness and deference. Not until they are around 6 years old do children understand that they should make requests politely and indirectly. A 5-year-old says, "I want the bicycle," or on good days, "I want the bicycle, please." But only after age 6 or so can children use forms like "Can I swing?" or "Please may I swing?" or "Can I have a turn on the swing?" (Bates, 1976; Garvey, 1975).

Even then, children's politeness is spotty. Politeness requires both linguistic and social skills. It requires that children not only know the polite language forms but also understand their listener's signals, status, and other aspects of the social situation. This understanding comes only with advanced cognitive development. In one study, 5- to 10-year-old children were studied as they interacted with adults (Axia and Baroni, 1985). Not until age 9 or so did children master polite requests to overcome difficulties in spontaneous interactions. Five-year-olds did not have the polite forms; 7-year-olds did not discriminate between situations in which politeness was required and those in which it was not.

> Every time I ask Peter to do something like put away his toys or eat his beans, he answers, very politely, ''No thank you.'' Then what am I supposed to say?

Solving social problems

Every day children face social problems—acquiring objects, entering interactions, seeking help, initiating friendships. Their ability to suggest solutions to these social problems increases in the school years. In one study (McCoy and Masters, 1985), for example, older elementary school children were able to suggest more useful and helpful strategies to cheer up or calm down fictional children experiencing sadness or anger than younger elementary school children were. In another study (Weiner and Handel, 1985), older elementary school children knew enough to say that they would withhold information that would make another person angry, such as the fact that the reason they hadn't shown up at a friend's house as invited was that they had decided to stay home and watch television (Rubin and Krasnor, 1986). Some psychologists have suggested that solving social problems improves in the school years because children's cognitive understanding increases (Spivack and Shure, 1974). To solve social problems, they have suggested, children need first to recognize the problem, then to generate alternative solutions to solve the problem, then to consider means to accomplish the solution, and finally to figure out consequences of the different alternatives. For this they must be able to take the other person's perspective.

What evidence do we have that this kind of cognitive understanding determines social behavior? Is there a connection between the level of understanding children demonstrate as they reflect about hypothetical situations and their social competence in real-life situations? When researchers in one study (Selman, Schorin, Stone, and Phelps, 1983) interviewed and observed girls in the second, third, fourth, and fifth grades, they found that the answers the girls gave to a dilemma about what a girl should do when her wishes and a friend's conflicted were related to the social interaction strategies they used in real groups. Immature strategies in these groups included grabbing and insulting; commands and tattling were more mature; suggestions and compromises were even more grown-up; and joking and making way for minority rights were the most mature of all. Girls with the lowest levels of reasoning in the social understanding dilemma used the fewest advanced strategies in their groups. Girls who reasoned at a high level of social understanding about hypothetical friendships also used relatively advanced strategies when they were in a group.

Research on socially isolated children, who do not interact with others in the classroom, supports this link between understanding and behavior. The solutions for social problems that these children offer have been found to be less relevant and flexible than those offered by more sociable children, at least in kindergarten (Rubin and Krasnor, 1986). Also, as we have already mentioned, popular children give better answers than rejected children to social problem-solving questions. But is this because advanced reasoning helps children know what to do in social situations, or are both reasoning and social behavior the consequences of children's general level of maturity? One way to find out would be to train children to reason about social problems and then to observe how they act in real problem situations. In one study (Shure and Spivack, 1980), children who were trained to solve hypothetical interpersonal problems were subsequently rated better adjusted by their teachers. But other researchers have not replicated these effects (Rubin and Krasnor, 1986). The strategies children need to solve their social problems are not necessarily the ones that are taught in such programs, and the strategies they are taught are not always used in their daily lives. Some social problems can be solved without thinking, and some cannot be solved even with it. The connection between children's cognitive understanding and their social skills clearly is imperfect. But the usefulness of training children in social problem-solving skills that can be translated into real actions has not yet been adequately explored by researchers. It is an area that is getting—and deserves—more attention.

The self

Middle childhood also brings with it rich new developments in children's feelings about themselves. Their self-images grow more complex and sophisticated.

Identity

In middle childhood, children's views of themselves change dramatically. They begin to recognize that they have unique personal qualities. They begin to feel a strong conscious sense of themselves as male or female and to project ahead to what they will do as grown women and men. To determine how the concept of the self changes in the school years, researchers have asked children to answer the question, "Who am I?" (Montemayor and Eisen, 1977).

> My name is Bruce C. I have brown eyes. I have brown hair. I have brown eyebrows. I'm 9 years old. I LOVE! sports. I have seven people in my family. I have great! eye site. I have lots! of friends. I live on 1923 Pinecrest Dr. I'm going on 10 in September. I'm a boy. I have a uncle that is almost 7 feet tall. My school is Pinecrest. My teacher is Mrs. V. I play Hockey! I'am almost the smartest boy in the class. I LOVE! food. I love freash air. I LOVE School (Montemayor and Eisen, 1977, p. 317).

Most children of 9 or 10 identified themselves as Bruce did, by the concrete facts of their existence. The 9- and 10-year-olds were likely to refer to their sex, age, name, territory, likes, dislikes, and physical self. But by age 11 many children, especially girls, began to emphasize their personality and relations with others. A sixth-grade girl of 11½ gave this answer to "Who am I?"

> My name is A. I'm a human being. I'm a girl. I'm a truthful person. I'm not pretty. I do so-so in my studies. I'm a very good cellist. I'm a very good pianist. I'm a little bit tall for my age. I like several boys. I like several girls. I'm old-fashioned. I play tennis. I am a very good swimmer. I try to be helpful. I'm always ready to be friends with anybody. Mostly I'm good, but I lose my temper. I'm not well-liked by some girls. I don't know if I'm liked by boys or not (Montemayor and Eisen, 1977, pp. 317–318).

These developments clearly parallel the changes we discussed in how children describe their classmates.

At the beginning of elementary school, most children do not differentiate their characteristics from those of others. They may love sports, but they do not think of themselves as athletes. They may say, "I ride a bike," but they do not say "I ride a bike better than my brother." Gradually, by selecting and integrating new discoveries about themselves and by using their newly developed cognitive skills, they bring into focus a picture of the self that is sharp and unique. They begin to appreciate their own **identity**.

To learn more about children's ideas of identity, researchers have asked them questions like, "What is the self?" and "What is the mind?" They also have used more indirect means, such as reading children stories like the one about 8-year-old Tom quoted earlier and then asking them questions about the children in the stories (Broughton, 1978; Secord and Peevers, 1974; Selman, 1980). Six-year-olds locate their "self" within their body, usually within their head. As one child replied, "I am the boss of myself. . .[because] my mouth told my arm and my arm does what my mouth tells it to do" (Selman, 1980, p. 95). At about the age of 8, children begin to understand the difference between self and body. They describe themselves largely in terms of what they do, but they also describe how well they do it compared to other children. In middle childhood comes the understanding that one is unique not solely for physical reasons but also for what one thinks and feels. As one 10-year-old put it:

> I am one of a kind. . . .There could be a person who looks like me, but no one who had every single detail I have. Never a person who thinks exactly like me (Broughton, 1978, p. 86).

Self-esteem

Although self-esteem builds over a lifetime, middle childhood is especially important because in this period children first develop more than vague, simple ideas about themselves and their worth.

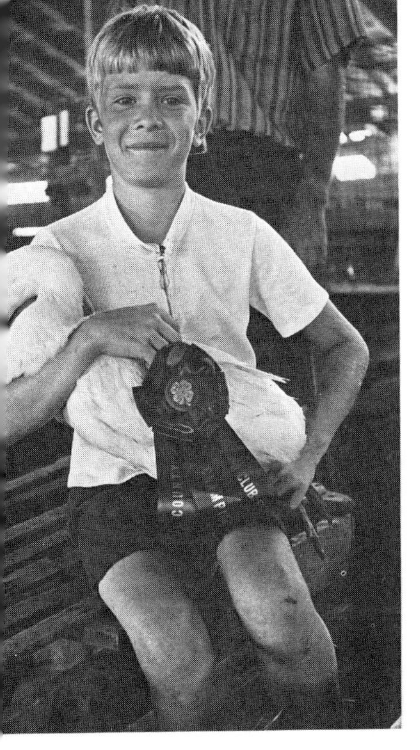

School-age children achieve a deeper and richer understanding not only of who they are but also of their self-worth. **Self-esteem** reflects children's mastery of developmental tasks, their performance on schoolwork, and their success in solving social problems. It also derives in part from other people's reactions to them and their achievements.

> My mother gets mad at me at least every day, and she tells me to get outside and find friends to play with because she doesn't "need a little friend around the house." I asked her once why she gets so mad at me, and she said it's because I need to be taken down a peg. Some days I want to run away, but I don't have anywhere to go.—Nancy, age 9

> Sam painted a wonderful pair of snakes, one lavender and one green, that we liked so much we framed it and hung it on the wall. When he walked into that room and saw his painting, the pride on his face was simply beautiful to see.

Self-esteem builds over a lifetime, of course, but the years of middle childhood are critical, for that is when children first go beyond vague, simple ideas of themselves and make more complex evaluations. Once formed, a child's self-esteem is difficult to change. Children with low self-esteem feel inadequate and are afraid of others and their rejection. Extremely low self-esteem is frequently accompanied by serious psychological problems.

Researchers have studied the development of children's self-esteem in middle childhood through extensive interviews with children and their parents. In one of the most detailed of these studies (Coopersmith, 1967), investigators found that boys (all the children in the study were boys) with high self-esteem were more independent and creative than those with lower self-esteem, more readily accepted in social groups as equals and leaders, more assertive, outspoken, and likely to express their opinions, and better at taking criticism. Their parents, in turn, were caring and attentive. They loved and respected their sons, set clear and reasonable limits for them, demonstrated their expectations and communicated the importance of social norms. They did not punish the boys harshly or maintain excessively high standards for their achievement in school. They were themselves self-assured, happily married, and socially active. In sum, children with high self-esteem are likely to have the same kind of warm, supportive, parents who, as we have seen before, are most likely to foster the development of social competence in their children.

As in other studies of parents' discipline and children's behavior, it is impossible to separate cause and effect in these correlations. But it is unlikely that children have high self-esteem simply because they have supportive parents. Children also have high self-esteem because they are successful in school and elsewhere. In one study, researchers traced the connections between boys' self-esteem and their success in school and work (Bachman and O'Malley, 1977). They found that boys with high grades in school later had higher self-esteem. In another study of third through ninth graders, achievement in school was found to influence students' estimations of their competence (Connell, 1981; Harter and Connell, 1982). This estimation of their competence in turn influenced their motivation to achieve. The higher the students' opinion of their academic abilities, the greater was their motivation to do well in school.

Continuity of personality

One of the most intriguing questions about human development concerns the continuity of social and personality traits. Does the irritable infant become a sullen school child? Does the predictable toddler become a predictable preteen? Does the popular preschooler become the college homecoming queen? Does the 4-year-old who never cries, even when her sister hugs her too tightly, later get through painful visits to the dentist without a fuss? Most of us—laypeople and psychologists alike—would be tempted to guess yes. Would we be right? The answer is hard to come by. Few researchers have followed their subjects from infancy into adulthood, and it is hard to follow the changing and tangled threads of personality across time. The scattered evidence we do have suggests that some aspects of personality are more stable than others. An adult's personality is never a wholesale continuation of the infant's or child's, but parts of early patterns surely do survive.

Followed from infancy into early childhood, a few temperamental traits seem to be quite stable. As we saw in Chapter 7, there is some stability in the "difficult" temperament pattern from infancy into adulthood. Data from the New York Longitudinal Study (Thomas and Chess, 1977) suggest that activity level, intensity, adaptability, and rhythmicity are most likely to be the stable traits (Table 13.1). Infants who moved often in their sleep were likely to be in perpetual motion at age 2. Infants who had regular eating and sleeping schedules as newborns ate and slept predictably at age 3. Infants who cried vigorously at wet diapers were likely to slam doors angrily at age 10. Infants who whimpered quietly when they were hungry later were likely to be stoic in the face of a scolding. Infants who adjusted quickly to new experiences like baths later adjusted easily to new experiences like nursery school. But although these temperamental traits were stable from year to year, they were not stable across the whole period from birth to age 5, and they were not stable enough to provide a basis for solid predictions in individual cases. When another researcher with a different sample (McDevitt, 1976) used the same categories of behavior as were used in the New York Longitudinal Study, he also found some stability between infancy and early childhood in activity level, intensity, adaptability, and, for girls, rhythmicity. But between infancy and middle childhood, the traits were much less stable. Clearly, the longer the interval between the measurements of traits, the less stable they are.

Table 13.1
Expression of temperament

Temperamental quality	At 2 months	At 10 years
1. ACTIVITY LEVEL[a]	Moves often in sleep. Wriggles when diaper is changed.	Plays ball and engages in other sports. Cannot sit still long enough to do homework.
	Does not move when being dressed or during sleep.	Likes chess and reading. Eats very slowly.
2. PHYSICAL RHYTHMS	Has been on four-hour feeding schedule since birth. Regular bowel movement.	Eats only at mealtimes. Sleeps the same amount of time each night.
	Awakes at a different time each morning. Size of feedings varies.	Food intake varies. Falls asleep at different time each night.
3. APPROACH AND AVOIDANCE	Smiles and licks washcloth. Has always liked bottle.	Went to camp happily. Loved to ski the first time.
	Rejected cereal the first time. Cries when stranger appears.	Severely homesick at camp during first days. Does not like new activities.
4. ADAPTABILITY	Was passive during first bath; now enjoys bathing. Smiles at nurse.	Likes camp, although homesick during first days. Learns enthusiastically.
	Still startled by sudden, sharp noise. Resists diapering.	Does not adjust well to new school or new teacher; comes home late for dinner even when punished.
5. INTENSITY OF REACTION	Cries when diapers are wet. Rejects food vigorously when satisfied.	Tears up an entire page of homework if one mistake is made. Slams door of room when teased by younger brother.
	Does not cry when diapers are wet. Whimpers instead of crying when hungry.	When a mistake is made in a model airplane, corrects it quietly. Does not comment when reprimanded.
6. THRESHOLD OF RESPONSIVENESS	Stops sucking on bottle when approached.	Rejects fatty foods. Adjusts shower until water is at exactly the right temperature.
	Is not startled by loud noises. Takes bottle and breast equally well.	Never complains when sick. Eats all foods.
7. USUAL MOOD	Smacks lips when first tasting new food. Smiles at parents.	Enjoys new accomplishments. Laughs aloud when reading a funny passage.
	Fusses after nursing. Cries when carriage is rocked.	Cries when he cannot solve a homework problem. Very "weepy" if he does not get enough sleep.

(continued next page)

Table 13.1 (*continued*)
Expression of temperament

Temperamental quality	At 2 months	At 10 years
8. DISTRACTIBILITY	Will stop crying for food if rocked. Stops fussing if given pacifier when diaper is being changed.	Needs absolute silence for homework. Has a hard time choosing a shirt in a store because they all appeal to him.
	Will not stop crying when diaper is changed. Fusses after eating, even if rocked.	Can read a book while television set is at high volume. Does chores on schedule.
9. ATTENTION SPAN AND PERSISTENCE	If soiled, continues to cry until changed. Repeatedly rejects water if he wants milk.	Reads for two hours before sleeping. Does homework carefully.
	Cries when awakened but stops almost immediately. Objects only mildly if cereal precedes bottle.	Gets up frequently from homework for a snack. Never finishes a book.

[a]High levels and positive behavior appear on white background; low levels and negative behavior on gray background.
SOURCE: Thomas, Chess, and Birch, 1970.

Across childhood and beyond, the greatest degree of stability has been found for two important qualities, aggressiveness and sociability. Aggression toward peers seems to be one of the most stable behavioral characteristics measured by researchers (Huesmann, Eron, Lefkowitz, and Walder, 1984; Kagan and Moss, 1962; Olweus, 1979). Boys, especially, who are aggressive as 5-year-olds turn out to be aggressive as 8-, 18-, and 30-year-olds as well. Children who are inclined to insult or hit other children are, in adulthood, more likely to be guilty of criminal acts, physical abuse of their spouse, speeding, drunk driving, and getting into fights. In the study by Rowell Huesmann and his associates, for example, among 8-year-olds considered by their peers to be *low* in aggression, 10 percent of the males and none of the females were convicted of crimes before age 30. Among 8-year-olds considered *highly* aggressive, 23 percent of the males and 6 percent of the females were guilty of crimes. Aggressiveness is stable, it should be noted, not only because of continuity in constitutional factors—genetic, hormonal, neurological—but because of continuity in environmental factors as well.

Sociability is another quality of personality that seems to be quite stable over time (Beckwith, 1979; Bronson, 1966; Kagan and Moss, 1962; Martin, 1964). Evidence from one longitudinal study (Waldrop and Halverson, 1975), for example, showed that children who at age 2½ were friendly and smiling, involved with and helpful towards other children, and who could cope with aggression from other children, at 7½ were likely to spend their time outside of school with other children, to feel socially at ease, choosing what and with whom they would play. In another longitudinal study (Schaefer and Bayley, 1963), boys who were friendly preschoolers became friendly and cooperative schoolchildren and, later, friendly and outgoing adolescents. The stability of sociability between early and middle childhood was lower for girls than for boys in this and one other study (Kagan and Moss, 1962). But girls who in *late* childhood were friendly, outgoing, and cooperative, tended to stay that way during adolescence.

lifespan focus

Lifelong personality

"She's got a million friends." "He's a total nerd." "I've always been shy." Statements like these refer to people's personalities. Lifespan psychologists have been interested in tracing the continuity of individuals' personalities over time. Is a nerd at 8 years a nerd at 80? Is a sociable 16-year-old a sociable 60-year-old?

Traditionally, psychologists studied individual personality *traits* like sociability or shyness and whether people continued to be high or low on these traits over time. More recently, psychologists (Mumford and Owens, 1984) have tried to study personality *types* as they change or remain the same over time. These psychologists assessed college freshmen, of about age 18, from a wide range of social backgrounds. The students were from the entering classes of 1968 through 1973 at a southwestern university. They answered questions about their parents' behavior, family size and interactions, their school and extracurricular experiences, leisure activities, feelings, social relationships, religious activities, independence, and achievement.

Then the researchers assessed these same students when they were seniors. In this assessment, students were asked about their reading habits, school and job experiences, leisure activities, social relationships, religious activities, and their evaluation of the university. A third assessment was made when the students were between 23 and 32 years old. It asked about the students' college and job experiences, marriage and family relations, leisure and religious activities, and reading habits.

From the data, the researchers classified over 80 percent of the students into a number of personality "types."

Among the types for the men were

Channeled Concrete Achievers—These men were a bit introverted, worked hard for material success, and were involved in their families and communities. They came from warm, supportive, upper-class families with traditional values and who supported achievement.

Upwardly Mobile Individuals—These men were independent and social leaders in college and afterwards. They were satisfied with and continued to be social leaders in their jobs. They came from warm, business-class families, and had been achievers and social leaders in high school.

Unrealistic Independents—These men did not achieve well at school, were unhappy with their jobs after college, and resisted merging with adult social institutions. They came from warm, supportive families that encouraged their independence.

Among the types for the women were

Unscathed Adjusters—These women were socially active in college, married six years or so after graduation, had happy marriages, and worked at social-service jobs. They came from warm, traditional families and had been encouraged to achieve in traditionally feminine areas.

Competent Nurturers—These women were bright and nurturant, socially active in college (although introverted in high school), and took social-service jobs after graduation. Their successful marriages, religious and community activities were the focus of their young adult lives. They came from warm, religious, upper-class families with socially active mothers.

These broad personality "types," the researchers discovered, were more stable over time than individual personality traits alone.

These personality traits are relatively stable. But it is important to remember that although *some* people change little from year to year, *others* seem in continual process of transformation (Block, 1971). Environmental, physical, and cultural changes influence how stable any individual's personality traits are across the years. People adapt their behavior to the circumstances in which they find themselves and so, in changing circumstances, their behavior changes, often markedly so, as they age.

Jean Macfarlane (1964), one of the psychologists who followed the development of the people in the Berkeley studies, noted that the personality characteristics of childhood often predict little about adult adjustment. According to her, half of the Berkeley sample became more stable adults than had been predicted, one-fifth became less stable, and less than one-third developed according to predictions. Macfarlane suggested that the researchers weighted troublesome aspects of children's personalities too heavily and weighted aspects related to adult stability too lightly. An awkward, shy, sometimes depressed girl who was a mediocre student matured into a happily married, enthusiastic mother and college teacher. A boy who broke rules and was expelled from school matured into a steady, understanding father with a good job. In some cases, the very experiences that seem to undermine personality development make people come to terms with themselves and actually enhance their development. The outlook for long-term personality development is often bright. It seems that as children develop, *if* the environment permits, they outgrow the maladaptive in their personalities and hold onto what is adaptive and healthy.

Summary

1. In middle childhood, the world is made up of family, school, and peers. School-age children feel deeply about their families. They need and want rules and restrictions, but the more controlling relationship between involved parents and younger, immature children gradually changes into a less controlling relationship in which both parents and child are responsible for the child's behavior.
2. School-age children form friendship cliques, with leaders and followers, in which they do such things together as watch television, go to the movies, play games, go to the beach, and talk. Girls' social networks are intensive. They play with one other girl, express intense feelings about her, and share experiences and fantasies with her. Boys have extensive social networks. They play noisily in a group and usually focus on a game like baseball or soccer. In the age-segregated elementary school, boys and girls rarely mix.
3. Popular children are confident, outgoing, active, friendly, and cooperative. They are rarely aggressive, get along with anyone, and know what to say and when to say it. Children who are socially inept, aggressive, disruptive, and act inappropriately are often rejected.
4. Cooperation, helping, and showing concern for others increase during early childhood but decline toward the end of middle childhood. Competitiveness, individualism, and generosity increase over the school years.
5. Almost all schoolchildren sometimes cheat or fib. They also sometimes hurt each other. Children learn aggression from parents who are aggressive, rejecting, and unloving and who punish the children for aggression but do not reward them for sharing or cooperating. Children's peers reinforce, elicit, model, and absorb aggression. When children watch a

lot of violent television shows, think that these shows are real, identify with aggressive television characters, are reinforced for their own aggressive behavior, and are objects of aggression, they are likely to act aggressively.

6. As they get older, children get better at understanding other people's feelings, moods, intentions, and motives. As their social understanding increases, their social competence increases, too.
7. The way that children see themselves changes in middle childhood as they begin to recognize that they have unique qualities. Children's self-esteem derives in part from others' opinions of them and in part from their mastery of developmental tasks, their performance in school, and their ability to deal with social situations.
8. Some temperamental traits—activity level, intensity, adaptability, and rhythmicity—seem somewhat stable from infancy to early childhood. Across childhood and beyond, the most stable qualities are aggressiveness and sociability.

Key terms

intensive peer relations
extensive peer relations
identity
self-esteem

Suggested readings

ASHER, STEVEN, and GOTTMAN, JOHN (Eds.). *The Development of Children's Friendships.* Cambridge: Cambridge University Press, 1981. A broad sampling of research into peer relations and their effect on children's development. The chapters on children without friends and how training can improve their social standing, and Robert Selman's chapter on children's capacities for friendship are of special interest.

DAMON, WILLIAM. *The Social World of the Child.* San Francisco: Jossey-Bass, 1977. A discussion of how children reorganize their ways of understanding social reality and deal with the principal social relations and regulations of their lives. The book is liberally sprinkled with actual conversations between children and adults.

HIGGINS, E. TORY, RUBLE, DIANE, and HARTUP, WILLARD (Eds.). *Social Cognition and Social Development.* New York: Cambridge University Press, 1983. An overview of current research and theory concerning social cognition and social behavior in children, with a focus on the developmental roots of social cognitive abilities.

RUBIN, KENNETH, and ROSS, HILDY (Eds.). *Peer Relationships and Social Skills in Childhood.* New York: Springer-Verlag, 1982. A collection of articles examining the development of peer relationships and social skills from infancy through early adolescence, by many of the most eminent researchers in the field.

PART SIX

Adolescence

chapter fourteen
Physical development

> I look at these kids I've known since they were pip-squeaks, and all of a sudden they're all grown up. Sam and his friends eat *incredible* quantities, their voices crack and embarrass them to death, and they need a bigger shoe size every two weeks. No more Saturday morning cartoons; the kids are *asleep*. The girls are getting curvy and attractive. Natalie comes over to our house now wearing lipstick and perfume, pocketbook over her shoulder. No more dollies. She wants to talk about dates.

Young people go through a series of biological and psychological changes at the end of childhood, as they enter **adolescence**. Their bodies visibly mature, their roles in society change, and the very content and complexity of their thoughts change as well. In this chapter, we begin our discussion of the complexities of development in adolescence by looking at the physical changes adolescence brings and at some of the implications of these changes for the way adolescents think and feel about themselves.

Puberty

The set of biological changes that mark the beginning of adolescence is called **puberty**. Puberty begins as increased levels of hormones enter the bloodstream, in response to signals from the hypothalamus region of the brain (Figure 14.1). Early in prenatal life, sex hormones and other chemicals in the

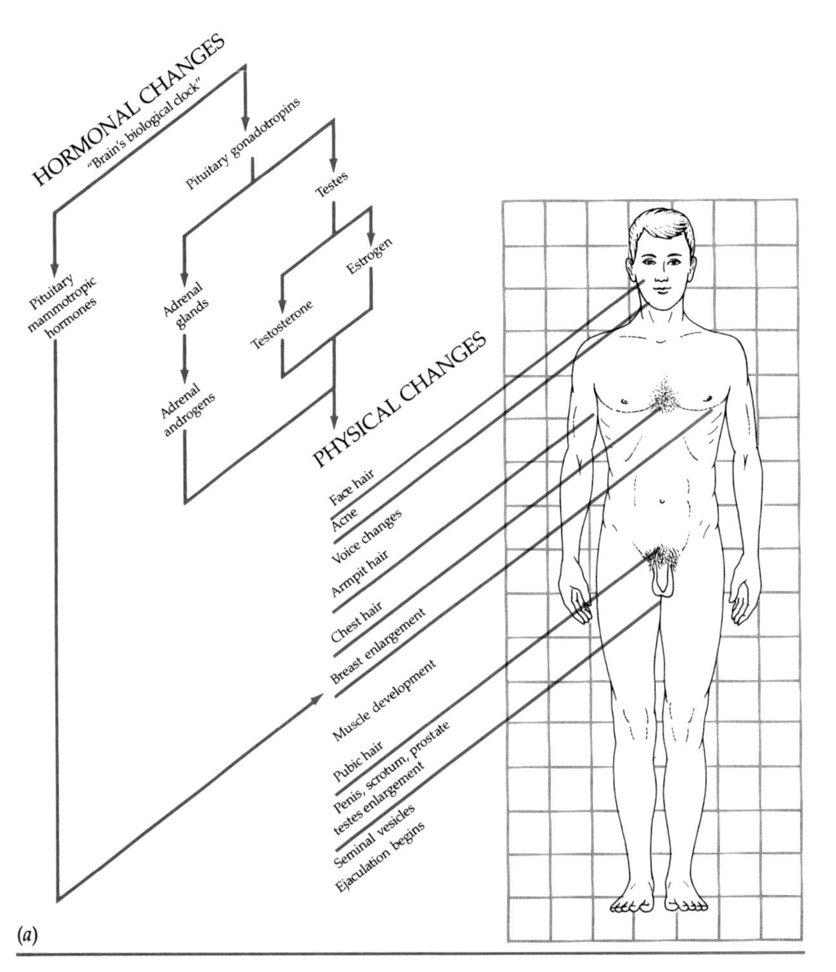

Figure 14.1 **Processes of sexual maturation at puberty in males and females.**

fetus and in the uterine environment program the hypothalamus of a male to develop differently from that of a female. At puberty, signals from these sexually differentiated parts of the hypothalamus communicate to the pituitary gland (also located in the brain) to increase its production of growth hormone and the hormones that stimulate the development of the gonads—ovaries and testes. The gonads then increase their production of sex hormones—estrogens and androgens—that turn the bodies of girls and boys into those of sexually mature young women and men. Throughout childhood, boys' and girls' bodies have produced roughly equal levels of estrogens and androgens. But with puberty, girls' bodies begin producing more estrogens than androgens, and boys' bodies begin producing more androgens than estrogens. As these sex hormones enter the bloodstream, they stimulate the development of particularly sensitive cells elsewhere in the body. In girls, breasts and genitals mature in response to estrogens; in boys, genitals mature in response to androgens.

The growth spurt

> Sam grew five inches in ninth grade. His body changed so fast that for a while he lost his physical coordination. He seemed to be bumping into things all the time and looked really gangly there for a while.

As the levels of growth hormone rise in the body, the steadily but slowly growing child seems suddenly to spurt up. It is this **growth spurt** that to the outside world visibly signals the beginning of puberty. The rate of growth may

PROCESSES AT PUBERTY

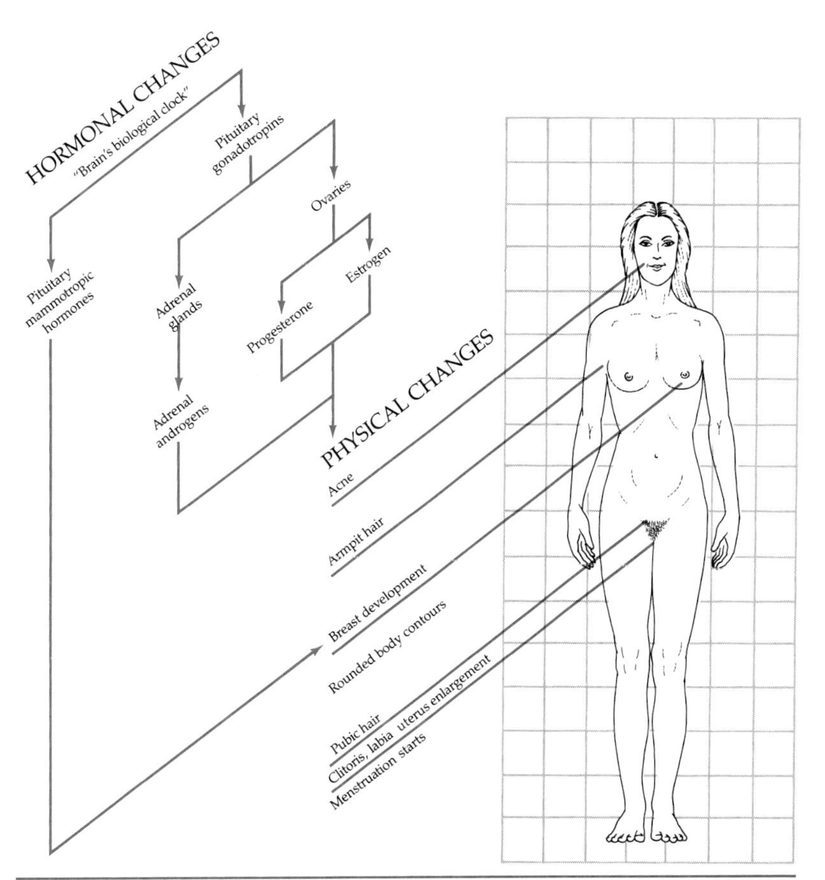

On the average, girls start their growth spurt two years befo boys do.

double, with young adolescents growing as much as five inches in their peak year. For boys, the growth spurt usually starts at 12 or 13 and peaks at about 14 years, then tapers off after 16, and finally stops at 18 or 19. For girls, the growth spurt starts at 10 or 11, peaks at 12 or 13, and stops at 17 or 18 (Bayer and Bayley, 1976; Faust, 1977; Stolze and Stolze, 1951; Tanner, 1970; see Figure 14.2). But there are wide individual differences in when the growth spurt occurs. One longitudinal study, for example, showed that of 94 girls, one began her growth spurt at the age of 7½, and one began hers at 13 (Faust, 1977).

The growth in height results from the final stages of bone maturation. It is at puberty that the epiphyses—the parts of the long bones made of cartilage—finally turn to bone. The muscles, too, lengthen and strengthen. Internal organs—lungs, heart, stomach, kidneys—grow to adult size and capacity. Boys particularly, in response to androgens, gain muscle mass, strength, and stamina. Their bodies grow more efficient at metabolizing lactic acid, the by-product of strenuous exercise. Girls, in response to estrogen, develop curves as breasts, hips, and buttocks grow larger and padded with fat.

But growth is not smooth and even. It is asynchronous. Legs grow before trunk and shoulders. Forearms grow before upper arms. Whereas growth in infancy and early childhood was from center to periphery—proximodistal—growth at puberty is just the opposite, from periphery to center—

Figure 14.2 These graphs illustrate the main aspects of physical change in adolescent girls and boys at puberty. Curves represent changes in height. Secondary sexual characteristics are shown by typical age of onset. Although individual variations in age are normal, the sequence of onset is essentially as shown (Tanner and Whitehouse, 1976; Tanner, 1978).

TIMING OF PHYSICAL CHANGES AT PUBERTY

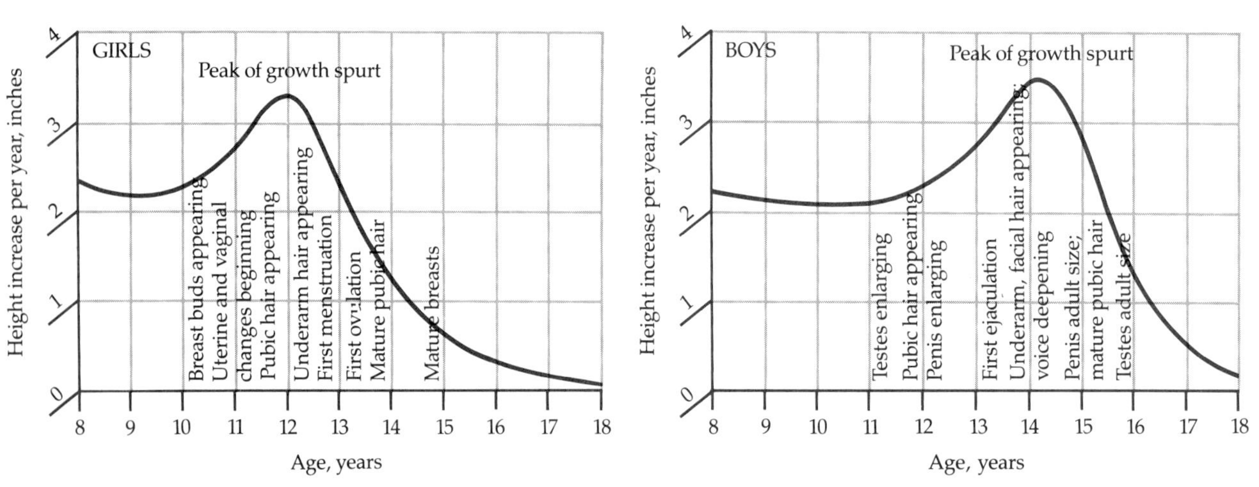

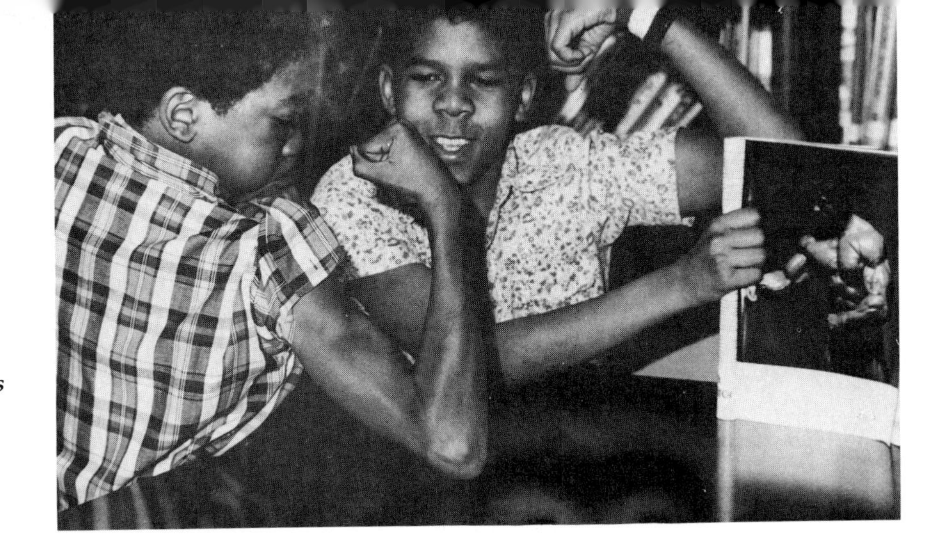

As their body image forms, adolescent boys constantly compare their physical growth and prowess against each other and with those of their ideal.

distoproximal. Adolescents in the midst of the uneven growth patterns of puberty sometimes look gawky and uncoordinated. Big feet and hands, long legs and arms poke out of pants and shirts. Jaws and ears look too big. At puberty, nose, lips, and ears reach their adult size before the upper part of the head grows. As the eyes grow in size, many adolescents become nearsighted. The sweat and oil glands of their skin make baby-fine complexions coarsen and break out.

Development of sexual characteristics

The most profound transformation of puberty is the development of sexual maturity, the ability to reproduce. The reproductive systems of boys and girls at adolescence enlarge and mature into those of sexually mature men and women. The primary sexual characteristics—genitals and associated structures—have been in existence since birth, but at puberty they become functional. Girls begin to menstruate and then to ovulate, boys to ejaculate semen.

At puberty, too, the **secondary sexual characteristics** develop (see Figure 14.2). Before his growth spurt begins, a boy's testes grow larger, downy pubic hair emerges, and the skin of the scrotum turns pinker and coarser. Then, just as growth hormone makes him taller, it makes his penis grow larger. The internal pathways for delivering sperm also develop, and after a year or two, the sperm-producing structures have matured and are producing live sperm. Nocturnal emissions, or wet dreams, may occur. Gradually, the numbers of sperm produced increase to the point that the young man is fertile and biologically capable of fathering a child. Because his hypothalamus signals for the fairly constant production of sex hormones, his production of sperm is also constant—many thousands every day.

It is not only the boy's reproductive system that develops at puberty. Fine hair appears on his upper lip, chin, and cheeks and under his arms. This fine hair then grows coarser and darker. The boy's larynx and vocal cords lengthen, deepening his voice. To his horror, his breasts may enlarge temporarily, a side effect of the high levels of hormones in his body. Eventually though, a boy's body takes on its distinctly masculine outline. The soft contours and plumpness many boys have disappear. Their faces get bonier and more muscular. Androgens make the boy's shoulders widen and his rib cage expand. By the end of puberty, he is likely to have beard and body hair, broad shoulders and chest, narrow hips, and a lean, muscular, angular form.

The parallel transformation for the girl is to take on the curves and padding of fat typical of the feminine outline. First, the girl sees breasts bud and fine pubic hair sprout. Even before she notices these outward signs, her uterus has begun to enlarge and her vaginal lining to thicken. Her hips and thighs grow heavier and wider, and her waist seems narrower. Estrogens cause her pelvis to widen, a development that will make childbearing easier. Her voice loses its girlish reediness. Her breasts continue to grow for several years, the area around the nipple—the areola—darkening and expanding. Her external genitals also enlarge, coarser pubic hair covers her mons and external labia, and underarm hair develops. A year or two after she has reached the peak of her growth spurt, her first menstrual period—**menarche**—appears. Menstruation usually begins only when a girl weighs over 100 pounds, presumably because at that weight her body has enough fat to store the estrogen needed for menstrual cycles. The first menstrual cycles may be irregular, and the girl's body may not begin to ovulate for a year or so after menstrual cycles have begun. Only when she ovulates does she become fertile, capable of conceiving a child.

By the end of puberty, boys have bigger and stronger muscles than girls, larger hearts and lungs, more blood and more oxygen capacity in their blood, and lower resting heart rates (Faust, 1977; Grinder, 1978; Malina, 1974; Tanner, 1970). With these physical advantages and their greater upper body strength, boys tend to excel in some sports, such as throwing and fielding balls, wrestling, and lifting weights. With their larger reserves of fat and greater flexibility, girls tend to excel at different sports, like gymnastics, distance running, and swimming. Although the *best* male athletes usually can outperform the *best* female athletes in a given sport (in professional basketball or marathon running, for example), the *best* female athletes can outperform the *majority* of males.

Puberty confers on males and females measurable differences in size, shape, and strength. But cultural attitudes intensify the effects of these biological differences. When gender-role expectations are such that females are pressured to be more delicate and weaker than males, adolescent girls actually may perform at reduced levels of strength after puberty. As attitudes toward the participation of women in high school and college athletics have changed, women in many fields have broken records and surpassed traditional expectations. In fact, female athletes have surpassed world records held by males 20 years ago (for example, in swimming). Old-fashioned fears that women should not engage in strenuous activity during their menstrual periods, that the menstrual cycle makes women's performance erratic, and that breasts and uterus make women too delicate to participate safely in sports have all proved unfounded. Surveys conducted at the Olympic Games show that women have set world records at all stages in the menstrual cycle (Grinder, 1978). The uterus is one of the most shock-resistant of all internal organs, and the external genitals of females are less exposed than those of males.

The trend toward earlier maturity

Both the growth spurt and the age of first menstruation have been occurring at younger and younger ages over the past century. This secular trend in maturation has been noted in Canada, the United States, Great Britain, Sweden, Japan, and other places with comprehensive statistics on growth (Moore, 1970). In the United States, for example, girls of a century ago first menstruated at 13 to 15 years; today the age is 11 to 13. Similarly for boys, the growth spurt now, on the average, begins two years earlier than it did at the turn of the century (Blizzard et al., 1974; Bullough, 1981; Hamburg, 1974; Tanner, 1962). Not only does the growth spurt start earlier, but adolescents also grow taller and heavier than they did in the past. Armor worn by medieval knights would fit a 10-year-old today, and the average height of male colonists was five feet. On average, for every ten years between 1910 and 1940, white adolescents grew one-half inch taller; for every ten years between 1940 and 1960, they grew one-quarter inch taller (Espenschade and Meleney, 1961).

Is this trend likely to continue? Are boys and girls going to mature ever earlier and grow ever larger—until they are sexually mature in kindergarten and as tall as giraffes in college? Some researchers have suggested that the timing of sexual development must have a limit (Tanner, 1965), and, in fact, the average age of menarche, 12.6 years, has not altered in the last 30 years (McAnarney and Greydanus, 1979). Puberty probably began earlier primarily because young people were eating better and suffering from fewer diseases than previously (Tanner, 1962). When girls eat well, they reach the critical body weight of 100 pounds earlier and so begin to mature sexually. It is likely that disease and poor nutrition can delay maturation and that good health and nutrition can stimulate it—a kind of biological "greenhouse effect"—but only so much.

Effects of physical change

Do the hormones that trigger the physical changes of puberty also trigger psychological changes? Do the physical changes themselves have psychological ramifications? Yes they do, but the effects are not as simple as one might expect.

Hormones and psychological states

When menarche begins, many adolescent girls undergo profound changes in their self-image (see Figure 14.3), in their attitudes toward others, and even in perceptual abilities. In one investigation (Diamond, Carey, and Back, 1983), for example, girls were found to encode unfamiliar faces less efficiently during puberty than either before or after puberty. What might cause this? It has nothing to do with changing schools, the research showed, and it has nothing to do with any age-related reorganization in knowledge of faces. It is possible that the hormones of puberty directly affect the brain's ability to recognize faces. It is also possible that the phenomenon is psychological. Perhaps in adolescence children become interested in different aspects of faces and so, through puberty, go through a period of less efficient encoding. For now the question remains open. Further research is necessary to explore the relations between hormonal and psychological changes during adolescence.

Preliminary findings from a study underway at the National Institute of Mental Health (Nottelmann and Susman, 1985) suggest that a low level of the sex hormone estradiol may be related to psychological adjustment problems in 9- to 14-year-old girls. In boys the link between hormones and behavior is even clearer and stronger. Boys with high adrenal androgen were more likely

MENARCHE AND SELF-IMAGE

Figure 14.3 ***The drawing on the left was done by a girl 12 years and 1 month old—before she began menstruating. She saw herself differently six months later, after she had begun to menstruate (Gardner, 1982, from Koff, Rierdan, and Silverstone, 1978).***

to have behavior problems—rebelliousness, talking back, fighting with classmates—or feelings of sadness and confusion. This finding fits with earlier research showing that when adolescents were given androgen, they became more aggressive (Wolstenholme and O'Connor, 1967).

The effect of hormones on adolescents' behavior are complicated by social and psychological factors, however. Cultural and family attitudes make a difference in whether hormones bring about psychological change. Attitudes toward menstruation and toward menarche, for example, affect adolescents' anxiety and physical discomfort around their menstrual periods.

> I have terrible cramps every month. No one can convince me that they're all in my head. It's the "curse."—Megan, age 15

> I almost never have cramps, and I do anything and everything when I have my period. My mother says just to do what I normally do. My friend's mother won't let her swim when she has her period.—Erica, age 15

In the United States, cultural attitudes toward menarche and menstruation are mixed. We have no formal ceremony or other means of recognizing the event, but we may surround the newly menstruating girl in a kind of pall.

> We prescribe no ritual; the girl continues on a round of school or work, but she is constantly confronted by a mysterious apprehensiveness in her parents and guardians. Her society has all the tensity of a room full of people who expect the latest arrival to throw a bomb (Mead, 1971, p. 180).

Although some adolescent girls have been exposed to a bit of education—films on menstruation, talks with their mothers, and the like—in one study of white, middle-class girls, most had little idea about what the inside of the body was like or how it functioned (Whisnant and Zegans, 1975). Although they had been told about showering and changing sanitary pads during their periods, their ideas about the purpose of menstruation were sketchy. Those who had not yet menstruated thought that when their first period came they would proudly announce it to family and friends. But when they did begin to menstruate, most grew secretive and told only their mothers.

> When I first got my period, I knew what it was, and so I wasn't scared. But I was embarrassed and asked my mother not to tell anyone. She was so proud "that her little girl had become a woman" that she crowed about it to everyone. I felt like crawling under the rug.—Sarah, age 15

Some grew closer to their mothers, even those who had not felt close before, and some focused more on their relationships with their fathers. In another study in which 350 adolescent girls were polled, most said that they believed menstruation caused physical pain, emotional upheaval, mood swings, and disruptions in behavior and relationships (Ruble and Brooks, 1977). The girl's age at the time, her knowledge and expectations, her personality, and the support she receives from her family all influence how she interprets menstruation. Girls who are well prepared both physically and psychologically and who begin menstruating at about the same time as most of their friends tend to feel that menstruation is a normal event (Ruble and Brooks-Gunn, 1982).

Body image

> Jason works out with weights, he's a vegetarian, and he runs 50 miles a week. He looks a lot healthier than he did as a young boy, but I worry

> sometimes that he's doing all this because underneath he feels that his body doesn't measure up somehow.

Just as menarche changes adolescents' images of themselves, so do other physical changes ushered in by puberty—particularly changes in appearance. Not a few adolescents are unhappy with the way they look; they have a poor **body image**. "Too fat" thought nearly half of all 12- to 17-year-olds surveyed in one national poll; "too short" thought nearly half the boys (Scanlon, 1975). "Too pimply" thought one-third of the girls and one-fifth of the boys at age 12; half the girls and two-thirds of the boys at age 17. Acne is only one of the complaints adolescents make about their appearance. Others dislike their posture (Hamburg, 1974). Asynchronous growth makes them feel awkward and ungainly. Many worry that they do not measure up, do not meet a tacit standard among their peers, and differ in some way from what girls or boys "should" look like. Boys want to see whether they measure up in height, endurance, muscle strength, and virility. Many worry that their penis is the wrong size, shape, or color; that their chests and chins are not hairy enough; that their voices are not deep enough. Girls, already uncomfortable because from age 11 to 13 they tower over the boys in their class, are concerned about measuring up in prettiness. Many of them worry that their breasts and genitals are the wrong size and shape, that their hair, teeth, hips, legs are not attractive enough. Those with chronic illnesses or handicaps can feel miserably self-conscious and different from "the other kids" at a time when being like them is nearly the most important thing in life (Magrab and Calcagno, 1978).

Adolescents who are attractive and who have a favorable body image are likely to have a generally favorable self-image and to be happier, more socially successful, and more pleased with themselves right into their adulthood than unattractive adolescents with poor body images (Berscheid, Walster, and Bohrnstedt, 1973; Jaquish and Savin-Williams, 1981). Adolescents' body images are based not only on what their bodies look like at present but on a lifelong accumulation of perceptions and feelings about their appearance.

> I was a chubby little girl, and so now every time I look in the mirror, I see *fat*. My grandmother tells me to stop dieting, that I look fine. But I don't believe her. Once a tub, always a tub.—Natalie, age 16

Even when their bodies have changed enormously at adolescence—the skinny becoming round, the round becoming skinny, the cute becoming plain, the plain becoming handsome—the things adolescents heard and saw about themselves in childhood do not allow them to perceive themselves objectively. Adolescents whose self-images in childhood were positive are more likely to approve of their new images than are those whose early self-images were negative and harsh.

How parents, teachers, and peers judge their appearance also contributes to adolescents' body images. During the rapid changes of early adolescence, when many lose the sense of what "normal" looks like for themselves, other people's opinions can be especially important, and those people's judgments usually are based on stereotypes of physical attractiveness. School principals, teachers, and teenagers all tend to think more highly of large, muscular, athletic boys than of smaller, weaker, less athletic boys, for example (Clarke and Clarke, 1961).

Sexuality

The physical changes of puberty make possible feelings of mature sexuality. As Freud was first to point out, sexuality is present since infancy, but present

lifespan focus

Early and late maturers

Some adolescents, as we have seen, reach puberty earlier and some later than the norm. Some 14-year-olds look and act like sexually mature men and women, but some are still children. Variation in the timing of puberty is normal. Researchers have studied whether this difference in timing has psychological effects.

In one early study, researchers picked out 16 of the earliest maturing and 16 of the latest maturing boys from the Berkeley Growth Study (Jones and Bayley, 1950). They observed the boys on the playing field, in a doctor's waiting room, talking with girls at a dance, and getting along with their classmates, and they talked to the boys' teachers and peers. They found that other adolescents thought that the **late maturers** were restless, immature, and bossy. Adults also considered them restless and immature, as well as tense, eager, energetic, and self-conscious. Many of the late maturers tried to compensate for their physical immaturity with bids for attention and bursts of activity; a few withdrew from others. In contrast, the **early maturers** were self-confident, calm, and considered mature by both adults and other adolescents.

The early maturing boy has social advantages in high school and is likely to be a leader and greatly admired.

in immature form. Infants and young children are no strangers to genital stimulation, exploration, or sexual arousal. In the preschool years, sexual games and fantasies are normal and widespread. In research by Alfred Kinsey and his associates, for example, 56 percent of boys and 30 percent of girls said that they masturbated (Elias and Gebhard, 1969). But mature sexuality is made up of more than genital exploration and stimulation. It is made up of more than the hormones, ova, and sperm that develop at puberty. Mature sexuality is also made up of mature *ideas* about sex. Most adolescents develop a fuller understanding about reproduction than they had as children, new interests in intimate sexual relationships, and new cultural expectations about the who, when, and where of sexual behavior.

Cultural expectations strongly influence sexual behavior during adolescence. As attitudes toward sexual behavior have changed in the last 50 years,

They engaged in little compensating or striving for status, though many were leaders in high school. When the boys made up stories about ambiguous pictures in the Thematic Apperception Test (TAT) (Mussen and Jones, 1957), the early maturers' stories were characterized by self-confidence, independence, and social maturity. The late maturers' were characterized by low self-esteem, dependence, rejection, conflict with their parents, and domination by others. The evidence suggests that early maturation confers social advantages, and late maturation, social disadvantages.

In a more recent and comprehensive study, 6000 adolescents were classified into early-, average-, and late-maturing groups (Duke et al., 1982). The adolescents, their parents, and their teachers were asked about the adolescents' school performance, intelligence, and related matters. Again, for late maturing boys, at every age beyond 12 years, disadvantages in development showed up. Compared to mid-maturing boys, late-maturing boys were less likely to intend to finish college and were less likely to be expected by their parents to do so. Their teachers considered late-maturing boys less intelligent and less academically able.

The effects of the timing of maturation on girls are similar, though more complicated than the effects on boys (Duke et al., 1982). Whereas early-maturing boys are viewed as athletic and as leaders, maturing early can present problems for the adolescent girl because of the ways that parents and peers react to the girl's potential for sexuality. More important than being early or late, for girls, is being "on time." Just as with menarche and physical appearance, it is very important to adolescents how others respond. Reaching puberty when most of their peers do tends to make adjustment easier (Faust, 1960; Greif and Ulman, 1982; Harper and Collins, 1972; Weatherly, 1964; Wilen and Peterson, 1980).

When the adolescents in the Berkeley Growth Study were observed at the age of 33, the early maturers still bore social advantages, still seemed poised and cooperative, were social leaders and professional successes. But they were also more likely to be rigid and tightly controlled, clinging in later life to the behavior that won them social acceptance in adolescence—and paying a psychological price for it. The late maturers were given to impulses, irritable, rebellious, and less successful and cooperative. Some continued to feel inferior and rejected. But, on the positive side, many had grown to be more flexible, perceptive, assertive, expressive, eager, uninhibited, insightful, and playful than the other men (Jones, 1957; Peskin, 1967). When puberty begins apparently does affect an individual's emotions and social development. But the outcomes are neither all good nor all bad.

so have adolescents' sexual habits (see Figure 14.4). For example, relatively few adolescents in one study reported feeling guilty about masturbating, a change from earlier generations (Sorenson, 1973). Moreover, the number of adolescents having sexual intercourse began to increase in the 1960s and 1970s and continues to increase in the 1980s (Dreyer, 1982). Adolescents today see the world as sexually active, even sexually preoccupied (Miller and Simon, 1980). They feel that social disapproval of sexual activity has weakened and that they must decide as individuals about the degree of sexual intimacy they find comfortable. Most adolescents date for some time without engaging in sexual intercourse, but eventually, after kissing, necking, and petting, most of them do have intercourse with their steady dates. Among 14-year-olds in the recent survey, 20 percent reported having had sexual intercourse (Search Institute, 1984). Among 16-year-olds, in another major survey, 45 percent of

Cultural expectations strongly influence sexual thoughts and behavior during adolescence.

the boys and 33 percent of the girls were sexually active. Among 19-year-olds, 80 percent of the boys and 70 percent of the girls were sexually active (Alan Guttmacher Institute, 1981).

The increase in sexual activity over the past two decades may be more a reflection of attempts to achieve personal identity and physical intimacy than a reflection of uncontrolled impulse gratification. For most adolescents, sex is not casual. But neither is sex an obsession. When 600 adolescents were asked to rank the importance of activities in their lives, they ranked sex last, after doing well in school, having friends of their own sex, having friends of the other sex, participating in sports, and being romantically involved (Haas, 1979). The adolescents who have sex differ somewhat from those who do not. They tend to have less conventional attitudes and values, to be less involved in conventional social institutions, to be more likely to use drugs or alcohol, and to have parents who are less controlling (Jessor and Jessor, 1975).

As sexual standards have relaxed, the old double standard—a stricter standard of sexual conduct for women than for men—has relaxed as well (Miller and Simon, 1980). More girls are having sex before marriage. By college, 74 percent of *both* young women and men have had sexual intercourse (Dreyer, 1982). But boys still are likely to have more sexual partners than girls. In one recent study, the majority of the adolescent girls had had sex with only one partner; the majority of boys had had sex with more than two partners (Haas, 1979). Most adolescent girls have sex first with someone they love and hope to marry. Boys are not so likely to have sex with a partner whom they love (Jessor, Costa, Jessor, and Donovan, 1983). The reason for this difference may be biological (boys have spontaneous erections; girls do not) or it may be social (boys brag about their exploits; girls do not). Whichever it is, the **sexual scripts**—norms about how to be sexual, norms about what, when, and with whom one does sexual things, and what they mean—differ for adolescent boys and girls (Gagnon and Simon, 1970). Adolescent boys usually proceed from masturbation to sex with a partner, and they must learn from girls about the interpersonal and the emotional aspects of sex. In contrast, adolescent girls usually begin with sex with a partner and must learn from boys how to focus on the physical aspects of sex. On dates, boys teach girls how to abandon themselves to sensual pleasure and to put aside manners and tidi-

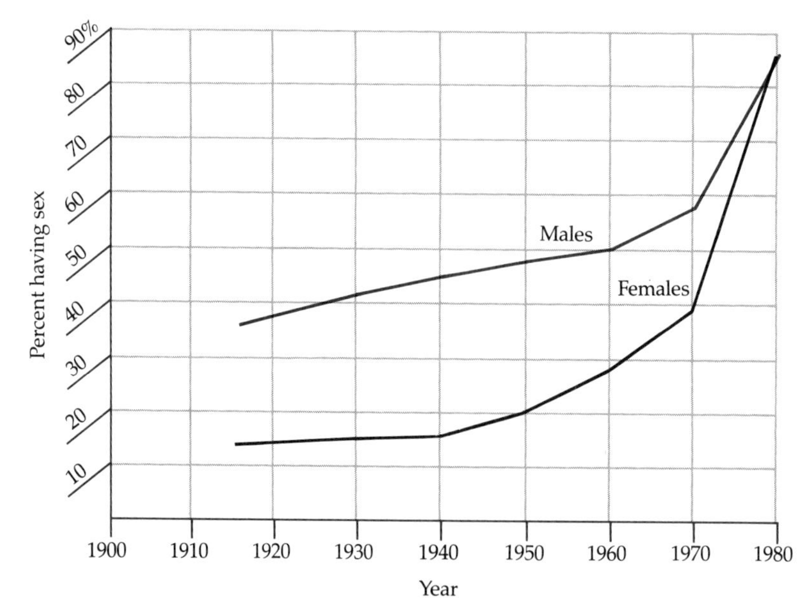

Figure 14.4 Rates of sexual intercourse for never married male and female college students, based on 35 published studies (Darling, Kallen, and Van Dusen, 1984), illustrating both the increase in premarital sex and the decline in the "double standard."

ness. Girls teach boys how to nurture and support. If they learn well, late in adolescence both boys and girls will combine elements of both sexes' scripts into a mature sexuality. In one survey, only 5 percent of the college students questioned said that they were not emotionally involved with their partner, compared with 45 percent of young adolescent boys (Miller and Simon, 1980).

Homosexuality

Most adolescents are sexually attracted to people of the other sex. But some adolescents wish for sexual relationships with members of their own sex. Although few adolescents have committed themselves to homosexuality, a large minority have thought about it. Among adolescents in one study, although they did not identify themselves as homosexual in orientation, 11 percent of the boys and 6 percent of the girls had had some homosexual experience (Sorensen, 1973). In a more recent study, 15 percent of boys and 10 percent of girls reported having had a homosexual experience (Dreyer, 1982). Forty percent of the adolescent boys and girls in the first study agreed that "If two boys and two girls want to have sex together, it's all right so long as they both want to do it." But 75 percent also said, "I'm sure I'd never want to." Typically, boys have their first homosexual experience with an older boy or a man at ages 11 or 12, and with increasing age, they are more likely to have homosexual experiences (Bell, Weinberg, and Hammersmith, 1981). Exclusively homosexual men have homosexual feelings as boys, and their homosexuality usually is well established by their late teens. Bisexual men do not have homosexual feelings as boys, although they do have homosexual experiences; their bisexuality usually emerges after age 19.

> I was *afraid* that I was gay when I was very young. I *knew* it when I was in my teens. But it took me until after college to come out.—Tim, age 25

Girls, in contrast, are most likely to have their first homosexual experience with another girl between the ages of 6 and 10.

In the past, the behavior and feelings of parents and peers were thought to cause a person's homosexuality. But these threads are probably just part of a larger and more complex biological, psychological, and social web. Biological factors probably predispose a person toward homosexuality. These biological factors then make the child act in certain ways that elicit certain reactions from parents and others. A boy who acts girlish, for example, might elicit hostility or rejection from his father and therefore would not want to identify with this masculine figure. Thus, sexual orientation may have its origins in very early hormonal events, and these may shape behavior and elicit reactions to that behavior. In one longitudinal study (Green, 1987; Green, Neuberg, Shapiro, and Finch, 1983), young boys referred to a clinic because their parents were worried about their preference for wearing girls' clothes, playing with girls' toys, and doing feminine things, were observed and tested. When dressed to conceal their sex, these boys threw a ball, walked, ran, and told a story in ways that were *neither* markedly feminine nor markedly masculine. In adulthood, 29 out of 43 turned out to be homosexual or bisexual. Only one factor in the boys' environments seemed to correlate with later homosexuality: the absence of the boy's natural father in the family. Why? It may be that although biology disposes a boy toward homosexuality, the presence of a strong father inhibits its expression. The pattern of feelings, attitudes, reactions, and thoughts that makes up any person's sexual orientation takes shape over the course of childhood, is reshaped during adolescence, and typically is confirmed in adulthood. The course of development, in this as in other areas, involves both biology and culture, nature and nurture.

Health problems

Because of the physical changes we have discussed, adolescence ushers in its own special health problems and hazards. Sexual activity brings teenage pregnancies and sexually transmitted diseases. Cigarette smoking, drug use, and drinking alcohol become fairly common. Violence, accidents, self-imposed starvation, and suicide take adolescents' lives (Schroeder, Teplin, and Schroeder, 1982).

Pregnancy, childbirth, and contraception

Mature enough physically to conceive and bear children, few adolescents are socially and emotionally mature enough to take good care of themselves when pregnant, to meet the intense demands of an infant, or to support themselves and their child financially and emotionally. When they do become pregnant, many adolescent girls do not receive adequate prenatal care. Only 40 percent of pregnant 15-year-olds get prenatal care in the first third of their pregnancies, half the number of those over 25 who seek prenatal care (Ventura, 1977). Over and above these problems of psychological maturity, pregnant adolescents face serious physical risks. It is a cruel irony that although adolescent girls can conceive, their bodies are too immature for problem-free pregnancy and delivery. The medical risks for a pregnant teenager are twice those of women who are pregnant in their 20s. Teenagers face the risks of toxemias of pregnancy, long labor, stillbirth, babies of low birth weight, and premature delivery. Recent studies show, however, that many of these complications can be avoided if the pregnant adolescent is identified early and given nutritional, social, prenatal, and perinatal care (Evrard and Gold, 1985).

When teenagers marry because the girl is pregnant, the marriage is likely to break up within six years (Alan Guttmacher Institute, 1976). But more often, the girl chooses to raise the child alone. Few teenagers who are about to become parents have married in the last decade, and out-of-wedlock births have increased (Alan Guttmacher Institute, 1981). Either way, the decision to go ahead and have the baby is often based on storybook notions of becoming a parent and living happily ever after. In the short run, having the baby may satisfy the adolescent's longing for a sense of adult worth. In the long run, when teenagers have babies, both teenage father and mother end up with less education, poorer economic opportunities, and, usually, more children than they want or can afford to support (Card and Wise, 1978). They feel sadder and more tense than those who wait to have their children (Brown, Adams, and Kellam, 1981).

Adoption is an alternative that few teenagers choose: 94 percent of the teenagers who give birth keep their babies. The frustration of lives pre-

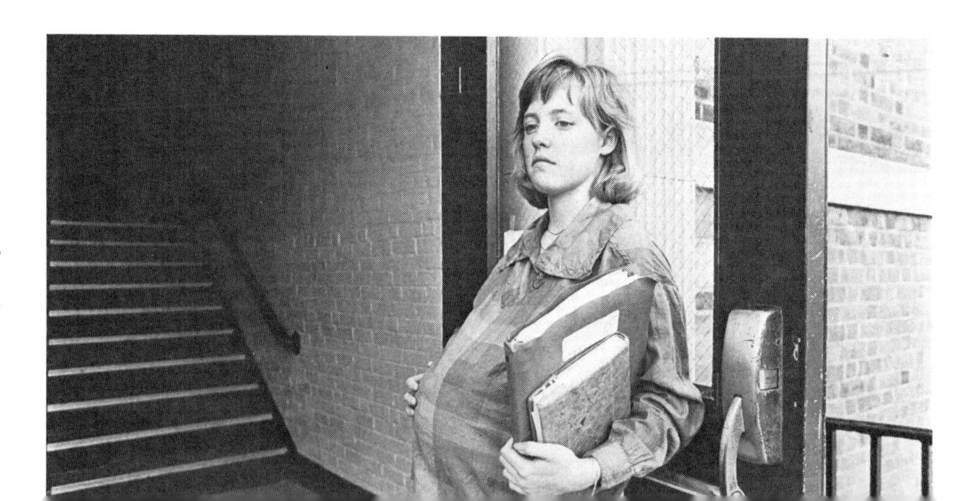

The cruel irony is that although adolescent girls can conceive, their bodies are not mature enough for problem-free pregnancies and childbirth.

maturely harnessed to the care and feeding of other utterly dependent human beings soon catches up with these young mothers and their children. Teenage parents have a high incidence of child abuse; their children in turn are more likely to develop behavior problems and to add to the numbers of those who will themselves later have difficulty being self-sufficient people and loving parents.

It is another cruel irony that although the majority of adolescents in this society are sexually active, they are given neither the information nor the support they need to make intelligent choices about their sexual activity. Parents are shy, teachers are distant, peers are uninformed, birth control is forbidden, illegal, or inaccessible, and sex education is scarce. Although teenagers can make abstract decisions about medical and psychological problems that are not different from adults' (Weithorn and Campbell, 1984), many people contend that adolescents are too immature to make practical decisions about their sexual activity or its consequences. The tug of war on this politically touchy issue is reflected in the contradictory judicial rulings about whether adolescents should have access to contraceptives, whether they are competent to consent to abortions, and whether their parents must give their consent as well. Some argue that requiring parents' consent for contraceptives will prevent many adolescents from using contraception and therefore will increase pregnancy rates. One study of 1200 teenagers showed, in fact, that 23 percent said that they would not use family planning if their parents had to know; 18 percent said that their parents did not know that they were using contraceptives (Torres, Forrest, and Eisman, 1980).

The United States is the only developed country in which teenage pregnancy rates are rising. Compared to 37 other developed nations, including Holland, France, and Canada, the rate of teenage pregnancy, abortion, and childbearing is highest by far in the United States. For every 1000 girls between 15 and 19 in the United States, 96 get pregnant, compared with 43 in France, 35 in Sweden, and 14 in Holland (Mall, 1985). Where attitudes toward sex are more liberal, where sex education is more thorough, and where contraceptives and health services are more readily available, the rates of teenage pregnancies, abortions, and childbirths are lower.

Many adults misguidedly fear that giving teenagers information about sex, contraception, or the right to abortion will increase their rates of sexual activity, pregnancy, abortion, and childbearing. Researchers have shown that sex education plus access to family planning does not prompt adolescents to begin having sex and does not increase pregnancies among sexually experienced adolescents; it does the reverse (Furstenberg, Moore, and Peterson 1985; Zelnick and Kim, 1982). Many high schools in the United States do not offer sex education, and of those that do, the majority do not include discussions of contraception. As a result, although over the past fifteen years increasing numbers of adolescents have used birth control, they still often use the least effective forms such as withdrawal and the rhythm method rather than pills, IUDs, condoms, diaphragms, or vaginal sponges (Mecklenburg and Thompson, 1983; Zelnick and Kantner, 1978).

One reason that adolescents are poor users of contraception is that many have only a vague grasp of the "facts of life." Young adolescents are too cognitively immature to understand the complexities of the unseen processes that make up sexual reproduction. Few know when during their menstrual cycles females are fertile. Some adolescent girls mistakenly believe that they cannot get pregnant, that they are sterile, that they are too young, or that it requires many sexual encounters to conceive. Many are too embarrassed to buy contraceptives or simply cannot find them. Some are too shy to touch their genitals. Some suspect that contraceptives are wrong or interfere with sexual pleasure and spontaneity. Of course, pregnancy is not always a result of ignorance or naïveté. Some adolescent girls want to get pregnant to bolster their self-esteem, to prove that they are women, to test their boyfriend's love, get attention or gain independence, or to assuage their loneliness; or they want to because their friends are doing it or they think that their families expect it of them (Chilman, 1979). Nevertheless, at least one study shows that most teenagers do not plan to get pregnant, nor are they happy when they find out that they are. When they are given instruction in contraception, they are less likely to get pregnant (Furstenberg, 1976).

Just where do teenagers get their information about sex? Most teenagers must piece together bits of sexual lore from three principal sources: their parents, their friends, and books. Their peers are teenagers' major source of information about sex. Teenagers talk together about intercourse, contraception, homosexuality, masturbation, prostitution, and ejaculation, and they are virtually the exclusive source of information for one another on petting (Thornberg, 1981). Parents impart their sexual attitudes to their children in more subtle and indirect ways. From the time their children are babies, parents express their attitudes toward sex in the way that they clean their children's genitals, teach them the culture's rules about modesty, and mislabel or ignore their sexual acts. When they are young and inquisitive, children may get special books from their parents about the "birds and the bees," many of which are literally that. But once the children are old enough to be interested in sex, most of the talk stops. Talk about intercourse and contraception is rare.

> The only thing my mother's ever told me about sex is, "You'll just know when it's time." Big help she is.—Megan, age 15

Parents are embarrassed, adolescents sense it, and so both parties withhold information about sex. In one study, many of the teenagers reported that their parents were so "uptight" about sex that they told them only what they thought they could stand hearing. A mere 21 percent asked their parents' advice about sex (Sorenson, 1973). Parents not only find it difficult to talk to their adolescent sons and daughters about sexual matters; they also overestimate what the adolescents know. They rarely go back over the rudimentary information that they gave their young children years before to clarify, explain, or correct the inevitable gaps and misimpressions. It is not surprising then that much of what adolescents know or believe about sex is inaccurate.

Parents are not only silent on the subject of sex; they also disapprove—not always silently—if their adolescent children are sexually active (LoPiccolo, 1973). Even parents who themselves had sexual intercourse before they were married tend to disapprove of it in their adolescent sons and daughters. In one study, although 30 percent of the mothers reported that they had had intercourse before marriage, only 3 percent approved of it in their daughters, and only 9 percent approved of it in their sons. Of the fathers who had had intercourse before marriage, only 10 percent approved of it in their daughters and 20 percent in their sons (Wake, 1969).

> My father is driving me crazy with all his talk about making sure that I keep a "good reputation." I nod and make noises like I agree with him, because I don't want to worry him, but boy is he out of it.—Erica, age 15

The adolescents who listen to their parents are most likely to remain virgins. One study of Canadian university students showed that half of those who were virgins said that their parents were the most important influence on their attitudes. But of those who had had intercourse, only 30 percent ranked their parents that highly. None of the students considered his or her parents major sources of factual information about sex (Barrett, 1980).

In families where parents and children do talk openly and fully about sex, where parents are openly affectionate with each other, and where parents explicitly forbid their daughter to have sex, adolescent daughters are less likely to be sexually active (Darling, 1979; Lenney, 1985; Lewis, 1973; Miller and Simon, 1974). Family discussion of sex is likely to increase the sexual activity of sons, but whether discussions with parents make them more responsible about matters like contraception remains an open question (Darling, 1979). In late adolescence especially, friends' attitudes may be more influential than parents' (Spanier, 1976). Emotionally intimate couples are likely to grow sexually intimate as well, no matter how their parents feel, and the likelihood grows the longer the relationship lasts.

Alcohol, drugs, and smoking

By the end of tenth grade, 90 percent of all adolescents have drunk alcohol (Figure 14.5). A recent survey showed that 40 percent of the high school juniors in Orange County, California, drank alcohol at least once a week (CBS News, June 1, 1984). In another survey, 41 percent of high school seniors had drunk five or more drinks in a row within the previous two weeks (Johnston, in Mervis, 1985). A sizable number of adolescents—as many as half a million in this country—are problem drinkers (Chafetz, 1974). Most adolescents do their first drinking at home and learn their drinking habits from their parents. Children are likely to drink moderately if their parents do, just as children are likely to drink excessively or to abstain altogether if their parents do (U.S. Public Health Service, 1974).

> I sometimes drink a beer with my dad when I visit him and we're watching a game together or something.—Jason, age 15
>
> I go drinking with the guys every weekend. We get a keg and party at the beach. My parents stay home and get sloshed on the hard stuff.—Brandon, age 17

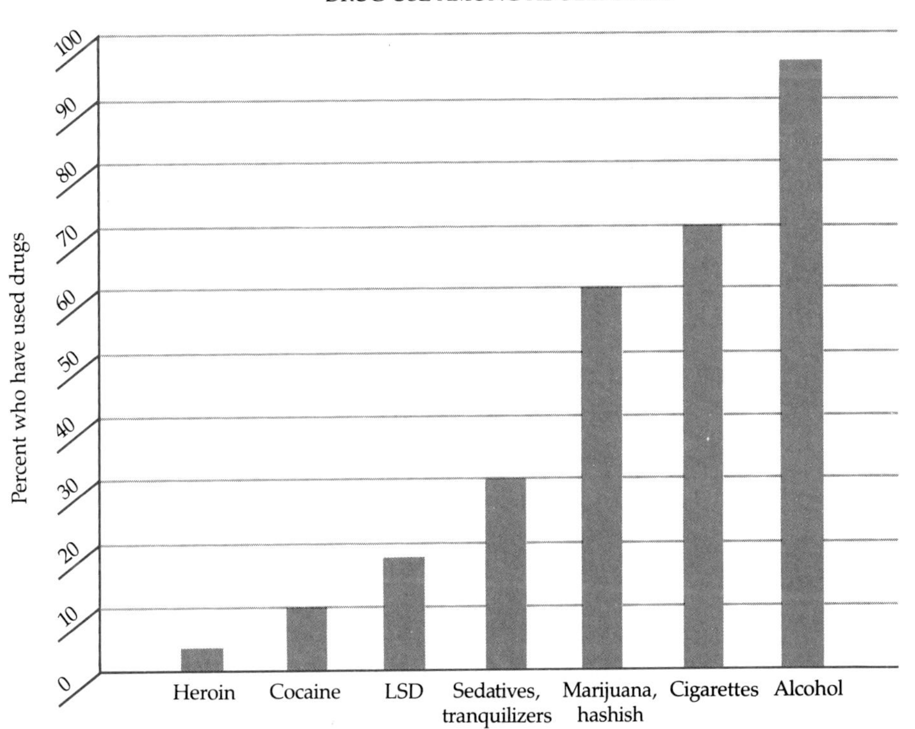

Figure 14.5 ***These percentages of adolescents graduating from high school have used the drugs indicated, without a doctor's order, at least once (Johnson, in Mervis, 1985; Johnson and Bachman, 1981).***

Adolescents who are most likely to drink too much are dependent and rebellious and have poor relations with others (Jones, 1968).

Many adolescents also use marijuana. Young adolescents are especially likely to smoke marijuana if their friends do (Brynner, 1980), and many begin to use drugs when they are most vulnerable to peer pressures—at ages 11 to 13 (see Figure 16.2). Recent surveys show that 40 percent of high school juniors and seniors have tried marijuana, 25 percent use it sometimes, and 5 percent use it every day (CBS News, June 1, 1984; Johnston, O'Malley, and Bachman, 1985). These figures are substantially lower than those of a decade ago, and the number of high school students who disapprove of marijuana smoking—85 percent—is substantially higher.

Although it is not true that marijuana dooms anyone to heroin addiction, it is true that teenagers who use one drug are likely to use several of them and to drink as well. In the last few years, the number of adolescents using marijuana, sedatives, and tranquilizers has decreased, but the number using heroin and other opiates has increased (Johnston, in Mervis, 1985). Twenty percent of adolescents have tried cocaine; 12 percent use it at least once a year; and 6 percent use it every month.

Just which teenagers are prone to using drugs? The profile is this: those most likely to smoke marijuana or use other drugs are not interested in school or academic achievement. They strongly value their independence and tolerate deviance. They do not care much for religion. Their parents are distant, unsupportive, uncontrolling, and express little disapproval of drugs. Their fathers are less likely to be well adjusted, to hold traditional values, or to be conventional and religious. Their friends strongly support drug use (Brook, Whiteman, Gordon, and Brook, 1984; Jessor, in Mervis, 1985; Stern, Northman, and Van Slyck, 1984; Tec, 1972).

One study of the factors that predict teenagers' drug use suggests that personality, peers, and family *all* operate (Brook, Whiteman, and Gordon, 1983). Teenagers with personality traits that make them likely drug users may do so even though their families and peers try to persuade them not to. Teenagers with poor family relationships may use drugs, even if their personal values and their peers' opinions are against drugs. Pressures from peers to use drugs can override personal antidrug values and good family relationships. An adolescent under pressure on all three fronts—a vulnerable personality, poor family relationships, and pressure from friends—is a prime candidate for using drugs.

What difference does it make if an adolescent does use drugs? Movies like *Reefer Madness* and antidrug campaigns over the last 20 years have implied that the first puff of marijuana smoke leads inevitably down the path of perdition and straight to heroin addiction. Such exaggerations were designed

Ninety percent of America's high school students drink at least occasionally with their friends or at home.

to scare people away from drugs. But drugs do cause real problems. They can damage the central nervous system and a person's abilities to pay attention and remember. Adolescents who smoke marijuana regularly may grow unresponsive, alienated, and passive. This passivity, in fact, makes marijuana smokers less likely to commit crimes than people who use other kinds of drugs. In this regard, one study has shown that as adolescents smoke more marijuana, the likelihood of their committing crimes drops (Gold and Reimer, 1975). Drugs and alcohol, however, are responsible for an estimated one-half of all the fatal accidents in which teenagers are involved. A recent study by the state of New York (1986) revealed that raising the drinking age from 18 to 19 in 1982 resulted in a 46 percent decrease in driving while intoxicated and a 21 percent decrease in crashes causing deaths and injuries among 18-year-olds.

By seventh grade, one-third of the adolescents in this country have tried smoking cigarettes. About 15 percent of high school seniors are regular smokers (Bachman, 1982).

> On a class trip in eighth grade, I sat in the back of the bus with a friend who had promised to teach me to smoke. The first few cigarettes tasted horrible. But then I got used to them. I liked the way I looked with a cigarette in my hand. I've been smoking ever since.—Sarah, age 16

Cigarettes, too, are harmful to health. Moreover, unlike marijuana, which can create psychological dependency but usually does not create physical addiction, the tobacco in cigarettes is physically addictive. Quitting a cigarette habit is a form of drug withdrawal that is painful and difficult. The best solution to cigarette addiction is never to start smoking in the first place. What then is the best way to keep children from starting to smoke cigarettes? In one recent study, researchers found that the most effective smoking prevention program for seventh graders was one taught by a fellow student and focused on things like appearance, bad breath, and exclusion from public areas (Murray, Johnson, Luepker and Mittelmark, 1984). (It worked only for those students who had not tried cigarettes.) More efforts like this may continue the downward trend in smoking among adolescents (Bachman, 1982).

Eating disorders

When 2000 high school students responded to a questionnaire (Kagan and Squires, 1984), 2 percent showed patterns of seriously disordered eating. Seven percent of the students (11 percent of the girls) were ''emotional eaters''—eating when they were depressed or anxious, for instance. Twenty percent went on binges at least once a week, 5 percent purged themselves afterward by vomiting, and 27 percent felt out of control about eating. In college, these numbers are likely to rise. For example, one-quarter of the women in introductory psychology and sociology courses at one midwestern college binged (Thompson, 1979). During freshman year in college, when students have to cope with separation from their families as well as adjust to a new academic environment, many of them gain weight. Few college men gain weight, because they unwind by exercising. But to cope with the pressures of classes and exams, college women drink, smoke, talk, and eat chocolate, milkshakes, burgers, and pizza (Little and Haar, 1985).

Finding themselves plumper than Brooke Shields, then, some girls may start on a spiral of dieting and gorging themselves. When adolescent girls with shaky self-images look in the mirror, they may see fat where no one else does and embark on a series of diets to control their ''overweight.'' Some diet strenuously and then, feeling starved and deprived, binge on thousands of calories in ''comfort foods'' like cookies, ice cream, cake, and doughnuts until they can stuff themselves no more. Then they make themselves vomit; some

Anorexia nervosa is a severe eating disorder, a form of self-starvation, which tends to afflict girls who have been good students and compliant daughters.

purge themselves with laxatives. This pattern of alternate starving, gorging, and purging is called **bulimia**. Bulimia takes both a physical and an emotional toll. The constant vomiting and purging play havoc with the digestive system, and the acidity of vomit corrodes tooth enamel. Many bulimics are secretive, ashamed, and depressed—some of them severely enough to try suicide (Cunningham, 1984).

A less common but more serious eating disorder is **anorexia nervosa**. Many adolescent girls are unhappy about their weight, but one in 300 starves herself and engages in bursts of compulsive exercise until she loses one-quarter or more of her body weight. Most anorexics are girls who are white, middle-class, affluent college students. They are pursuing to an exaggerated degree the thin body idealized in this society. Roughly half of anorexic women alternately binge and starve—the pattern of bulimia. Most are perfectionistic about their appearance, their schoolwork, and their family relationships. Laboring hard to be compliant daughters, excellent students, and beautiful girlfriends, and struggling against impending independence from their families all at the same time, they crumble (Bruch, 1969). Many feel out of control, without a sense of mastery over their own behavior, as they labor mightily to please others rather than themselves (Boskind-Lodahl, 1976). Their good grades are for their parents, their good looks for their boyfriends. With no firm sense of self-worth, anorexic women can set no bounds to their dieting. No matter how skinny they are, one look in the mirror convinces them that they are still ''too fat.'' Like many dieters, they grow preoccupied with food, reading recipes, preparing elaborate meals for *others*, counting calories. Without the necessary body weight, many anorexics stop menstruating. Starved and malnourished, some develop symptoms of dehydration, erratic heart rhythms, and other life-threatening conditions. Some anorexics must be hospitalized and force fed. One in five dies (Brody, 1982).

Physical maturity and fitness

Most of the health problems that we have discussed in this chapter affect a minority of adolescents. Most adolescents adjust to their new bodies and new capabilities with good health and good judgment. Some take advantage of their new capabilities to excel in sports and athletics. Gymnastics, tennis, basketball, and boxing, all of which demand flexibility, a slight body, quick reactions, speed, and strength, can be performed best by adolescents. On the average, female swimmers are at their fastest at age 13, male swimmers at about 18 (DeVries, 1980). In general, physical performance improves from early childhood straight through to the late teens, when it is at its height (DeVries, 1966). Then some physical functions—such as maximum oxygen consumption—begin to decline. For most, adolescence is a healthy, exciting time.

Summary

1. Puberty is the set of biological changes that transform children into sexually mature young adults. Puberty marks the beginning of adolescence, the stage of life characterized by social, emotional, and cognitive changes as well.
2. Puberty initiates a rapid growth spurt in height accompanied by the growth of muscles, internal organs, and bones. Secondary sex characteristics develop, primary sex organs and associated structures enlarge and become capable of reproduction, and adolescents' bodies take on the characteristic shapes of men and women.

3. A year or two after an adolescent girl's growth spurt has peaked, and once she has reached a critical body weight of about 100 pounds, her first menstrual period—menarche—begins.
4. The biological changes of puberty make males stronger than females, but cultural attitudes intensify these biological differences. Cultural attitudes also color girls' expectations for and experience of their menstrual cycles.
5. The changes and timing of puberty affect adolescents' behavior and self-image. Those who mature early may differ from those who mature late not only in physical traits but also in social, emotional, perceptual, and behavioral traits.
6. Sexual maturity and sexual activity mean that many teenage girls get pregnant. Most are not equipped physically, emotionally, or financially to meet the needs of an infant. The lack of sex education and the reticence of adults to discuss sex or contraception mean that many teenagers are ignorant of or have no access to effective means of preventing unwanted pregnancies. Many teenagers rely on the less effective methods like withdrawal or rhythm for preventing pregnancy. Many are frightened or shy of taking birth control pills or using other effective forms of contraception.
7. The health problems that afflict many adolescents include use of drugs like nicotine, alcohol, marijuana, and cocaine. Personality factors, peers, and family all influence whether adolescents use drugs.
8. The physical changes of puberty typically usher in a period when adolescents worry about their appearance. Eating disorders such as bulimia—overeating followed by purging—and anorexia nervosa—self-starvation—plague many young women who diet to extreme to meet a cultural ideal of thinness.

Key terms

adolescence
puberty
growth spurt
secondary sexual characteristics
menarche
body image
late maturers
early maturers
sexual scripts
withdrawal
rhythm method
bulimia
anorexia nervosa

Suggested readings

ALAN GUTTMACHER INSTITUTE. *Eleven Million Teenagers.* New York: 1976; and *Teenage Pregnancy: the Problem that Hasn't Gone Away.* New York: 1981. Excellent charts, statistics, and studies concerning the ongoing problems of teenage pregnancy, prenatal care, and parenthood.

BELL, ALAN F., WEINBERG, MARTIN S., and HAMMERSMITH, SUE K. *Sexual Preference.* Bloomington: Indiana University Press, 1981. A new perspective that stresses the deep-rooted, multifaceted nature of the choice to be heterosexual or homosexual, in the context of an in-depth survey of San Francisco homosexuals and lesbians.

BROOKS-GUNN, JEANNE, and PETERSEN, ANNE (Eds.). *Girls at Puberty: Biological and Psychological Perspectives.* New York: Plenum, 1982. Recent articles on the biological, social, and psychological components of puberty in young women.

BRUCH, HILDE. *Eating Disorders.* New York: Basic Books, 1973. Anorexia nervosa, obesity, and other eating disorders explained by a woman who has done extensive research in this field. She pays particular attention to their developmental roots.

COLES, ROBERT, and STOKES, GEOFFREY. *Sex and the American Teenager.* New York: Harper and Row, 1985. A thoughtful examination of teenage sexuality, which is not just a survey of sexual behavior but an explanation of how today's teenagers feel about what they do.

KAGAN, JEROME, and COLES, ROBERT (Eds.). *12 to 16: Early Adolescence.* New York: Norton, 1972. A collection of informative essays by experts in the fields of physical growth, sexuality, and cognitive changes. Thomas J. Cottle's "The Connections of Adolescence" and Tina DeVaron's "Growing Up" offer the personal insights of adolescents.

chapter fifteen

Cognitive development

Adolescence is the stage not only for dramatic physical changes but for dramatic cognitive changes as well. It is during adolescence that children become able to think about the possible, not just the actual, to discuss the world as it *might* be, not just as it *is*. It is during adolescence that children become able to reason abstractly and speculatively, to think of hypotheses and imagine their logical consequences, to discuss abstract topics like love, work, politics, religion, and the meaning of life—and to know what they're talking about. Words that were learned in early childhood—*love, country, parent*—take on new, symbolic meanings. Adolescents can construct theories about literature, philosophy, and morality. They can think in general terms and understand how history has affected current events and how current events will affect the future. Understanding this, they are likely to question the validity and goals of social institutions and human actions. They can imagine the world not only as it might be but as it *ought* to be, and many are idealists with strong ideas of how to make the world a better place. Adolescents can grapple with ideological goals—"No more nukes!" "Down with abortion!"—favor one political candidate over another, follow a spiritual leader, and think realistically about the work they may do as adults.

In this chapter, we will discuss the kinds of thinking that first become possible during adolescence. We will discuss what they mean, when they occur, and how they develop.

Scientific and logical thought

Although developmental psychologists had recognized for some time that the thinking processes of adolescents were different from those of younger children, these differences were not studied systematically until Piaget began his investigations. Piaget studied the development of adolescents' capacities for logical thought, by focusing primarily on their understanding of physical science.

To do so, Piaget made up many tasks, or problems, that involved scientific principles. Then he asked children of different ages to solve the problems. Children's and adolescents' ways of going about solving the problems revealed differences in their thought processes. In one task, for example, subjects were shown how to make a pendulum from a set of weights and a string. They then were shown how to change the weight, lengthen or shorten the string, adjust the height from which the weight on the pendulum was released, and push firmly or gently on the weight when it was released.

The development of scientific, logical thought means that adolescents can explore hypotheses and carry out experiments.

Finally they were asked to figure out what made the pendulum swing faster or slower. In another task devised by Piaget, children were shown four beakers, each containing a clear, odorless liquid, and a bottle labeled *g*, also containing a clear, odorless liquid. The child then was shown that by adding a few drops of *g*, some mixture of the other liquids turns yellow. The problem was to find out which mixture this was. With tasks such as these, Piaget probed the thinking of adolescents and compared it with that of younger children.

Adolescent thought had several distinctive qualities, Piaget discovered (Inhelder and Piaget, 1958). First of all, adolescents differed from younger children in their ability to go beyond here-and-now reality to the realm of the possible. In middle childhood, children approached Piaget's problems as concrete and practical, and they focused on what was there in front of them. But adolescents were more likely to examine the problem carefully to try to determine all the *possible* solutions and only then to try to discover the solution in that particular instance. An adolescent given the liquids in the beakers, for example, would realize that there were many possible solutions to the problem: liquid 1 + liquid 2 + *g*; liquid 1 + *g*; liquid 2 + *g*; and so on. A younger child would not consider all the possibilities. An adolescent asked by Piaget what people would do if there were no sun would consider the possibilities—that people would live underground, that they would eat cockroaches, that they would exploit other forms of energy, and so on. A younger child would flatly state that there *is* a sun or that it would be dark all the time.

Second, Piaget found that, based on this ability to think about the possible, adolescents could *hypothesize* and *deduce*. To solve the problems Piaget had set up, adolescents could apply the scientific method (see Chapter 2). They could:

- Analyze the problem.
- Hypothesize the correct solution.
- Deduce what empirical evidence would be necessary to prove the correctness of the hypothesis.
- Test the hypothesis by collecting the evidence.
- Come up with an alternative hypothesis if the evidence did not support the first hypothesis, and proceed as before.

An adolescent given the string and weights, for example, first might hypothesize that what really determined the speed of the pendulum's swing was its weight. She would test all the different weights and measure the speed of the swing. Finding that the weight did not affect the speed of the swing, she would then hypothesize that the length of the string determined the speed. She would vary the length of the string and measure the speed of swing. Finding that the length of the string did affect the speed, she would feel confident that her hypothesis was correct. But because she knew that she had to consider all the possibilities, she would also try pushing the weight harder or more gently to see whether these actions affected the speed of swing. Similarly, another adolescent given the beakers and liquids would understand that he had to take a systematic approach and test all the possible mixtures of all the liquids: First he would mix *g* with the contents of each beaker. Then he would systematically mix the contents of beaker 1 with *g* plus another liquid—liquid 1 + *g* + liquid 2, liquid 1 + *g* + liquid 3, liquid 1 + *g* + liquid 4. Then he would go through the same steps with the liquid in beaker 2, and then in beaker 3, then in beaker 4. Finally, he would systematically try all combinations of three liquids: liquid 1 + liquid 2 + liquid 3 + *g*, liquid 1 + liquid 2 + liquid 4 + *g*, liquid 1 + liquid 3 + liquid 4 + *g*, and liquid 2 + liquid 3 + liquid 4 + *g*. Many adolescents, Piaget found, were able to think like scientists, separating variables, generating and testing hypotheses, and weighing probabilities.

A third characteristic of adolescents' thinking discovered by Piaget was that, unlike younger children, they could judge the truth of the logical relation between propositions. During the concrete operational period, children could recognize the logic of a single proposition. They could formulate the proposition that there are the same number of cookies in two rows after one row had been spread out, and they could check this single proposition against concrete reality. Adolescents, Piaget found, could infer the logical relations between propositions, regardless of whether they were factually true. Suppose that an 8-year-old and a 16-year-old are told the following:

All suns are stars.
All stars are black.
Therefore, all suns are black.

The 8-year-old, knowing that the sun he sees is not black, declares the conclusion to be false. The child's reasoning is limited by what he knows to be true from concrete, active experience. He cannot suspend reality and listen to the logic underlying the conclusion. He cannot play with abstract ideas. But the adolescent can logically connect the three statements. *If* all suns are stars, and *if* all stars are black, then—*logically*—all suns must be black. He recognizes the logic of the propositions apart from their content and can combine, negate, and reverse even the most outrageous and blatantly false propositions.

Similarly, adolescents can figure out the answers to the following logical problems.

1. If this is Belgium, then this is Sunday.
   This is Belgium.
   Is this Sunday? [yes]
2. If this is Belgium, then this is not Sunday.
   This is Belgium.
   Is this Sunday? [no]
3. If this is Belgium, then this is Sunday.
   This is not Belgium.
   Is this Sunday? [can't say]

Children in middle childhood answer the first and second problems correctly but have trouble with the third because they cannot figure out the **inter-propositional logic** (Shapiro and O'Brien, 1970).

Piaget called this kind of scientific and logical thinking, which is first possible in adolescence, **formal operational thinking**. Whereas concrete operational thinking (discussed in Chapter 12) involves real, concrete objects, formal operational thinking involves abstract forms like propositions and hypotheses. A person with formal operational thinking realizes that logical forms and arguments have a life of their own regardless of what the person knows or sees. Because it involves thinking about propositions rather than about objects, formal operational thinking is said to be ''second-order'' thinking. Thinking about objects is ''first-order'' thinking.

Apart from Piaget's experiments, what evidence do we have about when and how logical thinking develops? Researchers have used Piaget's tasks and others like them to study the development of logical thinking in children and adolescents. No matter which task they have used, nearly all these researchers have found that between the ages of 11 and 15, children's thinking changes substantially (Neimark, 1975). In one study of several thousand adolescents, for example, researchers found that 14 percent of the 12-year-olds, 18 percent of the 13-year-olds, and 22 percent of the 14-year-olds succeeded on the pendulum task (Shayer, Kucheman, and Wylam, 1976; Shayer and Wylam, 1978). Logical thinking not only increased with age, this study showed, but

Formal operational thinking allows adolescents to form broad concepts and hypotheses, and these skills make for long conversations about themselves, society, love and sex, and the meaning of religion, justice, and life.

was tied closely to individual differences in intelligence. Whereas only 22 percent of the 14-year-olds from the general population succeeded on the pendulum test, 85 percent of the 14-year-olds from the top 8 percent in intelligence succeeded.

Levels of formal operational thought

Formal operational thinking does not appear all at once at the beginning of adolescence. According to Piaget, the first level of formal operations is thinking about alternatives, hypotheses, and possibilities. This kind of thinking usually begins in early adolescence (between 11 and 15 years). A second level, thinking *systematically* about possibilities, does not begin until later in adolescence. In the pendulum problem, for example, the early formal operational thinker tries out several possibilities. When he finds one that works, he stops. He does not test every possible variable systematically. The more mature formal operational thinker searches for what is *necessary* as well as what is *sufficient* to change the speed of the pendulum's swing, and therefore he tries out each variable and every combination of variables. Recent research suggests that there may be even more levels in the development of formal operational abilities than these two (Case, 1985).

As research in domains from physics to politics has shown, *many* older adolescents use formal operational thinking. But not *all* of them do so. Piaget (1972) himself pointed out that although most adolescents have the capacity for formal operational thinking, whether they actually use it depends on their experiences and their education in science and math. Children who have had formal science courses in school usually do better on Piaget's tasks than children who have not had such courses (McClosky, Caramazza, and Green, 1980). In fact, even preadolescent children can succeed on Piaget's tasks if they have been trained in scientific thinking. In one study, for example, 6- and 8-year-olds were tested on one of Piaget's tasks (Case, 1974). None of the children could solve the problem. Then the children were trained for four days on how to isolate variables and to test and keep track of possibilities in several problems that were similar to Piaget's task. When they were tested again on Piaget's task, 75 percent of the 8-year-olds did much better, and many solved it perfectly. Ordinarily, in our society, where the demand for this kind of formal thinking is rare before adolescence, young children do not succeed on tests of formal operations. The fact that they can be trained to do so supports the view that formal operational thinking depends on experience and education.

Formal operational thinking also depends on the situation. Even adolescents who use formal operational thinking *sometimes* do not use it *all* the time. Whether they use it depends on the demands of the task at hand. They are more likely to use formal operational thinking in more familiar tasks with simpler instructions and concrete objects (Surber and Grzesh, 1984). They are more likely to use it to solve physics problems at school than to solve personal problems at home. In adulthood, an auto mechanic may use formal operational thinking to figure out what is wrong with an engine but not to solve one of Piaget's tasks, whereas a scientist who uses formal operational thinking on a Piagetian task may not do so when his car breaks down.

Nor do *all* adolescents (or adults) use formal operational thinking even to solve Piaget's tasks (Neimark, 1975, 1982). In this sense, formal operational thinking is not universal. In nonliterate, technologically undeveloped cultures where there is little emphasis on science, adults generally do not use formal operational thinking as Piaget defined and measured it (for example, Cole, 1978). But to solve problems of everyday importance within their culture, even people from primitive cultures construct and test hypotheses. Members of the

lifespan focus

Magical thinking and formal operations

What role does chance play in your life? Can you control everything that happens to you? Is there a reason for everything that happens, even earthquakes, building collapses, and the winning of lotteries? It seems that the way that people answer questions like these tells a good deal about their cognitive development.

Piaget suggested that people's ideas about causality develop in conjunction with their logical abilities. Thus, he suggested, the preoperational child's ideas about causality and logic necessarily differ from those of the person who has developed concrete-operational or formal-operational thought. But Piaget's assumption was contradicted by findings from a study of women who belonged to a spiritual community (Lesser and Paisner, 1985). These women believed that random chance does not exist and that individuals control their destinies. On ideas about causality, the supernatural, and cognitive development, researchers compared the women from the spiritual community with other adults of the same age and education. Compared to the standard subjects, the women from the spiritual community had a stronger belief in the supernatural and a stronger tendency to believe that people control what happens to them.

Members of both groups could reason at the level of formal operations, but there were significant differences in the groups' ideas about why things happen. The standard subjects accepted the idea that there is some randomness in the universe—that luck determines who wins lotteries, who is born into poverty, or whether the train a person is riding derails. The standard subjects tended to assume a strong sense of personal responsibility in such areas of their lives as personal relationships, careers, and everyday decisions, and they sharply distinguished these areas from those they considered beyond human choice. In contrast, the women from the spiritual community believed that everything in the universe is fully determined. They believed in incarnation and in the idea that people choose the parents to whom they will be born. Asked what it would mean for the ceiling to fall as they were being interviewed during the course of the research, the women from the spiritual community said that the accident would have particular meaning for the people in the room; the standard subjects said that the accident would be due to something like faulty construction.

In some ways, the spiritual women's beliefs seemed almost *pre*operational—filled with the tendency to find messages in outside events or with a moral realism in which people are punished by events for their hostile feelings, for example. Their beliefs also seemed egocentric and magical—other qualities of preoperational thought. But despite these similarities to preoperational thought, the women's beliefs were clearly formal operational. Whereas preoperational thought is intuitive, the atypical women had sophisticated metacognitive awareness; they knew that they were thinking about their own thoughts.

The women from the spiritual community thus pose a challenge to Piagetian ideas about the linked development of logic and the understanding of causality. For although the women's ideas about causality were in some senses magical and seemingly preoperational, their level of reasoning was clearly sophisticated and formal operational. People's understanding of why things happen and logical relations may not, it seems, develop hand in hand. This study is interesting because it shows that there are discontinuities as well as continuities in lifelong development.

!Kung tribe of the Kalahari Desert, for example, use hypotheses in figuring out the most effective hunting strategies (Tulkin and Konner, 1973).

One researcher (Dulit, 1972) has suggested that even within our own culture, with its emphasis on science, some people use alternative patterns of thought that are also mature and effective for solving everyday problems. Some of them solve problems by applying standard solutions, which they have learned in other situations, rather than reasoning through each new problem systematically and formally. Other, artistic types leap intuitively to solutions, although they cannot explain their reasoning, as formal reasoners can.

Scientific, logical, formal thinking is an important cognitive development that usually occurs for the first time during adolescence. But whether a person uses it in a particular situation depends on the person's level of intelligence, training in scientific reasoning, and experience with similar problems, and on the importance, complexity, and familiarity of the situation itself. Although we often think of adolescent thinking as logical, scientific, and formal operational, clearly this kind of thinking occurs only in some of the people some of the time.

Information processing

With age and experience, adolescents and adults become increasingly systematic, scientific, logical, and abstract in their thinking. But just what allows them to move to higher levels of thinking? Some psychologists have suggested that improvements in information-processing abilities during adolescence account for these advances. They suggest that the main difference between children's and adolescents' reasoning is the completeness and thoroughness of their information processing. Children stop looking for possible solutions to problems, they suggest, because their memories become overloaded (Sternberg, 1977; Sternberg and Nigro, 1980). Adolescents can keep more information in short-term memory than children can and so are better at solving problems that draw on the ability to move back and forth from information just encoded to information previously stored. Children's memories may not hold enough information to allow them to exhaust the possibilities of a problem. Evidence to support this suggestion comes from a study in which researchers found that children's short-term memory for ratios increased with age (Case, 1985). Children between 9 and 11 years old could remember one ratio (for example, the number of yellow dots per green dot); those between 11 and 13 could remember two ratios (the number of yellow dots per green dot and the number of red dots per blue dot); those between 13 and 15 years, three ratios; and those 15 years and over, four ratios.

As well as having better memories, adolescents also select and plan their information-processing strategies better than children (Brown, 1975). Adolescents who use mnemonics and other cognitive strategies to organize information are likely to remember more than children who do not plan, and adolescents know better than younger children which information-processing strategies are helpful and select among them appropriately. Greater knowledge about the content of scientific and logical problems also enhances adolescents' reasoning and their abilities to solve formal and scientific problems.

Achievement in school

Our society places heavy emphasis on achievement and success, and therein for many Americans lies the value of abstract thinking, logical analysis, and effective writing. Well before adolescence, children understand the importance of achievement, fear failure, and feel nervous during tests (Feld, 1967). But the issue of achievement in adolescence is especially important because significant—and often irrevocable—educational and career choices are made during adolescence, and adolescents begin to comprehend the implications of these choices and of the possibilities before them. Adolescents have the cognitive capacities to understand and to explore systematically the long-term consequences of their educational and career choices.

> I don't know whether I should go to college right away or take a year off and work. If I work, I may lose some academic momentum, but I'll have a better idea of what I want to do with my life. If I go straight to college, when will I get another chance to take a year off?—Mike, age 17

School factors

How much adolescents know, how they think, and how well they write is determined to some extent by what they are taught in school, be that mathematics, creative writing, and computer programming, or gym, shop, and home economics.

> At the beginning of high school, we all got our course schedules for the first half-year from our homeroom teachers. We had all heard about "tracking," but those schedules were the first hard evidence of it. I was scheduled for French, algebra, history, and English composition every day, with music and art two or three times a week. But the guy sitting next to me was scheduled for remedial reading, algebra, shop, and study hall every day. Guess which one of us was on the college track.—Sam, age 16

In one study, 5000 adolescents between the ages of 13 and 17 were given 85 mathematics problems to solve (National Assessment of Educational Progress, 1983). How well they did on these problems was then analyzed in relation to a wide range of factors: the students' sex and race, their parents' levels of education and occupations, the amount and kinds of reading material in their homes; the students' year in school and grade point average; the geographic region and type of community where they lived; and the number of courses in high school algebra and geometry they had taken. Among these variables, the best predictor of the students' mathematics test scores was the number of years of high school algebra and geometry they had taken. The

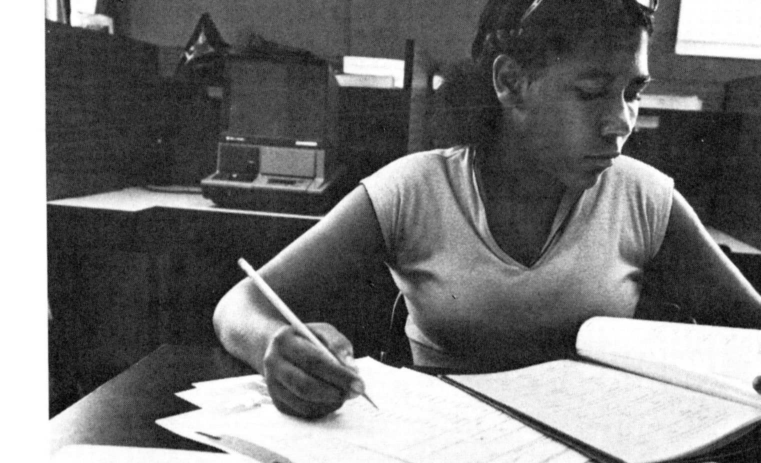

How much adolescents achieve in school depends on their own personal characteristics—including their abilities, motivation, and work habits—and on the characteristics of their families, such as social class.

more high school courses in mathematics that students had taken, the higher their scores on math achievement tests. This finding is consistent with findings from other tests (Jones, 1984), suggesting—no surprises here—that sheer amount of instruction in a subject is critical to what adolescents know.

So how much instruction are students getting in high school? In one study, in urban Chicago, high school students admitted that they paid attention to class instruction only about 40 percent of the time they were in class (Csikszentmihalyi, Larson, and Prescott, 1977). The rest of the time they daydreamed, whispered, did homework for other classes, and felt bored. Many said that they had trouble concentrating, felt self-conscious, and wanted to be elsewhere. Results from an extensive five-year study of high schools across the United States (Sizer, 1985) painted a similar picture. Interviewing many students and teachers, researchers found that high school students typically spend only five hours a day in classes. Most of that time is spent listening to teachers lecture, memorizing facts for tests (and promptly forgetting them), and earning credit by quietly going along with class routines. Little class time is spent in serious, rigorous, or original thinking. The researchers who conducted the study suggested that to improve the quality and quantity of instruction in high school so that students will achieve more, educators should:

- Insist that students know what of their schoolwork they must master and then clearly master it before they are allowed to graduate.
- Adjust achievement incentives to each individual student.
- Focus on teaching students how to think.
- Make curricula and school structure simpler and more flexible, and less specialized.

Personal factors

Of course things besides the quality and quantity of instruction received in school also affect how much an adolescent achieves. The adolescent's personal characteristics, for example, are an extremely important ingredient in this mix.

> Jason is smart. He remembers batting averages and TV commercials, and he's great at video games. But at the rate he's going, I'm not sure he's ever going to pass ninth grade. The teachers say that he rushes through his work, refuses help, and spends most of his time in class talking to the other kids or looking out the window. Now and then, he completes a project in a way that shows how bright and talented he is. That's what's so frustrating—he's *under*achieving.

Achievement motivation and standards

Underachievers are adolescents whose achievements in school and sometimes in other areas fall short of what could be expected of them, given their high level of abilities. Underachievers are likely to have IQs over 120 but school grades of Cs and below. Although children may not perform up to their potential in elementary school, underachievers usually are first identified in junior high, perhaps because that is when academic demands intensify. Underachievers shirk their homework, do not finish assignments, do not seem to want to do anything academic, and are socially and intellectually immature (McCall, 1986). They have little motivation to strive for success; they have a low **need for achievement**. Need for achievement is the motivation to perform well even in the absence of external rewards (McClelland et al., 1953).

A child's expectations about her success or failure in solving problems, as well as her teacher's expectations, strongly influence her performance in school.

Another aspect of achievement motivation that influences an adolescent's achievement is his or her own personal standards for achievement. For example, two students get a B− on a paper. One feels ashamed of herself for doing so poorly; the other feels proud of herself for doing so well. Their different reactions stem, in part, from their different standards for excellence.

Another aspect of achievement motivation is the adolescent's fear of failing—or succeeding—at tasks. Adolescents who feel anxious about taking tests or going for college interviews, for example, may be afraid of failing. Although some anxiety can help to focus attention and increase motivation and therefore improve performance in a task, too much anxiety is likely to impair performance. A high level of anxiety is especially harmful when a person is trying to learn something *new* or solve a *complex* problem (Spielberger, 1966). Consequently, fear of failing often interferes with school performance. In school and on tests, relaxed but alert attentiveness is most conducive to learning and achieving (Hill and Sarason, 1966). Conversely, an adolescent's fear of success also may undermine the need for achievement and lead to ambivalent or negative feelings about doing well in school or on tests.

Locus of control

Not only does an adolescent's motivation to succeed or fail affect his achievement, but so does his explanation of *why* he succeeded or failed. Adolescents may attribute their achievements to their own ability and effort, to luck, or to the difficulty of the task (Dweck and Wortman, 1980). Those who attribute their achievement to their own ability or effort are more likely to feel sure of their ability to achieve, to expect to succeed, and to do so.

> I did great on the history test, because I love history, and I really studied.—Nancy, age 15

Those who attribute their achievement to luck or to the difficulty of the task—factors beyond their personal control—are more likely to feel unsure of their ability to achieve, to expect to fail, and to do so.

> I got lucky last year and got all Bs. But the teacher liked me. I know I'm going to mess up this year.—Tim, age 15

Individuals who aim for a realistically high level of success and who feel personally responsible for their successes and failures are most likely to succeed.

Adolescents who feel that they are in control of their successes and failures are said to have an **internal locus of control**. Those who feel that the events in their lives are beyond their control have an **external locus of control**. To identify children's and adolescents' locus of control, psychologists have used the Intellectual Achievement Responsibility (IAR) questionnaire. The IAR poses questions like "When you get a high grade on a test, is it (*a*) because you studied hard?"—internal locus of control—or "(*b*) because the test was easy?"—external locus of control. Several researchers have found that children and adolescents who think that they are personally responsible for their successes, in fact, succeed most often. They do well in reading, math, and language, spend more time on homework, try longer to solve complex logical puzzles, and get higher grades (Crandall, Katkovsky, and Crandall, 1965; Franklin, 1963; McGhee and Crandall, 1968). Students who think that events are beyond their control are likely to get along poorly with their teachers and to attribute more negative qualities to both their teacher and themselves than students who think that they can control the events in their lives (Bryant, 1974).

When adolescents believe that nothing they do makes a difference, that they are not bright enough, that the work is too hard for them or that the teacher dislikes them, they expect to fail. Those who have failed over and over again may begin to believe that failing is inevitable and unavoidable. They sink into **learned helplessness**, an attitude that is likely to fuel the cycle of failures, and stop trying to achieve (Dweck and Licht, 1980). Adolescents who feel helpless are likely to dwell on their failures and the reasons for them. In contrast, those who expect to succeed prefer to move on to something that they can succeed in after they have failed (Diener and Dweck, 1978). In one study of the debilitating effects of learned helplessness, children were asked to do a task that was impossible. In the room with them was a person called a "failure experimenter." Later, the children were asked to perform other tasks with the "failure experimenter" present. The children identified as having learned helplessness failed on these tasks when the "failure experimenter" was present even though the tasks were well within the children's competence. But when the children with learned helplessness were taught how to take pride in their successes, they succeeded more often. The fact that this feeling of helplessness has been learned implies that it can, with training, be unlearned. When children are trained to connect their own efforts with their successes, they feel—and are—more in control (Dweck, 1976).

Gender

Some differences in achievement in school are related to gender. Boys are likely to achieve more—especially in the physical sciences—and the differences in achievement increase from middle childhood through late adolescence. But just why do these differences exist? In part, they may be the result of biologically based differences in spatial ability (discussed in Chapter 12). In part, the differences may be the result of differences in interests and experience. However, a recent study of the SAT scores of 40,000 high-achieving male and female adolescents showed that males scored higher than females in math, even when the comparison was limited to adolescents who had taken the same high school math courses and expressed the same degree of interest in math (Benbow and Stanley, 1983).

Socialization by parents, peers, and teachers no doubt contributes to the observed sex differences in achievement. Most adolescent girls get the message that they had better not do *too* well, that competition is aggressive and not feminine, and that achieving can threaten their relationships with boys (Bardwick and Douvan, 1971). What is more, girls are more likely than boys to have been socialized to feel that their failures are attributable to factors beyond their control, such as their lack of ability, and to be victims of learned helplessness (Dweck and Reppucci, 1973; Dweck and Licht, 1980; Nicholls, 1975). Adolescent girls are also more likely than boys to have few expectations of success, low standards of achievement, and considerable anxiety about failing (Stein and Bailey, 1973).

But socialization is unlikely to provide a complete explanation of gender differences in achievement. Among high-achieving adolescents, although girls achieve less than boys in science and math, they achieve more than boys in verbal tasks, especially writing (Benbow and Stanley, 1980; National Center for Educational Statistics, 1985). Is verbal achievement more acceptable for girls than achievement in science and math? Possibly. But other explanations are also possible. For one, the content of the material learned in science and math classes, on the one hand, and that learned in language and literature classes, on the other hand, may help explain the achievement differences between the sexes. In math classes, students are required to master new concepts as they move from arithmetic, to algebra, to geometry, to calculus, whereas in language classes, new concepts build on familiar skills, such as reading, spelling, and vocabulary. Because girls are more likely to fear failure, to feel that they cannot achieve, and to be debilitated by anxiety about their performance, they may be more likely to perform worse than boys in math and science, subjects that regularly call on students to master new concepts (Licht and Dweck, 1984).

Another explanation for boys' higher achievement rests on what happens in the classroom. As one recent study showed, boys dominate class discussion, call out answers more often, and get more of their teachers' attention than girls (Sadker and Sadker, 1985). Girls are encouraged to sit quietly and raise their hands, boys to assert themselves actively. Teachers often did things for girls but instructed boys how to do things for themselves. They tended to criticize boys' wrong answers, to help them to find the correct answer, and praise them for answering correctly. But when girls answered incorrectly, teachers tended to move on to another student.

No doubt many factors contribute to observed gender differences in achievement—aptitudes, abilities, interests, experience, socialization, motivation, and attention. These differences remain a focus of investigation—and concern—for researchers and educators. Most upsetting, despite attempts by feminists and equality-minded parents and teachers, the 1985 SAT figures showed girls falling further behind boys—in verbal as well as mathematical and scientific tests.

Family factors

Many of the differences in adolescents' achievement—including the differences related to gender—can be traced to family influences. Family factors are linked to adolescents' opportunities, abilities, and motivation to achieve.

The opportunities for achievement presented to an adolescent depend in part on the family's socioeconomic status and race. Poor and minority students are found in disproportionate numbers in the lower tracks in school (Featherman, 1980). Even the bright students from poor families are likely to achieve less than students from wealthier families if they know that they cannot afford to go to college. Adolescents from middle-class families consistently achieve more in school than those from poorer families (Garbarino and Asp, 1981). Despite attempts to close the gap in educational opportunities between social classes, social class remains a strong determining factor in an adolescent's school achievement (Featherman, 1980).

Adolescents from middle-class families are also likely to achieve more in school than adolescents from poorer families because of differences in their academic abilities. These differences stem from environmental and genetic sources. On the average, middle-class parents have higher IQ scores than parents from lower social classes, and their children are likely to inherit greater intellectual abilities. They also receive the advantages of living in an environment where the importance of learning and achievement is stressed, and good nutrition, health care, and lessons are provided (Featherman, 1980).

Middle-class parents also may instill stronger achievement motivation in their children. Adolescents from families in which parents expect and reward their children's achievement and autonomy have been found to have a higher need for achievement than adolescents whose parents are indifferent to their achievements (Rosen and D'Andrade, 1959; Winterbottom, 1958). In laboratory research, adolescents who have been exposed to a model who rewards himself or herself for excellence are likely to have high standards for achievement (Bandura, 1977).

> My parents have always told me if something is worth doing, it's worth doing well. Of course, I don't always like it when they tell me that. But I think it probably makes me think about trying to do a good job.—Natalie, age 15

Parents of adolescents who are underachievers are likely to be uninterested in education or in their children's school or to be interested but of the opinion that the school, not they, should make their children shape up (McCall, 1986).

Moral reasoning

Moral maturity

Another important change that adolescence brings is the development of greater moral maturity. What *is* moral maturity? Psychologists who subscribe to different theories of development would answer that question in different ways. Those who emphasize cognitive development might say that moral maturity is being able to make moral judgments that are just and fair. Those who accept a psychoanalytic explanation of human development would answer that moral maturity grows out of an internalized sense of guilt, which curbs impulses, lets a person accept blame, and urges the person to make good after doing wrong. Those who subscribe to social learning theory might answer that honest and upright actions characterize mature morality. But followers of each of these theories would agree that adolescence brings with it a higher level of moral maturity. The reasons that adolescents are more morally mature than younger children are many: their cognitive development allows them to think abstractly; their psychological development makes them question rules accepted by their parents and teachers; and their social experience exposes them to a variety of moral positions and moral dilemmas.

In adolescence thinking about moral issues like crime, punishment, and the law undergoes change. A 13-year-old boy might suggest violent, punitive ways to control crime:

> I think that I would . . . well, like if you murder somebody you would punish them with death or something like this. But I don't think that would help because they wouldn't learn their lesson. I think I would give them some kind of scare or something . . . these people who are in jail for about five years must still own the same grudge, then I would put them in for triple or double the time. I think they would learn their lesson then (Adelson, 1982, p. 8).

But later in adolescence, thinking moves away from this view of law as a way to suppress wayward behavior toward the view that the law protects and helps people. Older adolescents are more likely to believe that laws provide for harmony among people, that they make a nation ''a better place to live.'' Their ideas of justice are more benevolent. They are surer of themselves, more self-controlled, more consistently honest (or dishonest) than younger children. Compared to younger children, adolescents better understand their own reasons for acting as they do, sympathize more with others, and have a more highly developed sense of personal guilt and social injustice.

Adolescents who reach a mature level of moral reasoning make moral judgments based on their own personal values and standards, not on social conventions or the persuasion of authorities. They form their own ideas about moral issues—''Is segregation a moral system?'' ''Are the immigration laws of the United States fair?''—and reason according to principles of justice and social reciprocity. They acknowledge the rights of individuals to life and liberty as a higher moral good than even such hallowed principles as the interests of the majority.

According to Lawrence Kohlberg (1976), whose theory of moral development was discussed in Chapter 12, mature moral reasoning goes beyond the rules prescribed by social conventions; it is postconventional. But few adoles-

Many adolescents feel sympathetic distress and social guilt. These feelings compel them to embrace social causes and to worry about the future of humanity.

cents reach this level (Colby et al., 1983). Formal operational thought is *necessary* for postconventional moral reasoning, because postconventional reasoning requires abstract thought and the understanding that social conventions of right and wrong compete with equally plausible possibilities. But formal operational thinking alone does not *guarantee* postconventional moral reasoning. Most adolescents in the United States reason at the conventional level, the third and fourth of Kohlberg's stages of moral development. Some reason at even lower—preconventional—levels, Kohlberg's Stages 1 and 2.

Moral ideas and moral acts

What is the relation between moral reasoning and moral action? Some people may reason maturely on moral issues but act criminally. Others, whose level of moral judgment is low, never do anything illegal. Moral reasoning is far from an exact predictor of moral action.

In one study, researchers looked at the relation between moral thought and honesty in junior high school students. The students were interviewed about Kohlberg's moral dilemmas and also were tempted to cheat (Krebs and Kohlberg, 1973). Among the preconventional reasoners, three-fourths succumbed to temptation and cheated in at least one of the situations. Among the conventional reasoners, two-thirds cheated. Among the postconventional reasoners, only one-fifth cheated. In this study, the relation between moral reasoning and moral action was clear-cut. But in another study, the relation between more advanced moral reasoning and honest behavior held only when students were interviewed about the moral dilemmas *before* they had a chance to cheat (Krebs, 1967). Interviewed afterward, the more advanced reasoners cheated *more* than less advanced reasoners. Perhaps the interviews put the students on guard not to cheat. Perhaps they made the students think about moral issues and reinforced their tendencies to act honestly. In another study, this one of delinquent and nondelinquent adolescents, it was shown that the moral reasoning of adolescents who had committed antisocial acts was no different from that of those who had not committed antisocial acts (Hains and Ryan, 1983). But eleventh and twelfth graders who reasoned at higher levels on Kohlberg's dilemmas were rated by their teachers as more likely to take morally courageous actions (Gibbs et al., 1986). In sum, although there is a link between moral reasoning and moral behavior, it is tenuous. Moral and immoral actions depend on more than one's abstract judgment.

James Rest (1983) has proposed a model for integrating moral reasoning with moral actions. First, he suggests, people interpret a situation and surmise how their actions might affect others. They also interpret their own feelings in the situation. Once they have thought about alternative pos-

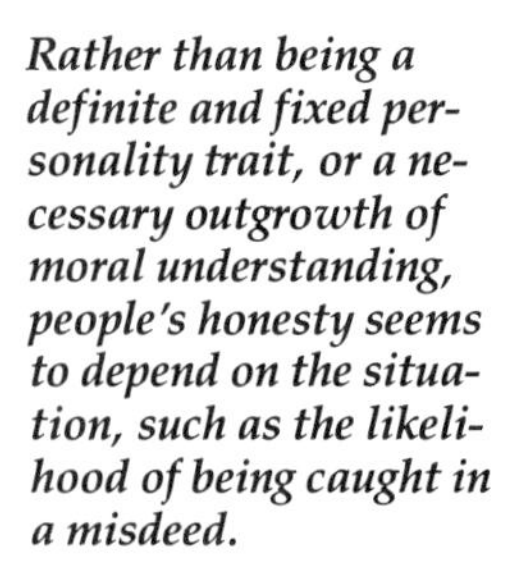
Rather than being a definite and fixed personality trait, or a necessary outgrowth of moral understanding, people's honesty seems to depend on the situation, such as the likelihood of being caught in a misdeed.

sibilities and how these might affect others and themselves, they judge and integrate these alternatives. Then people decide what *ought* to be done in the situation—Kohlberg's moral reasoning. Third, they decide what they actually are *going* to do. Often, values other than moral ones weigh heavily in a person's choice of action. Religious values, ambition, and self-interest are three such possibilities. There is often a gap between what people think they *should* do and what they *will* do. Finally, people carry out a plan of action, a process that involves planning a sequence of actions, surmounting obstacles, fatigue, and disappointment, resisting distractions and other temptations, and keeping the goal in mind. Recent research has shown that the likelihood of a person's acting morally depends on the stress in a particular situation, the cost to the person of acting morally, and the person's ability to assess the situation accurately and choose a practical solution (Haan, Aerts, and Cooper, 1985).

Adolescent egocentrism

Another outcome of the physical, cognitive, social, and psychological changes that take place during adolescence is **adolescent egocentrism** (Elkind, 1967, 1985; Inhelder and Piaget, 1958). Adolescent egocentrism is what makes adolescents think that they are more important and unusual than they really are. In their mind's eye they stand before an audience under a social spotlight. They imagine that the audience can see into their innermost thoughts and is hanging on their every word and deed. After hours fiddling with a hairstyle in front of a mirror, an adolescent girl may expect others to applaud her appearance. A single unsightly pimple can lead a boy to cringe in the certain knowledge that "everybody" will notice. To avoid going to school—and into locker rooms especially—where others might judge how they look, some adolescents develop school phobia (Rutter, 1980).

Whereas younger children can imagine the thoughts of existing people, adolescents can imagine the thoughts of hypothetical people—a second-order symbolic creation (Elkind, 1985). The problem is that at first adolescents fail to distinguish between the concerns of this **imaginary audience** and their own, personal concerns. Results from questionnaires given to hundreds of adolescents show that self-consciousness, which is presumably a reflection of heightened concern with an imaginary audience, is at its peak among 13-year-old girls and 15-year-old boys (Elkind and Bowen, 1979; Gray and Hudson, 1984).

As part of their egocentricism, young adolescents may imagine a **personal fable** for themselves. As if having dressed themselves in the kind of wonderful cape worn by fairy-tale characters, adolescents feel unique and indestructible. They believe that their thoughts and feelings are understood by no one, least of all their parents—"But Daddy, you just don't understand." They feel invincible, immune from death and disaster— "*I* won't get pregnant," "*I* won't crack up the car." They believe that their emotional experiences are new and unique. In failing to differentiate between what is new and thrilling to them and what is new in human experience, adolescents are likely to feel that no one has ever loved or hated as deeply as they—"But Mother, *you've* never been really in love."

What causes adolescent egocentrism? Originally it was suggested that adolescent egocentrism was the result of the onset of formal operational thinking, with its emphasis on speculation and abstraction (Elkind, 1967; Inhelder and Piaget, 1958). This suggestion was supported by studies showing that adolescent egocentrism peaked at the age that formal operational thinking began (Elkind and Bowen, 1979). But this is also the age that dra-

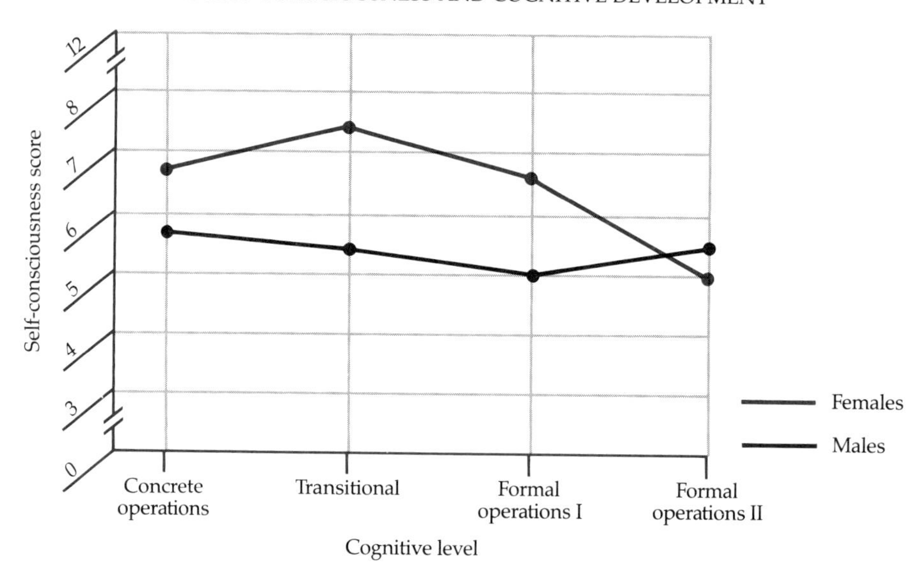

Figure 15.1 This graph shows the relation between measures of adolescent egocentrism and level of cognitive development. For girls, self-consciousness, which presumably reflects adolescent egocentrism, is most common during the transition between concrete and formal operations. For boys, self-consciousness is most common in the stage of concrete operations (Gray and Hudson, 1984).

matic physical changes brought on by puberty and dramatic social changes like heightened conformity to peers and conflict with parents take place. These changes also could lead to increased egocentrism. To test whether adolescent egocentrism is linked to formal operational thinking, researchers in one study gave adolescents tests of formal operational reasoning and logic and assessed their egocentrism (Gray and Hudson, 1984). It was expected that adolescents who were just making the transition from concrete to formal operations would be the most egocentrically self-conscious. The hypothesis was supported for girls but not boys (see Figure 15.1). Apparently, the ability to reason logically and abstractly in the physical domain does not, by itself, account for the increased subjectivity and self-consciousness of adolescence. Advances in social understanding may be even more relevant to the development of adolescent egocentrism (Lapsley and Murphy, 1985). The young adolescent is able for the first time to think about his own perspective, the perspective of another person, and the perspective of a third person observing the first two (Selman, 1980). With this ability, the adolescent can see himself as both actor and object and can imagine his interactions from the perspective of a generalized ''average'' member of the group observing him.

Fortunately for their *real* audience, adolescent self-consciousness diminishes after early adolescence. It is replaced, as we shall see in the next chapter, by new concerns, about identity and the future.

Summary

1. During adolescence, children can for the first time think abstractly, solve logical problems, and follow the scientific method. Piaget called this level of thinking *formal operational*. Formal operational thinking develops in most, but not all, adolescents by about age 15.
2. Academic achievement is predicted by factors in the school (like the number of courses offered), the family (such as its emphasis on education), and the person (for example, achievement motivation and learned helplessness).

3. No matter how it is defined, moral maturity does not appear until adolescence. Compared to children, adolescents better understand their reasons for acting as they do, can sympathize more with others, and have a more highly developed sense of personal guilt and social injustice.
4. Adolescents with formal operational thought often reason about moral dilemmas at a postconventional level. They base their moral judgments on their own personal values and standards, not on social conventions or the persuasion of authorities. They acknowledge the rights of individuals to life and liberty as a higher moral good than even such honored principles as the right of the majority. In the first stage of postconventional reasoning, people are oriented toward social contracts. In the second stage, they base their moral judgments on universal principles and unimpeachable ethics.
5. Moral reasoning is tenuously linked to moral actions; both moral and immoral actions depend on more than a person's awareness and abstract judgment.
6. Adolescents tend to think that they are more important and unusual than they really are and to imagine that they stand before an audience attentive to their every word and action. This is called *adolescent egocentrism*. Adolescent egocentrism may be a result of the increased cognitive and social understanding of adolescents, in which they can think not only about their own perspective but about that of others.

Key terms

interpropositional logic
formal operational thinking
need for achievement
internal locus of control
external locus of control
learned helplessness
adolescent egocentrism
imaginary audience
personal fable

Suggested readings

ERIKSON, ERIK. *Gandhi's Truth: On the Origins of Militant Nonviolence.* New York: Norton, 1969. A biography of the man who led India to independence through great leadership and passive resistance, written by one of the outstanding theorists in the field of development.

GILLIGAN, CAROL. *In a Different Voice: Psychological Theory and Woman's Development.* Cambridge, Mass.: Harvard University Press, 1982. An original theory of moral development in women, one which contrasts their moral and psychological orientations to those of men and is supported by data gathered in extensive interviews with women at various ages.

SIZER, THEODORE R. *Horace's Compromise: the Dilemma of the American High School.* Boston: Houghton Mifflin, 1985. Although Sizer's topic is the American high school, the larger subject of the book is universal—specifically, the ways in which society transmits education and cultural and social values to adolescents.

chapter sixteen
Social and emotional development

In some cultures, there is a clear demarcation between childhood and adulthood. The passage from one to the other, which in our society is long and complex, is quick. The onset of puberty is marked by public events and "rites of passage." Once the physically mature child has completed the rite, be it a circumcision, a ritual offering, a betrothal, or an elaborate ceremony, he or she is considered part of the adult world. In our culture, the gulf between puberty—physical maturity—and an adult social role—social maturity—yawns wider than ever. The period between them is almost as long as that between birth and puberty. Puberty occurs long before marriage, parenthood, or economic independence. For those seeking professional careers, 20 years may elapse between puberty and true economic independence.

The adolescent experience strongly reflects prevailing historical and cultural trends—in patterns of friendship and family life, opportunities for education, norms for leaving school and getting married, and so on. Because history and culture so thoroughly affect adolescents, it is important to remember that the information in this chapter is specific to contemporary adolescence in the United States. Views of adolescence and perhaps adolescent development itself change right along with adolescents' changing educational, economic, and sexual experiences, expectations, attitudes, fads, values, and norms.

Storm and stress

Adolescence is full of ups and downs as situations and moods shift from one hour to the next.

For some, adolescence today is a difficult, stormy time of life. For others, it is a relatively smooth transition between childhood and adulthood. Researchers were once divided into those, like G. Stanley Hall and Anna Freud, who characterized adolescence as a period of **storm and stress**—a stormy decade of emotional turmoil—and those, like Margaret Mead, who did not. One reason for these different views was probably that different observers based their conclusions on different groups of adolescents. Clinicians, seeing patients in psychotherapy, were likely to see adolescents who were experiencing a great deal of stress. Researchers who saw a more representative sample of "normal" adolescents would see less stress.

It is probably more accurate to say that some adolescents feel a great deal of stress, some very little, and the others a moderate amount. These differences stem in part from differences in individual temperaments and circumstances and in part from historical and cultural conditions. Although some adolescents experience more stress than others, almost all adolescents are subject to ups and downs in their moods.

> I was feeling on top of the world until I got that note in the cafeteria from my boy friend. He wants to break up, because he says he's feeling "too penned in." I went to the girls' room and cried until I had to be in English class.—Megan, age 15

> After school my father drove me to Boston for a college interview. I was really up on the way in, thinking about all the things I was gonna say to the interviewer. But when we got there, it was pouring rain, and I had to run about two blocks from the car to the interview office. I was so rattled that I got bummed out, and the interview was horrible. I *know* I won't get in.—Ted, age 17

One minute a teenager is basking in the glow of a remembered kiss, and the next he is made miserable by a teacher who tells him to stop day dreaming in class. One day, a teenager is feeling warm and friendly toward her family, and the next she is irritated by their demands. As he spends an evening alone, a teenager worries about whether anyone really likes him; the next afternoon, he is buoyed by his friends' infectious high spirits and highjinks.

In one recent study in Chicago (Csikszentmihalyi and Larson, 1984), 75 adolescents, 13 to 18 years old, were paged at random times during the day and night by "beepers" that they carried and were asked what they were doing, whom they were with, and how they were feeling. Within the bland and broad regularity of waking up, going to school, going to work, coming home, doing homework, and going to sleep, there were frequent and rapid changes of mood. From one hour to the next, adolescents went from elation to dejection, from excitement to boredom. For example, one boy in the study, Greg, feels happy as he walks home from school on Tuesday with a girl whom he is beginning to think of as his girlfriend. But he is already "going with" someone else, and as the week goes on this worries him. He wakes up happy on Wednesday, but chemistry class makes him feel down and confused. After class, Greg happily socializes with some friends and admires graffiti he has written: "*Q*: Are we not men? *A*:We are DEVO!" He cheerfully thinks about the song that these lines come from. But 45 minutes later, in typing class, he is emotionally unhinged. By midafternoon, Greg is very happy as he walks by himself to work, because he has just kissed his new girlfriend. Work and the rest of the night are dull. On Thursday, Greg feels guilty and "suicidal" when he thinks about not taking his old girlfriend to the prom. But later that night, he happily plans the costume to wear to the annual Senior Banquet on Saturday. Over the week, Greg alternates between bored unhappiness in class and happy excitement when he is with his friends. Although adults, too, alternate between boredom and excitement in their daily lives, it is the rapidity of the change and their extremes of dejection and exhilaration that mark instability of moods as characteristic of adolescence.

Adolescents in the Chicago study reported far wider variations in moods than adults did in a comparable beeper study (Larson, Csikszentmihalyi, and Graef, 1980). They were more likely than adults to feel euphoric—to see the world as a perfect place—and to be vulnerable to the pain and chaos of unexpected events. Not only were their moods more extreme, but these extreme moods were shorter-lived than adults'. Within 45 minutes of feeling dejection or exhilaration, Greg and the other adolescents studied had usually returned to a neutral mood. But adults were still feeling up or down two hours later. In one respect, the adolescents' moods did resemble the adults'. Their moods were equally predictable from what they were doing at the time. But adolescents move more quickly from one setting to another and get into more emotional situations. They also are more likely to feel overwhelmed—by the demands of a teacher or a parent, and by their own high expectations.

> I thought I would go insane in class today. Who *cares* about memorizing all the losing vice-presidential candidates?—Erica, age 15
>
> When I heard that we were going to have to do a term paper in English, I panicked. One of the girls burst into tears.—Peter, age 15

Although some individuals have greater difficulty than others, the stresses on most adolescents are real. Some of these stresses are physical—puberty—and others are environmental—rejection by peers, tough exams and bad marks, demands by parents. Even the most competent children may become unpredictable in adolescence. Over the course of adolescence, though, moods become somewhat more stable. In another beeper study which extended for one week during each of several years, adolescents circled words that reflected how they were feeling about themselves when the beeper sounded (Savin-Williams and Demo, 1984). In seventh grade, only 11 percent of the subjects had stable feelings of self-esteem over the week. By tenth grade, the percentage of adolescents with stable feelings over the week had increased to 41 percent.

Making transitions: early adolescence

The early part of adolescence, from 11 or 12 to 16 or so, is when the storms are stormiest and the stresses most stressful for most adolescents. For this is when the adolescent is making the transition out of childhood.

> I checked on Kristy one night at bedtime when she was 13. She'd fallen asleep with lipstick and heels on, her hair in rollers, and a stuffed animal under her arms. It reminded me of the time when she was 2 and fell asleep with a pacifier in her mouth and sunglasses on her nose. She was straddling two stages of life once again.

In early adolescence, changes occur on all fronts. The child changes *physically* (as we saw in Chapter 14) and acquires an adult shape and reproductive capacity. These changes and associated side effects (like acne) may create stresses. Worries about physical changes have been found to be most common during early adolescence (Berzonsky, 1982). During early adolescence, children also change *cognitively* (as we saw in Chapter 15) and begin to acquire adult reasoning abilities. This allows them to worry about social issues, rules, and fairness. It also allows teachers to increase their expectations of what children can achieve and to set more challenging assignments. During this period, adolescents are most likely to mention that one of their greatest worries is nuclear war. Researchers in Finland and Canada, for example, have found that almost twice as many 12-year-olds as 18-year-olds worry most about nuclear war (Goldberg et al., 1985; Solantaus, Rimpela, and Taipale, 1984). And during early adolescence, children change *socially*. During early adolescence, the desire for comformity to peers is likely to increase, creating in its wake concerns about popularity—"Do people like me?" "Am I too different?" In one study, for example, 11- and 15-year-olds reported that their greatest anxieties were over relationships with members of the other sex and rejection by their peers (Coleman, 1974).

During this period, too, issues of independence may become critical at home. Many young adolescents spend large amounts of time away from home—at school, with friends, or working at a part-time job after school—and resent their parents' requests and questions.

> When I asked Sam where he'd been this afternoon, he snapped, "Here comes the Spanish Inquisition again."

The young adolescent may feel grown up but continue to be treated like a child by parents and older brothers and sisters. It is during early adolescence that family disruptions boil over most often.

> Erica turned impossible on the eve of her thirteenth birthday. We have dinner with my parents—Erica and Mike's grandparents—every Sunday, and we've been doing this for years. All of a sudden, Erica didn't want to go. Why? She had to wash her hair. Couldn't she wash it before or after dinnertime? No.Why? Because she was supposed to wait by the phone for her boyfriend to call her then. I suggested that she call him and change the time of his call to her. Oh no, she couldn't do that. Girls don't call boys, *Mother.* Erica stormed into her bedroom, slammed the door, and put some hideous music on ear splittingly loud. I thought that my blood pressure would blow steam from my ears, I was so frustrated. When I walked by her bedroom door half an hour later, the phone cord was pulled underneath it. She was back "in conference," as my husband puts it. Lord, give us strength to survive the child's adolescence!

In a study of 70 middle-class adolescent boys, family disputes were most common in early adolescence and tended to quiet down after ninth grade

(Offer, Ostrov, and Howard, 1981). The 12- and 13-year-olds were the most rebellious, often bickering with their parents.

Self-esteem

Young adolescents are most likely to be plagued with feelings of unhappiness and self-doubt if they are under stress from several sources. In one study of young adolescents, 12-year-old girls who were going through physical and social changes at the same time—moving to a new school, beginning to date—had the lowest self-confidence and self-esteem (Simmons, Rosenberg, and Rosenberg, 1973; see Figure 16.1). In another study, seventh-grade boys who were going through puberty had lower self-esteem than those who had not yet reached puberty (Jaquish and Savin-Williams, 1981). Many young adolescents just entering junior high school go through a period of questioning their competence (Connell, 1981; Harter and Connell, 1982).

> Our sixth-grade teacher was trying to impress us with what a big change junior high was going to be. She said that she hoped that we would find our way, because it was a much bigger school than the one we were used to. She hoped that we would do well in our courses, because no one would be there to hold our hands the way they'd done up till then. By the end of her talk, I was so terrified of junior high that I nearly passed out.—Natalie, age 13

> I remember sitting in the cafeteria during my first months in junior high and watching the ninth graders. They seemed so nonchalant, so mature, so cool. The boys would casually lean against the wall. The girls would saunter by. And they'd all laugh and joke around. We seventh graders just clumped around our little tables, gawking at them. At the end of seventh grade, I was invited to a party where there were going to be some ninth-grade boys in addition to our regular group. The ninth graders all swooped into the party at once, and the atmosphere turned positively electric. They were *so* sophisticated and cool that when one of them asked me to dance I thought I would levitate through the ceiling.—Sarah, age 17

By later adolescence, the self-doubts of the 13-year-old have eased (McCarthy and Hoge, 1982; O'Malley and Bachman, 1983). Over time, more adolescents answer positively questions like, "I feel I am a person of worth, on an equal plane with others," "I am able to do things as well as most other

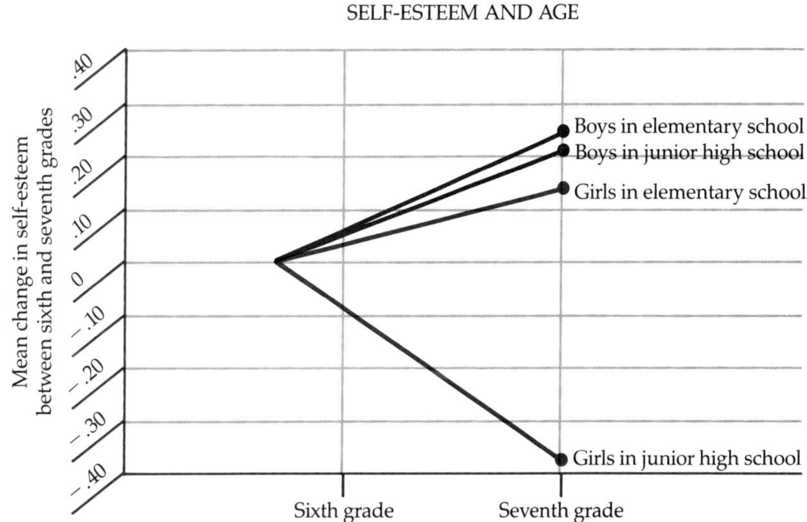

Figure 16.1 ***Boys gain some self-esteem between sixth and seventh grade; girls attending seventh grade in an elementary school do, too, although not as much. The self-esteem of girls attending seventh grade in a junior high school plummets (Simmons, Van Cleave, Blyth, and Bush, 1979).***

people," and "On the whole, I am satisfied with myself" (O'Malley and Bachman, 1983). Older adolescents are less moody, less irritable, and less self-conscious about their physical changes. Although they are still nestlings, they fly off on forays of their own and begin to set their sights on adulthood. They form plans and engage in reveries about what their lives will be like when they are grown up. They worry about leaving home (Berzonsky, 1982) and conflicts with their parents (Coleman, 1974).

The identity crisis

The most important psychological development in later adolescence—after age 16 or so—is the gradual appearance of a mature identity. A mature identity is both an inner sense of uniqueness and an outward statement of goals, values, and beliefs. Once the rough journey through early adolescence has calmed a bit, the task at hand is to answer the critical question, "Who am I?" If adolescents are to move toward an understanding of who they are, in fact and in fantasy, they must navigate new ways of getting along with family members and friends, new ways of feeling and acting sexually—whether in spite of or with the encouragement of peers, parents, and other authority figures—and new ways of conceptualizing themselves and their world.

An important focus of attention in adolescence is the "identity crisis," when young people reflect on their moral and sexual values, religious beliefs, and career options.

> When I grow up I want to be either a poet or a carpenter. I'll *never* be a businessman. My father's a businessman, and I think his values stink.—Tim, age 16

> Who am I? I am a girl, an American, a high school student. I am a Protestant, a Taurus, and a believer in human rights. I want to combine a career with a family life when I get married. I want to travel and see the world. I want to fall in love. I am moody, independent, a good listener, and an idealist.—Megan, age16

Erik Erikson (1968) theorized that adolescence is a crisis among crises, a period when old issues resurface from childhood and must be resolved once again and put into a new order in a newly emerging sense of self. Issues of sexuality, self-worth or self-doubt, industriousness or passivity, independence or dependence, social recognition or isolation that already have been resolved in infancy and childhood emerge in new forms. If infancy and childhood brought more trust and autonomy, then the adolescent will be more likely to feel autonomous and self-respecting and to believe that others can be loving and trustworthy. But if infancy and childhood brought mistrust, shame, or doubt, then the adolescent may have trouble separating from the family, loving others, trusting to the future, or respecting and believing in himself or herself. If childhood has brought a sense of industry and success in school, then the adolescent is likely to feel that work is rewarded. If it has not, the adolescent is likely to be left feeling inferior, thwarted, and frustrated. All these issues are again scrutinized, sorted, weighed, and tested with the new self-awareness of adolescence, as the adolescent tries to figure out and piece together who he or she is. Graduating from high school, going to college, and forming new relationships challenge adolescents' sense of self and precipitate an **identity crisis**. Resolving this identity crisis takes time—time to find and fit together all the pieces, to find the roles, the work, the attitudes, and the social connectedness that will let the adolescent take a place in adult society.

According to Erikson, adolescents enter a period of **psychosocial moratorium** when they have no task more important, no psychological debts more pressing, than those used in the service of finding themselves, as they experiment with roles and look for niches where they might comfortably fit. According to Erikson, adolescents must work out identity issues in four major areas: (1) career, (2) morality and religion, (3) political ideology, and (4) social roles, including gender roles. Adolescents devote attention to each area as they go through an identity crisis. Thus, they contend with questions such as Should I go to college? Where? Do I believe in God? Should I have sexual intercourse? With whom? Should I become politically active? Which job should I go after? Whom should I date? Should I take drugs?

> Sometimes I look in the mirror and say to myself, "Okay, who are you *really,* Natalie?" I pull my hair back and put on makeup and look very sophisticated. That's one me. I let my hair fall loosely on my shoulders, put on a Shetland sweater, and that's another me. I write poetry and stay up late to watch the stars and planets through my telescope; that's another me. I get really involved in my chemistry homework and think I'll be a doctor, or I want to be a translator, or I want to be a foreign correspondent. There are almost too many possibilities.—Natalie, age 17

What have researchers found out about the identity "crisis"? Do adolescents experience a crisis? Do their views of themselves change? There is some research support for Erikson's ideas. Some adolescents do consider alternative identities (Waterman, 1982) and try out being rebellious, studious, or detached. Nevertheless, not all adolescents experience a "crisis" or make major shifts in their self-concepts. In one study (Dusek and Flaherty, 1981), researchers repeatedly interviewed a group of adolescents over the period from age 11 to 18. They asked about the beginning of puberty and the intensification of sexual feelings, the changing relations between boys and girls, career choices, the emergence of idealistic thinking, and changes in intellectual performance and self-esteem. For the most part, they found that even important environmental events did not cause major changes in adolescents' identities. Adolescents' self-concepts formed slowly and steadily. The person who entered adolescence was essentially the same person who left it. Their research was supported by data from another longitudinal study (Savin-Williams and Demo, 1984), showing that adolescents' feelings about themselves were basically stable from year to year.

It is worth noting, though, that these data were collected during the middle to late 1970s, a time of relative political and cultural calm. In a more tumultuous era, like the 1960s, there is likely to be more marked change in an individual's identity. For example, during the Vietnam War era, a distinct identity status appeared—**alienated identity achievement** (Marcia, 1980). It appeared among adolescents who concluded that they did not want to pursue the careers offered by "straight" society, participate in a political system that waged an immoral war, or opt for traditional forms of marriage, family, and religion.

As with the experience of storm and stress, there are individual differences in the identity formation process. Although most adolescents have formed their identity by age 19 or 20, there are wide individual variations in the rate and endpoint of this process. Some adolescents emerge with an intact sense of identity, headed for some sort of work, committed to political views and having examined their religious beliefs. The changes that come with adolescence—changing schools, choosing a major, making job plans, getting along with friends and family—may cause stress, but these adolescents meet the challenges. Out of the many personal, social, cultural, and even historical strands from which they must weave a unified self-image, they emerge as sound, adequate, and successful young men and women. These adolescents have reached a stage of **identity achievement**.

> Junior high was hard for me because most of the other kids—boys, I mean—seemed so *silly.* I felt as though I was wasting a lot of time, and so

> I concentrated on my schoolwork. I wasn't sure what I wanted to be when I grew up, but I felt more serious somehow than most of the kids in my class. I still hung around with them, went to parties, talked on the phone, all that stuff. When I was in tenth grade, I had an English teacher who really turned me on to poetry and literature. He lent me books of his own and helped me pick out things to read that he thought I'd like. Anyhow, I pretty much decided then that I wanted to be either a writer or a teacher of literature when I grew up. I've had a few other ideas from time to time, but I always come back to that decision. Now I'm majoring in English at a good college, and I've already had a poem published.—Robbie, age 19

Other adolescents remain uncommitted, in a state of **identity diffusion**. They are willing to take the risk of letting go of their secure childhood dependencies and fantasies for an unknown future. They have trouble making decisions and commitments. They lack direction or interest in academic, political, or social questions. According to Erikson, identity diffusion is the major psychological risk of adolescence.

> I might major in history. But if I can get my psych grade up, I might major in psych. You know, I sort of like to read all those case histories. But I don't know what I'd actually do with a psych degree, so I might go into accounting. There's good money in accounting, and you can be your own boss. Well, I don't know.—Jason, age 18

Still other adolescents, actively looking for a way to commit themselves to meaningful work or political views, remain in the state of **identity moratorium** well into adulthood. They may think about religion and politics, they may question their parents' views, but they have not yet found anything with which to replace them. Although they may be preoccupied with questions, their questioning tends to remain unproductive and unresolved.

> I've been thinking about working for a political candidate. But I can't decide which one to work for. My father's the head of town Republicans, and I've always been a Republican too. But I might like to give the Democrats a chance. The Republicans haven't done too much for the poor, you know. But I can't see that much difference between the Republicans and the Democrats.—Nancy, age 18

Still other adolescents do not search for identity, never having felt a sense of crisis, and accept the identity that their parents have set for them. They are in a state of **identity foreclosure**. Asked about religious or political questions, they may say something about how "My family has always been. . . ."

> I'm a member of the Baptist Church. Our whole family goes there. My mother leads the choir, and I teach Sunday school to the little kids. I was baptized when I was 12, and that made my parents very happy. I hope to marry a guy from our youth group so we can get married in our church.—Nadine, age 19

Researchers have found that the proportion of adolescents with identity achievement increases from junior high school to the end of college, while the proportion with identity diffusion and moratorium decreases (Waterman, 1985).

Family relationships

> My parents don't understand me. They want me to stay home with them all the time and never go anywhere with my friends. I might as well be in prison.—Megan, age 15

One task facing both boys and girls in adolescence is to reshape their relationships with their parents. They must manage a fine balancing act, managing to lean far enough from their families to achieve some autonomy, but not leaning so far as to topple into isolation, rage, depression, or guilt. Most adolescents do manage this balancing act, moving gradually into more autonomous family relationships, but remaining strongly emotionally attached to their parents, brothers, and sisters even so. Among the 17-year-olds interviewed in one study, for example, over 80 percent said that they felt close to their mothers, and over 70 percent, close to their fathers (Greenberger, 1975). In another study, college students, asked to name their top five heroes and heroines, named their mothers and fathers above all others (Farley, in Stark, 1986). Mothers got six times as many first-place votes as anyone else, fathers twice as many. Many studies done in the 1960s and 1970s showed that adolescents, especially girls, generally agreed with their parents about politics, religion, education, and careers (Douvan and Adelson, 1966; Feather, 1980; Lerner et al., 1975; Offer, 1969, 1981).

Freedom and control

Adolescence is a time when parents' limits are thoroughly tested. Although the issues may vary from day to day and family to family, the underlying battle is to control what the adolescent thinks and does. In one study, college students described the typical relationship between parent and adolescent as like that between guard and prisoner or son-in-law and mother-in-law (Wish, Deutsch, and Kaplan, 1976). When the teenagers in the Chicago beeper study (Csikszentmihalyi and Larson, 1984) were with their families, their negative thoughts outnumbered their positive thoughts ten to one. The kinds of things that they reported included thoughts such as "How incompetent my mom is," "How pigheaded my mom and dad are," "How much I really don't like my sister's hair," and "Why my mother manipulates the conversation to get me to hate her."

Adolescents want freedom; parents want control. Whereas from their younger children parents once could demand and expect compliance, now they must explain their reasons and justify their positions. Adolescents are likely to argue, to press for advantage, and to test most of the limits that parents set. It is in early adolescence that children are most sensitive to any hint of parents' control. Irascible and eager to pounce on every perceived flaw in their parents, they are likely to feel their parents' praise or criticism as unbearable intrusions. Wanting autonomy but unskilled at getting it, young adolescents may withdraw into an unsociable, uncommunicative cold war. The more the parents defend their positions, the more the adolescents are likely to defend theirs. Although these arguments with parents may be tiring and unpleasant, they teach adolescents an important social skill. They learn that autonomy can be reached through peaceable compromise.

Dotting the battlefield, of course, are positive feelings and good family times—the night everyone sits around in the Jeep eating chili after they've been sledding all day; the time they sing the songs from *South Pacific* on the ride home from a cousin's wedding; the nights when the whole family watches a special television series together and talks about it afterward. These good, warm feelings are important anchors for adolescents, as they pick their sometimes rocky course through family relationships.

But parents and adolescents are in conflict in "all families some of the time and some families most of the time" (Montemayor, 1983). Because the psychological work of the period is to redefine the childish self into an adult self, conflict with parents is nearly inevitable. Adolescents feel the need to experience life on their *own* terms, not their parents'. Families pay a price for this breaking away. In a study of the problems associated with raising children at different ages, mothers said that there were more problems raising adolescents than any other age-group (Ballenski and Cook, 1982). They reported problems, particularly in disciplining their adolescent children and with the adolescent children's moodiness and desire for independence. Bad as it seems to the mothers, sometimes adolescents report even more problems than their mothers do (Montemayor, 1983). Why? Imagine a mother telling her daughter to "clean up your room. It looks like a tornado went through there." The daughter complies silently, seething inside. But her mother sees only the clothes hung up and the bed made—and does not register the episode as a conflict.

What kinds of things do families wrangle about? Usually, the sparks fly over the normal, everyday business of life such as schoolwork, social life, friends, chores around the house, disobeying rules, fighting with brothers and sisters, and looking presentable (Montemayor, 1983). One follow-up study of "Middletown," a typical Midwestern town first profiled in a classic sociological study of the 1920s (Lynd and Lynd, 1929), showed that adolescents in the 1980s were disagreeing with their parents about virtually the same matters as adolescents were in the 1920s—"the hours you get in at night" and "home duties" still led the list, half a century later (Caplow, Bahr, Chadwick, Hill, and Williamson, 1982). Although sex, drugs, religion, and politics are possible sources for rip-snorting arguments, in fact most adolescents and parents do *not* argue about these matters.

By late adolescence, as we have suggested, family conflict usually diminishes. Then, once adolescents reach the age of about 18 and leave home, the amount of stress in most families drops even further. In one study of college boys, researchers found that when the boys moved away from home, their relationships with their families improved (Sullivan and Sullivan, 1980). Their parents became more affectionate, communication improved, and parents and sons alike were happier with their relationships. If boys continued living at home, there was no improvement in family relations. Once the nest empties, and adolescents leave home, mothers report their marriages become happier and their own sense of well-being increases as well (Rubin, 1980).

Sons' and daughters' conflicts

The sometimes choppy path through adolescence differs for girls and boys. In childhood, it is boys who most often have problems. Boys mature more slowly than girls, they inhabit largely female-dominated environments—at home with Mother or in school with a female teacher—and they must conform to stricter gender-role expectations than girls. But come adolescence, it is girls who most often have problems, as they cope with the conflict between their need to remain dependent and their need to strike out on their own—problems that sometimes show up in early pregnancies, unstable marriages, and emotional distress (Werner and Smith, 1982).

In adolescence, boys are likely to begin to separate from their families by relying on their abilities to do things and their knowledge of the outside world. They gradually spend less time with their families and more time at school, at work, or engaged in hobbies and sports. Their identities are based on these activities and abilities (Montemayor, 1982). Girls are more likely to base their emerging identities on their ability to get along with other people, and so they cannot so clearly turn their backs on their families.

Nor can their parents so easily let them go. Parents give daughters less room to maneuver than they give sons. They keep girls in tighter check, worrying about all the things that might happen if they, the parents, were to relax. They worry about whether their daughters are safe on their own, and they worry about their sexual behavior and, worst of all, they worry about the

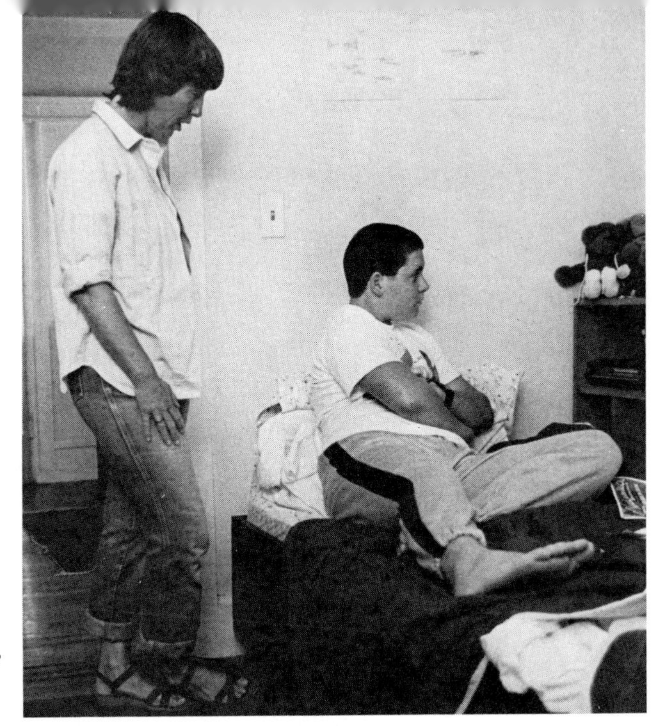

Parents and adolescents are in conflict in all families some of the time and in some families most of the time. For boys, a major source of conflict is keeping their rooms neat.

prospect of pregnancy. More tightly confined than boys, girls may have a harder time coming to terms with their independence. They want freedom of thought and expression—but without physically leaving the family (Coleman, 1974). Girls are expected to achieve independence but to maintain close emotional and physical ties with their families at the same time. They often feel ambivalent about asserting independence. If adolescent daughters do not feel close to their mothers, they feel sorry about it (Konopka, 1976), yet adolescent daughters want their parents to acknowledge that they are no longer children. Girls tend to have more conflicts with their parents, especially their mothers, than boys do. They get into more arguments, have more emotional outbursts, and get more threats from their parents. Their conflicts usually are about emotional issues (Kinloch, 1970; Montemayor, 1982).

> I had the chance to go camping with a bunch of kids in my class at school. And I really wanted to go. But my parents said that I couldn't go. "I thought you trusted me," I told them. "We *do* trust you, but it's not the kind of situation you should be in." I got so mad I stayed in my room all weekend.—Erica, age 14

The conflicts that boys have with their parents often involve issues like using the family car, "hanging out with the guys," going to church, having long hair, and doing chores around the house (Offer, Ostrov, and Howard, 1981). Boys also are more likely to feel crowded when they are with their parents and to resent being given orders. They assert themselves unambivalently toward their mothers (Steinberg, 1979), and they behave respectfully—even as they feel rebellious—toward their fathers (Montemayor, 1982).

> *Every* night, my mother makes me wash the dishes after dinner and take out the garbage. When I ask why I have to do it, she says she's "tired after a long day." Well, I have long days, too. How come she gets to sit around with my father and watch TV while I slave in the kitchen?—Ted, age 14

lifespan focus

Double trouble: overlapping crises

All families are vulnerable when children are going through adolescence, but they may be especially vulnerable when parents are experiencing a psychological crisis at the same time. Adolescents test their parents as they search for their own emerging identities, and they need parents who can weather the inevitable storms calmly, stably, and good-naturedly. Parents who are coping with the strains of a midlife crisis of their own may feel insecure and prone to depression themselves as they contend with the waning of their physical powers, with the dreams that will never be realized, and with the foreclosing of opportunity. When the psychological turmoil of the parents' mid-life transition gets heaped on top of the surliness, testing, and acting out of the child's adolescence, a family may be in "double trouble."

> I turned 40 less than a week before Sarah turned 15. For the first time ever, I felt miserably depressed by getting a year older. I looked in the mirror the next morning, and I started to cry because there was a middle-aged lady staring back at me. Then at breakfast, Sarah made some crack about "what was the point" of my putting on makeup before work. Some days I pull out of the driveway and swear I'll never come back.

Homemakers, whose children have been their major occupation, and parents who have turned to their children—and away from their spouses—for the satisfaction of emotional needs are most likely to suffer when their children become independent (Birnbaum, 1971; Lidz, 1969). They may feel quite threatened by a teenaged child's normal sexual interests.

> My father put his arm around my shoulders in the car the other day and said, "You'll always be my little princess." But I *won't* always be. That's the problem.—Kristy, age 17

When divorce breaks up the family, the adolescent's identity struggle can be especially severe (Wallerstein and Kelly, 1974, 1980). Some divorced parents act much like their own children both socially and sexually when they primp in front of the mirror on Saturday night, don their designer jeans, and pick up their dates in the family car. Adolescents may feel uncomfortable if they have to listen to parents' harangues about the former spouse's sexual problems or if they run into the parent's date as they stumble into the bathroom the next morning. Said one 15-year-old of her mother,

> I didn't like her going out with other men—I thought she was a tramp. I walked in on her when she had a man in bed—she should go out but not sleep with them (McLoughlin and Whitfield, 1984, p. 166).

In and of itself, divorce need not interfere with normal adolescent development. In fact, if families are riddled with conflict, adolescents may prefer to live with one rather than both parents (McLoughlin and Whitfield, 1984). Most feel that without the constant fighting, shouting, and tears, they can live and work more peacefully and that their schoolwork improves. Some adolescents want nothing more to do with one of their parents after a separation or a divorce. But most try to understand both their parents' situations and appreciate them for taking the time to explain their feelings and intentions and for continuing to provide support and security (Wallerstein and Kelly, 1974, 1980).

Parents' continuing influence

The sparks may fly, and the arguments may hurtle around the house. But adolescents continue to feel their parents' influence in this period, as they did earlier. Parents always have influenced their adolescent children and probably always will. When the inventor of the telegraph, Samuel F. B. Morse, attended Yale in 1810, he wrote home to his father that he wanted to be an artist. His horrified father wrote back:

> Dear Finley,
>
> I received your letter of the 22nd today by mail.
>
> On the subject of your future pursuits we will converse when I see you and when you get home. It will be best for you to form no plans. Your mama and I have been thinking and planning for you. I shall disclose to you our plan when I see you. Till then suspend your mind.
>
> Your affectionate father,
> J. Morse

Parents today may be less directive than the elder Morse, but their adolescent children nevertheless are sensitive to their influence.

Parents' styles of disciplining and communicating with their children tend to be stable as the children develop from childhood into adolescence. In one study, for example, mothers and fathers completed questionnaires about their child-rearing philosophies and methods (Roberts, Block, and Block, 1984). Their answers when their children were 3 years old were consistent with their answers ten years later. The degree of control parents exerted, their investment in their children, and their relative enjoyment of their children all tended to remain stable over time. When parents shifted their child-rearing methods, the shifts were appropriate for the changing needs of their growing children. Thus, all parents were likely to stress achievement more for their 13-year-olds than for their 3-year-olds and to express more affection physically with the 3- than with the 13-year-olds. Younger children were punished more physically; older children were punished by losing privileges. But despite these changes in specific actions, children were exposed to continuous, stable values and attitudes from their parents.

How adolescents think about themselves is related to their parents' attitudes and behavior. As we saw in earlier discussions of discipline (in Chapters 10 and 13), children are most likely to develop competence and

Parents' styles of disciplining and communicating with their children tend to be stable from early childhood into adolescence. Their styles may also have a bearing on the adolescents' exploration of identity issues.

confidence if their parents provide a balance of love and limits. Development in adolescence shows the same pattern. In one recent study, adolescents with high self-esteem perceived their parents as accepting and not overly harsh in making or enforcing rules (Litovsky and Dusek, 1985). Adolescents do best when their parents walk a fine line between giving too much support and too much freedom.

In another study (Cooper, Grotevant, and Condon, 1983), researchers first interviewed adolescents on their thoughts about their identities. Then they asked the adolescents and their families to make detailed plans for a hypothetical two-week vacation. Adolescents who reported that they had thought more about their identities and had explored more different identity possibilities were more likely to come from families in which, when planning their vacation, the father was willing to disagree openly with his wife, the mother was assertive, and both parents were willing to initiate compromises. These were families in which each member was clearly a separate individual with definite opinions and in which all members got along with humor, frankness, support, spontaneity, and vulnerability. As one father in such a family said as the family began planning their vacation (Cooper et al., 1983, pp. 54–55),

> I think probably what we all ought to do is decide the things that we want to do each one of us individually. And then maybe we'll be able to reconcile from that point. . . . Let's go ahead and take a few minutes to decide where we'd like to go and what we would like to do. And maybe we'll be able to work everything everybody wants to do in those fourteen days. Okay?

The mother of this family then said,

> "I think we all have good imaginations," and her husband commented, "I think that's kind of nice. I think we ought to be a rich gang."

In families in which members rarely disagreed or expressed individual opinions, adolescents explored few identity issues. as one such adolescent put it:

> I'm having a hard time deciding what to do. It would be easier if they would tell me what to do, but of course I don't want that.

Here is how her family discussed their vacation plans:

> MOTHER: Where shall we go?
> FATHER: Back to Spain.
> MOTHER: Back to Spain.
> JANET: Back to Spain.
> SISTER: Back to Spain.

For supporting the adolescent's quest for identity, the best type of family appears to be one in which members are connected but still individuals, not one in which the members are thoroughly enmeshed with each other, nor one in which they are indifferent or hostile to each other. Even though their rebelliousness may suggest that they would prefer it to be otherwise, adolescents do best when their parents are supportive, interested, and involved.

Conflict is nearly unavoidable in families with a teenage child, but a *moderate* amount of conflict between adolescents and parents sometimes may be constructive. By creating a certain degree of cognitive disturbance, moderate conflict may help adolescents to develop autonomous and advanced moral thought (Haan, Smith, and Block, 1968; Kohlberg, 1969). *Serious* conflict, a problem for an estimated 15 to 20 percent of all adolescents, does not support psychological growth (Montemayor, 1983). Instead, it is associated with a

Adolescents spend as much time as they can with each other and enjoy this time spent more than any other activity.

variety of problems for adolescents, such as leaving home and school, joining a religious cult, getting married or pregnant, becoming a juvenile delinquent, using drugs, or committing suicide. Which comes first, the family conflict or the other problems? It may be either. But whichever is the cause and whichever the effect, the connection between serious family conflict and serious social and psychological problems among adolescents is disturbing and real.

Peers versus parents

The lure of others their own age grows ever stronger throughout childhood, until adolescents are spending hours upon hours with their friends. With friends, adolescents can test their independence and aspects of their developing identities. They can test their sexual attractiveness. They can have a good time with others who like doing what they do (Hunter, 1984).

Time together

Adolescents spend their time in a variety of pursuits. In the Chicago beeper study, adolescents spent half of their waking hours with others their age, both in and out of school, and the time they spent with friends was the most enjoyable part of their lives.

> The other night eight of us went to the drive-in. But who could watch the movie? Five of us were in the backseat, and one guy had a leg cast that he slung over everybody's laps. So we left the drive-in and went to the diner. They're nice there. I had my usual—"Number 5 and a Coke"—and the other kids had whatever they wanted. Then we drove around some more, heard about a party at someone's house. But when we got there, no one was around except the kid's parents. We drove around some more, just hung around at the tennis courts near school, and got home early—about midnight.—Peter, age 16

With friends adolescents talked, joked, and hung out. They were spontaneous, open, and free of adult restraints. They felt excited, friendly, sociable, involved, and motivated. Boys especially felt more open, free, involved, strong, and active when they were with their friends and liked to spend time with them. Often adolescents described having fun with friends as "being rowdy"—getting to say anything, do everything. With friends, adolescents could drive cars from the *back* seat, have fights in school, act silly, and feel uncontrollably gleeful.

> The other night we all went "car surfing." You stand on the roof of a car and try to stay on while it goes around corners.—Sam, age 16

Adolescents also sometimes felt self-conscious, impatient, and angry with their friends. Even so, conflict with friends was much rarer than it was with family members.

Although adolescents spend a lot of time being *supervised* by adults—in school, at home, at work—they spend relatively little time actively *involved* with adults. Few adolescents in the beeper study reported spending time alone with adults. On the average, they reported spending only 8 percent of their time with their parents—usually their mother—and another 2 percent with other adults like bosses, teachers, or grandparents. When they did spend time with an adult they usually liked it, though. They loved it when their families played sports or games, and they also liked eating with their families. What they disliked was trying to read or do things privately when other family members were around.

In an interview study of the way that a different group of adolescents—those from Mormon Salt Lake City—spent their time and the amount of conflict they had with their parents, researchers found that these adolescents spent equal amounts of time with parents and friends (Montemayor, 1982). Clearly, there are differences in adolescents' activities depending on their families and their communities. But Mormon adolescents, too, did different kinds of things with their parents—work and chores—and their friends—play and recreation.

Spending time with friends makes adolescents happy and supports their social development. But adolescents in the Chicago study who spent the *most* time with their friends had wider mood swings and more problems in school. Adolescents who spent more time with their families, as boring or as conflict-ridden as that may be sometimes, adapted better to school and to the social system.

Conflict, conformity, and closeness

In the elementary school years, children usually move smoothly between the world of family and the world of friends. But in adolescence, the two worlds often collide. Because adolescence brings increased pressures to look, act, and think like peers, disagreements with parents may multiply. With their peers, adolescents may act outrageously—in their parents' eyes—and their parents may try to stop them.

> At the ripe old age of 12, Sarah suddenly began to wear heavy eye makeup and spiky hair, short, tight skirts, and high-top sneakers. We tried not to criticize, on the grounds that anything we objected to would only incite her further. But when she began criticizing the way I look—"Oh Mom, your hair is so *dowdy*!"—I drew the line.

Not surprisingly, adolescents feel closer to their peers than to their parents. Researchers have studied the development of conformity and closeness to peers and conflict with parents over the course of adolescence. Their findings are as expected: adolescents increasingly say that their friends are more important to them and more intimate than their parents. In a study of fourth, seventh, tenth graders, and college graduates (Hunter and Youniss, 1982), for example, fourth graders claimed to be more intimate with their parents than their friends. They claimed that their parents knew how they felt, talked things over with them, did things with them, and were more

Conforming to the ways of peers reaches an all-time high at age 13, but adolescents continue to affect uniformity of dress far later than this peak period.

enjoyable to talk to. Parents were perceived as more nurturant than friends. They helped their childen to solve problems, did things they needed, and the like. But by tenth grade, friends were perceived to be more intimate than parents and just as nurturant.

In one crucial way, adolescents' relationships with their friends differ from those with their parents: they want to look and act just like their friends and to belong to the group. They do not want to look and act just like their parents—at least not yet. In adolescence, conformity to one's peer group peaks; it is higher than it was in childhood and higher than it will be in adulthood. When one large sample of adolescents was asked with whom they most strongly identified, the majority—close to 60 percent of both boys and girls—said that they identified with people of their own generation (Sorensen, 1973). Adolescents conform to the values, advice, and judgments of their friends more and follow their parents' advice less than they did when they were younger (Bixenstine, DeCorte, and Bixenstine, 1976; Bowerman and Kinch, 1956; Devereaux, 1970; Utech and Hoving, 1969). In one study, for example, adolescents between 11 and 13 were more likely than younger or older subjects to change their answers to match their peers' (Costanzo and Shaw, 1966). In another study, adolescents were asked whether they would go along with their peers if the peers wanted them to do something prosocial, like helping the peer with schoolwork, or antisocial, like stealing candy (Berndt, 1979). Adolescents were most likely to conform to the prosocial suggestion at 11 to 12 years and the antisocial suggestion at 14 to 15 (see Figure 16.2). When the judgments of classmates and teachers are in conflict, adolescents are inclined to go along with their classmates, even when their suggestions are antisocial (Berenda, 1950; Berndt 1978b). Their friends' values strongly influence how adolescents behave (Siman, 1977). When it comes to antisocial behavior like smoking, skipping school, hurting someone, or committing crimes, peer values often override personal or parental values (Condry and Siman, 1974; Krosnick and Judd, 1982).

But adolescents are not mere pawns of their peers. Adolescents turn to peers or to parents for advice on different issues. In one study, perhaps somewhat outdated now, girls were likely to ask their parents about which job to take, whether to enter a beauty contest, and which boy to date (Brittain, 1963, 1966). But they were likely to ask their friends about what to wear to a football game, which subjects to take in school, and which dress to buy. Moreover, although peers' influence increases in adolescence overall, parents

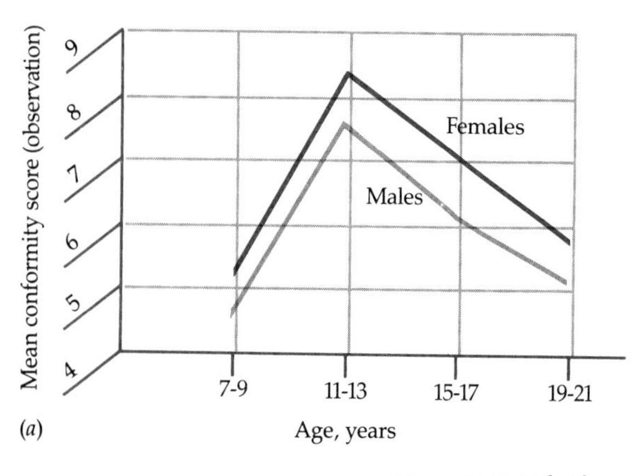

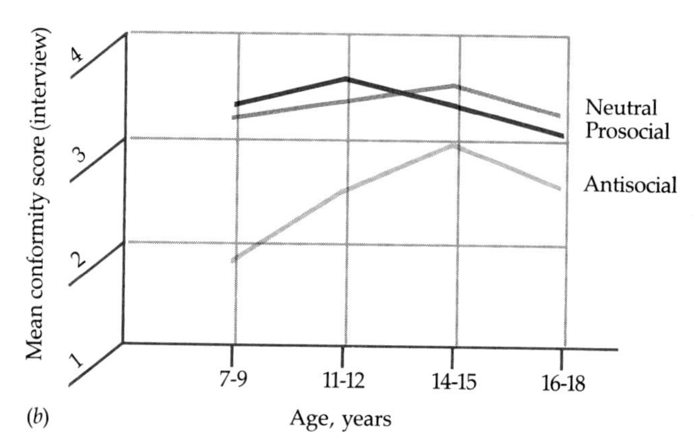

Figure 16.2 ***Whether conformity is measured* (a) *by observing children's behavior or* (b) *by asking children whether they would go along with a peer's suggestion, it seems to peak in early adolescence: at 11 to 14 for prosocial or neutral suggestions; at 14 to 15 for antisocial suggestions. Girls conform more than boys, and prosocial or neutral suggestions are more readily accepted than antisocial ones* (a: *Costanzo and Shaw, 1966;* b: *Berndt, 1979).***

are seen by their children as exerting more control over their behavior than friends do—telling them what to do and disagreeing with them—at *all* ages (Hunter and Youniss,1982). By age 18 or 19, most adolescents have undergone "a true growth in autonomy" and make their *own* decisions (Berndt, 1978b).

Peer relations

Cliques and crowds

Adolescents may conform in general to the peer culture, but they share the values of their immediate circle of friends even more. Young adolescents usually hang out in **cliques** of up to half a dozen members of their own sex (Csikszentmihalyi and Larson, 1984; Dunphy, 1963). In these cliques, young adolescents do things together—shop for clothes or records, go to concerts, do school projects, and share good times. Members of a clique usually are of the same age, race, and socioeconomic group. They share attitudes toward school, drugs, music, and clothes, spoken and unspoken norms: a private language, a "joke of the week," an agreement about partying every weekend. They share the attitudes of these friends both because they seek out others who think as they do and because they influence each other (Berndt, 1982; Douvan and Adelson, 1966).

Sometimes three or four cliques join into a larger, more loosely organized **crowd** that is organized by activity rather than by close friendship. A big party for the athletic crowd, for example, may include the girls' cheerleading and girls' basketball-playing cliques, which otherwise keep their distance from each other. Crowds serve several functions for adolescents and for adults. School officials, for example, may use crowd leaders to convey and carry out rules and expectations. Crowds therefore can help to maintain order and predictability in schools. Crowds and cliques also contribute to adolescents' sense of identity. Adolescents measure their own abilities against those of other clique members and mold their self-concepts according to the norms and attitudes of the crowd. Crowds and cliques give adolescents a social identity in the eyes of their schoolmates and mark social boundaries.

The names and nature of the cliques and crowds vary from school to school and community to community—"ropers" and "dopers," "kickers" and "potheads," "jocks" and "freaks," "preppies" and "sporties." But they have similarities nevertheless. In one widely quoted study of adolescent cliques and crowds in the 1950s, James Coleman (1961) found that regardless of the school, the leading crowds were likely to be male athletes and popular girls—jocks and cheerleaders, not brains. This finding has been confirmed in more recent investigations in the 1970s and 1980s (Eitzen, 1975; Thirer and Wright, 1985). Status within the crowd or clique depends, for boys, on physical maturity, fitness, and athletic ability (Savin-Williams, 1977). For girls, status is most likely to depend on physical attractiveness (Weisfeld, Bloch, and Ivers, 1984).

Friendship and understanding

Within cliques there may be closer pairs, their friendships based on similarities and shared interests. Friendships are oases where adolescents can learn about themselves and others. In adolescence, friends comfort each other when they are angry, sad, or in trouble (Csikszentmihalyi and Larson, 1984). As in childhood, adolescent girls' friendships are more intimate and more exclusive than boys' friendships. Girls talk about relationships, and boys talk about things to do (Johnson, 1983).

> My best friend and I promised to tell each other everything, including about sex.—Sarah, age 14

> When I get together with my friends, we usually listen to records or go to the movies. Sometimes we go shopping for stuff like clothes. Sometimes we shoot baskets or play a little soccer. Lots of times we just hang out.—Sam, age 14

But in adolescence friendship for both boys and girls is likely to be based on empathy, understanding, and self-disclosure (Bigelow, 1977). During adolescence, friendships mature into a fuller appreciation of idiosyncrasies and differences in personality or interests. Friends, adolescents come to realize, may be both dependent and independent at the same time. They trust each other, but they also "give each other a chance to breathe" (Selman and Selman, 1979). Friends are supportive and intimate, but they also act on their own as independent individuals. Adolescents want friends who are loyal and with whom they can identify and share worries. When they just start dating, girls want their girlfriends to remain loyal and supportive. Once they feel more secure about dating, girls' friendships may relax a bit (Douvan and Adelson, 1966).

Interestingly, the quality of adolescents' friendships seems to parallel the quality of their relationships with parents. Among high school girls who were asked to evaluate their relationship with a close girlfriend and with their mother, those who felt close to their mother also felt close to their girlfriend. Girls who had a poor relationship with their mother did not seem able to form a satisfying friendship, which might have compensated for their feelings of being unloved or out of touch with their mother (Gold and Yanof, 1985).

With the increasing intellectual maturity that adolescence brings, young men and women come to appreciate their friends more thoroughly because they understand better what makes them tick. With their abilities to think abstractly and to speculate about possibilities, as we discussed in Chapter 15, adolescents can draw inferences about how others think, feel, and strive, and they can use what they know in interpreting their own experiences. Their increasing powers to understand ambiguity, relative truths, and contradiction deepen adolescents' perceptions of others and themselves. They may understand that someone who *seems* snobbish is really just shy—an understanding of contradictions that younger children lack.

> My boyfriend acts very sophisticated and sarcastic when we're with other people. But when we're alone, he tells me things that he's never told anyone else before. Sometimes he even cries.—Kristy, age 16

They come to understand how one person's behavior changes from one setting to another, and they begin to perceive the unchanging core of the person's personality despite changing setting or circumstance. In so doing, adolescents come to understand what personality means, and they use this understanding in shaping their own emerging identities (Barenboim, 1977; Hill and Palmquist, 1978).

Dating

> My first date was when I was 12. I was fixed up with this really cute 15-year-old guy to go to my cousin's Sweet Sixteen party, a square dance. My mother took me shopping for a dress. It was pink, and I wore white shoes with little heels on them. I was kind of scared, and I worried that my date thought I was a nerd because I was so young. He was polite but didn't spend any time with me. But I had a good time dancing anyway.—Erica, age 16

Since the turn of the century, there has been a trend toward dating earlier and marrying later, effectively increasing the span during which adolescents date. The median age at which girls started dating was 16 in the 1920s, 13 in the 1970s, and even younger today (McCabe, 1984). Many girls now begin to date at age 12, boys a year or so later. Whether or not adolescents date, how they feel about dating, and their attitudes toward sex depend on both their physical and their psychological maturity. The adolescent who still has a child's body is likely to feel and act very differently on dates from the adolescent who is sexually mature. Whether, when, and how often adolescents date depend on how they look and think and on the norms prevailing within their family, their peer group, and their community. City adolescents have more sexual experience than country adolescents, for example (McCabe and Collins, 1981), and girls are more likely than boys to be influenced by their parents' wishes and by community standards (McCabe, 1984). Some adolescents do not feel comfortable with members of the other sex until late in adolescence (Douvan and Adelson, 1966). But most adolescents do begin to date at the socially approved age and feel left out if they do not.

Adolescents who begin dating at either extreme of the dating timetable may have problems. Girls who begin dating and then going steady early in adolescence may be hindered because they have not formulated autonomous personal or academic goals, yet girls who do not date by age 16 may have little idea of what boys find attractive (Douvan and Adelson, 1966; Douvan and Gold, 1966). These late daters are likely to be dependent on their families, self-absorbed, and insecure. They usually know that they are different from others their age, and they worry about it.

Dating serves several different functions for adolescents. It is plain fun. It helps adolescents learn to cooperate, consider others' feelings, take responsibility, and get along with others. It even teaches manners. Dating helps adolescents find their social niches, assess their popularity, and develop the social skills that they will need to form a mature, enduring relationship with a mate. It offers companionship and a pathway to achieving intimacy. On dates, adolescents experiment with sex. They reaffirm—or question—their deep personal convictions of their masculinity or femininity that took form during their early childhood. By testing whom they find physically attractive and who finds them attractive in turn, they shape a heterosexual or homosexual identity. Dating gives adolescents a series of temporary relationships within which

By dating, adolescents discover more about their masculinity or femininity, test their popularity, find their social niches, learn how to get along with others and have **fun.**

they can explore their sexual identity. Dating also gives adolescents a chance to test how closely they wish to conform to gender-role stereotypes. Boys experiment with finding a balance between acting macho and being vulnerable, and girls experiment with finding a balance between being emotional and dependent and being strong-willed and self-reliant.

Dates are attracted to each other for their friendliness, popularity, personality, intelligence, shared attitudes, and athletic ability and, perhaps most of all, their looks (Walster, Aronson, Abraham, and Roltman, 1966).

> I met Will at a party in ninth grade, took one look at him, my stomach went *boing,* and I fell in love with him. He's so gorgeous!—Sarah, age 16

> I like girls who are tall, thin, long-legged, smart, funny, and have a good personality.—Sam, age 16

> When someone tries to fix you up with a date, and they say, ''He's got a great personality,'' watch out. That always means he's ugly.—Natalie, age 16

Adolescent problems

Adolescence can be a difficult time and can bring with it a host of new problems—skipping school, running away from home, stealing, prostitution, and even suicide.

> Four kids in my class died this year. Two who were drunk got killed in a car accident. One overdosed on pills. And one got shot to death by another kid. It was so horrible around school for awhile that they brought in a psychologist to talk to everyone about death.—Peter, age 16

Delinquency

The term **juvenile delinquency** covers a multitude of crimes—murder, assault, theft, prostitution—and a multitude of acts considered wrong because they are committed by a minor—promiscuity, skipping school, running away from home. Most adolescents do things that they know are wrong. In an Illinois survey, for example, of the 3300 14- to 18-year-olds questioned, 73 percent admitted cheating in school, 47 percent admitted skipping school, 46 percent admitted getting drunk, and 23 percent said that they had bought

Delinquency has many different causes—neurological and cognitive abnormalities, destructive family relations, and pressure from delinquent peers.

liquor (Puntil, 1972). Fewer adolescents commit serious crimes. The most serious offenses adolescents are charged with usually are robbery, purse snatching, shoplifting, and pocket picking. Attacks on teachers and fellow students constitute a fairly new kind of serious offense by adolescents. Over 60,000 teachers were attacked by students in 1978 (O'Toole, 1978), and the federal government estimates that up to 280,000 students are attacked every month (U.S. Department of Health, Education, and Welfare and National Institute of Education, 1978). Even so, adolescents generally do not carry guns, injure the victims they rob, or inflict such heavy losses as adult criminals. Although boys are more likely to be delinquent than girls, the rate of delinquency among girls has been rising. For both boys and girls, delinquency peaks at around age 15 (see Figure 16.3).

What are the causes of juvenile delinquency? Neurological and psychological abnormalities are one cause. In a study of Michigan adolescents who had committed a crime, investigators found lower than average scores on tests of rhythm (a test of neurological integration, not musical ability), vision, expressive speech, writing, reading, arithmetic, and general intelligence (Brickman, McManus, Grapentine, and Alessi, 1984). The delinquent adolescents had trouble thinking abstractly, putting things in temporal order, and concentrating. The pattern of abnormalities led the investigators to conclude that these adolescents had not developed the cognitive control to manage their feelings or moods, and this lack of control plus learning problems may have made it difficult for them to deal with their feelings, to pay attention, or to control their actions. In another study (Chandler, 1973), delinquent boys were found to be less able than nondelinquent boys to take the perspective of another person. After being trained in social perspective taking, the boys, according to police records, committed fewer crimes over the next 18 months—or at least did not get caught at them. Interestingly enough, though, delinquents in yet another study were found to be able to reason morally at the same level as others their age (Hains and Ryan, 1983).

It has long been thought that children who think poorly of themselves misbehave more often. But does low self-esteem cause delinquency? In a recent three-year study (McCarthy and Hoge, 1985), students who started out low in self-esteem were no more likely to misbehave or commit crimes than students with high esteem. But delinquent students were more likely to *end up*

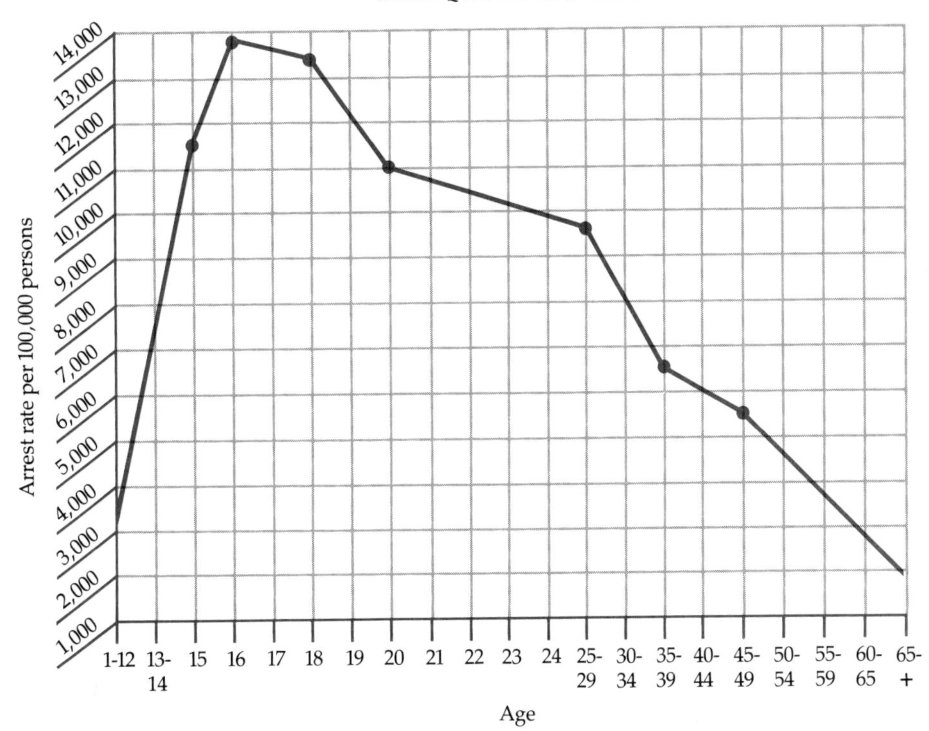

Figure 16.3 ***Whether one uses official arrest figures, as in this illustration (Sykes, 1980), or people's confidential reports of crimes committed (Gold and Reimer, 1975), the seriousness of delinquent acts appears to peak at about age 15. Boys commit more numerous and more serious delinquent acts than girls.***

with low self-esteem. It is possible that breaking rules resulted in condemnation by the people the adolescents cared about, and the condemnation made them think worse of themselves. It is also possible that the adolescents committed delinquent acts to get their friends' approval but reduced their self-esteem in the process. The psychological processes involved in delinquency are not fully understood and no doubt differ from one individual to another.

Destructive family relationships create a backdrop against which adolescents become delinquents.

> Every single kid, I don't care whether their parents are rich or poor, every kid who comes into the program [a residential drug treatment program] has some terrible story to tell. The parents are alcoholics or substance abusers. The kids have been molested, abused, farmed out, neglected. They've been victims of incest, or they've been beaten up. Sometimes the story isn't so blatant, but kids don't get into trouble just out of the blue.—Mike and Erica's father, a drug treatment counselor

In one study of Philadelphia gang members, it was found that the boys tended to be violent at home and had defied, hit, or abused their parents (Friedman, Mann, and Friedman, 1975). In another study of seventh- through tenth-grade boys in Eugene, Oregon, it was shown that delinquency was associated with how closely parents monitored and disciplined the boys' behavior (Patterson and Stouthamer-Loeber, 1984). If the parents supervised their sons and knew where they were, with whom they were spending time, and what they were doing, their sons were less likely to become delinquents. Eighty percent of the delinquent boys were inadequately supervised by their parents; only 10 percent of the nondelinquent boys were inadequately supervised.

Adolescents from every social class commit delinquent acts. But lower-class adolescents are more likely than middle- or upper-class adolescents to be caught by the police, to be charged with a crime, to be represented by a court-appointed lawyer, to go to prison, and to commit crimes as adults. Adolescents who have served time are likely to be stigmatized as troublemakers by the police as well as by people in the community, and people labeled troublemakers are likely to make a career out of crime (Chambliss, 1977). Thus, although delinquency may have various causes—neurological and cognitive abnormalities, destructive family relations, pressure from delinquent peers—the accident of birth into a lower-class family sometimes closes off an adolescent's avenues of escape from delinquency.

Suicide

Some adolescents look for a way out of their problems that is more final than running away from home or becoming members of a gang. They commit suicide.

> Jennifer was smart and sensitive. From the time she was a child, she took it to heart when she saw people acting cruelly or when someone she knew got hurt. When her father and I argued, the stricken look on her face was pathetic. Jennifer had a few close girlfriends, but she hadn't started dating actively. She spent a lot of time by herself, reading, listening to classical music—"Rock is for street people, not me"—and talked a lot to me or her grandmother. She was devastated when her grandmother died last year of cancer. For a while when she was in junior high, she saw a therapist because she seemed so depressed and lonely.

lifespan focus

Many pathways to maturity

When you meet an adult, you may be able to look back on that adult's past and see clearly how he or she became the person you see—whether that person is well or poorly adjusted, productive or destructive, happy or troubled. But when you meet a child, it is far more difficult to predict the kind of adult that child will develop into. Predicting a person's development is different from explaining it. One reason that prediction is so difficult is that there are many pathways to maturity. Two children from similar backgrounds may develop quite differently. Conversely, people who are similar as adults may have developed by quite different pathways.

Why do some people run into social and psychological problems while others, from essentially the same backgrounds, do not? In a 20-year study in which 700 Asian and Polynesian adolescents born in Hawaii in 1955 were followed until early adulthood, researcher Emmy Werner and a team of physicians, nurses, social workers, and a psychologist tried to answer this question (Werner and Smith, 1982).

Most of the adolescents in the study came from lower-class backgrounds. Many had been born prematurely or had had difficult births, had been raised in chronic poverty by parents with little education or serious mental health problems, and lived in unstable families. Not surprisingly, about 200 of the children did run into serious problems at some time before they turned 20. About one-third of the children went through an adolescence that was stormy and stressful, and one-fifth had developed serious problems by late adolescence. The boys with problems were aggressive and delinquent. The girls with problems got pregnant or had mental health problems. Many of the troubled adolescents of both sexes felt frustrated and not able to control their own fates.

But most of the adolescents studied—who had been exposed to the same serious stresses—did *not* develop problems. This resilient majority developed into mature, autonomous, and competent young adults who "worked well,

> But then she stopped therapy, got more active at school and with other kids her age. She had seemed lonely again lately, and I talked with her about going back into therapy. But she pooh-poohed the idea. I guess you always think back over the things you might have done, and I think that if we had insisted on more therapy, she might still be alive. We found her body on the first anniversary of her grandmother's death.

The problems of adolescence can loom so large and insurmountable, the person feels so bleak and despairing, that 14 in every 100,000 people between the ages of 15 and 24 kill themselves every year, and 100 times that many try. Only murder and car accidents kill more people in this age range, and many of these "accidents" actually may be suicides (*Monthly Vital Statistics Report*, 1979; Seiden, 1969). The rate of suicide among adolescents has increased 300 percent in the last 30 years (U.S. Department of Commerce, 1981). More girls attempt suicide than boys, but because boys use more violent methods—guns rather than pills—more boys succeed (Cantor, 1985; Petzel and Cline, 1978). Of the adolescents who attempt suicide, 90 percent are firstborn girls who are unusually close to their mothers. Of the adolescents who actually commit suicide, 75 percent are white, middle-class boys. They tend to be later-born children who keep their problems inside themselves and cannot accept help from others. Many feel pressured to achieve but unappreciated and unrecognized by their parents.

played well, loved well, and expected well." Compared to the troubled adolescents, many of the resilient adolescents had been firstborn children and had fewer brothers and sisters. They were described by their mothers as having been physically robust, easy to deal with, active, socially responsive, and good-natured as infants. It is likely that their temperamental characteristics allowed these infants to seek and hold the attention of their parents and other adults—grandparents, friends' parents, babysitters—and that this attention provided the children with important emotional support. Girls who took care of siblings while their mothers had worked outside the home developed reserves of autonomy and feelings of responsibility that further bolstered their development.

Whereas temperament and physical health seemed to have more effect on infants' development, environmental factors like family structure and an individual's own intellectual ability had more effect in early adolescence. By late adolescence, factors like self-esteem and feelings of control or helplessness most heavily affected whether the adolescents ran into trouble. It was this sense of control and faith in the effectiveness of their own actions that seemed most important to a good outcome, according to Werner. By late adolescence, many of the resilient individuals had positive self-concepts, were nurturant, had a sense of coherence to their lives, and were interested in "improving themselves"—that is, continuing to grow psychologically.

As they study how people mature across the lifespan, many developmental psychologists are finding that development is a series of choices among many possibilities. In one study of the quality of personal relationships with parents, peers, and spouses across the lifespan (Skolnick, in press), three quarters of those studied between infancy and adulthood developed discontinuously. For them, the quality of relationships in infancy was inconsistent with later stages of their lives. Only one quarter of those studied developed continuously, with their quality of relationships remaining the same from infancy onward. Lifespan psychologists recognize that people's choices shape the course of their development and that all of us could have been different from what we have become.

Suicide may seem like the only out to adolescents who feel hopeless, powerless, and without options. They have never had experience resolving serious problems. They do not know much about rising above their problems or just waiting them out.

> To my family and friends:
>
> I'm sorry it has to be this way. For some reason, I have set unattainable goals for myself. It hurts to live and life is full of so many disappointments and problems. . . . Please don't cry or feel badly. I know what I am doing and why I am doing it. I guess I never really found out what love or responsibility was.
>
> Bill
>
> I might also add that I had had in recent years no great desire to continue living. Saying goodbye to all of you who I was close to would only make things harder for me. Believe me, I tried to cope with my problems, but I couldn't (Jacobs, 1971).

The tendency for adolescents to see things as all black or all white, to narrow their focus into a kind of tunnel vision that shuts out all light, can make bad situations seem utterly hopeless. We do not know which factors make an adolescent turn to suicide, but we do know some of the things that generally can be ruled out: poverty, deprivation, crime, pregnancy, sexually transmitted diseases, and drugs (Wynne, 1978; Yankelovich, 1981). Recently it has been found that adolescents who committed suicide were more likely to have experienced stress before or at birth. In a study comparing suicide victims with a matched comparison group, it was found that 60 percent of the suicide group, compared to 12 percent of the comparison group, had been born of a pregnancy lacking in prenatal care, had experienced respiratory stress at birth, or had a mother who was chronically ill (Lipsitt, 1985). The causal chain between birth stress and adolescent suicide is not yet clear, but this finding does suggest that physical and psychological stresses that begin even before a child is born may culminate in suicide.

Some adolescents are so depressed by the death of a member of their family or a close friend that they try suicide (Tishler, McKenry, and Morgan, 1981). Others try suicide after they have been isolated for a long time, after they have failed repeatedly at things, or when they have been feeling unworthy, desperate, and hopeless. Although many suicides follow depression, hopelessness actually is the key. Some adolescents attempt suicide because they want out of a destructive family; others have problems with members of the other sex, problems at school, problems with brothers and sisters, or problems with friends (Tishler, McHenry, and Morgan, 1981). Whatever the specific cause—and adolescents themselves may not always know what has driven them to the brink—suicide is one increasingly common reaction to the problems of adolescence.

Work

Part-time jobs

> I have a job at the Boys' Club after school every day watching the 4- and 5-year-olds. It's hard to get my homework done when I've been running around after little kids all afternoon.—Nadine, age 16

Many adolescents hold part-time jobs. Although work teaches them to be responsible, punctual, dependable and self-reliant, the jobs also may interfere with their schoolwork and relationships with family and friends.

Many adolescents hold part-time jobs. In one recent month during a school year, for example, 43 percent of the 16- and 17-year-old boys and 37 percent of the girls in this country held jobs (Greenberger and Steinberg, 1986). Even one-sixth of the 14- and 15-year-olds were working or looking for work in 1983 (Nilsen, 1984). More adolescents hold part-time jobs today than at any other time in the last 40 years. Fully 80 percent of all high school students hold jobs at one time or another. In one short-term, longitudinal study of high school students in Orange County, California, the researchers set out to learn about the costs and benefits of working during adolescence (Greenberger and Steinberg, 1986). Work, they noted, is assumed to benefit adolescents' development and to teach them about taking responsibility, being on time, and being dependable. At work, people are assumed to cooperate and to interact socially. At work, barriers of age and class fall, and tolerance rises. But are these assumptions valid? Work, the researchers found, apparently did teach adolescents to be responsible, on time, dependable, and self-reliant. But it did not increase their cooperativeness, social interaction, or concern for others. In general, the more time that high school students worked beyond 14 hours a week, the more often they were absent from school, the less time they spent on homework or extracurricular activities, the lower their grades, and the less they enjoyed school. The more time they worked, the less time they spent doing things with their families and the less close they felt to their friends. Students who held jobs had more negative attitudes toward work than students who did not hold jobs, were more tolerant of unethical work practices, and used more drugs, alcohol, and cigarettes. Of course, it is not clear that part-time work itself *causes* these negative outcomes. Adolescents are not randomly assigned to work, and it may be that those adolescents who already smoke and don't care about school choose to or have to work. But it is clear that part-time work does not guarantee benefits to adolescents in psychological, social, or academic development. Moreover, the jobs that adolescents are likely to hold—in food services, manual labor, retail sales, cleaning, office work, and child care—tend to be repetitive, to provide little chance for learning, and to expose the adolescents to environmental, social, and personal stress.

Career choice

Whether or not they hold part-time jobs, adolescents are concerned about work, for adolescence is when initial decisions about occupation and careers must be made. Adolescents begin to think about their abilities and interests, to experiment with work roles, and to make vocational plans that are somewhat more realistic than younger children's fantasies about what they will be when they grow up. According to one theory (Super, 1967), from 14 to 18 adolescents first begin to narrow their choices according to their interests and values and seek out information about general categories of occupations and professions—for example, mental health or medicine. From 18 to 21 they specify their vocational interests within one general career category and seek information about specific occupations—for example, social worker, clinical psychologist, psychiatrist. In the real world, not all adolescents are so systematic and rational, however. Many flounder after high school, making repeated job shifts without a clear plan based on adequate information (Super, Kowalski, and Gotkin, 1967). Occupational choices are not based solely on adolescents' interests and values. For one thing, people's interests and values continue to develop past adolescence. For another, there are other factors that affect vocational choices. Another vocational theory (Holland,

1973) suggests that personality is important in the choice of an occupation. "Realistic" individuals prefer practical jobs; "sociable" individuals are interested in the helping professions; "enterprising" individuals look for power and status, and so on. These are the kinds of personality types assessed in inventories of interests that are used in vocational counseling.

Just as important as adolescents' interests and personalities, though, are other people. Parents and peers clearly influence adolescents' career choices. Adolescents from middle-class families with middle-class friends almost always enter middle-class occupations. Why? The reasons include family connections, family pressure, and family expectations. They include educational opportunities, information, and the leisure to explore career options. They also include the more specific example set by parents. In one study (Werts, 1968), for example, it was found that 40 percent of physicians' sons entered the field of medicine. This tendency for sons to take over the family business used to be even more common. Now, adolescents have a wider range of choices and fields than their parents had and are less likely to follow in their father's footsteps. But they still choose occupations of the same socioeconomic level.

Broader conditions in society, like the economy and the job market, also influence adolescents' vocational choices, and so do societal expectations. For example, in our society, expectations for men and women likely are responsible for the fact that adolescent girls' career plans center on a narrower range of occupations than boys' plans do (Marini and Greenberger, 1978). The women's movement has made it more likely for girls to say that they will have jobs when they grow up, but the jobs they aspire to—nurse, teacher, secretary—are as gender-typed as ever (Lueptow, 1984).

One reason for the strong influence of family, friends, and society on adolescents' career choices is the lack of vocational guidance at school. In theory, vocational classes and counseling in high school would seem to be good sources of career guidance, but in fact, they usually do not offer effective guidance (Lueptow, 1984; Sizer, 1985). Most high school guidance counselors have more urgent problems to solve—like getting the students through high school—and most vocational classes are out of date. As a result, adolescents may continue their naive fantasies about their intended occupations. In one study (Sarason, 1980), college seniors who were planning to be doctors or lawyers claimed that they had not considered careers in business because they didn't want to be small cogs in big wheels, to have to struggle to reach the top, or to become morally corrupt in the process. Students who were already in law school or medical school realized that these factors were also part of careers in law and medicine. Career development begins in adolescence, but it certainly doesn't end there.

Psychosocial maturity

Adolescents are faced with the task of learning the patterns of behavior that society requires of its functioning members. At the most basic level, they must learn that there are times for sleeping and eating, working and studying, relaxing and playing. At a higher level, they must learn how to bring pleasure and meaning to the things that they do. They must learn to narrow their sights and concentrate their energies on a limited course of development. One person cannot be a builder, a rock star, a playwright, *and* a dancer. Somehow adolescents must learn to listen to the outer voices— "Read the history chapter," "Let's go bike riding," "Give to the United Way"—and to the inner voices—"I don't care what anyone thinks," "I will never give in," "I love this." They must heed both social and personal values. They must develop constructive attitudes toward working—to know how to perform some kind of work, to persist and to resist distraction, to have standards for and to take pleasure from their work. They also need a clear sense of their own identity—a clear idea of themselves, of their goals and values—and must think well of

focus *Personhood for adolescents*

> Natalie is 17—going on 70. She's more mature than half the adults I meet.

> Mike got suspended from school for gathering names on a petition against the war in Nicaragua. He's so mad about his right of free speech being violated that he wants to take the school principal to court.

In the eyes of their parents, in the eyes of the state, and in the eyes of the law, adolescents are part children and part adults. They are entitled by law to some rights granted adults but not all. They are entitled to due process of law in delinquency proceedings but not to a jury trial. They have some rights of free speech at school but not the full rights guaranteed adults. They may choose on their own to have an abortion—even against their parents' wishes—but their parents may still "volunteer" them for admission to a mental hospital. Some advocates of minors' rights want to liberate adolescents and increase their powers of self-determination, whereas others want to protect them, and still others see no distinction between the interests of parents and the interests of children. The Supreme Court has ruled as though adolescents do not have the capacity to make mature, independent choices. In the words of Chief Justice Burger:

> Most children, even in adolescence, simply are not able to make sound judgments concerning many decisions, including their need for medical care or treatment. Parents can and must make those judgments (Melton, 1983, p. 100).

But do minors have the ability to make mature choices? In many areas, they do seem able to make mature decisions. For example, adolescents' reasoning is indistinguishable from that of adults when it comes to understanding the concepts of psychotherapy and mental disorders. Adolescents give the same arguments for and against abortion as adults do (Melton, 1983). If adolescents think like adults, perhaps they should be treated like adults. Some psychologists have suggested that the psychological benefits might be substantial (Melton, 1983). If adolescents had more legal rights, they suggest, they would feel more in control of their lives, more competent, less helpless, and would have an easier time developing into independent individuals. Giving children the right balance between freedom and control is essential for the state, just as it is for parents.

It is probable that as the concept of adolescence as a separate stage of development enters its second century, we will see it shifting again in response to social and historical factors. Will the concept of adolescence come full circle, and will adolescents be treated essentially like adults? The weight of the evidence on adolescent development appears to caution society against such a move. For although adolescents seem mature in many ways—with their formal operations and sexual maturity—they still need a protected time in which to form their identities. Without that time, many adolescents are likely to find themselves with identities that are foreclosed rather than integrated and fully resolved.

themselves. They must learn to function independently and to contribute to their society's well-being and survival. All this sounds like a tall order. Can any real-life adolescents fill it?

On a questionnaire designed to test for precisely these aspects of psychological and social maturity, Ellen Greenberger (1974, 1983) found that 1100 sixth, ninth, and twelfth graders advanced significantly over the course of a year in both autonomy and social responsibility. Other studies confirm steady progress toward autonomy among 12- to 17-year-olds and in social responsibility for boys at all ages and for girls until about age 14, when they reach a temporary plateau (Greenberger, 1982). But although girls are more likely than boys to say that they should contribute to their communities, at no point in adolescence does a clear majority of either sex endorse this position. Adolescents in this country are clearly more interested in growing independent than in taking social responsibility.

Summary

1. All adolescents go through ups and downs in their moods and feelings about themselves. Some adolescents feel great storm and stress; others find adolescence a fairly smooth transition to adulthood.
2. The early part of adolescence, from age 11 to 16, typically is when physical and psychological pressures are more intense, when family arguments are most heated, and when adolescents' feelings of self-doubt and unhappiness are acute.
3. By later adolescence, moodiness, irritability, and self-consciousness have eased, and the search for a personal identity intensifies, as old issues of personality resurface from childhood and must be put in order in an emerging sense of self. It takes adolescents time to resolve this identity crisis and to find the necessary roles, work, attitudes, and social connectedness.
4. Adolescents face the task of achieving autonomy from their families without removing themselves so far that they feel isolated, enraged, depressed, or guilty. They test their parents' limits and bristle at their parents' attempts to control their behavior. Conflict between adolescent and parents occurs in every family, although it usually diminishes when the adolescent reaches 18 or so.
5. Girls are more likely to have problems coping with the conflict between remaining emotionally close to their parents and breaking away than boys are. Girls argue with their parents most over emotional issues, whereas boys argue over practical issues like using the family car, going to church, or doing chores around the house.
6. Parents continue to influence their children during adolescence. The amount of emotional support they provide influences how successfully adolescent children resolve their identity crises.
7. Adolescents spend many happy, sociable, and exciting hours with their friends. Although they sometimes feel angry and self-conscious with their friends, they are in conflict with them less often and come to feel more intimate with them than with their parents. As they search out their own identities and test their parents' values against their friends', adolescents turn to peers and parents on different issues. By age 19, most adolescents conform less to both peers and parents and are more autonomous.

8. Not only is dating fun, but it helps adolescents learn to cooperate, consider others' feelings, take responsibility, get along with others, test their popularity, develop social skills and manners, and experiment sexually.
9. Few adolescents commit serious crimes. Those who do commit crimes are likely to suffer from neurological and psychological problems. Their family relationships may be destructive and violent, and their parents probably do not supervise them adequately.
10. When adolescents feel that their problems are hopeless, they may attempt suicide. In this country, more girls attempt suicide than boys, but more boys actually kill themselves.
11. Many adolescents hold part-time jobs. Work teaches adolescents to be responsible, dependable, punctual, and self-reliant. But it may interfere with school grades and participation and with an adolescent's family life.
12. The work of adolescence is to develop a coherent sense of personal identity and positive self-esteem, to be autonomous and socially responsible, to enjoy working, and to be capable of making mature decisions.

Key terms

storm and stress
identity crisis
psychosocial moratorium
alienated identity achievement
identity achievement
identity moratorium
identity diffusion
identity foreclosure
clique
crowd
juvenile delinquency

Suggested readings

COLEMAN, JAMES, and HUSÉN, TORSTEN. *Becoming Adult in a Changing Society.* Paris: Organisation for Economic Co-operation and Development, 1985. An excellent overview of issues and findings concerning changing conceptions of youth and the transition to adulthood.

CSIKSZENTMIHALYI, MIHALY, and LARSON, REED. *Being Adolescent.* New York: Basic Books, 1984. A rich source of information about adolescents' feelings and experiences.

ERIKSON, ERIK. *Identity: Youth and Crisis.* New York: Norton, 1968. Erikson's detailed description of the many paths identity formation can take, with his analysis of literary figures, historical circumstances, and the American scene.

GROTEVANT, HAROLD, and COOPER, CATHERINE (Eds.). *Adolescent Development in the Family.* San Francisco: Jossey-Bass, 1983. Alan Waterman, (Ed.). *Identity in Adolescence: Processes and Contents.* Two collections of brief articles by major researchers in the field of adolescent development (both in the New Directions for Child Development series), which address how identity formation relates to family interaction, religious thinking, political commitment, and vocational development.

LIPSITZ, JOAN. *Growing Up Forgotten.* Lexington, Mass.: D. C. Heath, 1977. A review of research on early adolescence, showing the extent to which we have neglected this age group. Schools, service agencies, and the juvenile justice system are examined and all found wanting.

MEAD, MARGARET. *Coming of Age in Samoa.* NewYork: Morrow, 1971 (originally published in 1928). A readable work that changed the way Americans looked at adolescence because it showed that not *all* adolescents experienced storm and stress.

STEINBERG, L. D. *Understanding Families with Young Adolescents.* Carrboro, N.C.: Center for Early Adolescence, 1980. An easy-to-read overview of the maturation of adolescents and the simultaneous maturation of parents. Ideas for coping with adolescent change in families are included.

PART SEVEN

Early and Middle Adulthood

chapter seventeen
Physical development

Adolescence shades gradually into adulthood. Some people think that adulthood begins at age 18, some at 21 with legal majority, and some think that it begins in the twenties or even early thirties, when formal education is done. Lifespan developmental psychologists usually establish the beginning of early adulthood at around age 20, the beginning of middle adulthood at age 40, and late adulthood at age 65 or so. The boundaries of early and middle adulthood are particularly flexible; in contrast, late adulthood often is more clearly marked—by retirement from work, the receipt of Social Security and other financial benefits, and other clear changes in social and economic status.

The adult body

The rate and extent of the physical changes in adulthood vary greatly from one person to another. They are an excellent example of a principle of development discovered by researchers into lifespan development. This principle is that as people get older, their development becomes more *differentiated*. In other words, there are more individual variations among older adults than among younger people. Most 5-year-olds, for example, are likely to be experiencing many of the same physical, social, and cognitive developments. But development among most 35-year-olds and, to a greater extent, among most 65-year-olds differs far more from one individual to the next. Throughout life, individuals differ widely in their energy levels, capacity for work, and general health. As people age, these individual differences get larger (Weg, 1983).

In addition, aging rarely progresses uniformly throughout a person's whole body. Some physical systems continue to function well, while others deteriorate. One 50-year-old may have a sturdy heart but brittle bones; another may have a weak heart but strong bones. Chronological age and physical age often diverge as well. Some people seem old at 40, and others seem robust at 80. What is more, the old-seeming 50-year-old who is suffering from stress or disease may recover and become the young-seeming 60-year-old.

Healthy adults can meet all of the demands of life, but professional athletes nearing middle age begin to notice a decline in speed and endurance.

By some measurements, early and middle adulthood are periods of gradual physical decline in many systems of the body. But for most adults, these years are the prime of life. Until most people are well into late adulthood, their bodies function well enough to meet virtually all of the demands of everyday life. For one thing, most organs of the body have so great a reserve capacity that it is only when young and middle-aged adults are under stress that they feel most of the physical changes associated with aging. For example, a woman pregnant at age 40 is likely to feel more tired than a woman pregnant at age 25, but otherwise the 40-year-old feels far from "old." For another thing, many people can accommodate to the gradual physical changes of adulthood. For example, a 45-year-old can buy reading glasses to accommodate to the farsightedness that is a normal part of aging for many people. Similarly, many couples adjust their sexual relationships to accommodate the gradual age-related changes of men's and women's reproductive systems.

We humans begin to age virtually from the moment of birth. Gradually, our bodies begin to deteriorate, to grow less efficient, and to slow down. This deterioration ordinarily is so well hidden from our view that most of us do not notice the physical signs of our aging until we are nearly middle aged. Imperceptibly, cells, organs, and internal structures of the body accumulate damage, and physical systems function in a less integrated way (Shock, 1977b). Thus middle-aged tennis players, joggers, and professional athletes notice that their endurance and speed are not what they were when they were in their teens and twenties. Forty-year-old boxers, with slowed reflexes and tiring muscles, resign their titles, and 40-year-old baseball players hang up their jerseys. People who at age 20 could drink cup after cup of coffee banish caffeine or risk sleeples nights at age 40.

This chapter describes the physical changes that take place in various systems of the body during early and middle adulthood. In the process, it also describes the differences between the effects of **primary**, or **normal aging** and **secondary aging** on the systems of the body. Primary aging is universal and unavoidable. It may be programmed into the genes. In contrast, secondary aging is neither universal nor unavoidable. Secondary aging is the product of disease, physical abuse, and physical disuse. For example, some wrinkling of the skin on people's face and hands is a result of normal aging, but most wrinkling is secondary aging, the result of exposure to damaging sunlight. Researchers are still trying to distinguish primary from secondary aging in many areas of physical function. Because the effects of secondary aging often can be avoided, averted, and reversed, knowing the difference between primary and secondary aging can mean the difference between health and disease, comfort and discomfort, use and abuse.

Appearance

During early adulthood, few people notice signs of aging in their appearance. Their skin is smooth and taut, their hair full and the same color that it has been for years. Some young adults do notice a few gray hairs, but usually physical deterioration is so gradual that people do not notice it until something jogs their attention. A mirror shows that crows' feet and gray hairs, balding and slack muscles have started in earnest. Most people are surprised by the first signs of aging.

> I was window shopping one day and saw a face reflected in the glass. I heard myself thinking, ''Who's that middle-aged woman?'' It was *me*. —Erica's mother at 39

The outward signs of aging in middle adulthood—gray hair, crow's-feet, slacker muscles—may come as blows to self-esteem.

Not only do healthy, energetic 40-year-olds think of themselves as young, but they live in a society that reveres youth. To them, the physical signs of aging come as an unwelcome surprise, and the negative stereotypes and attitudes of other people toward aging mean that it is felt by many people as a blow to self-confidence and self-esteem. In cultures where the aged are revered, the physical signs of aging may enhance self-confidence and self-esteem. But in cultures like our own, where aging is feared and reviled, seeing oneself age can feel like a punishment (Weg, 1983).

By middle age, many of the signs of aging have become unmistakable. But most people do not feel old just because they have turned 45 or 50. *Other* people's reactions to their aging appearance ultimately make them believe that they really are old. Being ignored or patronized or seeing television shows and magazines that caricature older adults as foolish and feeble of mind and body make people feel old. Few give in easily to this unpleasant feeling. Some embark on exercise programs and diets. Others spend their money on creams, dyes, cosmetics, plastic surgery, and hair transplants that will smooth, slim, firm, and cover the signs of age. Plastic surgery, once the province of women, now is sought increasingly by adult men as well (Kelly, 1977). Facelifts and skin tucks can freshen the appearance for five or ten years, by tightening or removing excess, slack skin (Wantz and Gay, 1981).

Hair

The baby-fine, silky hair of the infant is different from the thicker hair of the child, that of the child differs from that of the young adult, and that of the young adult differs from that of the older adult. The process of aging affects both the color and texture of people's hair.

> When I was a baby, my hair was straight, blonde, thin, and silky. By the time I was a teenager, it was wavy, brown, and thick. In my forties, it turned grayish and thin on top. Now that I'm in my fifties, my hair's turned gray all over and straight again.—Lainie's mother at 52

Whether hair turns gray at 30 or stays a natural black until 65 is largely determined by genes. People's hair may turn gray or white because their bodies lack a certain enzyme (Rossman, 1977) or because the cells in the hair follicles that produce pigment diminish or malfunction (Selmanowitz, Rizer, and Orentrich, 1977). Despite wide individual differences in the rate of change of hair color, graying hair is considered one of the most reliable signs of aging. By one reckoning (Damon et al., 1972), out of more than fifty physical indicators of aging, gray hair is the most reliable.

The aging process also shows itself in the changing texture of hair on various parts of the body. Hair follicles may produce fine, short, colorless **vellus hair** or coarse, longer, colored **terminal hair**. Men over 30 may notice that the fine, nearly invisible vellus hair in their ears has turned into tufts of dark terminal hair. But to a man's dismay, the thick, colored terminal hair on his scalp may give way to a spreading bald patch, sparsely covered with thin, colorless vellus hair. It is with a mixture of anxiety and resignation that a man watches himself go bald from his temples, across the top of his scalp, until only a fringe of terminal hair is left. Men inherit this kind of baldness, called male **pattern baldness**, on their sex chromosomes. Other forms of baldness appear after disease or injury. Although men, and not women, inherit pattern baldness, people of both sexes lose some hair from their scalps as they get older. The scalp hair of the average 50-year-old man and woman, for example, has thinned from 615 follicles per square centimeter to only 485 (Rockstein and Sussman, 1979). After menopause, women may find that their bodies' lower levels of sex hormones cause their pubic hair and the hair in their armpits to thin or to disappear.

Skin and nails

The smooth, dewy skin that we see on models in cosmetics advertisements gradually goes through a process of normal aging. Once past menopause, as hormone levels fall, women's skin ages relatively quickly. Men's skin seems to age less quickly, probably a result of their continually renewed beard growth. Past the age of 50 or so, the surface skin cells begin to replace themselves more slowly. Dead and damaged cells remain on the surface for a longer time, and the result is old-looking skin. In the deeper skin layers, elastic fibers and connective tissues stiffen up, and the skin becomes less pliable. As underlying fat deposits and supporting muscles shrink, the skin grows thinner, drier, droopier, and more wrinkled. Gone are the plump cheeks and dimpled knuckles of childhood. Said one 55-year-old woman:

> I noticed that the skin on my jaw and neck started getting looser when I was in my forties. When I'd complain about my "turkey wattles," my husband would kid me and say he loved me still, "and you don't look a day over 103." He really made me feel better.

Some wit once said that by the age of 50, people have the face they deserve. By then, wrinkles have etched a permanent expression—perhaps a sour, or smiling, or worried look—onto a person's face. Wrinkles on the face tend to form at an angle to the direction that the skin is pulled (Rossman, 1977). This tendency explains why laugh lines radiate around the eyes, frown lines stripe the forehead, and smile lines stretch from lips to nose.

The color, texture, and rate of growth of nails are affected not only by aging but by what people eat and drink, their general health, hormone levels, blood supply, and conditions in their environment. Aging itself slows the nails' rate of growth, so that the nails of the 80-year-old may grow 38 percent more slowly than those of young people (Selmanowitz, Rizer, and Orentrich, 1977). Aging also changes the nail bed and the blood vessels of the fingers.

Muscle and fat

All those gorgeous middle-aged women on television have inspired me to stick with my aerobics and jogging. I'm still nice and firm at 31. I'd give anything to look like Jane Fonda or Linda Evans in twenty years.—Natalie's mother

''Middle age spread'' is no laughing matter when it's *you* who's spreading! Since I turned 50, I've had to exercise twice as much to keep my muscles toned. My body seems to want to get flabby.—Sarah's mother

With age, the make-up of muscle cells changes, and the proportion of muscle to fat changes, too. Until people are about 39 years old, their muscles grow increasingly dense, but thereafter the muscles shrink and their fibers grow fewer in number and smaller in diameter (Bulke et al., 1979). Some of this shrinking of the muscles is normal aging, and some is secondary aging, a result of inactivity. The ''use it or lose it'' principle applies to aging muscles. Muscles atrophy when they are not used, and many adults are less active than they once were (Rockstein and Sussman, 1979). This inactivity probably is not healthy. Cross-sectional studies show that adults whose bodies are small and lean live longer than others (Weg, 1983). When people do not use their muscles, they weaken, a trend that may set in as early as age 30. Men aged 62 who worked as machinists, whose jobs required them to exert themselves strenuously, had the muscle strength and endurance of 22-year-olds (deVries, 1983). Researchers are not certain how much of the poorer muscle tone, strength, flexibility, and speed that they see among older people is a result of normal aging and how much is a result of a sedentary life, inactivity, and disuse.

With aging, as the muscle cells atrophy and die, fat cells increase. The net result is a body that is made up of relatively less muscle and more fat. For this reason, the 60-year-old who weighs the same as she did at age 25 is likely to have more body fat at 60. CAT (computerized axial tomography) scans, which are cross-sectional X rays of the body, show that until the age of 50, fat cells are found mainly between skin and muscles. But after that age, healthy men have fat cells in and between their muscles (Borkan et al., 1983).

Sensory system

The senses generally become less responsive to external stimulation as people get older, but most of these changes do not interfere with everyday functioning until late adulthood. For example, most people are at the peak of their sensory acuity at around age 5. Sensitivity to glare increases after adolescence, and visual acuity declines early in adulthood. But these changes are so slight as to go unnoticed by most adults. By middle age, many people need glasses to read or to perform other activities up close. But it is not until late adulthood that most people feel the deficit in their depth perception, their perception of blues, greens, and violets, or in their adjustment to changes in illumination (see Figure 17.1).

By middle adulthood, many people need glasses for reading and close work—an accommodation to normal aging that is easily made.

Hearing also declines slowly after the mid-twenties, although most people do not notice any change until middle adulthood or later. The ability to hear sounds at high frequencies is lost first, especially by men. Thus middle-aged men are less likely than middle-aged women to hear the sound of a doorbell or the high notes of a woman singing on the radio. Many speech sounds are high-pitched—*s, z, f, g,*—for example—and may eventually become inaudible. When this happens, older adults may find that they miss parts of conversations. In hearing loss as in so many other physical changes of aging, the differences among individuals are great. The differences in 65-year-olds' abilities to hear high-pitched sounds are 3½ times greater than those among 25-year-olds (Corso, 1977) (see Figure 17.2).

Throughout adulthood, the tastebuds are continually replaced and restored, and the sense of taste generally remains keen throughout most of adulthood. Only late in adulthood is there a drop-off for certain flavors. Sensitivity to touch and pain also remains keen throughout early and middle adulthood.

Cardiovascular system

The cardiovascular system is made up of the heart and the blood vessels—veins, arteries, and tiny capillaries—that carry blood throughout the body (see

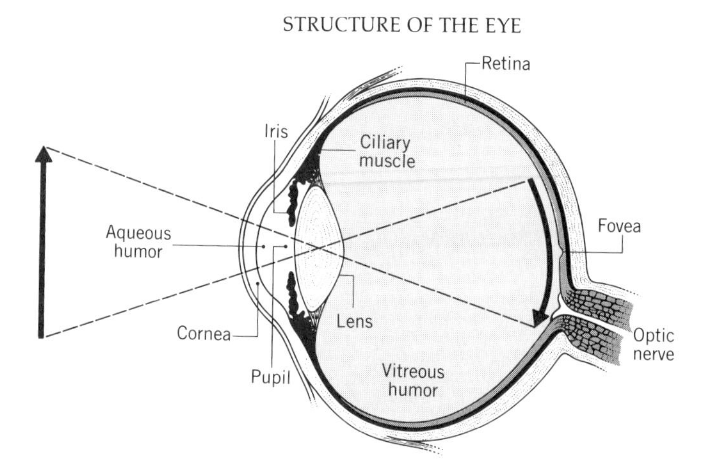

Figure 17.1 Clear vision requires that images be focused sharply on the fovea. It is tightly packed with color receptors. In adulthood, age-related changes in the structure of the eye are normal but may reduce visual acuity, color and depth perception, and sensitivity to changes in visual stimuli.

MAJOR STRUCTURES OF THE EAR

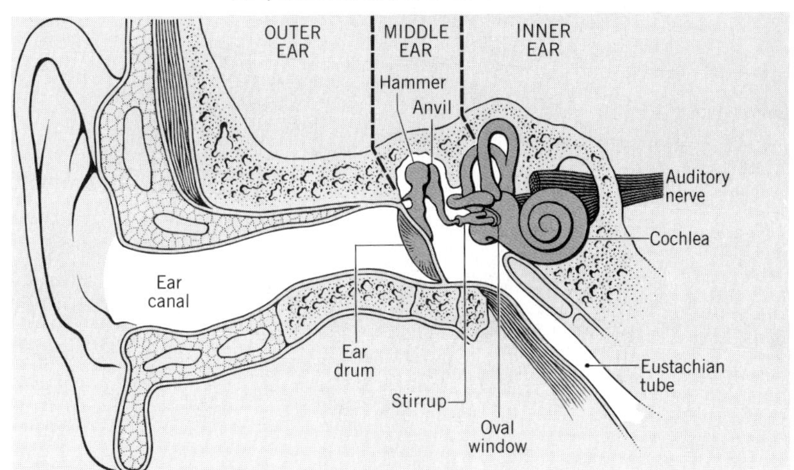

Figure 17.2 The age-related changes in hearing are virtually imperceptible to most young adults. But by middle age, people may notice that they have trouble hearing certain high-frequency tones as clearly as they once did.

Figure 17.3). As people get older, normal changes take place in their cardiovascular systems (see Chapter 20 for more information). Unlike bone and muscle cells, the cells within the cardiovascular system cannot divide and reproduce themselves, and so ultimately the system grows less efficient. Some of the deterioration of the cardiovascular system results from the normal wear and tear of aging, some results from disease, and some results from people's habits of diet and exercise.

Heart

From soon after conception until death, the human heart beats continuously. It rests for only part of a second between beats and may beat 3 *billion* times in the course of a lifetime (Rockstein and Sussman, 1979). At rest, the muscles of the heart may pump 75 gallons of blood an hour; it may pump 750 gallons during strenuous exertion. With age, the weakening muscle means that the heart pumps less blood with each pumping action and pumps less slowly (Kohn, 1977). **Cardiac output**, which is the measure of the blood volume the heart pumps in one minute, drops by 1 percent every year after age 20.

Respiratory system

The amount of air flowing into the young adult's lungs is 20 to 30 percent more than it is in the older adult. Past the age of 20, the amount of breath that a person can take into his or her lungs in a single breath begins to decline. Between 25 and 85, this vital capacity has dropped 40 percent. The total capacity of the lungs, however, does not drop with age. As a result, the lungs' residual volume, the amount of air that stays in the lungs after a person breathes out as much air as possible, increases over time and with increasing speed after age 45. The residual volume of air in a 20-year-old's lungs is 20 percent; in a 60-year-old it is 35 percent (Klocke, 1977).

Although the amount of air that people can take in and out with each breath begins to fall during the twenties, there are wide individual differences in people's lung capacity. For some 60-year-olds, maximal oxygen uptake, the greatest volume of air they can breathe in and out in 12 seconds, has fallen by 30 percent. But people who exercise regularly and former athletes have greater oxygen uptake than people who are sedentary (Klocke, 1977).

HEART

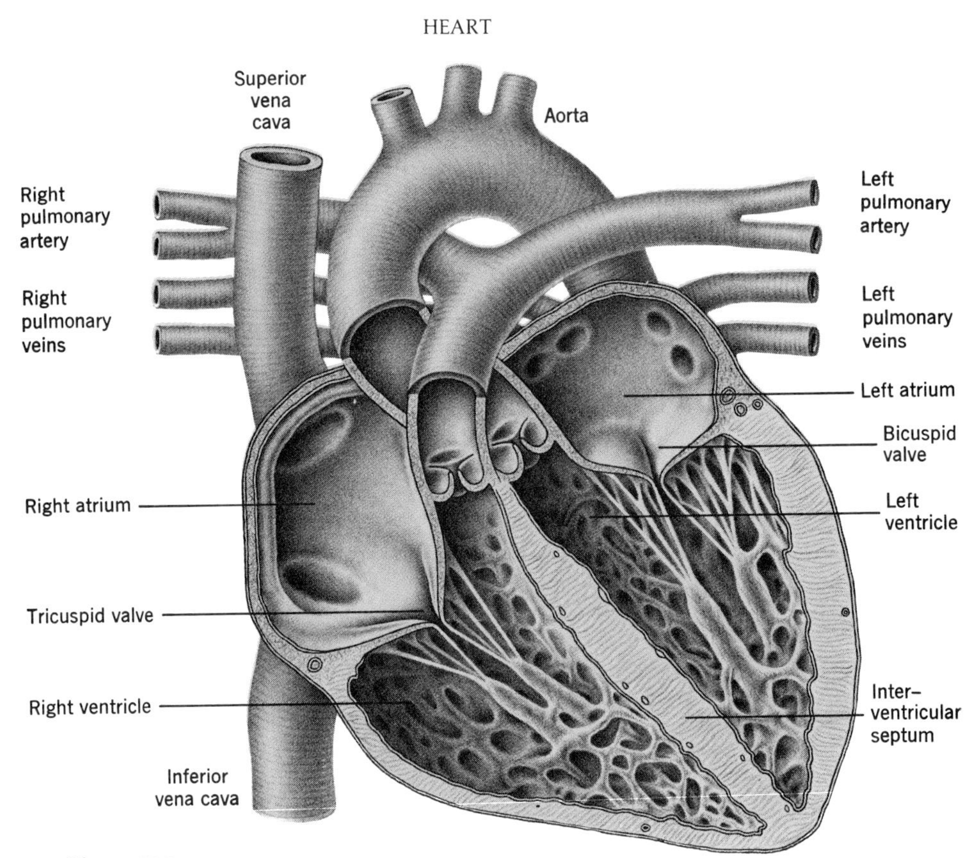

Figure 17.3 ***The cardiovascular system is made up of heart and blood vessels. Changes in this system are normal to aging. Many of the problems associated with the cardiovascular system are preventable.***

Exercising and maintaining a healthy diet can slow the normal aging process. This woman and her husband work out in a gym.

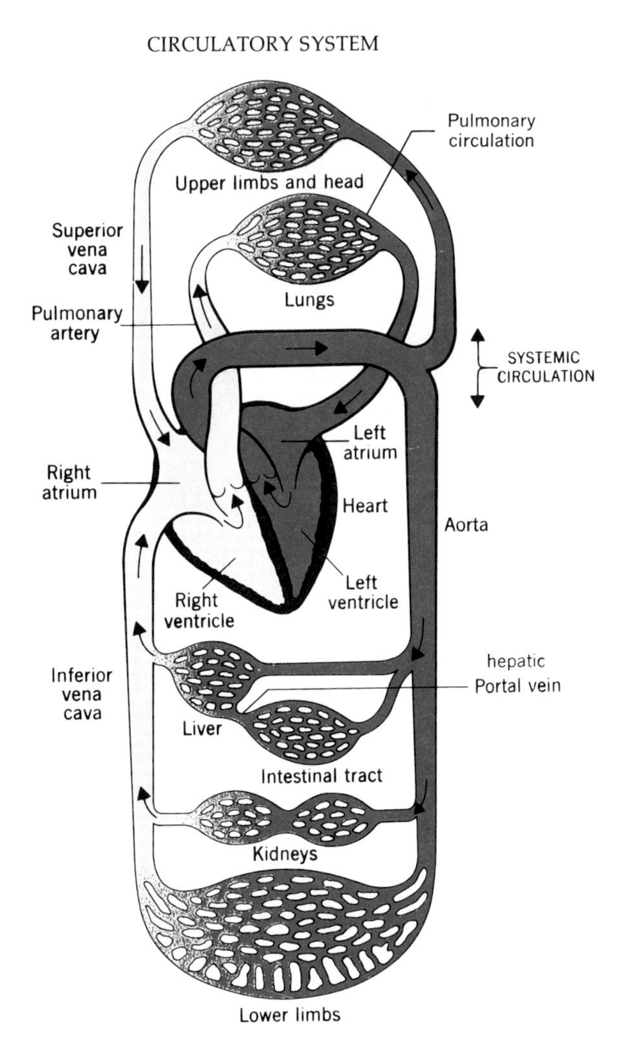

Endocrine system

The endocrine system is made up of several different glands that secrete hormones directly into the bloodstream (see Figure 17.4). The endocrine system is regulated by feedback loops and controlled by the pituitary gland. The endocrine glands regulate such vital functions as sexual reproduction, immunity to stress and disease, metabolism, growth, and even the aging of cells. Aging affects the endocrine system by slowing the secretion of hormones, by making cells in various parts of the body less responsive to hormones, by changing the chemical messengers that carry hormonal "messages" into cells, and by changing the levels of enzymes that respond to hormones (Marx, 1979). The endocrine system itself is so complex in its functioning that researchers are just beginning to understand how it ages.

Pituitary gland

The pituitary is the master gland of the endocrine system. It sits deep within the brain, nourished by blood, transmitting hormonal messages to cells throughout the body. Following signals from another part of the brain, the hypothalamus, the pituitary may actually set off the process of aging. Re-

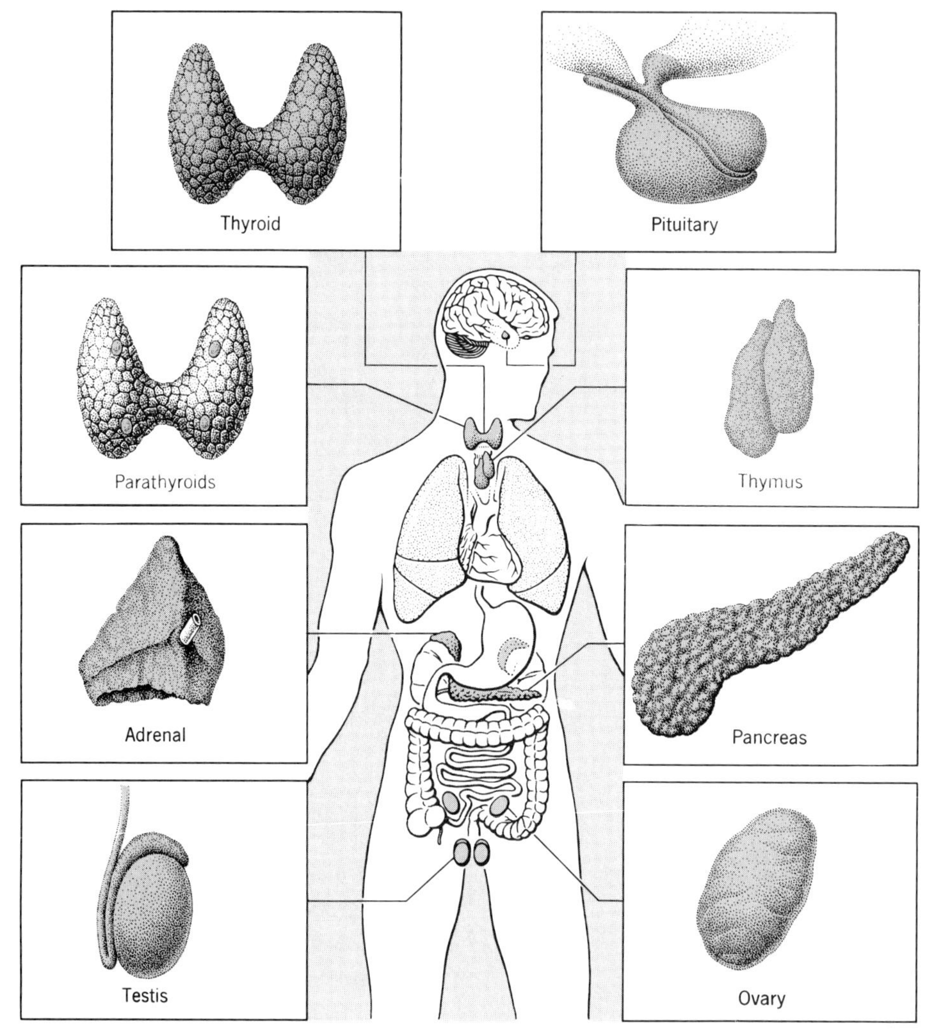

Figure 17.4 ***The endocrine system regulates sexual functioning, immunity to stress and disease, growth, metabolism, and the aging of cells.***

searchers have removed the pituitary of young rats, injected them with one third of its usual secretions, and have found that the rats do not age. Damage to kidneys slows, arteries repair themselves, and the rats' heart stays healthy (Walford, 1983).

Although the size and weight of the pituitary do not change with age, the blood supply to it gradually decreases after puberty. By the time a person is 60, the blood supply to the pituitary has decreased appreciably (Rockstein and Sussman, 1979). The amount of connective tissue increases, and the distribution of types of cells within the pituitary shifts.

The hormones that the pituitary secretes generally remain at adequate levels even into old age. One of these hormones is growth hormone (GH). It influences the metabolism, growth, and repair of cells. Women's blood levels of GH ordinarily are higher than men's, and people of either sex who are overweight have higher levels of GH than people of normal weight. After they are 40, however, women of normal weight have gradually declining levels of GH. But the pituitary continues to produce GH throughout the lifespan. Even in people between 63 and 99 years old, the pituitary responds to a sharp drop in blood sugar with just as efficient a production of GH as occurs in young adults (Andres and Tobin, 1977). Throughout adulthood, the pituitary also continues its steady production of thyroid-stimulating hormone (TSH) and ACTH, the hormone that stimulates the adrenal glands to respond to stress.

Parathyroid glands

At the base of the throat are the parathyroid glands. The hormones they secrete help the body to metabolize minerals such as calcium and phosphates (Rockstein and Sussman, 1979). The parathyroid glands grow heavier until men turn 30 and women turn 50. They do not shrink with age, and their output of hormones remains steady as well.

Thyroid gland

As people age, the cells within their thyroid gland undergo structural changes, and collagen fibers appear. Although the blood levels of thyroid hormones do not drop with age, the thyroid itself produces fewer hormones. These hormones regulate the body's metabolism of fats and carbohydrates, regulate the bones' absorption of calcium and phosphate, and stimulate the cells' use of oxygen. The thyroid continues to function normally in healthy older people, and when they have acute infections, their thyroid produces extra hormones (Andres and Tobin, 1977).

Pancreas

The pancreas is the gland that produces the hormone insulin. Without insulin, the body cannot metabolize the sugars in the diet. Insulin stimulates the carrying of blood sugar (glucose) to the cells of the body. When the metabolism of blood sugar goes awry, a person has diabetes. Some people have the kind of diabetes caused by an insufficient production of insulin, but most people whose diabetes develops in adulthood do produce enough insulin. In "maturity-onset" diabetes, the cells of the body have grown insensitive to insulin (Marx, 1979). But even in many older adults without diabetes, the pancreas is slow to release insulin in response to blood sugar and releases less insulin than it once did. Over half of the people older than 65 metabolize blood sugar less efficiently than they once did, although less than one in ten has diabetes (Rockstein and Sussman, 1979). The pancreas of an older adult also releases more inactive insulin. The net result of these changes in pancreas functions, in addition to the aging cells' tendency to absorb less blood sugar from the bloodstream, is longer and higher levels of blood sugar (Weg, 1983). Is the decline in the body's metabolism of blood sugar normal aging or not? Researchers do not know.

Adrenal glands

The adrenal glands sit above each kidney. They release the hormones epinephrine and norepinephrine into the bloodstream, which mediate the response to stress. They release the sex hormones androgens in both sexes. They mediate the kidneys' absorption of sodium and chloride and the body's metabolism of fat and carbohydrates.

The output of some adrenal hormones declines with age. For reasons that no one understands, the output of androgens declines with age. But the adrenal hormones that regulate the metabolism of fat and carbohydrates does not decline with age. It does, however, take the older body longer to metabolize these hormones than it once did—40 percent longer for the 75-year-old man than the 35-year-old man (Andres and Tobin, 1977). The hormones that

lifespan focus

Champion swimmers

Lifespan developmental psychologists have been interested in learning about the performance of top athletes as they get older. They have wanted to know whether their performance declines with age and, if so, whether the decline is the result of environmental or biological factors. Researchers therefore studied championship swimmers between the ages of 30 and 84 (Hartley and Hartley, 1984a). First, they wanted to compare differences among athletes from different cohorts with differences over time among athletes within the same cohort. Second, they wanted to compare age-related changes, if any, in performance in swimming races that were mainly *anaerobic*—that required a brief, explosive output of energy and that used available oxygen—with performance in races that were mainly *aerobic*—that required longer output of energy and

Training may have a stronger influence on swimmers' performance than normal aging. These men are competing in the Senior Olympics.

affect kidney function are stable in old age, and so the fluid balance of the healthy older person is normal (Rockstein and Sussman, 1979).

Thymus gland

The thymus gland sits in the upper chest cavity and plays a vital role in the immune system. The thymus produces T cells, those parts of the immune system responsible for rejecting foreign cells and tumors. The bone marrow produces B cells, those parts of the immune system responsible for producing antibodies to disease, but B cells need T cells to produce antigens (Mackinodan, 1977). The role the thymus gland plays in immunity has led some researchers to single it out as one of the most important regulators of the aging process (Walford, 1983).

The thymus gland grows throughout childhood and then shrinks. Cells in its cortex die and give way to fat and connective tissue. The thymus gradually produces less hormones as it shrinks, and the drop-off is pronounced in people between 25 and 45 years old (Walford, 1983). As the hormones drop, antibodies drop, too. The T cells respond more weakly to invaders to which used oxygen from the lungs. The researchers knew that aging is associated with less efficient use of oxygen in the muscles, a decrease in muscle size, and poorer coordination. Therefore, they predicted that physiological aging would take a greater toll in short, explosive races than in longer, aerobic races. Their findings supported this prediction. (However, other researchers used a different measurement of the swimmers' performance and found that performance declined more in the long-distance races than the sprints [Stones and Kozma, 1984]. The original researchers suggested that there was more absolute loss of speed in sprints and more relative loss of speed in endurance races and that changes in muscle would explain the former, changes in oxygen use would explain the latter [Hartley and Hartley, 1984b].)

The researchers studied swimmers both longitudinally and cross-sectionally. The longitudinal study showed that among the men, those in their thirties improved slightly over five years, but among older men, performance declined increasingly with age. For women, cohort effects also were significant, but performance did not drop significantly until women were in their seventies. The longitudinal data therefore suggest that cohort differences had a stronger effect on the swimmers' performances than physiological aging itself. The cross-sectional data also showed strong cohort effects among the men and the women. For both sexes, their speed in races declined as their age increased.

Thus these researchers suggest that cohort differences were more significant influences on the swimmers' performance than age changes were. For example, possibly adults go through some periods of their lives when they train less because of other demands on their time. Possibly they are not exposed to new training techniques, or chances to compete are fewer for older adults. Possibly, too, given the finding that there is less of a drop in speed from short to long races among the older swimmers than the young, older swimmers may train for aerobic fitness and may adopt race strategies that keep them within their aerobic limits. Although adults should expect their performance to drop off somewhat with age, the studies of championship swimmers suggest that the drop-off may not be so great as cross-sectional data have led people to expect. In the future, it will be interesting to see whether the long-term changes in performance decrease still more as older swimmers have more and more regular chances to train and to swim competitively.

they have not already been sensitized. With a weaker immune system, older people may come down with more infections, cancerous tumors, and diseases of the immune system, including late-onset diabetes and a form of anemia (Mackinodoan, 1977).

Reproductive system

The male reproductive system

Over time, the reproductive system of aging men functions less quickly and efficiently, but most functions remain well into extreme old age. Some of the decline in reproductive function stems from aging of the endocrine system, because hormones from the two systems are interdependent. For example, the pituitary hormones follicle-stimulating hormone (FSH) and luteinizing hormone (LH) regulate the hormone production of the testes. For sperm to mature through their many stages of development, FSH must be present. For the testes to produce testosterone, LH must be present. As the bloodstream fills with testosterone, the pituitary slows the production of LH.

Researchers are just beginning to understand the effects of hormones on human sexuality, but the research results on the effects of aging on hormone production and function are not always consistent. Some studies show LH levels but not FSH levels rising in men over 65; other studies show LH and FSH levels rising (Talbert, 1977). Some studies show testosterone levels falling in men over 60, but others suggest that this decline is not a part of normal aging. When a group of healthy older men were tested, their testosterone levels were no lower than, and in some cases were slightly higher than, the levels of young men (Marx, 1979). The likelihood is that hormone production falls off slightly with age, and the aging body takes longer to metabolize the hormones.

Men's testes produce sperm throughout adulthood, although the number does drop off gradually. The testes of adult men between the ages of 20 and 39 produce sperm in 90 percent of the sperm-producing tubules. The number of sperm-producing tubules drops off again during the forties to about 50 percent. After the age of 70 comes another decline, and the testes of men over 80 produce sperm in only 10 percent of the tubules. Some men ultimately produce no sperm at all, but half of the men tested between 80 and 90 years old still produced sperm (Talbert, 1977).

The prostate gland, which secretes fluid that mobilizes sperm and carries them out of the penis during ejaculation, starts to age during a man's forties. Muscle gives way to connective tissue, blood supply diminishes, and the prostate of most older men begins to grow. In a 70-year-old man, the prostate may be twice the size it was and may have to be removed if it impinges on the bladder (Rockstein and Sussman, 1979).

> When I went in for prostate surgery, I was 58 and pretty down. I felt all washed up. But once that nuisance was cured, things turned around for me. It took some doing, and I got my share of strange looks from my wife, but now we enjoy sexual activity pretty regularly. It keeps us young.—Tim's grandfather

The penis of a man in his thirties may show signs of age, as cells and blood vessels harden and grow less elastic. These changes in the blood vessels can interfere with the filling of erectile tissue in the penis with blood during sexual arousal, and erections of the penis may grow slower and less firm. Compared to that of a 20-year-old, it takes the penis of a 50-year-old six times longer to grow fully erect (Solnick and Birren, 1977). By the time a man is 60, the angle of his fully erect penis is 45 degrees, or half what it was when he was younger (Comfort, 1980). Much to his relief, these are the only changes in the reproductive system of the healthy older man.

The female reproductive system

The adult woman's reproductive system, like the man's, produces different levels of hormones from those it once did. The woman's pituitary gland, also like the man's, produces FSH and LH. FSH stimulates the ovaries, or egg-producing glands, to grow the follicles within which ova are nourished and mature. LH stimulates ovulation, the cyclical release of mature ova and follicle from an ovary. LH also stimulates the production of the hormones estrogen and progesterone. The ovaries and adrenal glands produce androgens as well, some of which are changed into estrogen. (Androgens are sometimes called ''male'' hormones, estrogen and progesterone ''female'' hormones. In truth, the endocrine and reproductive systems of both sexes produce both kinds of hormones. Only the ratios differ, with males producing more ''male hormones'' and females producing more ''female hormones.'')

All of the ova in a woman's ovaries develop to an immature state before birth, and the aging of the female reproductive system begins at birth. The number of immature ova in each ovary falls from an average of 700,000 at the birth of a baby girl, to half that amount when she first begins to menstruate, halves again between ages 18 and 24. By the time she is 40 years old, each of a woman's ovaries contains 11,000 immature ova (Rockstein and Sussman, 1979).

> When I was in my twenties, I looked ahead to menopause with horror. But twenty-five years and two children later, I felt no horror, just a twinge of regret and a lot of relief. For the first time in my life, I could enjoy sex without worrying about getting pregnant.—Kristy and Jason's mother

Fertility declines gradually once a woman is past her twenties, but it declines more quickly during her forties as menstrual periods get shorter and more irregular. These changes signal the ''change of life,'' or **climacteric**. Now her ovaries slowly stop functioning and, when she is about 50 years old, menstrual cycles grow longer and more erratic. Finally, menstruation stops altogether, and a woman has reached **menopause**. At that point, her body's production of estrogen and progesterone sharply declines and their levels in the bloodstream fall. Her body keeps on producing androgens, and because these are converted to estrogens, some estrogen remains in the bloodstream of the postmenopausal woman (Solnick and Corby, 1983). Two years after menopause, FSH levels are 18 times higher and LH levels three times higher than they were before the change of life (Hammond and Maxon, 1982).

Despite the widespread belief in this culture that virtually every woman suffers through menopause with hot flashes and night sweats, bursts of temper and crying, only one quarter of the women in a large sample said that they were uncomfortable during menopause (Corby and Solnick, 1980). These uncomfortable symptoms include head- and backache, nervous tension, and flashes of hot and cold. Rises in LH have been found to correspond to hot flashes, themselves caused by changes in the hypothalamus (Marx, 1979). Other symptoms may be associated with changing estrogen levels.

The cultural stereotype is that the menopausal woman suffers both psychologically and physically. In many cases, however, the truth is quite different.

Another consequence of falling estrogen levels may be changes in the vagina. The muscular walls of the vagina grow thin and less elastic. The walls also produce fewer normal secretions, so the vagina feels dryer. Women who find vaginal dryness a problem during sexual arousal can use sterile lubricating jelly, available at any drugstore. Some women take prescribed estrogen, and this therapy can stem not only the atrophy of the vagina, but other consequences of menopause, such as hot flashes and osteoporosis. But estrogen replacement also increases the risks of certain kinds of cancer and of gallbladder disease (Hammond and Maxon, 1982). Women who remain sexually active and regularly masturbate or have intercourse have less atrophy of their vagina and higher levels of estrogen.

Sexuality

Early and middle adulthood may well be the sexual prime of life. True, adolescents have sexual experiences and relationships. But for most people, the social and psychological developments of adulthood greatly enrich sexuality. For example, in young adulthood, many people live together in marriage or in other kinds of couples. During the first year that couples live together, their rates of sexual intercourse typically are the highest of their adult lives. After the first four years of marriage, the rate of intercourse declines, even among couples in their twenties (Udry, 1980). And as adulthood continues, the rate of intercourse declines gradually, the Kinsey reports showed (Kinsey et al., 1948, 1953). (See Figure 17.5.)

But it is likely that cohort differences will become apparent between today's baby boom generation, who will be the 40-year-olds of the 1990s, and the middle-aged adults who responded to Kinsey's questions. And these differences in sexual histories are likely to show up in sexual behavior during middle age. For example, many of the baby boomers have had more extensive sexual experience and more liberal sexual views than the older generation. Many also have fewer children and live in two-job families. Lifespan psychologists do not yet know whether these factors will push the coming generation of middle-aged adults toward more sex later in their lives than their predecessors had or toward less sex because of all the competing demands on their time and energy (Luria, Friedman, and Rose, 1987).

Sexual responsiveness slows and weakens with age, and each stage of the sexual response cycle is longer. Sixty-year-old men can hold their erections longer than younger men before ejaculating. They may not sense the inevitability of orgasm, and their orgasms may consist of fewer and weaker contractions. The refractory period, during which the previous physical changes of the sexual response cycle reverse themselves, gets longer, too. For older men, that means a longer time when they cannot get another erection. Women's sexual responses go through parallel changes. Signs of sexual stimulation—lubrication of the vagina, increase in size and color changes of breasts and clitoris—are weaker and take longer. The number and intensity of contractions in a woman's orgasm also drop.

Although none of the physiological changes in sexual responsiveness necessarily affects people's enjoyment of sex, the frequency of intercourse declines with age. Why? The answer is part biology and part sociology. Duke University researchers have found that when middle-aged couples stop having intercourse, typically the reason is the husband. In general, a man's physical functioning tends to relate significantly to his level of sexual interest and enjoyment, and men who feel unhealthy are likely to show lower levels of sexual enjoyment and interest than healthy men (Pfeiffer and Davis, 1972). The aging male's gradual decline in the sex hormone testosterone explains only part of his decline in sexual activity. Although a woman's level of testosterone does not decline slowly with age, more middle-aged women than men report declining interest in sex (Pfeiffer, Verwoerdt, and Davis, 1972). What accounts for this pattern? Possibly older men find it harder to admit to sexual decline than older women. Possibly some older women are happy to see sex winding down.

The psychological growth of early and middle adulthood can enrich sexuality for many people.

In the past, many couples gave up sex entirely at the first sign of a man's problems with erections. Sudden, irreversible erectile problems are not common in middle-aged men, but fears of "impotence" are. Among all men, the hormone testosterone must remain at a certain level in the bloodstream for erections in coitus to remain. Men whose testosterone falls well below that level may benefit from externally administered testosterone (Davidson, Camargo, and Smith, 1979). Still other men have erection difficulties that stem from problems with their blood vessels: blood apparently flows into the penis but flows out of it too soon—for reasons unknown (Karacan, Aslan, and Hershkowitz, 1983).

Older men can learn to work with rather than against their slowed responsiveness. Longer, more intense stimulation can help arousal. When sex with a familiar partner may have grown stale, older men sometimes turn to a new partner. The excitement and anxiety of a new liaison dramatically increase a man's level of arousal. Sometimes shifting to new techniques (turn-taking, mutual masturbation, oral–genital sex) or to a new use of erotic materials that spur fantasy can provide the necessary impetus to arousal. Older men

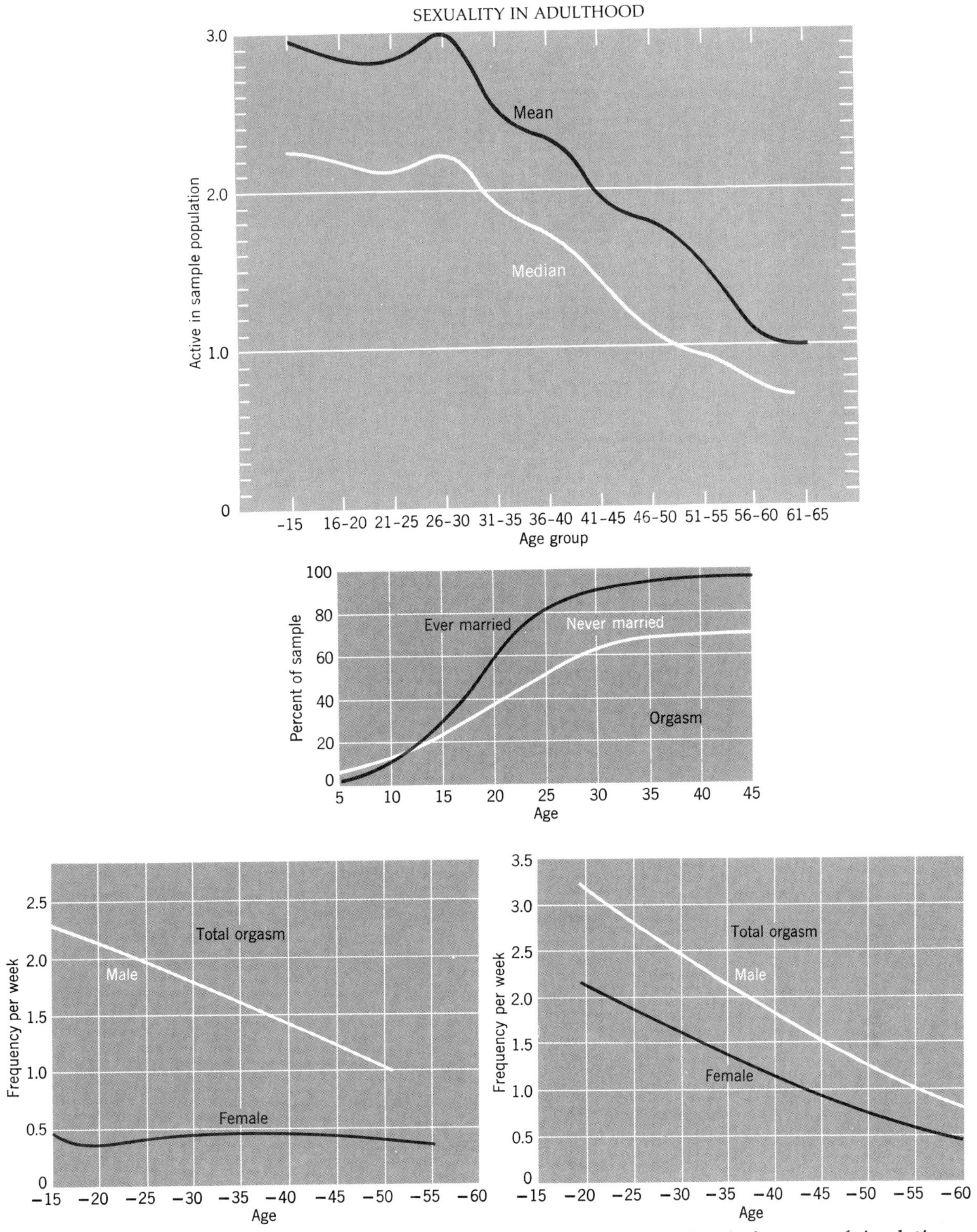

Figure 17.5 **(a)** ***The Kinsey reports showed that for males, the frequency of ejaculations was highest in the early years of adulthood and then waned gradually. Means are mathematical averages. Medians are midpoints between top and bottom statistical halves of samples. (Based on Kinsey et al., 1948).*** **(b)** ***The frequency of orgasms for women—whether from intercourse, masturbation, or another source—tended to rise from early to middle adulthood (Based on Kinsey et al., 1953).*** **(c)** ***The number of orgasms for males and females tended to decline during adulthood. Rates for single people appear on the left, those for married people on the right (Kinsey et al., 1953).***

vary tremendously in their individual sexual responsiveness. Their physical and mental health, their willingness to experiment, their interest in sexual relationships, and the frequency of their sexual activity as younger men (whether a genetic factor or the "use it or lose it" principle again) all affect an aging man's sexual responsiveness.

Summary

1. Aging begins at birth, as various systems of the body gradually deteriorate. The individual differences among adults in the physical changes associated with aging are greater than those among younger people. Individual differences in energy levels, capacity for work, and general health grow larger as people age. Aging rarely is uniform throughout an individual's body.
2. Young and early middle adulthood generally are considered the prime of life. Although measurable physical decline has set in, the body's large reserve capacity, the ability to meet virtually all everyday physical demands, the slowness of decline, and the ability to accommodate to the changes all tend to prevent adults from feeling significantly in decline until late in adulthood.
3. Primary aging is universal and unavoidable. But secondary aging is the product of disease, physical abuse, and physical disuses. It is neither universal nor unavoidable.
4. After age 39, muscle fibers shrink, and the ratio of fat to muscle increases.
5. Aging affects the sensory system, vision and hearing especially. With age, people gradually become less sensitive to external stimuli.
6. The cardiovascular and respiratory systems slowly decline during adulthood, although many of the changes are the result of secondary aging. For example, cohort differences have been shown to affect the performance of older athletes even more than age-related changes.
7. Aging affects the endocrine system by slowing the secretion of hormones, by making cells in various parts of the body less responsive to hormones, by changing the chemical messengers that carry hormonal "messages" into cells, and by changing the levels of enzymes that respond to hormones.
8. Early adulthood represents the peak years for many people's sexual relationships and activity. Over time, the reproductive system functions less quickly and efficiently, but most functions remain well into extreme old age. Sperm production gradually declines in men; women's menstrual cycles stop at menopause. Both sexes experience a decline in the production of some sex hormones. But the sexuality and sexual relationships of many adults remain healthy and flexible.

Key terms

primary (normal) aging
secondary aging
vellus hair
terminal hair
pattern baldness
cardiac output
climacteric
menopause

Suggested readings

FINCH, C.E., and SCHNEIDER, E.L., *Handbook of the Biology of Aging*. New York: Van Nostrand, 1985. Many up-to-date reviews of the most important work on the biology of aging, including chapters on aging and mortality, molecular biology, cell biology, physiology, neurobiology, and pathology.

KLINE, DONALD, and SCHIEBER, FRANK. "Vision and Aging." In J.E. Birren and K.W. Schaie (eds.), *Handbook of the Psychology of Aging*. New York: Van Nostrand, 1985. A comprehensive review of visual aging, including a discussion of age-related differences in light sensitivity and color vision, spatial and temporal resolution, visual information processing, and perceptual organization and flexibility.

OLSHO, L. WERNER, HARKINS, STEPHEN, and LENHARDT, MARTIN. "Aging and the Auditory System." In J.E. Birren and K.W. Schaie (eds.), *Handbook of the Psychology of Aging*. New York: Van Nostrand, 1985. A comprehensive review of auditory aging, including a discussion of age-related differences in the anatomy and physiology of the auditory system, performance on psychoacoustic and speech perception tasks, and auditory rehabilitation.

WELFORD, A.T., "Sensory, Perceptual, and Motor Processes in Older Adults." In J.E. Birren and R.B. Sloane (eds.), *Handbook of Mental Health and Aging*. Englewood-Cliffs, N.J.: Prentice-Hall, 1980. A review of age-related differences in sensory, perceptual, and motor processes, including a discussion of sensory function, random central nervous system activity, integration of perceptual and memory data, and factors involved in taking action.

chapter eighteen
Cognitive development

At any age, learning, thinking, and remembering depend upon the information-processing "system." This system largely reaches maturity by early childhood but becomes progressively more efficient throughout childhood. Through most of adulthood, the information processing system remains relatively stable, although, in some cases, it may continue to improve in efficiency. Adults process and learn new information; they remember and assess old information. They learn new skills and hone old skills. Although relatively few psychologists have studied cognitive change in early adulthood, there is general agreement in the field that cognition undergoes little change for the worse until past the age of 60 or 70 or so.

> When I was a kid, I seemed to remember everything. By the time I was in my forties, I was much more forgetful. But when I was a kid I knew a mere fraction of what I know today.—Jason's mother

> To keep up in my field, I have to read lots of specialized journal articles every month. I've been doing it for over 30 years now, and I have no trouble absorbing the information.—Erica and Mike's father

In this chapter, we first discuss the information-processing abilities of adults (some of the research in this chapter covers not only early and middle but later adulthood as well). Then we turn to long- and short-term memory and intellectual and reasoning abilities.

The information-processing system

Processing information means taking it in from the environment, manipulating, storing, classifying, and retrieving it from memory. Thus information processing includes everyday cognitive acts such as taking notes on the kind of car to buy, reading car reviews in magazines, listening to a salesperson describe the differences among models, remembering those differences, and thinking ahead to the pleasures of driving to the beach in your new car during upcoming vacation.

Studying information processing in adults

As we have seen in earlier chapters, learning and remembering rely on cognitive processes that transfer information. When we learn, we acquire information or skill from our experience. When we remember, we store and retrieve the information or skill that we have learned. Learning and remembering abilities change little during adulthood, at least until old age. Age itself is not a very good predictor of adults' abilities to learn or to remember. On experimental tasks, younger adults often do better than older adults. But the margin usually is slim, and the breadth of individual variation is great. Nearly always, some older adults outperform some of the younger adults. Moreover, there appear to be reasons for the age differences that occur in average performance that have nothing to do with age itself.

Nevertheless, when psychologists compare the cognitive performance of young and old adults on experimental tasks, they often find that the average scores of young adults are higher than those of many of the older adults. Why? These apparent age changes in cognition may be the result of two phenomena: cohort effects and research designs. Psychologists find cohort differences between young and older adults. One such difference is the fact that as a group, young adults today have more years of formal education than older adults. More young adults have finished high school and college. In fact, many of the young adults who participate in studies of cognitive development are college students. Because more highly educated people tend to perform

One reason that younger adults generally score higher on tests of cognitive skills than do older adults is that young adults are likely to have much more recent practice in the kinds of cognitive skills that the tests tap.

better on cognitive tests than less educated people, unless researchers control for the level of their subjects' education, the older adults are likely to test more poorly than the young adults.

In addition, young adults have much more recent practice in the kinds of cognitive skills that many researchers test than older adults—another cohort effect. College students regularly are called on to learn and memorize information and to exercise the very skills that experimental learning tasks tap into. Older adults are likely to be decades away from formal education and to be rusty on these cognitive skills and strategies. When people stop using certain cognitive skills, they may have trouble summoning them up when they want them later on—the mental equivalent of the "use it or lose it" principle that affects physical skills. In addition, if older adults expect themselves to learn or remember poorly, they may set in motion a self-fulfilling prophecy. By failing to try, they may fail to exercise their cognitive potential.

> After my kids were grown, I decided to go back to school and get a law degree. I had a really tough time with the preadmission standardized tests. It had been more than 20 years since I'd had to answer questions like those, and I spent days practicing so that I would do well.—Megan's mother

> I was 38 when I took the state examination to get my license as a psychologist. To prepare, I had to memorize an undergraduate-level textbook on development. It took me months to plow through that mass of information. That part of the professional requirements was much harder for me than writing my dissertation or seeing patients ever were.—Nancy's mother

Another cohort effect derives from the fact that the educational experiences of young adults may differ substantially from those of older adults, with resulting differences in their cognitive processes. Most young adults have been watching television and gathering information from it since they were in diapers. They may have reached a level of information from television that people once could reach only with many more years of formal education. In contrast, older people learned in a more disciplined atmosphere and may have reached a wider range of scholarship (Perlmutter, 1983).

A second phenomenon that produces apparent age differences in adults' cognition derives from the way that cognition typically is studied. Cognition often is studied in the laboratory, divorced from the everyday situations in which people actually learn and remember. As we saw in the research on very

A second reason that young adults generally test better than older adults has to do with the conditions under which cognition is studied—in a laboratory, divorced from the everyday situations in which older adults ordinarily learn and remember.

young subjects, the nature of the surroundings—a child's own house versus a strange laboratory, even one with a home-like playroom or kitchen—strongly influenced the children's cognitive skills and performance. Similarly, older adults often do not test as well as young adults in the laboratory when they are asked, for instance, to repeat a sequence of numbers or to learn a new task in 3 minutes. But what laboratory research like this does not reveal are the differences, if any, between young and older adults' abilities in everyday situations—to learn how to use a new microwave oven, word processor, or automatic teller machine or to adapt to a new bus schedule, bank statement, or income tax form.

> I've been teaching college-level English literature for about 15 years, and I find that every time I go back to Shakespeare, or Keats, or Dickens, to mention just a few of the true greats, I understand the work better and appreciate it more deeply. Instead of getting tired of things after the dozenth time, I actually know them better. And that makes me a better teacher, too.—Sam's father

> When I was 41, my boss asked me to learn how to use a computer with spreadsheets for financial data and a word processing program. I was afraid ahead of time that I wouldn't be able to do it. But it turned out to be a lot of fun. A couple of years later, we switched to a new computer system, and I had to work my way through new software "languages." It wasn't too hard at all.—Robbie's mother

Vital to people's cognitive functioning at any age is their familiarity with the material to be learned and their familiarity with related material (Perlmutter, 1983). In real life, people do not encounter information in isolation. They encounter information in context; they acquire and store new information within the context of old information. For example, over time a person learns in increments as he moves from manual typewriter, to electric typewriter, to word processor or from simple arithmetic, to algebra, to geometry, to higher mathematics.

In an attempt to make the demands of a laboratory study resemble the demands of everyday life, researchers (Waddell and Rogoff, 1981) supplied some of their subjects with objects in context. Groups of small models of objects such as cars, animals, furniture, household items, and people were shown to women of middle age—31 to 59 years old—and old age—65 to 85 years old. Some of the women saw a researcher arrange the objects in a

When daily life—as in this computer class—supplies adults with meaningful contexts for learning and remembering, older adults function as well as or better than younger adults.

landscape filled with mountains, houses, church, parking lot, and street. Others saw a researcher arrange the objects in a box containing various sized cubicles and the same mountains, houses, and other props. Later, the women were asked to reconstruct the display they had seen. It was found that the middle-aged women did much better than the older women at reconstructing the cubicle display—which stripped the objects of their meaningful context. But when the objects were in the meaningful landscape, women of all ages did equally well.

Daily life, of course, supplies people with meaningful contexts. Older people, with their many years of experience, in fact may find that they function as well as or better than younger adults in everyday situations that tap their memories and learning skills. The experienced teacher may recognize and solve the problem presented by an unruly child more skillfully than the novice, and the senior lawyer may understand the implications of a new law more quickly and thoroughly than the recent graduate. Their years of accumulating and of organizing information can make older adults practiced, skillful, learned, and wise.

Short-term memory

Human memory apparently operates at several different levels, as we have discussed in earlier chapters (see Figure 18.1). The shallowest level is **sensory memory**, which records information quite close to the time and form of the perception. These sights, sounds, odors, tastes, and touches are retained in sensory memory for about one second. Then sensory memories decay, unless they are processed to the next level, short-term memory. Information remains

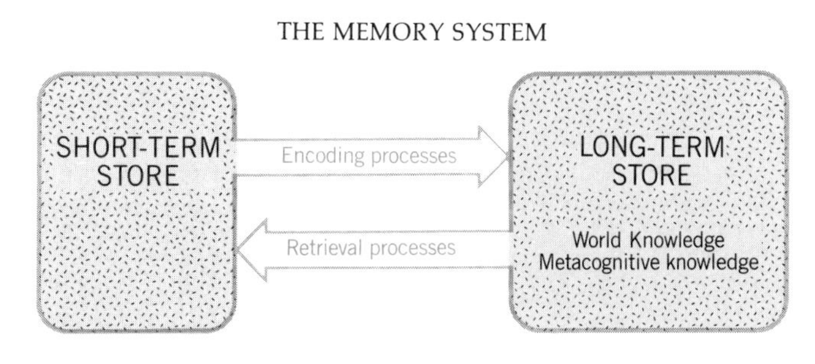

Figure 18.1 ***The memory consists of long- and short-term capacities and encoding and retrieval processes that move information between the two stores. The contents of memory include semantic and episodic memories and understanding of how memory itself works.***

in the limited storage of short-term memory for up to 15 seconds. While there, information is consciously attended to and processed. For example, a person may repeat a telephone number until he dials it. When he stops holding the number in conscious memory, it is lost. If he wants to remember the phone number permanently, he must encode it for later retrieval. By applying mnemonic strategies, such as organizing or imaging information, a person may transfer information to **long-term memory** and later recall it.

What happens to short-term memory during adulthood? In general, adults' short-term memory capacity does not change appreciably during most of adulthood, although there may be a slight decline after the age of 60 (Craik, 1977). Age differences are most likely to show up in people's abilities to manipulate information in short-term memory. Older adults do less well than younger adults when asked to repeat a string of numbers backwards. This effect may be the result of information processing abilities growing less flexible with age, a loss of information during reorganization, or a diminished ability to carry out the reorganization (Walsh, 1983).

> As I've gotten into my forties, I've noticed that my memory is a little bit slower and creakier than it used to be. When I have to add up a restaurant tab and figure in the tip, it takes me a bit longer, and I have to concentrate a bit harder than I used to. It's not anything that anyone else would notice. But I do.—Sam's father

Older adults also have more trouble than younger adults in dividing their attention between two tasks. Briefly interrupt a young woman while she is trying to keep a telephone number in short-term memory, and she is likely still to remember the number. But the same brief interruption is likely to make an older person forget the number altogether. Why? It may be that the contents of short-term memory grow more fragile and easily disrupted as people age (Arenberg, 1980).

> One thing I notice that's different now that I'm middle-aged is that when someone interrupts me, I lose my train of thought more easily. I have to ask, ''What was I saying?'' I never used to have to do that.—Sarah's father

Short-term memory also seems to slow down as people age. People may be able to store as much information as they once did, but they are slower at retrieving it. In experiments, these differences in time are measured in thousandths of seconds. A 35-year-old might need 100 milliseconds longer, and a 75-year-old might need 300 milliseconds longer than a 20-year-old to remember five numbers from short-term memory. For example, when people are given a list of numbers and asked later whether a certain number had been part of the original list, it takes older people longer than younger ones to remember (Anders, Fozard, and Lillyquist, 1972). (On this task, adults between 58 and 85 years old took longer to remember a number, even when the list had contained only one number.) When the list approached seven numbers—the limit of short-term memory capacity—those in their 30s needed more time than those in their 20s.

Long-term memory

> The other day Erica, my daughter, had her fourth birthday, and someone gave her a nightgown that she didn't like. She embarrassed me terribly and made me remember the embarrassment I caused my mother at my fifth birthday, when I critized the frilly underwear my aunt gave me.

> It's been 20 years at least since I needed them, but I still remember perfectly my best friends' telephone numbers from when we were teen-

> agers. My grandparents have been dead for years, but I still remember their telephone number and street address.—Sam's mother

People's memories of childhood, what they know about the world, and what they understand about people and things comprise information in long-term memory. When people encode information in short-term memory for longer storage, it enters long-term memory. When they want to retrieve the information from long-term memory, it returns to short-term memory. There it can be consciously manipulated once again. The capacity of long-term memory is assumed to be virtually unlimited.

Aging seems not to affect long-term memory adversely. Eighty-year-olds seem to remember information from long-term memory as effectively as 20-year-olds (Walsh, 1983). Even when people cannot retrieve information from long-term memory—''Oh dear, it's on the tip of my tongue, but I just can't remember it.''—it is nonetheless assumed to be in storage but not accessible. If the person were given the right cue, presumably the memory would surface.

Young and old adults also seem to organize information in similar ways. Young and old have **episodic memories**, which are memories of things that happen to us, linked to specific times and places, such as the mother's memory of her fifth birthday party or the 90-year-old man's memories of World War I.

> I was in the navy during the First World War when there was a terrible influenza epidemic. I remember seeing the bodies of the sailors who had died of influenza stacked like logs on the dock at New London, Connecticut.—Sam's great grandfather

Young and old alike also have **semantic memories**, which are organized facts and concepts, such as the names of all the states in the union, the names for family members and their relationships—Aunt Margaret; her brother-in-law, Uncle Mike; cousin Steven; your two sisters, and so on—and the knowledge that baboons, chimpanzees, and humans are all primates. Experimental work suggests that episodic memories, but not semantic memories, may be affected by aging (Craik and Simon, 1980).

Older adults are less efficient at encoding information into long-term memory as well as at retrieving it from long-term memory. Thus older adults generally are less efficient at *recalling* episodic memories. In experiments, when people are asked to remember a list of words, and then to recall it, the number of words recalled may fall off in 30-year-olds and continue to fall with increasing age. However, when people are given a cue to help their recall—perhaps a category name such as ''primates''—they can recall more words from a list, and age differences are less pronounced (Craik, 1977).

How, if at all, do the cohort effects we mentioned earlier affect these observed changes in memory? What about the level of people's formal education? Older adults with Ph.D.s as well as older adults with high-school diplomas showed the same decline when they were measured against people in their twenties with comparable levels of education (Perlmutter, 1978). But educational level does have some effect on recall. Young and older adults with more years of schooling could recall more words than others their age with less schooling. It should be noted that although there was an age decline in episodic memory of word lists, older adults with high-school diplomas as well as doctorates did better at answering semantic, factual history questions than young adults with comparable levels of education. In general, semantic memories about the world and metacognitive memories (about the cognitive system) increase during adulthood and may compensate for decline in more basic memory capacity.

Memory for pictures seems stronger in both older and young adults than memory for words. All age-groups are better at recalling a list of words that have been accompanied by line drawings than simply the list of words (Winograd, Smith, and Simon, 1982). In general, people's ability to recognize declines less than their ability to recall information. In experiments that test recognition, people may be shown groups of pictures or long lists of words, then later see a second array, and have to answer whether the pictures or words were in the original array. The second array serves as a memory cue and helps people to retrieve the information about the original array (Perlmutter, 1979). Some researchers have found few, if any, age differences in recognition memory but consistent age differences in recall. Recall (as you may recall from earlier chapters) is a more difficult process than recognition, because it is carried out in the absence of the original information.

The conditions under which information is encoded into and retrieved from memory as well as the characteristics of the people under study all seem to affect whether age differences in recognition appear. For example, researchers in one experiment found that older adults' IQ scores correlated closely with their recognition memories (Bowles and Poon, 1982). Among a group of people in their seventies, those with high verbal IQs recognized as many words as did college students with high verbal IQs. But among those with low verbal IQs, not only did both age-groups recognize fewer words than did those with high IQs, but the older group recognized many fewer words than did college students with low verbal IQs.

> When I was in college, it took me at most two or three times through my notes to memorize information. But now that I'm 44, it takes me twice that many times to memorize. It feels as if my memory isn't as "sticky" as it used to be.—Erica's mother

Memory strategies

All people remember better if they use memory strategies. (Recall our discussion of the advantages they afford to school-age children over preschoolers.) When older people have trouble encoding and retrieving information from memory, they may have failed to use the memory strategies that younger people use spontaneously. Once reminded to use these strategies, older people are capable of doing so (Perlmutter, 1983). Memory strategies elaborate on information or move it to a deeper level of cognitive processing, making it more likely that the information will be encoded and later retrievable. As we saw in earlier chapters, encoding strategies that move information to deeper levels of processing increase learning. These strategies include organizing information by category—remembering all the meat-eating versus all the plant-eating dinosaurs, for example. Verbal aids to memory, or mnemonic devices, also help—such as remembering the number of days in the month with "Thirty days hath September . . . " Creating a vivid mental image of items to be remembered and systematically rehearsing the information are other memory strategies that increase the depth of processing.

Older adults do not use these strategies spontaneously on memory tasks. It has been suggested that this failure is responsible for older adults' poorer recall and recognition memory. For example, in one experiment (Zacks, 1982), college students and 70-year-olds had to memorize a list of several categories of common words. The students were better at memorizing than the older adults. But when the students were prevented from using a memory strategy—when they were made to repeat each word aloud until the next word appeared—the students remembered only slightly more words than the 70-year-olds. When older and young adults are asked to memorize information by similar strategies, they can do so (Perlmutter and Mitchell, 1982).

By middle adulthood, people generally are less quick than they once were at encoding and recalling certain kinds of information. But they may compensate for this limitation because the understanding of how memory works improves throughout adulthood.

> I had to learn a new language when I was 49 and my company transferred me to Germany. The first week of class was agony. I couldn't remember new vocabulary words or verbs forms for the life of me. But after a couple of weeks, I remembered how to memorize things the way I did in college. When I took a shower, for instance, I'd conjugate German verbs, instead of planning for a business meeting. Once I got my memory back in the swing of things, learning German was a breeze.—Megan's father

In one study, young, middle-aged, and older women were asked to recall lists of words (Zivian and Darjes, 1983). Half of the middle-aged women were college students, and it was found that no matter their age, the college students recalled as many words as the 20-year-olds. The middle-aged women who were not students recalled many fewer words than the students, but more words than the group of women who were between 60 and 86. When everyday situations call on people to process information so that they can recall it at will, they practice memory strategies that they might otherwise abandon.

Metamemory

Middle-aged and older adults may compensate for limitations in the way they encode and recall information because their understanding of how memory works—*metamemory*—improves throughout adulthood. A number of questionnaire studies of metamemory knowledge show that with age, adults maintain or increase their understanding about the memory system (Perlmutter, 1985). In addition, middle-aged students understand metamemory better than middle-aged adults who are not in school (Zivian and Darjes, 1983).

Old and young adults were equally adept at the everyday metamemory process called "the feeling of knowing" (Lachman and Lachman, 1980). When they were asked questions that tapped general knowledge, such as "What was the former name of Mohammed Ali?" old and young were equally accurate in telling whether they definitely did or did not know the answers or could recognize the answer if told it.

If older adults understand the way memory works as well as younger adults do and are as good at assessing, or monitoring, their memory, why do older adults sometimes fail to use effective memory strategies? Perhaps they have less energy available for processing information and cannot or will not apply themselves to the demands of deeply processing information (Craik and Byrd, 1982). When older adults do not know that they will be tested on their recall of a list of words, they generally remember as many words as when they know a test is to follow. This tendency may indicate that they do not have the relevant information about memory strategies for list learning, which is, of course, not a very usual task. Also, when adults are shown distractor words as well as target words, older adults remember fewer of the distractors than younger adults. This tendency may result because young adults have a capacity *surplus*, which allows them to pick up extra information, even when they are not trying to do so intentionally (Perlmutter and Mitchell, 1982).

Even when older adults adopt strategies for memorizing, many of them still have trouble retrieving the information from memory. In one experiment, when subjects were forced to process information deeply, the young but not the old benefited (Simon et al., 1982). Under conditions that fostered shallow, moderate, or deep processing of a story about a family with financial problems, young adults remembered more main ideas than middle-aged or older adults. However, when the subjects were asked merely to read and remember the story, adults in all age-groups recalled equally well. It may be that al-

lifespan focus Expectations and competence

According to one stereotype, the older adults become, the less competent they become at many kinds of things—social, physical, and intellectual. Not only are older adults often considered less competent, but their supposed incompetence is attributed to internal, stable factors such as age and inability rather than to external, changeable factors such as luck. In contrast, young adults often are presumed to perform well socially, physically, and cognitively because of internal, stable factors. This unflattering portrait makes older adults seem less competent than younger adults, and it is a portrait painted by old and young alike. Expecting older people to be incompetent is likely to influence how people behave—both how older adults behave and how others behave toward them. Among young adults, for example, such diminished expectations often go hand in hand with learned helplessness.

To learn about the expectations of a group of young adults—students in an introductory psychology class—and older adults—members of a community senior citizens' group—researchers asked all of them to answer questions about hypothetical situations that involved themselves, others their age, or others of a different age (Lachman and McArthur, 1986). The questions tapped the subjects' expectations about how people would behave in the cognitive, physical, and social domains. For example, some of the men were asked whether someone might be so absent-minded that he bought two subscriptions to a magazine (cognitive domain). Some were asked whether a man might have broken a door by slamming it too hard (physical domain). Some were asked whether a child might have listened to an adult's stories for hours on end because the stories were so entertaining (social domain).

The researchers found that young and older adults painted something close to the unflattering stereotyped portrait of older people's cognitive and physical—but not social—abilities. But they also found, to their surprise, that older adults were more likely to be given credit for their good performance than to be blamed for their poor performance. Contrary to the stereotype, older adults were seen as physically quick but were not seen as particularly nurturant.

Why did the subjects have more favorable views of older adults' abilities than findings from earlier studies had reported? Unlike those in other studies, the subjects were told that the people in the hypothetical situations were real, and the situations were presented more realistically, too. The implications of this research are important, because when older people learn adaptive attributions, their performance actually improves. In the same way as people's expectations affect their physical health, their expectations affect their psychological health and their cognitive, social, and physical performance as well.

though older adults process information deeply, because they have encoded the information less specifically and distinctively than young adults, they cannot retrieve the information. If so, older adults need different or more general cues to recall processed information (Craik and Simon, 1980).

Learning skills

On some tasks, laboratory tasks for example, younger adults may learn and remember more efficiently than older adults, but at all ages, people *do* learn and remember new things throughout their daily lives. In fact, the complexity of learning in everyday life is so great that important factors in learning and memory may well be missing from current theories (Perlmutter and List,

1982). Some of the very factors that interfere with laboratory learning may enhance everyday learning. For instance, the proactive interference observed in the laboratory suggests that well-practiced habits may enhance older adults' learning of related information (Arenberg and Robertson-Tchabo, 1977). Middle-aged and older adults may use their cognitive habits as familiar roads that take them where they want to go and bypass the delays they faced when they were younger.

Intellectual abilities

Throughout the lifespan, people draw on their intellectual abilities. Everyday life calls upon people of every age to learn new facts, faces, figures, and relationships, to recognize and recall all kinds of information, trivial and important, small and large, and to solve problems, difficult and easy. The cognitive changes that characterize mature, adult learning and memory make some older people more adept than younger people at certain kinds of intellectual tasks, although this age advantage is not universal across either individuals or domains.

Several kinds of factors are related to adults' intellectual performance. People's educational levels are correlated with their responses on tests of classifying objects and their IQ test results, with those who have had more schooling scoring higher than those with less schooling. As we saw in earlier chapters, social circumstances are related to performance on intelligence tests as well, with those relegated to the margins of society scoring more poorly than the privileged.

Fluid intelligence remains stable until late adulthood and then declines. It is made up of basic intellectual processes such as abstract thinking, the speed of thinking, and the drawing of inferences.

The skills that make up what we usually call ''intelligence'' can be classified into two basic types, fluid and crystallized. Fluid intelligence corresponds to basic cognitive processes and is required for the identification and comprehension of relationships and the drawing of inferences out of that comprehension. In contrast, crystallized intelligence corresponds to acquired knowledge and intellectual skills and to all of a person's quantitative thinking, judgment, wisdom. Crystallized intelligence shapes the way that people apply their fluid intelligence within specific social and cultural contexts (Horn, 1982). Crystallized and fluid intelligence take different developmental paths during adulthood (see Figure 18.2).

Fluid intelligence

> When I was in my early fifties, I had a physical problem that took me to a neurologist. I was given a battery of tests. Some of them made me feel like a dope! I had to do speed calculations of simple numbers and remember lists of American presidents. I felt like I was back in fourth grade—and flunking!—Sam's grandmother
>
> My grandson, Jason, is 9 now. The other afternoon, he and I played with his Transformer. I had to figure out how the thing transformed from a robot to a truck and back. I got the hang of it right away, even faster than Jason did.—Jason's grandfather

Until late adulthood, test results show that fluid intelligence remains stable. Then, in late adulthood, fluid intelligence apparently declines an average of 4 points for every decade of old age—a small decline when one considers that within any given age-group people's scores vary by more than ten times that amount (Horn, 1982).

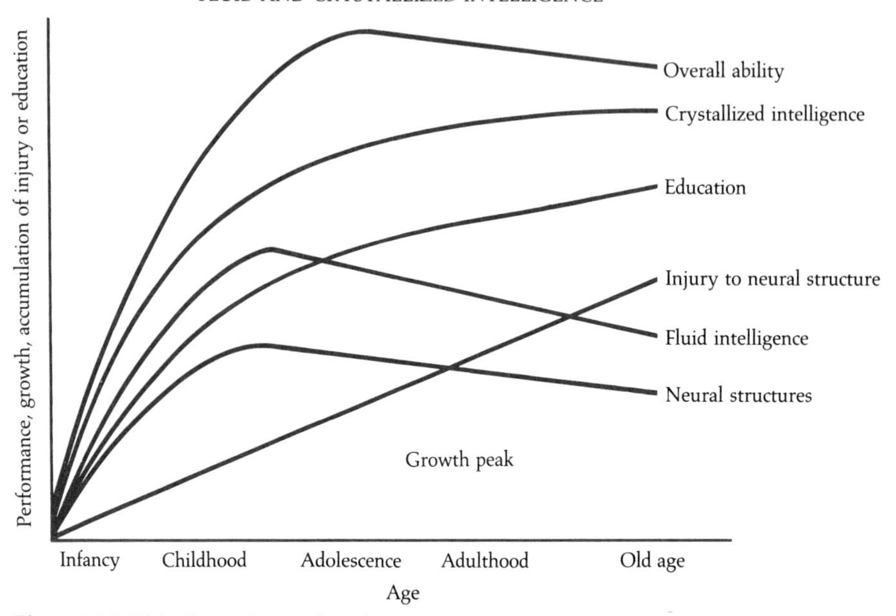

Figure 18.2 ***This chart shows the relation among fluid and crystallized intelligence, education, growth and decline of neural structures, injury to neural structures, and overall ability (Horn, 1970).***

Crystallized intelligence

> As a writer, I found that I didn't really hit my stride until I was in my late thirties. My command of the language, of forms of writing, and of my own personal style didn't mature until I was in my forties.—Sam's mother

> I work as a builder, and I know more today, at 43, about putting houses together than I knew when I first started out. Of course, when I was younger I *thought* I knew more. But that was because I didn't realize how much there was to learn. I bet I'll keep on getting better for a while to come.—Nadine's father

Until people are about 65 years old, their loss in fluid intelligence is countered by a comparable gain in crystallized intelligence. Crystallized intelligence appears to increase at least until the mid-sixties. On vocabulary tests, understanding analogies, and thinking divergently about putting common objects

Crystallized intelligence increases throughout middle adulthood and possibly longer. It is made up of the intellectual abilities that result from learning, such as knowledge of vocabulary and general information about the world.

to new uses, older people outperform younger ones. It stands to reason that older people have a wider knowledge base and a more organized, cohesive, accessible, and accurate store of knowledge than younger people do (Horn, 1982).

Reasoning and problem solving

The cognitive skills of reasoning and problem solving also remain essentially stable until late adulthood. In some respects, adults' reasoning abilities may grow progressively deeper over time. In Piaget's theory, cognitive maturity is reached during adolescence, with the acquisition of formal operations. Then during adulthood, suggested Piaget, there is a plateau. However, several developmental psychologists have suggested that formal operations may not be the ultimate stage of cognitive development.

Realistic understanding

It has been suggested, too, contrary to Piaget's suggestion, that most adults do not typically deal with abstractions that may require formal operational thought (Arlin, 1980). Instead, adults typically are faced with the concrete realities of everyday life, and these require more concrete thought. Thus, during adulthood, the increased application of experience, knowledge, and wisdom to everyday events may be adaptive (Neimark, 1982). Adults regularly and intelligently apply what they have learned from past experience to problems with uncertain outcomes, such as deciding whether to start a new business or back a certain candidate, which alternative to recommend to a client or patient, whether to hire or fire, whether to move to a new place and change jobs, and so on.

College students have less practice with decisions like these, and although they score well on IQ and problem-solving tests and on Piagetian tasks, they do not apply their cognitive abilities to everyday problems as effectively as older adults do. Asked to decide among alternatives that involve people's lives or money, college students do not exercise the same judgment or caution as older people (Tversky and Kahneman, 1981). They often abandon formal reasoning for simplistic guidelines and make decisions without considering all the relevant facts. As a consequence, their decisions often are unwise or faulty (Tversky and Kahneman, 1974). In short, when problems have an uncertain outcome—as most real-life problems do—college students' skills in formal, abstract reasoning do not stand up to those of older adults.

> When I was in my late thirties, I could feel my thinking about a lot of different issues beginning to take a new shape. Whereas once I had thought the most important thing was to get things right, to win, to compete and succeed, I began to think that the most important thing was for people to stay in touch and to solve their problems.—Natalie's mother

> I've been teaching college physics for 15 years. I'm less interested in grappling with abstract problems these days than in the social and political consequences of physics—nuclear power plants and "Star Wars" space systems, for instance. I've been spending more and more time lately volunteering information to our senator and debating on panels.—Peter's father

Adults are more expert at relating logic and abstractions to actions, emotions, issues of social responsibility, and personal relationships. As they do so, their thought grows more global and aimed at new goals. Adults may well be less interested in following a narrow path of reasoning to its logical conclusion than in following a path to broader moral and practical concerns (Labouvie-Vief, 1982).

Dialectical reasoning

Michael Basseches (1984) has proposed that some adults continue to develop intellectually. These adults, he suggests, have gone beyond formal thought to a more powerful way of understanding the world. They engage in **dialectical reasoning**, which is a way of making sense of the world. Adults who reason dialectically see the evolution of order in the universe as a continuous, ongoing process. They see the processes of finding and creating order in the universe as basic to human life and understanding but recognize the relativity and transience of the world. Although people who are dialectical reasoners are not likely to believe that they can include all of life's processes within a single, comprehensive ordering, they constantly seek to integrate and order their understanding of life, knowing that new syntheses will still be required.

> When I was younger, I used to argue with my kids all the time. I felt it was important to get my point across and to teach them the right way to see things. But these days I'm less inclined to assume that either I'm wrong *or* they're wrong. I try to hear to what they're telling me. They've learned important things from their own experiences, and I could stand to benefit from their experiences, too. It's a two-way street. My kids have listened to me and applied values from my day to their lives, and I listen to them and apply what they know to my own future.—Natalie's mother

Problem-solving skills

Adults of all ages must solve problems in their daily lives, and many do so skillfully. They must decide which route will get them to the store most quickly, which refrigerator is the best buy, which investment retirement account offers the best benefits, and which clothes to pack for a vacation. However, laboratory-based studies of adults' problem-solving abilities tap different kinds of skills. In these studies, researchers fairly consistently find young and middle-aged adults outperforming older adults in such tasks as searching efficiently for a correct array of stimuli, identifying and changing concepts, and reasoning verbally (Reese and Rodeheaver, 1985).

Forming concepts

As anyone who has gone shopping for the week's groceries, a new car, or an airline ticket can attest, when people organize their knowledge about certain

The ability to solve problems is required of people throughout the lifespan. Young and middle-aged adults tend to be more efficient than older adults at certain problem-solving skills, including searching for arrays of stimuli, reasoning verbally, and identifying and changing concepts.

subjects efficiently, they can solve problems quickly and easily. When they do not organize their knowledge or organize it inefficiently, they solve problems with great difficulty or not at all. Experimentally, the way adults form concepts can be studied if they are asked to figure out the rule by which an experimenter has divided geometrical forms. For instance, when adults are shown large and small, red and blue circles and squares, they then guess whether each form they see is an example of the governing concept. Feedback from the experimenter on the correctness of their answers helps them discover the rule. In one experiment, the rule had two dimensions. It took adults between 35 and 51 longer to discover the rule, and they made more wrong guesses than did younger adults (Wiersma and Klausmeier, 1965). On a similar task, adults between 18 and 22 were faster and more accurate than children 10 to 14 or adults 60 to 78 years old (West, Odom, and Aschkenasy, 1978).

One strategy for discovering rules in experimental situations like those just described is to form a hypothesis and to modify it according to the feedback the experimenter supplied after each guess. The general rule is to stick to a correct choice but switch to a new hypothesis after an incorrect choice. When the performance of school-age children, college students, and older adults was compared, the older adults did worst and broke the general rule more often than the others (Offenbach, 1974). The older adults switched even when their hypothesis had been correct or stuck with a hypothesis that was wrong, as if they were not applying the available feedback. Even when older adults' memories were jogged by placing results of previous guesses in front of them, their performance did not improve (Brinley, Jovick, and McLaughlin, 1974).

Expertness

Some adults know enough about certain areas of knowledge to rate as true experts. Their understanding of history, poetry, carpentry, human development, computers, organic chemistry, or other fields of knowledge is broad, deep, and expert. A person's expertness tends to be restricted to one or two areas, and so, for example, the person who has an expert memory for numerical information may have only an ordinary memory for verbal information (Hoyer, 1985). Although older adults remain experts in their fields, they are less likely than younger adults to develop proficiency in new areas or to draw

By adulthood, some people have become expert in an area of knowledge, and their expertise may become so well practiced that it is nearly automatic.

upon the cognitive abilities that do decline with age. For example, an older adult who has an expert understanding of English literature is not likely to take up the study of linguistics to broaden that understanding. A younger adult is much more likely to study linguistics as a way of increasing his or her understanding of literature and language. Older experts may understand their fields, but they may not always know what they know—a metacognitive process. For example, older radiologists may have trouble explaining errors in diagnosis to novices (Lesgold, 1983). Expertness may become so well practiced that it is nearly automatic. A musician or an athlete, for example, may have a nearly automatic command of the muscular, perceptual, and intellectual skills necessary for playing an instrument or a sport. An expert lecturer has mastered the script for lecturing. It is likely that certain age-related declines in abilities to pay attention and to shift attention affect older adults' abilities to formulate new rules and integrate vast quantities of new information in their fields of expertness. Even so, older doctors are likely to have mastered thousands of rules about alternative diagnoses, treatments, and prescriptions. In addition, older doctors also have mastered thousands of exceptions to those rules—what to do when a patient does not respond to a certain drug, surgical procedure, and so forth. In short, as adults gain in experience and competence, despite age-related declines in certain information-processing skills, they may gain truly expert understanding.

Summary

1. Learning and remembering abilities change little during adulthood, at least until old age. Vital to people's cognitive functioning at any age is their familiarity with the material to be learned and their familiarity with related material.
2. When psychologists compare the cognitive performance of young and old adults on experimental tasks, they often find that the young adults outperform many of the older adults. These changes in cognition may be the result of cohort differences rather than age changes.
3. Adults' short-term memory capacity and speed do not diminish appreciably until around the age of 60.
4. Young adults are more efficient than older adults at encoding information into long-term memory and retrieving it. Memory for pictures seems stronger in both older and young adults than memory for words. When older people have trouble encoding and retrieving information from memory, they may have failed to use the memory strategies that younger people use spontaneously.
5. It may be that young adults have a processing surplus, from which they can pick up incidental information or upon which to draw when learning conditions are difficult.
6. Until late adulthood, test results show that fluid intelligence remains stable. Crystallized intelligence appears to increase until the mid-sixties.
7. Reasoning and problem-solving skills do not seem to decline in adulthood and may actually be enhanced by the life experiences of middle and late adulthood.
8. Adults may use formal thought to advance their understanding of everyday life and its concrete realities. Some adults may go beyond formal operational thinking and apply dialectical reasoning to make sense of the world.

9. Cognitive growth during adulthood may be the application of experience, knowledge, and wisdom to everyday events.

Key terms

sensory memory	episodic memories	dialectical reasoning
long-term memory	semantic memories	

Suggested readings

BALTES, PAUL, DITTMANN-KOHLI, FREYA, and DIXON, R.A., "New Perspectives on the Development of Intelligence in Adulthood: Toward a Dual Process Conception and a Model of Selective Optimization with Compensation." In P.B. Baltes and O.G. Brim (eds.), *Life Span Development and Behavior* (vol. 6). New York: Academic Press, 1984. Theoretical discussions of the nature of intelligence in adulthood.

BASSECHES, MICHAEL, *Dialectic Thinking and Adult Development*. Norwood, N.J.: Ablex, 1984. A presentation of the author's somewhat idiosyncratic but interesting dialectic framework for conceptualizing age-related improvements in adult intelligence.

CHARNESS, NEIL. *Aging and Human Performance*. New York: Wiley, 1985. Reviews of research on age-related differences in many kinds of cognitive performance, including sensory and perceptual functioning, attention, memory, problem solving, and work behavior.

HORN, JOHN. "The Aging of Human Abilities." In B.B. Wolman (ed.), *Handbook of Developmental Psychology*, Englewood Cliffs, N.J.: Prentice-Hall, 1982. Review and methodological critique of research on age differences in people's performance or psychometric intelligence.

chapter nineteen
Social development

Adulthood is a time of change and development, no less than childhood is. Adulthood can be a time of great emotional and personal development, a time of important changes in people's roles. Far from being static, adulthood is a time of changes in people's roles and, therefore, in their emotions, motivation, and relations with others.

> When I was in college, I used to think that soon I'd be an adult and therefore I'd stop feeling confused about life. Well, it's 20 years later, and I know now that you never really stop changing. —Natalie's father, age 42

There are several different but related theories of dynamic emotional development in adulthood, and we will discuss them in this chapter. The theories—each in its own way—all stress the constant psychological and social reorganizations of adulthood, the stresses and dissatisfactions that result in the series of transformations that make up the course of an individual's adult development.

Stages of adult emotional development

According to Erik Erikson (1980, 1982), people face psychosocial crises in adulthood, as in adolescence, from the standpoint of individual personalities shaped since infancy (see Table 1.1). They are influenced, for example, by their sense of autonomy and identity.

Entering adulthood

The special developmental task of young adulthood is to resolve the tension between *intimacy* and *isolation*. In intimate relationships, young adults fuse their identities with those of other people. They learn how to commit themselves to relationships and how to sacrifice and compromise to maintain those relationships. Adults who successfully resolve this crisis have the ability to love.

Young adults face the developmental task of forming intimate friendships, that is, of fusing their identity with those of others.

> When I met my husband-to-be, I'd dated other men, but I'd never *really* opened myself up the way I did to him. Sometimes I got frightened because I felt so vulnerable with him. What if he betrayed me? But then I decided that it would be worse for me if I didn't stay open to him. I didn't want to be a closed-off person. What would that get me?—Mike and Erica's mother, age 39

Young adults who do not successfully resolve the crisis between intimacy and isolation remain psychosocially isolated, and their relationships are without spontaneity, warmth, or deep emotional exchanges. Some may fear intimacy as a loss of the self, a loss of identity. The sexual relations of people who remain isolated are likely to be problematic. When an adult, for example, separates sex from emotional involvement, he or she may not perceive a sexual partner as a whole person. The consequence may be a feeling of extreme isolation (Erikson and Hall, 1983).

> I've been close to settling down with a woman several times. But every time, I realized that the woman was wrong for me in some way. One of them said that I was lousy at "forming alliances," as she put it. Maybe so. But I'm not ready to be crowded by someone else's needs. I need space for *me*.—Natalie's uncle, age 34

Erikson's theory of emotional development at the beginning of adulthood has received some empirical support. In one sequential study (Whitbourne and Waterman, 1979), college students answered questions that reflected their development of trust, autonomy, initiative, industry, identity, and intimacy. Ten years later, they again answered questions. The data showed that they had grown in these qualities over time, especially in industry, identity, and intimacy, which Erikson theorized develop between childhood and early adulthood. In 1976, at the second testing, a second group of college students answered the questions as well, to provide a cohort comparison. College women in 1976 scored higher than college women in 1966 had scored on measures of industry (rather than inferiority). The researchers could not establish whether the change had resulted from changes in gender roles or changes in college admissions procedures.

Early adulthood

The seasons of a man's life

Like Erikson, psychologist Daniel Levinson was influenced by Freud's psychoanalytic theory. Levinson has suggested that his theory is consistent with and grows out of Erikson's. The theory of adult male development that he has proposed (Levinson, Darrow, Klein, Levinson, and McKee, 1978) focuses on the interaction of the self and society. Levinson interviewed 40 men—ten novelists, ten biologists, ten factory workers, and ten business executives—and although his work may apply to women, he did not interview any women and did not try to theorize about women's development.

The ''seasons of a man's life,'' as Levinson's theory has been called, are a series of adult stages, bridged by five-year-long transitions that merge with the stages before and after them (see Figure 19.1). The first ''season'' is the transition out of adolescence, at roughly the ages of 17 to 22. Next comes early adulthood, which extends from ages 17 to 45. Middle adulthood extends from 40 to 60, late adulthood from 60 on. Levinson also describes the possibility of late-late adulthood, from age 80 on, which may develop as life expectancy increases. The onset and length of any stage vary from one man to another and depend on particular psychological, social, and biological circumstances. In general, however, individuals vary by no more than two or three years in their progression through these stages.

The men whom Levinson interviewed were born between 1923 and 1934 and were 35 to 45 years old when first interviewed. They ranged in class background from working- to upper-class; in educational background from high school dropout to Ph.D. Some were Catholic, some Protestant, some Jewish. Five were black. Levinson interviewed each man for ten to 20 hours during a three-month period, interviewed them again two years later, administered a short form of a personality test (the Thematic Apperception Test), and interviewed most of the men's wives. Levinson also analyzed fictional and nonfictional accounts of other men's lives as presented in poems, plays, biographies, and novels.

Levinson interviewed no men older than 47, and so his ideas about development in late adulthood are speculative. Although Levinson suggests that the stages he found apply to men everywhere, those he interviewed all were from one place—the Northeast United States—and from one cohort. Only further research will tell whether men (and women) from other places and other times develop through the same ''seasons.''

According to Levinson, the seasons of a man's life ultimately form a pattern, or *life structure*. This structure is made up of alternating periods of stability—when developmental tasks are met and goals actively pursued—and flux or transition—when men question the direction of their lives, make important choices, and explore new avenues. This pattern of dynamic change is consistent with a dialectical view of adult development. In some cases,

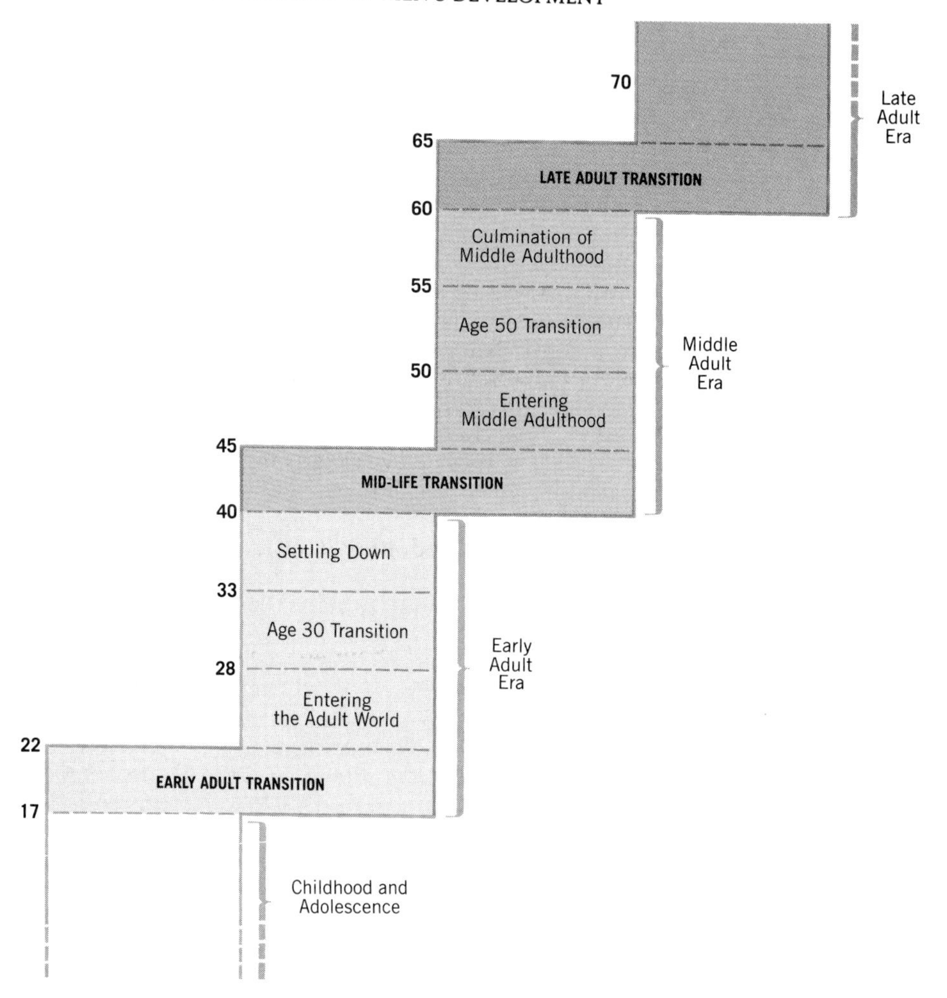

Figure 19.1 ***Adulthood for men is conceived of as stages of relative clan surrounded by more tumultuous transition periods (Levinson et al., 1978) and the Sterling Lord Agency.***

periods of flux were introduced by inner, psychological conflict. In some cases, external events from the social sphere introduced conflict or affected its resolution.

As young men make the transition out of adolescence, they once more grapple with the personality issue of autonomy. Autonomy, as Erikson suggests, is a concern of young children. Young adults also need to free themselves psychologically from their parents. According to Levinson, once they have done so, at about the age of 22, men enter a six-year period of relative stability when they enter the adult world. A young man now is a novice whose life dream is taking form. As he commits himself to a line of work and to collaboration with others, he aims both for stability and for keeping his options open. He also commits himself to both casual and intimate relationships with women. These resemble Erikson's descriptions of the tasks of identity and intimacy. Most of the men in the study married and had children during this stage. However, few had an intimate, nonsexual relationship with another man or with a woman other than their wife.

When the men reached their late twenties, they entered a less stable "age 30 transition." During the years between 28 and 33, the men noticed problems in their life structures and consequently faced new decisions about where their lives would lead. Once they negotiated this transition, they entered a stable ''settling down'' period. Having developed beyond the role of novice, men now applied themselves to their work, to earning prestige, assuming a place in society, and shaping what they considered to be a good life for themselves and their families. Near the end of young adulthood, men grappled once again with the issue of autonomy. Men distanced themselves from mentors who had advised and guided them along in their careers and tried to ''become their own man'' at work, in the family, and in the social world.

The decade birthdays, Levinson found, seemed to summon forth strong feelings: 30 with its striving for direction and 40 with its tough reappraisal. Fifty reopened some of the same questions that appeared at 30: how a man was to change the structure of his life to meet the personal and social needs that confronted him. If he dealt with his feelings of turmoil at age 40 by making few changes, 50 was likely to be stormy indeed. According to Levinson, few men can develop through middle adulthood without stumbling into one period of crisis. But once past the age 50 watershed, some men once again settled into a stable, fulfilling period. This relative calm lasted until the beginning of late adulthood, at about age 60.

The midlife transition

Between the ages of about 40 and 45, Levinson suggests, a man navigates the sometimes rocky transition to middle age. As he evaluates his life to this point, he is likely to feel that reality has not lived up to all he had hoped for. As he looks ahead, he is likely to feel that his life will never match his dreams. Disillusioned, he ponders the state of his work, his marriage and other personal relationships, his beliefs and the assumptions that have guided him to the brink of middle age. A man is likely to change the structure of his life at this point, to try and strike a better balance between his personal and social needs. Some men change radically. They leave a marriage or a job, they remarry, they move to a new place.

> I'd been practicing medicine in San Francisco since I got out of medical school. I've been married for 14 years, and my wife and I have two children. Three years ago, I was feeling desperate and trapped. I needed a change, and my wife wanted to go back to work now that our kids are in school. So we moved back to the East Coast. But this arrangement hasn't worked out for me. About six months ago, I met an old girlfriend

Some men draw strength from the uncertainties of mid-life by serving as mentors to younger adults.

at a convention, and I've decided to move in with her—back in San Francisco—and get back into medical practice there.—Jason and Kristy's father, age 42

Among the men whom Levinson interviewed (the physician quoted above was not among them), four out of five found the midlife transition a major crisis. **Midlife crisis** is a developmental state of physical and psychological distress that arises when developmental issues threaten to outstrip a person's resources for dealing with them (Cytrynbaum et al., 1980). For the men in Levinson's study, old conflicts resurfaced, and emotional turmoil tossed them about. For the first time, the men felt physical decline—waning energy and strength—and saw in themselves the first signs of aging. With a thud, they felt their own mortality. Out of this flux and turmoil, some of the men took new strength and new generative ability to act as mentors, advisers, and guides to younger adults.

But not all developmental psychologists are convinced of the existence of a midlife crisis. The research on the question of its existence has produced mixed results. Levinson, of course, found evidence of midlife crisis among the men he interviewed. George Vaillant (1977), however, who has collected longitudinal data on Harvard graduates, suggests that although some men do divorce, change jobs, and suffer depression at midlife, the frequency of these occurrences is essentially the same throughout adulthood. Vaillant argues that experiencing an actual crisis at midlife is the exception, not the rule. Yet the men in Vaillant's study were six to 12 years older than Levinson's subjects, and their histories therefore were quite different. Furthermore, Harvard graduates are far from a cross section of the general population. Cohort effects and the special nature of the sample may well have skewed the results upon which Vaillant has based his assertions.

In another set of studies, too, the California Intergenerational Studies (Clausen, 1981; Haan, 1981), there was little evidence of midlife crisis in the lives of middle-class men and women born either between 1920 and 1922 or 1928 and 1929. Most of the men were satisfied with their work at midlife. Men and women felt self-confident, insightful, resourceful in coping with stress, and both introspective and open to others. Finally, in a study that was designed specially to identify midlife crisis (McCrae and Costa, 1982), it was found to crop up in only a few men and at any age between 30 and 60. Among a different group, men who went through a midlife crisis before age 40 or after age 45 tended to suffer from long-term emotional problems. Thus the midlife crisis seems neither universal nor clear-cut. Perhaps it has appeared in some studies as an artifact of people in therapy (as in Roger Gould's work [1975, 1978], reported later) or of introspective middle-class people with strong needs for self-fulfillment. Perhaps, too, the normative "seasons" of life do lead to changes in people's identity and self-concept. But when these normative events—marriage, parenthood, entering and leaving jobs—are on schedule, they rarely lead to crisis, because people have had time to prepare themselves and to work through the associated problems, and their sense of the continuity and integrity of the life cycle remains intact. In other words, there may be a midlife transition even if not a crisis.

Middle adulthood

When men have passed through the midlife transition and have embarked on the stable period that begins middle adulthood, they are likely to feel more productive and satisfied than ever before. They may devote the years between the ages of roughly 45 and 50 to solidifying the life that they have chosen. Some men now develop qualities of compassion, judiciousness, and wisdom. Others stagnate and enter a period of decline.

According to Erikson, the developmental crisis of middle adulthood is between *generativity* and *self-absorption* or *stagnation*. Generativity is caring about the next generation and expressing this caring in any one of several ways. Perhaps the most obvious generative acts are "generating" children—bearing and caring for them. Other generative acts are teaching, guiding, nurturing, and acting as a mentor to the children and young adults of other people. Teachers, physicians and nurses, writers, artists, and people in many other lines of work express generativity. Some people express generativity in trying to better society. No matter how people express it, generativity is a personality strength of early adulthood that grows on the strengths of earlier periods of development: hope, will, purpose, competence, loyalty, and love.

Adults who do not develop a feeling of generativity are in danger of stagnating and growing self-absorbed. They may indulge themselves, as if they were their own children. Erikson has suggested that for generativity to develop, people must satisfy the drive to procreate. Those who do not have children, he suggests, must direct their feelings of generativity into other social outlets, or they risk eventually feeling frustrated, empty, and lost (Erikson and Hall, 1983).

Results of longitudinal studies support Erikson's theory of personality development in middle adulthood. Vaillant's (1977) study of men who graduated from Harvard showed that they had developed through the stages that Erikson describes. By the time they were 47 years old, the least successfully adapted men tended not to trust the world or themselves, to feel antagonistic to people in authority, to lack initiative, to feel insecure, to have trouble with intimacy and with taking responsibility for other adults. The most mature, well adapted men had satisfactorily resolved the psychosocial issues that had faced them through middle adulthood.

"Transformations"

Levinson was interviewing in the Northeast at roughly the same time as Roger Gould, (1975, 1978) a psychiatrist, was gathering cross-sectional data on southern California men and women between 16 and 60. Because Levinson's and Gould's findings are similar in many respects, lifespan developmental psychologists have questioned whether the findings reflect cohort effects. On this point, too, further research will have to provide the answers. Gould gave a questionnaire to over 500 white, middle-class Americans, men and women from 16 to 60 years old and from their responses observed that they tended to have the same kinds of problems at the same ages. He saw a pattern emerging of a series of predictable changes. Gould did not see adulthood as a period of stability, of unchanging emotions and motivation but of dynamic change and transformations.

In Gould's theory of development during adulthood, ambivalence characterizes late adolescence, the years between 16 and 22, when people are most concerned with forming an identity and gaining independence from their parents. In young adulthood, people have gained their independence and work to reach the goals they have set for themselves. Between 28 and 34, adults are again in flux, questioning their goals, their finances, and their marriages.

The adults were still questioning between the ages of 35 and 43, with a heightened sense of urgency. As they faced middle adulthood, they realized that their time was limited, that if they were going to make changes, they had to get on with them or forever lose their chance. This period was one of instability, emotional turmoil, and psychological pain. But then it opened onto the stability of middle age, the years between 43 and 50, when people felt happier with their marriage, their friends, and their finances. For many

lifespan focus Sources of women's well-being

What makes women feel satisfied with themselves as members of society? What makes them feel like masters of their own fate? From what sources do women derive pleasure and enjoyment? These were the questions asked by a group of researchers (Baruch, Barnett, and Rivers, 1983) interested in locating the sources of women's feelings of well-being. To get a balanced picture, they interviewed 300 women from many different roles—women who were parents and women who were not, women who were married, divorced, and never married, women in high-status and women in low-status jobs, homemakers and women who worked outside the home. They chose as subjects women between the ages of 35 and 55 on the assumption that these are the years when

The sources of adult women's well-being are marriage, motherhood, **and** *work outside the home.*

people, the feeling of acceptance pervaded the fifties, too, along with a sense of generativity, pleasure in personal relationships, marriage, family, and concern for society. But at the same time, people worried about their health and worried, too, that their time was running out. Thus to Gould, adulthood unfolded in a fairly predictable sequence of changes in emotion, satisfaction, and motivation.

Sex differences in emotional development

Some developmental psychologists have questioned whether the emotional development of men and women during adulthood is essentially the same. They have questioned, for example, whether women suffer midlife turmoil in the same way as some men apparently do. Researchers therefore have tried to trace the long-term connections between gender roles and emotional development to see whether and how the development of men and women typically differs.

One difference between women's and men's development turned up among people in their twenties. Many men, as you have seen, use their most women's lives have settled into a stable pattern. The women all were white and lived in a large town near Boston.

The researchers found that the women with the greatest sense of well-being were those who had taken on several roles. They had not confined themselves only to homemaking or to a career. Despite feelings of stress and strain, the adult women who were most satisfied with their lives had taken on the roles of marriage, motherhood, and outside work. For these women, although the stress was real, the rewards of their lives outweighed it. No single life pattern, the researchers found, was hard or easy. Each presented women with periods of greater or lesser stress. Thus a married, working woman who was trying to establish her career and mother two young children might find her twenties particularly stressful. But her middle age might be far less stressful as her children became more independent, her income rose, and her career stabilized. In contrast, a woman who stays home to raise her children may find her twenties easier than her middle years if she reenters the job market and tries to compete in the marketplace.

What were the main sources of stress in women's lives? Women at home with children, the researchers found, were more likely to feel overloaded—role strain—by midlife than were women who worked outside the home. In fact, homemaking often turned into "today's high-risk job." Divorce, the death of a husband, financial problems, and changing personal and social needs often left women trapped in their roles as homemakers. Women in low-status jobs also were likely to feel more stress than those in high-status jobs. Why? The doctor, for example, can work part-time when her children are young and can bring them to the office if her other child-care arrangements fall through. But a secretary cannot.

Women in marriages, and especially those in satisfying marriages, often reported them as important sources of pleasure in their lives. Marriage offered these women intimacy, sexual pleasure, and economic security. But divorced and unmarried women, all of whom worked outside the home, reported feelings of mastery strong enough to outweigh their problems. By the time the women in this study were in middle adulthood, many had come to terms with the less satisfying aspects of their lives and derived feelings of pleasure and mastery from the more satisfying aspects. Women were able to derive feelings of well-being from many different patterns of life.

twenties to build careers and families and thereby follow through on plans formed during adolescence and early adulthood. But researchers found that many women felt uncertain and dissatisfied during their twenties (Baruch, Barnett, and Rivers, 1983). The women were unhappy and anxious as they worried about their future and wrestled with problems and choices about their careers, marriages, and childbearing. The women in their twenties also reported experiencing a feeling of crisis when unexpected events befell them—divorce, car accident, job transfer, and so forth—and when normative life events happened to them *off* schedule—the early death of a parent, for example. The researchers suggested that for women in their twenties, there are fewer role models, more confusion, less clearly marked pathways for development than there are for men of the same age.

Erik Erikson has asserted that the stages of psychosocial development he describes are essentially alike for men and women. Although the specifics of the developmental tasks may vary from one society to the next and from one era to the next, they are essentially universal (Erikson and Hall, 1983). But Carol Gilligan (1982) suggests that the development of males and females differs significantly. Whereas Erikson and Levinson portray males as defining themselves by progressively separating from and growing independent in their relationships with others, Gilligan portrays females as locating their independence and identity *within* their relationships with others. Adult men subordinate personal relationships to their aspirations and their work. As men face the task of intimacy, they feel a conflict between the demands of their work and their personal relationships. Society generally pushes men to resolve the conflict—with its implications for their identities—in favor of work. But women generally are pushed to resolve the same conflict in favor of dependence, compassion, and gentleness. From girlhood on, they have been expected to be caring and generative. Contrary to the timetable proposed for men, many adolescent women work through issues of identity and intimacy at the same time, and many adult women consider the postponement of generativity until middle age to be contrary to their personal experience. Gilligan suggests that many of the misunderstandings between the sexes arise from such differences in socialization. Although people of both sexes use the same vocabulary, in fact they are talking about very different sets of experiences.

Findings from the California Intergenerational Studies suggest that people's emotional development during adulthood correlated with their gender roles during adolescence. By adolescence, the individuals had evolved certain styles of personality, and these styles developed in complexity during adulthood. Men and women who at age 50 were psychologically healthier than average had followed traditional gender roles as adolescents (Livson, 1981). That is, as adolescents the boys had prepared more for achievement in sports and careers, and the girls had prepared more for social and family roles. Men and women who had followed less traditional gender roles in adolescence had developed less smoothly. At age 40, many were not especially psychologically healthy. But during their forties, they improved, and by age 50, they were as successful as their more traditional age-mates.

Among the women who had followed traditional gender roles, all were married; one-third had divorced and remarried. During adolescence, they had been conventional, popular, and outgoing, and they had established firm identities for themselves. By age 40, these women were trusting, sympathetic, cooperative, outgoing, and capable of intimacy. They did not seem to be entering a period of midlife crisis. By age 50, their outgoingness had become more giving, their sociability more nurturing. Thus by middle age, the traditional women were generative, sociable, conforming. They functioned well in their roles as wives, mothers, friends, and took great satisfaction from their personal relationships.

Among the adult women in the California Intergenerational Studies who had followed less traditional gender roles, one-third had divorced, but not all had remarried. One-third worked full time. Compared to the traditional women, this group focused less on family and home, but their relationships with their children were much the same. During adolescence, the nontraditional women were ambitious, intellectual, and unconventional, and they established firm identities for themselves as well. By age 40, the less traditional women were functioning less smoothly. They were irritable, unpredictable, and their intellectual skills were no longer in the forefront. Women in this group did seem to be suffering a midlife crisis, and their identities were in flux. By age 50, they had resolved the crisis, and their untraditional identities were firm once again. Their intellectual skills, insight, and skepticism were flourishing. Compared to the traditional women, this group was more ambitious and independent, and found it easier to express their feelings spontaneously. They took great satisfaction from developing themselves.

What about the emotional development of traditional and untraditional men? During adolescence, traditional men controlled their impulses and expressions of feeling as they concentrated on achieving. By age 40, men from this group had become productive, dependable, competent, rational, and self-controlled. They also were expressing their feelings and learning how to be intimate. By age 50, these men were still productive, disciplined, and ambitious and perceived themselves realistically.

In contrast, during adolescence the nontraditional men were humorous, a bit impulsive, and expressed their feelings. They did not foreclose the issue of their personal identity. By age 40, their style was assertive, power-oriented, and exploitive. The men were angry, anxious, and neither trusting nor emotionally expressive. Like the nontraditional women in the study, the nontraditional men in adulthood were in conflict about their identities. By age 50, also like the nontraditional women, the men had resolved the conflict, had cast aside aspects of exaggerated masculinity, and were developing the capacity for intimacy. They once again were outgoing and could express their feelings.

Thus traditional men and women seemed to develop emotionally much as Erikson had suggested. The nontraditional men and women adhered less closely to Erikson's suggested timetable. The nontraditional men and women may have followed parallel developmental paths (Livson, 1981). As young adults, they slowed their development by suppressing their androgynous sides—men suppressing their feminine side, women suppressing their masculine side—and only felt free to express these aspects of themselves in middle age. The men and women in the California Intergenerational Studies were born in the 1920s, a time when gender role expectations were fairly inflexible. It is possible that men and women who grow up in periods when gender role expectations are more flexible experience fewer obstacles to healthy emotional development.

Personal relations

People universally seem to need emotional relationships with others. Women seem to need emotional relationships more strongly than men, younger women more strongly than older women, older men more strongly than younger men (Huyck, 1982; see Table 19.1). As we have seen, women traditionally are socialized to define themselves in terms of their relationships with others, men to define themselves in terms of their personal achievements. When the styles of relationship and attitudes toward intimacy of homosexual and heterosexual men and women were compared, it was found that a person's gender correlated more highly with attitudes toward intimacy than a person's sexual orientation (Peplau, 1981). Relationships between lesbians, in which both partners have been socialized to focus on long-term emotional commitment, are likely to be tender, warm, and caring—but also sometimes to falter under the weight of the partners' emotional overinvolvement (Nichols and Lieblum, 1983). In short, gender seems to have a strong bearing on how individuals relate to others.

> I tell my kids that you can have all the money in the world, but if you have no one to share it with, what have you got? *Nothing*.—Natalie's mother

Choosing a mate

By choosing a mate, people try to satisfy their needs for intimate connection with another person. Some intimate relationships are loving—whether passionate, companionate, or a mixture of both—and some intimate relationships are not loving. Not all married people nor all sexual partners feel love or

Table 19.1
Married people in the United States

	Age								
	18–19 Years	20–24 Years	25–29 Years	30–34 Years	35–44 Years	45–54 Years	55–64 Years	65–74 Years	>74 Years
Men	4.2%	28.6%	60.7%	74.7%	82.1%	86.2%	84.7%	83.0%	72.0%
Women	14.7%	44.2%	68.9%	77.1%	80.7%	78.6%	70.3%	50.1%	23.3%

SOURCE: U.S. Bureau of the Census, *Statistical Abstract of the United States, 1982–1983,* 103d ed., 1982.

intimacy. But in this society, marriage is idealized as a relationship between two people who feel intense romantic love for each other. In other societies, marriage is not expected to have a foundation in romantic love. Instead, marriage may be seen as a way to transfer property, to form family alliances, and to propagate children.

Romantic love flourishes in societies where people *expect* to fall in love. It also tends to flourish between people with strong similarities—in education, economic status, religion, race, and ethnic group. Although boundaries in our society are fairly fluid, most marriages are between people within the same social groups and do not cross racial, religious, or ethnic boundaries (Murstein, 1980). People also choose mates who are essentially their equals in physical attractiveness, intelligence, and sexual interest.

An **equity theory** has been proposed to account for why people choose the mates they do (Murstein, 1980). Each mate brings certain assets and liabilities to the relationship. Once partners have passed through the first blush of physical attraction, they evaluate the compatibility of these assets and liabilities. They also evaluate the compatibility of their attitudes and values. Partners who offer many assets and few liabilities choose each other; those who offer few assets and many liabilities settle for each other.

Marriage

Once people have chosen their mates, they begin the work of living together as partners in marriage. During the first year of marriage, many partners assume complementary roles: one passive and the other assertive, one nurturant and the other receptive, one dominant and the other submissive, and so forth.

> Sometimes I'm the boss, and sometimes my husband is the boss. I don't really know who's more powerful. But when things feel out of whack, we both know how to speak up and get things back in balance.—Erica and Mike's mother

> My mother always deferred to my father and said that the big decisions were up to him. She believed that women should complement their husbands. But she exerted more influence than she liked to let on—not that it was a problem for either of them. Somewhere along the line, they must have tacitly agreed to an official story and the slightly different private reality.—Sam's mother

Marriage partners also divide power between them. The findings from one study (Scanzioni and Scanzioni, 1976) showed three basic patterns to the power within marriages. In one pattern, the husband is the more powerful, and the wife acts in a complementary role. In a second pattern, the husband acts as a senior partner, the wife as junior partner. In a third pattern, the

After they have formed an intimate relationship, married people must work together as partners. In most marriages, the partners share power unequally; in some, the partners do cooperate and share equally.

husband and wife are equal partners. Typically, men and women wield power in different ways in marriage and other intimate relationships (Falbo and Peplau, 1980). Men are likely to wield power directly and to expect compliance. Women whose power is recognized within a relationship are likely to make requests and to bolster them with a statement about how important they feel the request to be. Women whose power is not recognized within a relationship are likely to resort to indirect means of wielding power, including hinting, withdrawing into silence and emotional coldness, and going ahead and doing as they wish. The results from other studies have shown other marriage patterns. In one study (Miller and Olson, 1978) that followed 1000 young adult couples for several years, nine patterns of marriage emerged. In three of them, wives held more power than husbands. Although most young adults say that a marriage of equals is their ideal, few actually live a marriage like that. Among the 1000 couples, although 80 percent said that their marriages were cooperative, with power equally shared, the researchers rated only 12 percent as cooperative. A similar discrepancy between how young adults described their marriages and how researchers rated them appeared in another study (Hill et al., 1970) in which the marriages of young adults, their parents, and their grandparents were compared.

Satisfaction with marriage

> Sam's father and I are very much alike. Who wouldn't be after nearly 20 years together? But we're also basically different in ways that work well for the marriage and for the family as a whole.—Sam's mother

Which factors affect individuals' satisfaction with their marriages? Neither age nor income seems to affect marital satisfaction, and couples with hefty incomes seem no more satisfied with marriage than couples with lean incomes (Spanier and Lewis, 1980). Among happily married couples, throughout adulthood, women report that intimacy and emotional security are the most important factors in the happiness of their marriages, whereas men report that loyalty and commitment to the future of the marriage are the most important factors to them (Reedy, Birren, and Schaie, 1981).

Data from the California Intergenerational Studies (Skolnick, 1981) suggest that happily married middle-aged partners like, admire, respect, and enjoy being with each other. Their personalities are similar. Certain personality traits in this happily married group correlate with traditional gender roles: nurturance in women and self-confidence in men. Members of both sexes also are socially mature and high achievers. Over the years, their marriage has gotten better, and they would choose the same partner all over again. Their marriage is personal and intimate, not merely a useful living arrangement. Happily married couples like these—who seem to embody the companionate ideal that young married couples wish for—may comprise only one marriage in five. The other marriages are likely to be more utilitarian, cemented by practical rather than emotional matters. In unhappy marriages in the California sample, bonds tended to be utilitarian, and the personalities of husband and wife were dissimilar.

In the California Intergenerational Studies, not only did husbands and wives rate their own levels of marital satisfaction, trained observers independently rated the satisfaction of both husbands and wives as well. These independent ratings correlated highly with the husbands' and wives' own ratings. Even more interesting, however, the correlations of these independent ratings were significantly higher than the correlations *between* husbands and wives. In other words, in many of the marriages, there were considerable discrepancies in partners' ratings of their satisfaction with their marriages but not in observers' ratings. Results from other studies confirm that the happier partner is likelier to be the husband than the wife (Douvan and Kulka, 1979; Veroff and Feld, 1970). Middle-aged wives were especially prone to disappointment with their marriages. In a study (Lowenthal, Thurnher, and Chiriboga, 1975) of San Francisco couples, for example, only 40 percent of the middle-aged women viewed their husbands favorably, compared to 80 percent of older and younger wives. The middle-aged wives complained most about poor communication in their marriages, a complaint echoed by older wives in another study (Stinnett, Collins, and Montgomery, 1970).

Results from cross-sectional studies tell another part of the story of how happy adults are with their marriages. They suggest that people are happy with their marriages in the early years. Satisfaction is lowest after the birth of the first child and highest when the children leave home. But cross-sectional results, which show satisfaction high in later years of a marriage, may obscure the tendency for the least happy marriages to dissolve in divorce or in the partners' denials that they have devoted decades to an unhappy relationship (Huyck, 1982). Cohort effects and other factors relevant to the partners' ages also may affect the outcome of cross-sectional data. The traditional pattern of marital satisfaction did not appear in one study of middle-class couples (Spanier and Lewis, 1980). The pattern did not appear in the California Intergenerational Study, where couples married for 16 to 18 years and with adolescent children still at home reported as much satisfaction with their marriages as they had reported ten years earlier (Skolnick, 1981).

Results from a study described earlier (Baruch, Barnett, and Rivers, 1983) in which adult women aged 35 to 55 reported their satisfaction with their marriages tells yet another part of the story of marital satisfaction. Among this group of over 200 women, those who took the most pleasure—happiness, satisfaction, plus optimism—from their marriages were mothers. Those who took the least pleasure had no children and did not work outside their homes.

> My mother used to tell me that the secret of a good marriage was always to give more than 50 percent. "It comes back to you," she'd say. Now that I'm married, I think she was right. If I counted all the compromises I've made in my marriage, I'd get depressed. We both compromise—without counting.—Natalie's mother

Marriage itself does seem to make people happier overall. Married adults report being happier than unmarried adults (Hutchinson, 1975). Married adults are also healthier and less afflicted with mental illness than unmarried adults (Traupmann, Eckels, and Hatfield, 1982). Of course, we cannot tell from these data whether marriage makes people happier and healthier or whether people who are happier and healthier are more likely than others to marry. In general, findings from cross-sectional and longitudinal studies are consistent about the personality traits that correlate with happy marriages. These traits include emotional maturity, self-esteem, adaptability, consideration, and the ability to express affection. It is not surprising that people who are well adjusted socially and emotionally would succeed at marriage.

Divorce

Today, divorce in the United States ends four marriages in ten. Although the averages suggest that divorce happens after about six years of marriage, to a man of 31 and a woman of 29, the specifics say that nearly 10,000 marriages of people over 65 years old will end in divorce. Not only does partners' satisfaction with their marriage influence whether they ultimately seek divorce, but external pressures do, too: critical career decisions, physical attraction to someone outside the marriage, and others. In an age in which marriage is no longer the central cultural event in the passage into adulthood, divorce is more common than it was in times past (Furstenberg, 1982).

Even though divorce is relatively common today (see Table 19.2), it remains a painful process. Divorce may free people from a burdensome relationship, but it also is likely to make people feel more anxious, guilty, incompetent, depressed, lonely, lower in self-esteem, and more likely to drink alcohol and smoke marijuana than either marriage or remarriage (Cargan and Melko, 1982). Many people feel a sense of personal failure when their marriages end. As a group, divorced people have the highest rates of emotional disturbance, accidental death, and death from cancer, cardiovascular disease, and cirrhosis of the liver (Brody, 1983b).

Remarriage

Most people who divorce remarry, and they do so within three years on average (see Figure 19.2). When over 200 recently divorced people from Pennsylvania were studied (Furstenberg, 1982), two-thirds said that they were unlikely to remarry. But within three years, nearly half of them had remarried. When middle-aged adults divorce, over 80 percent of the men and 75 percent of the women remarry (Glick, 1980). Among older adults, 95 percent of marriages are remarriages, although most of the earlier marriages end in death, not divorce (Troll, Miller, and Atchley, 1979).

> I married my second wife because we both were lonely and wanted to make a home together. Life is easier when you're not alone.—Tim's grandfather, age 56

Table 19.2
Divorced people in the United States

	Age								
	18–19 Years	20–24 Years	25–29 Years	30–34 Years	35–44 Years	45–54 Years	55–64 Years	65–74 Years	>74 Years
Men	0.1%	1.8%	5.4%	8.8%	8.9%	7.3%	6.1%	3.9%	2.5%
Women	0.7%	3.8%	8.8%	11.6%	11.6%	9.9%	7.1%	4.4%	2.3%

SOURCE: U.S. Bureau of the Census, *Statistical Abstract of the United States, 1982–1983,* 103d ed., 1982.

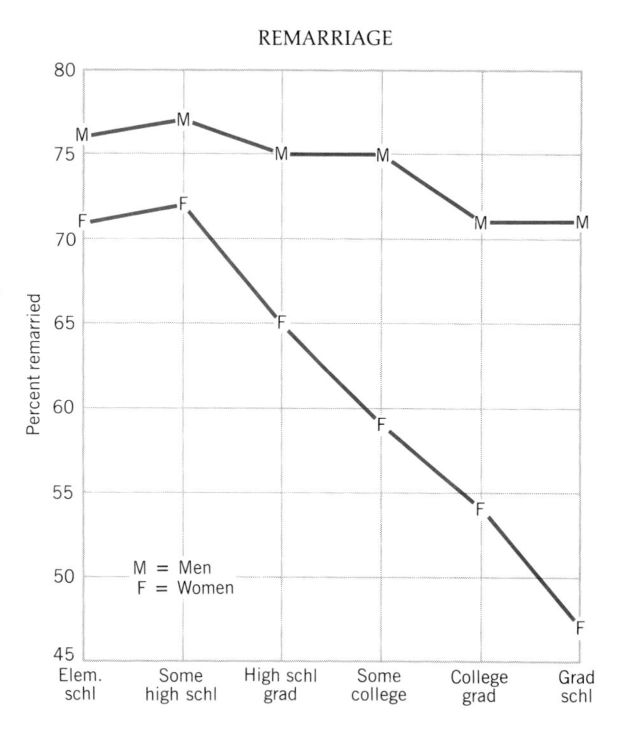

Figure 19.2 ***This graph shows the relations among remarriage, sex, and education. More men than women remarry at every educational level. The more education women have had, the less likely they are to remarry (U.S. Bureau of the Census, 1970*** **Census of Population).**

> After my first marriage broke up, I was shellshocked. I moved 2000 miles across the country and buried myself in my work. I dated now and then, never very seriously. I think that I was nursing my wounds. It took a long time for me to fall in love with my second wife and an even longer time for us to want to try marriage again. But it's working out well. We're more alike than my first wife and I were. We do more things together. We have a sense of being companions that I think will last us for a long time.—Robbie's father, age 38

Most people who divorce eventually remarry. With the children from former and current marriages, they form a "blended" family.

Remarriages differ from first marriages in several respects (Furstenberg, 1982). People form expectations about marriage and bring these expectations to later marriages. Many people use an earlier marriage as a basis for comparison and try not to repeat what they perceive as their earlier mistakes. People in remarriages also are likely to be parents, many with young children. Family relationships among stepparents, stepchildren, and the sometimes unavoidable contact with a former mate can make remarriages complex. When people remarry, they are older, more mature, and have greater experience and, in some cases, social status. When men remarry, they tend to look for women who are younger than they are, and women tend to look for older men with good income (Glick, 1980). Thus remarriages put people in different marriage cohorts, and their relationships are subject to different cultural and social events (Furstenberg, 1982).

People who remarry are likely to have lowered their expectations and to look for companionship, long-lasting affection, and admiration (McKain, 1969). Among the remarried adults from Pennsylvania (Furstenberg, 1982), most said that their later marriage was more egalitarian—decisions and household jobs were more often shared—and was characterized by better communication, trust, and goodwill than their first marriage had been. Most later marriages seem to be about as satisfying, happy, and worrisome as first marriages. Many remarried women report that they are very happy. Many remarried men say either that they are very happy or very unhappy; few say that they feel somewhere in between (Huyck, 1982).

Other relationships

Adults have personal relationships with brothers and sisters and with friends, relationships that may (or may not) be deeply meaningful.

Brothers and sisters

My sister and I were at each other's throats when we were kids. But now that we're both grown up and married, we get along fine. I can actually confide more in my closest friends than in my sister, but my sister and I have shared some important things, too—Sam's mother

My younger brother and I are not especially close. We get together every few months for a meal, but our lives are very different. He's gay and lives in the city. I'm a suburban father of two young boys. But we love each other, and he knows that he can come to me for help.—Tim's father

The relationship with brothers and sisters generally lasts longer than the relationship with parents—from the birth until the death of one sibling. It is a relationship in which much is shared—genetic inheritance, culture, and many early experiences. It is a relationship of relatively equal power and freedom. It is an unearned relationship—conferred by birth alone (Cicirelli, 1982).

The relationship between brothers and sisters is likely to be most intense during childhood and to diminish somewhat once children leave home. Most people stay in touch with their siblings, and many report intense feelings even in adulthood. Most also report that they get along well with their siblings, feel that their siblings are interested in them. However, most adult siblings do not talk intimately together or discuss important decisions about their lives. Adults are especially likely to retain intense feelings for their grown siblings when, as children, they have lost a parent through death, divorce, or abandonment or when they have perceived their parents as especially weak or hostile. When the family social system has been either unpredictable or absent, brothers and sisters may turn to each other for support (Bank and Kahn, 1982).

Researchers have found several typical patterns of relationship among adult siblings (Cicirelli, 1982). Some are essentially apathetic. They may get together at family rituals—holidays, weddings, funerals, and the like—and are unlikely to stay in close touch once their parents die. It is unusual for apathetic siblings to fall completely out of touch, however. Some siblings, as we have said, are close. Sisters are more likely than brothers to be very close in adulthood, and cross-sex relationships are likely to be closer than relationships between brothers (Adams, 1968). Although most people say that they have felt rivalry toward a brother or sister during childhood (Ross and Milgram, 1982), sibling rivalry in adulthood is reported by a scant 2 percent of middle-aged adults (Cicirelli, 1981).

Friends

Unlike marriages and family relationships, friendships are not bound by law or clear social customs. They therefore tend to be more fragile than these other personal relationships among adults. Friendship is a highly subjective term, given to widely various definitions, rarely a predictable guide to actual behavior (Huyck, 1982). In one study (Cohen, Cook, and Rajkowski, 1980), for example, adults who lived in hotels in Manhattan listed their "friends" as well as their daily contacts with other people. It turned out that many of the adults saw their friends seldom and considered them neither intimate nor important. They saw their "nonfriends" often and considered some of them both intimate and important.

People need not live near one another to feel friendship. When people have shared a long and intense relationship—during childhood, college, military service, or the like—their friendships may regain intimacy at each new contact (Hess, 1972). Many middle-class people sustain long-distance friendships, whether by letter, phone call, or periodic visits.

Friends tend to be roughly similar in interests, attitudes, social backgrounds, age, sex, race, and religion. They tend to be similar because friendship seems to offer people ego support, self-acceptance, and validation (Tesch, 1983; Hess, 1972). Friends affirm one another's identity. They also provide intellectual stimulation, practical help, and plain fun. Friends help each other to learn new roles—how to be a parent or a marriage partner—and monitor each other's behavior (Hess, 1972).

When friendships flicker or die, reported a sample of young adult readers of *Psychology Today* (Parlee, 1979), they do so because one friend moves away, marries, has a child, divorces, gets involved with someone the other friend dislikes, betrays the other, holds conflicting views on an important subject, borrows money, or suddenly has a significantly different income or job level. People's social roles tend to influence the people they choose and keep as friends. Parents are likely to be friends with other parents, singles with other singles, and these tendencies are especially marked among older adults. Older married people see more of their friends than others their age, older people without children see more of their friends than older people with children, and older people with jobs make new friends more than older people who have retired (Riley and Foner, 1968).

> When I was in my teens and twenties, I had lots of male and female friends. Some of them are still my friends, and we try to stay in touch. But all of my close friends are women. I don't really confide in any men—except my husband. He's absolutely my best friend. But it's still very important for me to have a few women friends to talk with intimately.—Sam's mother

> My college roomate is still my closest friend, twenty years later. We live close enough to see each other every few months. We like to talk about our work, and movies, and what's going on in our lives. The friendship is important to both of us.—Mike and Erica's father

Do the friendships of men differ from those of women? People have many different opinions on the subject. Some say that men form truer and deeper friendships than women (Tiger, 1969), and others say just the reverse (Booth and Hess, 1974). Men's friendships often revolve around activities and experiences, women's around emotional support, assistance, and intimacy (Weiss and Lowenthal, 1975; Wright, 1982). When strains appear in a friendship, women are more likely to discuss the source directly in the hope of eliminating them. Men are likely to end the friendship or to tolerate the strain. Do men hide their feelings from their friends while women express them? A study (Hacker, 1981) of single, middle-class, Catholic adults showed that the men confided in their same-sex friends about as much as the women confided in theirs (see Table 19.3). It may be that friendships differ among conventional and unconventional men and women. In a study (Bell, 1981) of upper middle-class Australians and Americans in their thirties, those who were unconventional revealed themselves to their friends and focused on the emotional aspects of friendship. They were likelier to form friendships with members of the other sex, and some of these friendships were sexual. In contrast, conventional women's friendships were more personal, intimate, and emotional than

Table 19.3
Self-disclosure among friends

	Same-Sex Pairs		Cross-Sex Pairs	
	Women	Men	Women	Men
Reveal strengths and weaknesses	77%	86%	50%	62%
Reveal only weaknesses	18%	0	33%	0%
Reveal only strengths	0	9%	0	31%
Reveal neither	5%	5%	17%	7%

SOURCE: H. M. Hacker, "Blabbermouths and Clams: Sex Differences in Self-Disclosure in Same-Sex and Cross-Sex Friendship Dyads," *Psychology of Women Quarterly*, 5 (1981), p. 393.

those of conventional men. The friendships of conventional men revolved around sharing activities and were low in self-disclosure. The friendships of both conventional men and women were not likely to cross gender lines. Although the adults were not rated on feminine and masculine personality traits, the unconventional adults had more in common and were probably more androgynous than the conventional adults.

Family

The family can be viewed as a social system, established when two adults marry, added to by the births of children, subtracted from by the departure of various family members, dissolved when marriage partners divorce or die (Hill and Mattessich, 1979). Within this system, a change in any one part affects all the other parts. Thus when a first child is born, husband and wife assume the roles of father and mother. When a second child is born, new relationships form.

The length of the cycle of family development has stretched in recent years. On the average, adults marry later, have children later, and have longer periods without children at the beginning and end of marriage (Glick, 1977). Over time, family members respond to one another's needs. The nature of their responses are affected in part by the historical era in which they live. Social meanings attached to birth, sexual maturity, age, and death all change from one era and one culture to another (Elder, 1978). These social meanings shape the way families view the "right" times for children to enter and finish school, to leave home, and for adults to marry, to become financially independent, have children, retire from work, and the like.

Being a parent

Most couples today have children, although they have fewer children than couples did in the 1960s and earlier decades. Many women also hold jobs and postpone having their first child until they are in their late twenties or thirties. It has been suggested (Rollins and Galligan, 1978) that children powerfully affect the degree of satisfaction married people feel. They affect not only how well parents perform as husband and wife, but how parents evaluate one another as husband and wife, and the sacrifices each makes for the sake of the family. People generally find parenthood to be both stressful and satisfying, and nearly all say that they would have children again if they had the chance (Yankelovich, 1981).

Parents say that the satisfactions come as feelings of joy, fun, pride, fulfillment, and maturity. Many women feel that parenthood enhances their sense of personal worth; many men feel that it enhances their self-esteem (Bell and Harper, 1977). Many mothers feel a strong link between their effectiveness as caregivers to their children and their self-esteem. This link does not appear so often or so strongly among fathers, although many fathers today actively take part in caregiving to their children. Among fathers who do little routine caretaking of their infants and young children, the link between self-esteem and even a clumsy attempt at feeding or dressing a child is essentially nonexistent (Lamb and Easterbrooks, 1981).

Children change a marriage. Even among couples who had divided chores relatively equally before they had children, the division of labor takes on a more traditional cast after children are born (Cowan et al., 1978). When new mothers feel overburdened by household tasks plus childcare, conflict within the marriage may increase and satisfaction may decrease. Parents say that the stresses of parenthood come as restrictions on their personal and financial freedom. Children tend to intrude upon parents' time to talk together, to act and express their feelings for each other spontaneously, and to have sex. As many as one quarter of mothers with children under the age of 5 occasionally take tranquilizers for depression or anxiety (Brown and Harris, 1978). Part of the pressure on mothers comes when they work to make the family system operate smoothly, only to run into anger and resistance from their children and little help from their husbands. Feeling stressed and boxed in, mothers in these situations may feel low self-esteem and depression (Maccoby, 1980).

> When Sam was 4 years old, he went through a phase when he was defiant and fresh. He refused to do what he was told, ordered me around, threatened me—''I'll blow you up!''—and stuck out his tongue at me 50 times a day. One day at lunch, he snapped his fingers and commanded, ''Milk!'' My husband just tuned all of this out. I got furious. I told my husband that we had to do something because the situation was intolerable. We spent a whole afternoon, while Sam

Families grow and develop through fairly predictable stages.

> napped, discussing how to handle the problem. I really don't know what I would have done if we hadn't talked.

Not only may the division of labor within a marriage affect parents' feelings of satisfaction, but other factors affect it, too. A mother's age is one such factor. Younger mothers tend to feel less satisfaction with parenthood than older mothers. In a study of mothers between the ages of 16 and 38 who had one child and who did not hold outside jobs, the older mothers reported more satisfaction with parenthood. This effect was especially prominent in mothers of infants who had been born prematurely (Ragozini et al., 1982).

> When Erica was about 6 months old, we were feeling stretched to our limits. I guess life is a series of lowered expectations. It was harder for us to get used to the second baby than it had been to get used to having just one. I know that lots of people say that one more doesn't make much of a difference, but that wasn't our experience. It probably has a lot to do with how easy the kids are.—Tim and Willy's mother

The birth of a second child requires further adjustments within a family. Parents must again shift the balance of household and childcare tasks as well as their needs as married partners, jobholders, and friends. A study (Kreppner, Paulsen, and Schuetze, 1982) that followed 16 couples for two years after the birth of their second child showed how parents readjusted. Some traded roles, standing in for one another as the situation demanded. Some fathers took over more responsibility for the first child, while the mothers established intimate ties with the new babies. Some fathers took over more of the household tasks, while mothers took over most of the childcare. Once the babies were out of infancy, the parents redefined their roles, seeing themselves as parents of children to whom they responded as individuals and attending somewhat more to their own needs and wishes as well.

Another study (Baruch, Barnett, and Rivers, 1983), which focused on mothers at midlife, presented two kinds of relationships with children. **Autonomous mothers** took satisfaction in watching their children behave as individuals, felt proud of them, liked them, and enjoyed doing things with them and watching them grow up. **Coupled mothers** took satisfaction in the way their children completed their own sense of personal identity, felt that their children gave meaning to their (the mothers') lives, and made them feel needed, special, and irreplaceable. Mothers of both kinds took great satisfaction from their children, but autonomous mothers tended to have higher self-esteem, to feel more in control of their lives, and to complain less of depression or anxiety than coupled mothers.

The empty nest

> The nest may be empty, but our lives certainly aren't empty. We're still parents, even though our kids are on their own. We still see them and love them. But it's lovely not to have to consult a child's wishes about what to eat or watch on television, not to have to close the bedroom door, not to have to do two or three loads of laundry every day.—Nancy's mother, age 44

When the last child leaves home, parents face an ''empty nest.'' The question is: do they face the empty nest sorrowfully or joyously? It was once presumed that all parents grieved when their children left the nest and that mothers grieved more than fathers (Troll, Miller, and Atchley, 1979). After her years of concentrated childcare, a middle-aged mother facing menopause and the flight of her children was thought to be headed for depression. But research (Glenn, 1975; Neugarten, 1970) showed that the ''empty nest syndrome'' was more fiction than fact. Few women had trouble coping with their freedom from the daily stresses of parenthood. Women who did not hold outside jobs seemed to adapt to their freedom even more easily than women who did hold jobs. The minority of women who reacted to the empty nest with unhappiness tended to have histories of emotional problems (Lowenthal and Chiriboga, 1972). In general, an empty nest means that parents have more time for one another and less spend less time cleaning, shopping, cooking. The styles of marriages once children leave home are varied. Some husbands and wives have the time and the privacy for a second honeymoon and renewed intimacy. Others develop separate interests. Still others, who have stayed together ''for the sake of the children,'' part company (Troll, Miller, and Atchley, 1979).

Like so many other normative life events, when the nest empties ''on schedule''—when it is expected and prepared for—the transition is easier for family members than when it empties precipitously, too early, or too late. Caring for frail elderly parents is another normative life event the timing of which can make it more or less difficult.

Caring for frail parents

Parents and their grown children are likely to cherish not only their independence but their interdependence as well. In most families, the bonds between parents and grown children are relatively close, maintained by affection, assistance, and fairly frequent contact. But children do not have to feel warmth for their parents to stay in touch with them (Clausen, Mussen, and Kuypers, 1981) or to offer them assistance. When aging parents face declining health or income, their adult children may have to assume considerable responsibility for them. When the adult children are also caring for their own children, they may be caught in what has been called (Oppenheimer, 1981) a **life cycle squeeze**. Responsibility for aging parents (and in some cases, grandparents, too) plus maturing children can take a great emotional and financial toll. Many middle-aged women who report feeling greatly stressed by caring for their aging parents also have children whose development is ''off schedule'' (Hagestad, 1982)—a son in his late twenties and still without work or a divorced daughter who has moved back to her parents' home.

> My poor mother is running around trying to take care of my sister and her two young kids, who moved back home a few years ago when her marriage fell apart, plus her own widowed mother who's got a heart condition, plus my father's mother, who is healthy but nearly 90 years old. Every once in a while my mother asks, ''So where's the gold in the 'golden years'?''—Mike and Erica's father

Middle-aged men are more likely to feel the stress in financial terms.

When elderly parents live in another city, a fairly common phenomenon among middle-class families, the squeeze can grow tighter still. Many middle-aged children make long-distance arrangements for their parents' care and spend their weekends traveling. They are under great stress. The stress can be severe enough to destroy marriages, careers, and bank accounts (Collins, 1983b).

Work

> At Sam and Natalie's nursery school, the kids' choices of what they would be when they grew up were recorded on a long sheet of paper. ''Laboratory worker,'' said the son of a biologist. ''Doctor,'' said Natalie. ''Skyscraper builder,'' said Sam.

> I've struggled ever since I was in college to "find myself" and the kind of work I wanted to do. I went part of the way through graduate school and quit to get married. When Kristy and Jason were in school, I worked for an advertising agency writing copy. After I was divorced, I got my real estate license. Everyone I know has made some commitment to a job or a career. But not me. I feel so at sea. It's really awful.

Work plays an important a part in the development of a person's sense of self. The adult's ability to be productive develops upon the foundation of a sense of industry which, as we have seen, is laid down during middle childhood (Erikson, 1968). So important is productive work to people's psychosocial identities that when adult men and women were asked whether they would keep on working even if they did not need the income, the vast majority said yes (Pfeiffer and Davis, 1974; Renwick and Lawler, 1978).

Choosing work

In the best of all possible worlds, people would know and find the job or career that best suited their interests, skills, and personal qualities. But many people end up working where they do by reason of accident, luck, sex, social class, and proximity. Among a group of readers of *Psychology Today* magazine, of whom most were professionals and executives, 40 percent reported that chance had determined their present occupations, and only one-fourth reported that they deliberately had chosen their occupations (Renwick and Lawler, 1978).

People are most likely to find occupations that suit their interests, skills, and personal qualities if they can identify both those interests, skills, and qualities and the occupations that tap them. But when young adults are making career decisions, they may not understand the everyday realities and requirements of occupations. They may well operate according to stereotypes and casual sources of information, with the result that many adults ignore occupations that might suit them well or prepare at length for occupations that suit them badly (Anastasi, 1976).

focus Vocational tests

Vocational tests exist that can help steer people toward occupations suitable to their skills and interests. One such test is the Strong-Campbell Interest Inventory (Campbell, 1974). It contains hundreds of questions about a person's likes, dislikes, and preferences. The answers create a profile of test takers' general interests and the kind of working environment—farm or office, for example—that would be most suitable.

Answers to the Strong-Campbell test questions cluster around themes described by researcher J. L. Holland (1976). According to Holland, the interests of people in particular occupations tend to cluster around certain basic themes. Thus people show more or less interest in the realistic, investigative, artistic, social, enterprising, and conventional. Farmers are likely to score high on realistic and conventional themes and low on social and artistic themes. These themes, Holland suggests, correspond to important aspects of people's personalities. When people's personalities match the themes of a particular occupation, they are likely to feel satisfied and to stick with their work and to achieve well.

Career development

In their first five years of working, people may change jobs more often than in later years. Early adulthood is a period of initial commitment to a line of work.

During adulthood, some people develop along with their careers. So closely bound are some people's personal and work identities that it has been suggested (Vaillant, 1977) that a stage of *career consolidation* should be inserted into Erikson's model of adult development. As we have seen, many studies of adult development show that for men especially, a period of career consolidation precedes a period of generativity.

Some people follow an orderly occupational course and others, a less orderly one, over many years (Super, 1957). The orderly careers can be described as a series of stages, beginning with adolescence. During the earliest, *crystallization stage*, adolescents explore many fields and try to match their career opportunities with their personal needs, interests, values, and skills. Some adolescents picture themselves in a future occupation, but few have any real understanding of what the occupation actually entails. During the *specification stage*, people find out more about specific occupations, and their knowledge further shapes the direction they take. This stage corresponds to the undergraduate college years and is a transitional period devoted to job training.

By early adulthood and the early twenties, people make an initial commitment to an occupation. During the *implementation stage*, people may get further training, whether in school or on the job, and they may continue to shift jobs in search of the one that best suits them. Some new workers find mentors, established older workers who serve as guides and advisers to them, smoothing their way to advancement and promotion. Mentors may be important to people in trades, business, and most professions. They may be nearly essential in some occupations. In the longitudinal study (Vaillant, 1977) of Harvard graduates, for example, those men whose careers had been relatively unsuccessful had not had mentors during their twenties and thirties. Many women who have succeeded into the highest corporate ranks also had mentors.

During the first five years or so of people's working lives, they tend to change jobs more often than in later years. But up to one in five workers changes jobs every year, and most of these are voluntary changes. Among the men in the California Intergenerational Studies (Clausen, 1981), for example, many changed jobs during their thirties and forties. Top executives and those who were upwardly mobile were more likely than other men to change jobs. Blue-collar workers rarely reported changing jobs once they were in their thirties.

When workers reach their mid-twenties, they enter the *stabilization stage*. For about a decade, they buckle down to establishing their positions within their chosen fields. By the mid-thirties, a *consolidation stage* begins, when the experienced and knowledgeable worker advances as far as possible and consolidates the gains he or she has made. The consolidation stage ends with retirement from the work force.

Personality not only influences the occupations people choose but their success in those occupations as well. The most successful adult men in the California Intergenerational Studies, for example, had been ambitious, productive, dependable, and self-disciplined in early adolescence. They also scored lowest on personality traits such as anxiety, fearfulness, punitiveness, withdrawal when frustrated, feelings of victimization, and the tendency to complicate simple situations. Middle-class men who had been working-class as adolescents were more dependable, considerate, likable, sympathetic, and warm than those who had been middle-class as adolescents. The upwardly mobile men remained warm and sympathetic as adults, developed their intellectual skills, were conventional and unlikely to test limits.

By the time working people are in their mid-thirties, most become experienced and knowledgeable and consolidate their gains.

Data on women from the California Intergenerational Studies (Stroud, 1981) also reflect the interaction of personalities and occupations. Women at midlife who were committed to their careers had not conformed to the traditional feminine role as adolescents. They had chosen higher education at a time when that choice was not the norm for women, their intellectual interests had centered on "masculine" fields, such as science, or they had committed themselves deeply to "feminine" fields, such as creative writing. By the age of 40, the women were ambitious, rational, insightful, and warm—an androgynous mixture. They were deeply committed to and satisfied with their careers. However, the women's commitment to their work was typically less extensive than men's commitment to their work. The occupational course of most of the women had not been orderly, and most had developed their commitment to work during their thirties.

In contrast, women who worked primarily to keep busy or to earn money were less satisfied with their work than either the work-committed women or full-time homemakers. The working women who had been assertive as adolescents and had gone to college at midlife remained highly assertive, independent, but cool. They tended to have low self-esteem and low morale. The women who had not been to college and were not committed to their work were less warm still, power-oriented, rebellious, and without deep intellectual interests. Those women who were full-time homemakers had conformed to the conventional feminine role as adolescents. Full-time homemakers who had attended college had the highest self-esteem and morale of any of the women studied. They were happy, nurturant, submissive, and eager to please. However, college graduates who had left their jobs to become full-time homemakers tended to be depressed, hostile, angry, critical of their children, and dissatisfied with the worlds of work and family. The full-time homemakers who had not attended college were ambitious, warm, nurturant, insightful, but critical of themselves.

Occupation also may affect personality. One reason why it does is because a person's working environment can be considered a stimulus that endures over long periods of time (Garfinkel, 1982). Sociologist Melvin Kohn (1980) has suggested that a job's complexity, pressures, closeness of supervision—and not the job's status—influence a personality. People whose jobs are complex—requiring independent thought and judgment—and challenging are likely to respond by developing increasing intellectual flexibility. The complex-

Many adults benefit from job counseling and retraining, both to enter and to reenter the job market.

ity of a job may affect personality characteristics such as self-esteem, anxiety, responsiveness to change, moral standards, authoritarianism, intellectual quality of leisure pursuits, and degree of alienation. Conversely, repetitive, fast-paced, mindless work can be deadening, alienating, and injurious to health (Geyer, 1972).

> I took my job with a general practice law firm because I could make an excellent living there. But I hate everything else about it—my partners, my clients, the deals. Sunday nights are almost unbearable. When I make it to lunchtime on Mondays, I reward myself with a cigar.—Tim and Willy's father

> I've been teaching English literature to undergraduates for over fifteen years, and I still love it. I love what I teach, and I like the students. They do seem to get younger every year, I notice. The one thing that I hate about my job is grading papers. That actually gets more painful every year.—Sam's father

No matter the job, when a person finds work creative, productive, or a way to achieve, that person is likely to take pride and satisfaction from that work (Garfinkel, 1982). The stockbroker who feels that his job is superfluous, that he cannot make meaningful decisions, and that his customers are on a hopeless chase after money derives no sense of achievement from his work. But the supermarket clerk who takes pride in knowing the prices of thousands of items, in her speed and accuracy in ringing them up, and in her unfailing cheerfulness toward all customers gets to work 45 minutes early every day.

Summary

1. According to Erikson, the main developmental task of young adulthood is to resolve the tension between intimacy and isolation. The developmental crisis of middle adulthood is between generativity and self-absorption or stagnation.

2. According to Levinson, the seasons of a man's life ultimately form a pattern, or life structure, with alternating periods of stability—when developmental tasks are met and goals actively pursued—and flux or transition—when men question the direction of their lives, make important choices, and explore new avenues.
3. Midlife crisis is a developmental state of physical and psychological distress that arises when developmental issues threaten to outstrip a person's resources for dealing with them. Research into whether a midlife crisis exists has produced mixed results. Adulthood probably is not a period of stability, but a period of continual emotional and motivational change.
4. Erik Erikson has asserted that the stages of psychosocial development he describes are essentially alike for men and women. But whereas Carol Gilligan asserts that relationships form the core of girls' and women's lives, Erikson and Levinson portray males as defining themselves by growing independent in their relationships. Gender seems to have a strong bearing on how individuals relate to others.
5. By choosing a mate, people try to satisfy their needs for intimate connection with another person. Among happily married couples, throughout adulthood, women report that intimacy and emotional security are the most important factors in the happiness of their marriages, whereas men report that loyalty and commitment to the future of the marriage are the most important factors to them.
6. Not only does partners' satisfaction with their marriage influence whether they ultimately seek divorce, but external pressures do, too: critical career decisions, physical attraction to someone outside the marriage, and the like. Many people feel a sense of personal failure when their marriages end. Most people who divorce remarry.
7. Adults have personal relationships with brothers and sisters and with friends, relationships that may or may not be deeply meaningful. Adults are especially likely to retain intense feelings for their grown siblings when, as children, they have lost a parent through death, divorce, or abandonment or when they have perceived their parents as especially weak or hostile. Friendships are not bound by law or clear social customs and therefore tend to be fragile.
8. Children affect not only how well parents peform as husband and wife, but how parents evaluate one another as husband and wife, and the sacrifices each makes for the sake of the family. People generally find parenthood to be both stressful and satisfying. Many women feel that parenthood enhances their sense of personal worth; many men feel that it enhances their self-esteem.
9. Like many other normative life events, when the nest empties "on schedule"—when it is expected and prepared for—the transition is easier for family members than when it empties precipitously, too early, or too late. Most adults report positive transitions from the nest emptying.
10. Responsibility for aging parents (and in some cases, grandparents, too) plus maturing children can take a great emotional and financial toll on people in middle adulthood.
11. Work plays an important part in the development of a person's sense of self. Personality influences the occupations people choose and their success in those occupations. Occupation also may affect personality. People whose jobs are complex—requiring independent thought and judgment—and challenging are likely to become increasingly intellectually flexible.

Key Terms

midlife crisis
equity theory
autonomous mothers
coupled mothers
life cycle squeeze

Suggested readings

ERIKSON, ERIK. *Dimensions of a New Identity*. New York: Norton, 1974. Erikson's theory of the stages of lifelong development.

GOULD, ROGER. *Transformations: Growth and Change in Adulthood*. New York: Simon & Schuster, 1978. Gould's research on predictable adult life stages.

LEVINSON, DANIEL. *The Seasons of a Man's Life*. New York: Knopf, 1978. Research and theory of adult development that suggests specific age-linked phases that underline personal crises, emotional states, attitudes, and behavior.

PART EIGHT

Late Adulthood

BIOLOGICAL PERSPECTIVES ON AGING
- Theories of Aging

PHYSICAL CHANGES
- Skin
- Skeletal System
- Cardiovascular System
- Respiratory System
- Gastrointestinal System
- Excretory System
- Reproductive System
- Nervous System

DISEASE AND AGING
- Acute Illness and Injuries
- Chronic Illness
- Health and Expectations

HABITS AND HEALTH
- Smoking
- Diet
- Lifespan Focus: Obesity
- Exercise

chapter twenty *Physical development*

Biological perspectives on aging

The nightly television news camera was focused on a charming, alert, and attractive woman as she celebrated her 103rd birthday. What was her secret for such a long life, asked the reporter. Was it something special that she ate or drank? Did she avoid doing certain things? ''No,'' she answered with a smile, ''nothing special. I just thank God every day.'' At the age of 103, this woman probably was pushing the upper limit of the human lifespan. The claims of some super-old people to have survived to the ages of 137 or 150 have never been authenticated. Although researchers are working to push back the apparent biological limit to the human lifespan, and although someday people may survive well into their hundreds, to date the oldest humans have lived to the age of about 110.

Most of us, of course, would like to live to be exceptionally old, so long as we are healthy and productive. In this chapter, we will discuss both the longest lifespan that researchers believe that people one day *may* enjoy and the average lifespan they actually *do* enjoy. We also will discuss several different theories of the biology of aging. We will look at the effects of both primary, or normal aging, and of secondary aging, in the hope of understanding which aspects of aging are inevitable parts of the human condition and which aspects are the avoidable costs of illness, physical abuse, and physical disuse.

Theories of aging

Which factors determine the biological limits on the human lifespan? What exactly makes people age? These are just two of the fundamental questions that motivate researchers. We know that even though people escape the ravages of chronic disease, genetic, and environmental hazards, all people ultimately age and die. Just what mechanisms cause the human body to falter and die? Theories of aging tend to explain this seemingly inevitable process in terms of either genetic programming or wear and tear.

Programmed theories

My grandmother was widowed at 78, but she was still healthy and able to take care of herself. She lived in her own house until she was 92. Then she fell and broke her hip. It was hard for her to get around after that, and so she went to an extended-care facility where she had her own apartment, but meals and medical care were provided. She did quite well for several years, even though she walked with a walker and needed a hearing aid. She was alert and cheerful, spent time with her friends and family, and liked to play cards and knit things. When she was 99, she seemed to weaken. There was no clear illness that anyone could diagnose; she just turned inward and quiet. She died within a few months.—Mike

In some cases, people die from ''aging itself.'' When one pathologist conducted autopsies on 200 people who had died past the age of 85, he could find no clear cause of death in one quarter of them (Bishop, 1983). There was no heart or artery disease, no pneumonia, no infection, no sign of accident or any other generally accepted cause of death. These old people had died of old age itself. According to one school of thought, people die of ''age itself'' because their aging and death are programmed into their genes.

As we have seen, human development before birth is directed by the molecules of DNA contained within the cells of the body. **Programmed theories** of aging suggest that aging is under the direction of the DNA in human genes. As various genes turn on and off over the course of human development, a person goes through a succession of predictable biological events—childhood's growth in height and weight, puberty, menopause, and the aging of cells and organs. Programmed theories of aging are supported by results of several lines of research.

Premature aging. One line of research that lends support to theories of genetically programmed aging focuses on people who age prematurely. People who inherit **progeria**, for example, a rare disease that turns young children into aged old men and women, may begin to age during infancy. Soon their hair has turned gray, their faces have wizened, their joints have stiffened, their bodies look gaunt, and their skin looks dry and mottled. When cells from a progeric child's body are cultured in a laboratory, they divide with the diminished vigor of an old person's cells. The cardiovascular system of a child with progeria resembles that normally found in someone 75 years old. Most children with progeria die from a heart attack or a stroke between the ages of 7 and 27 (Selmanowitz, Rizer, and Orentreich, 1977). In support of the suggestion that progeria is a genetic disorder is the fact that the disease has been known to strike more than once in a single family. But no one fully understands what progeria actually is. It is not simply rapid aging, because children with progeria escape certain symptoms of old age, such as cataracts, diabetes, and mental slowness (Walford, 1983).

People with **Werner's syndrome** show signs of abnormal aging in their teens or twenties. Unlike progerics, they develop cataracts and diabetes. By their thirties, people with Werner's syndrome resemble progerics, and their lifespan is shortened by 20 to 30 years. Werner's syndrome is inherited when both parents carry a recessive gene for the disorder.

Down's syndrome, which we discussed in Chapter 3, is another genetic disorder that causes premature aging. People with Down's syndrome prematurely get gray hair and show symptoms of an aging nervous system, endocrine system, and immune system. In their thirties and forties, people with Down's syndrome undergo the same structural deterioration of the brain as is found in people with **Alzheimer's disease**, a form of dementia (Matsuyama and Jarvik, 1980). They usually do not live beyond the age of 50. In fact, some cases of Alzheimer's disease have been traced to the same chromosome implicated in Down's syndrome.

Research on premature aging suggests that aging is governed by the actions of many different genes. This line of research eventually may reveal the biological causes of both premature and normal aging.

The Hayflick limit. Leonard Hayflick (1977), a prominent researcher in the field of aging, found that when he put cells from human embryos into a laboratory dish, they divided only up to a certain limit and then stopped. Cells from some individuals divided 40 times, and some divided 60 times, but on average, human embryonic cells divide 50 times. When Hayflick treated cells from adults in the same way, he found that they divided only between 14 and 29 times. Hayflick had discovered that the individual human cell has a limited lifespan, the **Hayflick limit**.

Further research answered some questions about aging but raised others. Why did the cells from one 87-year-old divide 29 times but those from one 26-year-old divide only 20 times? In other words, why did chronological age and cell division seem *not* to correlate? When cells from a 9-year-old boy with progeria were studied, they divided only twice. Those from individuals with Werner's syndrome divided between 2 and 10 times. Cancer cells defy the Hayflick limit altogether. In fact, it is cancer cells' tendency to divide without limit that makes them so dangerous to the body's healthy cells. Research on other species showed that cells from each species seem to have their own

Hayflick limit. Cells from the short-lived mouse divide 15 times; cells from the long-lived Galapagos turtle divide 130 times.

The Hayflick limit shows up in the laboratory dish, but it does not always show up so neatly in a natural environment. Perhaps the Hayflick limit is affected by some unknown aspect of laboratory cell culture (Daniel, 1977). When researchers have tried to study the behavior of cells in living animals, they have found exceptions to the rule. Skin cells and mammary gland cells grafted from one living animal to another have divided many more times than the Hayflick rule would suggest they should. Thus although cells in living individuals have a limited lifespan, the workings of that limit are not yet entirely understood.

> When I told a friend that my grandmothers both had lived to their late eighties, she congratulated me on my ''good genes.''—Natalie

Programmed theories of aging receive support from the facts that each species has a different maximum lifespan, identical twins have similar lifespans, and children whose parents and grandparents have long lifespans tend to live longer than children whose parents have shorter lifespans. Statistically, someone whose four grandparents survive to the age of 80 is likely to live four years longer than someone whose grandparents do not live to the age of 60 (Jones, 1956). Someone whose four grandparents live to the age of 90 is likely to live seven years longer. A mother's longevity may affect a child's longevity more than a father's (Abbott et al., 1974).

Programmed deterioration. If several genes direct the aging process, how do they do so? Several possibilities are under investigation. One possibility is that certain genes are activated at particular points in the lifespan, and these genes trigger gradual aging and death. A second possibility is that ''young'' genes are turned off or overpowered by ''aging'' genes during middle age. Finally, it is possible that the genes that make people youthful somehow change in middle age so that thereafter they make people age. The third possibility may explain the action of the hormone estrogen on women's aging (Rockstein and Sussman, 1979). Before menopause, estrogen not only is bound up with women's reproductive functioning, but it also protects women from the same degeneration of the arteries and high blood pressure seen in men the same age. Men are not protected by estrogen. Yet once past menopause, women's estrogen levels drop, and in a few years, a woman's risk of getting high blood pressure and other forms of cardiovascular disease catch up to a man's.

Another possible explanation for the programmed deterioration of aging lies with the hypothalamus. This part of the brain, as we have seen, sends neurotransmitters to the pituitary gland. The pituitary, in turn, signals for the release of certain hormones. If some kind of biological signal causes these neurotransmitters in the hypothalamus to decrease, the pituitary may preside over a hormonal imbalance. This imbalance may set off the deterioration that makes up the aging process.

> Now that I'm 72, my doctor insists that I get flu shots every year. He says that I may feel great, but my immune system isn't what it used to be. As he says, ''It's the very young and the old who come down with pneumonia after a bout of flu, and there's no sense in taking a risk.''—Sam's grandmother

One of the most exciting research areas is focused on the role that the immune system plays in the aging process. Perhaps the deterioration of aging results because the immune system is somehow programmed to grow less efficient and even to attack the body (Walford et al., 1981). The aging immune system grows less efficient at fighting off attacking microorganisms, foreign proteins, and at checking the abnormal growth of cancer cells. Roy Walford (1983) has noted that the signs of aging mirror the signs of a body rejecting a transplanted organ—rejection being another function of the intact immune system.

Wear-and-tear theories

Some theorists propose that aging is a result of long-term, accumulated damage to systems of the body. These **wear-and-tear theories** propose that in the course of living, the body suffers functional damage that it repairs ever less effectively. Each of the several wear-and-tear theories is focused on particular forms of physical damage. The focus may be on damage to DNA, connective tissue, or cell compounds and membranes.

DNA repair. Just as some theories of programmed aging focus on the genetic program within body cells, the **DNA repair theory** of aging focuses on the long-term damage to DNA. DNA, a complex molecule within the cells, may sustain damage from many different sources. The ultraviolet light in sunlight, for example, can damage DNA. In the laboratory, damage to skin cells is repaired five times more quickly in species with long lifespans, such as humans, elephants, and cows, than in short-lived rats and mice (Sinex, 1977). DNA repair generally is quicker in species with longer lifespans. One common kind of damage to DNA takes place when one strand of the molecule's two laddered spirals breaks. Particles within the cell function to recognize and remove the broken portion and to repair the break by replicating proteins from the parallel section of the undamaged spiral. The cells of older people are more vulnerable to damage from radiation and less efficient at repairing themselves (Staiano-Coico et al., 1983).

Although cells may repair themselves, the DNA-repair theory suggests that the rate of repair is always behind the rate of damage. As damage accumulates, function is impaired, especially in the nerve and muscle cells that cannot divide. Over time, for instance, the function of the heart muscle may become impaired as damaged DNA accumulates. Some researchers have suggested that the rate of DNA repair is under the direction of the genes themselves (Walford, 1983), making this theory a combination of wear-and-tear and programmed theories.

Cross-linkage. If you take a piece of moist and supple animal hide and tan it with harsh chemicals, you end up with a tough and dry piece of leather. The tanning process causes the cells in the hide to bind together and rigidify. The **cross-linkage theory** proposes that living cells age in a similar way. When stable bonds, or cross-links, form between cells, the cells grow rigid and lose some of their function. Cross-links are formed during cell metabolism and appear in both DNA and in connective tissue. As the number of cross-links in the body increases, tissue loses its suppleness. It has been suggested that cross-links within the body's connective tissue account for much of the deterioration in the aging body's functioning (Kohn, 1971).

Free radicals. The **free-radical theory** is another wear-and-tear theory of aging. It was proposed by researcher Denham Harman (1968). Free radicals are not terrorists on the loose. They are chemicals that are inherently unstable, because they have a free electron. This instability means that the chemicals will bond easily with other chemicals. For example, some free radicals bond with unsaturated fats in cell membranes, damaging the cells. Free radicals also can damage chromosomes. The process of metabolism creates some free radicals, and others are present in the environment. Some free radicals combine with unsaturated fats and form lipids that break down into aldehydes (Walford, 1983). Aldehydes form cross-links between molecules. Thus two wear-and-tear theories of aging, the cross-linkage and the free-radical theories, are interrelated.

Free radicals multiply in ever greater numbers every time they react with a molecule (Rockstein and Sussman, 1979). Although the body can protect itself against free radicals with certain scavenger molecules, it cannot protect itself against all of them. With time, the number of free radicals builds up, and it shows up in the signs of aging.

Lipofuscin. Lipofuscins, the inert brown pigments that accumulate in many cells of the body, are made up of protein, carbohydrate, and fats. They increase in the cells of the aging body. They may be created during metabolism. As they fill cells, lipofuscins interfere with function. Some old people have cells full of lipofuscins, but others have little. Some young adults also have lots of lipofuscins (Sanadi, 1977). Thus although the presence of lipofuscins often is correlated with aging, its role in the aging process remains unspecified.

Before we have a unified theory of aging, the theory will have to answer

- Why some people age prematurely
- Why cells divide only a limited number of times
- Why the rate of DNA repair correlates with a species' maximum lifespan
- Why the immune system grows less effective with age
- Why connective tissue deteriorates with age
- Why chemical changes occur in aging cells
- Why most animals use about the same amount of energy per pound of body weight over the course of their lifespans (Walford, 1983).

At present, no single theory of aging—programmed theory or wear-and-tear theory—can answer all of these intriguing questions.

Physical changes

Normal aging tends to turn the skin paler and to increase "age spots." Secondary aging—the effects of sun, wind, and exposure—causes wrinkling and, in increasing numbers of people, skin cancer.

Many of the physical changes associated with aging have been underway since early or mid adulthood. But because of the body's great reserve capacity and because most of the changes are so gradual, most of the physical changes of aging do not require people to make significant adjustments in their daily lives until they are in their seventies or beyond.

> When I asked my 6-year-old grandson what "old" means, he said, "old means being sick."—Alyssa's grandfather

Many people, young as well as old, assume that illness is an inevitable part of old age. But it need not be so. For example, evidence from the Duke Longitudinal Studies (Maddox and Douglass, 1974), which traced the development of 300 men and women from North Carolina for a period of 15 years, showed that the health of more than half of the subjects remained stable and the health of one-fifth actually improved. Because statistics on illness among older people usually state the incidence for people over 65, people may assume that 65 is a watershed after which health begins to crumble. That is a misleading picture. The truth is that when the statistics are broken down, most people in their sixties and early seventies are healthy. Most people do not have major health problems until they pass the age of 75.

Skin

The older adult's skin may grow paler, as the pigment-containing cells, or **melanocytes**, decrease. In some people, especially fair-skinned Caucasians, the melanocytes that remain may grow larger and form clusters. These brownish freckles appear on the back of the hands, forearms, and face. People call them "age spots" and "liver spots" or, more formally, "pigment plaques" and "lentigines."

Besides the processes of normal aging in the skin, secondary aging takes its toll through exposure to sunlight, wind, abrasion, and exposure. Skin ages in direct proportion to the frequency, intensity, and length of its exposure (Selmanowitz, Rizer, and Orentreich, 1977). In fact, environmental damage so commonly affects people's skin that researchers cannot always separate its effects from those of normal aging (Rockstein and Sussman, 1979). But they do know that light-complected people—redheads and blonds, those with blue and green eyes—have skin that is most vulnerable to the damaging effects of sunlight. Blacks and other dark-complected people are less vulnerable. Sunlight stimulates the skin to protect itself by darkening; it produces the dark pigment, **melanin**. But sunlight also harms the skin's ability to renew itself, because it interferes with its ability to produce DNA and to synthesize protein. The price of an attractive, "healthy" tan, unfortunately, eventually is thin and wrinkled skin (Wantz and Gay, 1981). For some people, the price is even steeper: skin cancer. Skin cancer, which appears mainly in people over 50, usually is a result of exposure to sunlight. It is not a normal part of the aging process. Older people sometimes have other complaints about their skin. Because older skin is dry, it may chap and itch. Itching can be aggravated by a vitamin A deficiency or the drying effects of soap, and it may be caused by drug side effects or certain degenerative diseases.

What should people do if they want to slow down the effects of secondary aging on their skin? Besides eating well and staying well, they should treat their skin kindly—protect it from blasts of wind, of hot and cold air, and, perhaps most important, from sunlight. The bronze skin that sunbathers long for eventually turns into the wrinkled, leathery, sometimes cancerous skin that they, as older people, regret.

Skeletal system

Skeleton

Normal aging causes the bones to thin and the spine to shrink somewhat, but no one really knows where the effects of normal aging stop and those of secondary aging begin. For example, many people have long believed and cross-sectional studies of the population have supported the belief that people shrink several inches as they age. One study (Stoudt et al., 1965) showed that the average 75-year-old American man was 3.3 inches shorter than the average 35-year-old man and the average 75-year-old American woman 2.7 inches shorter than the average 35-year-old woman. But longitudinal studies show that succeeding generations of Americans have been growing taller—an average of 1.2 inches over 40 years of young men entering the army, for example (Rossman, 1977). Therefore, the estimates of shrinkage with age have been revised, and it is now estimated that men shrink about 1 inch and women 2 inches as their posture changes and as the cartilage disks between the vertebra in their spine thin and lose some of their water content.

Past the age of 30 or so, the bones also begin to lose the calcium of which they are made. Women tend to lose more calcium than men. The results of one study (Garn, 1975) showed that women between 55 and 85 years old lost 25 percent of the density of their bones, men half that amount. Although the skeletal bones grow thinner, they can still repair fractures. Bone fractures in older people heal less quickly than in younger people, however (Tonna, 1977).

In the scientific community, there is some disagreement about the dividing line between normal aging of the bones and more pronounced pitting and brittleness of the bones, a condition known as **osteoporosis**. Osteoporosis causes the bones to break more easily than normal. Found four times more

Osteoporosis is associated with aging and causes the bones to be brittle and break easily. It may cause the upper spine to curve. Normal aging, as well as inactivity and a diet low in calcium, may be causes of osteoporosis.

often in postmenopausal women than in men, osteoporosis may cause the upper spine to curve and give older women a characteristic "dowager's hump." Fully 25 percent of postmenopausal women have suffered at least one broken bone by the time they are 65 years old, and osteoporosis is a major factor in these accidents, many of which result in severe disability and even death (Hogue, 1982a; Sterns, Barrett, and Alexander, 1985). What causes osteoporosis? Among the proposed causes are lower levels of the hormone estrogen, inactivity, lactase (an enzyme necessary for the digestion of milk) deficiency, a low-calcium diet, and increased parathyroid secretions in response to low blood levels of calcium (Tonna, 1977). Some people think that the normal American diet, which is high in animal protein, causes the bones to shed calcium in order to metabolize the protein (Whitaker, 1985). They suggest that people cut down on the amount of animal foods they eat and increase the amount of cereals and grains, vegetables and fruit. What else helps to prevent osteoporosis? Some doctors prescribe estrogen replacement therapy, because it can prevent bone loss. Others suggest that older people keep their intake of calcium, vitamins D and K, and flouride high and to engage regularly in some form of weight-bearing exercise, such as walking, dancing, or playing sports (Weg, 1983).

> My doctor tells me to talk a walk every day and to eat some kind of dairy product, for the calcium, at every meal. That's not such a tough prescription to fill, and it ought to be good protection against osteoporosis—Sarah's grandmother at 64

> Six months after I had had Sam, when I was 35, I limped to the orthopedist's office with a painful, bruised knee. He told me I had osteoarthritis and that I must stay off my feet as much as possible. But how was I supposed to take care of the baby—which meant running up and down stairs 100 times a day—and still stay off my feet? The knee has gotten worse as I've gotten older, just as the doctor said it would. But how does anyone really prevent normal wear and tear?—Sam's mother

Osteoarthritis is a painful degeneration of the joints that is related to a person's heredity, hormones, and diet. Osteoarthritis is a result of wear and tear on the skeletal system, for as the connective tissues between bones thins with age, the ends of the bones may rub together and grow painfully inflamed. Weight-bearing joints like the ankle, knee, and hip, which take the most pounding, are most liable to developing osteoarthritis. Osteoarthritis sometimes strikes young adults; it grows increasingly common and severe in people up to about 80 years old. Once past 80, few people develop a new case of osteoarthritis (Tonna, 1977).

Teeth

One of the happy consequences of getting older is the greater resistance of one's teeth to decay. Look, grandchild, no cavities! Past the age of 17, few people get tooth decay in their sound teeth (Jackson and Burch, 1969), because the tooth surface accumulates deposits that resist acid and the tooth enamel is harder and more impermeable. Aging causes the pulp found in the center of the teeth to lose cells; about half are gone by the age of 70. Blood vessels inside the teeth also decrease with age, and fibers increase in number. But healthy human teeth are so durable that they might not wear down for 200 years, if they were not lost to decay, infections of the roots, and gum disease (Tonna, 1977).

> My father came home from a dentist's visit all upset because he'd found out he had gum disease. He tried to make a joke of it and said, "My

teeth are healthy, but my gums have to come out.'' But he knew he was in for some painful and expensive treatment.—Sarah

Periodontal disease is the bacterial infection of gums, bone, and other tissue surounding the teeth. It makes the gums swell, bleed easily, and pull away from the teeth. As the gums pull away, the infection invades deeper layers of tissue, bone is lost, and teeth may loosen, fall out, or require removal. Periodontal disease crops up in young as well as older people, although its incidence increases with age. It is not considered a normal part of aging. Many people with periodontal disease also have osteoporosis (Weg, 1983). People whose bodies are young for their age are less likely than those whose bodies are old for their age to develop periodontal disease (Tonna, 1977). Not only heredity, but poor diet and dental hygiene, an ill-fitting bite, and disease all may contribute to the development of periodontal disease.

Cardiovascular system

The young heart looks red, but the normal aging heart is covered with fat and looks brownish, because lipofuscin accumulates on it. The degree of this brown coloring does not seem to relate to the incidence of heart disease, however (Kohn, 1977). The muscle fibers of the aging heart are surrounded by collagen that grows progressively stiffer, more inert, and less soluble. As more collagen accumulates in and around muscle fibers and valves, the valves thicken and the relative proportion of muscle drops. One cross-sectional study showed (Davies and Pomerance, 1972), for example, that in people younger than 50, the heart had 2½ times more muscle than collagen. But in people older than 75, the heart had 1⅓ times more collagen than muscle. The aging heart also grows larger with age. When healthy men between 23 and 76 years old were followed for ten years, their heart size increased slightly over that period. The greatest increases in size were found among men who had put on weight (Ensor et al., 1983).

The aging heart also beats less quickly, even under conditions of maximum exertion. When people perform strenuous exertion, their heart beats more quickly to pump necessary oxygen to the muscles. A young person's heart may beat up to 200 times a minute, but the heart rate of a person between 70 and 90 years old is only 125 beats a minute. Because the aging heart contracts more slowly and relaxes more slowly after each contraction, the net result may be a slowed heart rate during even maximum exertion (Shock, 1977b).

Arteries

One of the most exciting fields in science today is concerned with identifying the causes and cures for secondary aging of the cardiovascular system. One area of disagreement is what normal aging of the arteries consists of. As the arteries age, the composition and texture of their walls change. The **elastin** fibers thin and break; some die. In the process, calcium attaches itself to the elastin fibers, and the artery walls become inelastic and inflexible. Collagen also builds up in the artery walls, and several different kinds of large fat molecules, or lipids, stick to the walls. These processes are not in dispute, but the extent to which they are the normal result of aging is in dispute. It is likely that the changes in elastin are the result of normal aging. But the build-up of collagen, calcium, and fats—conditions characteristic of inflammation—is likely to be the result of damage to the artery walls (Kohn, 1977).

When artery walls stiffen, they do not readily expand when they are filled with blood. As a result, a person's blood pressure may go up, although many older people have blood pressure that is healthily low. With age, smaller arteries in the cardiovascular system grow less efficient at carrying blood. In combination with the heart's weaker pumping of arterial blood, the less efficient peripheral vessels can mean a reduced blood supply to kidneys and certain other internal organs as well as to feet and hands.

> My 90-year-old great grandfather has to walk with a cane now because he has so little feeling in his feet. He complains that his feet are always cold and numb. The doctor tells him it's because his heart is getting less blood to his extremities and encourages him to take a brisk walk every day.—Sam

The brain, heart, and skeletal muscles usually suffer least from this reduction in arterial blood supply (Kohn, 1977). As someone once aptly said, people are as old as their arteries.

Respiratory system

The lungs, trachea, and bronchi age in ways that make breathing less powerful and efficient. Yet many of the signs of deterioration once considered part of the normal aging of the respiratory system now are considered the result of such environmental factors as air pollution, cigarette smoking, and inactivity. Understanding how to avoid damaging the respiratory system, as well as the capacity of the lungs to repair themselves, can keep the deterioration of the system to a minimum.

One of the reasons that older people breathe more shallowly than younger people is because the aging rib cage is rigid, and the muscles involved in moving the chest during breathing are less elastic. Some older people also stoop as a result of osteoporosis or other spine conditions. For all of these reasons, when older people breathe, their chests expand less and they take in less air than younger people do. Whereas the expanded rib cage contributes 40 percent of a young adult's lung volume, it contributes only 30 percent of a 70-year-old's (Rizzato and Marazzini, 1970).

The lungs change inside and out. They change from the pink of youth to gray, studded with black patches of inhaled carbon particles. Inside, cartilage in the trachea and bronchi turn rigid and bony. Tubes leading to the lungs' air sacs, the **alveoli**, get larger as the alveoli themselves get more narrow and shallow. Once past the age of 40, the alveoli diminish in these ways, and by age 80, over 80 percent of them have diminished (Klocke, 1977). Although the number of alveoli does not necessarily change with age, their surface area diminishes. As collagen is increasing in the arteries, it is decreasing in the alveoli. Similarly, as elastin is decreasing in the arteries, it is increasing in the alveoli. With less collagen, the lungs are less elastic, and the alveoli expand less with every breath. Whereas young adults have four times as much collagen as elastin in their lungs, 70-year-olds have only twice the collagen. The arteries and veins that supply the lungs thicken, stiffen, fill with fatty deposits, and deliver less blood. Although the amount of oxygen in the alveoli does not drop with age, the lungs become less efficient at exchanging gases—turning carbon dioxide and other gases to oxygen—and so the arterial blood carries less oxygen. As a result of these changes, some older people complain that they often feel short of breath.

Just as much of the decrease in oxygen uptake results from secondary aging, so do many other lung problems found among older people. Long-time smokers, for example, may develop emphysema, a debilitating condition in which the walls between the alveoli are destroyed. So much of the lungs' surface area is dead and so little oxygen is breathed in that people with emphysema have trouble exerting themselves. So little oxygen may travel through their bloodstream that emphysema can cause people to become confused, disoriented, or unconscious (Wantz and Gay, 1981). Smoking and,

Many of the signs of physical deterioration, once considered to be normal aging, are now understood to be the results of inactivity, cigarette smoking, poor diet, and other controllable factors.

to a lesser extent, air pollution cause lung cancer, the incidence of which increases with age. Lung cancer used to strike more men than women, but women have taken up cigarette smoking in such great numbers that lung cancer has surpassed breast cancer as a killer of women.

Gastrointestinal system

The gastrointestinal system is like other systems in the aging body in the sense that it is subject to a certain predictable degree of deterioration in the course of normal aging, and it is also subject to deterioration as a result of secondary aging—poor diet, lack of exercise, and the like. Although the laxative and antacid industries would have us believe that every older person in America is at the mercy of a touchy digestive system, in reality their gastrointestinal system need not trouble older people.

True, the digestive system slows down and grows less efficient with age. Food normally is moved through the esophagus and into the stomach by waves of muscular contractions. In young adults, nearly every swallow is followed by effective contractions. But in people as old as 90, only half of the swallows are followed by effective contractions. The other half are followed by sporadic, ineffective contractions that draw the esophagus into a corkscrew shape (Soergel, Zboralske, and Amberg, 1964). Among young people, the muscle at the bottom of the esophagus relaxes after each swallow so that food can pass, but in older people, this muscle relaxes only half the time, and so food remains in the esophagus. Researchers suggest that normal aging weakens the contractions of the muscles in the esophagus but that the other forms of slowing in the digestive system result from disease (Bhanthumnavin and Schuster, 1977).

The stomach digests food by secreting hydrochloric acid, and this secretion falls off after the age of 50, especially in men. By the time they have turned 70, 23 percent of men and 28 percent of women produce no acid. Most of them have a chronic inflammation of the lining of the stomach, atrophic gastritis, that results from chronic use of aspirin or alcohol or repeated injury to the stomach lining by the bile salts produced during digestion or by gamma globulin, a chemical that the immune system produces (Bhanthumnavin and Schuster, 1977).

Food passes from the stomach into the intestines. The normal intestines of an older person secrete the same enzymes as those of a younger person, although the amount of secretion begins to taper off around the age of 30 (Rockstein and Sussman, 1979). Most older people's intestines are efficient at absorbing nutrients from their food, despite the normal atrophy of intestinal muscles and mucous membranes. The smooth muscle in the wall of the large intestine grows smaller and less powerful with age. Only after the age of 80 does normal aging keep a person from absorbing molecules of one simple sugar (Bhanthumnavin and Schuster, 1977).

The liver is another organ in the gastrointestinal system, and it has great reserve capacity. It has been estimated that people need only 20 percent of their liver (Bhanthumnavin and Schuster, 1977). Once people turn 50 to 60 years old, their liver may shrink, its cells may change, and it may produce enzymes less efficiently and in less concentrated form than it did when they were younger. As the liver grows less efficient, older people's abilities to metabolize certain drugs may decline, and so physicians may have to reduce the amounts of drugs they prescribe for older patients. The gallbladder stores bile produced by the liver and releases it during digestion. With age, the gallbladder's walls shrink and thicken, but it functions efficiently in most older people.

Many of the disorders of the digestive system that afflict older people result from problems of diet and inactivity. Gallstones, for example, which grow increasingly common with age, may develop when the bile within the gallbladder contains too much cholesterol (Bhanthumnavin and Schuster, 1977). Gallstones may be treated surgically or with drugs. Constipation also grows more common with age, although it certainly is not inevitable. Constipation is less likely to be the result of normal aging of the digestive system than of drinking too little fluid, eating a diet low in fibrous foods, and exercising too little. When these habits are compounded by dependence on laxatives, people may develop irregular pouches in the walls of the large intestine that then become blocked, infected, and painful. This condition, diverticulitis, is found in few young adults but in 40 percent of those over 70. It has been called a disease of modern society (Bhanthumnavin and Schuster, 1977). Many of the problems of the digestive system that plague older people are found in young adults, too, though usually in less severe form. In more than half the cases, the problem is of psychological or emotional origin (Rockstein and Sussman, 1979).

Excretory system

Once the gastrointestinal system has digested and absorbed the nutrients from the food a person has taken in, the waste matter is processed by the organs of the excretory system—the kidneys, bladder, and the blood vessels and ducts that supply them. Although the kidney begins to shrink at about age 30 and by age 90 it is only two-thirds its former size (Goldman, 1977), the kidney nevertheless works fairly efficiently for most older people. The arteries that supply blood to the kidneys age like arteries elsewhere in the body; they thicken and stiffen. People with high blood pressure may have arteries in which these changes are pronounced (Goldman, 1977). With age, the bladder shrinks, although it does not undergo any obvious structural change with age (Goldman, 1977). The smaller capacity is a nuisance to those people over 65—70 percent of men and 61 percent of women—who wake up in the middle of the night to urinate.

Reproductive system

> When my 79-year-old grandfather remarried 11 months after he was widowed, he confided that one reason he wanted to marry again was because he missed having sex. He thought it wasn't healthy to go without sex for too long, no matter how old you are.—Sam's mother

Although the reproductive system responds less quickly and vigorously in middle and late adulthood, people can and do adapt to these changes. As they learn how to stimulate and satisfy themselves sexually, they find that sex remains deeply pleasurable well into their eighties and nineties. People may be frightened by illness or by their suspicions that sex in old age is somehow irrelevant or dirty, but reasonably healthy old people can remain sexually happy virtually all of their adult lives.

Sexuality

For men, health is especially important in determining the experience of sexuality in middle age and beyond. For women, it is the presence of a husband or stable partner. Today, by the age of 65, half of all married women are widowed. One might expect masturbation to increase among this group. But although formerly married women do masturbate more than married women of their age, they do not masturbate as often as their married sisters have intercourse. Many women depend on a relationship with a man to provide sexual cues. Masturbation does not provide a social relationship, an element that many older women find necessary. The sex lives of 200 healthy old people between the ages of 80 and 100, living in retirement in California, have been studied (Bretschneider and McCoy, 1983). The participants were taking no medications known to affect sexual interest or functioning. More men than women were still married at the time of the study. More women then men had never been married or had been divorced. More of the men than the women still masturbated, and more men than women who masturbated found it enjoyable. Many more women (70 percent) than men (38 percent) did *not* have intercourse. Of course, many more women (75 percent) than men (47 percent) had no regular sexual partner. About twice as many men (76 percent) as women (39 percent) who had intercourse enjoyed it. There had been no difference between men and women in frequency or enjoyment of intercourse when they were young.

Men more than women liked touching and daydreaming and thought sex was important. But both sexes reported that sex was less important than it once had been. Even from their eighties to their nineties, some people reported that they less often enjoyed touching and caressing without intercourse. For men and women, the importance of sex in youth and old age

Older adults can and do learn to adjust to the physical changes of their reproductive systems. Sexuality can be a source of pleasure well into old age.

correlated highly, a confirmation of the generalization that frequent, enjoyable sex early in life goes along with frequent, enjoyable sex late in life.

In sum, the evidence on sex and aging suggests, first, that sexuality can continue throughout life, even though sexual response slows with age. Second, human sexual behavior is largely independent of reproduction. Third, substantial declines in sex hormones are associated with declines in sexual interest and activity. But not all loss of sexual interest and erection is determined by hormones alone. At every age, women express less interest in sex than men. In women, social factors like social class, attitudes, and a sharing sexual partner probably avert or mask such declines. Last, and most important, at all ages and in both sexes, sexual interest has a wide range. At all ages, some adults are sexually active and interested. Health, the availability of a partner, and genetic and temperamental factors all play a role in the maintenance of sexual interest and activity.

Nervous system

The central nervous system is made up of the brain and spinal cord. Its aging is crucial, because through nerve impluses to the muscles and to internal organs and through brain signals to the endocrine system, the nervous system coordinates and controls many of the body's vital functions. One of the frightening myths about aging says that we are destined to grow confused and ''senile'' as our brain ages. But confusion, memory loss, and ''senility'' are not normal courses of aging; they are the symptoms of disorders.

> My body tells me it's getting old through lots of small annoyances. But my head still seems clear. I don't find my thinking cloudy or any slower than it was when I was younger. I do have more trouble remembering little things, like new telephone numbers. But that started happening when I was only 40—78-year-old woman

> I've been practicing family medicine for 50 years. I've decided to retire this year, when I turn 75, because I don't want to make ''the big mistake'' with any of my patients. I can still keep up with what's going on in the field. I still pore through the journals and don't have any trouble remembering what I read. But I want to retire *before* I have to, not after. Then I think I'll spend more time politicking for the local medical society.—Sam's grandfather

Only in the recent past have researchers been able to study the aging of the nervous system in living people. Before that they had had to study the nervous system in human corpses and in animals. But today, great strides in understanding the normal functions of the nervous system are being made as researchers use techniques such as CAT scans, in which computers generate cross-sectional X rays of brain tissue, as well as PET (positron emission tomography) scans, which take pictures of brain cells as they metabolize glucose. Researchers also use nuclear magnetic resonance (NMR). With this technique, a person's body is surrounded by a magnetic field and, as radio waves bounce off living tissue, they produce images of the brain's biochemical functioning.

Autonomic nervous system

The autonomic nervous system regulates the internal organs—heart rate, breathing, blood pressure, temperature, digestion, excretion, and response to stress (see Figure 20.1). Age seems to slow and weaken some of these automatic functions, as neurotransmitters weaken or are metabolized differently and as neurons change in structure and die. But the core of the autonomic nervous system undergoes no marked deterioration with age, and only slight changes in function appear (Everitt and Huang, 1980). The development of higher blood pressure over time may be one effect of normal aging in the autonomic nervous system. Similarly, when older people are subject to rapid changes in surrounding temperature, their ability to regulate body temperature is less efficient than it once was. After exercising, it takes the pulse and respiration rates longer to return to their resting rates. Said one 72-year-old:

> I still jog a couple of miles four or five times a week, and I've been doing that for nearly 40 years now. I may be a little bit slower around the track, and I may not push that pulse rate quite as high as I did, but I can still keep up with my grandson. So don't count me out yet!

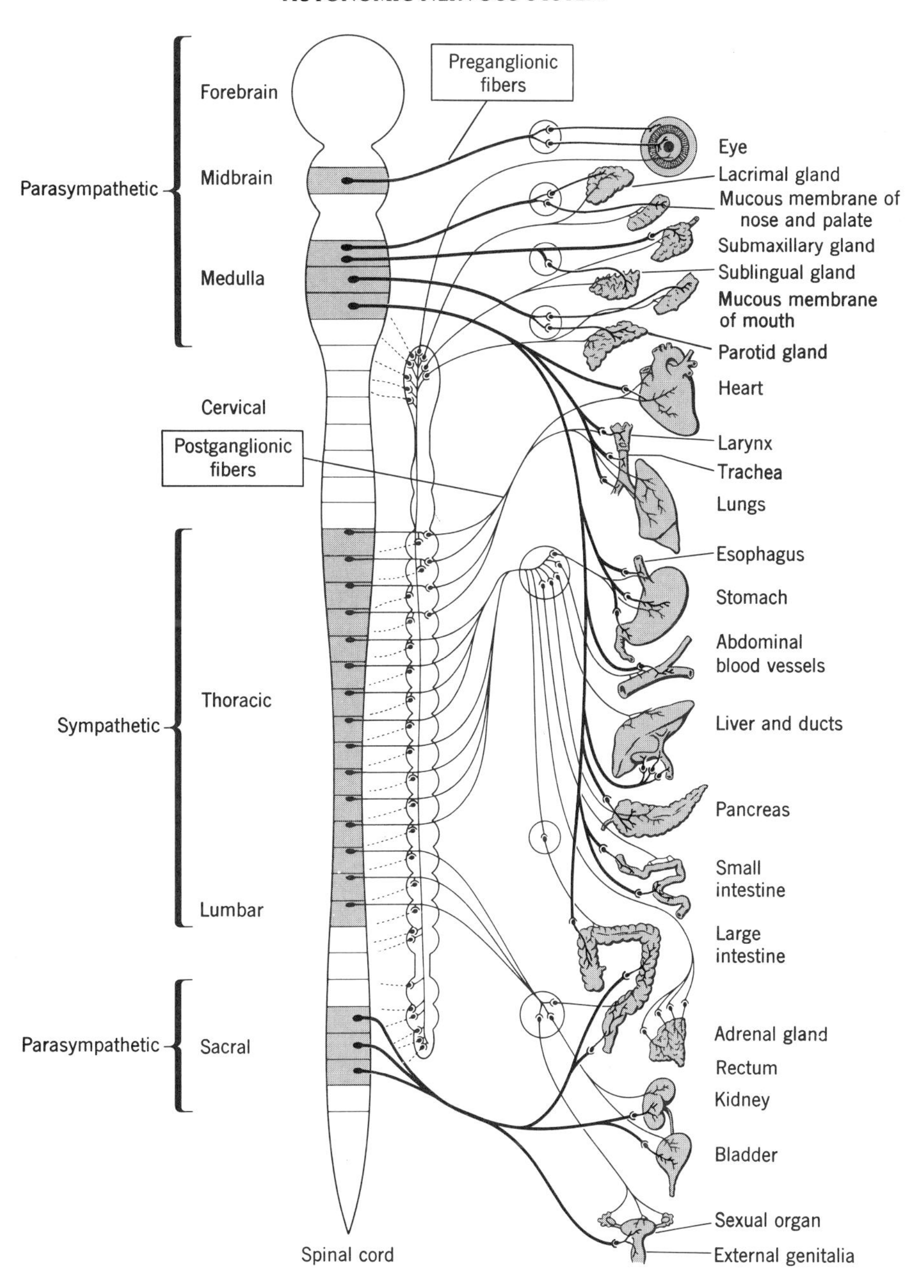

Figure 20.1 The autonomic nervous system controls the body's internal organs.

Sleep

When Erica was young, she always woke up with the birds, just after dawn. Since I woke up then, too, we had lovely times together in the morning when she visited me. By early afternoon, we'd both been awake for 7 or 8 hours, and we'd take a nap together. As I've gotten older, sleep seems to get lighter and less refreshing.—Erica's 66-year-old grandmother

My grandchildren are teenagers now, and good Lord, do they sleep! Until noon, if you let them. I'm lucky if I can get six hours at a stretch.—Natalie's grandmother

The long, unbroken, refreshing sleep of the young seems to elude many older people, much to their regret. Although many older people still sleep 7 to 8 hours in every 24, their sleep is usually broken. Nighttime sleep is usually an alternation of sleeping and waking, for a total of 5 to 6 broken hours of sleep. A daytime nap of an hour or two rounds out the total. Men in their fifties seem to have more disturbed sleep than women (Webb, 1982).

Older people sleep restlessly for several apparent reasons. As we have seen, the aging bladder signals the need to urinate during the night. Older people also may suffer from sleep apnea, a temporary cessation of breathing. Sleep apnea is harmless, but as the blood fills with carbon dioxide, the sleeper wakes up and starts breathing again. Apnea, in short, makes for choppy sleeping. Harmless but annoying muscle twitches and jerks can also interrupt sleep. One researcher (Hubbard, 1982) has estimated that nearly 30 to 60 percent of people over 58 suffer either from choppy (apnea) or jerky (myoclonus) sleep—or both. Physical fitness seems to have an effect on both of these conditions. When a group of healthy 85- to 94-year-olds from Vilcabamba, Ecuador, were studied, they showed a far lower incidence of both sleep apena and myoclonus (Okudaira et al., 1983). It is likely that the Ecuadorians were especially fit because they lived at a high altitude and had strong, healthy lungs, because they exercised strenously, and because they took no sleeping medicine. Used for more than a few nights running, sleeping medicines have the unpleasant tendency to rebound and *prevent* people from sleeping. People grow tolerant of the drugs, grow physically dependent on them, and suffer from ''drug-dependency insomnia'' (Butler and Lewis, 1982). In sum, at least some of the restlessness of older people's sleep may not be a result of normal aging. Depression, drugs, drinking alcohol or caffeine, poor physical fitness, and inactivity all may interfere with the amount or quality of older people's sleep.

Central nervous system

What happens to the aging brain in healthy people? Answering that question has been difficult for researchers, in part because the technology for studying the living brain and spinal cord had been lacking and in part, as we have seen, because it was difficult to separate the effects of normal aging from those of secondary aging. Diet, drugs, alcohol, the health of the cardiovascular system, cancer, and organic brain disease all affect the anatomy of the brain. It seems to lose weight, to shrink so that there is more room between it and the skull, and its interior ventricles get larger. Some estimates have pegged the loss of brain mass at 5 percent by age 70, 10 percent by age 80, and 20 percent by age 90. But when only the brains of people with normal brain function were tested, the shrinkage is less (Adams, 1980). The evidence on the enlargement of the ventricles is similarly contradictory. In one study, CAT scans showed that the ventricles gradually get larger after the age of 10 and, between 70 and 90, rapidly get larger (Barron, Jacobs, Kinkel, 1976). But another CAT scan study showed no parallel increases in ventricular size (Brody and Vijayashankar, 1977). Yet a third CAT scan study showed that over time, in some people the increases in ventricular size actually were reversed (Jernigan et al., 1980).

Neurons

It is likely that from infancy onward, day by day over the course of a lifetime, many neurons die. But the human brain has so great an excess capacity of neurons that their gradual loss does not affect how well people function. When neurons die, they are not replaced. They seem to die in only certain regions of the brain. For instance, neurons in one layer of the cerebellum die off fairly rapidly once a person turns 60 years old. The cerebellum lies at the base of the brain, near the top of the spinal cord, and coordinates the voluntary muscles and physical balance. Neurons also die in a few regions of the cerebral cortex, the "gray matter" covering the two hemispheres of the brain (Bondareff, 1980; see Figure 20.2).

Not only do some neurons die during the normal course of aging, but they also lose some of their connecting branches, or dendrites and axons, that send and receive impulses from other neurons (see Figure 20.3). Researchers have found that people with normal brain function who died between the ages of 68 and 92 had more and longer dendrites than people who died between the ages of 45 and 55 (Buell and Coleman, 1979). Only in the brain of adults with organic brain disease had large numbers of dendrites died, and researchers suggest that serious, irreversible loss of connections between neurons is a symptom of disease.

The chemicals that spurt between the neurons, the neurotransmitters, either make them fire or inhibit their firing. Studies of aging rats have shown weaker concentrations of some neurotransmitters and neurons that are slower to take up available neurotransmitters (Bondareff, 1980). Researchers are beginning to understand the roles neurotransmitters play in diseased brains—the role of dopamine in Parkinson's disease and acetylcholine in Alzheimer's disease, for example—and in normal aging brains, too.

Neurons in the aging brain also begin to show other characteristic changes, but these changes are far more common in older people with organic brain disease. For example, in the brain of a person over 60, *vacuoles* may appear. Vacuoles are thick granules surrounded by fluid, and although they are most common in diseased brains, they show up in three quarters of the brains checked from people over 80. Similarly, *senile plaques*, collections of debris in the brain, appear most often in the diseased brain. But a few senile plaques also appear in the brain of a person over 90 (Adams, 1980). Finally, *neurofibrillary tangles*, tangled clumps of double-helical strands of protein, appear in older brains, although no one quite understands their effect (Bondareff, 1980).

Just as some of the reflexes seen in the newborn infant disappear during the normal course of infancy, so do some of the reflexes of the adult disappear during the normal course of later life. These unconditioned reflexes, such as the jerking motion made by a tapped knee or ankle, are governed by nerve impulses in the spinal cord. Many people over 60 find that some of their reflexes are weak or absent. In one study, 15 percent of a group of people between 70 and 80 years old did not have the knee jerk reflex, and 70 percent did not have the ankle jerk reflex. In a group of those up to 90 years old, most people had no jerk reflexes (Rockstein and Sussman, 1979).

Neuroglia

Neuroglia are the delicate, branching connective tissue that fills the spaces between neurons in the brain and spinal cord, supporting and binding them together. An increase in their number is thought to result not from

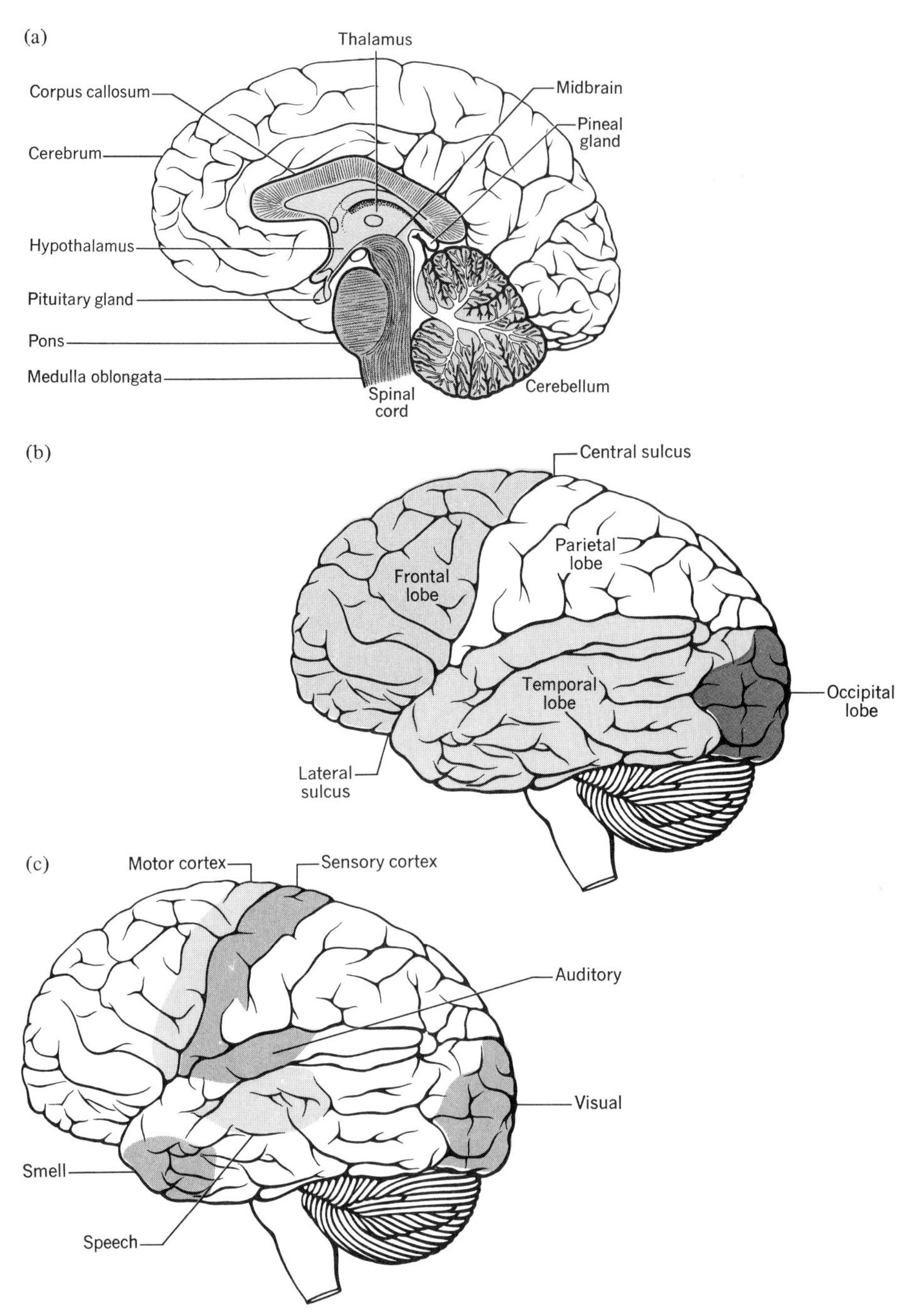

Figure 20.2 **(a)** ***The major parts of the human brain;*** **(b)** ***left hemisphere of the cerebrum;*** **(c)** ***cortex and gray matter that covers it.***

normal aging but as a response to the deterioration of neurons or dendrites. The bodies of **astrocytes**, star-shaped neuroglia, get larger in the aging brain, although their fibers do not enlarge. Astrocytes prevent neurotransmitters from building up between nerve cells, and it has been suggested that the enlargement of the astrocytes is associated with this function (Bondareff, 1980).

One of the older techniques for studying brain activity, the electroencephalogram (EEG), picks up the electrical impulses between nerve cells and

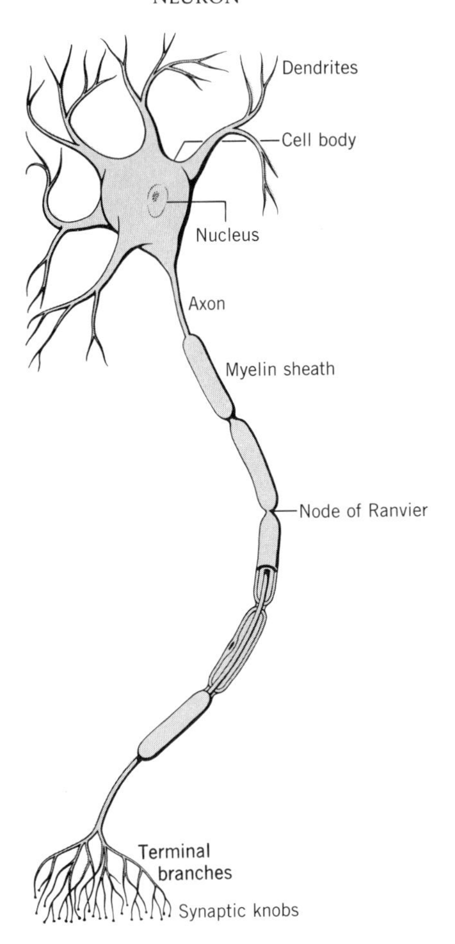

Figure 20.3 ***This shows the major parts of a neuron, or nerve cell.***

charts them. Rapid brain waves correspond to alertness and mental activity; slower brain waves correspond to slower, more meditative thought; the slowest brain waves appear during deep sleep. Certain changing patterns of electrical activity are found in the aging brain. Some are associated with changes in function, and some are not. For example, one change that is not associated with any impairment is a burst of high-voltage slow brain waves from localized regions. This pattern shows up in healthy people of about 50. In healthy 65-year-olds, the bursts show up in the EEGs of 30 to 50 percent of those tested.

Healthy adults younger than 75 have the same amount of slow brain waves as younger people. But one-fifth of the adults tested who were older than 75 had EEGs with an increase in slow brain waves. The slower waves may result when blood flow to the brain slows, perhaps as a symptom of aging heart and arteries. When EEGs show slow waves emanating from the whole brain, a corresponding drop in intelligence test scores usually occurs (Marsh and Thompson, 1977).

With age, alpha brain waves decline in amplitude and speed, especially at the low end of their frequency range. Alpha waves correspond to relaxed wakefulness or meditation. For younger people, alpha waves are classified as falling in the range of 8 to 13 cycles per second (Hz). For older people, some researchers include 7 Hz in the range of alpha waves. But alpha waves do not slow in all older people. In one study, nearly one in four people tested between the ages of 60 and 80 showed an *increase* in the frequency of alpha waves. In people whose alpha waves do diminish, they are replaced by slower alpha waves. In healthy older adults, slowed alpha waves may be caused by reduced blood supply to the brain, an effect that is especially pronounced when people are under stress. One longitudinal study showed that over 12 years, for adults who were highly educated and high on the socioeconomic scale, slowed alpha waves corresponded with lower intelligence test scores. But this effect was not found in people from other socioeconomic groups (Wang, Obrist, and Busse, 1970).

Beta waves, which correspond to full alertness, are the fastest brain waves (14 to 40 Hz). Normal aging causes no slowing of beta waves: alert older people can think just as quickly as alert younger people. Some researchers have found beta waves to increase in people 60 to 70 years old. Once past 80, however, a person's beta waves do seem to diminish (Marsh and Thompson, 1977).

> I find it shocking when young people assume that someone my age would stop reading, thinking, caring about world events. I still read two newspapers every day, several books and magazines a week, keep up with current events, and carry on heated debates with my friends about what goes on in this world.—Denny's grandfather at 69

Periodically as a person enters REM sleep, the brain produces waves that are a mixture of alpha waves, a few very slow theta waves (4 to 7 Hz), punctuated by bursts of beta waves. REM sleep gets shorter as people age and drops off markedly in men in their fifties (Webb, 1982). Until about age 80, however, the *proportion* of REM to nonREM sleep remains the same. After 80, the proportion of REM sleep declines. Because REM sleep is usually the time when people dream, it is likely that aging brings a decrease in dreams. Deep, stage 4 sleep, when the brain produces big, slow delta waves (1 to 3 Hz), also decreases with age. Yet because healthy, undepressed older adults show the same proportions of deep sleep as younger adults do, the decrease in deep sleep found in many adults may be, as we have said, an effect of depression or other disorders and not an effect of normal aging.

Disease and aging

Even with the best of care not everyone remains healthy into old age. Some people struggle with chronic and acute illnesses. Both kinds of illnesses can cut into the quality and length of life.

Disease is a form of secondary aging, not an inevitable part of aging. What is more, the ill effects of disease, physical abuse, and physical disuse often compound each other. Diseases may prevent people from using parts of their body, may lower their resistance to other diseases, and may limit their activity. Thus the man who suffers from arthritis in his hands may stop using them, and so his muscles and joints deteriorate further. Soon he cannot write a letter, tie his shoelaces, or cut his own food. The woman who is overweight does further damage to her heart and blood vessels by eating fatty foods, failing to exercise, and smoking cigarettes. Over time, she finds that she cannot climb a flight of stairs or walk a city block without gasping for air and suffering chest pains, and so she grows more inactive, more overweight, and more incapacitated.

Acute illness and injuries

> My 4-year-old granddaughter fell one afternoon while she was playing and injured a tendon in her leg. She was carried into the house, her knee was bandaged, and she sat with her leg up on a pillow for a few hours.

> Four hours later she stood up and walked on her leg. She was completely recovered. Compare that to the tendon I injured when I turned my ankle. I, too, was helped indoors, bandaged, and sat with my leg up, applying heat and cold alternately. What my granddaughter's 4-year-old body took half a day to repair, my 76-year-old body took 6 weeks to repair.—Mike and Erica's grandmother

> When I was young, even the worst colds I caught were over in five to seven days tops. But ever since I reached 80, I sometimes get colds that my system can't shake. They last for a month, two months. They frighten me.—Peter's grandmother

Once past childhood, people generally come down with ever fewer **acute illnesses**, illnesses that last only a relatively short time. Throughout adolescence, adulthood, and old age, the incidence of infections, respiratory illnesses, and digestive illnesses falls steadily. Although old people usually have fewer bouts of acute illness each year than they did when they were younger, each illness tends to be more severe (See Table 20.1).

Not only are acute illnesses more severe during old age, they are also likely to be complicated by certain aspects of normal aging (Hickey, 1980). Older people respond to treatment for acute illnesses more slowly and weakly than younger people do, and as we have seen, their acute illnesses are more likely to develop serious complications. Certain aspects of normal aging can make an acute illness difficult to diagnose. For example, an older person's temperature may rise only slightly, and the sense of pain is dulled enough to make its source difficult to pinpoint. Many older people also suffer from one or more long-term health problems, and their presence may complicate the diagnosis and treatment of acute illnesses.

Chronic illness

> My husband has diabetes. The doctor tells him to take his medicine, eat right, and get some exercise. He's pretty good about following the doctor's orders, but he still has trouble with the circulation in his feet. His eyesight is affected, too.—Nadine's mother

Table 20.1
Acute illness

	Infective and Parasitic	Respiratory		Digestive System	Injuries
		Upper	Other		
Total	24.6%	57.0%	59.2%	11.4%	33.4%
Male	23.4	50.9	52.9	11.2	39.0
Female	25.7	62.7	65.0	11.6	28.1
By age					
Under 6 years	56.3	127.6	73.5	18.6	35.2
6 to 16 years	42.5	78.0	81.5	13.4	40.4
17 to 44 years	21.3	53.3	60.0	13.0	38.6
45 to 64 years	11.2	31.2	46.0	6.3	23.0
Over 64 years	7.2	27.9	32.4	5.6	19.0

SOURCE: U.S. Bureau of the Census, 1982.

This table shows the percentage of those across the lifespan who had acute illnesses in 1980. Acute illnesses either require medical attention or restrict activity for one day or more.

> Both of my parents had heart attacks, and my father had a series of strokes, so I've inherited a double dose of heart disease. Even though I take care of myself—I quit smoking ten years ago, and I watch my weight—my arteries keep on getting worse. It's gotten so bad that it hurts me too much to walk to the bus stop. I don't want to live out my seventies as a burden on my children.—Jason's grandmother

Chronic illnesses are those that last for long periods of time, resist cure, and tend to worsen over time. Diabetes, arthritis, emphysema, and heart disease are just a few of the chronic illnesses that plague older people. Some researchers estimate that among people older than 65, as many as 85 percent have one or more chronic illnesses. Twenty percent have two chronic illnesses, and 33 percent have three or more (Hickey, 1980). Over the course of the lifespan, both the incidence and severity of chronic disease increase (See Table 20.2).

Evidence gathered during longitudinal research on Californians, the Intergenerational Studies (Eichorn et al., 1981), showed that for subjects in their forties, the balance shifted from a predominance of acute to chronic illnesses. Before their forties, people tended to suffer more from short-lived diseases such as colds and flu, but after their forties, they tended to suffer more from chronic illness such as arthritis and associated problems such as pain and stiffness. Past the age of 50, they are more likely to suffer from chronic bronchitis, emphysema, high blood pressure, and arthritis. The most widespread chronic illnesses among older people in the United States are cardiovascular disease, cancer, osteoarthritis, and osteoporosis (Wantz and Gay, 1981). Osteoarthritis and osteoporosis cause disability; cardiovascular disease, cancer, respiratory disease, and diabetes are the most common causes of death.

Cardiovascular disease

New findings about what causes and cures cardiovascular disease roll off the presses every day. One group claims that diseases of the heart and blood vessels are inherited; another group claims that they result from the diet and exercise habits of an affluent society. One group claims that cardiovascular disease is best treated by surgery; another group claims that surgery is rarely useful. One group claims that cholesterol in the diet clogs the arteries; another claims that cholesterol is hogwash. What no one disputes is the fact that cardiovascular disease kills more men over 45 and more women over 65 than any other cause of death (U.S. Bureau of the Census, 1982).

Table 20.2
Activity limitation

	No Limitation on Activity		Some Limitation		Limitation in Major Activity	
	1970	1980	1970	1980	1970	1980
Under 45 years	94.7%	93.2%	5.3%	6.8%	3.3%	4.2%
45 to 64 years	80.5	76.1	19.5	23.9	15.7	18.8
Over 64 years	57.7	54.8	42.3	45.2	37.0	39.0

SOURCE: U.S. Bureau of the Census, 1982.

These figures show the percentage of people whose physical activities were limited in the years 1978 to 1980. The number of limitations increases with age, although the large majority of older adults remain able-bodied.

Nearly all older people show some degree of the stiff, inelastic artery walls called *arteriosclerosis*, or "hardening of the arteries." Many also show some degree of *atherosclerosis*, the build-up of hard, yellow, fat deposits on the artery walls. These deposits are made of collagen, lipids, and other substances. As they increase in size, they fill the artery so that less and less blood can flow through it. Even children may have some degree of atherosclerosis. It may be that a small amount of atherosclerosis is a normal part of aging, and greater amounts comprise cardiovascular disease (Rockstein and Sussman, 1977).

The major risk factors associated with cardiovascular disease include a high level of blood cholesterol (cholesterol is made up of several kinds of lipids); cigarette smoking; and high blood pressure. The secondary risk factors associated with cardiovascular disease include lack of exercise; a high level of blood triglycerides (another lipid); heredity; diabetes; overweight; stress; a diet high in animal (saturated) fats; and a hard-driving, hurrying, competitive, Type A personality (DeBakey et al., 1984). The more of these risk factors a person has, the greater the person's risk of suffering a heart attack, stroke, or other complications of cardiovascular disease. Most people whose arteries are substantially blocked by fatty particles also have high blood pressure, or *hypertension*, and in turn, hypertension tends to speed up the build-up of fatty particles. Hypertension is considered the result of disease, not a normal part of aging. It is associated with many different factors. Heredity contributes to hypertension, as do kidney abnormalities, stress, overweight, and a diet high in salt.

The fatty deposits on the artery walls prevent blood from passing and eventually may starve muscles of the oxygen they need to survive. When the heart muscle itself is starved for oxygen, a condition called *ischemic heart disease* is created. Twelve percent of women over 65 and 10 percent of men have ischemic heart disease (Rockstein and Sussman, 1979). When the heart does not have oxygen, a heart attack follows. The severity of the attack depends on the location and extent of cell death.

When their brain's supply of blood is cut off, people suffer from strokes, also called *cerebrovascular accidents*. A massive stroke kills, and in America massive strokes are the third leading cause of death, just behind heart disease and cancer. Depending on which part of the brain has been starved of oxygen, less extensive strokes may paralyze the muscles and interfere with memory and speech. An accumulation of transient, small strokes can produce a brain disorder that resembles dementia.

Health and expectations

> A 90-year-old woman was wheeled into the hospital emergency room with a broken hip. After a young surgeon examined her, he told her enthusiastically, "We can replace that hip joint, and you'll be as good as new." The woman wasn't buying it. "But I'm 90 years old, Doctor. *Ninety* years old."
>
> A 103-year-old woman made an appointment with an orthopedist because her knee was causing her pain. The doctor examined her and suggested that the best course of action was to do nothing. "After all, you're 103 years old," he said. She replied, "Yes, but my other knee doesn't hurt, and it's 103 years old, too."

Expectations play an important role in determining the health of old people. People's stereotypes and expectations, the effects of acute and chronic illness, and the effects of normal aging combine to affect the quality and the length of life. Thus in assessing older people's overall level of health, one must take into

There is a consistent relation between how healthy people **expect** *themselves to be and how healthy they actually* **are**. *Older adults also tend to be quite accurate at assessing their own health.*

account not only the actual physical condition and level of functioning, but also what older people expect of themselves and what others expect of them (Hickey, 1980). When others expect too little of the old, and when they subscribe to a stereotype of aging people as inevitably in poor health, they tend to underestimate and undertreat the problems of the old.

> My great grandfather was fuming at the dinner table. Numbness in his feet was making walking increasingly difficult. ''The doctor tells me that I'm too old to have surgery on my feet. He told me I should expect to 'slow down' at my age.''—Sam

Generally, there exists a consistent relation between how healthy a person is and how healthy that person expects him- or herself to be. When older people or those who take care of them accept the idea that to be old is to be disabled, the older people function below their physical capacity.

How well do older people's assessments of their health jibe with the assessments of their doctors? When the subjects in the Duke Longitudinal Study were asked to rate their own health, the ratings of 60 percent agreed with the ratings of their doctors. One quarter of the older subjects rated their health as better than their doctors did, and half that number rated their health as worse than their doctors did. The older people's ratings of their health were better predictors of their health at their next examination than their doctors' ratings were (Maddox and Douglass, 1974). Older people's ratings of their own health also have been found to be good predictors of death from ischemic heart disease (Siegler and Costa, 1985). People who considered their health poor were twice as likely to die from heart disease as those who considered their health excellent. On the whole, people are quite accurate at evaluating their health.

Habits and health

There may not be any way for people to avoid the effects of normal aging, but there are many ways for them to avoid the effects of secondary aging. It has been estimated that *half* of the people in this country die prematurely because their habits are unhealthy (U.S. Public Health Service, 1979). Evidence from longitudinal studies shows that smoking, diet, exercise, stress, and the quality of the environment all affect health and longevity. In the Duke Longitudinal Study, for example, the amount of exercise people got correlated most highly with their level of health (Palmore, 1974). Older people who exercised regularly were less likely to sicken or die prematurely than those who exercised infrequently. People who smoked and who were overweight or underweight also were ill more often. The healthiest older subjects were nonsmokers of normal weight who exercised regularly.

When children know about and follow a healthy way of life, their chances of remaining healthy well into old age increase. But adults who trade in harmful habits for healthy ones also do themselves substantial good. For example, adults who stop smoking dramatically reduce their risk of heart attack and lung cancer. Adults who exercise and watch their weight and diet can reverse cardiovascular damage. Evidence from the Framingham Heart Study, a longitudinal investigation of cardiovascular disease, suggests that people are aware of the risk factors and are doing something about them. When the health of parents 20 years ago was compared with the health of their children today, the children were found to have lower blood pressure and lower blood cholesterol and to smoke less than their parents did at a comparable age (Riley and Bond, 1981). Those data imply that the children will become adults with a lower incidence of heart disease than their parents.

Deaths from cardiovascular disease in the United States have been declining in the past few decades. They took a sharp drop in the 1970s (Levy and Moscowitz, 1982). Just as certain habits—among them a diet rich in animal products, more sedentary jobs, widespread cigarette smoking—caused the incidence of cardiovascular disease to rise during the first half of this century, a reversal of these habits today seems to be diminishing the incidence of cardiovascular disease. It is likely that many different factors are contributing to the decline, including better diet, more exercise, less smoking, better drugs to control high blood pressure, and better medical care for people with cardiovascular disease.

Smoking

> In my experience, once my patients have reached 40 "pack years," I can expect to see them getting lung cancer. A "pack year" is smoking one pack of cigarettes a day for a year, or the equivalent.—Sam's grandfather

As the clinical experience of this family doctor has shown, lung cancer often follows cigarette smoking. Smoking is the cause of three quarters of all cases of lung cancer, a form of cancer that is curable in only a small minority of cases. Not only do smokers have ten times the rate of lung cancer as nonsmokers, but they also have three to five times the rate of cancer of the mouth and tongue, three times as much cancer of the larynx, and twice as much bladder cancer (U.S. Public Health Service, 1979). In the United States, one-fifth of the deaths from cancer can be traced directly to smoking, and another one-third of the deaths can be traced indirectly to smoking.

Smoking also accelerates damage to the heart and vascular system. Twice the number of smokers as nonsmokers die of heart disease, and the more cigarettes people smoke, the more likely they are to have a heart attack. Tobacco smoke contains carbon monoxide, a poison that seems to increase the formation of plaques and the deposit of fats in the arteries. Tobacco smoke also contains the stimulant nicotine, another poison. Nicotine boosts the heart's demand for oxygen and may contribute to the crushing chest pain of angina, a symptom of the heart's oxygen starvation. Smokers are also likely to have reduced lung capacity, ephysema, chronic bronchitis, a greater vulnerability to microorganisms and lung debris, and an increased risk of developing stomach ulcers.

When people of any age stop smoking, their increased risks of developing these conditions eventually return to normal levels.

Diet

> As I've gotten older, I notice that my tastes have changed. I seem to be adding a lot of sugar to my food. I even found myself putting sugar in the tuna salad. It takes more seasoning for me to be able to taste things.—Peter's grandmother

> We used to make fun of my great grandfather's diet. He doesn't like red meat, just fish and chicken. He never eats sweets. In the morning he drinks a cup of hot water with lemon in it, not tea or coffee. He's always been thin and a great walker. Nobody's making fun of him now. He's 96 and still going strong.—Alyssa

As evidence from many different quarters filters in, the influence of diet on aging is increasingly well understood. It is possible that good diet can to some extent slow both normal and secondary aging. Many different studies suggest that the rate of aging is affected by people's lifelong eating patterns (Guigoz and Munro, 1985; Schlenker et al., 1973). We form food preferences, and our eating patterns take shape when we are children (see Lifespan Focus: Obesity). These preferences and patterns in turn reflect those of our parents, our ethnic and religious background, social class (Davis and Randall, 1983), and even the region of the country where we live.

What, when, how, and with whom we eat also reflect broad changes in culture and technology. Today, for example, families are less likely to eat together, more likely to eat processed, restaurant, and fast food, and more likely to eat a wider variety of foods than families did before World War II. Since 1900, Americans have been eating more fat and sugar. Today Americans also eat more protein from animal sources.

These changes in the typical American diet tend to hasten aging and contribute to diseases such as high blood pressure, cancer, diabetes, and heart disease. One longitudinal study of 4000 middle-aged men showed that when they ate a diet that reduced the level of cholesterol in their blood, they also had fewer heart attacks and less heart disease (Kolata, 1984). A comparison of the American and Japanese diets shows differences in both diet and in the incidence of certain diseases. The Japanese eat less fat, less beef, more salt, and more pickled foods. Their rates of breast, colon, and intestinal cancer are lower, but their rate of stomach cancer is higher. When Japanese move to the United States and adopt its diet, their rates of breast, colon, and intestinal cancer rise, and their rate of stomach cancer drops. It is likely that a diet both low in fat and high in natural fiber prevents the development of certain diseases (see Table 20.4). A diet low in salt—without the salty processed foods, salty snacks, and table salt that Americans love—is also likely to reduce the incidence of hypertension in aging adults. Recent evidence suggests that Americans are beginning to eat more healthily. The per capita consumption of animal fats, dairy products, and eggs has declined, while the consumption of fats from vegetable sources has increased (Weg, 1983).

A diet low in fat and high in natural fiber is also likely to prevent obesity.

Table 20.4
Dietary goals for the United States

	Current Diet	Dietary Goals
Saturated fat	16%	10%
Polyunsaturated and unsaturated fat	26	20
Protein	12	12
Starches	22	43
Sugar	24	19

SOURCE: U.S. Senate, Select committee on Nutrition and Human Needs, "Dietary Goals for the United States" (Washington, D.C.: U.S. Government Printing Office, 1977).

Americans who reduce the amount of fat in their diet to 30 percent or less, reduce the amount of sugar, and increase starches are likely to improve their health.

As people get older, they tend to be less active and therefore to need fewer calories per pound of body weight. Yet when older people are physically active, they actually burn more calories than younger people do to perform the same physical task (Hickey, 1980).

As people age, they tend to take in fewer calories, especially calories from fats (Elahi et al., 1983). Older people's needs for certain nutrients remain unclear in many instances. But because aging brings a decline in digestive secretions, enzymes, and the absorption of nutrients from the intestine, older people not only metabolize nutrients differently from younger people but may need more of certain vitamins and minerals (Butler and Lewis, 1982). Poor

Vigorous exercise slows or even reverses many of the symptoms of aging and the diseases associated with it. These two men are finishing the Boston Marathon.

lifespan focus — *Obesity*

Chunky toddlers may well turn into fat children and, later, adults, and fatness can cause psychological as well as physical ills. Five percent of American schoolchildren are obese, weighing more than 85 pounds at age 9, compared to the norm of 66 pounds. The prevalence of obesity has increased by more than 50 percent in the past 15 to 20 years and even more so for black children (Dietz, 1986). Forty percent of the children who are obese at age 7 become obese adults (Epstein, 1986).

Why are there so many fat people? There are many reasons (see Table 20.3). For one thing, parents pass on both their body type and their eating habits to their children (Stunkard, 1986). They begin to do this in infancy by overfeeding their babies. By the time a child is 6 years old, overeating has become a habit that is hard to break.

According to one hypothesis, the first year or two of life is a critical period during which the infant's diet affects how many fat cells form. Malnourished infants develop few fat cells; overfed infants, many (Winick, 1975). Overfed infants then are permanently predisposed to obesity, turning into children and adults who crave more food and gain weight more easily than those with fewer fat cells (Harding, 1971). But some evidence suggests that fat babies are not necessarily doomed to be fat adults. One researcher (Roche, 1981) has challenged this hypothesis with data from animal experiments, which suggest that the number of fat cells is *not* fixed in the early years. Furthermore, in humans, he points out, measures of obesity in infancy are not

For many people, inherited predisposition plus environmental factors make obesity a lifelong problem.

Table 20.3
Causes of overweight

- *Body build.* People inherit a body build that is lean or heavy.
- *Activity level.* They also inherit a characteristic level of activity that burns off more or fewer calories.
- *Taste preference.* Infants are born with taste preferences. Some have a "sweet tooth" before they even have teeth. They prefer to drink sugary water more than other infants do.
- *Basal metabolism.* Even at rest, people's rates of using the calories they take in varies by as much as 400 to 500 calories a day.
- *Endorphins.* Overweight people may produce too many endorphins, with the effect that they crave food all the time.
- *Food consumed.* People often have been urged by their parents to "clean their plates" and thereby overeat.
- *Sugar and fat.* People may know that candy and cookies are sugary and fattening, but they may not know about the sugar hidden in processed foods like peanut butter, ketchup, and cold cereals. Foods low in sugar and fats—fresh vegetables, fish, poultry, and lean meats, and, for infants, breast milk—rarely cause obesity.
- *Stress and anxiety.* Anxiety, loneliness, boredom, and stress can all cause people to overeat.

significantly correlated with measures of obesity in adulthood. In addition, factors besides the number of fat cells have been shown to influence one's weight. The *size* of fat cells, for example, also influences how much a person weighs (Brownell, in Turkington, 1984).

Another factor that contributes to leanness or obesity is a person's general body metabolism, the rate at which calories are burned. How efficiently people burn calories is more important in determining obesity than simply the number of calories they take in. Lean people burn the extra calories they take in without turning them into deposits of fat. Some governing mechanism in their brain makes their body "waste" extra calories. But the metabolism of obese people makes their bodies thrifty about burning extra calories; they store those calories as fat (Bennett and Gurin, 1982). Related to this, perhaps, is another factor: activity level. Quiet, placid babies who eat only moderately are fatter than thin, tense babies who cry often and eat a lot (Mayer, 1975). Later, sitting passively while watching television and munching aggravates the tendency to obesity (Dietz, 1986). Taste preferences also may contribute to obesity. From the very beginning, it has been found, some infants prefer sweet flavors more than other infants do (Milstein, 1978). These infants are thought to be likely candidates for later overweight.

Obesity affects more than an individual's appearance. In adulthood, obesity poses health risks like diabetes and high blood pressure. In childhood, the risks are mainly to mental health. Overweight children often feel rejected by others. When one researcher (Mayer, 1975) showed girls a picture of a girl a short distance away from a group of other girls, thin girls usually interpreted the picture to mean that the girl was walking *toward* the group; fat girls interpreted the picture to mean that the girl had been *left out* of the group. In fact, overweight children, especially girls, may be left out of social activities. One study of college admissions showed that, if applicants were interviewed in person, obese girls had only one-third the chance of acceptance at their chosen college as thin girls (Canning and Mayer, 1966). Many other studies support the suggestion that both children and adults hold negative stereotypes about, keep farther away from, and discriminate against obese people (Jarvie, Lahey, Graziano, and Framer, 1983).

nutrition can cause older people to seem mentally confused, depressed, and unable to learn.

When older people are undernourished, the reason is less likely to be ignorance than poverty, stress, loneliness, and immobility. Poverty makes it difficult for people (of any age) to buy enough nourishing food, and immobility makes it difficult for people to get hold of nourishing food. Older people who are lonely, depressed, and anxious may have little appetite and little desire to prepare well-balanced meals. Older people who live alone eat a less adequate and less varied diet, prepare fewer of their foods, and less often eat an evening meal than older people who live with someone else (Davis and Randall, 1983).

Exercise

Many Americans can drive everywhere, sit down on the job, and sit in front of a television set for recreation. A study of 76 older women showed that their rates of physical activity ranged from sedentary to very sedentary, with an average level of activity only one-third that of college students (LaPorte et al., 1983). As women and men get older, they become less active. Women become sedentary before men, and people with less education become sedentary before college graduates (Ostrow, 1980). Many older Americans look askance at physical exercise. As one wit said, whenever the urge to exercise came over him, he'd lie down until it passed. Older Americans not only think that they do not need exercise, they also think that it endangers their health, that they are incapable of vigorous activity, and that slight exercise from time to time is healthy. Half of all Americans surveyed for the National Health Survey said that they did not exercise, and the reason they gave was that they did not need to (U.S. Bureau of the Census, 1982). But they were wrong. All that sitting around simply is not healthy.

As we have seen, some of the healthiest older people are those who exercise vigorously and regularly. What many people take to be normal aging—muscle weakness, stiff joints, mental deterioration—is likely to be secondary aging, a result of inactivity. When older people are inactive, their muscles shrink and lose endurance. When young men in excellent physical condition were confined to bed rest for three weeks, their heart grew one-fourth less efficient, and their breathing capacity and oxygen consumption declined by one-third (Saltin et al., 1968).

Exercise reduces the risk of heart disease. A study of 17,000 Harvard graduates showed that those who exercised regularly had only half the incidence of heart disease and many fewer heart attacks than those who did not exercise reguarly (U.S. Public Health Service, 1979). Exercise increases the number of red blood cells and blood volume, improves the condition of blood vessels in the heart, improves blood circulation, and reduces the levels of cholesterol and other harmful fats in the blood (Wantz and Gay, 1981). A program of six months of regular exercise may also reduce high blood pressure (Boyer and Kasch, 1970), reduce pain from angina, and reverse other effects of cardiovascular disease. Exercise makes people slimmer, makes their joints more supple, improves the respiratory system, and increases bone strength and density. Infrequent, mild exercise does not have the same benefits to health as regular, vigorous exercise. For the greatest benefit, older people need to exercise large muscle masses by swimming, running, jogging, or walking quickly.

Older people who are physically active generally say that they feel better and more energetic and need less sleep. Many say that they feel better about themselves and suffer less from mild depression and anxiety (U.S. Public Health Service, 1979). Others says that their memory is better, their reaction times swifter, and their attention sharper (Wiswell, 1980). When a group of ten elderly men who suffered from high anxiety were tested with exercise, tranquilizers, and a placebo, exercise proved more effective than the tranquilizer, an effect corroborated by measures of the men's muscle tension (deVries and Adams, 1972).

Improvements in health habits are likely to translate into significant gains in longevity and quality of life for aging Americans. The statistics on longevity are quite clear: if people escape the dangers of premature death, they can live out a lifespan that comes close to the biological maximum for the human species (see Chapter 23). By keeping themselves healthy and fit, adults stand an excellent of chance of surviving to an advanced old age.

> I retired from my job as manager of a hardware store when I was 75, but after a few years of idleness I wanted to get back to work. Gardening and babysitting the grandchildren weren't enough to keep me busy. When I was 78, I found a job as a stockroom clerk for a big mail order firm. They would have kept me, but my eyesight got too bad for me to be able to drive myself to work. I retired for good at 86.—Sam's (other) great grandfather

The best news is that health statistics already have begun to reflect many of the promising changes in people's habits.

Summary

1. Programmed theories of aging ascribe the aging process to genetic programming. As various genes turn on and off over the course of human development, people go through a predictable series of biological events, including the aging of cells and organs. Programmed theories of aging are supported by research on causes of premature aging. The Hayflick limit is the maximum number of replications of individual human cells.
2. To account for the deterioration that accompanies aging, some people have suggested that certain genes trigger aging and death; that "young" genes are overpowered by "aging" genes during middle age; and that genes change during middle age. Others suggest that the hypothalamus or the immune system directs the programmed deterioration.
3. Wear-and-tear theories of aging propose that over the lifespan, the body suffers functional damage that it repairs increasingly less effectively. The DNA repair theory locates the source of damage within the DNA molecule itself. The cross-linkage theory suggests that stable links form between cells, which turn them rigid and impair function. The free-radical theory suggests that certain unstable chemical compounds build up in the body and cause deterioration.
4. Changes in appearance often are the first signs of aging. Nails grow more slowly, and skin that has been exposed to sunlight and other environmental hazards looks wrinkled, lined, and weathered.
5. In normal aging, the bones thin and the spine shrinks somewhat. Much of the bone loss and brittleness seen in older adults results from inactivity and poor diet. Teeth grow more resistant to decay, but surrounding gums and bone may be infected.
6. In the cardiovascular system, the aging heart pumps more slowly and pumps less blood with each beat. The artery walls thicken and grow less elastic—arteriosclerosis—and large fat molecules stick to the walls—atherosclerosis—gradually blocking more and more blood flow. De-

creased arterial blood flow is usually least pronounced in the brain, heart, and skeletal muscles and most pronounced in the kidneys, other internal organs, and the extremities. The causes of cardiovascular disease are not fully understood, but much of it probably is due to consuming a high-fat diet, not exercising, and smoking cigarettes. Insufficient blood to the heart can cause heart attacks; insufficient blood to the brain can cause strokes.

7. The aging of the respiratory system makes breathing less powerful and efficient. Older people also breathe more shallowly because the rib cage grows more rigid and the muscles that move the chest during breathing grow less elastic. As the lungs age, they turn from pink to gray, they grow less elastic, the surface area of the alveoli shrinks, and the lungs are less efficient at cleansing the blood. The lungs' residual volume increases with age, along with a person's maximal oxygen uptake. Much of the aging of the respiratory system is a result of secondary aging—including breathing polluted air and cigarette smoke and inactivity.
8. The aging digestive system also slows and grows less efficient. Contractions that move food along the digestive tract grow less effective as muscles weaken, the stomach's production of hydrochloric acid diminishes, the intestines secrete fewer enzymes, and the smooth muscle in the wall of the large intestine grows smaller and weaker. After the age of 80, normal aging keeps a person from absorbing molecules of one simple sugar. The liver may also grow less efficient with age, and its ability to metabolize certain drugs diminishes. But liver and gallbladder continue to function well in most older people. Many of the disorders of the digestive system that afflict older people result from a poor diet and from inactivity.
9. In the excretory system, the aging kidneys and the blood vessels that supply them shrink, but kidney function remains efficient in healthy older people. The number of urinary tract infections also increases with age.
10. During late adulthood, sexual responsiveness is slower and weaker than it once was, and many older people have sex less often. But old age need not mean the end to pleasurable sexual activity.
11. The aging brain loses weight, shrinks, and its interior ventricles enlarge. Neurons in certain regions of the brain die but cause no apparent loss of brain function; they also lose some of their dendrites and axons. Lipofuscin builds up, and some unconditioned reflexes disappear. Brain wave activity during both sleep and waking changes. Sleep itself may grow shorter, lighter, more restless, and more likely to be broken by apnea and muscle jerks.
12. Age weakens some autonomic functions, as neurotransmitters weaken or are metabolized differently and as neurons change in structure and die. But the core of the autonomic nervous system undergoes no marked deterioration with age, and only slight changes in function appear.
13. Acute illnesses tend to grow less frequent but more severe among older people. Chronic, incurable illnesses such as diabetes, arthritis, emphysema, and cardiovascular disease grow more common in old age.
14. Certain habits are known to influence the quality and length of life. These include smoking tobacco, quality of diet, and amount of physical exercise.

Key terms

programmed theories
progeria
Werner's syndrome
Alzheimer's disease
Hayflick limit
wear-and-tear theories
DNA repair theory
cross-linkage theory
free-radical theory
melanocytes
melanin
osteoporosis
osteoarthritis
elastin
alveoli
astrocytes
acute illnesses
chronic illnesses

Suggested readings

HICKEY, T. *Health and Aging*. Monterey, CA: Brooks-Cole, 1980. A review of current and probable future conditions of health and aging, including health assessment, the social-psychological, political, and economic contexts of health and aging, and long-term care.

SIEGLER, ILENE, and COSTA, PAUL. "Health Behavior Relationships." In J.E. Birren and K.W. Schaie (eds.), *Handbook of the Psychology of Aging*. New York: Van Nostrand, 1985. A review of findings about the effects of health on behavior and the effects of behavior on health.

WANTZ, M.S., and GAY, J.E., *The Aging Process: A Health Perspective*. Winthrop, 1981. A discussion of many aspects of health and aging, including psychological aging, physiological aging, disease, and enhancing health.

chapter twenty-one

Cognitive development

Normal decline in cognition

In late adulthood, there appears to be a decline in some cognitive abilities. But unless this decline is associated with disease, it is slight and affects only some individuals. Other cognitive abilities remain remarkably vital into very old age. After looking at this nonpathological decline in cognitive ability, we will turn to the decline associated with disease and, finally, we will discuss the possible growth in cognition and wisdom that some older adults display.

Information processing skills

Learning, thinking, and remembering are basic information processing skills that probably become less efficient in late adulthood. But developmental psychologists still know relatively little about the rate of this decline or its point of onset. They know relatively little because most of the relevant research has been cross-sectional in design, comparing the abilities of college students with older adults. Cross-sectional research has shown age differences in abilities associated with conditioned learning, verbal learning, and learning specific cognitive skills and strategies.

It is important to remember, however, that older people adapt to changes in their cognitive functioning. The person who knows it takes longer at age 60 than age 30 to master information allots a longer learning period. The person who forgets easily learns to write things down.

> I've gotten so bad at remembering certain things that I've made it a habit to write them down—things to pick up at the supermarket, ideas I want to bring to work in the morning, reminders to mail my grandson a birthday card. If I don't write them down, they're often gone for good.—Natalie's grandmother, age 71

Age changes in conditioning

The pattern of age changes demonstrated experimentally differs for classical and operant conditioning. In one experiment in classical conditioning, adults were asked to press a key on the right side of a board when a light appeared on that side and a key on the left when a light appeared on the left side (Kimble and Pennypacker, 1963). With each appearance of the light, a puff of air blew at the adult subject's eyes. When the subject blinked at the appearance of the light, in the absence of the puff of air, they were conditioned. Older adults took longer to condition than younger adults, and the older adults' blink was also less robust.

Experimental results suggest that age differences in classical conditioning set in during the sixties, as a result of changes in the central nervous system that weaken involuntary responses (Gendreau and Suboski, 1971; Schonfield, 1980). The decline in sensory acuteness that affects older people may make stimuli such as puffs of air feel less unpleasant to them than to younger people. Age changes may also appear because it takes older people longer than younger people to register and respond to stimuli (Perlmutter and List, 1982).

Although there has been little research on whether older people grow less responsive to operant conditioning, it is known that older people can be conditioned. One 82-year-old man with a heart condition was keeping to his bed and neglecting his medicine, exercise, and food (Dapcich-Miura and Hovell, 1979). He was placed on a system of token reinforcements whereby every time he took a short walk without being reminded, drank orange juice, or took his medicine, he earned tokens that could be exchanged for privileges, such as dinner at a restaurant. Within a few weeks, the conditioning had succeeded to the point that the man was taking walks several times a day and regularly taking medicine and juice. His chest pains also had subsided.

Even among older people with organic brain disease, operant conditioning can succeed. Prompts and reinforcement have succeeded in getting nursing-home patients as old as 90 to bathe themselves (Rinke et al., 1978). Operant conditioning has increased social interaction, appetite, and exercise and reduced incontinence among patients with brain disease (Schonfield, 1980). However, operant conditioning is not always successful with older adults, in part because the rewards that are effective in changing younger adults' behavior are less effective with older adults (Perlmutter and List, 1982).

Age changes in verbal learning

In verbal learning tasks, people are expected to repeat specific information presented by a researcher. The trials usually are structured as serial-learning or paired-associate tasks. In serial-learning tasks, people watch a list of words appear one after another on a screen and try to memorize them. In paired-associate tasks, people are given pairs of words, such as ''coat–roof,'' and later are expected to supply the second member of the pair when given the first.

Results from cross-sectional studies show that although older adults do not do as well as younger adults at verbal learning, there is usually overlap between the scores of the two age-groups. Some older adults do as well as younger adults. Cohort differences probably exaggerate the observed age differences.

In one longitudinal investigation of a group of 60- to 74-year-olds who had been tested 40 years before, scores on learning and remembering declined significantly in late adulthood, but scores on vocabulary stayed high (Gilbert, 1973). In another longitudinal study of six cohorts, well educated, upper-middle class men born between 1885 and 1932 and originally tested between the ages of 32 and 75 were tested again after an interval of eight years (Arenberg and Robertson-Tchabo, 1977). Among the men younger than 60, there was little change in learning abilities. Among those older than 60, the change was marked. At both intervals, the men took paired-associate and serial-learning tests, with similar results on the two kinds of tests. On the paired-associate test, the younger cohort, who had been in their thirties on the original test, improved slightly. Middle-aged men, who had been in their forties and early fifties on the original test, declined slightly. Older men, who had been 55 and older on the original test, declined markedly, with the steepest decline among those who were 69 to 76 on the original test. Differences among individuals were smallest among the youngest group and largest among the oldest. Thus in verbal learning, a decline appears to set in after the age of 60.

Encoding ability does seem to decline with age. Even when they have cues, older adults recall fewer words than younger adults, leading researchers to infer that older adults encoded less information to begin with. When older and younger adults do learn (encode) the same amount of information, however, these age differences in recall diminish (Perlmutter, 1983). It may be that older adults require more time to encode information. They require more time to register information visually and to integrate it. Older adults may encode only the general features of information and skip certain details that might help them to distinguish one bit of information from another. Slowing of this nature can affect both recognition and recall. For example, when older and younger adults were given half a second to remember each picture and word, older adults later could not recognize many of them. Given a whole second, the age difference declined significantly (Waugh and Barr, 1982).

Why does this decline set in? Perhaps slowing of perceptual processes accounts for the decline (Birren, 1974). Slowed perceptual processes themselves probably are the result of older adults' reduced abilities or willingness to concentrate on intellectual tasks (Horn, 1982). Even when older poeple are given extra time on these intellectual tasks, they score lower in fluid intelligence than younger people do.

Age changes in learning cognitive skills

Although few researchers have made the attempt, and although the overall results have been mixed, some older adults have been successfully taught certain cognitive skills. In one experiment, older adults were trained in reasoning inductively—forming general rules out of bits of information—and later transferred their improved ability to another inductive reasoning task (Labouvie-Vief and Gonda, 1976). Older adults can also form new concepts, if more slowly than younger people. Taught a specific, systematic strategy for testing hypotheses, older adults' skills at forming concepts improves noticeably and remains better for at least one year (Sterns and Sanders, 1980).

Older adults can also be trained in perspective taking and describing things for others—referential communication. When older adults went through six weeks of training in discussing and role-playing problems, their social understanding improved (Zaks and Labouvie-Vief, 1980). Compared to a control group and a group who had only discussed problems, the more thoroughly trained group was better at taking another person's perspective. They were also better at describing objects, such as geometric figures, so that another person could pick out a similar object from an array working solely from the verbal description.

Influences on learning

Age differences in learning may be influenced by the pacing of the learning task and by the motivation, caution, and distractibility of the subject. For these reasons, plus others we have already mentioned, it has been difficult for researchers to determine whether and to what extent the observed decline in learning is a result of aging.

Pacing

As we have seen, older people need more time than younger people to learn, to extract visual information, and to process information (Salthouse and Somberg, 1982). Thus the pace at which laboratory experiments are run strongly influences older adults' performance.

Older adults often recall fewer words on laboratory tests than younger adults do. But such deficits tend to be small and, in the everyday world, virtually unnoticeable.

The pacing of paired-associate learning tasks is varied either during the study period—the length of time the pairs of words appear together—or during the testing period—the interval between when the first word in each pair appears and the appearance of the two words together. Older adults' performance improves when either period is increased. Younger adults also improve when they have more time, but older adults improve more. For example, when people search for a paired word, they first must register the word presented, then scan their memory for its partner, and produce it before the pair appears on the screen. When the testing period lasts only 1½ seconds, older adults perform far less well than younger adults. But when the testing period lasts 3 seconds, older adults improve far more than younger adults do (Canestrari, 1963). When both the testing and the study periods are increased, older adults do best of all.

When men had to learn complex light patterns, older men could adapt to all but the most rapid pacing (Perone and Baron, 1982). It took the older men longer to learn the sequences, but by the fifth time through, the older men did as well as the younger men. Once the men had mastered a complex sequence, 67- to 75-year-olds performed as well as 18- to 20-year-olds, so long as they had 1 second or more to produce the sequence. Given only half a second, the older men's performance declined. People as young as 40 begin to perform less well on paired-associate learning tasks when the pacing is rapid (Monge and Hultsch, 1971). In the real world, people in their thirties begin to leave rapidly paced jobs where they must perform under great time pressure (Welford, 1958).

Motivation

People's (and animals') motivation invariably affects their performance, both in the laboratory and outside of it. Some researchers have assumed that older adults are poorly motivated to learn lists or pairs of words. In classical conditioning experiments, older adults' sensory deficits may make them underaroused and undermotivated. Other researchers assume that older adults who volunteer for research are highly motivated and interested. In fact, it has even been suggested that older adults may be so highly motivated to perform well on intentional learning experiments that they grow too aroused and anxious and perform poorly (Botwinick, 1978).

Several studies have shown higher measures of arousal during learning tasks among older adults than among those younger than 50, whether the measurement of arousal has been free fatty acids in the blood (Eisdorfer, 1968) or heart rate and skin conductance (Furchgott and Busemeyer, 1976). To test the hypothesis that older adults are overly aroused in learning experiments, researchers (Eisdorfer, Nowlin, and Wilkie, 1970) gave half of their subjects a drug that lowers autonomic nervous system response. This group had lower levels of free fatty acids in their blood and did better on learning tasks than the other group, supporting the hypothesis that some older adults grow too aroused to do their best on learning experiments.

The practical implications of this research are important. If older adults find themselves in situations that make them feel anxious or prone to failure, they may not perform up to their abilities (Schonfeld, 1980). In one experiment, women who were in college and women between the ages of 60 and 74 who had graduated from college carried out detailed work with pencil and paper (Bellucci and Hoyer, 1975). Some of the women from each age-group worked near an experimenter who remained silent, and some worked near an experimenter who praised them for doing better than most their age, no matter how well they actually were doing. Young and older women who were praised did better than those who were not. But according to how they later rewarded themselves for their work—as measured by how many trading stamps they awarded themselves—it became clear that the older women who had not been praised thought less well of their work than the younger women who had not been praised. In short, these older women's self-confidence was lower.

Caution

Older adults often are so cautious about their responses in learning situations that they not only respond slowly, but they take longer to master information, and may seem to have learned less than they have. On learning experiments, older adults' errors are largely errors of omission (Arenberg and Robertson, Tchabo, 1977). Many older adults prefer to omit an answer rather than give a wrong answer. When older adults in one experiment (Leech and White, 1971) were paid for all their answers, right and wrong, their caution was sidestepped, and their errors of omission were reduced. Because their rate of responding increased, the subjects needed fewer trials to learn than older adults who were paid for correct answers only.

Distractibility

People whose attention is easily distracted from a learning task have trouble learning. If older adults are especially vulnerable to distraction by irrelevant information or by intruding thoughts, they learn less efficiently (Rabbitt, 1965; Schonfield, 1980). It has been suggested that in the information processing system of older adults, all information persists longer, and therefore has the potential for distraction.

To test the distractibility hypothesis, college students and adults between 63 and 77 years old looked for certain letters in a visual display (Madden, 1983). The older adults made more mistakes than the younger adults. After four days of using the same target letters, new letters were substituted. But some of the old letters were used as distractors. In both age-groups, seeing a familiar letter increased reaction speed, but by about the same amount. Thus it is likely that older adults have fewer problems with distraction than with focusing their attention to begin with.

Interference

As you will recall, when previously learned material interferes with the learning of new material, the phenomenon is called *proactive interference.* When given lists of commonly associated words—hot–cold; blossom–flower—older adults have more trouble than younger adults in forming new associations (Lair, Moon, and Kausler, 1969). In contrast, *retroactive interference* describes what happens when the learning of new material interferes with already learned material. It has been nearly impossible to tell whether there are age changes in retroactive interference because so many other factors influence performance (Arenberg and Robertson-Tchabo, 1977). However, in one study, young adults and older adults performed nearly perfectly to start with, and the older adults had more problems with proactive interference. The younger adults found only recall affected by proactive interference, but the older adults found both recall and recognition affected by it.

Intellectual abilities

What explains the observed decline in fluid intelligence—the ability to manipulate abstract symbols—during late adulthood? Fluid intelligence may decline because people pay less attention to incidental features of the environment. They pay less attention than younger people do to aspects of their surroundings that are not immediately relevant to the task at hand. Younger people can recall such incidentals moments or hours later, if they emerge as relevant. If older adults have only a limited cognitive capacity to concentrate on intellectual tasks, or if younger adults have a surplus capacity, differences in fluid intelligence may be the result. About half of the decline in fluid intelligence associated with age changes can be attributed to older adults' trouble with organizing information at the encoding stage, with focusing their attention, and with the metacognitive process of forming expectations about a task (Horn, 1982).

The decline in fluid intelligence can be offset with training. When people between 60 and 80 years old were given training and practice in skills related to fluid intelligence, such as reasoning inductively, figure relations, and the attention and memory tasks of perceptual discrimination, paying selective attention, switching attention, and concentrating, their performance improved (Baltes and Willis, 1982).

As you will recall from earlier chapters, the relevance of intelligence test scores to everyday life has been questioned for a number of reasons. Yet because developmental psychologists are still without a test that accurately measures and predicts how people will apply their intellectual abilities to problems of everyday life, they rely on standard IQ scores to assess adults' intelligence (Salthouse, 1982).

Age changes in intelligence test scores

When adults' performance on various subtests of the standard intelligence tests are charted, the picture is mixed. One view suggests that IQ test performance is steady from early adulthood to middle age and then declines. Yet another view suggests that during adulthood, scores on some subtests rise as others fall, raising the distinct possibility that the same overall IQ may mean different things at different ages. Finally, results from cross-sectional and longitudinal tests give still different views of adult intelligence.

Cross-sectional and longitudinal studies of intelligence

Results of cross-sectional and longitudinal IQ tests scores present different and conflicting pictures of what happens to intelligence across adulthood (see Figure 21.1). From cross-sectional results, we get a depressing picture of intelligence falling with advancing age. Tests conducted by the army and other organizations earlier in this century showed intelligence declining during people's twenties or at age 40 (Schaie, 1979; Yerkes, 1921). When cross-sectional test results were analyzed by social class, it looked as if verbal scores declined little. Nonverbal and psychomotor functions did decline, especially when speed was a factor. The decline was modest during middle age, but after age 70 scores dropped markedly. This pattern held for men and women, blacks and whites, all social classes, and even for people in institutions (Botwinick, 1977).

From longitudinal test results, we get a less depressing picture. These show increasing scores until age 60. For example, the army tests from the First World War (Yerkes, 1921) show verbal scores rising until age 50 and performance scores, tapping manipulation of objects, remaining essentially steady. Among people in their sixties, some verbal scores continued to rise, although arithmetic scores declined (Owens, 1966). On tests from the California Intergenerational Studies (Eichorn, Hunt, and Honzik, 1981), overall IQ increased slightly between the ages of 18 and 36 to 48, although performance IQ dropped somewhat during some people's mid-thirties. The differences among individuals were great, and some had large changes in IQ during the period between early and middle adulthood. Some lost ground; others gained. Many of those whose IQ increased significantly had traveled extensively outside of

CROSS-SECTIONAL AND LONGITUDINAL SCORES IN VERBAL IQ

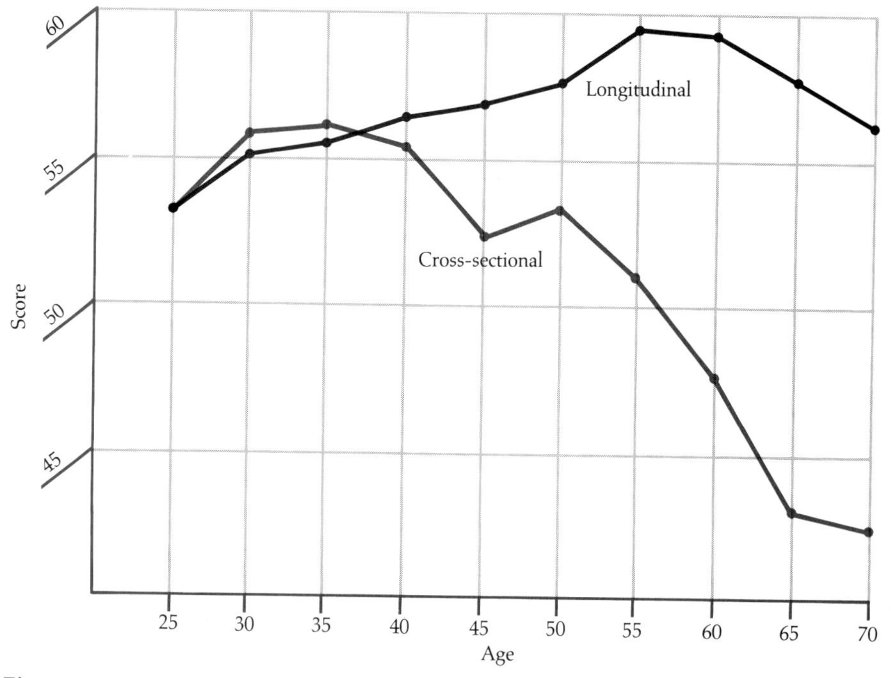

Figure 21.1 ***Although the cross-sectional data showed intellectual decline in adulthood, the longitudinal data showed something quite different—steady increases until late adulthood and then only a slight decline from the peak (Schaie and Strother, 1968).***

the United States and had married someone with an IQ at adolescence at least 10 points above their own. Many of those whose IQ decreased significantly had serious health problems and drank heavily.

In the Duke Longitudinal Study (Eisdorfer and Wilkie, 1974), which followed adults for ten years, results showed a small decline in IQ among adults in their sixties when first tested. Most of this decline was in performance IQ. Among adults in their seventies when first tested, the declines in verbal and performance IQ were marked.

What explains the discrepancies between the results of cross-sectional and longitudinal studies? First, the later age of decline in IQ in longitudinal studies may be a result of certain subjects dropping out. Some people died, and they may have experienced a pronounced decline in intelligence before death, a fairly common phenomenon (discussed in the next section). Some people who scored poorly did not return for later testing. Over the long term, the participants tended to be those with higher scores (Botwinick, 1977). In addition, cohort differences help to account for the discrepancy. Younger people have more years of schooling and the benefits of greater exposure to radio, television, and sophisticated technology. Over the long term, results from each new series of standardized IQ tests show higher IQ peaks at successively later ages (Schaie, 1983). For example, the 1916 Standford-Binet scores showed adult intelligence peaking at age 16. In 1930, the peak was age 20; in 1939, age 20 to 24; and by the mid-1950s, the peak was between ages 25 and 35.

In one study of adult intelligence that used a cross-sequential design (Schaie, 1979, 1983; Schaie and Herzog, 1983), the researchers found that different generations of people had different levels of ability at the same chronological age. The test results showed that younger cohorts performed best on tests of inductive reasoning, word comprehension, and spatial ability. Older cohorts did best on tests of word fluency. In interpreting these data, Warner Schaie (1983) has suggested that many older adults continue to function at the level they reached as young adults, but that this level is not always most appropriate for succeeding in today's society. Thus in some areas, older adults are not deficient, but out of date.

Cohort effects have exaggerated the decline in intellectual abilities among older adults. Even so, results from longitudinal studies and this cross-sequential study show that there are small but statistically significant declines in certain intellectual abilities among people in their fifities. The decline becomes noticeable during their sixties, although it may have almost no effect on everyday life. Test results may show that a 30-year-old could produce 40 words beginning with the letter ''s'' in three minutes but at age 70 could produce only 36 words. In everyday life, this decline is likely to have no real effect.

Terminal drop

Results from longitudinal studies of intelligence in adulthood have shown that people only a few years or a few months from death experience a marked decline in IQ known as **terminal drop** (Berkowitz, 1965; Kleemeier, 1962). Markedly declining verbal abilities have proved to be better predictors of mortality than chronological age (Blum, Clark, and Jarvik, 1973). Perhaps cardiovascular disease is responsible for terminal drop (Birren, 1968). Some age differences in cognition may stem from biological changes associated with aging (Albert and Kaplan, 1980; Butters, 1980). Older adults' abilities to learn and remember may diminish because wear and tear, cell structure, or other biochemical changes affect the central nervous system. These abilities may diminish in the face of depression, degenerative illness, or the sharp decline in cognition that often predeces death.

> When we visited my grandmother the autumn before she died, she complained that she couldn't remember street names any more. She asked which school Erica was going to be a student at, but she couldn't remember the word *student*.—Mike and Erica's mother

The connections between IQ and physical health among older adults are complex. For example, in the Duke Longitudinal Study, those with high blood pressure declined significantly in intellectual abilities over a ten-year period. However, terminal drop was only sometimes connected with high blood pressure or other symptoms of cardiovascular disease. Adults without cardiovascular disease but with acute illnesses, such as urinary tract infection,

Older adults generally learn to adapt to changes in their physical, sensory, and cognitive functioning.

and chronic illnesses, such as diabetes and emphysema, showed greater terminal drop (Wilkie and Eisdorfer, 1974a,b). It seems that cardiovascular disease as well as other degenerative diseases impair intellectual abilities.

In sum, although the nature of terminal drop is not yet fully understood, natural aging does not have to mean cognitive decline. People who remain healthy can expect little change for the worse in their intellectual abilities (Jarvik, 1973).

Psychomotor performance

Adaptations to changes in physical, sensory, and cognitive functioning are forms of learning in their own right. In everyday matters, older adults adapt to many kinds of change and learn to cope with the problems these changes may cause. As muscular control deteriorates and older people walk more clumsily, for example, they rely for feedback on sensory cues like their heels or soles touching the ground. As they walk, younger people rely on their anticipation of muscular performance. Although older people's muscular control slows, they compensate quite well. Old typists are just as quick as young typists (Salthouse, 1983). Although their simple reaction time has slowed, older typists compensate by anticipating longer strings of letters. Old people can play piano and golf, drive a car, and do many things by relying on experience and practice when cognitive control declines. Older people adapt to changes in their sensory functioning. The person who grows hard of hearing adapts by paying closer attention to other people's faces and the context of spoken information.

Reasoning and problem solving

Older people do far better at solving everyday problems than results of laboratory procedures might lead one to expect. Abilities that do not show up in the laboratory may well show up in real life, and abilities that seem to have been lost may quickly reappear after brief training (Denney, 1979). Retirement and widowhood, for example, require new forms of social learning from older adults. They must adapt to new social roles and learn new social skills. As people face retirement, they rehearse for their new role: quizzing other retirees, gathering information from the media, joining formal retirement-planning programs sponsored by businesses and the Social Security Administration. Adults in their sixties learn just as much as younger adults about pension benefits, financial, legal, and recreational planning, medical benefits and health care, nutrition, and exercise (Perlmutter and List, 1982).

> I'm planning on retiring within the year, and it's forced me to take a completely new perspective on things. What shall I keep or give away from my office? What's the best way to handle finances now? How will my wife and I manage our longer hours together? Will we crowd each other? Now that I'm retiring, I feel the urge to pass along what I've learned to my grandchildren all the more strongly. Retiring is a major readjustment, that's for sure.—Sam's grandfather, age 74

Retired adults who later switch careers again draw heavily on new learning and memory. Many older people, especially those with years of formal education, read regularly. Many also use television and radio as sources of information, even those who did not begin to watch television until middle age (Parker and Paisley, 1966). Some older people go to continuing education classes, Elder hostels, and other types of formal education.

However, older adults generally perform more poorly than younger adults on certain problem-solving tasks. It is difficult to pinpoint the reasons for this decline, because most research has not been able to separate the effect of various cognitive processes on adults' abilities to solve problems (Rabbitt, 1977). The decline may result from changes in memory, speed of processing, organization of information, or other factors that affect performance. For example, when older adults are in good physical condition, their physical and intellectual performance is enhanced (Perlmutter and List, 1982). A group of older adults who undertook physical training also improved in attention span and on simple cognitive tasks (Ohlsson, 1976). Others have improved their grasp of complex nonverbal relations (Elsayed, Ismail, and Young, 1980). Probably, several factors in combination affect adults' problem-solving skills.

Formal operations in late adulthood

In earlier chapters, we discussed Piaget's theory of cognitive development at length. We saw that the most mature level of cognition, formal operations, begins during adolescence and continues to develop throughout adulthood. It is heavily influenced by formal education and cultural factors as well.

How does performance on Piagetian tasks of cognitive development fare during late adulthood? In most cases, older adults do not do as well as younger adults on these tasks. Cognitive development ultimately seems to reverse itself, as the more advanced cognitive skills decline first (Storck, Looft, and Hooper, 1972). In one experiment focused on formal operations, college women outperformed older women (Overton and Clayton, 1976). Women in their sixties and seventies were less able to solve problems that required formal operations, such as determining the speed of a swinging pendulum. Intelligence test scores correlated with the women's scores on the tests of formal operations. In other research, it has been found that intelligence and level of education were better predictors than age of formal operations (Hawley and Kelley, 1973). For example, older scientists outperformed younger scientists on tasks of formal operations, and only among older nonscientists did formal operations decline (Sabatini and Labouvie-Vief, 1979).

On tests of older adults' concrete-operational thought, especially their understanding of the principles of conservation, results have been contradictory. Some results suggest no changes among older adults (Eisner, 1973). Others suggest that older adults have no trouble with simple problems of conservation, such as conservation of number, but do have trouble on more difficult problems (Papalia, 1972). Older adults revert to more functional classifications of objects, a change to what Piaget considered concrete operations.

Age differences have been found on many Piagetian tasks, but most of the relevant research fails to provide information on logical functioning in

Late adulthood calls for new learning and remembering in many situations.

The contexts within which older adults apply their cognitive skills tend to be quite important to the performance of these adults.

everyday life and confounds age with such factors as chronic disease, brain damage, and the effects of institutionalization. Cohort differences that are left unspecified are likely to have affected the findings. Another problem is that the Piagetian tasks were designed for use with children. Their repetitive questions may bore or irritate older adults into giving childlike answers.

But older adults' underlying knowledge may not be fully tapped on Piagetian tasks. When older adults have been trained in feedback techniques, they have succeeded in solving conservation tasks (Hornblum and Overton, 1976). When they have seen a skill modeled, they have solved classification tasks (Denney, 1974). It may be that adults' poor performance on cognitive tasks is a result of the tasks being stripped of meaningful context or content.

It seems that developmental psychologists still have much to learn about the application of adults' intellectual skills in everyday life (Neimark, 1982). Most tests of formal operations have a scientific and academic slant, and the findings they produce therefore reflect a temporary historical orientation toward abstract, logical thinking, rather than a biologically guided course of cognitive development (Labouvie-Vief, 1982). In addition, the contexts within which adults apply their intellectual skills may be extremely important. Psychologists are still learning about the effect of these contexts on intellectual competence. Finally, it is likely that people's health, level of education, and cultural background are powerfully related to intellectual performance. Some researchers (Labouvie-Vief and Chandler, 1978) have suggested that, contrary to the results of test scores and laboratory experiments, intellectual competence bears little relation to people's age. Instead, the changes in people's social, economic, and intellectual situations that typically accompany retirement may cause the sharp declines in intelligence test scores that crop up among people in their late sixties and seventies.

> When I was 68 and my wife was 66, we both were hospitalized for several weeks with hepatitis. When I finally got back to my medical practice, I found that I had lost my working knowledge of prescription drugs. It was a real chore to relearn all of that information.—Sam's grandfather

Forming categories

In the game of Twenty Questions, one person has 20 tries to guess the object someone else is thinking of. The person can ask any question answerable by yes or no, but the secret to winning is to ask questions that rule out (or in) whole categories of items at a time. Asking "Is it alive?" is a better question than "Is it a dog?" In laboratory investigations of people's problem-solving strategies and abilities, researchers may ask subjects to play a version of Twenty Questions. In one version, adults see colored pictures of 42 common objects. By asking as few questions as possible answerable by yes or no, they try to discover which object an experimenter has in mind.

When adults between 30 and 90 years old played this game, the number of questions required to discover the object increased with age (Denny and

Denny, 1982). Whereas two-thirds of the questions asked by people in their thirties embraced classes of objects: ''Is it a tool?'' ''Can you ride in it?'' only half of the questions asked by those in their fifties, and one quarter of those from people in their eighties did so. The results of this cross-sectional study suggest that more than age change was at work. The educational level of the adults correlated with the *type* of question asked, and so cohort differences were probably affecting performance. However, educational level did not correlate with the *number* of questions asked. Age alone accounted for this increase. It is unlikely that the decline in constraint-seeking questions reflected cognitive decline. When they were reminded of this strategy, older adults quickly put it to use (Denney, 1979).

In a longitudinal study that tapped similar abilities, subjects were presented with a problem scenario in which they were to discover the poisoned food (Arenberg, 1982). From a menu of two drinks, two meats, two vegetables, and two desserts, people chose one from each category. If the choices included a poisoned food, the experimenter said ''died.'' If the choices were safe, the experimenter said ''lived.'' In a more complicated version of the game, two foods were poisoned, and the meal was safe unless one or two were chosen from the menu. When the subjects knew which food(s) had been poisoned, they told the experimenter. The subjects were white men between 24 and 87 years old and either employed or retired managers, professionals, and scientists. Analyzed cross-sectionally, the results showed that men in their twenties had the most correct solutions. Those in their eighties had the fewest correct solutions. The largest decline came between the sixties and seventies. Analyzed longitudinally, the results showed that the men's performance improved over a period of seven years. Only among the oldest age-group, men who were in their seventies when the study began, did performance decline, and the decline correlated with age. The researcher, David Arenberg, ruled out memory deficits as the reason for this decline. He has suggested that older people suffer from having too much information to sift through when they try to solve complex problems. They have trouble going over the possible choices, planning, and executing their next choice. As people age, they grow less efficient at organizing the elements of a problem and at holding and manipulating more than one idea at a time (Hebb, 1978).

Pathological decline in cognition

A minority of older adults have organic brain disorders that interfere with their intellectual abilities. Their pathological decline in cognition may be extreme. Most **organic brain disorders** produce similar symptoms, although the symptoms may be slight or profound, and their progression may be slow or rapid. The symptoms that are generally found in organic brain disorders include:

1. *Impaired memory*—failure to register, retain, or recall information
2. *Impaired intellect*—difficulty understanding facts or ideas, doing arithmetic, or learning
3. *Impaired judgment*—difficulty understanding personal situations, planning, or deciding
4. *Impaired orientation*—confusion about time and subsequent confusion about place and others' identity
5. *Exaggerated or shallow emotions*—excessive reactions such as temper tantrums, or apathy, or emotions shifting rapidly for no apparent reason (Butler and Lewis, 1982).

lifespan focus

Classifying objects

When people classify objects, they search for similarities among them. These may be similarities of function, of appearance—shape, color, size, substance—or of abstractions like "cities," "mammals," and "stages of cognitive development." Piaget demonstrated preschool children's limited ability to classify objects by using cutouts of rectangles, triangles, and arcs that were red, blue, and yellow. He found that 2- and 3-year-old children arranged the cutouts in a line or big circle, or they made a house or a wagon with them. But they paid no attention at all to the shapes or colors of the cutouts. Four- to 6-year-olds began to pair the cutouts, sometimes by shape and sometimes by color. But as they arranged the cutouts in, say, a line, they were distracted from one property to another and set down a few blue rectangles, then a yellow one, then two yellow arcs, and finally a red and a blue arc. They had begun to sort the cutouts but were still easily distracted. Only at about age 7, Piaget found, could the children carry out their plan to arrange all the cutouts by shape, by color, or by both.

Today in the United States, where many youngsters watch "Sesame Street" and go to nursery school, preschool children are more likely to know something about classification. In one study (Denney, 1972), for example, about half of the 2-year-olds could sort cardboard cutouts of four different shapes, two sizes, and four colors. Two-thirds of the 4-year-olds sorted the cutouts correctly or nearly so, according to two of the three properties. In another study (Rosch et al., 1976), children were asked to sort pictures of familiar pieces of clothing, furniture, vehicles, and people's faces into categories. One kind of category was described in terms of the basic, functional properties of the items, such as "things people put on their bodies." The other kind of category was described in more abstract, general terms, such as "clothing." Preschool children were able to sort the objects into the basic, functional categories. But only half of the kindergartners could sort the objects into general categories. Even preschoolers, it appears, can classify

People with organic brain disorders do not necessarily display all five symptoms, and some symptoms may be more marked than others. People with reversible symptoms are said to suffer from **acute brain dysfunction**. People with irreversible symptoms suffer from **chronic brain dysfunction**.

Acute brain dysfunction

People who suffer from acute brain dysfunction may be confused, in a stupor, have fluctuating levels of awareness, or delirious (Butler and Lewis, 1982). Some 10 to 20 percent of these people suffer **delirium**, a disorder that may appear swiftly or slowly and that disturbs brain metabolism. Delirium usually has more than a single cause (Sloane, 1980). People with acute brain dysfunctions may show some or all of the organic symptoms listed before as well as hallucinations, delusions, fever, tremors, rapid heartbeat, abnormal brain waves, sweating, flushed face, dilated pupils, and high blood pressure.

Just as the symptoms of acute brain dysfunction are variable, so are its causes. The possible causes include medicine, malnutrition, brain tumor, liver and cardiovascular disease, stroke, fever, emphysema, and acute alcoholism. When diagnosed correctly and when the underlying cause is treated, acute brain dysfunction lasts less than a week. Treatment may consist of administering oxygen, blood, blood sugar, or fluids or treating infection or body temperature. Without proper diagnosis and treatment, people may develop chronic brain dysfunction or die. Nearly half of all patients with acute brain objects—but only if they are familiar enough and are described in simple, specific enough terms.

Adults over 60 have been found to classify objects according to function more often than younger adults (Cicirelli, 1976). Shown pictures of 50 common objects and asked to sort them, adults younger than 60 sorted them into abstract, thematic categories such as "appliances" or "kitchen utensils." Older adults sorted the objects into taxonomic, functional categories, classifying the frying pan with the stove and the match with the pipe. Among adults in their sixties, fewer than 15 percent of the classifications were functional; among those in their eighties, 30 percent were functional. In this study, educational level did not correlate with differences in classification strategies. But when other researchers compared college students with alumni in their seventies, people in both groups used the same classification strategies (Laurence and Arrowood, 1982). However, older veterans hospital patients used significantly fewer abstract classifications than the other two groups, perhaps because their neurological, intellectual, and social competence all had declined.

Cross-cultural research has shown a correlation between educational level and the use of abstract classifications. In the 1920s, Soviet psychologist Alexander Luria asked peasants from a remote region to sort common objects. Eighty percent of the peasants, who had had little formal education, sorted the objects according to function: log with saw, pail (for watering) with horse, and the like. In contrast, younger people who had been to school for several years sorted the objects abstractly into "tools," "animals," and the like.

In our society, why do older adults increasingly group objects by function? The reason is probably not cognitive decline. It simply may be a cohort difference and not age decline, or it may be related to something about everyday situations. In most everyday situations, objects' functions and not their abstract properties are what count. Thus when older adults classify objects functionally, their behavior may be an adaptive demonstration of intelligence (Kogan, 1982).

dysfunction die, either from the underlying cause or from exhaustion (Butler and Lewis, 1982). More than half recover completely. Those with the best chances of recovering have suffered from social or environmental problems—for example, isolation leading to malnutrition or alcoholism—that can be reversed. Those whose symptoms have arisen from drug toxicity also may recover.

Chronic brain dysfunction

Chronic brain dysfunction severely impairs the cognitive functioning of about 5 percent of those over 65 and mildly to moderately impairs the cognitive functioning of another 10 percent (Coyle, Price, and DeLong, 1983). However, these statistics are somewhat misleading, because chronic brain dysfunction usually develops among those in their seventies and eighties.

In some cases, people who seem to be suffering from an untreatable, chronic brain dysfunction actually have a treatable condition. In one group of 60 patients thought to have chronic brain dysfunction, nearly one-third improved when treated for chronic drug toxicity, liver failure, or thyroid condition (Freemon, 1976).

Alzheimer's disease

About half of the older adults with chronic brain dysfunction suffer from Alzheimer's disease, a condition that causes severe cognitive disability (U.S.

Public Health Service, 1980). Named after the German neurologist who identified the physical changes in the brain that comprise the disease, Alzheimer's is often used to describe the degenerative dementia that afflicts both middle-aged and older adults (Bartus et al., 1982; Coyle, Price, and DeLong, 1983). Alzheimer's disease is more common among women than men and has a hereditary component. A genetic marker for some cases of Alzheimer's has been located on chromosome 21. Close relatives of Alzheimer's patients are somewhat more likely than others to develop the disease. Research into the cause of Alzheimer's disease presently takes several different approaches. Some researchers believe that the cause is a slow virus.

It is known for certain that Alzheimer's patients lack the neurons that supply a critical neurotransmitter, acetylcholine (Coyle, Price, and DeLong, 1983). When the brains of Alzheimer's patients are autopsied, they show acetylcholine levels that are 60 to 90 percent lower than normal. The neurons that supply acetylcholine are gradually destroyed, eventually clumping into tangles. The brains of Alzheimer's patients are abnormally shrunken and show great atrophy of the dendrites, senile plaques, and vacuoles. When acetylcholine does not enter neurons, cognitive functioning is disrupted, learning and memory are impaired, and amnesia may follow (Longo, 1966; Olton and Feustle, 1981). Alzheimer's disease is diagnosed by a process of eliminating other possible causes of brain dysfunction. Patients undergo thorough physical, neurological, and psychiatric examination. They may have CAT scans, EEGs, spinal taps, and comprehensive blood tests. Psychological tests can help to rule out depression and other functional disorders. PET scans help in the process of diagnosis, because the brains of people with Alzheimer's use glucose more slowly than normal (Roach, 1983).

Alzheimer's disease has a slow onset. In its early stages, family members may not realize than anything is wrong. They may misinterpret symptoms of the disease.

> As I look back on it, I realize that some things my wife did in the early stages were symptoms of the disease. At the time, they baffled me. She'd forget things that were so obvious that I thought she was being sneaky. I worried that maybe she had a drinking problem that I didn't know about or that she was cheating about her tennis scores. We didn't know at the time that her memory was failing.

At first, people with Alzheimer's disease have trouble with short-term memory. They have trouble writing checks, adding up a restaurant check, reading

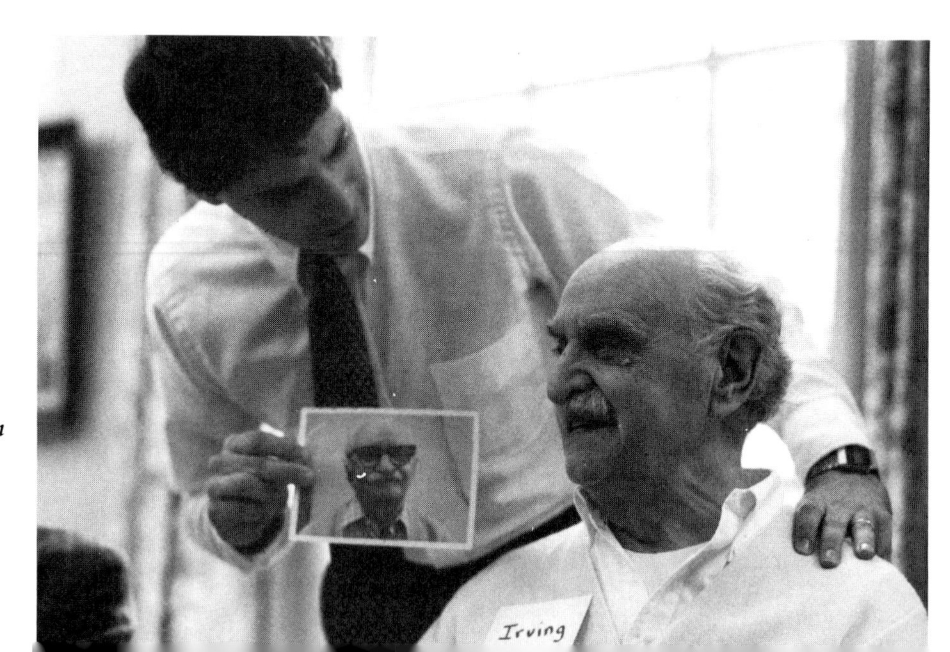

This man, who has Alzheimer's disease, wears a name tag and is shown a photograph of his face to prompt his memory.

the dial of a watch. Recall memory fails before recognition. It can help Alzheimer's patients if family members substitute recognition for recall by labeling things around the house—"spoons," "your glasses," "Turn off the oven." Higher functions deteriorate first. As the disease progresses, people are increasingly disabled. They cannot care for themselves and require skilled nursing. The disease cuts life expectancy in half. Some people live up to ten years after the appearance of symptoms or about five years after diagnosis (Zarit, 1980).

Presently, there is no cure for Alzheimer's disease. Treatment consists of keeping patients functioning and comfortable as long as possible. Most researchers working on developing a cure rely on the knowledge that acetylcholine is disrupted. Another clue has been provided by the fact that in the brains of younger adults with Alzheimer's disease, those sometimes diagnosed as having *presenile* dementia, the area of the cortex that supplies the neurotransmitter noradrenaline degenerates. But the same degree of degeneration is not found in the brains of those with later onset *senile* dementia (Coyle, Price, and DeLong, 1983).

Multi-infarct dementia

When blood clots repeatedly cut off the blood supply to the brain, a person is diagnosed as having **multi-infarct dementia**. Because some people with multi-infarct dementia hallucinate, they may be misdiagnosed as paranoic when in fact they have a functional brain disorder (Butler and Lewis, 1980). Multi-infarct dementia accounts for about 20 percent of the cases of chronic brain dysfunction. It is also present in about 10 percent of Alzheimer's patients. More men than women suffer from it, perhaps because women are protected by the hormone estrogen from developing cardiovascular disease, a contributing factor in multi-infarct dementia. Most people have the first symptoms of multi-infarct dementia in their mid-sixties, although some have symptoms as early as age 50. Early symptoms include dizziness or headaches; about half of the people experience a sudden episode of confusion (Butler and Lewis, 1982). Some people hallucinate or grow delirious. Some have spotty memory loss. Memory and lucidity may return, and only sensitive tests will show cognitive impairment (Sloane, 1980). Insight remains essentially intact until the disease has progressed. Treatment of multi-infarct dementia consists of treating the underlying hypertension and cardiovascular disease. Patients may live for 15 years or longer. Most die from stroke, heart disease, or pneumonia.

Parkinson's disease

> One night I tried to stand up from my chair in the living room. But my body was frozen. I couldn't make my arms and legs work. I thought it was strange, but I didn't see a doctor until it happened again. When I told him what was happening, he examined me and diagnosed Parkinson's disease.—Ted's grandmother, age 71

Found primarily among adults in their sixties and older, **Parkinson's disease** begins with slowed movement, stooping, and a shortened stride. Twice as many men as women are afflicted with Parkinson's disease. The facial expression of emotion is impaired, and the voice grows toneless. Tremors appear that affect the function of fingers and arms, eyelids, and tongue. Some Parkinson's sufferers can temporarily still their tremors (Bootzin and Acocella, 1980). Parkinson's disease also affects people's abilities to concentrate and remember. Some patients become severely depressed.

Parkinson's disease is characterized by a deficiency of the neurotransmitter dopamine. The brains of Parkinson's victims show a loss of dopamine-producing cells in the substantia nigra (Wyatt and Young, 1983). Some patients are treated with a drug that the body uses to make dopamine, Levadopa (L-dopa). This drug reduces or eliminates symptoms and allows some people to function relatively well for a time.

Possible growth in cognition

On the one hand, test results show that older adults' problem-solving abilities, fluid intelligence, and other forms of reasoning decline. On the other hand, many older adults function extremely well in the everyday world. Although the experimental evidence would lead one to expect otherwise, older adults work as judges and legislators, executives and managers, teachers and writers, scientists and artists. It may be more useful to think of intelligence in adulthood, especially late adulthood, as something besides what tests measure. Some people remain creative, and others develop great wisdom in their old age. Jean Piaget and Sigmund Freud, Chagall and Picasso, Vladimir Horowitz and Igor Stravinsky, Ronald Reagan and Charles De Gaulle, and many others have been productive in very old age. Let us look at creativity and wisdom specifically.

Creativity

Creativity, as we mentioned in Chapter 12, is an aspect of intelligence, the novel combining of elements, a divergent leap of thought. Although creativity requires a certain level of intelligence, intelligence alone does not guarantee its presence. Researchers who study creativity in adults may focus on the products of creative minds, or they may focus on the personality—the thought, motivation, and personality of creative people. They may also focus on the creative process and use tests to reveal the originality, fluency, flexibility, and unusual responses of people across the lifespan (Kogan, 1973).

From research that focuses on people's creative output, it appears that some people reach a creative height early in middle age and then decline. One study of scientists showed that their creative contributions peaked during the late thirties. But this research was restricted to only some kinds of creative achievements. When total creative output is considered, people in some fields show no decline in creativity. For example, among creative people who lived to the age of 80 or beyond, productivity remained through middle age in most professions and well into old age (Dennis, 1968). Chamber-music composers reached their peak of productivity in their thirties, whereas architects, playwrights, poets, opera composers, and most scientists reached their peak in their forties. Novelists were most productive in their fifties, historians, philosophers, and inventors in their sixites. Mathematicians were productive from age 30 to 69. Scholars remained productive into their seventies. Scientists declined in productivity, and those in the arts showed the sharpest declines of all. Another analysis (Simonton, 1984) of the relation between age and productive output suggests that although most people reach their peak of productivity in their forties, it is a waning of enthusiasm that accounts for the decline in productivity, and the quality of output remains essentially stable throughout adulthood. Some old people say that the passions and enthusiasm they felt as younger people continue to inspire them in their work. People who love what they do, be that playing cello, reading novels, or running marathons, are likely to pursue their passions as long as they are able. Thus for some people, old age is a period of expansion.

But what about the creative process itself? Research shows that people usually show a decline in divergent thinking, an aspect of intelligence which has been linked to creative thought. A study of teachers from Southern Cali-

Creativity is a lifelong cognitive process. Pablo Picasso, shown here when he was a mere 80 years old, remained productive into his nineties.

fornia between the ages of 20 and 83 showed that some teachers showed a lessened interest in complexity and reduced divergent thinking, phenomena that the researchers suggest correspond to an age-related decline in creativity (Alpaugh and Birren 1977).

Some researchers have studied the creative process by comparing highly educated, active women between the ages of 25 and 74 who had never been highly creative with professional artists and writers of comparable ages (Crosson and Robertson-Tchabo, 1983). Among the noncreative women, test results showed that the older women showed a declining preference for complexity, much as Alpaugh and Birren had found. But no such age difference appeared among the creative women. Cohort effects did not show up among the creative women either. The researchers interpreted these data to suggest that traits, preferences, and abilities that people consider important will tend to be retained into old age. They also suggest that the continuous exercise of skills means that they are unlikely to decline with age.

James Birren (1985) has suggested that older people must remain in touch with their inspiring passions and keep them up to date so that they continue to spark their divergent thinking. But older people can do this only if they can look back on important issues and resolve important conflicts from their past. When old age is fraught with unresolved conflicts, inspiration dies, divergent thought declines, and creativity withers. When old age is a time for cognitive reorganization, for new understanding of the self and the world, creativity has room to flourish.

Wisdom

People in many cultures regard wisdom as a venerable characteristic that ripens with age. But cultural definitions of wisdom and the path that leads to it vary widely (Clayton and Birren, 1980). For example, in western cultures, wisdom is considered a mixture of intellect, feeling, and intuition. From the Judeo-Christian tradition comes the idea that wisdom takes time to develop. Yet not all mature people are wise. Wisdom is reached through learning from one's parents, formal schooling, or divine gift. The emphasis on what comprises wisdom may differ from one cultural tradition to another, but most

lifespan focus

Lifelong creativity

What really happens to people's creativity over the course of the lifespan? On the one hand, there are the poets who create in a white heat during their young adulthood and whose output then cools dramatically. On the other hand, there are the scholars and musicians who are still productive in their eighties and nineties. On the one hand, the great physicist Sir Isaac Newton said that he had his best creative ideas when he was in his twenties. On the other hand, the cellist Pablo Casals and the artist Grandma Moses were both going strong in their nineties.

To try to answer this question, lifespan psychologists have studied the lifetime careers of creative people in the arts, the sciences, and in scholarship. They have found, first of all, that from the time a person begins his or her creative career, productivity increases rapidly, levels off at a peak productive age, and then slowly declines. Dean Simonton (1983) suggests that creativity is a two-step process. The first step consists of the formation of the creative person's ideas; the second consists of the translation of these creative ideas into paintings, poems, scholarly articles, scientific investigations, or the other established forms of communication within a discipline. The important point is that there often is a lag between the two steps. Although people have their greatest number of creative ideas—step 1—in their early twenties, they may not translate them into creative works—step 2—for another twenty years.

In each field, the lag between the creative idea and its realization tends to differ. The lag is longer in scholarship than in the sciences, for example. Scholars may have 35 years and artists 14 years between the peak of their inspirations and the peak of their creative output. In contrast, mathematicians and poets are likely to have an especially short lag—only about 8 years—between their peaks of inspiration and productivity. But in no field does creative potential ever get used up altogether. Even the 90-year-old creator has a store of new inspirations as well as creative productions to come.

seem to agree that wisdom entails understanding life's purpose, that it takes time and study for wisdom to develop, and that wisdom is reflected in behavior (see Table 21.1)

Erik Erikson (1982) described wisdom as a knowledgeable and detached concern with life in the face of death. He has suggested that old people grow wise as they resolve the conflict between despair over impending death and integrity over a meaningful life. Lawrence Kohlberg (1973) suggested that only when people move from the principled morality that occasionally accompanies formal thought do they begin to develop wisdom. Wisdom only develops, suggested Kohlberg, when people have lived for a long time according to these advanced moral principles.

How do Americans perceive wisdom? When well-educated men and women of various ages were asked to rate several qualities that might correspond to wisdom, people from all age groups agreed that wisdom is made up of reflectiveness, feeling, and intellectual qualities (Clayton and Birren, 1980). They also agreed that wisdom develops over time. College students and middle-aged adults thought that wisdom was more tightly bound to old age than the older adults did. The older adults thought that the old are no wiser than people of other ages. Older adults also considered empathy and understanding more important to the emergence of wisdom than age or experience.

Wisdom draws from several different domains. People who are wise are *expert* in certain forms of abstract and concrete thinking. People who have the quality of wisdom are likely to be able to penetrate the surface of a problem

Table 21.1
Wisdom

Although wisdom is a concept that most people understand, it is difficult to describe exactly what it is. Someone who is wise might be thought of as:

Having a good understanding of life and people
Understanding life's paradoxes and contradictions
Having learned much from his or her own experiences
Being very knowledgeable
Thinking clearly and carefully
Being highly observant, aware, and perceptive
Being open-minded, empathetic, and reflective
Being able to see things within a broad context
Being able to consider all points of view
Being able to consider all options in a situation
Having good common sense
Having intuition, insight, and foresight
Being able to predict how things will turn out
Being nonjudgmental about other people
Being interesting to talk with
Having worthwhile things to say
Being a source of good advice
Knowing when to give advice

SOURCES: Clayton and Birren (1980), Holliday and Chandler (1986), Meacham (1983), Sternberg (1985).

The definition of wisdom changes from one culture to another. But generally, people are considered wise when they have lived for many years and have developed a certain expertness, pragmatic understanding, and knowledge of the behavior of others.

and go to its heart. They have a *pragmatic* understanding of problems, crises, and their solutions. Generally, people are wise about only some things and some areas of knowledge. Wisdom tends to flourish within specific *contexts*. People who are wise acknowledge and deal with the inevitable *uncertainties*, ambiguities, and complexity of problems and tasks, and their thinking takes into account *relative* possibilities, alternatives, and outcomes, differences between people's levels of understanding and competence, and the like (Dittmann-Kohli and Baltes, 1984). It may be that the values of a technological society—productivity and problem solving—are inconsistent with values necessary for the development of wisdom—reflection and the search for meaning in life. If we understood more about what fosters wisdom, our society might contain more people whose wisdom enriched us all.

Summary

1. In late adulthood, there appears to be a normal decline in cognitive ability. But unless this decline is associated with disease, it is slight and affects only some individuals. Learning, thinking, and remembering are information processing skills that may decline in late adulthood. Age changes in conditioning, verbal learning, and learning certain cognitive skills all have been demonstrated experimentally.
2. Age differences in learning may be influenced by the pacing of the learning task and by the motivation, caution, and distractibility of the subject.
3. Cross-sectional IQ test results show intelligence falling with advancing age from early adulthood on. Longitudinal test results show increasing scores until age 60. Cross-sequential research shows that different generations of people have different levels of ability at the same chronological age. Cohort effects have exaggerated the decline in intellectual abilities among older adults. Even so, there are small declines in certain intellectual abilities among people in their fifties and older.
4. People only a few years or a few months from death experience a marked decline in IQ known as terminal drop.
5. Older adults generally perform more poorly than younger adults on certain problem-solving tasks, such as Piagetian tasks. The decline may result from changes in memory, speed of processing, organization of information, or other factors that affect performance.
6. Abilities that do not show up in the laboratory may well show up in real life, and abilities that seem to have been lost may quickly reappear after brief training. Intellectual competence may bear relatively little relation to people's age. Instead, the changes in people's social, economic, and intellectual situations that typically accompany retirement may cause much of the decline in intelligence test scores that crop up among some people in their late sixties and seventies.
7. In general, older adults differ from younger adults in classifying objects, forming and testing hypotheses, and solving problems. As people age, they grow less efficient at organizing the elements of a problem and at holding and manipulating more than one idea at a time.
8. A minority of older adults have organic brain disorders that interfere with their intellectual abilities. Their pathological decline in cognition may be extreme. People whose symptoms are reversible suffer from

acute brain dysfunction. Common chronic, irreversible brain disorders include Alzheimer's disease, Parkinson's disease, and multi-infarct dementia.

9. Creativity is one aspect of intelligence that may remain vital into very old age. Wisdom, which is made up of feeling, understanding, and intuition, also may flourish in people of advanced age.

Key terms

terminal drop
organic brain disorders
acute brain dysfunction
chronic brain dysfunction
delirium
multi-infarct dementia
Parkinson's disease

Suggested readings

BUTLER, ROBERT, and LEWIS, M.I., *Aging and Mental Health*. St Louis: Mosby, 1983. A discussion of mental health and aging, including a consideration of functional and organic brain dysfunctions.

CLAYTON, V., and BIRREN, J.E., "The Development of Wisdom across the Life Span: A Re-examination of an Ancient Topic." In P.B. Baltes and O.G. Brim (eds.), *Life Span Development and Behavior*. New York: Academic Press, 1980. The authors provide a historical perspective on the quality of wisdom and summarize recent work that defines this elusive concept.

SIMONTON, D.K., *Genius, Creativity, and Leadership*. Cambridge, Mass.: Harvard University Press, 1984. An examination of historical data for evidence of factors contributing to genius, creativity, and leadership and a discussion of what these data indicate about the relationship of age to genius, creativity, and leadership.

ZARIT, STEPHEN. *Aging and Mental Disorders*. New York: Free Press, 1980. A discussion of the mental disorders associated with aging, including cognitive dysfunctions.

chapter twenty-two
Social development

Personality and emotional development

What happens to personality as adults get older? Do they change markedly, or do they stay much the same? This is a central area of disagreement among researchers who study late adulthood. But in this as in other areas, the nature of the findings tends to vary with the method of the research. Cross-sectional studies often tell a different story from longitudinal studies about the development of personality late in the lifespan.

Stability and change in personality

> I've always loved being with people, ever since I was a girl. After I was widowed, I was doing fine on my own, and I wondered whether I could mesh with someone else's habits at such a late date. But I wanted the companionship, so I remarried when I was 70.—Sam's great grandmother

Some researchers believe that people feel, think, and react fairly consistently over the course of adulthood because their personality traits are fairly stable (Epstein, 1980). Thus the young adult who was sociable at parties also is likely to be sociable and to seek people out in his old age. The woman who was apprehensive before getting married at age 40 is also likely to be apprehensive before she retires in her seventies. No matter how personality is tested, lifespan psychologists find that it remains quite stable during adulthood (McCrae and Costa, 1982; Mortimer, Finch, and Kumka, 1982). For example, longitudinal studies have shown that among West Germans between the ages of 60 and 79, level of activity and attitudes toward life remained stable, and there was no increase in rigidity (Thomae, 1980). Thus there is no such thing as an "old" personality. Instead, people's personality traits give clues about their possible *ranges* of responses. They do not allow others to predict with certainty what anyone will do in a specific situation.

Studies of **normative stability** show that personality is stable during adulthood. Normative stability describes the tendency for an individual's personality to remain stable in relation to members of his or her age-group even if it changes in relation to the population as a whole. Thus the 30-year-old who is more outgoing than other 30-year-olds is likely to be more outgoing at 75 than others her age, although she may not be as sociable as she was at age 40 or as other 40-year-olds.

When people's lives are continuous and essentially unchanging, their self-concepts and, therefore, their personalities are not likely to change. When people's personalities do change, the change seems to be a response to life events rather than to aging itself (Moss and Susman, 1980).

> After my father died, my mother's personality seemed to take on lots of sharp edges. She was irritable and bossy. She wanted endless attention from us kids. But she was so clearly at loose ends, and lonely, and miserable that it was hard to stay angry with her for long. Then, about two years after he died, she brightened up. Her old personality emerged, although she seemed a little less exuberant than she had been. It's been nice to see her get back on her feet again.—Natalie's mother

A sudden crisis—a spouse or child becoming ill or dying—may change an adult's personality. Among the majority of people—those who remain healthy and keep up community ties—personality does not change markedly from young to late adulthood (Thomae, 1980).

When psychologists study the development of adults' personalities over time, they stress the difference between the stereotype and the reality. Different groups of people hold different stereotypes of old age. For example,

American graduate students hold the stereotype of old age as an unpleasant mixture of stubbornness, touchiness, bossiness, and complaining. Some German people hold the stereotype that older adults are inactive and withdrawn, and textbooks in German elementary schools show old people as dependent, incompetent, and passive. Some psychiatrists hold the stereotype of old people as rigid, extreme, and irritable (Thomae, 1980). Stereotyped ideas probably are generalizations based on isolated cases.

But stereotypes themselves have the power to create expectations, and some old people meet these expectations. Although most old people do not fit the stereotype, some may feel pressured into acting out the ''old'' role. Also, some older adults *believe* that their personalities have changed more than they actually have. But longitudinal studies suggest that older adults' memories of what their personalities were like are hazy. Self-reports of personality traits show that middle-aged adults do not recall accurately what they were like as adolescents. They recall themselves more harshly than they did in their adolescent self-reports (Woodruff and Birren, 1972).

To assess the stability of personality during adulthood, Paul Costa and Robert McCrae (1980) evolved three dimensions of personality out of a study of 2000 adult men. The first dimension is *neuroticism*. It consists of anxiety, depression, self-consciousness, vulnerability, impulsiveness, and hostility. The second dimension is *extraversion–introversion*. It consists of attachment, assertiveness, outgoingess, excitement seeking, positive feelings, and activity. The third dimension is *openness*. It consists of ideas, feelings, fantasy, esthetics, actions, and values.

Costa and McCrae investigated how these personality dimensions change over time. Over a period of ten years, no changes appeared in men's levels of neuroticism. Men high in neuroticism tended to complain about their health, to smoke heavily, to have problems with drinking, sex, and finances, and to feel dissatisfied with life. Other longitudinal studies also have shown stability in the behavior associated with neuroticism among adult men and women (Douglas and Arenberg, 1978; Siegler, George, and Okun, 1979).

The behavior associated with extraversion also was quite stable over a ten-year period (see Chapter 13, ''Continuity of Personality''). Men who were high in extraversion tended to value power and humanitarian concerns and were happier and more satisfied with life. They chose jobs that involved working with people. Over the ten-year period the only change toward greater introversion among these men was a slight decline in independence. But other longitudinal studies have painted a less clear-cut picture. One study showed no changes in extraversion in either men or women (Siegler, George, and

People's personalities tend to be fairly stable during adulthood. The person who was shy at 40 is likely to be shy at 75, and the person who was outgoing at 40 is likely to be outgoing at 75, too.

Okun, 1979). Another study showed that over a 30-year period between middle and old age, men grew somewhat more introverted (Leon et al., in press).

Studies of openness showed no long-term changes but some cross-sectional differences. Men who were very open tended to score high on measures of esthetic and theoretical values, low on measures of religious and economic values. Highly open men were more likely to work as psychologists, psychiatrists, and ministers than as bankers, veterinarians, or undertakers. They tended to have IQ scores above average, to change jobs often, and to get themselves involved in lawsuits. Their lives were uneventful, and they experienced both pleasant and unpleasant events intensely. Other longitudinal studies have shown no changes in openness among men or women (Siegler, George, and Okun, 1979).

> I don't care so much about going out as I used to. I have my television programs. I used to love going to the movies and restaurants, but I don't care for that any more.—Kristy and Jason's grandmother at 77

Some researchers, in contrast, have found consistent age-related changes in introversion in late adulthood. They have found that as people get older, they generally turn more inward and focus increasingly on inner concerns. As they turn inward, they also tend to become more cautious and conforming (Reedy, 1983). They may gradually cut many of their social ties. For example, they may retire from work and give up active control of their children. Then, in the social roles that they do maintain, they may become more passive and marginal.

Generally, results from cross-sectional studies have not agreed on other age-related changes in adult personality. One study may show an age-related change in some trait, but then another study comes along and shows no change in the trait. One sequential-cohort study, which focused on a period of seven years during adults' lives, showed that although most personality traits were stable, when people did change, it was in the direction of greater excitability and humanitarian concern (Schaie and Parham, 1976). One woman reported that her seventies were fairly serene but that her eighties, to her surprise, were passionate, intense, and bursting with "hot conviction" (Scott-Maxwell, 1979).

People's self-concepts, their organized, coherent, and integrated pattern of beliefs about themselves, include both self-esteem and self-image. In a longitudinal study of younger adult men, self-image was found to have four dimensions: feelings of well-being, activity, interpersonal qualities, and unconventionality. The fact that several of these qualities resemble McCrae and Costa's (1983) three personality dimensions suggests that self-concept and personality are related.

Among West German men and women in their late sixties, self-concept remained positive (Bergler, 1968). They saw themselves as active, self-controlled, and competent. People with a positive self-concept are likely to incorporate some of those positive feelings into their concepts of how older people act. Among older people who deny to themselves that they are old—70-year-olds who see themselves as middle aged—this positive self-image is likely to be reflected in better adjustment and higher morale than in people who see themselves as old (Turner, 1979). Older people who think of themselves as middle aged are also likely to be in better health, more active, feel better about their work, and not to have lost a spouse or been seriously ill. Life events seem to shape self-concept just as they shape other aspects of personality.

Because self-concept is vulnerable to life events, aging itself has no one clear effect on it (Thomae, 1980). Health, finances, social involvement, social class, sex, housing conditions, and marriage—in other words, social, economic, and biological factors—all affect people's self-concepts. Widowhood, as we shall see later in this chapter, may have a powerful effect on psychological development during late adulthood.

Satisfaction with life

If, as some research suggests, adults' personalities are essentially stable, so long as the events in their lives are stable, too, it stands to reason that their feelings of satisfaction with life would remain stable over time as well. Do the data bear out this assumption? Yes. One investigator who studied over 6000 people between the ages of 4 and 99, at home, at school, at work, and at play, concluded that satisfaction is the same at all ages (Cameron, 1975). He measured satisfaction in terms of expressed happiness, sadness, and neutral moods. Mood, he found, was determined more by social class, sex, and factors in the immediate surroundings than by age.

Similar conclusions emerged from a review of research into older adults' feelings of well-being and satisfaction (Larson, 1978). Older adults' responses to questions like "I am just as happy now as when I was younger" and "I sometimes feel that life isn't worth living" showed that age had little bearing on feelings of satisfaction. Health was the most important factor. Money, social class, social interaction, marital status, housing, and even transportation all were factors in older adults' feelings of satisfaction or dissatisfaction with their lives. When other researchers studied the moods of older adults for 18 months, they found no age-related changes in happiness (Kozma and Stones, 1983). But again, factors such as health, housing, activity, the events of life, and marriage all influenced older people's feelings of happiness. For older adults who lived in cities, happiness with housing arrangements was the most significant factor. For older adults who lived in rural areas, health was the most significant factor in their feelings of overall happiness. The Duke Longitudinal Study showed that satisfaction with life was associated with high levels of activity. Compared to adults who had disengaged from life, more adults who were active and engaged reported feeling more satisfied than average (Maddox, 1970).

Not only life events, but personality traits themselves influence older adults' feelings of satisfaction. For example, men in the McCrae and Costa (1983) study of personality dimensions who were high in neuroticism were more likely than others to feel dissatisfied and unhappy. Men who were high

Research suggests that older people's moods and satisfaction with life are determined less by age than by social class, sex, and aspects of the immediate situation.

in extraversion were more likely to feel satisfied and happy. These associations held true no matter how psychologically mature the men were rated to be.

Does the essential nature of satisfaction itself change over time? Does the 75-year-old who says "I'm happy" mean the same thing as the 35-year-old? One researcher suggests that older adults' feelings are as intense as ever, although the quality of those feelings may change (Schulz, 1982). Feelings in later years may be more ambivalent. Love may feel more bittersweet the fourth time around. Feelings also may last longer among older adults, in part because physical regulatory systems respond more slowly with age. Compared to adolescents and young adults, older adults may go through fewer mood changes in a given day.

Emotional development in late adulthood

In earlier chapters, we have traced the psychosocial crises that, according to Erikson, confront individuals at every stage of life, from infancy through adulthood. The last achievement in this cycle of lifelong emotional development is a sense of *ego integrity*, a sense that one's life has been whole and meaningful.

> I'm an old man now, and I won't be around forever. Sure I'd have done some things differently if I could have. Who wouldn't? But I've had a wonderful life, and I tell my kids to remember that when I'm gone. Life's blessings, that's what they should remember.—Robbie's great grandfather, age 91

Older adults who successfully resolve the conflict between feelings of integrity and *despair* emerge with a sense of wisdom, which is one of the true strengths of old age. Older adults may despair over the courses their lives have taken, they may fear death, and they may wish they could turn back the clock and do things differently. Some despair is a measure of health; pervasive despair is not. Out of the psychosocial identity that formed during adolescence, the older adult has the chance to build an existential identity in which he or she squarely faces the prospect of death, understands and accepts its inevitability, and goes on living fully even so (Erikson and Erikson, 1981).

Family and personal relationships

The personal relationships special to late adulthood are a mixture of the bitter—widowhood—and the sweet—spouses, children and grandchildren, brothers and sisters, and friends.

As we saw in Chapter 19, the traditional pattern is for satisfaction with marriage to remain high in the early years, to drop when children are born, and to rise when the children leave home. But are older married couples more satisfied with their marriages *because* their children have left home? The answer: probably not. Older married couples without children report the same feelings as older married couples with grown children: feelings of satisfaction, harmony, and absence of stress (Troll, Miller, and Atchley, 1979). Older adults who report greater than average satisfaction with their marriages are likely to have an active sexual relationship, engaging in sex more than once a week (Busse and Eisdorfer, 1970). The partners are likely to have similar levels of intelligence, and the husband is likely to be older than the wife. Many older adults feel that their happy marriage is the emotional center of their lives. It brings comfort, support, and intimacy that increases over time. Happy marriages among older adults also are likely to be egalitarian and cooperative, and traditional gender roles and power relationships are likely to be transcended (Troll, Miller, and Atchley, 1979).

Widowhood

> I've had to adjust to this awful business of being a widow. First I had to adjust to the loneliness and to missing my husband. Since then I've had to get used to socializing with other widows. I had a happy marriage. Some people say that makes being a widow harder. Some say it makes it easier. I don't really know. But you do have to go on.—Natalie's grandmother

More women are widowed than men. In this country, there are 11 million widows and 2 million widowers (U.S. Bureau of the Census, 1982; see Table 22.1). Of those who lose a spouse between the ages of 50 and 70, only 5 percent of women but most men remarry (Troll, Miller, and Atchley, 1979)

The mourning process (discussed in detail in Chapter 23) is a period of bereavement and, eventually, of slow and painful adjustment to a new role. Several different factors affect how severely widowhood disrupts the survivor's life (Lopata, 1975). These factors include the survivor's earlier dependence on the spouse who has died; the couple's earlier involvement in family, community, and work; the survivor's economic and social resources; and the survivor's acceptance of widowhood or attempts to move out of it.

> After my grandmother died, my grandfather was a lost soul. After the funeral, someone asked him whether he had food for breakfast the next day at his house. He wasn't sure, he said. My aunt told him how to make orange juice, tea, and toast. My poor grandfather seemed overwhelmed.—Nancy's mother

The little research available on widowers suggests that they have more trouble adjusting than widows do (Bernardo, 1970). Widowers may be more lonely and isolated than widows (Lopata, 1980). They are likely to have depended on their wives to keep ties to family and friends. They are not likely to have a close friend outside of their marriage. They are suddenly faced with keeping house, facing the many household chores that they have never mastered.

Like so many life events, if widowhood is perceived as "on time," it is much easier for the widow to adjust to. Many young widows, for example, have a much more difficult adjustment than older widows. Older widows, especially those who have seen a spouse through a final illness, have prepared themselves somewhat for widowhood. They are likely to have rehearsed the psychological loss (Treas, 1983).

Most of what we know about widows applies to older women who have had marriages with traditional gender roles. Many such widows devote them-

Table 22.1
Widows

	Age				
	35–44 Years	**45–54 Years**	**55–64 Years**	**65–74 Years**	**>75 Years**
Men	0.5%	1.4%	4.1%	8.2%	22.1%
Women	2.2	6.8	18.4	40.1	68.2

SOURCE: U.S. Bureau of the Census, *Statistical Abstract of the United States*, 103d ed., 1982.

Friends, brothers, and sisters may become increasingly important to one another during late adulthood and may act as substitute parents and spouses.

selves to their families. But when women who are now young and middle-aged working adults are widowed, the pattern is likely to change. Many of these women will be widows who are high school and college educated, with professional and technical backgrounds (Neugarten and Brown-Rezanka, 1978). Most of these women will have been accustomed to working, to handling finances, and will have had fewer children than earlier generations of widows. Thus they will be equipped with work identities in addition to their identities as wives and mothers. Therefore they also may be better equipped to handle the decade or more of widowhood that the statistics say they can expect during late adulthood.

Brothers, sisters, and friends

> My sister and I have grown very close over the last few years. She lost her husband a few years ago. When I'm widowed, we may live together.—Sam's great grandmother

Among older adults, relationships with brothers and sisters may deepen, and feelings of closeness may intensify—even though overt communication such as visits, letters, and phone calls may become less frequent. Among adults who have never married or who are widowed, feelings toward and contact with siblings may be especially intense (Troll, Miller, and Atchley, 1979). Whereas brothers and sisters may need little help from each other in early adulthood, and help only during crises during middle adulthood, they may offer a reliable source of help in late adulthood and become substitute parents or spouses (Cicirelli, 1982).

Friendships in late adulthood seem to depend on people's other roles in life. Thus friends tend to be from similar social classes and to be of the same sex (Riley and Foner, 1968), but older adults who are married see more of their friends than do older adults who are widowed or single. Older adults without children also see more of their friends than do older adults with children. Older adults who have retired are less likely to make new friends than those who are still working.

Being grandparents

Many grandparents today are middle-aged, not older adults. Several generations in which people had a few children, closely spaced in age, relatively early in their lives has produced a large crop of grandparents in their forties and fifties. Many of these grandparents have no children of their own still at home. Among earlier cohorts of grandparents, larger families were common, and many of the oldest children left home and produced grandchildren who were about the same ages as their aunts and uncles.

The longer life expectancies of older adults also means that now four, and sometimes even five, generations may be alive within a single family. Grandparents thus are not always the oldest living generation within a family. Today about 40 percent of older adults are great grandparents (Troll, Miller, and Atchley, 1979). It is quite possible that soon we will see families with living great-great grandparents. But it is still too early to predict that trend. The many women today who are waiting to have children until their thirties are effectively postponing the age of grandparenthood for their parents and, eventually, for themselves as well.

Divorce also may be changing the experience of grandparenthood. Not surprisingly, the results of one study of divorce showed that grandparents whose children had custody of the grandchildren saw more of their grandchildren than did grandparents whose children did not have custody (Matthews and Sprey, 1984). The fact that most young adults remarry within three

lifespan focus Never-married older adults

Most older adults are married or have been married at some time in their lives. Only about 5 percent of older adults have never married, but that percentage translates to over one million people in the United States. Research shows that compared to others, married people generally are more satisfied with their lives and have better health in old age. But what is the truth about older adults who have never married? Are they unhappy, isolated, maladjusted—as stereotypes suggest? Is there a true "social type" of never-married older adults, and if so, what is it?

To counter the myths and learn the answers to these questions, Robert Rubinstein (1987) studied data on never-married older adults. What he found in many cases did not confirm the stereotyped view. First, the data did not support the view that never-married older adults were isolated. For example, most had lived with relatives—parents, brothers and sisters—as adults. Many reported having close friends. Some of the men had dated and, having outlived their parents and others and having passed by various problems, felt free to think about marrying for the first time in their lives. But although few were isolated, many never-married adults reported that they felt lonely some of the time.

Second, people who have never married do not, of course, suffer the pain of losing a husband or wife. But because many have lived with family members, they are likely to suffer deeply the pain of losing a parent or sibling or the pain of family problems. For example, one 72-year-old man had lived with his parents as an adult. When they divorced, he lived with his father until his late middle age. This man reported still feeling deep emotional pain from his parents' divorce.

The patterns of relationships among the never-married men and women were too varied and individualistic for any "social type" to emerge. Like others their age, people who have never married are affected by many different social and cultural factors as well as by their personalities. Some are homosexual. Some have lived in places where there were no available marriage partners. Some have chosen not to marry in order to pursue a career or to fulfill family responsibilities such as caring for aging parents. Most have been influenced by more than one of these factors. In short, the individual differences among never-married adults are no less pronounced than they are among other older adults—which is pronounced indeed.

years of a divorce means that many young children today have four sets of grandparents—their own two sets of biological grandparents, the parents of their stepparent, and the parents of their noncustodial parent's new spouse. It is likely that all of these grandparents touch their grandchildren's lives to a greater or lesser extent. No one yet knows how this complicated abundance of grandparents will affect relationships between grandchildren and grandparents (Cherlin, 1983).

Most grandparents take pleasure in the role, although about one-third of those in one classic study (Neugarten and Weinstein, 1964) said that they were disappointed in some way. Many grandmothers say that they are not eager to babysit and resent it when they are asked (Cohler and Grunebaum, 1981). Most of the grandparents studied between 50 and 79 enjoyed the role. Most of those over 80 and those in their forties felt neutral (Troll, 1980a). Grandchildren give older adults opportunities to express physical affection and intimacy and even to act silly when they play together. Even though many grandparents say that they take pleasure from their grandchildren, few grandparents' feelings of satisfaction with their lives derive from their involvement with their grandchildren (Wood and Robertson, 1976).

People's satisfaction with their roles as grandparents sometimes varies according to their gender roles. In one study of white Midwestern grandparents, for example, the grandmothers expressed more satisfaction with their role than the grandfathers did (Thomas, 1986). But grandmothers and grandfathers both reported that having relatively high levels of responsibility for young grandchildren contributed to their feelings of satisfaction. The grandmothers may have felt satisfied because taking care of their grandchildren drew on their childrearing skills. The grandfathers may have felt satisfied because their grandchildren offered them a chance to be nurturant and intimate. Often when people become grandparents, they become more androgynous (Feldman, Biringen, and Nash, 1981). Grandmothers become more independent, grandfathers more tender and compassionate.

Most grandparents enjoy their role. Grandchildren offer grandfathers the chance to be nurturant and intimate, tender and compassionate.

> My parents have changed so much since becoming grandparents it's hard to believe. When we were young kids, my mother stayed home and always accepted that as her role. My father worked long hours—he'd leave at 6:30 in the morning and get home at 8 at night—to support us. When they reached their mid-sixties, my father couldn't wait to retire. He's taking art classes and jogging, and he adores babysitting for my kids. The minute my younger sister left for college, my mother took a full-time job as a bookkeeper, and even though she could retire she is horrified at the thought of it. Now she's the one who works all the time. She uses her vacations to spend extra time with her grandchildren.—Mike and Erica's mother

Because grandparents have no psychologically or socially prescribed function, they are free to relate to their grandchildren as they wish. In one early study of middle-class grandparents, the researchers described five different styles of relationships (Neugarten and Weinstein, 1964). One style was *formal*, in which grandparents did not involve themselves in childrearing but were interested in their grandchildren and gave them special treats and indulgences. Most formal grandparents were older adults, in their sixties and seventies. A second style was *fun-seeking*, in which grandparents interacted informally with their grandchildren and spent free time with them. A third style was *distant*, in which the remote grandparents were kindly, had contact with their grandchildren only on holidays and birthdays, and then dropped out of their grandchildren's everyday lives. Some grandmothers became *surrogate parents*, usually in families in which mothers worked and grandmothers took over as main caregivers to the children. A few grandfathers became *reservoirs of family wisdom*, teaching skills and dominating the parents in the family.

> My mother's parents take care of us a couple of days a week because both of our parents have to work late. My grandfather's retired, so he can pick us up at school. He drives us home and stays with us until late afternoon when my grandmother comes to our house straight from her job and fixes supper for us. They let us watch TV and give us baths. We get to stay up a little later with them—but not much on school nights.—Mike

After studying 300 grandparents, one researcher described five dimensions to the role (Kivnick, 1982). Although all of the grandparents studied knew about all of the dimensions, they varied individually in the stress they gave to each dimension. Some grandparents *spoiled* their grandchildren. Some considered their activities with their grandchildren of *central* important to their lives and enriched their identities. Some played the role of *valued elder*, and some felt a sense of personal *immortality* through their young descendants. Finally, some grandparents relived their own early experiences with their grandchildren and became *reinvolved with their personal past*.

Grandchildren usually enjoy their grandparents, and many have relationships free of the frictions that can mar the relationships between parents and children. Most grandchildren are closest to their grandparents until they are about 10 years old. Preschool-age children enjoy grandparents' special treats and presents. School-age children enjoy their grandparents' entering into their play (Kahana and Kahana, 1970). When young adult grandchildren were asked how they felt about their grandparents, nearly all said that they felt warmly toward them, liked spending time with them, and would help them. They said that they expected little of their grandparents except warm feelings (Robertson, 1976). Yet the relationship between grandparent and child leaves a lasting mark. The relationships that children develop with their grandparents affect how they relate to their own grandchildren half a century later (Kivnick, 1982).

Relations among generations

> I see my folks three or four times a week. I do lots of errands for them, and they return the favor. We get together for supper about once a week, or we go shopping or visiting together. I see my grown children several times a month, too.—Sam's grandmother, age 64

In this country, each generation within a family usually prefers to live in its own household. But relations between the generations—middle-aged children, their parents, and their children—are close and frequent enough to make the American family a **modified extended family**. In a modified extended family, the generations do not live together, but they have strong bonds of affection and aid (Litwack, 1960). Many older adults, for example, see and hear from their children often (Bengston and Treas, 1980). Results from many studies show that the majority of older adults see their children every day. Married daughters are more likely than married sons to have contact with their elderly parents. Widows are more likely to have contact than married people.

Most elderly parents and their middle-aged children say that they feel close to each other. In a study of adults from Boston, for example, the elderly parents who were healthy, were financially secure, and felt positively about aging had warm relations with their children (Johnson and Bursk, 1977). When elderly parents' and their children's reports of the degree of warmth in their relationships disagreed, usually the elderly parents thought the relationship was warmer than the children did. Similarly, other research has shown that when elderly parents and middle-aged children rated each other on trust, fairness, understanding, affection, and respect, each group gave high ratings to the other. But the elderly parents consistently rated their relationships more positively than their children did (Bengston and Treas, 1980).

Elderly parents report valuing their children's respect and affection even more than their help (Treas, 1983). Middle-aged children often help their parents. They may offer their elderly parents help in dealing with bureaucrats in government agencies, in finding out about housing, pensions, and medical care, in filling out forms, and in offering advice for problems (Sussman, 1976). Elderly parents often return the favor, offering money, babysitting services, help in preparing meals, and transportation. One study has shown that more elderly parents help their children than receive help themselves (Riley and Foner, 1968).

Middle-aged children and elderly parents often try to influence one another, too. Many studies have shown that both elderly and middle-aged parents and their children try to influence each other in matters of diet, health, politics, childrearing, and virtually every aspect of life (Hagestad, 1982). Parents' influence on their children is felt for a lifetime. In a study of three-generational families in Detroit, adults of all ages were asked to describe a man and a woman whom they knew (Troll, 1980b). Most of them, including adults in their eighties, described their own parents.

Work and leisure

Some older adults continue to work through their seventies and eighties. During the 1800s, most older adults worked as long as they were physically able. Those who lived in farming communities always had chores to do. But today, fewer than 20 percent of older men and fewer than 10 percent of older women still work. The many older adults who no longer work must adjust to a quite different life, likely to be filled with more leisure time.

Retirement

The retirement age of 65 was chosen arbitrarily by the government as the time when people could begin to collect full retirement benefits. That age gradually has come to be considered the time when people enter old age (Neugarten and Hall, 1980). The law today provides that no one can be forced to retire until age 70.

> The last few years before my husband and I retired from our high school teaching jobs went by slowly. We were so eager to get off the treadmill and start doing all the things we've always wanted to do. We travel a lot now, and I garden. George putters around the house. We feel very free now. It's lovely.—Nancy's grandmother

Some people continue to work long past the age of 65. But an increasing number of people have been retiring after they turn 62 and accepting lower retirement benefits. Many of those who retire before age 62 are either in good physical *and* financial health or in poor physical *and* financial health. Most fall into the latter category and die within a few years of retiring (McConnell, 1983). A survey of automobile industry workers showed that those who could count on a retirement income that would let them live comfortably retired in their early sixties (Barfield and Morgan, 1975). Results from longitudinal studies show that workers who retired early were likely not only to be in poor health but to feel positively about retirement itself (Palmore, George, and Fillenbaum, 1982). Some who retired early had been dissatisfied with their work, and their work had been central to their lives (McConnell, 1983). Workers who retired at the expected time were likely to have suffered from physical and psychological job stresses, to have worked under poor—noisy, smelly, extremely hot or cold—conditions, and to have been without power to make decisions (Quinn, 1978). Workers who retired late tended to be those with higher than average education and job status and those who worked for themselves.

Most people who are retired say that they want to work at least part-time. This man does volunteer work at a geriatric center.

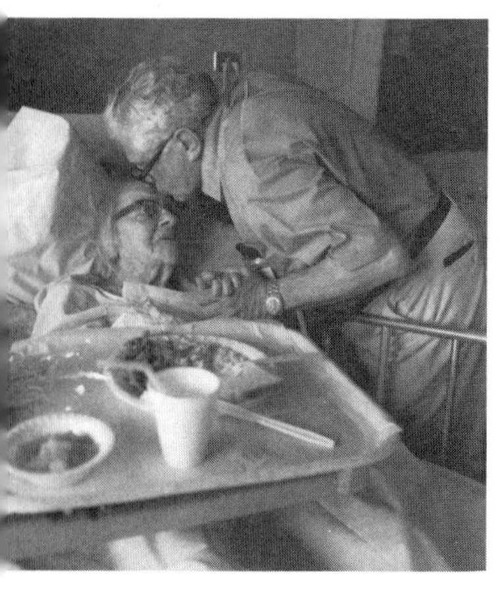

Although today many retirees continue to be paid, people did not easily accept the idea of paid retirement. For example, in a 1950 a poll of older steel workers, most said that paid retirement should be given only to disabled older workers. But ten years later, only one-fourth of the steelworkers polled felt that way (Ash, 1966). In fact, most working people today think of retirement as a just reward and a deserved rest.

> When my grandfather died, my grandmother had a small income. She was too proud to accept any money. But my mother dropped off grocer-

ies at her house every week, and when she had to go into the hospital, my parents paid the bills.—Mike and Erica's mother

For most people who retire, income drops substantially. Although stopping work may lower expenses, usually there is still a net loss in income following retirement. Many older people are poor. But the proportion of older people who are poor has not changed in the last ten or 20 years, in contrast to the experience of younger people, among whom more have become poor. Older women are more likely to live in poverty than older men. One reason is that women traditionally have been paid less than men or have not held jobs at all, and so their retirement income is low.

Seventy-five percent of retirees say that they want to work. Most workers over 55 say that they want to work at least until they are 65. Eighty percent want to work part time after retirement (Harris, 1981). Longitudinal study results show that those workers who said before retirement that they wanted to keep on working were as good as their word and were working more hours than those who had said that they wanted a leisurely retirement (Palmore, George, and Fillenbaum, 1982).

Discrimination against older workers

One reason that older adults retire is because many people discriminate against them. Most people—those over 65 included—believe that older adults are less efficient and adaptable, slower, and weaker workers than younger adults. A national poll showed that young adults considered older adults as warm and friendly but as lacking in qualities necessary in the workplace (Harris, 1975). These qualities included efficiency, flexibility, alertness, and physical ability. But older adults considered themselves, except in physical abilities, as capable as younger adults considered themselves to be. The problem for the older adult, however, is that younger adults often can hire and fire older adults. The same negative sterotypes of older adults that might contribute to discrimination against older workers surfaced in a survey of *Harvard Business Review* readers (Rosen, 1978). For example, these readers would fire older computer programmers but retain younger ones. Other studies show that older workers are hired last and laid off first, and if they find another job, it usually carries less pay and prestige. Older workers also are more likely to be passed over for promotions and less likely to get the schooling and retraining that younger workers get to improve their job skills. A longitudinal study of workers between the ages of 55 and 64 showed that their incomes actually dropped during the period, compared to a 12 percent gain among younger workers (Wanner and McDonald, 1983).

Older workers fall prey to harmful stereotypes, such as being slower, weaker, and less adaptable than younger workers. These stereotypes are false. Older workers, in fact, work more steadily and are more committed to their work than are younger people.

The stereotypes have little if any truth to them. Cross-sectional studies show that older workers generally are not less productive than younger workers. When there is any decline in productivity, it is slight (Foner and Schwab, 1981). Other studies have shown that older workers are actually more productive. When work output is tallied against job experience, the age difference in productivity disappears (Schwab and Heneman, 1977). Only when we have results from longitudinal studies will researchers be able to separate out the possible effects of differences in attitude and education. For now, it seems that for the kinds of work studied cross-sectionally—clerical and factory production work—older workers are about as productive as younger workers.

Older workers outshine younger workers in some areas. Older workers work more steadily day in and day out, are absent less, and are more committed to their work than younger workers (Foner and Schwab, 1981). Older workers also can be innovative and flexible. Because many younger workers are trying to learn about company politics and to get ahead in their careers, they often are inhibited from proposing the kinds of radical innovations that older workers may come up with (Schrank and Waring, 1983).

Many older workers cannot find jobs or feel that they have been let go or demoted solely because of their age. Although the unemployment figures for older workers are lower than those for younger workers, the figures may be misleading. Because many older workers finally become too discouraged to look for new jobs, they are not included in the unemployment figures (Schram and Osten, 1978). Thus many of these older workers have retired only reluctantly.

Adjusting to retirement

For many people, retirement is a big change. They go from steady work that gives meaning to their lives and gives them a sense of well-being to a life without this central core. How do retired people adjust? The research results are mixed. Some people call retirement an affliction, crisis, trauma, or stressful transition (Foner and Schwab, 1981). But many studies suggest that most retired people feel satisfied, still feel useful, and still have their sense of identity. Most retired people are neither depressed, declining, nor dying.

Some researchers suggest that retirement has either no effect on health or improves it (Atchley, 1976). In an examination of longitudinal data begun by the Veterans Administration in 1963, researchers found that the men who said that their health had improved after retirement were likely to have worked at stressful jobs, to have had serious health problems before they retired, and to have retired for reasons related to their health. But the men's health itself was not likely to have improved; only their *perceptions* had changed after retirement. The statistics in this study were comparable to those in other studies in which men had been asked about the effect of retirement on their health. Thus half of the men said that retirement had not affected their health. Just over one-third said that it had improved their health. The rest of the men were uncertain.

In sum, retirement is good for many people, has little effect on others, and is bad for others (McConnell, 1983). People's personality traits, their attitudes, and their finances all contribute to feelings about retirement. Some people do experience retirement as a stressful loss of a powerful role that has been central to their adult lives. These people are likely to have held high-status, relatively influential jobs. But high work status in and of itself does not doom anyone to an unhappy retirement. For example, over half of the people who had retired from high-status jobs at several large corporations considered retirement to be the best years of their lives (Kimmel, Price, and Walker, 1978).

People who feel positively about their retirement are especially likely to be financially comfortable, to be relatively healthy, to have planned how to occupy themselves, to have retired voluntarily, and not to have considered their work central to their lives (McConnell, 1983). Planning for retirement can be difficult. How much money can and should be set aside? People do not know how long they will live after retiring. They have no crystal ball for reading the future of the economy or their health. One economist has suggested that without Social Security or private pensions, people would have to save 20 percent of their lifetime incomes to live comfortably during their retirement (Schulz, 1980).

Among men who have retired, those who are married are more likely to be satisfied with their retirement than those who are divorced, separated, or widowed (Beck, 1982). In one study of working-class men and women who had retired and then worked part time as park gardeners and maintenance workers, after ten weeks of working, they felt happier and healthier (Soumerai and Avorn, 1983). The improvement probably stemmed from increased income, feelings of usefulness, the ability to exert some control over the environment, and physical exercise.

The suddenness of retirement can make it stressful for some people. To make retirement more gradual, one corporation offers workers the chance to take a three-month leave with full benefits. If the workers want to return after the three months, they may do so. About half of the workers choose to return. Another option the corporation offers is gradually to cut back workers' work hours and salaries (Clendinen, 1983). When the transition is eased between a life of constant work and a life of leisure, people may experience less stress.

Leisure

> Now that we're retired we do more of the things we've always liked. We play golf during the week as well as on weekends. We go out to dinner all the time. We go to movies and plays. We travel here and there. Last year we took our grandsons to Disney World, and next year we're all going to Florida for two weeks. My husband plays cards with his gang.—Tim and Willy's grandmother, age 70

When people retire, their free time increases, and they are likely to engage in more leisure activities. Leisure activities are those that people do not *have to* engage in. Many retired people spend time with family. Others pursue hobbies and special interests. "Leisure" describes people's attitudes toward activities rather than activities themselves. Thus what one person does for leisure another might consider obligatory, and vice versa. The leisure activities that people find most pleasurable change over the course of the lifespan and reflect variations in factors such as age, sex, education, personality, and socioeconomic level. The goals of any leisure activity include (Gordon, Gaitz, and Scott, 1976):

1. Relaxation, derived from daydreaming, sleeping, and resting.
2. Diversion, derived from changing pace and doing light reading or television watching, engaging in hobbies, talking with friends, playing games, going to parties, and watching sports events.
3. Developmental pleasures, derived from exercising, athletics, doing serious reading, attending museums, clubs, interest groups, learning to play a musical instrument, traveling, and playing "educational" games.
4. Creative pleasures, derived from artistic, musical, and literary pursuits, volunteer work, and serious discussions.
5. Sensual pleasures, derived not just from sexual activity but also from dancing, highly competitive sports, religious experiences, and other activities that are extremely pleasurable and joyous.

Retirement offers older people more free time, so they are likely to engage in more leisure activities than younger adults do. Not surprisingly, these older people choose activities that complement their values and roles.

A study of 1500 adults from Houston, Texas, showed that older adults spent more time relaxing and in solitary activities than younger adults did (Gordon, Gaitz, and Scott, 1976). Older men also spent time cooking. Older adults spent less time than younger adults in events away from home that were exciting, escapist, physically demanding, or required sensorimotor skills. These events included dancing and drinking, exercising, going to movies, hunting and target shooting, traveling, reading books, playing musical instruments, singing, drawing, and painting.

People choose leisure activities that complement their roles and values (Rapoport and Rapoport, 1975). For many older adults, children have grown and left home, disposable incomes have increased, and so they have more time to travel, to pursue their own interests, and to spend evenings out. People from lower socioeconomic groups are more likely to engage in leisure activities devoted to relaxation, diversion, and sensual pleasure (Gordon, Gaitz, and Scott, 1976). People from higher socioeconomic groups are more likely to choose developmental, creative, and sensual pleasures. Once older adults retire, not only socioeconomic status but their health are likely to influence their choice of leisure activities. In a study some years ago of 5000 retired adults, who were asked what they had done the day before, most had watched television, visited, read, napped, and daydreamed (Riley and Foner, 1968).

Whether retired people are generally satisfied with their lives or not is best predicted by their level of social involvement. In the Duke Longitudinal Study, for example, the most satisfied adults were also those who were quite active. Most of these people kept up the same levels of social activity for several years and cut back only in the face of illness. Some of the adults in the study were socially isolated, and most of these had disengaged relatively early in adulthood (Maddox, 1970). Today among older adults, women usually have less free time than men because for those who are married, their traditional gender roles require household responsibilities that do not go away in old age.

Education

In the future, retired people are likely to have higher incomes and more education than today's retired people do. Formal education is becoming increasingly popular for older adults. They are taking traditional college programs, special college programs for ''mature'' students, community adult education courses, correspondence courses, and even televised courses.

Increasing numbers of older adults are pursuing formal education, which helps them to function well cognitively and socially and offers a chance for personal growth and satisfaction.

> I tried to reach my mother on the phone, but no one answered. Where was she? At Spanish class. She takes an evening class at a local high school.—Megan's mother

One program especially for adults 60 and over is Elderhostel, in which a network of colleges in all 50 states offer short-term courses. In some Elderhostel programs, students room and board in colleges for three weeks while they take courses from regular faculty members. Many Elderhostel students are female, retired, and financially comfortable and had gone to college already. But among the 10 percent who had never gone to college before, intellectual growth was greatest. Students in their seventies and eighties also seemed to gain more than those in their sixties (*AARP News Bulletin*, 1983d).

Three general attitudes toward education in later life can be identified (Sterns and Saunders, 1980). One attitude is that education is a good way to fill leisure time. Another attitude is that education helps older adults to participate fully in society, to function well, and to compensate for minor cognitive and physical problems. Finally, education is sometimes held to offer personal growth and satisfaction to older adults.

The few studies available show that older adults do as well as or better than younger students, although the differences among individuals are great (Kasworm, 1980). But older adults face certain problems when they go back to school, especially when they take regular college courses. Widely held views that older people are incompetent, dependent, and not productive hamper their self-esteem and academic performance. Some older adults benefit from techniques such as courses in which they have the chance early on to succeed at certain tasks. Tests that draw on recognition rather than recall, for example, can reinforce older students' sense of their effectiveness (Rebok and Offermann, 1983).

> When I was in college, the older students used to drive me crazy. They would ask so many questions, they took it all so seriously, and sometimes they kept the whole class there after the period had ended while the teacher answered some question.—Nancy's mother

Community involvement

Older adults participate in community activities of religious, political, legal, economic, and social nature. Although it once was believed that older adults

Older adults for whom religion is important tend to be well adjusted and satisfied with their lives, although psychologists do not know which is the cause and which the effect.

routinely disengaged from their communities, it is now realized that for most older adults, the greater their social involvement, the greater their satisfaction with life.

Religion

In particular, religious involvement correlates with feelings of satisfaction about one's life. Thus in a study of older adults who lived in a rural part of New York state, religious attitudes, beliefs, and practices were the best predictors of people's feelings of meaning and purpose to their lives (Telis-Nayak, 1982). It is hard to answer whether people who are well-adjusted engage in religious activities or whether religious activities make people well-adjusted. But results from several studies do show a connection.

We know little about developmental trends in religious feeling during adulthood. But we do know that throughout adulthood, more people participate in religious activities than all other voluntary community activities put together (Moberg, 1968). Attendance at religious activities generally is stable across adulthood, although lower than in childhood (Hammond, 1969). Among older adults, however, problems of health, followed in significance by problems of transportation and finances, interfere with religious participation away from home (Moberg, 1968). But older adults may compensate by increasing their private religious activities, a sign that religious interest does not always decline along with religious participation (Ainlay and Smith, 1984). For example, older adults may watch services on television, listen to them on radio, read the Bible, pray privately, believe in immortality, believe that the world needs religion more than economic security, and report that religion is personally meaningful for them (Hammond, 1969). One-third of a group of older adults who were poor, black, inner-city residents said that religion had become more important to them during their lives even though their religious participation had not changed (Heisel and Faulkner, 1982). Older and middle-aged adults are more likely than younger adults to believe that religion may provide solutions to world and national problems.

Some older adults report that religion becomes more meaningful to them over time (Moberg, 1968). Data from cross-sectional studies show that 85 percent of those over 65 consider religion very important in their lives (Riley and Foner, 1968). Data from longitudinal studies show that one group of gifted older adults was more interested in religion later than they had been earlier in adulthood (Marshall and Oden, 1962). But among older Anglo and Mexican-American adults from Texas, religious participation declined (Markides, 1983). Whereas Mexican-American women and men and Anglo women reported an increase in religious feelings over a period of four years, the Anglo men reported no such increase.

Politics

Many people believe that we become more conservative as we age. Are they right? Not entirely, it seems. Although it is true that people with relatively little education are conservative, those with more education are more liberal. But cohort experiences also influence political outlook. Those who begin voting when the country is moving toward conservatism are likely to remain on the conservative side over the long run. The same holds true for those who begin voting when the country is moving toward liberalism. Further, older adults are more likely to favor government support for medical care—a liberal position—than younger adults. Older adults, of course, see it as in their own interest to take that political position. Like people of any age, older adults vote for their own self-interest. It is also probably true that older adults have not become more conservative with age but that, at least for a while, younger adults were less conservative, and so the older group seemed more conservative by comparison.

Although some older adults are political liberals, they form a minority. As might be expected, personality characteristics often are associated with political outlooks. For example, data from the California Intergenerational Study have shown that the personality characteristics associated with later political outlooks were already evident in adolescence. Thus people who were liberals in adulthood had been independent, unconventional, rebellious, objective, and interested in philosophical questions ever since early adolescence (Mussen and Haan, 1981). Those who were conservatives in adulthood had been dependent, submissive, in need of reassurance, and interested in moral questions ever since early adolescence. They were not especially introspective and did not feel comfortable with uncertainty. The researchers suggest that the political attitudes of later adulthood may have roots in childhood experiences. Thus parents' styles of childrearing combine with individuals' social and economic histories and broader social, historical, and economic events to shape political outlook in late adulthood.

As a group, older adults are quite politically active (Hudson and Binstock, 1976). Adults who are politically active in middle age tend to remain so in old age. They typically do not retire from politics at age 65. Many political leaders are themselves in their fifties or beyond. In many cases, the more important the political office, the older the officeholder. Older adults also vote in greater numbers than people from other age-groups. The more educated the older adult, the more likely he or she is to vote (Wolfinger and Rosenstone, 1980). Older men are also more likely to vote than older women, even when education levels are taken into account. Older women who hold jobs are no more likely to vote than other women. It is likely that this is a cohort effect—a reflection of some older women's beliefs that they should not vote or participate actively in politics—and it is likely to change in coming generations.

Some studies show that once adults reach their seventies, they begin voting less. This trend is more pronounced among women than men and is generally interpreted to reflect health problems (Milbrath and Goel, 1977). But when factors such as education, marital status, and sex are held constant, the rate of voter participation continues to rise among even the oldest adults (Wolfinger and Rosenstone, 1980).

Legal and economic issues

Laws may deal people in or out of roles, status, rights, or privileges. In the 1800s, age began to be used as a way to deal people out of certain rights and privileges (Cain, 1976). One effect of undermining the negative stereotype of older adults may be to undermine the support programs that some of them need. Yet one effect of these programs automatically going into effect on a person's sixty-fifth birthday is to perpetuate the negative stereotype of the aging person and to reinforce discrimination. As we have seen, laws now forbid discrimination against people on the basis of age. But negative attitudes toward older workers are widespread even so. Other issues of discrimination and older adults have been raised. For example, is it fair to make older adults pass vision tests to renew their drivers' licenses? Is it fair for older adults to pay only half fare on public transportation? It has been suggested that only the needy old be eligible for economic, medical, and social support programs. It also has been suggested that the marker of old age be moved from 65 to 75, an age at which people are more likely to be physically frail and vulnerable.

Although many older adults live in poverty, their proportion to older adults as a group is about the same as the proportion of poor in other age-groups (Pear, 1984). Over the past few decades, the financial position of older adults has improved greatly (Schulz, 1980). Social Security benefits have increased along with private pension benefits, the institution of public health programs and food stamps, and laws to grant relief from property taxes. As a result, many older Americans can anticipate a comfortable retirement.

Compared to younger adults, older adults spend more of their income on goods, housing and household functions, and medical care and less on furniture, transportation, personal care, clothes, and leisure activities (Kreps, 1976). Compared to older adults who are retired, those who are still working spend less on housing, meals at home, or medical care and slightly more on running their households, fuel, and utilities (McConnel and Deljavan, 1983). Older working adults also spend more on cars and eating out. Retired people change their consumption patterns to adjust to their reduced incomes, and few use savings for daily living expenses. When retired people face a substantial drop in income, and especially when they have not expected that drop, they are likely to feel deprived (Espenshade and Braun, 1983).

Summary

1. Throughout adulthood, an individual's personality is quite stable, because both personality traits and self-concept are stable. When an individual's personality does change during adulthood, it is usually in response to changes in life events. Adults' satisfaction with their lives generally is stable over the long term, too.
2. Introversion is the one personality trait consistently found to increase with age. Some studies also have shown age-related changes in conformity, social responsibility, excitability, and humanitarian concern.
3. Older adults who successfully resolve the conflict between ego integrity and despair emerge with a sense of wisdom.
4. More women are widowed than men. The severity of the adjustment to widowhood depends on the survivor's earlier dependence on his or her spouse; earlier social involvements; economic and social resources; and the survivor's personal acceptance of widowhood. When widowhood is perceived as "on time," the widow's adjustment is likely to be easier than when it happens suddenly.
5. Relationships of brothers and sisters may deepen during late adulthood, especially among those who are not married. Similarly, older adults who are not married see more of their friends than those who are married.
6. Although most grandparents take pleasure in the role, few say that their feelings of satisfaction with life derive from their involvement with their grandchildren. Different people adopt different styles of interacting with their grandchildren. Some grandparents are more formal, intimate, and involved with their grandchildren than others.
7. In this country, each generation within a family generally prefers to live in its own household. But relationships among the generations are close enough to make many American families into modified extended families. Parents and children each value and try to influence the other.
8. Retirement poses older adults with the need to adjust to new psychological, social, and economic circumstances. Many older adults are poor. Even so, most retired people report feeling satisfied, useful, and an intact sense of personal identity.

9. Although many retired workers say that they would rather work, many are forced to retire. Discrimination against older workers—erroneously based on stereotypes of older people as inflexible, slow, and inefficient—also makes it difficult for many older adults to find or keep jobs.
10. Retirement increases most people's amounts of available free time. Older people who feel satisfied with their lives are likely to remain socially active and engaged. Religious involvement also correlates with many older people's feelings of satisfaction with their lives.
11. Although some older people become more politically conservative as they age, not all do. Personality traits later associated with political liberalism and conservatism have been evident in some people since adolescence.

Key terms

normative stability
modified extended family

Suggested readings

BENGSTON, VERN, REEDY, MARGARET, and GORDON, CHAD. "Aging and Self Conceptions: Personality Processes and Social Contexts." In J.E. Birren and K.W. Schaie (eds.), *Handbook of the Psychology of Aging*. New York: Van Nostrand, 1985. Discussion of aging and self-concept including cognitive, affective, and conative components.

BINSTOCK, R.H., and SHANAS, E. *Handbook of Aging and the Social Sciences*. New York: Van Nostrand, 1985. A book of current reviews of the most important social-science research on aging. It includes chapters on the social aspects of aging, aging and social structures, social systems, and social interventions.

BURRUS-BAMMEL, LEI L., and BAMMEL, GENE. "Leisure and Recreation." In J.E. Birren and K.W. Schaie (eds.), *Handbook of the Psychology of Aging*. New York: Van Nostrand, 1985. A review of theories and research on leisure in adulthood and aging.

KAHANA, B. "Social Behavior and Aging." In B.B. Wolman (ed.), *Handbook of Developmental Psychology*. Englewood Cliffs, NJ: Prentice-Hall, 1982. A discussion of self-concept, attitudes, and social behavior among older adults.

chapter twenty-three
Death and dying

Longevity

Humans are probably the only species to try to slow down the aging process and push back the limits of human **longevity**, or length of life. What characteristics influence longevity? The length of prenatal development, age of sexual maturity, size of brain and body, rate of metabolism, and body temperature all have been correlated with longevity, although none is perfectly related. Among mammals, the most consistent correlations are among longevity, brain weight, and body size (Walford, 1983). Tortoises may live to a ripe old 150, humans to 114, whales to 80, elephants to 70, horses to 62.

Researchers also have found a correlation between an animal's rate of metabolism and its longevity. Each species burns calories at a fairly constant rate, and the faster this metabolism, the shorter the lifespan of the species. Mice burn calories relatively rapidly and live for only a few years; elephants burn calories relatively slowly and live for up to 70 years. But there is an exception to this rule—the human being. Compared to our mammal cousins, we humans burn as many as four times the calories for every pound we weigh.

A correlation also has been found among the longevity of a species, the characteristic length of its prenatal development, and the length of the young's development before reaching sexual maturity. In general, the briefer the prenatal development and the briefer the period of sexual immaturity, the shorter the lifespan of a species. But we humans prove the exception to this rule as well. We are longer lived than might be expected for a species whose prenatal development lasts only 9 months. Horses develop prenatally for 11 months, and apes and cattle, which also have a 9-month gestation period, live briefer lives than humans do. But human children remain sexually immature and dependent on their parents for a relatively long time, and this factor may help to explain the long human lifespan (Rockstein, Chesky, and Sussman, 1977). During their period of dependence on adults, the young of many species learn the skills they will need to reproduce sexually, to rear their young successfully, and to maintain the survival of their species. The especially long period of dependence found in humans, during which children learn a complex array of skills necessary for survival, may contribute to the longevity of our species.

Maximum lifespan

If each of us were favored by health, inheritance, accident of birth, and pure luck, we would survive to between 110 and 120 years old (see Figure 23.1). The oldest human being whose great longevity could be verified was a Japanese man who reached the age of 114. A California woman lived to be 113 years and 215 days old, and a New York woman lived to be 113 years and 114 days old (Freeman, 1982; Walford, 1983) One hundred and fourteen years seems to be the **maximum lifespan** for humans. The maximum lifespan is defined as the oldest age to which any members of a species survive. Each species has its own maximum lifespan, and it is the product of biological characteristics unique to the species (Cutler, 1981; see Table 23.1).

People in other times and from other places have claimed to survive longer than 114, but their claims have not been verifiable. Claims of extreme longevity have emanated from remote mountain communities in the Soviet Union and Ecuador. But when researchers have combed through church, state, and other records, they find that people who have claimed to be 127 or 130 are, in fact, younger than that (Bennett and Garson, 1986). The record for human longevity remains 114 years.

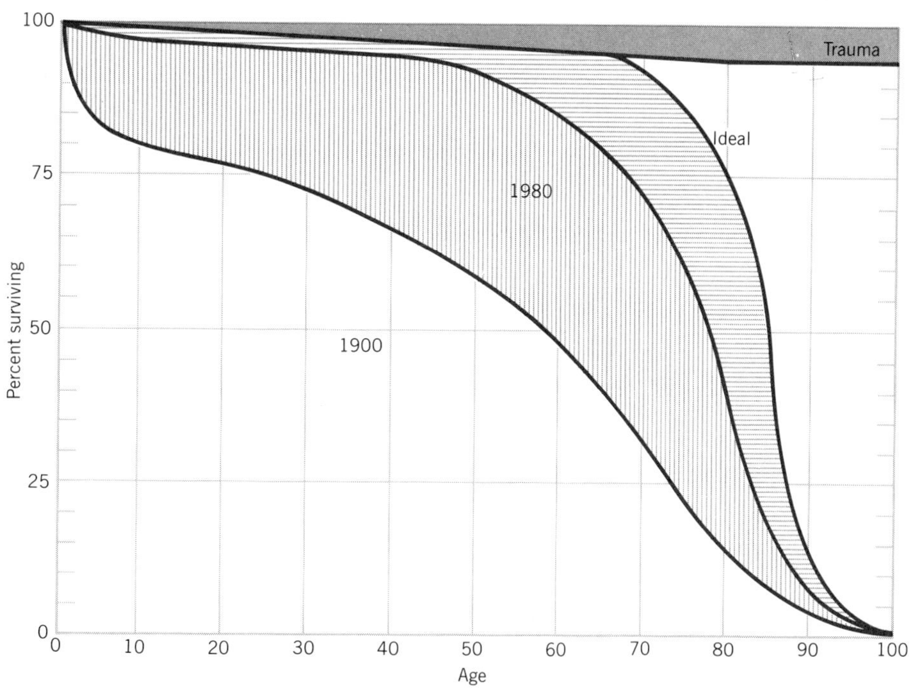

Figure 23.1 As premature deaths from disease are eliminated, more people live lives that approach the maximum human lifespan (Fries and Crapo, 1981). From **Vitality and Aging** *by James Fries and Lawrence M. Crapo. Copyright © 1981 W.H. Freeman & Co. Reprinted with permission.*

For the last 200 years, the average human lifespan has been increasing. Premature deaths resulting from infection, disease, and malnutrition have been declining steadily.

The maximum human lifespan has not changed for eons. It probably has remained the same since the human species evolved into its present form. Although the maximum lifespan of modern humans has remained at between 110 and 120 for hundreds of generations, the *actual* lifespan has only recently begun to approach that upper limit.

Average lifespan

Early humans had shorter lives than modern humans because they were more vulnerable to disease, malnutrition, the elements, and predators. For the last two hundred years or so, the **average lifespan**—the average length of time that members of a species actually survive—has been increasing. Until the mid-1800s, as people got food that was both more abundant and more nourishing, they became stronger, healthier, and more resistant to disease. The average lifespan lengthened. After the mid-1800s, better sanitation contributed to a longer average lifespan. Once people understood the connection between microorganisms and infection, they curbed the contamination of food and water and the spread of disease and infection. More recently, medical advances such as widespread immunization, antiseptic surgery, and the discovery of vitamins and antibiotics have further extended the average human lifespan. Today, as researchers begin to understand the effects of nutrition, exercise, and habits of living on longevity, the average human lifespan may be lengthened still further.

As premature deaths from infection, disease, and malnutrition have been reduced steadily over the years, the average length of time a person can expect to live, a measure called **life expectancy**, has increased. In 1900, the average life expectancy was 47 years. Eighty years later, life expectancy was 73 years. Although more people reach old age today than in the past, those who

Table 23.1
Maximum lifespan

Species	Years
Tortoise	150.0
Human being	114.0
Whale	80.0
Indian elephant	70.0
Horse	62.0
Great apes	
Gorilla	50.0
Chimpanzee	50.0
Orangutan	50.0
Old World monkeys	
Baboon	40.0
Macaque	40.0
Gibbon	35.0
Brown bear	36.8
New World monkeys	
Spider monkey	35.0
Squirrel monkey	21.0
Dog	34.0
Cat	30.0
Cattle	30.0
Swine	27.0
Sheep	20.0
Goat	18.0
European rabbit	13.0
Guinea pig	7.5
Golden hamster	4.0
Mouse	3.5

SOURCE: M. Rockstein, J. A. Chesky, and M. L. Sussman, 1977, S. L. Washburn, 1981, R. L. Walford, 1983.

reached age 65 in 1900 could expect an average of 13 more years of life. Those who reached age 65 in 1979 could expect an average of 16.6 years of life—not a great increase (U.S. Bureau of the Census, 1982).

Because average lifespan is strongly influenced both by environmental factors such as diet, weight, and exercise, smoking and drinking, disease and infection, availability and quality of medical care and by constitutional factors such as genetic inheritance and sex, the average lifespan varies within different population groups. For example, females (human and in other species) generally live longer than males. From conception onward, more females survive, they resist infection and disease more effectively, they have fewer accidents, and they are less vulnerable to the effects of physical deprivation. At least until they reach menopause, women have hormones that seem to protect them from heart disease, today's greatest killer in our society. Changes in living habits have modified the life expectancy of the two sexes. In 1900, the numbers of men and women over 75 were equal (Walford, 1983). But as men smoked more cigarettes and suffered more alcoholism and accidents than women, and as reliable contraception, legal abortion, and better hygiene prevented ''childbed fever,'' the life expectancy for females rose higher than that for males. Until the 1980s, women also had lower rates of lung cancer than men, but the increased numbers of women smokers have raised their death rate from lung cancer as high as men's.

Both environmental and constitutional factors modify the life expectancy of blacks and whites in this country. In general, whites live longer than blacks, although the difference has declined since 1900 (Butler and Lewis, 1982; see Table 23.2). Today, a newborn black baby's life expectancy is shorter than a white baby's. Once people reach the age of 65, however, the advantage to whites dwindles to only about one year. By the age of 85, black men and women have the advantage over whites in life expectancy. The environmental factors associated with poverty, such as poor nutrition, sanitation, and medical care, contribute to the lower life expectancy of blacks in this country, to the even lower life expectancy of Hispanics, and to the still lower life expectancy of Native Americans, who can expect to survive for only 45 years after birth (Block, 1979). But in all societies, once people reach the age of 65, the rest of their expected lifespan is about the same length (Fries and Crapo, 1981). In other words, once people survive to old age and evade the environmental hazards that kill people prematurely, they tend to live about the same length of time.

Death

Although the average lifespan comes increasingly close to the maximum lifespan in some societies and for some individuals, inevitably death visits the young as well as the old. Infants die from congenital defects, from malnutrition, and from disease and infection. Children die from accidents, poisons, and acute diseases. Adolescents die in accidents and by suicide, and adults die from chronic and acute diseases. The age of the person who has died and the circumstances surrounding the death usually have a profound effect on the feelings of both the dying person and the survivors. As we will see in this chapter, children understand death differently from older people, and a child's death is interpreted differently from an older person's. Across the lifespan, people have different attitudes towards and expectations about death. Whereas the death of an infant or a child perhaps inevitably seems premature, the death of someone very old may be accepted as the timely end to a long and full life.

Definitions of death

We speak of ''death'' as a single event, but in fact today there are several different categories of death. Historically, death has been said to occur when a person stopped breathing. As breathing stops, the heart stops beating, oxy-

Table 23.2
Life expectancy in the United States

Age	White Men	White Women	Black Men	Black Women
Birth	70.6	78.0	64.0	72.7
Age 65	14.2	18.7	13.3	17.2
Age 85	5.5	7.0	6.8	9.2

SOURCE: U.S. Bureau of the Census, 1982.

gen no longer moves through the bloodstream, brain cells die, and death has arrived. Even today, many people all across the lifespan do, of course, die in this way. But many people are given emergency care that reverses this sequence. They are given mouth-to-mouth respiration, oxygen, electric shocks, adrenaline. Tubes are threaded into an airway, and machines keep them breathing. People can be kept for years on respirators and other life-sustaining equipment. But is a person truly alive who cannot breathe independently and who would die without a respirator? Is a person truly alive whose breathing continues but who cannot eat, respond, or think?

Questions like these have become more and more pressing as technology has blurred the once clear boundaries between life and death. As this boundary has been blurred, the definitions of death itself have had to change as well. For most people, **clinical death** marks the end of life. Clinical death occurs when spontaneous breathing and heartbeat end. In some cases, the decision is made ahead of time not to extend life beyond the point of clinical death.

> When Sam's grandfather was dying of Alzheimer's disease, the family and the doctors decided not to try to save him if he suffered a heart attack or other life-threatening medical emergency. On his medical chart was written in large letters, "DO NOT RESUSCITATE."—Sam's father

For other people, **brain death** technically marks the end of life. Brain death takes place when the higher centers of the brain stop functioning, leaving a person comatose. First the cortex dies; it controls thought, memory, and voluntary actions. Then the midbrain dies; it controls reflexes. Some people who are brain dead may survive in a vegetative state for years, so long as they are fed intravenously. Ultimately, the brain stem dies, and breathing and heartbeat stop. Karen Quinlan was a famous example of the dilemma that certain families and doctors face. Her parents sued to have their brain dead, comatose daughter taken off a respirator, assuming that she would soon die peacefully. But Karen survived for ten years.

The anguish and moral ambiguity surrounding death under circumstances such as these led a committee at the Harvard Medical School to recommend that brain death be considered the death of a person. This recommendation is widely followed today. Brain death is defined as

1. Total unresponsiveness to any stimuli, even the most painful;
2. Lack of movement for one hour and lack of breathing for 3 minutes when a respirator is removed;
3. Absence of reflexes and activity in the brain stem;
4. Absence of activity in the cortex, as indicated by a flat electroencephalogram (EEG) (Jeffko, 1979).

For a person to be considered brain dead, these criteria must be met twice in 24 hours, and the person must not be suffering from hypothermia—a body temperature of less than 90 degrees—or from the effects of a drug overdose, which can stop breathing. Hypothermia and drug overdose can mimic brain death, but they are reversible conditions. These criteria for brain death help doctors and family members decide when to remove a person from a respirator.

> A teenager at Erica's high school was killed when he drove his car into a telephone pole. He was brain dead when they got him to the hospital, but they kept him on a respirator so that they could ask his parents to donate his heart, liver, kidneys, and other organs to other people. They consented. They said it would mean their son's death wasn't completely in vain.—Erica's mother

focus *Life and death issues*

In 1982, Baby ''Jane Doe'' was born in New York with a severe spinal defect. Her parents and pediatrician decided not to operate on the infant, for surgery could only prolong a blighted existence. They would feed and care for her otherwise and let nature take its course. But an advocate for the baby sued the parents in court to provide surgery, and this position was supported by the federal government. Ultimately, the court ruled in favor of the parents.

The Baby ''Jane Doe'' case is but one among many. Should an infant's life be preserved under all conditions? How much, if at all, should the quality of that life weigh in decisions about medical treatment? Who is to decide—the family, the doctor, or the courts?

How does a doctor's moral reasoning influence his or her decisions about treating infants like Baby Jane Doe, who are severely defective or terminally ill? To assess levels of moral reasoning, one research group (Candee, Sheehan, Cook, Husted, and Bargen, 1982) interviewed medical residents and other pediatricians practicing in several hospitals about the care they would give to newborn infants. They asked, for example, about the treatment a physician would recommend for an infant born with brain damage, no nerve function in its legs, and no bladder or bowel control. What would the physician do, they also asked, if this severely defective infant developed a life-threatening infection like meningitis or if its heart stopped beating? Would the physician aggressively treat the meningitis or try to resuscitate the failed heart?

The medical residents were more likely to prescribe treatment than the teaching and practicing physicians were. Moreover, among the medical residents, those who engaged in the highest levels of principled moral reasoning were less likely to treat defective infants with a poor prognosis when the families did not want such treatment. Among teaching and practicing physicians, either there was no correlation between moral reasoning and actual treatment, or the relation was just the reverse. The norms within the particular institution an infant was in also had a bearing on whether the physicians did or did not actively treat the infant. Clearly, the decision about whether and how to treat a severely defective infant is complicated. It draws not only on physicians' levels of moral reasoning but on their levels of experience, on medical issues, on families' attitudes and wishes, and on social norms. Parallel issues are also arising in the medical treatment of terminally ill adults.

Not all cases of brain death are so clear-cut. In some cases, the cortex is dead, but the midbrain and brain stem continue to function. Many doctors and other people have suggested that the Harvard criteria for brain death are too narrow and that **cerebral death** be considered sufficient to establish brain death. Cerebral death is the death of the cortex. As one neurosurgeon (Sweet, 1978) has argued, studies of over 2000 people have shown that no one ever recovers after the cortex stops functioning and its electrical activity stops. (The only exceptions are people in a coma brought on by drugs.) The criteria for establishing cerebral death include taking tracings from two different EEG machines six hours apart.

Many states accept total brain death as the legal end of life, but none currently accept cerebral death as the legal standard. The circumstances surrounding many deaths remain a gray area, raising troubling legal and moral questions for family members, people in medicine, and those who are dying.

The conditions of death

Just as the definition of death has shifted to reflect changes in medical care for the dying, the conditions surrounding death have been changing. Both the when and the where of death have changed markedly since the beginning of this century.

Age of death

An 81-year-old woman was telling her 78-year-old sister about the recent death of a neighbor. "It was such a shame. She was so young, only 64, with so many good years ahead of her."—Peter's mother

Improvements in diet and effective cures for many of the acute diseases that once killed many children gradually have shifted the average age of death upwards. In 1900 in the United States, for example, the death rate among infants younger than 1 year old was 180 per thousand boys and 145 per thousand girls. Today, the death of an infant or child is much more unusual; it is likely to be considered a tragic exception or aberration. In 1981, the death rate among infants younger than 1 year old was 13 per thousand boys and 10 per thousand girls (U.S. Census Bureau, 1982–1983; see Table 23.3).

As the age of death in this country has shifted upward, the form of the final illness also has changed. Whereas in earlier centuries death suddenly struck infants and children, today death is more likely to follow after a chronic, lingering disease associated with secondary aging. Thus the sudden deaths from acute illness such as pneumonia, diarrhea, diphtheria, polio, influenza, and others have been replaced by the slower degeneration associated with heart disease, cancer, diabetes, and other chronic illnesses.

The age of death has been shifting upward, and the form of final illness has shifted from a sudden illness to a chronic, lingering illness associated with secondary aging.

With these changes in the typical age at death and the typical last illness, the stamp of death on the typical family pattern has changed as well. Not only are parents today less likely to lose an infant or young child to death, but young children today are less likely to lose a parent to death than in earlier

Table 23.3
Death rate in the United States

	1900		1950		1981	
	Men	Women	Men	Women	Men	Women
All ages	17.9*	16.5	11.1	8.2	9.6	7.8
>1 year	179.1	145.4	37.3	28.6	13.5	10.3
1 to 4	20.5	19.1	15.2	12.7	0.7	0.5
5 to 14	3.8	3.9	0.7	0.5	0.4	0.2
15 to 24	5.9	5.8	1.7	0.9	1.6	0.6
25 to 34	8.2	8.2	2.2	1.4	2.0	0.8
35 to 44	10.7	9.8	4.3	3.0	3.0	1.6
45 to 54	15.7	14.2	10.7	6.4	7.6	4.1
55 to 64	28.7	25.8	24.0	14.0	17.8	9.4
65 to 74	59.3	53.6	49.3	33.3	40.1	21.2
75 to 84	128.3	118.8	104.3	84.0	85.3	51.9
>84	268.8	255.2	216.4	192.0	182.4	140.9

*per 1000 people

SOURCE: U.S. Bureau of the Census. *Statistical Abstract of the United States, 1982–83.* Figures for 1900 from U.S. National Center for Health Statistics, U.S. Public Health Service.

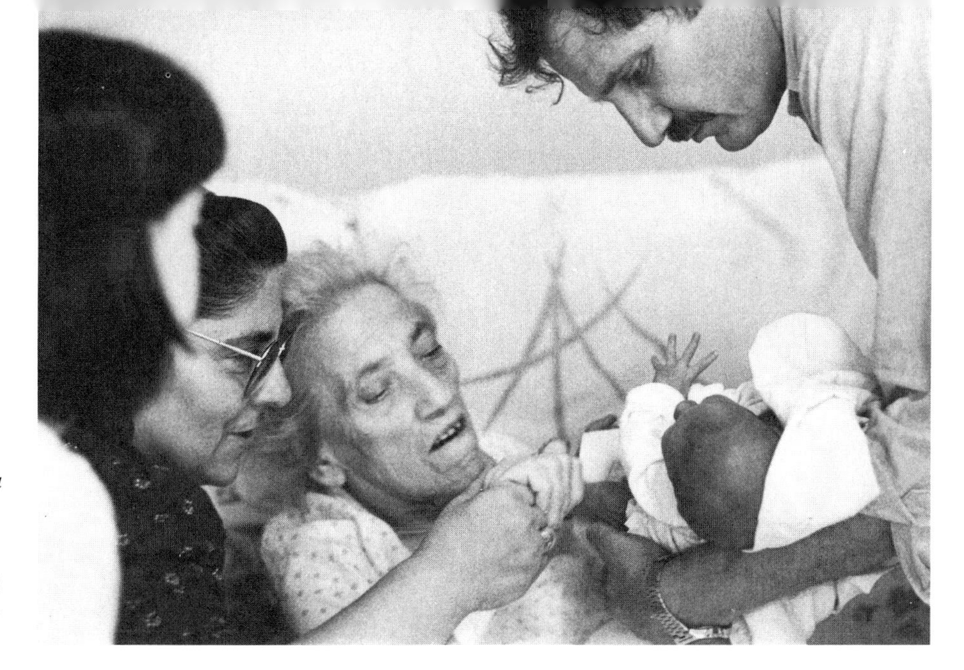

In a hospice, the life of a dying person is not prolonged artificially, but the person is kept as comfortable and treated with as much dignity as possible.

periods. Historical studies show, for example, that in American Quaker families during the early 1800s, it was not uncommon for a parent to die ten years before the youngest child was ready to leave home (Wells, 1973). In the early 1900s, a parent died on the average one year *after* the youngest child left home. Today, a child is likely to have two parents living for more than a dozen years after the child leaves home.

> When I listened to my grandparents talk about their childhoods, I used to get confused by the stories of the great grandmothers who died and whose husbands remarried and of the great grandmothers who died and whose unmarried sister would move into the family to take care of the children. But now I realize that the pattern really was different for families in those days.—Sam, age 18

Today fewer parents mourn the loss of children than do children mourn the loss of parents. Today few young children lose parents. Most people grow up knowing at least some of their grandparents. Fewer grandparents mourn the loss of grandchildren than do grandchildren mourn the loss of grand- and great grandparents.

Place of death

In a trend that parallels the removal of childbirth from private homes to hospitals in the United States since the early 1900s, death, too, has been moved out of people's homes and into hospitals and other institutions. There also has been a further change in attitudes that has led some people to favor "natural," medically supervised birth in a more home-like setting than the typical hospital offers and medically supervised, "natural" death at home or in home-like settings.

As the nature of the typical final illness in our society has changed from acute to chronic, and as the ability of medicine to prolong life has grown, more and more people have entered a hospital for their last illness—and their death. Until medical science grew so powerful during this century, doctors were likelier to try to comfort the dying than to cure them. Most people died at home. But today, medicine can work cures and prolong life in so many cases that more than 80 percent of Americans die in a hospital, a nursing home for the aged, or other institution (Bok, 1978). The thrust in a hospital usually is toward prolonging life, sometimes even against the wishes of a dying person and family members. When this prolongation of life becomes artificial and undignified, it turns into what has been criticized as a ''degradation ceremony'' (Shneidman, 1980). To avoid this kind of degradation, some people choose to die at home or in a **hospice**. A hospice is part of a hospital or an independent institution where the life of a dying person is not prolonged but is made as comfortable and as dignified as possible.

Since the first hospice was founded in London in 1948, the hospice movement has become international, and the number of hospices in the United States continues to grow. Hospices generally coordinate home and hospice care. Families are encouraged to visit, to spend time with the dying person, to offer emotional and physical support. Paid staff members and volunteers spend time with those who are dying, offering affection, attention, and physical contact (Burns and Carney, 1985). Doctors prescribe analgesics that dull the pain but not the senses. Experience has shown that there is little reason to be concerned that dying people might grow addicted to painkillers. Hospices not only help to provide people with an **appropriate death**—the kind of death a dying person chooses when that person is allowed to make the choice—but they also make dying less expensive. Hospice care does away with expensive medical care, enlists the aid of volunteers and family members, and keeps people at home for an average of two weeks longer than if they were getting standard medical care (Schulz, 1978). In a hospice, many dying people can accept death calmly, without anguish, and free from the fear of a painful, draining, lingering last ordeal (Bok, 1978). The hospice movement exemplifies new attitudes toward death.

Attitudes toward death

People's attitudes toward death are shaped by both social and psychological forces. Thus people within certain social groups share certain attitudes toward death, individual people form different ideas about the prospect of their own death, and people at different points in cognitive development have different ways of understanding death.

Historically, particular strategies of reconciliation have been embraced by people in certain social groups. When older people in this society were asked what comforted them as they thought about death, one-third said that they took comfort in their religious faith, 40 percent took comfort in their achievements, and one quarter took comfort from the love of friends and family. Data from people in the United States who knew that their death was near suggest that they relied on few strategies of reconciliation. Data from the National Hospice Study (Kastenbaum, Kastenbaum, and Morris, in preparation), in which a national sample of 390 terminally ill adults were asked to describe their greatest sources of strength in adversity, show that very few—3 percent—drew strength from the prospect of an afterlife. Even fewer—2 percent—took strength from the thought that their children and grandchildren would survive and carry on the dying person's name and tradition. But over 50 percent said that they drew strength from the support of family members and friends.

Social attitudes toward death have been shaped by many of the changes that we have discussed: an increasingly longer average life expectancy; the shift from death following an acute illness that might strike at any point during the lifespan to death following chronic illness late in adulthood; and the shift from death at home, in the care of family members, to death in an institution, in the care of medical professionals. Thus many people in our society today think of decline and death as the natural province of the old. In an era when relatively few children die, a child's death violates widely held expectations. A child's death is perceived as disruptive and ''unscheduled,'' just as an old person's death is perceived as ''scheduled'' (Glaser and Strauss, 1966; 1967).

Children of all ages think about death. Their comprehension of it follows the same path of cognitive development as do other areas of comprehension.

> "Why did Nana die?" Mikey asked me. "Nana was old," I explained. "When people get old and sick, they die."

The death of an older person causes grief and sorrow, but it also confirms people's expectations and, in so doing, provides a measure of psychological security (Kastenbaum 1985). Death is "playing the game" according to the rules that people want to believe in. By ascribing death to the old, younger people can create the comfortable illusion that they are at a safe distance from death (Feifel, 1977). But by considering death the province of old people, people of all ages run certain psychological risks. When young people distance themselves from the psychological reality of their own death, they postpone coming to terms with their own inevitable decline and sense of loss. Unless people can cross this psychological barrier, they will find it difficult to develop an integrated sense of their own identity, encompassing their entire lifespan (Kastenbaum, 1985).

As Robert Kastenbaum (1985) suggests, people's attitudes toward those who are dying run in one of two different directions. Either the dying are segregated, socially isolated, and demeaned as hopelessly old and impaired, or they are imbued with inspirational qualities such as wisdom, saintliness, and heightened or deepened spiritual wisdom. Either they are thought to have a less legitimate claim on social resources than younger people, or they are thought too pure for this imperfect earthly existence. Neither attitude helps anyone to see old people, dying people, or old and dying people as distinct individuals.

Developmental change in concepts of death

> Natalie's grandfather died suddenly when she was 5. Right after he died, Natalie cried and seemed sad for a few days. She asked when Grandpa was coming to visit. It was hard for her to grasp that Grandpa wasn't coming back. For a few months, she liked to pretend to talk to him on the telephone. But six months after his death, she almost never mentioned him anymore.

Many parents and teachers long assumed that children cannot, do not, and should not understand death (Kastenbaum, 1985), but the empirical evidence suggests that their assumption is unwarranted. People's understanding of death follows the same path of cognitive development as we have seen in other areas of understanding. Even very young children show an awareness of death and its relation to other matters (Bluebond-Langner, 1977; Rosemeir and Minsel, 1982). Infants', children's, adolescents', and adults' levels of awareness of death are consistent with their stages of cognitive development.

During their first years, infants begin to grasp ideas about object permanence, about the disappearance, transformation, separation, and loss of objects and people. As infants develop an understanding of object permanence, they develop a complementary understanding of *im*permanence, change, destruction, and disappearance. Out of infants' understanding that people and things disappear from their immediate experience may grow an understanding of "the death of the object." It will be some time, however, before young children distinguish sufficiently between themselves and their surroundings to distinguish between the "death of the object" and "the death of the self" (Kastenbaum and Aisenberg, 1976). Retrospective reports imply that even very young children may deeply feel someone's death.

Children of all ages think about death (Anthony, 1972; Bluebond-Langner, 1976). Findings from one influential early study (Nagy, 1948) of how children think about death suggest that their ideas develop through two characteristic stages. Preschool children are likely to think that death is temporary.

> When Sam was 3 or 4, he started asking questions about death—who died, when they died, why they died. One day he asked, "How you get out of dead?"
>
> I overheard Erica playing one morning. She had a "bad robot" who killed people and a "good robot doctor" who could "unkill" them.

During this stage of understanding, which corresponds to the stage of preoperational logic described by Piaget, children may think that death means a shrinking rather than an end to life. "Dead people don't get thirsty—well, maybe just a little!" a child at this stage might reason. During the next stage, as they enter the stage of concrete operations, children understand that death is the end to life, but they do not yet understand that it is inevitable and universal. Some children between the ages of 5 and 9 think of death as a person (Anthony, 1972). At around the age of 10, children grasp the idea of death as inevitable and universal, something that will happen to them.

In children who are healthy, the emerging understanding of death reflects the stages of cognitive development postulated by Piaget. But children who are terminally ill learn by experience. They learn from changes in their own condition, from the responses—or lack thereof—of parents, nurses, and doctors, and they learn especially from observing other terminally ill children. They have a systematic understanding of death well beyond that of healthy children their own age (Bluebond-Langner, 1977). Even those terminally ill children who might be considered too young to understand do, in fact, grasp the finality, inevitability, and totality of death. Regardless of their age, terminally ill children see death and dying as mutilating processes that bring on separation and a loss of identity.

The research is sketchy on adolescents' understanding of death. During adolescence, young people try to make sense out of many aspects of their experience. Typically, they reach adolescence in our society without much help in thinking about or coping with death (Lannetto, 1980). Because adolescents are at a transition point between one major role and status and another, they are vulnerable to thoughts of death and suicide (Maris, 1981). In this vulnerability, they resemble adults on the verge of old age, another transition point leading to discontinuity and redefinitions of the self.

Death and the sense of the future

As some adults approach old age, they think more about death in general and about their own death in particular. Some reshape their thinking gradually and some suddenly, perhaps in response to crises.

> I used to be happy-go-lucky, but around the time I was 40 things took a nosedive. My father was operated on for lung cancer, and then he retired. I'd gone into medicine with soaring ideals, but that winter I saw a thousand too many kids with ear infections and runny noses. My job started to feel overwhelming and dull at the same time, a rotten combination. Mikey and Erica were all wrapped up in their friends. I felt drained and bleak. Sometimes I found myself crying for no reason or lost in depression, worrying about whether I was going to get cancer like my father.—Mike and Erica's mother

Children and adults share the perspective that their own death will happen in the future. Research suggests that as some people enter middle age, they begin to think of their lives not in terms of how many years they have lived, but in terms of how many years remain to them (Neugarten, Crotty, and Tobin, 1964). With a foreshortened sense of their future, some older people reorganize their lives and how they spend their time (Kalish, 1976). When people were asked what they would do if they knew that they would die in six months, those over 60 years old were less likely than younger people to say that they would markedly change the way they were living (Kalish and Reynolds, 1976). The implication is that those over 60 already had reorganized their lives with their death in mind.

Other research, however, suggests that older people do not necessarily have a foreshortened sense of the future (Chappell, 1975). Chronological age—the number of birthdays elapsed—has not proved a particularly useful correlate to these people's orientation to death (Kastenbaum, 1985). Older adults' feelings about the future seem to depend more on the degree of control they feel they exert over their environment than on the number of years they have lived or expect to live (Chang, 1979; Schulz and Hanusa, 1977).

Studies of older adults show that there are certain other correlates between people's perspectives on the future and their attitudes toward death. Among older women in good health, for example, those who are more anxious about their own death are also more possessive of time and disturbed by its quick passage than those who are less anxious about death (Bascue, 1973; Durlak, 1972). Those older women who reported the greatest sense of meaning to their lives were least anxious about death, and those women who reported the greatest anxiety about death were most sensitive to the rapid passage of time (Quinn and Resnikoff, in press).

Fear of death

We human beings probably are the only creatures who understand that we will die, and so we probably are also the only creatures who fear death. It has been suggested that people's unconscious, unexamined fear of their own death influences all that they do (Becker, 1973). Specifically, adults say that they fear the dissolution of their body, the end of experience, pain, facing the unknown, and the grief of their families (Diggory and Rothman, 1961). Young adults have been found to say that they fear all of these things except the dissolution of their body (Shneidman, 1971).

Although some middle-aged people have been found to fear death more than either younger or older adults (Schulz, 1978), it has more often been found that the fear of death tends to decrease over the course of adulthood

(Kalish and Reynolds, 1976). When over 1000 adults in the Los Angeles area were polled, only 4 percent admitted to being ''very afraid'' of death, and two thirds said that they were ''not at all afraid'' (Bengston, Cuellar, and Ragan, 1977). It is plausible that older people are less likely to fear death because they value life less than they once did, a consequence of illness, financial problems, the loss of social roles, and the awareness of death. Older people also may feel that they have lived their allotted number of years and may feel psychologically prepared to die (Kalish, 1976).

Death and the sense of the past

Out of the fear of death comes the human urge to understand and reconcile ourselves to its inevitability. As Erikson (1982) suggests, the last stage of psychosocial development requires that people integrate the prospect of their death with the achievements of their past. The final developmental task is to reconcile one's struggle against death and despair with the fulfilling sense that one's life has been meaningful. Thus people attempt to integrate past and future: they turn backward to go forward. Robert Butler (1975) suggests that as people face the prospect of their death, they embark on a process of **life review**. They reflect on their past and review their memories. When people think that only a few years remain to them, they are more likely to talk about the past with others and to consider their memories their most important possession (Marshall, 1975). People who are satisfied with their lives, who remember more turning points in their past, who think that past events have turned out well, and who feel in control of their lives are more likely to reminisce with others. Those who are less satisfied with their lives and the direction of their past are more likely to reminisce by themselves. Reviewing one's life may lead to anxiety, guilt, depression, and despair, or it may lead to resolutions of old conflicts, insight into the past, and a sense that one's life has been meaningful. Once people have reviewed their lives, they tend to move on to other things and to reminisce less (Butler, 1975).

Reviewing the past also can serve other needs. It may lead people to create a continuous sense of themselves, it may enhance their self-esteem, but it may also lead them to try to impress other people (Marhsall, 1980). Reviewing one's life may be a defensive process that does not improve one's ability to adapt. Even those who have reviewed their lives may not handle stress well. A study of older people who were entering homes for the aged showed that about the same proportion, one-third, of those who had avoided thinking of

The last psychological task is to integrate the prospect of one's death with the achievements of one's past. Many older adults embark on a life review that churns up depression, despair, and often resolution, insight, and contentment.

the past, of those who had completed a life review, and those who were in the process of a life review could not handle stress well. It may be that those who successfully review their lives and fashion a meaningful idea of themselves possess special psychological skills and resources (Lieberman and Tobin, 1983).

Dying

Except for those who die suddenly, people go through a psychological process of anticipating their own death. This process has been called the **dying trajectory** and is made up of the emotional states that a dying person traverses (Glaser and Strauss, 1967). How old a person is affects this emotional progression, as does the nature of the person's last illness. The emotional progression begins when a person understands that he or she is dying.

> My grandmother was found to have inoperable bone cancer when she was 76 years old. She was alert enough to ask what was wrong with her. "I can take the news," she insisted. She told the family and the doctors that she wanted "no heroic measures" taken to keep her alive. We agreed to that. She was fed until she couldn't eat and then given fluids until she couldn't drink. For the last ten days of her life, she was unconscious. She died peacefully, as she had wished, at 5 o'clock in the morning.—Sam's mother

When people with terminal cancer were asked whether it was right to tell people that they are dying, 80 percent said that it was, and the same proportion of healthy people agreed (Hinton, 1967). In short, most people want to be told that they are dying. When people are not told that they are dying, they lose the chance to set their affairs in order, to make decisions about when and whether to go to hospital or hospice, to have surgery, and to speak to and see certain people. They also do not tolerate physical pain or surgery as well nor benefit from therapy as fully as people who have been told truthfully that they are dying (Bok, 1978). Some people may realize only unconsciously that they are dying and may alternate between awareness and unawareness (Schulz, 1978). When people are not told of their impending death, they may play out a "ritual drama of mutual pretense" with family, friends, doctors, and nurses in which everyone understands that death is imminent but everyone acts as if it were not (Glaser and Strauss, 1965).

Even when people are not told in so many words that they are dying, they learn the truth from the way that people react—or fail to react—to them (Glaser and Strauss, 1965; Kübler-Ross, 1969). People suddenly grow reluctant to talk about the dying person's condition. People's facial expressions change, and they begin to avoid the dying person. Nurses take longer to answer the calls of people who are dying than of those who are severely ill but expected to recover (Kastenbaum and Aisenberg, 1976). A troubling possibility is that as people withdraw, they may set in motion a self-fulfilling prophecy (Kastenbaum, 1985). Withdrawing care and social contact from people who are sick is likely to make them sicker.

Stages of dying

Elisabeth Kübler-Ross (1969), a physician who has worked with dying people, has theorized the existence of five stages in the dying process. In the first stage, nearly everyone feels *denial and isolation*. People briefly refuse to admit to themselves that they are really dying.

> I went to an orthopedist because my knee was bothering me, expecting nothing more serious than a diagnosis of arthritis or something like that.

> But the doctor said my X rays showed a tumor on the bone. My mind went blank. I stopped being able to understand what he was saying to me. I kept telling myself to calm down and try to listen, but I was in a daze. I had to call him on the phone later that day when I felt I could absorb what he said—"cancer."—Robbie's father

The brief stage of denial gives way to *anger and resentment* when people rail against the unfairness of their plight and bitterly envy those who are healthy. The rage against death may be turned against a doctor, friend, or relative.

In the third stage, people *bargain* over their fate in an attempt to postpone death. People may promise their body to science if only the doctor helps them, or they may promise eternal devotion if only the Lord helps them. Beyond bargaining is *depression*, a time of preparing for the loss of everything and everyone beloved. And beyond depression is *acceptance*, a calm contemplation and expectation of death.

This theory of dying's stages may be neater than people's actual experience proves to be. People may go through some of the stages that Kübler-Ross describes, but many neither go through certain of the stages nor go through them in the proposed sequence (Shneidman, 1973). Some people have been found to withdraw when they learn that they are dying and remain inactive until the end, whereas others have been found to grow more active and to make new friends (Kastenbaum and Weisman, 1972). Kübler-Ross's theory also has been criticized on the grounds that it does not make allowances for individual differences in emotional responses to dying, for the nature of a person's illness, for sex, ethnic group, cognitive style, level of development, or the environment of the dying person (Kastenbaum, 1975).

The timing of death

Some people seem able to control the timing of their death, at least to some extent. A Catholic priest who worked among Alaskan Indians noted that many dying people spent several days planning, reviewing their life, and praying for their family members. Then they called for the priest to administer sacraments and died quietly a few hours afterwards (Trelease, 1975). The postponement of death raises several intriguing questions. How do people postpone their own death? Is dying a continuous extension of the aging process, or is it a discontinuous development? Do people have premonitions of death, and are there warning signs of death that others can learn to recognize? Some older people do not show a pattern of gradual decline before death. Instead, some who are frail but not apparently terminally ill suffer sudden falls, recover quickly, but then quickly relapse and die (Kastenbaum, 1985)—a less expected and more discontinuous pattern of decline. Other people, as we have discussed in earlier chapters, show a pronounced drop in cognitive ability and die soon thereafter.

Grieving

Just as people from different social groups have different attitudes toward dying, they also have different customs for disposing of the dead and for managing the grief of the survivors (Bowlby, 1980). Death tears a hole in the social fabric, and social customs help to mend it. In some parts of the United States and England, social customs such as wearing black and observing a specified period of mourning have fallen into disuse. When mourning customs disappear, survivors may have no clear ways of expressing or establishing psychological and social limits to their grief (Gorer, 1965; Parkes, 1972).

Grief may be the most profound psychological trauma. But it has important psychological functions as well.

The process of grieving

> I had been very close to my mother's mother. She died when I was 18—more than 20 years ago. I still miss her. Sometimes I dream about her, and I can hear her voice and see her so clearly in my dreams that it hurts when I wake up and realize she's not here anymore.—Natalie's mother

> My husband died after 11 gruesome years of Alzheimer's disease. He died by inches. Sometimes I thought his death would be a relief to us all. In a way, it was a release. But as sick as he was, he wasn't *gone* until he died. And when all of a sudden he was gone, it hurt unbelievably.—Sam's grandmother

When a beloved person dies, the survivors experience a feeling of desolation known as **bereavement**. Bereavement has been called the most severe psychological trauma that most people ever feel (Parkes and Weiss, 1983). During the period of acute grief, people suffer both psychologically and physically. The throat feels tight and choked, breath is short, the stomach feels empty, the person sighs, feels weak, and has pangs of intense distress. Sleeping, eating, and engaging in everyday activities are impossible. These feelings come intermittently, last for 20 minutes to an hour, and then subside (Lindemann, 1944).

The uses of grief

> Our older son died of leukemia when he was 6. He had been sick for 3 years. Even though we knew that he was going to die, and we were with him when he died, and even though he died peacefully, I wasn't always sure that I could live with the pain of his death. Before he died, there were times when I cried so hard that I fell down. My throat would close. I felt like I was choking. A year after he died, I cried less often. I slept more. I had more energy. Things gradually felt more like normal for longer stretches of time, and I could function more normally. But the pain still hits me sometimes. I don't think that will ever go away completely.—Megan's mother

Although acute grief is traumatic, it is considered a normal response to the loss of someone beloved. The pain of grief may have important psychological uses. For example, Sigmund Freud (1917) suggested that by grieving, people try to free themselves from their emotional ties to the person they have lost. Grief reflects the inner struggle between the wish to be free and the wish to cling to the dead person. As people grieve, Freud noted, they follow their memories of the dead person and in the process work through many of their feelings. In this view, grief is useful in helping survivors to detach themselves emotionally from the dead and in reestablishing their lives.

John Bowlby and Murray Parkes (1970) explain the use of grief as an adaptation in which people try to reunite themselves with the dead person. The signs of grief are like other signs of distress following separation from a loved one: anxious searching, crying, attacking anyone who interferes. These distress signals may well succeed in reuniting people who have not been separated by death. The distress signals also succeed in drawing others near to offer help and to protect a group member who has been rendered vulnerable (Averill, 1968). The signs of acute grief also signal the existence of an extraordinary situation in which ordinary social rules are suspended. People lost in grief often are treated as if they are ill, and they are allowed to do things that otherwise would be considered socially unacceptable (Parkes and Weiss, 1983).

focus Therapy for the bereaved

Many grieving people begin recovering within a year after a death, although it may take two to three more years before they reorganize their new identity. Therapy sometimes helps people who fall into a pattern of abnormal grief to complete the process of grieving. Those who benefit from this kind of therapy need help in ending their grieving, not in establishing it (Parkes and Weiss, 1983). Widows and widowers whose marriages were difficult and who are in danger of falling into a pattern of conflicted grieving, for example, may be good candidates for this kind of counseling. Bereavement counseling helps people to live more productively and to cope with the stress they feel in the wake of a death (Shneidman, 1973).

Most such counseling follows principles of individual or group therapy, in which people are helped to express their feelings and to develop a personal relationship with their therapist. People may be led back and helped to experience their loss, to understand it intellectually and emotionally, to deal with their ambivalent feelings, and ultimately, it is hoped, to complete the grieving process (Horn, 1974). In the Widow-to-Widow program, widows who have completed the reorganization stage of grieving get in touch with new widows, and provide sympathy, advice, reassurance, and practical as well as emotional support (Silverman et al., 1974). At St. Christopher's Hospice, in London, the staff visit all survivors who rate high on the Bereavement Risk Index. Therapists visit about ten days following a death, and since the counseling program has begun, the numbers of suicides among survivors have dropped markedly (Parkes and Weiss, 1983).

The stages of grieving

Grief for a young person's death may be more intense than grief for an older person.

Grieving is a process of adapting to the loss of someone beloved. As they move through this process, grieving people feel numbness, yearning, disorganization and despair, and reorganization (Parkes, 1972). Although these feelings tend to unfold in sequence, they also tend to blur, and the boundaries between them are not sharp.

People who have just learned of a death may feel *shock, numbness,* and little real feeling. They may even disbelieve the news or insist that someone has made a mistake. Eventually, this protective numbness gives way to *yearning,* when people are flooded by the feelings of acute grief. During this second stage of the grieving process, people are tense, extremely aroused, restless, and in continual search for the dead person. They may call out the dead person's name or go to places associated with the dead person. In this stage of vigilant yearning, people prime themselves to expect to see or hear the dead person, interpreting sights and sounds as the dead person's image and voice. Many grieving people feel the presence of the dead person, a feeling that may be an actual hallucination (Rees, 1970). People dwell on their memories of the dead person, sometimes wishing ''if only'' they had done something differently, the person would still be alive. Feelings of guilt and anger, blaming self and others, alternate.

Yearning is followed by *depression,* a disorganized stage of grief when people may grow lethargic, passive, and defeated. Anger has subsided, and panicky feelings emerge. Acute attacks of grief become less frequent. People may want to run away. Widows are more likely to go through this stage than widowers (Parkes and Weiss, 1983).

For grieving to be completed, people must enter a stage of *reorganization.* They must accept the reality of their loss intellectually and must believe that the world still makes sense. Without this intellectual acceptance, grieving

lifespan focus

Sudden infant death syndrome

In the United States, some 5000 to 10,000 babies die suddenly and unexpectedly each year after suffering few, if any symptoms. These babies go to sleep, apparently well, and never wake up. They are victims of **sudden infant death syndrome (SIDS)**, or crib death—every new parent's nightmare. Found mostly in 2- to 6-month-olds, SIDS may strike as early as at two weeks but rarely occurs after one year. SIDS is the major cause of infant death after the first month. Researchers have proposed many hypotheses about the causes of SIDS, but no one yet knows the precise cause.

Infants who die of sudden infant death syndrome (SIDS) pose especially difficult problems of grieving for their parents. Parents' immediate reactions to SIDS include shock, disbelief, and rage. Usually, the infant has seemed healthy or suffering from nothing more than a cold, and the death is unanticipated. In addition, because many people do not know how to deal with the death of an infant, grieving parents may meet with a wall of silence. Friends may avoid talking about the death, they may try to minimize the loss with insensitive remarks such as, "Oh, you'll be all right. You have another child at home" (Helmrath and Sternitz, 1978). Some people hint that the parents may be to blame for the death (Callahan, Brasted, and Granados, 1983). Parents' grieving for an infant who has died of SIDS is also complicated by the fact that no one yet understands what causes this syndrome, and it is possible for parents to accuse one another of neglect.

Elisabeth Kübler-Ross (Kübler-Ross and Goleman, 1976) has found that parents who lose children to accidents react first with shock and then with a deep rage that must be vented for the grieving process to complete itself. When parents are sedated as soon as they hear the news of a child's accidental death, Kübler-Ross suggests, they may be denied the chance to express the necessary grief. When they do not have a chance to see the child's body and to take leave of it, they find it difficult to accept the reality of the death. Kübler-Ross described one mother who had not viewed her child's body and who, five years later, still turned down the covers on the child's bed every night. When parents can return to the hospital where the child died about a month after the death and can ask questions that they were too benumbed to think of at the time, they can absorb the details of the death in a way that can make the death real to them.

people remain depressed and continually expect new losses. Those who grieve also must accept their loss emotionally and eventually come to feel more pleasure than pain as they remember the dead person. They need to reorganize their identity so that it is independent of the dead person's identity.

Grieving for young and old

The intensity of people's grief varies according to nature of their relationship and the age of the person who has died. People who lose an elderly parent feel less devastated and disrupted emotionally than people who lose a spouse or a child (Owen, Fulton, and Markuson, 1983; Sanders, 1979). Part of the reason that people grieve less for an elderly parent is that thinking about, preparing for, and rehearsing a parent's death is considered acceptable in our society. This preparation helps younger adults to adapt to and prepare emotionally for their own death (Moss and Moss, 1980).

Under social conditions in which children's deaths were common occurrences, older people were more accustomed to these deaths than are people in societies, such as our own, in which children's deaths are felt as cruel and exceptional. Parents and grandparents grieve acutely for the child they have lost as well as for their image of what the child might have become (Callahan, Brasted, and Granados, 1983). Ultimately, most parents do complete their grieving for a child, many of them leaning more heavily on their roles as a worker and spouse in compensation. But the death of some children can disrupt an already shaky marriage. In other cases, bereaved parents are so fearful of bearing another loss that they decide against having other children.

The death of an older husband or wife is generally experienced as less painful than the death of a child or of a young husband or wife. Older people may adjust more easily to the death of a spouse for several reasons (Parkes and Weiss, 1983): they may rely on and take more comfort from traditional mourning customs than younger people do, they may have had more experience with grieving, and they have had more time to prepare themselves psychologically for their loss.

Summary

1. Some characteristics correlated with the longevity of a species include length of prenatal development, age of sexual maturity, size of brain and body, rate of metabolism, and body temperature.
2. The maximum lifespan for humans is between 110 and 120 years and has remained at that limit for hundreds of generations. Although some claims have been made of people living longer than that, the claims have not been authenticated. Only recently has the actual human lifespan begun to approximate the maximum lifespan.
3. For the past two centuries, the average human lifespan has been increasing, a result of improvements in diet, sanitation, and medical care. As premature deaths from infection, disease, and malnutrition have decreased, life expectancy has increased. The average lifespan of different population groups varies according to biological and environmental factors.
4. Clinical death takes place when spontaneous breathing and heartbeat end. Brain death takes place when the higher centers of the brain stop functioning, leaving a person comatose. Cerebral death is the death of the cortex.
5. Whereas in earlier centuries, death suddenly struck many infants and children, today death is more likely to follow after a chronic, lingering disease associated with secondary aging.
6. Since the early 1900s, fewer people die at home, and more die in hospitals and other institutions. Changing attitudes toward death have led some people to favor medically supervised, ''natural'' death at home or in a hospice.
7. People's attitudes toward death are shaped by both social and psychological forces. People in various societies adopt certain strategies to reconcile themselves to the idea of death. Many people in our society today think of decline and death as the natural province of the old. The dying may be isolated and considered hopelessly old and impaired, or they may be considered wise, saintly, and spiritual.
8. Infants', children's, adolescents', and adults' levels of awareness of death are consistent with their stages of cognitive development. Older adults' feelings about the future seem to depend more on the degree of control they feel they exert over their environment than on the number of

years they have lived or expect to live. The final developmental task is to review one's life and to reconcile one's struggle against death and despair with the fulfilling sense that one's life has been meaningful.

9. People go through a psychological process of anticipating their own death, called the dying trajectory.
10. Elisabeth Kübler-Ross has theorized the existence of five stages in the dying process: denial, anger, bargaining, depression, and acceptance.
11. When a beloved person dies, the survivors experience a feeling of desolation known as bereavement, perhaps the most severe psychological trauma that most people ever feel. The pain of grief may have important psychological uses.
12. Grieving is a process of adapting to the loss of someone beloved. As they move through this process, grieving people feel numbness, yearning, disorganization and despair, and reorganization. The intensity of people's grief varies according to the nature of their relationship and the age of the person who has died. The death of an older parent, husband, or wife is generally experienced as less painful than the death of a child or of a young husband or wife.

Key terms

longevity
maximum lifespan
average lifespan
life expectancy
clinical death
brain death
cerebral death
hospice
appropriate death
life review
dying trajectory
bereavement
sudden infant death syndrome (SIDS)

Suggested readings

KALISH, R.A. "The Social Context of Death and Dying." In R.H. Binstock and E. Shanas (eds.), *Handbook of Aging and the Social Sciences*, New York: Van Nostrand, 1985. A discussion of attitudes toward death and dying, the dying process, and grief, bereavement, and mourning.

KASTENBAUM, ROBERT. "Dying and Death: A Life Span Approach." In J.E. Birren and K.W. Schaie (eds.), *Handbook of the Psychology of Aging*. New York: Van Nostrand, 1985. A discussion of expectations about growth and decline and about people's orientation to death throughout the lifespan.

KÜBLER-ROSS, ELISABETH. *On Death and Dying*. New York: Harper and Row, 1973. The author's findings about the stages that generally occur during the process of dying.

KÜBLER-ROSS, ELISABETH. *Death: The Final Stage of Growth*. Englewood Cliffs, NJ: Prentice-Hall, 1975. A discussion of the death-denying tendencies in our society.

WALFORD, ROY. *Maximum Life Span*. New York: Norton, 1983. Current thinking about the lifespan as well as Walford's research and perspective on lengthening the lifespan.

Epilogue

We have come full circle, from cradle to grave, from one end of the human lifespan to the other. We have seen that development is not the sole province of the young, as psychologists once believed. People do not stop developing when they emerge from adolescence; adulthood is not all loss of abilities and decay. Much of the decline associated with aging, in fact, turns out to be the symptoms of disease, of disuse, and of abuse. The implications of this discovery by lifespan developmental psychologists are profound. Now that this message has begun to reach more people, their health habits have begun to improve, and the mortality statistics reflect the improvement. For just one example, now that people know that diet, exercise, smoking, and stress affect the development of cardiovascular disease, many people have modified their habits and are living healthier, longer lives.

We also have seen that development—in the sense of both growth and decay—is a lifelong process. Normal aging is a lifelong process. Decay, for example, is present when cells die in the infant's cortex. Growth is present when the 70-year-old starts a business or develops true wisdom about human behavior. Longitudinal research tells us that intelligence may continue to increase well into adulthood, and an understanding of the dialectical complexities of experience seems to be the exclusive province of the aged. Throughout life, people change and develop in response to events in the physical, social, and historical world.

We have seen that from birth to death, there is both continuity and change in people's development. We are the complex sum totals of all our individual yesterdays, our todays, and our expectations about tomorrow. We are the sum totals of the personality, the body, the intelligence that we were born with and, over time, have become.

Our early experiences seem to influence our later development in several different ways. Some early experiences leave an indelible mark on all later development. A birth defect or an extraordinary musical talent, for example, marks all later developments in one's life. In contrast, some early experiences leave few traces or none. Some people, as we have seen, are resilient enough to develop adaptively despite the most unpromising early experiences—premature birth, physical or mental abuse, an impoverished or broken family. The short- and long-term effects of still other early experiences vary according to where in their development people are when the experiences take place. For example, parents' divorces affect the young child, the adolescent, the young adult, and the older adult differently. To an adolescent, divorce might affect identity formation, but it might affect a young adult's decision to marry. Finally, some long-term effects appear only under certain circumstances, almost as if they had lain dormant for years. For example, adults are likely to be strongly influenced in the way they behave to their children by the way that their own parents behaved to them.

The questions that remain for researchers to answer are vital questions for everyone. Throughout the lifespan, what is the dividing line between normal aging and the results of abuse, neglect, or disease? How much *can* people change? And what are the limits of possibility at each stage of the lifespan, from conception until death?

Glossary

A *not* B phenomenon The inability of the child, until 12 months of age, to follow the movements of an object when it is first hidden under cover A, then put, while the child watches, under cover B. The child looks for the object under A.

accommodation In Piaget's theory, the process of adjusting existing ways of thinking, of reworking schemes, to encompass new information, ideas, or objects. In the visual system, the adjustment of the lens of the eye to shifting planes of focus.

acquired immune deficiency syndrome (AIDS) A sexually transmitted disease that impairs the immune system and makes the patient susceptible to opportunistic infections. It may cause dementia and is so far always terminal.

acuity Sharpness of the visual system, which in infancy depends partly on physiological maturity.

acute brain dysfunction An organic brain disorder with reversible symptoms (compare chronic brain dysfunction).

acute illnesses Illnesses that last a relatively short time; most prevalent in childhood (compare chronic illnesses).

adaptation In Piaget's theory, the mental processes by which people extend and modify their thinking. They do so in two ways, by assimilation and by accommodation.

adolescence The biological and psychological changes that individuals go through at the end of childhood.

adolescent egocentrism The tendency of adolescents to think that they are more important and unusual than they actually are.

affordances Properties of objects that afford or allow for certain activities.

afterbirth The placenta and the parts of the fetal membranes that are delivered after the birth of the baby, in the third and final stage of labor.

alienated achievement An identity status of adolescents during the Vietnam War era, in which they rejected certain "straight" social conventions.

alternative reproduction Recently developed methods for treating infertility, such as in vitro fertilization, ovum transfer, and surrogate mothering.

alveoli Tiny air sacs of the lungs walled by a very thin epithelial membrane; they are surrounded by capillaries and are the site of respiration.

Alzheimer's disease A chronic brain dysfunction that causes, first, the loss of higher cognitive abilities, then the loss of basic abilities, and, finally, death.

ambivalent-insecure attachment Relationship of an infant to his or her caregiver in which the child is angry and resistant.

amniocentesis A prenatal medical procedure for diagnosing genetic abnormalities. A small amount of amniotic fluid is withdrawn from the pregnant woman by hollow needle. The chemical contents of the fluid reveal some diseases directly; cells from the fetus are cultured and studied for chromosomal anomalies.

amnion A thin, membranous, fluid-filled sac surrounding the developing embryo and fetus.

anal stage In Freud's theory, the psychosexual stage of toddlerhood, from 12 to 36 months, during which the child receives pleasure through stimulation of the anus and pays much attention to elimination.

animism The belief that all things are living and endowed with intentions, consciousness, and feelings.

anorexia nervosa A severe disorder in which the person, usually an adolescent girl, eats very little and loses 25 percent or more of original weight. The girl has an intense fear of becoming obese, feeling fat even when emaciated.

Apgar score A rating of the neonate's physical well-being, ranging from 0 to 10, determined by grading the infant's heart rate, breathing effort, muscle tone, reflex irritability, and color at 0, 1, and 2 points—poor, fair, or good.

appropriate death The kind of death a person chooses when the person is allowed to choose.

apnea A transient interruption of breathing.

areola Pigmented ring surrounding the nipple.

arteriosclerosis Stiffening and thickening of the arterial walls.

artificial insemination A procedure in which either the husband's or a donor's sperm are introduced into the vagina via syringe rather than through sexual intercourse.

assimilation In Piaget's theory, the process of adjusting new information and objects to make them fit existing ways of thinking or scheming.

astrocytes Glial cells that prevent the build-up of neurotransmitters in between neurons.

asynchrony Want of coincidence in time; the condition of two or more events not happening together.

astigmatism A defect of vision owing to corneal irregularity.

atherosclerosis Loss of elasticity of arterial walls plus deposits of fat that decrease the blood flow.

atropic gastritis Chronic inflammation of the stomach lining, which results from repeated injury and not from normal aging.

attachment An affectionate, close, and enduring relationship between two persons who have shared many experiences; the tie to their parents felt by children 6 months and older.

attention deficit disorder See hyperactivity.

authoritarian parent A parent who rears the child to believe that his or her often arbitrary rules are law, that misconduct by the child will be punished. The authoritarian parent is detached and seldom praises the child.

authoritative parent A parent who sets firm limits and provides direction for the child but at the same time is willing to listen to the child's objections and make compromises.

autism A rare but devastating childhood psychosis characterized by profound aloneness and an indifference to communication and social relationships.

autonomous mothers Mothers who take pleasure in seeing their children as independent individuals.

autosome A chromosome other than a sex chromosome. In the human being, any one chromosome of the 22 pairs that carry most of the genetic code for physical size, eye color, hair color, nose structure, chin structure, intelligence, and so on.

average lifespan Average length of time that members of a species survive.

avoidant-insecure attachment Relationship of an infant to his or her caregiver in which the infant tends to ignore, turn away from, or avoid the caregiver, even after a brief separation.

axon A long thin fiber that may extend a considerable distance from the nerve cell body and that conducts impulses away from it.

babbling Extending repetitions of consonant-vowel combinations beginning when babies are about 4 months old; vocal play that exercises the speech musculature and gives babies practice in making sounds.

baby talk The special language form used by adults when speaking to infants. The voice is usually high in register but has an exaggerated range of pitch and loudness. The adult chooses simple phrases and nonsense syllables and repeats them in a singsong fashion.

behaviorism A school of psychology that originated with John Watson and his proposal that only observable behavior should be studied, without reference to consciousness or mental processes.

bereavement The feelings of desolation survivors feel after someone's death, perhaps the most severe psychological trauma people can feel.

blastocyst A hollow, fluid-filled sphere whose wall is one cell thick except for the heap of large cells at one side forming the embryonic disk.

blastula A hollow, fluid-filled sphere made up of more than 100 cells; its wall is uniformly one cell thick, but the cells on one side are larger.

bonding The formation of an enduring attachment, such as to one's child, especially in the period immediately following birth.

brain death The loss of function of the higher centers of the brain.

bulimia A disorder in which the individual indulges in secret and episodic eating binges, which often end in self-induced vomiting.

canalized Relatively invulnerable to environmental effects.

cardiac output The amount of blood the heart pumps in 1 minute.

case study Research that collects detailed historical or biographical information on a specific individual.

catch-up growth Rapid physical growth during the first 5 months by an infant whose size was restricted by a small womb or by maternal malnutrition; or by a child who has been restored to health after illness or malnutrition.

cephalocaudal (development) A pattern of physical growth and motor development progressing from head to foot; in human beings, followed prenatally and throughout childhood.

cerebellum The region of the brain that coordinates movement and sensory input.

cerebral cortex The outer layer of gray matter covering the brain; responsible for complex information and conscious thoughts.

cerebral death Death of the cortex of the brain.

cerebrovascular accidents The cutting off of the brain's blood supply, which results in strokes.

cervix The very small outer or lower opening of the uterus; opens to the vagina.

child abuse The flagrant mistreatment or neglect of children, which federal and state laws now require doctors, clinic and hospital personnel, teachers, social workers, police officers, coroners, and other professionals to report. Defined by federal law as "the physical or mental injury, sexual abuse, negligent treatment or maltreatment of a child under the age of 18 by a person who is responsible for the child's welfare."

chorion The outermost of the membranes surrounding the developing embryo and fetus; villi from the chorion become part of the placenta.

chorionic villus biopsy A prenatal chromosome test in which a sample of fetal tissue, from the chorion, is taken and tested for abnormalities.

chromosomes Microscopic threadlike structures, consisting of a linear string of genes; chromosomes are the transmitters of inheritance and are present in the nucleus of every cell.

chronic brain dysfunction Organic brain disorder with irreversible symptoms (compare acute brain dysfunction).

chronic grieving An abnormal pattern of bereavement in which people feel intense grief, anxiety, and depression long after they could be expected to resolve those feelings.

chronic illnesses Illnesses that last a relatively long time, resist cure, and tend to get worse (compare acute illness).

circular reaction In Piaget's theory, an action that is repeated because it pleases.

classical conditioning A basic form of learning in which a neutral stimulus, through association with a physiologically significant stimulus, eventually comes to evoke the response made spontaneously to the significant stimulus.

climacteric In women, the decline in ovarian function that results in menopause; in (some) men, a similar downward trend in sexual function.

clinical death The point at which spontaneous breathing and heartbeat stop.

clique A closely knit, small circle of persons who tend to exclude others.

closed classroom A self-contained classroom.

cochlea In the inner ear, a bony, fluid-filled canal, coiled like a snail, containing within it a smaller, membranous, fluid-filled spiral passage wherein lie the nerve endings essential for hearing.

cohort A group of individuals born at the same time or during the same historical period.

cohort differences Differences between groups of people (cohorts) born at different points in time; for example, education differences between young and old adults.

cohort–sequential research design A type of research that includes elements of both longitudinal and cross-sectional designs, in which several different-aged samples (cohorts) are studied for a period of time.

colostrum A yellowish protein-rich fluid that is secreted from the mother's breasts a few days before a baby's birth and for two to four days after.

conception The formation of a viable zygote through the union of ovum and sperm, being the first step in the development of a new human being. The act of becoming pregnant.

concordance In genetic studies of twins, their similarity in one or more traits or in a diagnosis.

concrete operational thinking In Piaget's theory, the mental ability of children from ages 7 to 11 by which they can think logically about physical objects and their relations. Concrete operations include conservation, reversibility, seriation, and classification.

conflicted grief An abnormal pattern of bereavement in which people at first seem to accept a spouse's death but later cannot accept their loss.

conservation In Piaget's theory, the ability to recognize that important properties of a given quantity of matter, such as number, volume, and weight, remain constant despite changes in shape, length, or position.

conventional moral reasoning Kohlberg's intermediate level of moral reasoning, in which conforming to social conventions, or standards, such as pleasing others by being a ''nice guy'' and maintaining ''law and order'' for its own sake, are the primary moral values.

convergent thinking Thinking that has as its goal the selection of a single, acceptable, conventional solution to a problem.

corpus callosum The band of myelinated tissue that connects the two hemispheres of the brain.

correlation The degree and direction of relationship between two variables, which may be positive, with both increasing or decreasing at the same time, or negative, with one increasing and the other decreasing.

coupled mothers Mothers who take pleasure in how their children give meaning to their lives and to their sense of identity.

couvade A ritual of cultures in many parts of the world in which the father takes to his bed when his child is born and complains that he has suffered the pains of childbirth.

critical period An interval during which certain physical or psychological growth must occur if development is to proceed normally and during which the individual is vulnerable to pertinent harmful events and open to beneficial effects.

cross-linkage theory A theory of aging proposing that the stable bonds between cells become rigid and lose function.

cross-over The exchange of portions from strands of homologous chromosomes that have broken at the same level during meiosis.

cross-sectional study A research method in which groups of children of different ages but similar in other important ways—educational level, socioeconomic status, proportion of males to females—are compared at some specific point in time, usually for a specific aspect of behavior. The groups represent different age levels.

crowd A loosely organized group of three or four cliques of adolescents centered on some activity rather than on close friendship.

crowning During childbirth, the first appearance of the newborn's head.

data Evidence systematically collected during scientific studies.

day-care center An institutional setting in which preschool children are cared for when their parents are working or otherwise unavailable. Day-care centers usually are open 5 days a week, all day, all year round.

decentration In Piaget's theory, the ability to focus or the process of focusing on more than one dimension, such as length and height, simultaneously; a concrete operation.

delirium A disorder that disturbs brain metabolism (see acute brain dysfunction).

dendrites Any of the usually short and branched extensions of a nerve cell that conduct impulses toward the cell body.

deoxyribonucleic acid (DNA) A complex chemical substance whose molecules constitute genes and are able to replicate themselves.

depth of focus The distance that an object can be moved without a viewer perceiving a change in sharpness.

diabetes Illness in which the pancreas does not manufacture sufficient insulin, which is necessary to metabolize sugars and regulate blood sugar.

dialectical reasoning A kind of thought that develops in adulthood in which people constantly try to order their understanding of life, knowing that new syntheses will be needed in the future.

dialectical view A conception of human development—and especially such processes as communication, language, and problem solving—as a constant process of thesis, antithesis, and synthesis.

diethylstilbestrol (DES) A hormone formerly prescribed to pregnant women to prevent miscarriages. It was found to cause abnormalities and disease in the sex organs of their children when they reached their teens.

divergent thinking Thinking that goes in different directions and searches for new ideas and a number of solutions to a problem.

diverticuitsis Irregular pouches along the intestinal walls that become inflamed and painful.

dizygotic (DZ) twins Birth partners who have developed from two ova fertilized at the same time by two sperm. The genotypes of these twins are as different as those of any siblings.

DNA repair theory A theory of aging proposing that aging results from long-term damage to the DNA in cells.

dominant gene In a pattern of genetic inheritance, the one of a pair of genes that determines the trait and suppresses expression of the other.

double-blind study Research in which neither subjects nor researchers know which subjects are in a treatment group and which are in a control group. Double-blind studies may prevent bias from coloring research results.

Down's syndrome A disorder caused by an extra chromosome 21; the affected child is mentally retarded and has a number of distinctive physical signs.

dwarfism Abnormally short stature and immature physical development found in children under severe emotional stress.

dying trajectory The psychological process of anticipating one's death.

dyscalcula A learning disability that affects calculating skills.

dysgraphia A learning disability that affects writing skills.

dyslexia A pattern of difficulties in reading and spelling that appears in people of normal intelligence. They see letters backward on the page, because they process visual information unusually slowly and because letters persist in their mind's eye.

early maturers Young people who undergo the physical changes of adolescence a year or two before the majority of their age-mates.

ecological approach A view of human behavior that looks at how people are affected throughout their lives by the changing environments within which they live and grow.

ectoderm The outermost of the three primary cell layers of the embryo from which develop the skin, sensory organs, and nervous system.

ego In Freud's theory, the predominantly conscious part of the personality that acts on the reality principle and makes decisions while restraining the id and taking the strictures of the superego into account.

egocentrism One's inability to take another person's point of view; this imbues and colors thinking in early childhood.

elastin A protein that makes up some of the arterial walls.

Electra conflict The female counterpart of the Oedipal conflict in psychoanalytic theory. In the phallic stage, at about age 4, little girls have an erotic attachment for father and are antagonistic toward mother. These feelings are usually repressed.

embryo In the human being, the developing individual from 2 to 8 weeks following conception, during which time basic body structures and organ systems are forming.

embryonic stage The period extending from the 2nd to the 8th week after conception, during which the organs of the conceptus are differentiated and rudimentary anatomy becomes evident.

empathy The capacity to experience vicariously another person's emotional state.

emphysema Reduction of the lungs' surface area as walls between alveoli (air sacs) are destroyed.

encoding Putting impressions into short-term memory.

endoderm The innermost of the three primary cell layers of the embryo from which develop the lining of the digestive tract, the salivary glands, pancreas, liver, and the respiratory system.

endogenous smiles Spontaneous smiles made by the sleeping or drowsy infant during the first 2 weeks of life; they come when the baby relaxes after internal neurophysiological arousal.

endometriosis Growth of uterine lining tissue outside the uterus, a common cause of infertility in women.

environment All the conditions, circumstances, and influences that surround an organism.

epidural anesthesia Medication injected under the mother's skin at the bottom of the spine to relieve the pain of childbirth; it numbs the body between chest and knees. Epidural anesthesia may slow the contractions of labor and the newborn's motor abilities.

episodic memories Memories of events tied to specific times and places, which may be affected by aging (compare semantic memories).

equity theory A theory of personal relationships according to which people are presumed to evaluate the compatibility of the assets and liabilites that each contributes to the relationship.

ethology The observational study of how animals behave in their natural surroundings.

exogenous smiles Smiles made by an infant, beginning at about 1 month of age, stimulated by something in the external world, such as a nodding head speaking with a high voice.

expression In heredity, the display of genetic characteristics in an individual's phenotype.

extensive peer relations The distribution of social interactions and friendships across a number of other children, not just a single, close friend. See intensive peer relations.
external locus of control The belief that success and failure are beyond one's own personal control.
external morality In Piaget's theory, the code of ethics of school-age children that has its origin outside of them. The children obey rules in the belief that the rules have been laid down by authority figures and that they will be punished if they do not obey them.
extrinsic motivation Inducement to do things to get a prize or promised reward.
extroverted A personality that is friendly and outgoing.
failure to thrive Want of attaining expected height, weight, and behavior by a baby for no evident organic reason; usually caused by emotional neglect and inattention of parents.
fallopian tubes The pair of tubes that conduct the egg from the ovary to the uterus, having at its upper end a funnel-shaped expansion to capture the egg; fertilization occurs in the fallopian tubes.
fetal alcohol syndrome A congenital condition of infants whose mothers drink excessively during pregnancy. The infants have a small head, widely spaced eyes, a flat nose, an underdeveloped upper jaw, and are retarded in mental and motor development.
fetal stage The period, extending from the 9th week after conception until delivery, during which the conceptus grows large and its organs and muscles begin to function.
fetology A new field of medicine that treats problems of the fetus before birth.
fetus The developing unborn child from approximately 8 weeks after conception until birth.
fissures Folds in the brain that increase its surface area.
fontanel One of the six spaces, covered by membrane, between the bones of the fetal or young skull.
formal operation In Piaget's theory, the mental ability, supposedly attained in adolescence, by which the individual can think logically about abstractions, can speculate, and can consider the future, that is, what might and what ought to be. Formal operations include hypothetico-deductive, inductive, and reflective thinking and interpropositional logic.
fovea The central part of the retina in which focused visual images are formed. In infants the fovea is not distinct.
free radical theory A theory of aging proposing that certain unstable chemicals (free radicals) bond and damage cells and chromosomes.
full-term baby An infant born in the normal range of 259 to 293 days after the first day of mother's last menstrual period.
gender roles The behaviors and attitudes a society considers acceptable for males and females.
gender identity The deeply ingrained sense of being a boy or girl, man or woman. After age 3, gender identity resists change.
gene The basic unit for the transmission of hereditary characteristics; a portion of DNA in a fixed linear position on a chromosome.
genital herpes Sexually transmitted viral infection that causes genital sores.
genital stage In Freud's theory, the last stage of psychosexual development in which the individual has mature sexual relations with the opposite sex.
genotype The individual's genetic makeup, the totality of the genes inherited from the parent cells. The genes are present in each cell of the developing individual, beginning with the fertilized egg.
gentle birth A method of childbirth to give the newborn a welcoming introduction ot the world, advocated by Frederick Leboyer. Lights are dimmed, sounds muffled; the baby is put immediately on the mother's stomach and then into a warm bath.
germinal stage The period from conception until about 14 days later when the many-celled conceptus has become firmly implanted in the wall of the uterus.
giantism Abnormally tall stature.
gonorrhea Sexually transmitted inflammation of the genital mucous membranes caused by the gonococcus; found in both males and females. Sometimes there are no symptoms.
growth curve A graphic representation of the relative growth of an individual or population during successive periods similar in length.
growth spurt In adolescence, the sharp and rapid increase in height and other physical growth that continues for about three years.
guilt A feeling of responsibility or remorse for having violated some ethical, moral, or religious principle.
habituate To become used to something by frequent repetition or prolonged exposure.
Hayflick limit The finite number of times a cell will divide.
Head Start A national preschool program designed to provide children from poor families with the advantages usually afforded middle-class children, including good nutrition, health care, and educational experiences.
holophrase Single-word utterance that 12- to 18-month-old children combine with gesture and intonation to accomplish the work of a whole sentence.
homologous (chromosomes) A pair of chromosomes that are generally similar in size and shape, one having come from the male parent, one from the female parent.
horizontal décalage In Piaget's theory, the notion that the development of similar cognitive abilities is not simultaneous; children do not acquire these abilities all at the same time.
hospice Institution devoted to making the life of dying people as comfortable and dignified as possible.

hydrocephaly An abnormal increase in the amount of cerebrospinal fluid within the cranium, especially in infancy, that causes the cerebral ventricles to expand; the skull, in particular the forehead, to enlarge; and the brain to atrophy.

hyperactivity A disorder in which the child cannot remain still or pay attention in situations that demand it. Other symptoms are distractibility, little tolerance for frustration, readily aroused emotions, and problems in learning, listening, and completing tasks. Now more commonly called attention deficit disorder.

hypertension High blood pressure.

hypothesis A guess or prediction about the world that may be tested empirically.

id In Freud's theory, the unconscious aspect of the personality, which is made up of impulses and is governed by the pleasure principle. The id, present at birth, wants whatever satisfies and gratifies and wants it immediately.

identity One's unique psychological picture of oneself.

identity achievement The stage of psychosocial development reached by adolescents who have a sense of identity, have career goals, and are committed to religious and political views.

identity crisis In Erikson's life-cycle theory, the core crisis of adolescence; the urgent need for greater self-understanding and self-definition.

identity diffusion In Erikson's terms, the resolution of the identity crisis of individuals who do not commit themselves to a cause or life course.

identity foreclosure The stage of psychosocial development reached by adolescents who accept the identity set for them by their parents.

imaginary audience Hypothetical people whom adolescents imagine scrutinize them.

implantation The process by which the trophoblast grows into the uterine wall.

imprinting The rapid, innate learning of very young birds and perhaps other animals, within a limited critical period of time, to follow and form a continuing filial attachment to the first large moving object seen.

incubator An apparatus for the housing of premature or sick babies.

infertility Inability to conceive or impregnate, diagnosed after a year of unprotected intercourse, and treatable in many cases.

information processing A theory that describes the flow of information into and out of memory; stresses quantitative mental capacities rather than qualitative advances.

intelligence quotient (IQ) Originally, an individual's intelligence defined as mental age, as determined by a standard test, divided by chronological age and multiplied by 100. Now determined by the standard deviation of the distribution of MA scores at a given age level and the deviation of the individual's score from the mean of this distribution.

intensive peer relations The expression of intense feelings and sharing of experiences and fantasies with just one or two other children. See extensive peer relations.

internal locus of control The belief that success and failure are within one's own personal control.

internal morality In Piaget's theory, the more advanced and mature form of moral reasoning in which the individual recognizes that rules are formulated through reasoning and discussion among equals.

interpropositional logic The ability to judge the truth of the logical relationship of propositions; a formal operation.

interval scale An arrangement of quantitative data on a scale in which divisions are equally spaced e.g., inches, pounds.

interventions Deliberate attempts to intercede in a person's life and development.

introverted A personality that is shy, withdrawn, and perhaps anxious.

intuition In Piaget's theory, the preconceptual or prelogical thought of children from about 4 to 6 or 7 years of age, when children reason by guesses, not logic.

invariants Unchanging properties of objects and events.

in vitro fertilization Literally "in glass," laboratory fertilization of a human ovum.

ischemic heart disease Oxygen starvation of the heart muscle.

juvenile delinquency Crimes and the antisocial acts considered wrong because they are committed by minors, like running away from home, promiscuity, and truancy.

keratoses Wart-like skin growths, associated with aging.

kwashiorkor Severe malnutrition of children 2 to 4 years old whose diet consists mostly of carbohydrates with very little protein. Symptoms are generalized swelling of the stomach, face, and legs; anemia; thinning hair, which changes color; and skin lesions.

labor The three-stage physiological process that results in the delivery and birth of an infant.

language play Rhyming, saying nonsense words, and playing with words.

lanugo A soft, fine hair that covers the fetus's body during the 5th and 6th months of prenatal development; some may persist on parts of the newborn's body for a few weeks after birth.

laparoscopy Examination of the abdominal cavity through a very small incision, now usually with the help of a fiber-optic illuminator.

late maturers Young people who undergo the physical changes of adolescence one or more years after the majority of their age-mates.

latency period In Freud's theory of psychosexual development, the years from age 7 to 11 during which the child has few sexual interests.

lateralization In the brain, the division of functions between the left and right hemispheres. The left hemisphere appears to control verbal skills, the right hemisphere visual skills and imagery.

learned helplessness An attitude of individuals that they simply cannot surmount failure, which tends to perpetuate it.

learning disability A problem that a child has in one or more of the basic processes necessary for using and understanding language and numbers.

life cycle squeeze Stress that afflicts people when they are overloaded by demands at their stage in the life cycle; for instance, when middle-aged adults must care for aging parents and adolescent children.

life expectancy The average amount of time a person can expect to survive.

life review Reflection on the past and attempts to integrate it with the future as a person faces the end of life.

lifespan approach The school of psychology that studies human development from birth to death; a perspective on human development, which considers growth and decline from conception to death.

longevity Length of life.

longitudinal research design A research method in which the same group of children is studied over an extended period of time to determine how they change with age.

long-term memory Storage of information at the deepest level of cognitive processing.

macrosystem In the ecological view of development, the broad institutional patterns in a person's culture that affect his or her development.

magico-phenomenistic thinking In Piaget's theory, the infant's faulty linking of an inappropriate action and a desired end, which seems magical to the child because he or she does not really understand means and ends.

marasmus Severe malnutrition of infants under 1 year of age who are deprived of necessary proteins and calories. They gain no weight, hardly grow, lose their muscles, and become emaciated.

masking The complete or partial obscuring of one sensory process with another.

maturation The developmental acquiring of skills by normal children through physiological growth, probably chiefly neural growth, rather than through learning.

maximum lifespan The greatest age to which any member of a species survives.

mean A value intermediate between two extremes.

meconium Tarry waste that accumulates in the bowel during fetal life and is discharged shortly after birth.

meiosis A process of cell division unique to the gamete cells, leaving each sperm or ovum with 23 single chromosomes instead of 23 pairs.

melanin Dark skin pigment.

melanocytes Skin cells that contain melanin.

menarche The first menstrual period; the establishment of menstruation.

menopause The end of menstrual periods.

mesoderm The middle of the three primary germ layers of the embryo from which the skeleton muscles, kidneys, and circulatory system develop.

mesosystem In the ecological view of development, the network of ties among major settings in a person's life, such as family, school, friends, church, and camp, which affect the person's development.

metabolism The chemical processes by which nutrients are broken down, releasing energy and wastes, and by which small molecules are built up into new living matter, by consuming energy.

metamemory The awareness of the phenomenon of memory and an understanding of how it works and how to aid it.

microanalysis In studies of social communications, a method that allows very close examination, such as stopping frames of films and videotapes or running them in slow motion.

microcephaly Abnormal smallness of head and brain area, usually causing mental retardation.

microsystem In the ecological view of development, the network of ties between a person and his or her immediate setting, such as a school or office.

midlife crisis A stressful period, hypothesized to occur as a person enters middle age and faces difficulties in psychological and physical development.

mitosis The usual process of cell division in which the chromosomes split and duplicate within the nucleus, then separate, with one member of each duplicate going to one of the two daughter cells formed through division of the cytoplasm.

moderate mental retardation An IQ score of 40 to 55, usually accompanied by physical disabilities and neurological dysfunctions.

modified extended family Family form in which units of different generations live separately but retain emotional ties.

monozygotic (MZ) twins Birth partners who have developed from a single fertilized egg and thus have the same genotype.

Moro (embracing) reflex The newborn's normal reflexive response to a loud sound or a sudden drop of head and neck: the infant flings arms and legs out, fans its hands, and then convulsively brings arms toward the middle of the body.

morula A solid ball of 12 to 16 cells formed 60 hours after fertilization through several cell divisions; resembles a mulberry.

motherese The special language form used by adults when speaking to immature speakers. It consists of short simple sentences, frequent repetition, and emphasis on the here and now.

multi-infarct dementia Chronic brain dysfunction resulting when blood clots repeatedly cut off blood supply to the brain.

mutation A sudden permanent change in hereditary material, physical in a chromosome, biochemical in a gene.

myelin A fatty covering on the axons of nerve cells, which increases the speed of nerve impulses.

myelination The accretion of myelin on axons.
naturalistic observation A research method in which records of people's or animals' spontaneous behavior are made in the natural setting with minimal intrusion by the observer.
nature In developmental psychology, the genetic and physiological factors that can affect development of the individual.
need for achievement An individual's motivation to strive for excellence.
negative identity A negative self-concept and low self-esteem.
neobehaviorist More recent versions of behaviorist theory.
neo-Freudian theories More recent versions of Freud's theory of psychosexual development. They emphasize the social and cultural factors in the development of personality.
neonatology A specialization in medicine directed to the medical care of the infant during the newborn period.
neo-Piagetian theories More recent versions of Piaget's theory of cognitive development in childhood.
neurofibrillary tangles Tangled bunches of spiral strands of protein that appear in older brains.
neuroglia Delicate, branching connective tissue that fills the spaces between neurons, supporting and binding them together.
neurons Nerve cells.
neurotransmitters Chemicals in the brain that carry a nerve impulse across the synapse between neurons or keep it from crossing.
Nissl substance Material surrounding neurons that often appears just when a new function emerges.
norm A single value or a range of values constituting the empirically established average or standard performance on a test under specified conditions.
normative stability The tendency for a person to change in some way over time but to remain in the same relation to members of his or her cohort.
nursery school A preschool that prepares children for kindergarten, usually with periods of group activities and shorter hours than day care. See day-care center.
nurture In developmental psychology, the factors in the environment that can affect development of the individual.
object permanence The understanding that objects continue to exist, even though they are out of sight or disappear from view; develops gradually between 6 and 18 months of age. A term used by Piaget.
Oedipal conflict In Freud's theory, the erotic attachment little boys in the phallic stage have for mother and their fear that father will find them out. These feelings are usually repressed.
ontogeny The development of the individual.
oocytes Immature egg cells.
open class A class in which choice of activities is flexible and decisions about where to be and what to do are often made by children, individually and in small groups, rather than by the teacher.
open classroom A schoolroom in which spatial arrangements are flexible and physical equipment, including desks and walls, are moveable.
operant conditioning A learning process described by B. F. Skinner in which the individual responds again in a way rewarded in the past or refrains from activity earlier punished.
operant learning A form of training in which reward and punishment are used to shape behavior.
operation In Piaget's theory, a basic and logical mental manipulation and transformation of information.
oral stage In Freud's theory, the psychosexual stage from birth to 12 months, during which the mouth of the infant is the source of sustenance and pleasure.
organic brain disorders Pathological brain conditions, usually accompanied by impaired memory, intellect, judgment, orientation, and emotions.
organization In Piaget's theory, the process of combining and integrating perceptions and thoughts that continues throughout a lifetime. In information processing theory, cognitive strategy for memorizing, grouping items into more general classes, clusters, or chunks.
ossification The process of bone formation in which, beginning in the center of each prospective bone, cartilage cells are replaced by bone cells.
osteoarthritis Painful degeneration of the joints, a result of heredity, hormones, diet, and wear and tear on the ends of the bones.
osteoporosis Loss of calcium from bones, resulting in brittle, easily broken bones; accelerated by estrogen decline following menopause.
ova Human egg cells.
ovary The typically paired female reproductive organ that releases ova and in vertebrates, female sex hormones; in the adult human female an oval, flattened body suspended from a ligament.
overextension Children's generalized use of a given word for several objects that share a particular characteristic but have others that are dissimilar, such as "fly" for flies, specks of dust, all small insects, crumbs of bread.
overregularize To extend the rules for the inflections denoting past tense and plural to verbs with irregular past tenses and nouns with irregular plurals.
ovulate To release a maturing ovum from the ovary into a fallopian tube; it occurs once about every 28 days from puberty until menopause.
ovum transfer A procedure for treating infertility whereby an ovum is surgically removed from a woman's ovary, fertilized in vitro, and then reintroduced into her own or a surrogate mother's uterus for gestation.
oxytoxin A hormone secreted together with vasopressin by the posterior lobe of the pituitary; stimulates both the uterine muscles to contract, initiating

the birth process, and the breasts to eject milk. Also synthesized and used in obstetrics to induce labor and control postnatal hemorrhage.

palmar grasp A newborn's normal reflexive response when a finger or thin object is pressed into the palm; the muscles in the hand flex so tightly that those in the upper arm and forearm flex too, and the infant can briefly suspend his or her weight.

parallel play Play in which two children are side by side using similar toys in similar ways, but their activities are not related.

parity The number of children previously borne by a mother; can affect the course of a child's prenatal development, birth, and infancy.

Parkinson's disease A chronic brain dysfunction caused by a deficiency in the cells that produce the neurotransmitter dopamine.

pattern baldness An inherited form of baldness, in which men lose hair from their temples and top of scalp.

pelvic inflammatory disease (PID) Sexually transmitted infection of the female's reproductive organs; may cause infertility.

periodontal disease Gum disease.

permissive parent A parent who gives children as much freedom as possible and allows them to do virtually what they want. The permissive parent does not state or enforce rules and gives children few responsibilities.

personal fable Adolescents' belief in their personal uniqueness and indestructibility.

personality The unique and consistent way in which an individual behaves and approaches the world. The complex of traits distinguishing one person from all others.

phallic stage In Freud's theory of psychosexual development, the period in which the genital area gives physiological pleasure and is of interest to the preschool child.

phenotype All the observable characteristics of the individual that depend on how the environment has affected expression of the genotype.

phenylketonuria (PKU) An inherited inability to produce a liver enzyme that metabolizes phenylalanine, which is contained in most protein foods; the build-up of phenylalanine and phenylpyruvic acid in body fluids damages the brain and causes restlessness and agitation. A special diet will prevent much of the retardation.

phocomelia The condition of having extremely short limbs so that feet and hands arise close to the trunk; the principal deformity of children whose mothers took the tranquilizer Thalidomide in the 1960s.

phoneme The smallest unit of sound that signals a difference in meaning in a particular language.

phylogeny The evolution of a race or genetically related group as distinguished from the development of the individual.

placebo An inert substance given to a naive subject who believes it to be medication; used to distinguish effects with a physiological cause from those with a psychological cause.

placenta The disk-shaped structure rich in blood vessels by which the circulation of the embryo and then the fetus interlocks with that of the mother, allowing food and oxygen to diffuse into the conceptus and carrying away its body wastes.

pneumograph A device for recording chest movements during respiration.

polygenic trait A characteristic that is produced by the equal and cumulative effects of several genes, for example, skin color, body shape, intelligence, and memory.

postconventional moral reasoning Kohlberg's highest level of moral reasoning in which the person chooses and follows universal principles, realizing that some of the rules of society can be broken.

postterm babies Infants born later than normal, more than 293 days after the first day of the mother's last menstrual period.

practice effects Distortions of the behavior of subjects in long-term research by their becoming used to testing. The results of the research may be affected.

preconventional moral reasoning Kohlberg's first level of moral reasoning in which the child obeys rules to avoid punishment and to secure a fair exchange with others.

preoperational period In Piaget's theory, the second stage in the development of logical thinking, lasting from approximately 2 to 7 years of age, in which the child begins to employ mental symbols, to engage in symbolic play, and to use words. Thought remains egocentric in nature, and the child focuses on the striking states and conditions of objects and events, ignoring others. The child cannot think by operations, cannot logically manipulate information.

preterm babies Infants born earlier than normal, fewer than 259 days after the first day of the mother's last menstrual period.

primary (biological) drive A physiological need of the individual, such as for food or sleep.

primary (normal) aging The universal, inevitable, possibly genetically programmed processes of aging.

primitive streak A thickening caused by the growth of ectodermal cells; gives rise to the mesoderm and ultimately to the brain, the spinal cord, all the nerves and sensory organs, and the skin.

profound mental retardation The condition of having no testable IQ and severe physical disabilities.

progeria A disease that causes premature aging.

programmed theories (of aging) Theories that suggest aging to be under the direction of the DNA in human genes (compare wear and tear theories).

prosocial Cooperative, helpful, and socially integrative.

proximodistal development A pattern of physical

growth and motor development progressing from the spine toward the extremities; in human beings, followed prenatally and throughout childhood.

pseudodialogue A conversation between mother and infant in which the mother pauses for an imagined response from her baby.

psychosexual development The development of human sexual wishes and behavior in stages from infancy to maturity.

psychosocial development The social, cultural, and sexual aspects of development, first described by Erik Erikson.

psychosocial moratorium In Erikson's life-cycle theory, a period of time in adolescence for establishing identity, for finding a role, the work, the attitude, and the sense of social connectedness that will allow a person to assume a place in adult society.

puberty The period when the reproductive system matures and the secondary sex characteristics develop.

punishment In operant learning theory, the unpleasant consequences that get people and other animals to decrease behavior.

quickening The first motion of a fetus in the uterus felt by the mother, usually between the 4th and 5th month.

reaction range The extent to which the expression of a genotype in a phenotype is affected by the environment.

recall A form of memory in which one must retrieve information seen or heard before without prompts from the external environment.

recessive gene In a pattern of genetic inheritance, a gene that must be paired with an identical one in order to determine a trait.

recognition Perception that an object, item, or image is one that has been seen or heard before; the simplest form of memory.

reflex An unlearned, involuntary response of a part of the body to an external stimulus.

rehearsal A cognitive strategy for remembering that consists of repeating information until it is fixed in memory.

reinforcement In operant conditioning, an instance in which a response is followed by a favorable consequence, increasing the likelihood that the response will be made again.

reliability In psychometrics, the consistency of a test as a measuring device; for example, if all the items on a test do a good job of predicting the total score.

replicability Repeatability of research results if a study is done again.

representative sample A group of subjects who accurately represent the composition of the entire population in which a researcher is interested.

restraining effect The maternal influences that keep a fetus with a genotype for large size small enough to develop in and be born from the uterus of a small mother.

retina The part of the eye that transforms light into nerve signals to the brain.

reversibility In Piaget's theory, the principle that one manipulation of an object can be reversed by applying the opposite.

rhythm method A method of preventing pregnancy that relies on abstinence from sexual intercourse during a female's fertile periods.

ribonucleic acid (RNA) A complex chemical substance that is able to use the information in DNA molecules to assemble proteins.

ribosomes The part of cells containing RNA and producing proteins necessary for the growth and survival of cells.

Ritalin A stimulant that is administered to hyperactive children and enables them to focus longer on tasks.

ritual play Repetitive, rhythmic exchanges or turns, found in preschoolers' play.

robust Strong (e.g., research results).

rooming in Procedure in which a newborn infant remains in the mother's hospital or birthing center room during the daytime, from soon after birth until they go home.

rooting reflex A newborn's normal reflexive response to a nipple on its cheek; the infant turns its head and starts to suck.

rubella (German measles) A mild form of measles; if contracted by a woman during the embryonic period of her pregnancy, may cause blindness, deafness, heart malformations, and mental retardation in her child.

saccadic eye movements (saccades) Small sideways movements of the eyes from one point of fixation to another. Saccadic eye movements move the eyes to objects of focus during visual scanning and reading.

scheme In Piaget's theory, a basic unit of knowledge, which may be an observable pattern of action or an image of an object.

schizophrenia The most common and severe psychosis, which may show itself in adolescence. The symptoms are jumbled thoughts and speech, flat or inappropriate emotions, hallucinations and delusions, loss of contact with reality.

scientific method A method for generating hypotheses and empirically testing them; extraneous factors are systematically eliminated, and factors under study are rigorously controlled by the investigator.

scoliosis Lateral curvature of the spine.

secondary aging Aging associated with disease, physical abuse, and physical disuse; not inevitable or universal, as primary aging is.

secondary sex characteristics Bodily and physiological signs of maleness and femaleness, other than the sex organs, which emerge during pubescence and indicate physical maturity—breast development, pubic and facial hair, voice changes.

secure attachment A relationship of an infant to his or her caregiver in which there is a good balance between exploration and proximity seeking.

self-concept The individual's sense of his or her identity as a person, beginning with the infant's discovery of parts of the body and then between 1 and 2 years of age the child's recognition of selfhood, of differentiation from others. Later self-concepts include the qualities and traits that individuals think are characteristic of themselves.

self-esteem One's high or low opinion of oneself.

self-righting tendency A regulative mechanism promoting normal prenatal development, overcoming all but the most adverse circumstances.

semantic memories Organized memories of facts and concepts, largely unaffected by aging (compare episodic memories).

sensitive period An interval of time during which a very young animal is supposedly predisposed to learn particular behavior, according to ethologists.

sensorimotor period In Piaget's theory, the period of cognitive growth from birth to approximately 2 years of age during which knowledge is acquired through sensory perceptions and motor skills. Babies show their intelligence through ingenious actions and ways of handling objects.

senile plaques Collections of debris found most commonly in diseased brains.

sensory memory The shallowest level of memory, at which sensory traces (sights, sounds, odors, etc.) are held in memory for about 1 second.

set goal The degree of proximity to mother required by a particular child. The child's natural predisposition to seek proximity is modified by his or her past experiences with mother.

severe mental retardation The condition of having an IQ score between 25 and 39 and related physical disabilities.

sex constancy The understanding that one cannot change sex at will or over time. Children only gradually acquire a notion of sex constancy.

sex-linked trait A trait inherited through genes located on the sex chromosomes.

sexual scripts Social norms about how people are to act sexually and about when and with whom to do sexual things.

shape constancy The tendency for an object viewed from different angles, which distort the retinal images, to be perceived as having a constant shape.

short-term memory Storage of information for a few seconds to a minute.

sickle-cell anemia An inherited blood abnormality in which the defective hemoglobin molecules crystallize and stick together to form elongated bundles of rods when oxygen level is low. The red blood cells sickle, blocking capillaries and bringing fever and pain.

small for dates Babies who are born at full term but who have low birth weights.

social conventions Rules of conduct, attitude, or behavior in a society, culture, or subculture, such as table manners, a style of dress, a form of address.

social referencing The act of seeking cues from others about how to behave in unfamiliar situations.

social understanding The capacity to understand and communicate with the people in one's social world, to interpret their emotions and intentions, and to apply social rules and norms.

sociobiology The school of thought that posits a genetic basis to social behavior.

sperm (spermatozoa) Male reproductive cells.

standard deviation A statistical index of the variability within a distribution. It is the square root of the average of the squared deviation from the mean.

statistical significance The magnitude of difference that has little probability of occurring by chance alone and therefore should be accepted as real.

stepping reflex Reflex in which the young infant seems to walk.

storm and stress Translation of the German phrase, *Sturm und Drang*, used by early psychologists to characterize the presumed disruptions and extremism of adolescence. Now widely regarded as a stereotyped view, it survives nevertheless.

stranger anxiety The baby's wariness of unfamiliar people, beginning when he or she is 4 to 6 months old and peaking at the end of the first year.

strange situation A standard laboratory method for assessing the attachment of children 12 to 18 months old to their mother. The child is observed in a playroom with mother and a stranger, alone with the stranger, reunited with mother, alone in the room, then with the stranger again, then reunited with mother.

structured class A traditional class in which children spend most of their time on lessons, usually in the "basics," the three Rs. They are generally in their seats at desks or tables arranged in neat rows, receiving instruction from the teacher. Subject matter is from a set curriculum established for the class as a whole.

structured observation Research conducted with subjects in situations set up by a researcher.

sucking reflex A newborn's reflexive response to anything that touches its lips; the infant immediately starts to suck.

sudden infant death syndrome (SIDS) The sudden and unexplained death of apparently healthy babies who are usually between 2 and 4 months old. They stop breathing during sleep.

superego In Freud's theory, the conscience that develops through identification with the parent; the imposition of civilization and culture as filtered through parents' rules and the regulations of adult society.

surrogate mother In animal experiments, an artificial mother (e.g., one made of wire or cloth) to test the infant's reactions to having only some of its needs of mother fulfilled. In human beings, a woman who agrees by contract to bear a child for a biological father and an adoptive mother and to give them the infant after birth.

symbol A sound, written word, object, action, or drawing that stands for or signifies something else.

symbolic play The pretending and make-believe of young children.

synapse The very small gap between the axon terminals of one neuron and a dendrite or cell body of another across which the nerve impulse passes.

syntax Rules of a specific language by which words and morphemes are combined to form larger units such as clauses and sentences.

syphilis A sexually transmitted disease caused by a spirochete; found in males and females. The first symptom is a hard sore in the genital region. Untreated, syphilis has a clinical course of three stages continued over many years.

telegraphic sentences The two-word and three-word sentences of young children, which contain only the most necessary words.

temperament A basic and apparently largely inherited behavior pattern of an infant or older person; the natural disposition of an individual.

teratogen An external agent that may cross the barrier of the placenta and harm the developing embryo or fetus.

teratology The study of serious malformations and deviations from the normal in animals and plants. Now applied to human beings with the goal of understanding the causes of abnormalities and anticipating risks in prenatal development.

terminal drop A marked decline on IQ that precedes death.

terminal hair Coarse, colored hair.

terrible twos An age at which children test their autonomy in the world; their bad behavior consists of negativism and temper tantrums, their good of growing independence and mastery of their surroundings.

testis The typically paired male reproductive organ, which is located in the scrotum and in which sperm develop after puberty.

test-tube baby A baby born as a result of in vitro fertilization.

tonic neck reflex The newborn's normal reflexive response when placed on its back; the infant turns its head to one side and assumes a fencing position, extends arm and leg on this side, bends opposite limbs, and arches body away from the direction faced.

totipotent The ability of a cell formed during the first 60 hours of cell division after fertilization to develop into a complete person if separated from the zygote.

toxemia (of pregnancy) A disorder that usually has its onset in the last trimester; symptoms are persistent nausea and vomiting or, in preeclampsia, hypertension, sudden and rapid increase in weight through retention of water in the tissues, and albumin in the urine.

transductive reasoning In Piaget's theory, preoperational children's logic in which they reason from the particular to the particular without generalization.

tympanum The eardrum.

Type A Behavior pattern of people who act hurriedly, aggressively, intensely, and competitively.

Type B Behavior pattern of people who act more slowly than Type A people and are relatively free of pressure.

ultrasound A wave phenomenon of the same physical nature as sound but with frequencies above the range of human hearing. Can be used to picture a fairly detailed outline of the fetus.

umbilical cord The organ connecting fetus to pregnant mother through which nutrients, waste, and other substances are transferred between their bloodstreams.

unanticipated grief An abnormal pattern of grieving in which people who have not had enough time to prepare psychologically for someone else's death may have trouble recovering from the grieving process.

underextension Children's tendency to use a word too narrowly, such as applying the word *cat* only to the family cat and not to cats in the alley, on television, or in picture books.

uterus An organ in female mammals, resembling, when not pregnant, a flattened pear in size and shape; contains and nourishes the young during development previous to birth.

vacuoles Thick granules surrounded by fluid, found in diseased brains and associated with aging.

validity The ability of a test or other instrument to measure what it is supposed to measure.

vellus hair Fine, short, colorless hair.

vernix A white, pasty substance made up of the fetus's shed dead-skin cells and fatty substances from oil glands; serves as a protective coating.

villus On the surface of the chorion, a rootlike extension that may grow into the lining of the uterine wall during implantation and rupture the small blood vessels it meets.

visual cliff A laboratory structure for testing infants' depth perceptions; it consists of a glass-topped table with a middle runway, on one side of which there appears to be a drop.

wear and tear theories Theories of aging suggesting that as the body suffers functional damage, it repairs it increasingly less effectively (compare programmed theories).

Werner's syndrome A disease that causes rapid aging of people in their twenties and thirties.

withdrawal An ineffective method of preventing pregnancy that relies on the male withdrawing from the female's vagina before ejaculating.

X chromosome A sex chromosome that, when combined with another X chromosome, carries the genetic information necessary for the development of a female.

Y chromosome A sex chromosome that, when combined with an X chromosome, carries the genetic information necessary for the development of a male.

zygote A single cell formed through the union of two gametes, a sperm and an ovum; in time the human zygote will develop into a human being.

References

ABER, L. The transition from attachment to autonomy concerns in early development: The influence of stage-specific parenting competencies. *Infant Behavior and Development*, 1986, *9* (Special Issue: Abstracts of papers presented at the Fifth International Conference on Infant Studies), 16.

ABIKOFF, H., GITTELMAN-KLEIN, R., AND KLEIN, D. F. Classroom observation code for hyperactive children: A replication of validity. *Journal of Consulting and Clinical Psychology*, 1980, *48*, 555–565.

ABBOTT, M., MURPHY, E., BOLLING, D., AND ABBEY, H. The familial component in longevity—A study of offspring of nonagenarians. II: Preliminary analysis of the completed study. *Johns Hopkins Medical Journal*, 1974, *134*, 1–16.

ABRAMOVITCH, R., CORTER, C., PEPLER, D., AND STANHOPE, L. Sibling and peer interaction: A final follow-up and a comparison. *Child Development*, 1986, *57*, 217–229.

______, PEPLER, D., AND CORTER, C. Patterns of sibling interaction among preschool-age children. In M. Lamb and B. Sutton-Smith (Eds.), *Sibling relationships: Their nature and significance across the lifespan*. Hillsdale, N.J.: Lawrence Erlbaum, 1982.

ABRAVANEL, E., AND GINGOLD, H. Learning via observation during the second year of life. *Developmental Psychology*, 1985, *21*, 614–623.

ACREDOLO, L. P., AND HAKE, J. L. Infant perception. In B. B. Wolman and G. Strickler (Eds.), *Handbook of developmental psychology*. Englewood Cliffs, N.J.: Prentice-Hall, 1982.

ADAMS, B. N. *Kinship in an urban setting*. Chicago: Markham, 1968.

ADAMS, G. R., AND JONES, R. M. Female adolescents' identity development: Age comparisons and perceived child-rearing experience. *Developmental Psychology*, 1983, *19*, 249–256.

ADAMS, J. F. Earlier menarche, greater height and weight: A stimulation-stress factor hypothesis. *Genetic Psychology Monographs*, 1981, *104*, 3–22.

ADAMS, R. D. Morphological aspects of aging in the human nervous system. In J. E. Birren and R. B. Sloane (Eds.), *Handbook of mental health and aging*. Englewood Cliffs, N.J.: Prentice-Hall, 1980, pp. 149–160.

ADAMS, R. E., AND PASSMAN, R. H. Effects of visual and auditory aspects of mothers and strangers on the play and exploration of children. *Developmental Psychology*, 1979, *15*, 269–274.

ADAMS, R. J., AND MAURER, D. *A demonstration of color perception in the newborn*. Paper presented at the meetings of the Society for Research in Child Development, Detroit, April 1983.

______, AND ______. *The use of habituation to study newborns' color vision*. Paper presented at the Fourth International Conference on Infant Studies, New York, April 1984.

ADELSON, J. The political imagination of the young adolescent. In J. Kagan and R. Coles (Eds.), *12 to 16: Early adolescence*. New York: Norton, 1972.

______, GREEN, B., AND O'NEIL, R. P. The growth of the idea of law in adolescence. *Developmental Psychology*, 1969, *1*, 327–332.

AHLGREN, A. Sex differences in the correlates of cooperative and competitive school attitudes. *Developmental Psychology*, 1983, *19*, 881–888.

AHRENTZEN, S. The environmental and social context of distraction in the classroom. In A. E. Osterberg, C. P. Tiernan, and R. A. Findlay (Eds.), *Design research interactions*. Ames, Iowa: Environmental Design Research Association, 1981.

AIELLO, J. R., NICOSIA, G., AND THOMPSON, D. E. Physiological, social, and behavioral consequences of crowding on children and adolescents. *Child Development*, 1979, *50*, 195-202.

AINLAY, S. C., AND SMITH, D. R. Aging and religious participation. *Journal of Gerontology*, 1984, *39*, 357–363.

AINSLIE, R. C., AND ANDERSON, C. W. Day care children's relationship to their mothers and caregivers: An inquiry into the conditions for the development of attachment. In R. C. Ainslie (Ed.), *The child and the day care setting*. New York. Praeger, 1984.

AINSWORTH, M. D. S. *Infancy in Uganda: Infant care and the growth of love*. Baltimore: Johns Hopkins University Press, 1967.

______. The development of infant-mother attachment. In B. M. Caldwell and H. N. Ricciuti (Eds.), *Review of child development research*, Volume 3. Chicago: University of Chicago Press, 1973.

______, BLEHAR, M., WATERS, E., AND WALL, S. *Patterns of attachment: Observations in the strange situation and at home*. Hillsdale, N.J.: Lawrence Erlbaum, 1978.

ALAN GUTTMACHER INSTITUTE. *Eleven million teenagers*. New York: Alan Guttmacher Institute, 1976.

______. *Teenage pregnancy: The problem that hasn't gone away*. New York: Alan Guttmacher Institute, 1981.

ALBERT, M. S., AND KAPLAN, E. Organic implications of neuropsychological deficits in the elderly. In L. W. Poon, J. L. Fozard, L. S. Cermak, D. Arenberg, and L. W. Thompson (Eds.), *New directions in memory and aging*. Hillsdale, N.J.: Lawrence Erlbaum, 1980, pp. 403–432.

ALLEN, J. *Visual acuity development in human infants up to 6 months of age*. Unpublished doctoral dissertation, University of Washington, 1978.

ALPAUGH, P. K., AND BIRREN, J. E. Variables affecting creative contributions across the adult life span. *Human Development*, 1977, *20*, 240–248.

ALTUS, W. D. Birth order and its sequelae. *Science*, 1966, *151*, 44–49.

AMBROSE, J. A. The development of the smiling response in early infancy. In B. M. Foss (Ed.), *Determinants of infant behavior*. New York: Wiley, 1961.

ANASTASI, A. *Psychological testing* (2nd ed.). New York: Macmillan, 1976.

ANDERS, T. R., FOZARD, J. L. AND LILLYQUIST, T. D. Effects of age upon retrieval from short term memory, from 20–68 years of age. *Developmental Psychology*, 1972, *6*, 214–217.

ANDERSON, D. R., AND LEVIN, S. R. Young children's attention to *Sesame Street*. *Child Development*. 1976, *47*, 806–811.

ANDERSON, G. C. *Crying in newborn infants: Physiology and developmental implications*. Paper presented at the International Conference on Infant Studies, New York, April 1984.

ANDRES, D., AND WALTERS, R. H. Modification of delay of punishment effects through cognitive restructuring. In *Proceedings of the 78th Annual Convention of the American Psychological Association*, 1970.

ANDRES, R., AND TOBIN, J. D. Endocrine system. In C. E. Finch and L. Hayflick (Eds.), *Handbook of the biology of aging*. New York: Van Nostrand Reinhold, 1977, pp. 357–378.

ANGLIN, J. M. *The growth of word meaning*. Research Monograph 63. Cambridge, Mass.: MIT Press, 1970.

______. The child's first terms of reference. In S. Ehrlich and E. Tulving (Eds.), *Bulletin de Psychologie* (Special Issue on Semantic Memory), 1975.

______. *Word, object, and conceptual development*. New York: Norton, 1977.

ANNIS, L. F. *The child before birth*. Ithaca, N.Y.: Cornell University Press, 1978.

ANTHONY, S. *The discovery of death in childhood and after*. New York: Basic Books, 1972.

APGAR, V., AND BECK, J. *Is my baby all right?* New York: Pocket Books, 1974.

APPEL, Y. H. Developmental differences in children's perception of maternal socialization behavior. *Child Development*, 1977, *48*, 1689–1693.

Appleton, T., Clifton, R., and Goldberg, S. The development of behavioral competence in infancy. In F. D. Horowitz (Ed.), *Review of child development research*, Volume 4. Chicago: University of Chicago Press, 1975.

Archer, S. L. Identity and the choice of social roles. In A. S. Waterman (Ed.), *Identity in adolescence: Processes and contents. New Directions for child development*, No. 30. San Francisco: Jossey-Bass, 1985.

Arenberg, D. Comments on the processes that account for memory declines with age. In L. W. Poon, J. L. Fozard, L. S. Cermak, D. Arenberg, and L. W. Thompson (Eds.), *New directions in memory and aging*. Hillsdale, N.J.: Lawrence Erlbaum, 1980, pp. 67–71.

______. Changes with age in problem solving. In F. I. M. Craik and S. Trehub (Eds.), *Aging and cognitive processes*. New York: Plenum, 1982, pp. 221–236.

______, and Robertson-Tchabo, E. A. Learning and aging. In J. E. Birren and K. W. Schaie (Eds.), *Handbook of the psychology of aging*. New York: Van Nostrand Reinhold, 1977, pp. 421–449.

Arend, R., Gove, F. L., and Sroufe, L. A. Continuity of individual adaptation from infancy to kindergarten: A predictive study of ego-resiliency and curiosity in preschoolers. *Child Development* 1979, *50*, 950–959.

Ariés, P. Centuries of childhood: A social history of family life. New York: Vintage Books, 1962.

______. *Western attitudes toward death*. Baltimore: Johns Hopkins University Press, 1974.

Arlin, P. K. Adolescent and adult thought: A search for structures. Paper presented at meetings of the Piaget Society. Philadelphia, June 1980.

Armitage, S. E., Baldwin, B. A., and Vince, N. A. The fetal sound environment of sheep. *Science*, 1980, *208*, 1173–1174.

Ash, P. Pre-retirement counseling. *Gerontologist*, 1966, *6*, 97–99.

Asher, S., and Renshaw, P. Children without friends: Social knowledge and social skill training. In S. R. Asher and J. M. Gottman (Eds.), *The development of children's friendships*. New York: Cambridge University Press, 1981.

Asher, S. T. Topic interest and children's reading comprehension. In R. J. Spiro, B. C. Bruce, and W. F. Brewer (Eds.), *Theoretical issues in reading comprehension*. Hillsdale, N.J.: Lawrence Erlbaum, 1980.

Ashmead, D. H., and Perlmutter, M. Infant memory in everyday life. In M. Perlmutter (Ed.), *New directions in child development: Children's memory*. San Francisco: Jossey-Bass, 1980.

Aslin, R. N., Pisoni, D. B., and Jusczyk, P. W. Auditory development and speech perception in infancy. In P. H. Mussen (Ed.), *Handbook of child psychology*, Volume 2. New York: Wiley, 1983.

______, and Sinnott, J. M. *Frequency discrimination of pure tones in human infants*. Paper presented at the International Conference on Infant Studies, New York, April 1984.

Atchley, R. C. *The sociology of retirement*. New York: Halstead Press, 1976.

Averill, J. R. Grief: Its nature and significance. *Psychological Bulletin*, 1968, *6*, 721–748.

Axelrod, J. Neurotransmitters. *Scientific American*, June 1974, *230*, 59–71.

Axia, G., and Baroni, M. R. Linguistic politeness at different age levels. *Child Development*, 1985, 56, 918–927.

Bach, M. J., and Underwood, B. J. Developmental changes in memory attributes. *Journal of Educational Psychology*, 1970, *61*, 292–296.

Bachman, J. G. *The American high school student: A profile based on survey data*. Paper presented in Berkeley, June 1982.

______, and O'Malley, P. M. Self-esteem in young men: A longitudinal analysis of the impact of educational and occupational attainment. *Journal of Personality and Social Psychology*, 1977, *35*, 365–380.

Baer, P. E., and Goldfarb, G. E. A developmental study of verbal conditioning in children. *Psychological Reports*, 1962, *10*, 175–181.

Bailey, D. A. Exercise, fitness and physical education for the growing child. *Canadian Journal of Public Health*, 1973, *64*, 421–430.

Baillargeon, R. *Reasoning about hidden obstacles: Object permanence in the six-month-old infant*. Paper presented at the International Conference on Infant Studies, New York, April 1984.

______. Young infants' representation of the properties of hidden objects. *Infant Behavior and Development*, 1986, *9* (Special Issue: Abstracts of papers presented at the Fifth International Conference on Infant Studies), 15.

Bakan, D. *The duality of human existence*. Chicago: Rand McNally, 1966.

Baker, A. J. L., and Aber, L. Reconceptualizing security of attachment in toddlers: Theoretical and methodological issues. *Infant Behavior and Development*, 1986, *9* (Special Issue: Abstracts of papers presented at the Fifth International Conference on Infant Studies), 16.

Bakwin, H. Sleep-walking in twins. *Lancet*, 1970, *2*, 446–447.

______. Car-sickness in twins. *Developmental Medicine and Child Neurology*, 1971, *13*, 310–312. (a)

______. Constipation in twins. *American Journal of Diseases of Children*. 1971, *121*, 179–181. (b)

______. Nail-biting in twins. *Developmental Medicine and Child Neurology*, 1971, *13*, 304–307. (c)

______. Enuresis in twins. *American Journal of Diseases of Children*, 1971, *121*, 222–225. (d)

Ballenski, C. B., and Cook, A. S. Mothers' perceptions of their competence in managing selected parenting tasks. *Family Relations*, 1982, *31*, 489–494.

Baltes, P. B. Lifespan developmental psychology: Observations on history and theory revisited. In R. Lerner (Ed.), *Developmental psychology: Historical and philosophical perspectives*. Hillsdale, N.J.: Lawrence Erlbaum, 1983, pp. 81–90.

______, and Reese, H. W. The lifespan perspective in developmental psychology. In M. H. Bornstein and M. E. Lamb (Eds.), *Developmental psychology: An advanced textbook*, Hillsdale, N.J.: Lawrence Erlbaum, 1984.

______, and Willis, S. L. Plasticity and enhancement of intellectual functioning in old age. In F. I. M. Craik and S. Trehub (Eds.), *Aging and cognitive processes*. New York: Plenum, 1982, pp. 353–389.

Bandura, A. The stormy decade: Fact or fiction? *Psychology in the Schools*, 1964, *1*, 224–231.

______. *Principles of behavior modification*. New York: Holt, Rinehart & Winston, 1969. (a)

______. Social-learning theory of identificatory process. In D. A. Goslin (Ed.), *Handbook of socialization theory and research*. Chicago: Rand McNally, 1969. (b)

______. Self-referent thought: A developmental analysis of self-efficacy. In J. H. Flavell and L. Ross (Eds.), *Social cognitive development: Frontiers and possible futures*. Cambridge, England: Cambridge University Press, 1981.

______, and Huston, A. C. Identification as a process of incidental learning. *Journal of Abnormal and Social Psychology*, 1961, *63*, 311–318.

______, Ross, D. M., and Ross, S. A. Transmission of aggression through imitation of aggressive models. *Journal of Abnormal and Social Psychology*, 1961, *63*, 575–582.

______, and Ross, S. A. Imitation of film-mediated aggressive models. *Journal of Abnormal and Social Psychology*, 1963, *66*, 3–11.

Bane, M. J., and Jencks, C. Five myths about your I.Q. *Harper's Magazine*, February 1973, 28–40.

Bank, S., and Kahn, M. D. Sisterhood-brotherhood is powerful: Sibling subsystems and family therapy. *Family Process*, 1975, *14*, 311–337.

______, and ______. Intense sibling loyalties. In M. E. Lamb and B. Sutton-Smith (Eds.), *Sibling relationships: Their nature and signifi-*

cance across the lifespan. Hillsdale, N.J.: Lawrence Erlbaum, 1982, pp. 251–266.

BANKS, M. S., AND SALAPATEK, P. Infant pattern vision: A new approach based on the contrast sensitivity function. *Journal of Experimental Child Psychology*, 1981, *31*, 1–45.

———, AND ———. Infant visual perception. In P. H. Mussen (Ed.), *Handbook of child psychology*, Volume 2. New York: Wiley, 1983.

BARDWICK, J. M., AND DOUVAN, E. Ambivalence: The socialization of women. In V. Gornick and B. K. Moran (Eds.), *Women in sexist society*. New York: Basic Books, 1971.

BARENBOIM, C. Developmental changes in the interpersonal cognitive system from middle childhood to adolescence. *Child Development*, 1977, *48*, 1467–1474.

———. The development of person perception in childhood and adolescence: From behavioral comparisons to psychological constructs to psychological comparisons. *Child Development*, 1981, *52*, 129–144.

BARFIELD, R. E., AND MORGAN, J. N. *Early retirement: The decision and the experience and a second look*. Ann Arbor, Mich.: University of Michigan Press, 1975.

BARGLOW, P., VAUGHN, B. E., AND MOLITOR, N. Effects of maternal absence due to employment on the quality of infant–mother attachment in a low risk sample. *Child Development*, in press.

BARNETT, M. A., HOWARD, J. A., MELTON, E. M., AND DINO, G. A. Effect of inducing sadness about self or other on helping behavior in high- and low-empathic children. *Child Development*, 1982, *53*, 920–923.

BARON, S. A., JACOBS, L., AND KINKLE, W. R. Changes in size of normal lateral ventricles during aging determined by computerized tomography. *Neurology*, 1976, *26*, 1011–1013.

BARRET, D. E., AND RADKE-YARROW, M. Prosocial behavior, social inferential ability, and assertiveness in children. *Child Development*, 1977, *48*, 475–481.

BARRETT, F. M. Sexual experience, birth control usage, and sex education of unmarried Canadian university students: Changes between 1968 and 1978. *Archives of Sexual Behavior*, 1980, *9*, 367–390.

BARRY, H., BACON, M. K., AND CHILD, I. L. A cross-cultural survey of some sex differences in socialization. *Journal of Abnormal and Social Psychology*, 1957, *55*, 327–332.

———, AND PAXSON, L. M. Infancy and early childhood: Cross-cultural codes. *Ethnology*, 1971, *10*, 467–508.

BARTLETT, J. C., AND SANTROCK, J. W. Affect-dependent episodic memory in young children. *Child Development*, 1979, *50*, 513–518.

BARTUS, R. T., DEAN, R. L., III, BEER, B. AND LIPPA, A. S. The cholinergic hypothesis of geriatric memory dysfunction. *Science*, 1982, *217*, 408–417.

BARUCH, G., BARNETT, R., AND RIVERS, C. *Life prints: New patterns of love and work for today's women*. New York: McGraw-Hill, 1983.

BASCUE, L. *A study of the relationship of time orientation and time attitudes to death anxiety in elderly people*. Doctoral dissertation, University of Michigan, Ann Arbor, Mich. 1973. University Microfilms No. 7318234.

BASKETT, L. M. Ordinal position differences in children's family interactions. *Developmental Psychology*, 1984, *20*, 1026–1031.

———, AND JOHNSON, S. M. The young child's interactions with parents versus siblings: A behavioral analysis. *Child Development*, *1982*, *53*, 643–650.

BASSECHES, M. *Dialectical thinking and adult development*. Norwood, N.J.: Ablex, 1984.

BATES, E. *Language and context: The acquisition of pragmatics*. New York: Academic Press, 1976.

———, MACWHINNEY, B., BASELLI, C., DEVESCOVI, A., NATALE, F., AND VENZA, V. A cross-linguistic study of the development of sentence interpretation strategies. *Child Development*, 1984, *55*, 341–354.

BATES, J. E. The concept of difficult temperament. *Merrill-Palmer Quarterly*, 1980, *25*, 299–319.

———, AND BAYLES, K. Objective and subjective components in mothers' perceptions of their children from age 6 months to 3 years. *Merrill-Palmer Quarterly*, 1984, *30*, 111–130.

———, FREELAND, C. A., AND LOUNSBURY, M. L. Measurement of infant difficultness. *Child Development*, 1979, *50*, 794–803.

BAUMRIND, D. Current patterns of parental authority. *Developmental Psychology Monographs*, 1971, *4* (1, Part 2).

———. Early socialization and adolescent competence. In S. E. Dragastin and G. H. Elder (Eds.), *Adolescence in the life cycle*. New York: Wiley, 1975.

———. *Sex related socialization effects*. Paper presented at the biennial meeting of the Society for Research in Child Development, San Francisco, March 1979.

———, AND BLACK, A. E. Socialization practices associated with dimensions of competence in preschool boys and girls. *Child Development*, 1967, *38*, 291–327.

BAY, E. Ontogeny of stable speech areas in the human brain. In E. H. Lenneberg and E. Lenneberg (Eds.), *Foundations of language development*, Volume 2. New York: Academic Press, 1975.

BAYER, L. M., AND BAYLEY, N. *Growth diagnosis* (2nd ed.). Chicago: University of Chicago Press, 1976. (First edition 1959.)

BAYLEY, N. Mental growth during the first three years: A developmental study of 61 children by repeated tests. *Genetic Psychology Monographs*, 1933, *14*, 1–92.

———. Learning in adulthood: The role of intelligence. In H. J. Klausmeier and C. W. Harris (Eds.), *Analysis of concept learning*. New York: Academic Press, 1966.

———. Behavioral correlates of mental growth: Birth to thirty-six years. *American Psychologist*, 1968, *23*, 1–17.

———. The development of motor abilities during the first three years. In M. C. Jones, N. Bayley, J. W. McFarlane, and M. P. Honzik (Eds.), *The course of human development*. Waltham, Mass.: Xerox College Publishing, 1971.

BEATTIE, W. M., JR. Aging and the social services. In R. H. Binstock and E. Shanas (Eds.), *Handbook of aging and the social sciences*. New York: Van Nostrand Reinhold, 1976, pp. 619–642.

BECK, S. H. Adjustment to and satisfaction with retirement. *Journal of Gerontology*, 1982, *37*, 616–624.

BECKER, E. *The denial of death*. New York: Free Press, 1973.

BECKER, M. H. The health belief model and sick role behavior. In M. H. Becker (Ed.), *The health belief model and personal health behavior*. Thorofare, N.J.: Charles B. Slack, 1974.

BECKER, W. C. Consequences of different kinds of parental discipline. In M. L. Hoffman and L. W. Hoffman (Eds.), *Review of child development research*, Volume 1. New York: Russell Sage Foundation, 1964.

BECKWITH, L. Relationships between infants' social behavior and their mothers' behavior. *Child Development*, 1972, *43*, 397–411.

———. Prediction of emotional and social behavior. In J. D. Osofsky (Ed.), *Handbook of infant development*. New York: Wiley, 1979.

BEE, H. L., DISBROW, M. A., JOHNSON-CROWLEY, N., AND BARNARD, K. *Parent–child interactions during teaching in abusing and non-abusing families*. Paper presented at the meeting of the Society for Research in Child Development, Boston, 1981.

BELL, A., WEINBERG, M., AND HAMMERSMITH, S. K. *Sexual preference*. Indianapolis: Indiana University Press (Alfred C. Kinsey Institute for Sex Research Publication), 1981.

BELL, R. Q., AND HARPER, L. V. *Child effects on adults*. Hillsdale, N.J.: Lawrence Erlbaum, 1977.

BELL, R. R. Friendships of women and of men. *Psychology of Women Quarterly*, 1981, *5*, 402–417.

BELL, S. M., AND AINSWORTH, M. D. S. Infant crying and maternal responsiveness. *Child Development*, 1972, *43*, 1171–1190.

BELLUCCI, G., AND HOYER, W. J. Feedback effects on the performance and self-reinforcing behavior of elderly and young adult women. *Journal of Gerontology*, 1975, *30*, 456–460.

BELSKY, J. Experimenting with the family in the newborn period. *Child Development*, 1985, *56*, 407–414.

_____, Gilstrap, B., and Rovine, M. The Pennsylvania Infant and Family Development Project, I: Stability and change in mother-infant and father-infant interaction in a family setting at one, three, and nine months. *Child Development*, 1984, *55*, 692–705.

_____, GOODE, M. K., AND MOST, R. K. Maternal stimulation and infant exploratory competence: Cross-sectional, correlational and experimental analyses. *Child Development*, 1980, *51*, 1168–1178.

_____, AND ISABELLA, R. Maternal, infant, and social-contextual determinants of attachment security. In J. Belsky and T. Nezworkski (Eds.), *Clinical implications of attachment*. Hillsdale, N.J.: Lawrence Erlbaum, in press.

_____, ROVINE, M., AND TAYLOR, D. The Pennsylvania Infant and Family Development Project, III: The origins of individual differences in infant-mother attachment: Maternal and infant contributions. *Child Development*, 1984, *55*, 718–728.

BEM, S. L., MARTYNE, W., AND WATSON, C. Sex typing and androgyny: Further explorations of the expressive domain. *Journal of Personality and Social Psychology*, 1976, *34*, 1016–1023.

BENBOW, C. P., AND STANLEY, J. C. Sex differences in mathematical ability: Fact or artifact. *Science*, 1980, *210*, 1262.

BENGSTON, V. L., CUELLAR, J. E., AND RAGAN, P. K., Stratum contrasts and similarities in attitudes toward death. *Journal of Gerontology*, 1977, *32*, 76–88.

_____, AND TREAS, J. The changing family context of mental health and aging. In J. E. Birren and R. B. Sloane (Eds.), *Handbook of mental health and aging*. Englewood Cliffs, N.J.: Prentice-Hall, 1980, pp. 400–428.

BENN, R. K. Factors associated with security of attachment in dual career families. Paper presented at the biennial meetings of the Society for Research in Child Development, Toronto, 1985.

BENNETT, N. *Teaching styles and pupil progress*. London: Open Books, 1976.

BENNETT, N. G., AND GARSON, L. K. Extraordinary longevity in the Soviet Union: Fact or artifact? *The Gerontologist*, *26*, 1986, no. 4.

BENNETT, W., AND GURIN, J. *The dieter's dilemma: Eating less and weighing more*. New York: Basic Books, 1982.

BERENBAUM, S. A., AND RESNICK, S. Somatic androgyny and cognitive abilities. *Developmental Psychology*, 1982, *18*, 418–423.

BERENDA, R. W. *The influence of the group on the judgments of children*. New York: King's Crown Press, 1950.

BERENTHAL, B., AND PROFFITT, D. The extraction of structure from motion: Implementation of basic processing constraints. *Infant Behavior and Development*, 1986, *9* (Special Issue: Abstracts of papers presented at the Fifth International Conference on Infant Studies), 36.

BERGLER, R. Selbstbild und Alter. In R. Schubert (Ed.), *Berich I. Kongress Deutsche Gesellschaft für Gerontologie*. Darmstadt: Steinkopff, 1968, pp. 156–159.

BERKOWITZ, B. Changes in intellect with age: IV. Changes in achievement and survival in older people. *Journal of Genetic Psychology*, 1965, *107*, 3–14.

BERLIN, B., AND KAY, P. *Basic color terms: Their universality and evolution*. Berkeley: University of California Press, 1969.

BERLYNE, D. E. Novelty and curiosity as determinants of exploratory behavior. *British Journal of Psychology*, 1950, *41*, 68–80.

BERMAN, J. L., AND FORD, R. Intelligence quotients and intelligence loss in patients with phenylketonuria and some variant states. *Journal of Pediatrics*, 1970, *77*, 764–770.

BERMAN, P. W., AND GOODMAN, V. Age and sex differences in children's responses to babies: Effects of adults' caretaking requests and instructions. *Child Development*, 1984, *55*, 1071–1077.

BERNAL, J. Crying during the first 10 days of life and maternal responses. *Developmental Medicine and Child Neurology*, 1972, *14* 362–372.

_____, AND RICHARDS, M. P. M. *The effects of bottle and breast feeding on infant development*. Paper presented at the annual conference of the Society for Psychosomatic Research, London, November 1969.

BERNARDO, F. M. Survivorship and social isolation: The case of the aged widower. *Family Coordinator*, 1970, *19*, 11–25.

BERNDT, T. J. *Children's conceptions of friendship and the behavior expected of friends*. Paper presented at the annual meeting of the American Psychological Association, Toronto, August 1978. (a)

_____. *Developmental changes in conformity to parents and peers*. Paper presented at the annual meeting of the American Psychological Association, Toronto, August 1978. (b)

_____. Developmental changes in conformity to peers and parents. *Developmental Psychology*, 1979, *15*, 608–616.

_____. The features and effects of friendship in early adolescence. *Child Development*, 1982, *53*, 1447–1460.

_____. Correlates and causes of sociometric status in childhood: A commentary on six current studies of popular, rejected, and neglected children. *Merrill-Palmer Quarterly*, 1983, *29*, 439–448.

_____, AND BERNDT, E. G. Children's use of motives and intentionality in person perception and moral judgment. *Child Development*, 1975, *46*, 904–912.

_____, AND BULLEIT, T. N. Effects of sibling relationships on preschoolers' behavior at home and at school. *Developmental Psychology*, 1985, *21*, 761–767.

_____, AND HOYLE, S. G. Stability and change in childhood and adolescent friendships. *Developmental Psychology*, 1985, *21*, 1007–1015.

BERNSTEIN, R. M. The development of the self-system during adolescence. *Journal of Genetic Psychology*, 1980, *136*, 231–245.

BERSCHEID, E., WALSTER, E., AND BOHRENSTEDT, G. Body image. *Psychology Today*, July 1972, 57–66.

_____, WALSTER, E., AND BOHRNSTEDT, G. Body image: The happy American body. *Psychology Today*, November 1973, 119–131.

BERSHAD, S., RUBINSTEIN, A., PATERNITI, J. R., LE, N., POLIACK, S. C., HELLER, B., GINSBERG, H. N., FLEISCHMAJER, R., AND BROWN, W. V. Changes in plasma lipids and lipoproteins during isotretinoin therapy for acne. *New England Journal of Medicine*, October 17, 1985, *313*, 981–985.

BERTENTHAL, B. I., AND FISCHER, K. W. Development of self-recognition in the infant. *Developmental Psychology*, 1978, *14*, 44–50.

BERTOFF. A. Tolstoy, Vygotsky, and the making of meaning. *College Composition and Communication*, 1978, *29*, 249–255.

BERTONCINI, J., AND MEHLER, J. Syllables as units in infant speech perception. *Infant Behavior and Development*, 1981, *4*, 247–260.

BERZONSKY, M. D. Inter and intraindividual differences in adolescent storm and stress: A life span developmental view. *Journal of Early Adolescence*, 1982, *2*, 211–217.

BEST, D. L., WILLIAMS, J. E., CLOUD, J. M., DAVIS, S. W., ROBERTSON, L. S., EDWARDS, J. R., GILES, H., AND FOWLES, J. Development of sex-trait stereotypes among young children in the United States, England, and Ireland. *Child Development*, 1977, *48*, 1375–1384.

BHANTHUMNAVIN, K., AND SCHUSTER, M. M. Aging and gastrointestinal function. In C. E. Finch and L. Hayflick (Eds.), *Handbook of the biology of aging*. New York: Van Nostrand Reinhold, 1977, pp. 709–723.

BHATIA, V. P., KATIYAR, G. P., AND AGARWAL, K. N. Effect of intrauterine nutritional deprivation on neuromotor behavior of the newborn. *Acta Paediatrica Scandinavica*, 1979, *68*, 561–566.

BIBRING, G. L., DWYER, T. F., HUNTINGTON, D. S., AND VALENSTEIN, A. F. A study of the psychological processes in pregnancy and of the earliest mother-child relationship. *Psychoanalytic Study of the Child*, 1961, *16*, 9–24.

BIGELOW, B. J. Children's friendship expectations: A cognitive-developmental study. *Child Development*, 1977, *48*, 246–250.

BIJOU, S. W., AND BAER, P. M. *Child development*, Volume 1. A systematic and empirical theory. New York: Appleton-Century-Crofts, 1961.

BILLER, H. B. The father and personality development: Paternal deprivation and sex-role development. In M. E. Lamb (Ed.), *The role of the father in child development*. New York: Wiley, 1976.

BIRCH, L. L., MARLIN, D. W., AND ROTTER, J. Eating as the "means" activity in a contingency: Effects on young children's food preference. *Child Development*, 1984, *55*, 431–439.

BIRNBAUM, J. A. *Life patterns, personality style and self esteem in gifted family oriented and career committed women*. Unpublished doctoral dissertation, University of Michigan, 1971.

BIRREN, J. E. *Creativity in the second half of life*. Paper presented at the American Psychological Association, Los Angeles, Calif., August 27, 1985.

______. Increment and decrement in the intellectual status of the aged. *Psychiatric Research Reports*, 1968, *23*, 207–214. (chap. 9)

______. Psychophysiology and speed of response. *American Psychologist*, 1974, *29*, 808–815.

______, AND SCHAIE, K. W. (Eds.), *Handbook of the psychology of aging* (2nd ed.). New York: Van Nostrand Reinhold, 1985.

______, AND D. S. WOODRUFF. Human development over the life span through education. In P. B. Baltes and K. W. Schaie (Eds.), *Life-span developmental psychology: Personality and socialization*. New York: Academic Press, 1973, pp. 305–337.

BISHOP, J. E. Medical science helps prolong the life span and the active years. *Wall Street Journal*, February 24, 1983.

BIXENSTINE, V. E., DE CORTE, M. S., AND BIXENSTINE, B. A. Conformity to peer-sponsored misconduct at four age levels. *Developmental Psychology*, 1976, *12*, 226–236.

BJORK, E. L., AND CUMMINGS, E. M. *The "A, not B" search error in Piaget's theory of object permanence: Fact or artifact?* Paper presented at the meeting of the Psychonomic Society, Phoenix, November 1979.

BJORKLUND, D. F., AND ZEMAN, B. R. Children's organization and metamemory awareness in their recall of familiar information. *Child Development*, 1982, *53*, 799–810.

BLACKBURNE-STOVER, G., BELENKY, M. F., AND GILLIGAN, C. Moral development and reconstructive memory: Recalling a decision to terminate an unplanned pregnancy. *Development Psychology*, 1982, *18*, 862–870.

BLACKSTOCK, E. G. Cerebral asymmetry and the development of infantile autism. *Journal of Autism and Childhood Schizophrenia*, 1978, *8*, 339–353.

BLEWITT, P. Dog versus collie: Vocabulary in speech to young children. *Developmental Psychology*, 1983, *19*, 602–609.

BLOCK, J. *Lives through time*. Berkeley: Bancroft Books, 1971.

BLOCK, J. H. Issues, problems, and pitfalls in assessing sex differences: A critical review of *The psychology of sex differences*. *Merrill-Palmer Quarterly*, 1976, *22*, 283–308.

______. Differential premises arising from differential socialization of the sexes: Some conjectures. *Child Development*, 1983, *54*, 1335–1354.

BLOCK, M. R. Exiled Americans: The plight of the Indian aged in the United States. In D. E. Gelfand and A. J. Kutzik (Eds.), *Ethnicity and aging: Theory, research, and policy*. New York: Springer, 1979, pp. 184–192.

BLOOM, K., RUSSELL, A., AND DAVIS, S. Conversational turn taking: Verbal quality of adult affects vocal quality of infant. *Infant Behavior and Development*, 1986, *9* (Special Issue: Abstracts of papers presented at the Fifth International Conference on Infant Studies), 39.

BLOOM, L. *Language development: Form and function in emerging grammars*. Cambridge, Mass.: MIT Press, 1970.

______, HOOD, L., AND LIGHTBOWN, P. Imitation in language development: If, when, and why. *Cognitive Psychology*, 1974, *6*, 380–428.

BLOS, P. *The young adolescent: Clinical studies*. New York: Free Press, 1974.

BLOTNER, R., AND BEARISON, D. J. Developmental consistencies in socio-moral knowledge: Justice reasoning and altruistic behavior. *Merrill-Palmer Quarterly*, 1984, *30*, 349–367.

BLUEBOND-LANGNER, M. Field research on children's and adults' views of death. Unpublished fieldnotes, 1976.

______. Meanings of death to children. In H. Feifel (Ed.), *New meanings of death*. New York: McGraw-Hill, 1977, pp. 47–66.

BLUM, J. E., CLARK, E. T. AND JARVIK, L. F. The New York State Psychiatric Institute Study of Aging Twins. In L. F. Jarvik, C. Eisdorfer, and J. E. Blum (Eds.), *Intellectual functioning in adults: Psychological and biological influences*. New York: Springer, 1973, pp. 13–20.

BLURTON JONES, N. Categories of child-child interaction. In N. Blurton Jones (Ed.), *Ethological studies of child behaviour*. Cambridge, England: Cambridge University Press, 1972.

BOGATZ, G. A., AND BALL, S. *The second year of Sesame Street: A continuing evaluation*. Princeton, N.J.: Educational Testing Service, 1971.

BOHANNON, J. N., III, AND MARQUIS, A. L. Children's control of adult speech. *Child Development*, 1977, *48*, 1002–1008.

BOK, S. Lying: Moral choice in public and private life. New York: Pantheon, 1978.

BOLES, D. B. X-linkage of spatial ability: A critical review. *Child Development*, 1980, *51*, 625–635.

BONDAREFF, W. Neurobiology of aging. In J. E. Birren and R. B. Sloane (Eds.), *Handbook of mental health and aging*. Englewood Cliffs, N.J.: Prentice-Hall, 1980, pp. 75–99.

BOOTH, A., AND HESS, E. Cross-sex friendships. *Journal of Marriage and the Family*, 1974, *36*, 38–47.

BOOTZIN, R. R., AND ACOCELLA, J. R. *Abnormal psychology* (3d ed.). New York: Random House, 1980.

BORKAN, G. A., HULTS, D. E., GERZOF, S. G., ROBBINS, A. H., AND SILBERT, C. K. Age changes in body composition revealed by computed tomography. *Journal of Gerontology*, 1983, *38*, 673–677.

BORKE, H. Interpersonal perception of young children: Egocentrism or empathy? *Development Psychology*, 1971, *5*, 263–269.

______. Piaget's mountains revisited: Changes in the egocentric landscape. *Development Psychology*, 1975, *11*, 240–243.

BORNSTEIN, M. H. Qualities of color vision in infancy. *Journal of Experimental Child Psychology*, 1975, *19*, 401–419.

______. *Infant attention and caregiver stimulation*. Paper presented at the International Conference on Infant Studies, New York, April 1984.

______. How infant and mother jointly contribute to developing cognitive competence in the child. *Proceedings of the National Academy of Sciences of the United States of America*, 1985, *82*, 7470–7473. (a)

______. Human infant color vision and color preception. *Infant Behavior and Development*, 1985, *8*, 109–113. (b)

______, AND SIGMAN, M. D. Continuity in mental development from infancy. *Child Development*, 1986, *57*, 251–274.

BOSKIND-LODAHL, M. Cinderella's stepsisters: A feminist perspective on anorexia nervosa and bulimia. *Signs*, 1976, 2, 315–320.

BOTWINICK, J. *Aging and behavior*. New York: Springer, 1973.

______. Intellectual abilities. In J. E. Birren and K. W. Schaie (Eds.), *Handbook of the psychology of aging*. New York: Van Nostrand Reinhold, 1977, pp. 580–605.

BOWER, T. G. R., BROUGHTON, J. M., AND MOORE, M. K. Infant responses to approaching objects: An indicator of response to distal variables. *Perception and Psychophysics*, 1970, *9*, 193–196.

______, AND WISHART, J. G. The effects of motor skill on object permanence. *Cognition*, 1972, *1* 165–172.

BOWERMAN, C. E., AND KINCH, J. W. Changes in family and peer orientation of children between the fourth and tenth grades. *Social Forces*, 1956, *37*, 206–211.

BOWLBY, J. *Loss*. New York: Basic Books, 1980.

______. *Maternal care and mental health*. World Health Organization Monograph 2. Geneva: World Health Organization, 1951.

______. *Attachment and loss*, Volume 1. *Attachment*. New York: Basic Books, 1969.

______. *Attachment and loss*, Volume 2. *Separation*. New York: Basic Books, 1973.

______, AND PARKES, C. M. Separation and loss. In E. J. Anthony and C. Koupernik (Eds.), *The child in his family*. Volume 1. New York: Wiley, 1970.

BOWLES, N. L., AND POON, L. W. An analysis of the effect of aging on recognition memory. *Journal of Gerontology, 37*, 1982, 212–219.

BRACKBILL, Y. Extinction of the smiling response in infants as a function of reinforcement schedule. *Child Development*, 1958, *29*, 115–124.

______. Obstetrical medication and infant behavior. In J. D. Osofsky (Ed.), *Handbook of infant development*. New York: Wiley, 1979.

BRADLEY, R. H., AND CALDWELL, B. M. 174 children: A study of the relationship between home environment and cognitive development during the first 5 years. In A. W. Gottfried (Ed.), *Home environment and early cognitive development*. San Francisco: Academic Press, 1984.

______, CALDWELL, B. M., AND ELARDO, R. Home environment, social status, and mental test performance. *Journal of Educational Psychology*, 1977, *69*, 697–701.

______, AND CASEY, P. M. Home environments of low SES nonorganic failure-to-thrive infants. *Merrill-Palmer Quarterly*, 1984, *30*, 393–402.

BRADLEY-JOHNSON, S., JOHNSON, C. M., SHANAHAN, R. M., RICKERT, V. C., AND TARDONA, D. R. Effects of token reinforcement on WISC-R performance of black and white, low socioeconomic second graders. *Behavioral Assessment*, 1984, *6*, 365–373.

BRADSHAW, D. L., CAMPOS, J. J., AND KLINNERT, M. D. Emotional expressions as determinants of infants' immediate and delayed responses to prohibitions. *Infant Behavior and Development*, 1986. *9* (Special Issue: Abstracts of papers presented at the Fifth International Conference on Infant Studies), 46.

BRADY, J. E., NEWCOMB, A. F., AND HARTUP, W. W. Context and companions' behavior as determinants of cooperation and competition in school-age children. *Journal of Experimental Child Psychology*, 1983, *36*, 396–412.

BRAINE, M. D. S., HEIMER, C. B., WORTIS, H., AND FREEDMAN, A. M. Factors associated with impairment of the early development of prematures. *Monographs of the Society for Research in Child Development*, 1966, *31* (4, Serial No. 106).

______, AND RUMAIN, B. Logical reasoning. In P. H. Mussen (Ed.), *Handbook of child psychology*, Volume 3. New York: Wiley, 1983.

BRAINERD, C. J. Piaget's theory of intelligence. Englewood Cliffs, N.J.: Prentice-Hall, 1978.

______. Varieties of strategy training in Piagetian concept learning. In M. Pressley and J. L. Levin (Eds.), *Cognitive strategy research: Educational applications*. New York: Springer-Verlag, 1983.

BRANIGAN, G. Some reasons why successive single word utterances are not. *Journal of Child Language*, 1979, *6*, 443–458.

BRASK, B. H. The need for hospital beds for psychotic children. *Ugeskrift for Laeger*, 1967, *129*, 1559–1570.

BRAZELTON, T. B. Psychophysiologic reactions in the neonate: II. Effects of maternal medication on the neonate and his behavior. *Journal of Pediatrics*, 1961, *58*, 513–518.

______. *Infants and mothers*. New York: Delacorte Press/Seymour Lawrence, 1969.

______, TRONICK, E., ADAMSON, L., ALS, H., AND WISE, S. Early mother-infant reciprocity. In *Parent-infant interaction*, Ciba Foundation Symposium 33. Amsterdam: Associated Scientific Publishers, 1975.

BRENNAN, W. M., AMES, E. W., AND MOORE, R. W. Age differences in infants' attention to patterns of different complexities. *Science*, 1965, *151*, 1354–1356.

BRENT, L. Radiations and other physical agents. In J. G. Wilson and F. C. Fraser (Eds.), *Handbook of teratology*. New York: Plenum, 1977.

BRETSCHNEIDER, J., AND MCCOY, N. Sexual attitudes and behavior among healthy 80–102-year-olds. Paper read at Sixth World Congress of Sexology, Washington, D.C., May 22–27, 1983.

BRICKMAN, A.S., MCMANUS, M., GRAPETINE, W. L., AND ALESSI, N. Neuropsychological assessment of seriously delinquent adolescents. *Journal of the American Academy of Child Psychiatry*, 1984, *23*, 453–457.

BRIDGES, K. M. B. *The social and emotional development of the pre-school child*. London: Routledge, 1931.

BRINLEY, J. F., JOVICK, T. J., AND MCLAUGHLIN, L. M. Age, reasoning, and memory. *Journal of Gerontology*, 1974, *29*, 182–189.

BRITTAIN, C. Adolescent choices and parent-peer cross-pressures. *American Sociological Review*, 1963, *28*, 385–391.

______. Age and sex of siblings and conformity toward parents versus peers in adolescence. *Child Development*, 1966, *37*, 709–714.

BRITTON, J. H., AND BRITTON, J. O. *Personality changes in aging: A longitudinal study of community residents*. New York: Springer, 1972.

BRODY, E. B., AND BRODY, N. *Intelligence: Nature, determinants and consequences*. New York: Academic Press, 1976.

BRODY, E. M., JOHNSEN, P. T., FULCOMER, M. C. AND LAND, A. M. Women's changing roles and help to elderly parents: Attitudes of three generations. *Journal of Gerontology*, *38*, 1983, 597–607.

BRODY, G. H., STONEMAN, Z., MACKINNON, C. E., AND MACKINNON, R. Role relationships and behavior between preschool-aged and school-aged sibling pairs. *Development Psychology*, 1985, *21*, 124–129.

BRODY, H., AND VIJAYASHANKAR, N. Anatomical changes in the nervous system. In C. E. Finch and L. Hayflick (Eds.), *Handbook of the biology of aging*. New York: Van Nostrand Reinhold, 1977, pp. 241–261

BRODY, J. E. TV violence cited as bad influence. *New York Times*, December 7, 1975, p. 20.

______. Therapy helps teen-age girls having anorexia nervosa. *New York Times*, July 14, 1982, Section 4, p. 20.

______. Heart attacks and behavior: Early signs are found. *New York Times*, February 14, 1984, Section C, p. 1

BRODZINSKY, D. M. The role of conceptual tempo and stimulus characteristics in children's humor development. *Developmental Psychology*, 1975, *2*, 843–850.

BRONFENBRENNER, U., ALVAREZ, W. F., AND HENDERSON, C. R., JR. *Working and watching: Maternal employment status and parents' perceptions of their three-year old children*. Unpublished manuscript, Cornell University, 1983.

______, AND CROUTER, A. C. Work and family through time and space. In S. Kamerman and C. D. Hayes (Eds.), *Families that work: Children in a changing world*. Washington, D.C.: National Academy Press, 1982.

BRONSON, G. W. Infants' reactions to unfamiliar persons and novel objects. *Monographs of the Society for Research in Child Development*, 1972, *37* (3, Serial No. 148).

______. Aversive reactons to strangers: A dual process interpretation. *Child Development*, 1978, *49*, 495–499.

BRONSON, W. C. Central orientation: A study of behavior organization from childhood to adolescence. *Child Development*, 1966, *37*, 125–155.

______. Mother–toddler interaction: A perspective on studying the development of competence. *Merrill-Palmer Quarterly*, 1974, *20*, 275–301.

BROOK, J. S., WHITEMAN, M., AND GORDON, A. S. Stages of drug use in adolescence: Personality, peer, and family correlates. *Development Psychology*, 1983, *19*, 269–277.

______, WHITEMAN, M., GORDON, A. S., AND BROOK, D. W. Identification with paternal attributes and its relationship to the son's personality and drug use. *Child Development*, 1984, *20*, 1111–1119.

BROOKS, J., AND LEWIS, M. Infants responses to strangers: Midget, adult, and child. *Child Development*, 1976, *47*, 323–332.

BROPHY, J. E., AND EVERTSON, C. M. *Learning from teaching: A developmental perspective*. Boston: Allyn & Bacon, 1976.

BROUGHTON, R. Sleep disorders: Disorders of arousal. *Science*, 1968, *159*, 1070–1078.

______, AND GAUSTAUT, H. Recent sleep research in enuresis nocturia, sleep walking, sleep terrors, and confusional arousals. In P. Levin and W. Loella (Eds.), *Sleep 1974*. Basel: Karger, 1975.

BROWN, A. L. The development of memory: Knowing, knowing about knowing, and knowing how to know. In H. W. Reese (Ed.), *Advances in child development and behavior*, Volume 10. New York: Academic Press, 1975.

______, BRANSFORD, J. D., FERRARA, R. A., AND CAMPIONE, J. C. Learning, remembering, and understanding. In P. H. Mussen (Ed.), *Handbook of child psychology*, Volume 3. New York: Wiley, 1983.

______, AND DAY, J. D. Macrorules for summarizing texts: The development of expertise. *Journal of Verbal Learning and Verbal Behavior*, 1983, 22(1), 1–14.

______, AND PALINCSAR, A. S. Reciprocal teaching of comprehension strategies: A natural history of one program for enhancing learning. In J. Borkowski and J. D. Day (Eds.), *Intelligence and cognition in special children: Comparative studies of giftedness, mental retardation, and learning disabilities*. New York: Ablex, in press.

______, SMILEY, S. S., DAY, J. D., TOWNSEND, M. A., AND LAWTON, S. C. Intrusion of a thematic idea in children's comprehension and retention of stories. *Child Development*, 1977, *48*, 1454–1466.

BROWN, E. *Personal communication*, University of Chicago, 1981.

BROWN, G. W., AND HARRIS, T. O. *Social origins of depression*. New York: Free Press, 1978.

BROWN, H., ADAMS, R. G., AND KELLAM, S. G. A longitudinal study of teenage motherhood and symptoms of distress: Woodlawn Community Epidemiological Project. In R. Simmons (Ed.), *Research in community and mental health*, Volume 2. Greenwich, Conn.: JAI Press, 1981.

BROWN, J., BAKEMAN, R., SNYDER, P., FREDRICKSON, W., MORGAN, S., AND HEPLER, R. Interactions of black inner-city mothers with their newborn infants. *Child Development*, 1975, *46*, 677–686.

BROWN, R. How shall a thing be called? *Psychological Review*, 1958, 65, 14-21.

______. *A first language: The early stages*. Cambridge, Mass.: Harvard University Press, 1973.

______, AND HANLON, C. Derivational complexity and order of acquisition in child speech. In J. R. Hayes (Ed.), *Cognition and the development of language*. New York: Wiley, 1970.

BRUCEFORS, A., JOHANNESSON, I., KARLBERG, P., KLACKENBERG-LARSSON, I., LICHENSTEIN, H., AND SVENBERG, I. Trends in development of abilities related to somatic growth. *Human Development*, 1974, *17*, 152–159.

BRUCH, H. Obesity in adolescence. In H. Caplan and F. Lebovici (Eds.), *Adolescence: Psychosocial perspectives*. New York: Basic Books, 1969.

BRUNER, J. S. The course of cognitive growth. *American Psychologist*, 1964, *19*, 1–15.

______. From communication to language: A psychological perspective. In I. Markova (Ed.), *The social context of language*. New York: Wiley, 1978.

______. The acquisition of pragmatic commitments. In R. M. Golinkoff (Ed.), *The transition from prelinguistic to linguistic communication*. Hillsdale, N.J.: Lawrence Erlbaum, 1983.

BRUNK, M. A., AND HENGGELER, S. W. Child influences on adult controls: An experimental investigation. *Developmental Psychology*, 1984, *20*, 1074–1081.

BRYAN, J. H. Children's cooperation and helping behaviors. In E. M. Hetherington (Ed.), *Review of child development research*, Volume 5. Chicago: University of Chicago Press, 1975.

BRYANT, B. K. Locus of control related to teacher–child interperceptual experiences. *Child Development*, 1974, *45*, 157–164.

______. Sibling relationships in middle childhood. In M. Lamb and B. Sutton-Smith (Eds.), *Sibling relationships: Their nature and significance across the lifespan*. Hillsdale, N.J.: Lawrence Erlbaum, 1982.

BUCKHALT, J. A., MAHONEY, G. J., AND PARIS, S. C. Efficiency at self-generated elaborations by EMR and nonretarded children. *American Journal of Mental Deficiency*, 1976, *81*, 93–96.

BUELL, S. J., AND COLEMAN, P. D. Dendritic growth in aged human brain and failure of growth in senile dementia. *Science*, 1976, *206*, 854–856.

BUGENTAL, D. B., CAPORAEL, L., AND SHENNUM, W. A. Experimentally produced child uncontrollability: Effects on the potency of adult communication patterns. *Child Development*, 1980, *51*, 520–528.

______, AND SHENNUM, W. A. *Adult attributions as moderators of the effects of shy vs. assertive children*. Paper presented at the meeting of the Society for Research in Child Development, Boston, April 1981.

______, AND ______. "Difficult" children as elicitors and targets of adult communication patterns: An attributional-behavioral transactional analysis. *Monographs of the Society for Research in Child Development*, 1984, *49* (1, Serial No. 205).

BÜHLER, C. *The child and his family*. London: Harper & Bros., 1939.

______. *Der menschliche Lebenslauf als psychologisches Problem*. Leipzig: Hirzel, 1929.

BULCKE, J. A., TERMOTE, J. L., PALMERS, Y. AND CROLLA, D. Computed tomography of the human skeletal muscular system. *Neuroradiology*, 1979, *17*, 127–136.

BULLOCK, M. Animism in childhood thinking: A new look at an old question. *Developmental Psychology*, 1985, *21*, 217–225.

BULLOUGH, V. L. Age at menarche: A misunderstanding. *Science*, 1981, *213*, 365–366.

BURNS, N., AND CARNEY, K. The caring aspect of hospice. In L. F. Paradis (Ed.), *Hospice handbook: A guide for managers and planners*. Md.: Aspen Systems Corp., 1985.

BURTON, R. V. Generality of honesty reconsidered. *Psychological Review*, 1963, *70*, 481–499.

______. Honesty and dishonesty. In T. Lickona (Ed.), *Moral development and behavior: Theory, research and social issues*. New York: Holt, Rinehart & Winston, 1976.

BURTON, W. *Helps to education*. Boston: Crosby and Nichols, 1863.

BUSS, D. M. Predicting parent-child interactions from children's activity level. *Development Psychology* 1981, *17*, 59–65.

BUSS, R. R., YUSSEN, S. R., MATHEWS, S. R., II, MILLER, G. E., AND REMBOLD, K. L. Development of children's use of a story schema to retrieve information. *Developmental Psychology*, 1983, *19*, 22–28.

BUSSE, E. W., AND EISDORFER, C. Two thousand years of married life. In E. Palmore (Ed.), *Normal aging: Reports from the Duke Longitudinal Study, 1955–1969*. Durham, N.C.: Duke University Press, 1970, pp. 266–269.

BUTLER, N. R., AND GOLDSTEIN, H. Smoking in pregnancy and subsequent child development. *British Medical Journal*, 1973, *4*, 573–575.

BUTLER, R. N. *Why survive? Being old in America*. New York: Harper & Row, 1975. (chaps 1, 16)

______, AND LEWIS, M. I. *Aging and mental health* (3d ed). St. Louis: C. V. Mosby, 1982.

BUTTERFIELD, E. C., AND BELMONT, J. M. Assessing and improving the executive cognitive functions of mentally retarded people. In I. Bialer and M. Sternlicht (Eds.), *Psychological issues in mentally retarded people*. Chicago: Aldine, 1977.

BUTTERFIELD, P. M. Women "at risk" for parenting disorder perceive emotions in infant pictures differently. *Infant Behavior and Development*, 1986, *9* (Special Issue: Abstracts of papers presented at the Fifth International Conference on Infant Studies), 56.

BUTTERS, N. Potential contributions of neuropsychology to our understanding of the memory capacities of the elderly. In L. W. Poon, J. L. Fozard, L. S. Cermak, D. Arenberg, and L. W. Thompson (Eds.), *New directions in memory and aging*. Hillsdale, N.J.: Lawrence Erlbaum, 1980, pp. 451–460.

BUTTERWORTH, G. *Infancy and epistemology: An evaluation of Piaget's theory*. Brighton, England: Harvester, 1981.

CAIN, L. D. Aging and the law. In R. H. Binstock and E. Shanas (Eds.), *Handbook of aging and the social sciences*. New York: Van Nostrand Reinhold, 1976, pp. 342–368.

CALDWELL, B. M., HERSHER, L., LIPTON, E. L., RICHMOND, J. B., STERN, G. A., EDDY, E., DRACHMAN, R., AND ROTHMAN, A. Mother-infant interaction in monomatric and polymatric families. *American Journal of Orthopsychiatry*, 1963, *33*, 653–664.

CALLAHAN, E. J., BRASTED, W. S., AND GRANADOS, J. L. Fetal loss and sudden infant death: Grieving and adjustment for families. In E. J. Callahan and K. A. McCluskey (Eds.), *Life-span developmental psychology: Nonnormative events*. New York: Academic Press, 1983, pp. 145–166.

CAMERON, P. Mood as an indicant of happiness: Age, sex, social class, and situational differences. *Journal of Gerontology*, 1975, *30*, 216–224.

CAMERON, R. Problem-solving inefficiency and conceptual tempo: A task analysis of underlying factors. *Child Development*, 1984, *55*, 2031–2041.

CAMPBELL, D. P. *Handbook for the Strong-Campbell Interest Inventory*. Stanford, Calif.: Stanford University Press, 1974.

CAMPBELL, S. B., AND PAULAUKAS, S. Peer relations in hyperactive children. *Journal of Child Psychology and Psychiatry*, 1979, *20*, 233–246.

CAMPOS, J. J., LANGER, A., AND KROWITZ, A. Cardiac responses on the visual cliff in prelocomotor human infants. *Science*, 1970, *170*, 196–197.

———, AND STENBERG, C. R. Perception, appraisal and emotion: The onset of social referencing. In M. E. Lamb and L. R. Sherrod (Eds.), *Infant social cognition*. Hillsdale, N.J.: Lawrence Erlbaum, 1981.

CAMRAS, L. A. Facial expressions used by children in a conflict situation. *Child Development, 1977, 48*, 1431–1435.

CANDEE, D., SHEENAN, T. J., COOK, C. D., HUSTED, S. D. R., AND BARGEN, M. Moral reasoning and decisions in dilemmas of neonatal care. *Pediatric Research*, 9182, *16*, 846–850.

CANESTRARI, R. Paced and self-paced learning in young and elderly adults. *Journal of Gerontology*, 1963, *18*, 165–168.

Cantor, P. These teenagers feel that they have no options. *People Weekly*, February 18, 1985, 84–87.

CAPLAN, P. J., MACPHERSON, G. M., AND TOBIN, P. Do sex-related differences in spatial abilities exist? *American Psychologist*, 1985, *40*, 786–799.

CAPLOW, T., BAHR, H. M., CHADWICK, B. A., HILL, R., AND WILLIAMSON, M. H. *Middletown families*. Minneapolis: University of Minnesota Press, 1982.

CARD, J., AND WISE, L. Teenage mothers and teenage fathers: The impact of early childbearing on the parents' personal and professional lives. *Family Planning Perspectives*, 1978, *10*, 199–205.

CAREW, J. Experience and the development of intelligence in young children. *Monographs of the Society for Research in Child Development*, 1980, *45* (1–2, Serial No. 183).

CAREY, S. Semantic development: State of the art. In E. Wanner and L. R. Gleitman (Eds.), *Language acquisition: State of the art*. New York: Cambridge University Press, 1982.

———. *Conceptual change in childhood*. Cambridge, Mass.: MIT Press, 1985. (a)

———. Constraints on semantic development. In J. Mehler and R. Fox (Eds.), *Neonate cognition: Beyond the blooming buzzing confusion*. Hillsdale, N.J.: Lawrence Erlbaum Associates, 1985. (b)

CARGAN, L., AND MELKO, M. *Singles: Myths and realities*. Beverly Hills, Calif.: Sage, 1982.

CARON, A. J., CARON, R. F., CALDWELL, R., AND WEISS, S. Infant perception of the structural properties of the face. *Developmental Psychology*, 1973, *9*, 385–399.

CARON, R. F., CARON, A. J., AND MYERS, R. Do infants see emotional expressions in static faces? *Child Development*, 1985, *56*, 1552–1560.

———, AND MYERS, R. *Do infants perceive emotions in static faces?* Paper presented at the International Conference on Infant Studies, New York, April 1984.

CARPENTER, C. J. Activity structure and play: Implications for socialization. In M. B. Liss (Ed.), *Social and cognitive skills: Sex roles and children's play*. New York: Academic Press, 1983.

CARR, S. J., DABBS, J. M., JR., AND CARR, T. S. Mother-infant attachment: The importance of the mother's visual field. *Child Development*, 1975, *46*, 331–338.

CARUS, F. A. *Psychologie*. Leipzig: Barth & Kummer, 1808.

CASE, R. Structures and strictures: Some functional limitations on the course of cognitive growth. *Cognitive Psychology*, 1974, *6*, 544–573.

———, KURLAND, D. M., AND GOLDBERG, J. Operational efficiency and the growth of short-term memory span. *Journal of Experimental Child Psychology*, 1982, *33*, 386–404.

CASSIDY, J., AND MAIN, M. *Quality of attachment from infancy to early childhood: Security is stable but behavior changes*. Paper presented at the International Conference on Infant Studies, New York, April 1984.

CATTELL, J. M. Mental tests and measurements. *Mind*, 1890, *15*, 373.

CAUGHILL, R. E. Supportive care and age of the dying patient. In R. E. Caughill (Ed.), *The dying patient*, Boston: Little, Brown, 1976, pp. 191–224.

CAVANAUGH, J. C., AND PERLMUTTER, M. Metamemory: A critical examination. *Child Development*, 1982, *53*, 11–28.

CAVANAUGH, P. J., AND DAVIDSON, M. L. The secondary circular reaction and response elicitation in the operant learning of six-month-old infants. *Developmental Psychology*, 1977, *13*, 371–376.

CECI, S. J., AND BRONFENBRENNER, U. "Don't forget to take the cupcakes out of the oven": Prospective memory, strategic time-monitoring and context. *Child Development*, 1985, *56*, 152–164.

CECI, S. J., AND TISHMAN, J. Hyperactivity and incidental memory: Evidence for attentional diffusion. *Child Development*, 1984, *55*, 2192–2203.

CEDARBAUM, S. Personal communication, 1976. Cited by C. B. Kopp and A. H. Parmelee, Prenatal and perinatal influences on infant behavior. In J. D. Osofsky (Ed.), *Handbook of infant development*. New York: Wiley, 1979.

CHAFETZ, M. E. Alcoholism: Drug dependency problem No. 1. *Journal of Drug Issues*, 1974, *4*, 64–68.

CHAMBLISS, W. J. The saints and the roughnecks. In T. J. Cottle (Ed.), *Readings in adolescent psychology: Contemporary perspectives* New York: Harper & Row, 1977.

CHANDLER, M. J. Egocentrism and antisocial behavior: The assessment and training of social perspective-taking skills. *Developmental Psychology*, 1973, *9*, 326–332.

CHANG, B. L. *The relationship of generalized expectancies and situational control of daily activities to morale of the institutionalized aged*. San Francisco: Gerontological Society, 1979.

CHAPMAN, M., AND ZAHN-WAXLER, C. Young children's compliance and noncompliance to parental discipline in a natural setting. *International Journal of Behavior Development*, 1982, *5*, 8194.

CHAPPELL, N. L. Awareness of death in the disengagement theory: A conceptualization and an empirical investigation. *Omega*, 1975, *6*, pp. 325–344.

CHARLESWORTH, R., AND HARTUP, W. W. Positive social reinforcement in the nursery school peer group. *Child Development*, 1967, *38*, 993–1002.

CHASE-LANSDALE, P. L. Effects of maternal employment on mother-infant and father-infant attachment. Unpublished doctoral dissertation, University of Michigan, 1981. *Dissertation Abstracts International*, 1982, *42*, 2562.

CHEN, E. Twins reared apart: A living lab. *New York Times Magazine*, December 9, 1979, p. 110.

CHERLIN, A. A sense of history: Recent research on aging and the family. In M. W. Riley, B. B. Hess, and K. Bond (Eds.), *Aging in society: Selected reviews of recent research*. Hillsdale, N.J.: Lawrence Erlbaum, 1983, pp. 5–24.

CHESNICK, M., MENYUK, P., LIEBERGOTT, J., FERRIER, L., AND STRAND, K. *Who leads whom?* Paper presented at the meeting of the Society for Research in Child Development, Detroit, April 1983.

CHESTER, N. L. Pregnancy and the new parenthood: Twin experiences of change. Paper presented at the meeting of the Eastern Psychological Association. Philadelphia, 1979.

CHI, M. T. Short-term memory limitations in children: Capacity or processing deficits? *Memory and Cognition*, 1976, *4*, 559–572.

CHI, M. T. Knowledge structures and memory development. In R. S. Siegler (Ed.), *Children's thinking: What develops?* Hillsdale, N.J.: Lawrence Erlbaum, 1978.

______. Changing conception of sources of memory development. *Human Development*, 1985, *28*, 50–56.

CHILMAN, C. *Adolescent sexuality in a changing American society: Social and psychological perspectives*: Washington, D.C.: Public Health Service, National Institute of Mental Health, 1979.

CHOMSKY, C. *The acquisition of syntax in children from 5 to 10*. Research Monograph 57. Cambridge, Mass.: MIT Press, 1969.

CHOMSKY, N. *Syntactic structures*. The Hague: Mouton, 1957.

______. *Aspects of the theory of syntax*. Cambridge, Mass.: MIT Press, 1965.

______. *Language and mind*. New York: Harcourt Brace Jovanovich, 1968.

______. *Reflections of language*. New York: Pantheon, 1975.

CHUGANI, H. T., AND PHELPS, M. E. Maturational changes in cerebral function in infants determined by FGG positron emission tomography. *Science*, February 1986,*231*, 840–843.

CIALDINI, R. B., AND KENRICK, D. T. Altruism as hedonism: A social development perspective on the relationship of negative mood state and helping. *Journal of Personality and Social Psychology*, 1976, *34*, 907–914.

CICIRELLI, V. G. Concept learning of young children as a function of sibling relationships to the teacher. *Child Development*, 1972, *43*, 282–287.

______. Effects of sibling structure and interaction on children's categorization style. *Developmental Psychology*, 1973, *9*, 132–139.

______. Categorization behavior in aging subjects. *Journal of Gerontology*, 1976, *31*, 676–680.

______. Children's school grades and sibling structure. *Psychological Reports*, 1977, *41*, 1055–1058.

______. Effects of sibling presence on mother-child interaction. *Developmental Psychology*, 1978, *14*, 315–316.

______. Interpersonal relationships of siblings in the middle part of the lifespan. Paper presented at the Biennial Meeting of the Society for Research in Child Development. Boston, April 1981.

______. Sibling influence throughout the lifespan. In M. E. Lamb and B. Sutton-Smith (Eds.), *Sibling relationships: Their nature and significance across the lifespan*. Hillsdale, N.J.: Lawrence Erlbaum, 1982, pp. 267–284.

CLARK, E. On the acquisition of the meaning of *before* and *after*. *Journal of Verbal Learning and Verbal Behavior, 1971*, 10, 266–275.

______. First language acquisition. In J. Morton and J. C. Marshall (Eds.), *Psycholinguistics Series 1: Developmental and pathological*. London: Paul Elek, 1977.

______. Meanings and concepts. In P. H. Mussen (Ed.), *Handbook of child psychology*, Volume 3. New York: Wiley, 1983.

CLARK, H., AND CLARK, E. *Psychology and language: An introduction to psycholinguistics*. New York: Harcourt Brace Jovanovich, 1977.

CLARK, M., AND LESLIE, C. Why kids get fat: A new study shows obesity is in the genes. *Newsweek*, February 3, 1986, p. 61.

CLARK, P. M., AND KRIGE, P. D. *A study of interaction between infant peers: An analysis of infant-infant interaction in twins*. Paper presented at the second Joint Congress of the South African Psychology Association and the Psychological Institute of South Africa, Potchefstroom, September 1979.

CLARKE, H. H., AND CLARKE, D. H. Social status and mental health of boys as related to their maturity, structural, and strength characteristics. *Research Quarterly*, 1961, *32*, 326–334.

CLARKE-STEWART, A. *Child care in the family: A review of research and some propositions for policy*. New York: Academic Press, 1977.

______. Predicting child development from day care forms and features: The Chicago study. In D. Phillips (Ed.), *Predictors of quality childcare*. NAEYC Research Monograph Series 1, in press.

______, FRIEDMAN, S., AND KOCH, J. *Child development: A topical approach*. New York: Wiley, 1985.

CLARKE-STEWART, K. A. Interactions between mothers and their young children: Characteristics and consequences. *Monographs of the Society for Research in Child Development*, 1973, *38* (6–7, Serial No. 153).

______. And daddy makes three: The father's impact on mother and young child. *Child Development*, 1978, *49*, 466–478.(a)

______. Recasting the lone stranger. In J. Glick and K. A. Clarke-Stewart (Eds.), *The development of social understanding*. New York: Gardner Press, 1978.(b)

______. The father's contribution to child development. In F. A. Pedersen (Ed.), *The father-infant relationship: Observational studies in a family context*. New York: Praeger Special Studies, 1980.

______. *Daycare*. Cambridge, Mass.: Harvard University Press, 1982.

______. Daycare: A new context for research and development. In M. Perlmutter (Ed.), *Parent-child interaction and parent-child relations in child development. Minnesota Symposia on Child Psychology*, Volume 17. Hillsdale, N.J.: Lawrence Erlbaum, 1984.

______, AND APFEL, N. Evaluating parental effects on child development. In L. S. Shulman (Ed.), *Review of research in education*, Volume 6. Itasca, Ill.: Peacock, 1979.

______, AND FEIN, G. G. Early childhood programs. In P. H. Mussen (Ed.), *Handbook of child psychology*, Volume 2. New York: Wiley, 1983.

______, AND HEVEY, C. M. Longitudinal relations in repeated observations of mother-child interaction from 1 to 2 ½ years. *Developmental Psychology*, 1981, *17*, 127–145.

CLAUSEN, J. A. Men's occupational careers in the middle years. In D. H. Eichorn, J. A. Clausen, N. Haan, M. P. Honzik, and P. Mussen (Eds.), *Present and past in middle life*. New York: Academic Press, 1981, pp. 321–351.

______, MUSSEN, P. H., AND KUYPERS, J. Involvement, warmth, and parent-child resemblances in three generations. In D. H. Eichorn, J. A. Clausen, N. Haan, M. P. Honzik, and P. Mussen (Eds.), *Present and past in middle life*. New York: Academic Press, 1981, pp. 299–319.

CLAYTON, V. P., AND BIRREN, J. E. The development of wisdom across the life span: A reexamination of an ancient topic. In P. B. Baltes and O. G. Brim, Jr. (Eds.), *Life-span development and behavior*. Volume New York: Academic Press, 1980, pp. 103–135.

CLEMENTS, D. H., AND GULLO, D. F. Effects of computer programming on young children's cognition. *Journal of Educational Psychology*, 1984, *76*, 1051–1058.

CLENDINEN, D. Testing the waters before retirement. *The New York Times*, October 27, 1983, Cl+.

COHEN, C. I., COOK, P. AND RAJKOWSKI, H. What's in a friend? Paper presented at the 33d Annual Scientific Meeting of the Gerontological Society. San Diego, November 1980.

COHEN, L. B., AND STRAUSS, M. S. Concept acquisition in the human infant. *Child Development*, 1979, *50*, 419–424.

COHEN, L. J., AND CAMPOS, J. J. Father, mother, and stranger as elicitors of attachment behaviors in infancy. *Developmental Psychology*, 1974, *10*, 146–154.

COHEN, S., EVANS, G. W., KRANTZ, D. S., AND STOKOLS, D. Physiological, motivational, and cognitive effects of aircraft noise in children. *American Psychologist*, 1980, *35*, 231–243.

COHEN, S., AND WEINSTEIN, N. D. Nonauditory effects of noise. In G. W. Evans (Ed.), *Environmental stress*. New York: Cambridge University Press, 1982.

COHLER, B. J., AND GRUNEBAUM, H. U. *Mothers, grandmothers, and daughters*. New York: Wiley-Interscience, 1981.

COHN, J. F. AND TRONICK, E. Z. Three-month-old infants' reaction to simulated maternal depression. *Child Development*, 1983, *54*, 185–193.

COIE, J. D., AND KUPERSMIDT, J. B. A behavioral analysis of emerging social status in boys' groups. *Child Development*, 1983, *54*, 1400–1416.

COLBY, A., AND DAMON, W. Listening to a different voice: A review of Gilligan's *In a Different Voice. Merrill-Palmer Quarterly*, 1983, *29*, 473–481.

———, Kohlberg, L., Gibbs, J., and Lieberman, M. A longitudinal study of moral judgment. *Monographs of the Society for Research in Child Development*, 1983, *48* (1, Serial No. 200).

COLE, M. How education affects the mind. *Human Nature*, 1978, *1*, 50–58.

———, Gay, J., Glick, J. A., and Sharp, D. W. *The cultural context of learning and thinking: An exploration in experimental anthropology*. New York: Basic Books, 1971.

———, AND SCRIBNER, S. Cross-cultural studies of memory and cognition. In R. V. Kail, Jr., and J. W. Hagen (Eds.), *Perspectives on the development of memory and cognition*. Hillsdale, N.J.: Lawrence Erlbaum, 1977.

COLEMAN, J. C. *The adolescent society*. Glencoe, Ill.: Free Press, 1961.

———. *Relationships in adolescence*. London: Routledge, 1974.

———. Friendship and the peer group in adolescence. In J. Adelson (Ed.), *Handbook of adolescent psychology*. New York: Wiley, 1980.

COLEMAN, M. A report on the autistic syndromes. In M. Rutter and E. Schopler (Eds.), *Autism: A reappraisal of concepts and treatment*. New York: Plenum, 1978.

COLLINS, G. Long distance care of elderly relatives a growing problem. *The New York Times*, December 29, 1983, A1+.

———. Changes in a marriage when a baby is born. *New York Times*, January 6, 1985, p. 12.

COLLINS, W. A. Social antecedents, cognitive processing, and comprehension of social portrayals on television. In E. T. Higgins, D. N. Ruble, and W. W. Hartup (Eds.), *Social cognition and social development: A sociocultural perspective*. Cambridge, England: Cambridge University Press, 1983.

COMFORT, A. Sexuality in later life. In J. E. Birren and R. B. Sloane (Eds.), *Handbook of mental health and aging*. Englewood Cliffs, N.J.: Prentice-Hall, 1980, pp. 885–892.

COMMONS, M. L., RICHARDS, F. A., AND KUHN, D. Systematic and metasystematic reasoning: A case for levels of reasoning beyond Piaget's stage of formal operations. *Child Development*, 1982, *53*, 1058–1069.

CONDON, W. S. AND SANDERS, L. W. Neonate movement is synchronized with adult speech: Interactional participation and language acquisition. *Science*, 1974, *183*, 99–101.

CONDRY, J., AND SIMAN, M. L. Characteristics of peer-adult-oriented children. *Journal of Marriage and the Family*, 1974, *36*, 543–554.

CONEL, J. L. *The postnatal development of the human cerebral cortex* (8 vols.). Cambridge, Mass.: Harvard University Press, 1939–1967.

CONNELL, J. P. *A model of the relationship among children's self-rated cognitions, affects and academic achievement*. Unpublished doctoral dissertation, University of Denver, 1981.

CONNERS, C. K. *Food additives and hyperactive children*. New York: Plenum, 1980.

COOPER, C. R. *Collaboration in children: Dyadic interaction skills in problem solving*. Paper presented at the meeting of the Society for Research in Child Development, New Orleans, March 1977.

COOPER, C. R., AYERS-LOPEZ, S., AND MARQUIS, A. Children's discourse during peer learning in experimental and naturalistic situations. *Discourse Processes*, 1982, *5*, 177–191.

———, GROTEVANT, H. D., AND CONDON, S. M. Individuality and connectedness in the family as a context for adolescent identity formation and role-taking skill. *New Directions for Child Development*, 1983, *22*, 43–59.

COOPERSMITH, S. *The antecedents of self-esteem*. San Francisco: W. H. Freeman, 1967.

CORDES, C. Attention disorder crime link dispelled. *APA Monitor*, November 1985, *16*, 16.

COREA, G. *The mother machine*. New York: Harper & Row, 1986.

CORTER, C., TREHUB, S., BOUKYDIS, C., FORD, L., CELHOFFER, L., AND MINDE, K. Nurses' judgments of the attractiveness of premature infants. *Infant Behavior and Development*, 1978, *1*, 373–380.

COSGROVE, J. M., AND PATTERSON, C. J. Adequacy of young speakers' encoding in response to listener feedback. *Psychological Reports*, 1979, *45*, 15–18.

COSTANZO, P. R., AND SHAW, M. E. Conformity as a function of age level. *Child Development*, 1966, *37*, 967–975.

CORBY, N., AND SOLNICK, R. L. Psychosocial and physiological influences on sexuality in the older adult. in J. E. Birren and R. B. Sloane (Eds.), *Handbook of mental health and aging*. Englewood Cliffs, N.J.: Prentice-Hall, 1980, pp. 893–921.

COWAN, P., COWAN, C., COIE, J. AND COIE L. In L. Newman and W. Miller (Eds.), *The first child and family formation*. Durham, N.C.: University of North Carolina Press, 1978.

COYLE, J. T., PRICE D. L., AND DELONG. M. R. Alzheimer's Disease: A disorder of cortical cholinergic innervation. *Science*, 1983, *219*, 1184–1190.

———. Age differences in human memory. In J. E. Birren and K. W. Schaie (Eds.), *Handbook of the psychology of aging*. New York: Van Nostrand Reinhold, 1977, pp. 384–420

———, AND BYRD, M. Aging and cognitive deficits: The role of attentional resources. In F. I. M. Craik and S. Trehub (Eds.), *Aging and the cognitive processes*. New York: Plenum, 1982, pp. 191–212.

———, AND RABINOWITZ, J. C. Age differences in the acquisition and use of verbal information. In H. Bouma and D. G. Bouwhuis (Eds.), *Attention and performance*, Volume 10. Hillsdale, N.J.: Lawrence Erlbaum, 1984.

———, AND SIMON, E. Age differences in memory: The roles of attention and depth of processing. In L. W. Poon, J. L. Fozard, L. S. Cermak, D. Arenberg, and L. W. Thompson (Eds.), *New directions in memory and aging*. Hillsdale, N.J.: Lawrence Erlbaum, 1980, pp. 95–112.

CRANDALL, V. C., KATKOVSKY, W., AND CRANDALL, V. J. Children's beliefs in their control of reinforcements in intellectual academic achievement behaviors. *Child Development*, 1965, *36*, 91–109.

CRANO, W. D. Causal analyses of the effects of the socioeconomic status and initial intellectual endowment on patterns of cognitive development and academic achievement. In D. R. Green (Ed.), *The aptitude-achievement distinction*. Monterey, Calif.: California Test Bureau, 1974.

———, KENNY, J., AND CAMPBELL, D. T. Does intelligence cause achievement? A cross-lagged panel analysis. *Journal of Educational Psychology*, 1972, *63*, 258–275.

CRAVIOTO, J., BIRCH, H. G., DELICARDIE, E., ROSALES, L., AND VEGA, L. The ecology of growth and development in a Mexican preindustrial community. Report I: Method and findings from birth to one month of age. *Monographs of the Society for Research in Child Development*, 1969, *34* (5, Serial No. 129).

CRAWFORD, C. It was a living nightmare. *TV Guide*. March 2, 1985, 36–38.

CROCKENBERG, S. Infant irritability, mother responsiveness, and social support influences on the security of infant-mother attachment. *Child Development*, 1981, *52*, 857–869.

———. Early mother and infant antecedents of Bayley Scale performance at 21 months. *Developmental Psychology*, 1983, *19*, 727–730.

———. *Toddlers reactions to maternal anger*. Paper presented at the International Conference on Infant Studies, New York, April 1984.

———. Are temperamental differences in babies associated with predictable differences in care giving? In J. V. Lerner and R. M. Lerner (Eds.), *New directions for child development: Temperament and social interaction during infancy and childhood*. San Francisco: Jossey-Bass, 1986.

———, AND ACREDOLO, C. Infant temperament ratings: A function of infants, or mothers, or both. *Infant Behavior and Development*, 1983, *6*, 61–72.

CROSSON, C. W., AND ROBERTSON-TCHABO, E. A. Age and preference for complexity among manifestly creative women. *Human Development*, 1983, *26*, 149–155.

CSIKSZENTMIHALYI, M., AND LARSON, R. *Being adolescent: Conflict and growth in the teenage years*. New York: Basic Books, 1984.

______, LARSON, R., AND PRESCOTT, S. The ecology of adolescent activity and experience. *Journal of Youth and Adolescence*, 1977, *6*, 281–294.

CUMMINGS, E. M., AND BJORK, E. L. The search behavior of 12 to 14 month-old infants on a five-choice invisible displacement hiding task. *Infant Behavior and Development*, 1981, *4*, 47–60.

______, IANNOTTI, R. J., AND ZAHN-WAXLER, C. Influence of conflict between adults on the emotions and aggression of young children. *Developmental Psychology*, 1985, *21*, 495–507.

CUNNINGHAM, S. Bulimia's cycle shames patient, tests therapists. *APA Monitor*, January 1984, *15*, 16.

CURTISS, S. *Genie: A psycholinguistic study of a modern-day wild child*. New York: Academic Press, 1977.

CUTLER, R. G. Life-span extension. In J. L. McGaugh and S. B. Kiesler (Eds.), *Aging: Biology and behavior*. New York: Academic Press, 1981, pp. 31–76.

CYTRYNBAUM, S., BLUM, L., PATRICK, R., STEIN, J., AND WILK, C. Midlife development: A personality and social systems perspective. In L. W. Poon (Ed.), *Aging in the 1980s: Psychological issues*. Washington, D.C.: American Psychological Association, 1980.

DALE, N. *Early pretend play in the family*. Unpublished doctoral thesis, University of Cambridge, 1983.

DALE, P. *Language development: Structure and function* (2nd ed.). New York: Holt, Rinehart & Winston, 1976.

DAMON, A., SELTZER, C. C., STOUDT, H. W., AND BELL, B. Age and physique in healthy white veterans at Boston. *Aging and Human Development*, 1972, *3*, 202–208.

DAMON, W. *The social world of the child*. San Francisco: Jossey-Bass, 1977.

DAMON, W., AND HART, D. The development of self-understanding from infancy through adolescence. *Child Development*, 1982, *53*, 841–864.

DAMON, W., AND KILLEN, M. Peer interaction and the process of change in children's moral reasoning. *Merrill Palmer Quarterly*, 1982, *28*, 347–367.

DAN, A. J. *Patterns of behavioral and mood variation in men and women: Variability and the menstrual cycle*. Unpublished doctoral dissertation, University of Chicago (Committee on Human Development), 1976.

D'ANGELO, R. *Families of sand: A report concerning the flight of adolescents from their families*. Columbus, Ohio: State University School of Social Work, 1974.

DANIELS, C. W. "Cell Longevity: *In Vivo*." In C. E. Finch and L. Hayflick (Eds.), *Handbook of the Biology of Aging*. New York: Van Nostrand Reinhold, 1977, pp. 122–158.

DANIELS, D., AND PLOMIN, R. Origins of individual differences in infant shyness. *Developmental Psychology*, 1985, *21*, 118–121.

DANIELS, P., AND WEINGARTEN, K. *Sooner or later: The timing of parenthood in adult lives*. New York: Norton, 1982.

DAPCICH-MIURA, E., AND HOVELL, M. F. Contingency management of adherence to a complex medical regimen in an elderly heart patient. *Behavior Therapy*, 1979, *10*, 193–201.

DARLING, C. A. Parental influence on love, sexual behavior, and sexual satisfaction. Unpublished doctoral dissertation, Michigan State University, 1979.

______, KALLEN, D. J., AND VAN DUSEN, J. E. Sex in transition, 1900–1980. *Journal of Youth and Adolescence*, 1984, *13*, 385–399.

DASEN, P. Are cognitive processes universal? A contribution to cross-cultural Piagetian psychology. In N. Warren (Ed.), *Studies in cross-cultural psychology*, Volume 1. London: Academic Press, 1977.

DAVIDSON, E. S., YASUNA, A., AND TOWER, A. The effects of television cartoons on sex role stereotyping in young girls. *Child Development*, 1979, *50*, 597–600.

DAVIDSON, J. M., CAMARGO, C. A., AND SMITH, E. R. Effects of androgen on sexual behavior in hypogonadal men. *Journal of Clinical Endocrinology and Metabolism*, 1979, *48*, 955–958.

DAVIDSON, J. R. Post-partum mood change in Jamaican women: A description and discussion on its significance. *British Journal of Psychiatry*, 1972, *121*, 659–663.

DAVIES, M. J., AND POMERANCE, A. Quantitative study of aging changes in the human sinoatrial node and internodal tracts. *British Heart Journal*, 1972, *34*, 150–152.

DAVIS, M. A., AND RANDALL, E. Social change and food habits of the elderly. In M. W. Riley, B. B. Hess, and K. Bond (Eds.), *Aging in society: Selected reviews of recent research*. Hillsdale, N.J.: Lawrence Erlbaum, 1983, pp. 199–218.

DAWSON, C. D. Cerebral lateralization in individuals diagnosed as autistic in early childhood. *Brain and Language*, 1982, *15*, 353–368.

DAY, B., AND BRICE, R. Academic achievement, self-concept development, and behavior patterns of six-year-old children in open classrooms. *Elementary School Journal*, 1977, *78*, 132–139.

DAY, R. H., AND MCKENZIE, B. E. Perceptual shape constancy in early infancy, *Perception*. 1973, *2*, 315–320.

______, AND ______. Infant perception of the invariant size of approaching and receding objects. *Developmental Psychology*, 1981, *17*, 670–677.

DAYTON, G. O., JR., JONES, M. H., AIU, P., RAWSON, R. A., STEELE, B., AND ROSE, M. Developmental study of coordinated eye movements in the human infant: I. Visual acuity in the newborn human: A study based on induced optokinetic nystagmus recorded by electro-oculography. *Archives of Ophthalmology*, 1964, *71*, 865–870.

DEBAKEY, M. E., GOTTO, A. M., JR., SCOTT, L. W., AND FOREYT, J. P. *The living heart diet*. New York: Raven Press/ Simon & Schuster, 1984, pp. 12–27.

DE BOYSSON-BARDIES, B., SAGART, L., AND DURAND, C. Discernible differences in the babbling of infants according to target language. *Journal of Child Language*, 1984, *11*, 1–15.

DECARIE, T. G. *The infant's reaction to strangers*. New York: International Universities Press, 1974.

DECKER, S. N., AND DEFRIES, J. C. Cognitive ability profiles in families of reading disabled children. *Developmental Medicine and Child Neurology*, 1981, *23*, 217–227.

DEFRIES, J. C., AND BAKER, L. A. Parental contributions to longitudinal stability of cognitive measures in the Colorado Family Reading Study. *Child Development*, 1983, *54*, 388–395.

______, PLOMIN, R., AND LABUDA, M. C. Genetic stability of cognitive development from childhood to adulthood. *Developmental Psychology*, 1987, *23*, 4–12.

DELFINI, L., BERNAL, M., AND ROSEN, P. Comparison of deviant and normal boys in home settings. In E. J. Mash, L. A. Hamerlynck, and L. C. Handy (Eds.), *Behavior modification and families*, Volume 1. *Theory and research*. New York: Brunner/Mazel, 1976.

DELISSOVOY, V. Child care by adolescent parents. *Children Today*, 1973, *14*, 22.

DELOACHE, J. S. Naturalistic studies of memory for object location in very young children. *New directions for child development: Children's memory*. San Francisco: Jossey-Bass, 1980.

______, AND BROWN, A. L. Very young children's memory for the location of objects in a large-scale environment. *Child Development*, 1983, *54*, 888–897.

______, CASSIDY, D. J., BROWN, A. L. Precursors of mnemonic strategies in very young children's memory. *Child Development*, 1985, *56*, 125–137.

Demographic yearbook 1983, 35th issue. New York: United Nations, Statistical Office, 1983.

DEMOS, E. V. Facial expressions of infants and toddlers: A descriptive analysis. In T. Field and A. Fogel (Eds.), *Emotion and early interaction*. Hillsdale, N.J.: Lawrence Erlbaum, 1982.

DEMOS, J., AND DEMOS. V. Adolescence in historical perspective. *Journal of Marriage and the Family*, 1969, *31*, 632–638.

DEMYER, M. K., BARTON, S., ALPERN, G. D., KIMBERLIN, C., ALLEN, J., YANG, E., AND STEELE, R. The measured intelligence of autistic children: A follow-up study. *Journal of Autism and Childhood Schizophrenia*, 1974, 4, 42–60.

DENIER, C. A., AND SERBIN, L. A. *Play with male-preferred toys: Effects on visual-spatial performance*. Paper presented at the annual meeting of the American Psychological Association, Toronto, 1978.

DENNEY, N. W. A developmental study of free classification in children. *Child Development* 1972, *43*, 1161–1170.

———. Classification abilities in the elderly. *Journal of Gerontology*, 1974, *29*, 309.

———. Problem solving in later adulthood: Intervention research. In P. B. Baltes and O. G. Brim, Jr. (Eds.), *Life-span development and behavior*, Volume 2. New York: Academic Press, 1979, pp. 38–66.

———, AND DENNEY, D. R. The relationship between classification and questioning strategies among adults. *Journal of Gerontology*, 1982, *37*, 190–196.

DENNIS, W. Infant development under conditions of restricted practice and of minimum social stimulation: A preliminary report. *Pedagogical Seminary*, 1938, *53*, 149–158.

———. Infant development under conditions of restricted practice and of minimum social stimulation. *Genetic Psychology Monographs*, 1941, *23*, 143–189.

———. Causes of retardation among institutional children: Iran. *Journal of Genetic Psychology*, 1960, *96*, 47–59.

DENNIS, W. Creative productivity between the ages of 20 and 80 years. In B. L. Neugarten (Ed.), Middle age and aging. Chicago: University of Chicago Press, 1968, pp. 106–114

DEUTSCH, F. Female preschoolers' perceptions of affective responses and interpersonal behavior in videotaped episodes. *Developmental Psychology*, 1974, *10*, 733–740.

DEUTSCH, M. New vitality in follow-ups of Head Start youngsters. *The Brown University Human Development Letter*, June 1985, *1*, 6.

DEVEREAUX, E. C. The role of the peer group experience in moral development. In J. P. Hill (Ed.), *Minnesota symposia on child psychology*, Volume 4. Minneapolis: University of Minnesota Press, 1970.

DE VILLIERS, J. G., AND DE VILLIERS, P. A. A cross-sectional study of the development of grammatical morphemes in child speech. *Journal of Psycholinguistic Research*, 1973, *2*, 267–278.

DE VILLIERS, P. A., AND DE VILLIERS, J. G. *Early language*. Cambridge, Mass.: Harvard University Press, 1979.

DEVRIES, H. A. *Physiology of exercise* (1st ed., 3rd ed.). Dubuque, Iowa: W. C. Brown, 1966, 1980.

———. Physiology of exercise and aging. In D. S. Woodruff and J. E. Birren (Eds.), *Aging: Scientific perspectives and social issues* (2nd ed.). Monterey, Calif.: Brooks/Cole, 1983, pp. 285–304.

———, AND ADAMS, G. M. Electromyographic comparison of single doses of exercise and meprobamate as to effects on muscle relaxation. *American Journal of Physical Medicine*, 1972, *51*, 130–141.

DIAMOND, R., CAREY, S., AND BACK, K. J. Genetic influences on the development of spatial skills during early adolescence. *Cognition*, 1983, *13*, 167–185.

DIBBLE, U., AND STRAUS, M. Some social structure determinants of inconsistency between attitudes and behavior: The case of family violence. *Journal of Marriage and the Family*, 1980, *42*, 71–80.

DIENER, C. I., AND DWECK, C. S. An analysis of learned helplessness: Continuous changes in performance, strategy and achievement cognitions following failure. *Journal of Personality and Social Psychology*, 1978, *36*, 451–462.

DIENSTBIER, R. A., HILLMAN, D., LEHNHOFFM, J., HILLMAN, J., AND VALKENAAR, M. C. An emotion-attribution approach to moral behavior: Interfacing cognitive and avoidance theories of moral development. *Psychological Review*, 1975, *82*, 299–315.

DIGGORY, J., AND ROTHMAN, D. Values destroyed by death. *Journal of Abnormal and Social Psychology*, 1961, *61*, 205–210.

DITTMANN-KOHLI, F., AND BALTES, P. B. Toward a neofunctionalist conception of adult intellectual development. In C. Alexander and E. Langer (Eds.), *Beyond formal operations: Alternative endpoints to human development*, unpublished ms., 1984.

DODGE, K. A. Social cognition and children's aggressive behavior. *Child Development*, 1980, *51*, 162–170.

———. Behavioral antecedents of peer social status. *Child Development*, 1983, *54*, 1386–1399.

———, AND FRAME, C. L. Social cognitive biases and deficits in aggressive boys. *Child Development*, 1982, *53*, 620–635.

DODSON, J. C., KUSHIDA, E., WILLIAMSON, M., AND FRIEDMAN, E. G. Intellectual performance of 36 phenylketonuria patients and their nonaffected siblings. *Pediatrics*, 1976, *58*, 53–58.

DOLLARD, J., DOOB, L., MILLER, N. E., MOWRER, O. H., AND SEARS, R. R. *Frustration and aggression*. New Haven, Conn.: Yale University Press, 1939.

DONALDSON, M. Children's reasoning. In M. Donaldson, R. Graine, and C. Pratt (Eds.), *Early childhood development and education: Readings in psychology*. Oxford: Basil Blackwell, 1983.

———, AND BALFOUR, G. Less is more: A study of language comprehension in children. *British Journal of Psychology*, 1968, *59*, 461–472.

DONOVAN, W. L., AND LEAVITT, L. A. *Effects of experimentally manipulated attributions of infant cries of maternal learned helplessness*. Paper presented at the International Conference on Infant Studies, New York, April 1984.

———, LEAVITT, L. A., AND BALLING, J. D. Maternal physiological response to infant signals. *Psychophysiology*, 1978, *15*, 68–74.

DORE, J. Conditions for the acquisition of speech acts. In I. Markova (Ed.), *The social context of language*. New York: Wiley, 1978.

DORR, A. *Children's reports of what they learn from daily viewing*. Paper presented at the biennial meeting of the Society for Research in Child Development, San Francisco, March 1979.

DOUGLAS, K., AND ARENBERG, D. Age changes, cohort differences, and cultural change on the Guilford-Zimmerman Temperament Survey. *Journal of Gerontology*, 1978, *33*, 737–747.

DOUGLAS, V. I. Higher mental processes in hyperactive children. In R. M. Knights and D. J. Bakker (Eds.), *Treatment of hyperactive and learning disordered children*. Baltimore, Md.: University Park Press, 1980.

DOUVAN, E., AND ADELSON, J. *The adolescent experience*. New York: Wiley, 1966.

———, AND GOLD, M. Modal patterns in American adolescence. In L. W. Hoffman and M. L. Hoffman (Eds.), *Review of child development research*, Volume 2. New York: Russell Sage Foundation, 1966.

———, AND KULKA, R. The American family: A twenty-year view. In J. E. Gullahorn (Ed.), *Psychology and women: In transition*. New York: Wiley, 1979, pp. 83–93.

DOVE, A. Taking the chitling test. *Newsweek*, July 15, 1968, pp. 51–52.

DREYER, P. H. Sexuality during adolescence. In B. J. Wolman (Ed.), *Handbook of developmental psychology*. Englewood Cliffs, N.J.: Prentice-Hall, 1982.

DREYFUS-BRISAC, C. Neurophysiological studies in human premature and full-term newborns. *Biological Psychiatry*, 1975, *10*, 485–496.

DUKE, P. M., CARLSMITH, J. M., JENNINGS, D., MARTIN, J. A., DORNBUSCH, S. M., GROSS, R. T., AND SIEGEL-GORELICK, B. Educational correlates of early and late sexual maturation in adolescence. *Journal of Pediatrics*, 1982, *100*, 633–637.

DUNCAN, D., SCHUMAN, H., AND DUNCAN, B. *Social change in a metropolitan community*. New York: Russell Sage Foundation, 1973.

DUNN, J. Sibling relationships in early childhood. *Child Development*, 1983, *54*, 787–811.

———, AND KENDRICK, C. Studying temperament and parent–child interaction: Comparison of interview and direct observation. *Developmental Medicine and Child Neurology*, 1980, *22*, 484–496.

———, AND ———. The speech of two- and three-year-olds to

infant siblings: "Baby talk" and the context of communication. *Journal of Child Language*, 1982, *9*, 579–595.(a)

______, AND ______ *Siblings: Love, envy, and understanding*. Cambridge, Mass.: Harvard University Press, 1982.(b)

______, AND MACNAMEE, R. The reaction of firstborn children to the birth of a sibling: Mothers' reports. *Journal of Child Psychology and Psychiatry*, 1981, *22*, 1–18.

______, PLOMIN, R., AND DANIELS, D. Consistency and change in mother's behavior toward young siblings. *Child Development*, 1986, *57*, 348–356.

DUNPHY, D. C. The social structure of urban adolescent peer groups. *Sociometry*, 1963, *26*, 230–246.

DURKIN, D. *Children who read early*. New York: Teachers College, Columbia University, 1966.

DURLAK, J. Relationship between attitudes toward life and death among elderly women. *Developmental Psychology*, 1972, *8*, 146.

DUSEK, J. B., AND FLAHERTY, J. F. The development of the self-concept during the adolescent years. *Monographs of the Society for Research in Child Development*, 1981, *46* (4, Serial No. 191).

DWECK, C. S. Children's interpretation and evaluative feedback: The effect of social cues on learned helplessness. *Merrill-Palmer Quarterly*, 1976, *22*, 105–109.

______. Social-cognitive processes in children's friendships. In S. R. Asher and J. M. Gottman (Eds.), *The development of children's friendships*. New York: Cambridge University Press, 1981.

______, AND BUSH, E. S. Sex differences in learned helplessness: I. Differential debilitation with peer and adult evaluators. *Developmental Psychology*, 1976, *12*, 147–156.

______, DAVIDSON, W., NELSON, S., AND ENNA, B. Sex differences in learned helplessness: II. The contingencies of evaluation feedback in the classroom: III. An experimental analysis *Developmental Psychology*, 1979, *14*, 268–276.

______, AND GOETZ, F. E. Attributions and learned helplessness. In J. H. Harvey, W. Ickes, and R. F. Kidd (Eds.), *New directions in attribution research*, Volume 2. Hillsdale: N.J.: Lawrence Erlbaum 1977.

______, AND LICHT, B. Learned helplessness and intellectual achievement. In J. Garber and M. Seligman (Eds.), *Human helplessness*. New York: Academic Press, 1980.

______, AND REPPUCCI, N. D. Learned helplessness and reinforcement responsibility in children. *Journal of Personality and Social Psychology*, 1973, *25*, 109–116.

______, AND WORTMAN, C. Achievement, text anxiety, and learned helplessness: Adaptive and maladaptive cognitions. In H. Krohne and L. Laux (Eds.), *Achievement, stress and anxiety*. Washington, D.C.: Hemisphere, 1980.

EASTERBROOKS, M. A., AND GOLDBERG, W. A. Toddler development in the family: Impact of father involvement and parenting characteristics. *Child Development*, 1984, *55*, 740–752.

EDELMAN, R. Cell-mediated immune function in malnutrition. In R. M. Suskind (Ed.), *Malnutrition and the immune response*. New York: Raven Press, 1977.

______, SUSKIND, R., SIRISINHA, S., AND OLSON, R. E. Mechanisms of defective cutaneous hypersensitivity in children with protein-calorie malnutrition. *Lancet*, 1973, *1*, 506–508.

EDWARDS, C. P. The comparative study of the development of moral judgment and reasoning. In R. H. Munroe, R. L. Munroe, and B. Whiting (Eds.), *Handbook of cross-cultural human development*. New York: Garland, 1981.

EDWARDS, J. R. Social class differences and the identity of sex in children's speech. *Journal of Child Language*, 1979, *6*, 121–127.

EGELAND, B., AND FARBER, E. A. Infant-mother attachment: Factors related to its development and changes over time. *Child Development*, 1984, *55*, 753–771.

EGELAND B., AND SROUFE, L. A. Attachment and early maltreatment. *Child Development*, 1981, *52*, 44–52.

EIBL-EIBESFELDT, I. The expressive behavior of the deaf- and blind-born. In M. von Cranach and I. Vine (Eds.), *Social communication and movement*. New York: Academic Press, 1973.

EICHORN, D. H., CLAUSEN, J. A., HAAN, N., HONZIK, M. P. AND MUSSEN, P. H. *Present and past in middle life*. New York: Academic Press, 1981.

______, HUNT, J. V., AND HONZIK, M. P. Experience, personality, and IQ: adolescence to middle age. In D. H. Eichorn, J. A. Clausen, N. Haan, M. P. Honzik, and P. H. Mussen (Eds.), *Present and past in middle life*. New York: Academic Press, 1981, pp. 89–116.

EIMAS, P. D., SIQUELAND, E. R., JUZCZYK, P., AND VIGORITO, J. Speech perception in early infancy. *Science*, 1971, *171*, 303–306.

EISDORFER, C. Arousal and performance experiments in verbal learning and a tentative theory. In G. A. Talland (Ed.), *Human aging and behavior*. New York: Academic Press, 1968, pp. 189–216.

______, NOWLIN, J., AND WILKIE, F. Improvement of learning in the aged by modification of autonomic nervous system activity. *Science*, 1970, *170*, 1327–1329.

______, AND STOTSKY, B. A. Intervention, treatment, and rehabilitation of psychiatric disorders In J. E. Birren and K. W. Schaie (Eds.), *Handbook of the psychology of aging*. New York: Van Nostrand Reinhold, 1977, pp. 724–748.

______, AND WILKIE, F. Intellectual changes. In E. Palmore (Ed.), *Normal aging: II*. Durham, N.C.: Duke University Press, 1974, pp. 95–102.

EISENBERG, L. The fathers of autistic children. *American Journal of Orthopsychiatry*, 1957, *27*, 715–724.

EISENBERG, N., LENNON, R., AND ROTH, K. Prosocial development: A longitudinal study. *Developmental Psychology*, 1983, *19*, 846–855.

EISENBERG, N., MURRAY, E., AND HITE, T. Children's reasoning regarding sex-typed toy choices. *Child Development*, 1982, *53*, 81–86.

EISENBERG, R. B. Auditory behavior in the human neonate: I. Methodologic problems and the logical design of research procedures. *Journal of Auditory Research*, 1965, *5*, 159–177.

______. *Auditory competence in early life: The roots of communicative behavior*. Baltimore, Md.: University Park Press, 1976.

______. Stimulus significance as a determinant of infant responses to sound. In E. B. Thoman (Ed.), *Origins of the infant's social responses*. Hillsdale, N.J.: Lawrence Erlbaum, 1979.

EISNER, D. The effect of chronic organic brain syndrome upon concrete and formal operations in elderly men. Unpublished manuscript. William Patterson College of New Jersey, 1973.

EITZEN, D. S. Athletics in the status system of male adolescents: A replication of Coleman's *The adolescent society*. *Adolescence*, 1975, *10*, 267–276.

ELAHI, V. K., ELAHI, D., ANDRES, R., TOBIN, J. D., BUTLER, M. G., AND NORRIS, A. H. A longitudinal study of nutritional intake in men *Journal of Gerontology*, 1983, *38*, 162–180.

ELDER, G. H., JR. *Children of the great depression*. Chicago: University of Chicago Press, 1974.

______. Family history and the life course. In T. K. Hareven (ed.), *Transitions: The family and the life course in historical perspective*. New York: Academic Press, 1978, pp. 17–64.

ELIAS, J., AND GEBHARD, P. Sexuality and sexual learning in childhood. *Phi Beta Kappan*, 1969, *50*, 401–405.

ELIAS, M. F. *Nursing and night walking in the first two years*. Paper presented at the International Conference on Infant Studies, New York, April 1984.

ELKIND, D. Egocentrism in adolescence. *Child Development*, 1967, *38*, 1025–1034.

______. Perceptual development in children. In I. L. Janis (Ed.), *Current trends in psychology*. Los Altos, Calif.: Kaufmann, 1977.

______. *The child and society*. New York: Oxford University Press, 1979.

______. *The hurried child: Growing up too fast too soon*. Reading, Mass.: Addison-Wesley, 1981.

______. Egocentrism redux. *Developmental Review*, 1985, *5*, 218–226.

______, AND BOWEN, R. Imaginary audience behavior in children and adolescents. *Developmental Psychology*, 1979, *15*, 38–44.

ELLIS, S., ROGOFF, B., AND CROMER, C. C. Age segregation in children's social interactions. *Developmental Psychology*, 1981, *17*, 399–407.

ELSAYED, M., ISMAIL, A. H., AND YOUNG, R. S. Intellectual differences of adult men related to age and physical fitness before and after an exercise program. *Journal of Gerontology*, 1980, *35*, 383–387.

EMDE, R. N., GAENSBAUER, T., AND HARMON, R. Emotional expression in infancy: A biohavioral study. In *Psychological issues*, Monograph 37. New York: International Universities Press, 1976.

EMMERICH, W., GOLDMAN, K. S., KIRSCH, B., AND SHARABANY, R. Evidence for a transitional phase in the development of gender constancy. *Child Development*, 1977, *48*, 930–936.

ENRIGHT, R. D., LAPSLEY, D. K., AND LEVY, V. M., JR. Moral education strategies. In M. Pressley and J. R. Levin (Eds.), *Cognitive strategy research: Educational application*. New York: Springer-Verlag, 1983.

ENSOR, R. E., FLEG, J. L., KIM, Y. C., DELEON, E. F., AND GOLDMAN, S. M. Longitudinal chest X-ray changes in normal men. *Journal of Gerontology*, 1983, *38*, 307–314.

EPSTEIN, S. Traits are alive and well. In D. Magnusson and N. S. Endler (Eds.), *Personality at the crossroads: Current issues in interactional psychology*. Hillsdale, N.J.: Lawrence Erlbaum, 1979.

ERICSON, A., KALLEN, B., AND WESTERHOLM, P. Cigarette smoking as an etiologic factor in cleft lip and palate. *American Journal of Obstetrics and Gynecology*, 1979, *35*, 348–351.

ERIKSON, E. H. *Identity: Youth and crisis*. New York: Norton, 1968.

———. *Identity and the life cycle*. Reissue. New York: Norton, 1980a.

———. *The life cycle completed*. New York: Norton, 1982.

———, AND ERIKSON, J. M. On generativity and identity: From a conversation with Erik and Joan Erikson. *Harvard Educational Review*, 1981, *51*, 249–269.

———, interviewed by E. Hall. A conversation with Erik Erikson. *Psychology Today*, June 1983, *17*, 22–30.

ERLENMEYER-KIMLING, L., AND JARVIK, L. F. Genetics and intelligence: A review. *Science*, 1963, *142*, 1477–1479.

ERON, L. D., HUESMANN, R., BRICE, P., FISCHER, P., AND MERMELSTEIN, R. Age trends in the development of aggression, sex typing, and related television habits. *Developmental Psychology*, 1983, *19*, 71–77.

———, WALDER, L. O., AND LEFKOWITZ, M. M. *Learning of aggression in children*. Boston: Little, Brown, 1971.

ERVIN-TRIPP, S. Discourse agreement: How children answer questions. In J. R. Hayes (Ed.), *Cognition and the development of language*. New York: Wiley, 1970.

ESPENSHADE, A. S., AND MELENEY, H. E. Motor performances of adolescent boys and girls today in comparison with those of twenty years ago. *Research Quarterly*, 1961, *32*, 186–189.

ESPENSHADE, T. J., AND BRAUN, R. E. Economic aspects of an aging population and the material well-being of older persons. In M. W. Riley, B. B. Hess, and K. Bond (Eds.), *Aging in society: Selected reviews of recent research*. Hillsdale, N.J.: Lawrence Erlbaum, 1983, pp. 25–51.

ETAUGH, C. Effects of maternal employment on children: A review of recent research. *Merrill-Palmer Quarterly*, 1974, *20*, 71–98.

ETZEL, B. C., AND GEWIRTZ, J. L. Experimental modification of caretaker-maintained high rate operant crying in a 6- and a 20-week-old infant: Extinction of crying with reinforcement of eye contact and smiling. *Journal of Experimental Child Psychology*, 1967, *5*, 303–317.

EVANS, D. R., NEWCOMBE, R. G., AND CAMPBELL, H. Maternal smoking habits and congenital malformations: A population study. *British Medical Journal*, 1979, *2*, 171–173.

EVANS, R. *Jean Piaget: The man and his ideas*. New York: E. P. Dutton, 1973.

EVERITT, A. V., AND HUANG, C. Y. The hypothalamus, neuroendocrine, and autonomic nervous systems in aging. In J. E. Birren and R. B. Sloane (Eds.), *Handbook of mental health and aging*. Englewood Cliffs, N.J.: Prentice-Hall, 1980, pp. 100–133.

FAGAN, J. F., AND SINGER, L. T. Infant recognition memory as a measure of intelligence. In L. P. Lipsitt and C. K. Rovee-Collier (Eds.), *Advances in infancy research*, Volume 2. Norwood, N.J.: Ablex, 1983.

FAGOT, B. I. The influence of sex of child on parental reactions to toddler children. *Child Development*, 1978, *49*, 459–465.

———. The construction of gender during the child's first two years. *Infant Behavior and Development*, 1986, *9* (Special Issue: Abstracts of papers presented at the Fifth International Conference on Infant Studies), 116.

———, AND LEINBACH, M. D. Play styles in early childhood: Social consequences for boys and girls. In M. B. Liss (Ed.), *Social and cognitive skills: Sex roles and children's play*. New York: Academic Press, 1983.

———, AND LITTMAN, I. Relation of preschool sex-typing to intellectual performance in elementary school. *Psychological Reports*, 1976, *39*, 699–704.

FAIRWEATHER, H. Sex differences in cognition: A function of maturation rate? *Science*, 1976, *192*, 572–573.

FALBO, T., AND PEPLAU, L. A. Power strategies in intimate relationships. *Journal of Personality and Social Psychology*, 1980, *38*, 618–628.

FANTZ, R. L. The origins of form perception. *Scientific American*, 1961, *204*, 66–72.

———. Pattern vision in newborn infants. *Science*, 1963, *140*, 296–297.

———. Visual perception from birth as shown by pattern selectivity. *Annals of the New York Academy of Science*, 1965, *118*, 793–814.

———, AND NEVIS, S. The predictive value of changes in visual preferences in early infancy. In J. Hallmuth (Ed.), *Exceptional infant*, Volume 1. Seattle: Special Child Publications, 1967.

FARBER, S. L. *Identical twins reared apart: A reanalysis*. New York: Basic Books, 1981.

FARKAS-BARGETON, E., AND DIEBLER, M. F. A topographical study of enzyme maturation in human cerebral neocortex: A histochemical and biochemical study. In M. A. B. Brazier and H. Petsche (Eds.), *Architectonics of the cerebral cortex*. New York: Raven Press, 1978.

FARNHAM-DIGGORY, S. *Learning disabilities*. Cambridge, Mass.: Harvard University Press, 1978.

———. Why reading? Because it's there. *Developmental Review*, 1984, *4*, 62–71.

FAUST, M. S. Developmental maturity as a determinant in prestige of adolescent girls. *Child Development*, 1960, *31*, 173–184.

———. Somatic development of adolescent girls. *Monographs of the Society for Research in Child Development*, 1977, *42* (1, Serial No. 169).

FEATHER, N. T. Values in adolescence. In J. Adelson (Ed.), *Handbook of adolescent psychology*. New York: Wiley, 1980.

FEATHERMAN, D. Schooling and occupational careers. Constancy and change in worldly success. In O. Brim and J. Kagan (Eds.), *Constancy and change in human development*. Cambridge, Mass.: Harvard University Press, 1980.

FEIFEL, H. The function of attitudes toward death. In *Death and dying: Attitudes of patient and doctor*. New York: Group for the Advancement of Psychiatry, 1965, pp. 632–641.

———. Death in contemporary America. In H. Feifel (Ed.), *New meanings of death*. New York: McGraw-Hill, 1977, pp. 4–11.

FEIN, G., JOHNSON, D., KOSSON, N., STORK, L., AND WASSERMAN, L. Sex stereotypes and preferences in the toy choices of 20-month-old boys and girls. *Developmental Psychology*, 1975, *11*, 527–528.

FEINGOLD, B. F. *Why your child is hyperactive*. New York: Random House, 1975.

FEINMAN, S., AND LEWIS, M. Social referencing at ten months: A second-order effect on infants' responses to strangers. *Child Development*, 1983, *54*, 878–887.

FELD, S. Longitudinal study of the origins of achievement strivings. *Journal of Personality and Social Psychology*, 1967, *7*, 408–414.

FELDMAN, S. S., BIRINGEN, Z. C., AND NASH, S. C. Fluctuations of sex-related self-attributions as a function of stage of the family life cycle. *Developmental Psychology*, 1981, *17*, 24–35.

______, AND INGHAM, M. E. Attachment behavior: A validation study in two age groups. *Child Development*, 1975, *46*, 319–330.

FERGUSON, C. A. Baby talk in six languages. *American Anthropologist*, 1964, *66*, 103–114.

FERNALD, A. *Four-month-olds prefer to listen to "motherese."* Paper presented at the meeting of the Society for Research in Child Development, Boston, April 1981.

______. *Acoustic determinants of infant preference for "motherese."* Unpublished doctoral dissertation, University of Oregon, 1982.

______. The perceptual and affective salience of mother's speech to infants. In I. Feagans, C. Garvey, and R. Golinkoff (Eds.), *The origins and growth of communication*. Norwood, N.J.: Ablex, 1984.

______. Four-month-old infants prefer to listen to motherese. *Infant Behavior and Development*, 1985, *8*, 181–195.

______, AND SIMON, T. Expanded intonation contours in mothers' speech to newborns. *Development Psychology*, 1984, *20*, 104–113.

FESHBACH, S. The catharsis hypothesis and some consequences of interaction with aggressive and neutral play objects. *Journal of Personality*, 1956, *24*, 449–462.

______. Aggression. In P. H. Mussen (Ed.), *Carmichael's manual of child psychology*, Volume 2. New York: Wiley, 1970.

FIELD, T. Interaction behaviors of primary versus secondary caretaker fathers. *Developmental Psychology*, 1978, *14*, 183–184.

______, DE STAFANO, L., AND KOEWLER, J. H., III. Fantasy play of toddlers and preschoolers. *Developmental Psychology*, 1982, *18*, 503–508.

______, SANDBERG, D., GARCIA, R., VEGA-LAHR, N., GOLDSTEIN, S., AND GUY, L. Pregnancy problems, postpartum depression, and early mother-infant interactions. *Developmental Psychology*, 1985, *21*, 1152–1156.

FILLMORE, C. J. The case for case. In E. Bach and R. T. Harms (Eds.), *Universals of linguistic theory*. New York: Holt, Rinehart & Winston, 1968.

FINCHER, J. *Human intelligence*. New York: G. P. Putnam & Sons, 1976.

FIRTH, R. *Elements of social organization* (3d ed.). London: Tavistock, 1961.

FISCHER, K. W. A Theory of Cognitive Development: The control and construction of hierarchies of skills. *Psychological Review*, 1980, *87*, 477–531.

______, AND CANFIELD, R. L. The ambiguity of stage and structure in behavior: Person and environment in the development of psychological structures. In I. Levin (Ed.), *Stage and structure: Reopening the debate*, Norwood, N.J.: Ablex, 1986.

______, AND SILVERN, L. Stages and individual differences in cognitive development. *Annual Review of Psychology*, 1985, *36*, 613–648.

FIVUSH, R. Children's long-term memory for a novel event: An explanatory study. *Merrill-Palmer Quarterly*, 1984, *30*, 303–316.

FLAPAN, D. *Children's understanding of social interaction*. New York: Teachers College Press, Columbia University, 1968.

FLAVELL, J. H. Developmental studies of mediated memory. In H. W. Reese and L. P. Lipsitt (Eds.), *Advances in child development and behavior*, Volume 5. New York: Academic Press, 1970.

______. Stage-related properties of cognitive development. *Cognitive Psychology*, 1978, *2*, 421–453.

______, BEACH, D. H., AND CHINSKY, J. M. Spontaneous verbal rehearsal in memory tasks as a function of age. *Child Development*, 1966, *37*, 283–299.

______, BOTKIN, P. T., FRY, C. L., WRIGHT, J. W., AND JARVIS, P. E. *The development of role-taking and communication skills in children*. New York: Wiley, 1968.

______, EVERETT, B. A., CROFT, K., AND FLAVELL, E. R. Young children's knowledge about visual perception: Further evidence for the Level 1–Level 2 distinction. *Developmental Psychology*, 1981, *17*, 99–103.

______, SHIPSTEAD, S. G., AND CROFT, K. Young children's knowledge about visual perception: Hiding objects from others. *Child Development*, 1978, *49*, 1208–1211.

FLEMING, A. T. New frontiers in conception. *New York Times Magazine*, July 20, 1980.

FOGELMAN, K. Smoking in pregnancy and subsequent development of the child. *Child: Care, Health and Development*, 1980, *6*, 233–249.

FOLEY, M., AND JOHNSON, M. K. Confusions between memories for performed and imagined actions: A developmental comparison. *Child Development*, 1985, *56*, 1145–1155.

FOLGER, J. P., AND CHAPMAN, R. S. A pragmatic analysis of spontaneous imitations. *Journal of Child Language*, 1978, *5*, 25–38.

FOLSTEIN, S., AND RUTTER, M. Genetic influences and infantile autism. *Nature*, 1977, *265*, 726–728.

FONER, A., AND SCHWAB, K. *Aging and retirement*. Monterey, Calif.: Brooks/Cole, 1981.

FORMAN, G. E., AND HILL, F. *Constructive play applying Piaget in the preschool*. Monterey, Calif.: Brooks/Cole, 1980.

FORREST, D. L., AND WALLER, T. G. *Meta-memory and meta-cognitive aspects of decoding in reading*: Paper presented at the meeting of the American Educational Research Association, Los Angeles, April 1981.

FOURCIN, A. J. Acoustic patterns and speech acquisition. In N. Waterson and C. Snow (Eds.), *The development of communication*. New York: Wiley, 1978.

FOX, N., KAGAN, J., AND WEISKOPF, S. The growth of memory during infancy. *Genetic Psychology Monographs*, 1979, *99*, 91–130.

FRAIBERG, S. Blind infants and their mothers: An examination of the sign system. In M. Lewis and L. A. Rosenblum (Eds.), *The effect of the infant on its caregiver*. New York: Wiley, 1974.

______. The development of human attachments in infants blind from birth. *Merrill-Palmer Quarterly*, 1975, *21*, 315–334.

______. *Every child's birthright. In defense of mothering*. New York: Basic Books, 1977. (a)

______. *Insights from the blind*. New York: Basic Books, 1977. (b)

FRANKEL, K. *Origins of mother-toddler problem solving interactions*. Paper presented at the International Conference on Infant Studies, New York, April 1984.

FRANKENBERG, W. K., AND DODDS, J. B. The Denver Developmental Screening Test. *Journal of Pediatrics*, 1967, *71*, 181–191.

FRANKLIN, R. D. Youth's expectancies about internal versus external control of reinforcement. *Dissertation Abstracts*, 1963, *24*, 1684.

FRAZIER, T. M., DAVIS, G. H., GOLDSTEIN, H., AND GOLDBERG, I. Cigarette smoking: A prospective study. *American Journal of Obstetrics and Gynecology*, 1961, *81*, 988–996.

FREEMAN, J. T. The old, old, very old Charlie Smith. *Gerontologist*, 1982, *22*, 532–536.

FREEMAN, N. H. *Strategies of representation in young children: Analysis of spatial skill and drawing processes*. London: Academic Press, 1980.

______, LLOYD, S., AND SINHA, C. G. Infant search tasks reveal early concepts of containment and canonical usage of objects. *Cognition*, 1980, *8*, 243–262.

FREEMON, F. R. Evaluation of patients with progressive intellectual deterioration. *Archives of Neurology*, 1976, *33*, 658–659.

FREMGEN, A., AND FAY, D. Overextensions in production and comprehension: A methodological clarification. *Journal of Child Language*, 1980, *7*, 205–211.

FRIES, J. F., AND CRAPO, L. M. *Vitality and aging*. San Francisco: W. H. Freeman, 1981.

FREUD, A. Adolescence. In J. F. Rosenblith, W. Alinsmith, and J. P. William (Eds.), *The causes of behavior* (3rd ed.). Boston: Allyn & Bacon, 1972.

FREUD, S. Three contributions to the sexual theory (translated by A. A. Brill). *Nervous and Mental Disease Monograph Series*, *7*, 1910.

______. *An autobiographical study* (translated by J. Strachey). London: Hogarth, 1935.

______. Mourning and melancholia. In *Collected papers of Sigmund Freud*, Volume 4. London: Hogarth Press, 1953, pp. 152–170 (orig. pub. 1917).

______. *Instincts and their vicissitudes* (Volume 14 of the Standard Edition). London: Hogarth, 1968. (Originally published 1915)

______. *An outline of psychoanalysis* (translated by J. Strachey). New York: Norton, 1970.

FREUH, T., AND MCGHEE, P. E. Traditional sex role development and amount of time spent watching television. *Developmental Psychology*, 1975, *11*, 109.

FRIEDLANDER, B. Z., JACOBS, A. C., DAVID, B. B., AND WETSTONE, H. S. Time sampling analysis of infant's natural language environments—the home. *Child Development*, 1972, *43*, 730–740.

FRIEDMAN, C. J., MAHN, F., AND FRIEDMAN, A. S. A profile of juvenile street gang members. *Adolescence*, 1975, *10*, 563–607.

FRIEDMAN, M., AND ROSENMAN, R. *Type A behavior and your heart*. New York: Knopf, 1974.

FRIEDMAN, S. L., AND JACOBS, B. S. Sex differences in neonates' behavioral responsiveness to repeated auditory stimulation. *Infant Behavior and Development*, 1981, *4*, 175–183.

FRIEDRICH, L. K., AND STEIN, A. H. Aggressive and prosocial television programs and the natural behavior of preschool children. *Monographs of the Society for Research in Child Development*, 1973, *38* (4, Serial No. 151).

FRIEDRICH, O. What do babies know? *Time*, August 15, 1983, pp. 52–59. Reprinted in *Annual Editions Human Development 86/87*. Guilford, Conn.: Dushkin, 1986, pp. 62–67.

FRISCH, H. L. Sex stereotypes in adult-infant play. *Child Development*, 1977, *48*, 1671–1675.

FRODI, A. M., BRIDGES, L., SHONK, S., AND GREENE, L. Responsiveness to infant crying: Effects of perceived infant temperament. *Infant Behavior and Development*, 1986, *9* (Special Issue: Abstracts of papers presented at the Fifth International Conference on Infant Studies), 131.

FUCHS, V. R. *How we live: An economic perspective on americans from birth to death*. Cambridge, Mass.: Havard University Press, 1983.

FURCHGOTT, E., AND BUSEMEYER, J. K. Heart rate and skin conductance during cognitive processes as a function of age. Paper presented at the Meeting of the Gerontological Society, 1976.

FURMAN, E. *A child's parent dies*. New Haven: Yale University Press, 1974.

FURMAN, W., AND BIERMAN, K. L. Children's conceptions of friendship: A multimethod study of developmental changes. *Developmental Psychology*, 1984, *20*, 925–931.

______, RAHE, D. F., AND HARTUP, W. W. Rehabilitation of socially withdrawn preschool children through mixed-age and same-age socialization. *Child Development*, 1979, *50*, 915–922.

FURROW, D., NELSON, K., AND BENEDICT, H. Mother's speech to children and syntactic development: Some simple relationships. *Journal of Child Language*, 1979, *6*, 423–442.

FURSTENBERG, F. F. *Unplanned parenthood*. New York: Free Press, 1976.

______. Conjugal succession: Reentering marriage after divorce. In P. B. Baltes and O. G. Brim, Jr. (Eds.), *Life-span development and behavior*, Volume 4. New York: Academic Press, 1982, pp. 107–146.

______, MOORE, K. A., PETERSON, J. L. Sex education and sexual experience among adolescents. *American Journal of Public Health*, 1985, *75*, 1331–1332.

FURTH, H. G. *Thinking without language: Psychological implications of deafness*. Englewood Cliffs, N.J.: Prentice-Hall, 1966.

GADBERRY, S. Effects of restricting first graders' TV-viewing on leisure time use, IQ change, and cognitive style. *Journal of Applied Developmental Psychology*, 1980, *1*, 45–57.

GAGNON, J. H., AND SIMON, W. *Sexual conduct: The social origins of human sexuality*. Chicago: Aldine, 1973.

GAMER, E., THOMAS, J., AND KENDALL, D. Determinants of friendship across the life span. In M. F. Rebelsky (Ed.), *Life: The continuous process*. New York: Knopf, 1975.

GANCHROW, J. R., STEINER, J. E., AND DAHER, M. Neonatal facial expressions in response to different qualities and intensities of gustatory stimuli. *Infant Behavior and Development*, 1983, *6*, 189–200.

GARBARINO, J., AND ASP, C. *Successful schools and competent students*. Lexington, Mass.: Lexington Books, 1981.

______, SEBES, J., AND SCHELLENBACH, C. Families at risk for destructive parent-child relations in adolescence. *Child Development*, 1984, *55*, 174–183.

GARDNER, H. *Frames of mind: The theory of multiple intelligences*. New York: Basic Books, 1983.

GARDNER, J. A. *Sesame Street* and sex role stereotypes. *Women*, 1970, *1*, 42.

GARDNER, L. Deprivation dwarfism. *Scientific American*, 1972, *227*, 76–82.

GARFINKEL, R. By the sweat of your brow. In T. M. Field, A. Huston, H. C. Quay, L. Troll, and G. E. Finley (Eds.), *Review of human development*. New York: Wiley-Interscience, 1982, pp. 500–507.

GARMEZY, N., AND NUECHTERLEIN, K. H. Invulnerable children: The fact and fiction of competence and disadvantage. *American Journal of Orthopsychiatry*, 1972, *77*, 328–329.

GARN, S. M. Bone loss and aging. In R. Goldman, M. Rockstein, and M. Sussman (Eds.), *The physiology and pathology of human aging*. New York: Academic Press, 1975, pp. 39–57.

GARNICA, O. K. Some prosodic and paralinquistic features of speech directed to young children. In C. E. Snow and C. A. Ferguson (Eds.), *Talking to children: Language input and acquisition*. Cambridge, England: Cambridge University Press, 1977.

GARVEY, C. Some properties of social play. *Merrill-Palmer Quarterly*, 1974, *20*, 163–180.

______. Requests and responses in children's speech. *Journal of Child Language*, 1975, *2*, 41–63.

______. *Children's talk*. Cambridge, Mass.: Harvard University Press, 1984.

______, AND HOGAN, R. Social speech and social interaction: Egocentrism revisited. *Child Development*, 1973, *44*, 562–568.

GAYL, I. E., ROBERTS, J. O., AND WERNER, J. S. Linear systems analysis of infant visual pattern preferences. *Journal of Experimental Child Psychology*, 1983, *35*, 30–45.

GEFFEN, G., AND SEXTON, M. A. The development of auditory strategies of attention. *Developmental Psychology*, 1978, *14*, 11–17.

GELMAN, R. Conservation acquisition: A problem of learning to attend to relevant attributes. *Journal of Experimental Child Psychology*, 1969, *7*, 167–187.

______, AND BAILLARGEON R. A review of some Piagetian concepts. In P. H. Mussen (Ed.), *Handbook of child psychology*, Volume 3, New York: Wiley, 1983.

GENDREAU, P., AND SUBOSKI, M. D. Intelligence and age in discrimination conditioning of eyelid response. *Journal of Experimental Psychology*, 1971, *89*, 379–382.

GEORGAKAS, D. *The Methuselah factors: Strategies for a long and vigorous life*. New York: Simon & Schuster, 1980.

GERBNER, G. Violence in television drama: Trends and symbolic functions. In G. A. Comstock and E. A. Rubenstein (Eds.), *Television and social behavior*, Volume 1, *Media content and control*. Washington, D.C.: U.S. Government Printing Office, 1972.

______, GROSS, L., ELERY, M., JACKSON-BEECK, M., JEFFRIES-FOX, S., AND SIGNORIELLI, N. *Violence profile #8: Trends in network television drama and viewer conceptions of social reality, 1967–1976*. Philadelphia: Annenberg School of Communications, University of Pennsylvania, 1977.

GESCHWIND, N. Specialization of the human brain. *Scientific American*, 1979, *241*, 180–201.

GESELL, A., AND AMATRUDA, C. *Development diagnosis*. New York: Paul B. Hoeber, 1941.

GETZELS, J. W., AND CSIKSZENTMIHALYI, M. *The creative vision: A longitudinal study of problem finding in art*. New York: Wiley, 1976.

______, AND JACKSON, P. W. *Creativity and intelligence*. New York: Wiley, 1962.

GEWIRTZ, J. L. The course of infant smiling in four childrearing environments in Israel. In B. M. Foss (Ed.), *Determinants of infant behavior*, Volume 3. New York: Wiley, 1965.

______. Contingent maternal responding can increase infant nondistress crying, in press.

______, AND BOYD, E. F. Does maternal responding imply reduced infant crying? A critique of the 1972 Bell and Ainsworth report. *Child Development*, 1977, *48*, 1200–1207.

GEYER, R. F. *Bibliography alienation* (2nd ed.). Amsterdam: Netherlands Universities' Joint Social Research Centre, 1972.

GHISELLI, E. E. *The validity of occupational aptitude tests*. New York: Wiley, 1966.

GIAMBRA, L. M. A factor analytic study of daydreaming, imaginal process, and temperament: A replication on an adult male lifespan sample. *Journal of Gerontology*, 1977, *32*, 675–680.

GIBBONS, J. L., JOHNSON, M. O., MCDONOUGH, P. M., AND REZNICK, J. S. Behavioral inhibition in infants and children. *Infant Behavior and Development*, 1986, *9* (Special Issue: Abstracts of papers presented at the Fifth International Conference on Infant Studies), 139.

GIBBS, J. C., CLARK, P. M., JOSEPH, J. A., GREEN, J. L., GOODRICK, T. S., AND MAKOWSKI, D. G. Relations between moral judgment, moral courage, and field independence. *Child Development*, 1986, *57*, 185–193.

GIBSON, E. J. *Principles of perceptual learning and development*. New York: Appleton-Century-Crofts, 1969.

______, AND SPELKE, E. S. The development of perception. In P. H. Mussen (Ed.), *Handbook of child psychology*, Volume 3. New York: Wiley, 1983.

______, AND WALK, R. D. The "visual cliff." *Scientific American*, 1960, *202*, 64–71.

GIBSON, J. J. *The senses considered as perceptual systems*. Boston: Houghton-Mifflin, 1966.

GILBERT, J. G. Thirty-five year follow-up study of intellectual functioning. *Journal of Gerontology*, 1973, *28*, 68–72.

GILL, N. E., WHITE, M. A., AND ANDERSON, G. C. *Transitional newborn infants in a hospital nursery: Time and behavior between first oral cue and first sustained cry*. Paper presented at the International Conference on Infant Studies, New York, April 1984.

GILLBERG, C., RASSMUSSEN, P., AND WAHLSTROM, J. Minor neurodevelopmental disorders in children born to older mothers. *Developmental Medicine and Child Neurology*, 1982, *24*, 437–447.

GILLIGAN, C. *In a different voice: Psychological theory and women's development*. Cambridge, Mass.: Harvard University Press, 1982.

GLASER, B. G, AND STRAUSS, A. L. *Awareness of dying*. Chicago: Aldine, 1966.

______, AND ______. *Time for dying*. Chicago: Aldine, 1967.

GLEASON, J. B. Do children imitate? *Proceedings of the International Conference on Oral Education of the Deaf*, 1967, *2*, 1441–1448.

GLEITMAN, L. R., NEWPORT, E. L., AND GLEITMAN, H. The current status of the motherese hypothesis. *Journal of Child Language*, 1984, *11*, 43–79.

______, AND WANNER, B. Language acquisition: The state of the state of the art. In E. Wanner and L. R. Gleitman (Eds.), *Language acquisition: The state of the art*. Cambridge, England: Cambridge University Press, 1982.

GLENN, N. D. Psychological well-being in the postparental stage: Some evidence from national surveys. *Journal of Marriage and the Family*, 1975, *37*, 105–110.

GLENN, S. M., AND CUNNINGHAM, C. C. What do babies listen to most? A developmental study of auditory preferences in nonhandicapped infants and infants with Down's syndrome. *Developmental Psychology*, 1983, *19*, 332–337.

GLICK, I. O., WEISS, R. S., AND PARKES, C. M. *The first year of bereavement*. New York: Wiley, 1974.

GLICK, P. C. Updating the life cycle of the family. *Journal of Marriage and the Family*, 1977, *39*, 5–13.

______. Remarriage: Some recent changes and variations. *Journal of Family Issues*, 1980, *1*, 455–478.

GLOGER-TIPPELT, G. A process model of the pregnancy course. *Human Development*, 1983, *26*, 134–148.

GLUCKSBERG, S., KRAUSS, R. M., AND HIGGINS, E. T. The development of referential communication skills. In F. D. Horowitz, (Ed.), *Review of child development research*, Volume 4. Chicago: University of Chicago Press, 1975.

GNEPP, J. Children's social sensitivity: Inferring emotions from conflicting cues. *Developmental Psychology*, 1983, *19*, 805–814.

GOODARD, H. H. *The Kallikak family: A study in the heredity of feeblemindedness*. New York: Macmillan, 1912.

GOLD, M. AND REIMER, D. J. Changing patterns of delinquent behavior among Americans 13 through 16 years old: 1967–72. *Crime and Delinquency Literature*, 1975, *7*, 483–517.

GOLD, M., AND YANOF, D. S. Mothers, daughters, and girlfriends. *Journal of Personality and Social Psychology*, 1985, *49*, 654–659.

GOLDBERG, R. J. *Maternal time use and preschool performance*. Paper presented at the meeting of the Society for Research in Child Development, New Orleans, March 1977.

GOLDBERG, S., AND DIVITTO, B. A. *Born too soon: Preterm birth and early development*. San Francisco: W. H. Freeman, 1983.

______, LACOMBE, S., LEVINSON, D., PARKER, R., ROSS, C., AND SOMMERS, F. *Thinking about the threat of nuclear war: Relevance to mental health*. Unpublished manuscript, 1985.

______, PERLMUTTER, M., AND MYERS, N. Recall of related and unrelated lists by two-year-olds. *Journal of Experimental Child Psychology*, 1974, *18*, 1–8.

GOLDBERG, W. A., AND EASTERBROOKS, M. A. Role of marital quality in toddler development. *Developmental Psychology*, 1984, *20*, 504–514.

______, MICHAELS, G. Y., AND LAMB, M. E. Husbands' and wives' adjustment to pregnancy and first parenthood. *Journal of Family Issues*, 1985, *6*, 483–503.

GOLDFARB, W. Effects of psychological deprivation in infancy and subsequent stimulation. *American Journal of Psychiatry*, 1945, *102*, 18–33.

GOLDIN, P. C. A review of children's reports of parent behaviors. *Psychological Bulletin*, 1969, *71*, 222–236.

GOLDIN-MEADOW, S., AND FELDMAN, H. The development of a language-like communication without a language model. *Science* 1977, *197*, 401–403.

______, AND MYLANDER, C. Gestural communication in deaf children: The effects and noneffects of parental input on early language development. *Monographs of the Society for Research in Child Development*, 1984, *49*, (3-4, Serial No. 207).

GOLDMAN, B. D., AND ROSS, H. S. Social skills in action: An analysis of early peer games. In J. Glick and K. A. Clarke-Stewart (Eds.), *The development of social understanding*. New York: Gardner Press, 1978.

GOLDMAN, R. Aging of the excretory system: Kidney and bladder. In C. E. Finch and L. Hayflick (Eds.), *Handbook of the biology of aging*. New York: Van Nostrand Reinhold, 1977, pp. 409–431.

GOLDSCHNEIDER, C. *Population, modernization, and social structure*. Boston: Little, Brown, 1971.

GOLDSMITH, H., BRADSHAW, D. L., AND RIESER-DANNER, L. A. Temperament as a potential developmental influence on attachment. In J. V. Lerner and R. M. Lerner (Eds.), *Temperament and social interaction during infancy and childhood*. New Directions for Child Development, No. 31. San Francisco: Jossey-Bass, 1986.

GOLDSMITH, H. H. Genetic influences on personality from infancy to adulthood. *Child Development*, 1983, *54*, 331–355.

GOLDSTEIN, H. Factors influencing the height of seven-year-old children: Results from the National Child Development Study. *Human Biology*, 1971, *43*, 92–111.

GOLEMAN, S. Studies of children as witnesses find surprising accuracy. *New York Times*, November 6, 1984, Section C, p. 4.

GOLOMB, C. *Young children's sculpture and drawing: A study in representation development.* Cambridge, Mass.: Harvard University Press, 1974.

GOODENOUGH, F. L. *Anger in young children.* Minneapolis: University of Minnesota Press, 1931.

GOODNOW, J. *Children's drawings.* Cambridge, Mass.: Harvard University Press, 1977.

GORDON, C., GAITZ, C. M., AND SCOTT, J. Leisure and lives: Personal expressivity across the life span. In R. H. Binstock and E. Shanas (Eds.), *Handbook of aging and the social sciences.* New York: Van Nostrand Reinhold, 1976, pp. 310–341.

GORDON, R. E., AND GORDON, K. K. Social factors in prevention of postpartum emotional problems. *Obstetrics and Gynecology,* 1960, *15,* 433–438.

———, KAPOSTINS, E. E., AND GORDON, K. K. Factors in postpartum emotional adjustment. *Obstetrics and Gynecology,* 1965, *25,* 158–166.

GOREN, C., SARTY, M., AND WU, P. Visual following and pattern discrimination of face-like stimuli by newborn infants. *Pediatrics,* 1975, *56,* 544–549.

GORER, G. *Death, grief, and mourning.* New York: Doubleday, 1965.

GOTTESMAN, I. I. Heritability of personality: A demonstration. *Psychology Monographs,* 1963, *77,* No. 572.

GOTTFRIED, A. W., ROSE, S. A., AND BRIDGER, W. H. Crossmodal transfer in human infants. *Child Development,* 1977, *48,* 118–123.

GOTTMAN, J. M. How children become friends. *Monographs of the Society for Research in Child Development,* 1983, *48* (2, Serial No. 201).

———, GONSO, J., AND RASMUSSEN, B. Social interaction, social competence and friendship in children. *Child Development,* 1975, *46,* 709–718.

———, AND PARKHURST, J. T. *Developing may not always be improving: A developmental study of children's best friendships.* Paper presented at the meetings of the Society for Research in Child Development, New Orleans, March 1977.

———, AND ———. A developmental theory of friendship and acquaintanceship processes. In W. A. Collins (Ed.), *Development of cognition, affect, and social relations. The Minnesota Symposia on Child Psychology,* Volume 13. Hillsdale, N.J.: Lawrence Erlbaum, 1980.

GOULD, R. L. Adult life stages: Growth toward self-tolerance. *Psychology Today,* 8 (February 1975), 74–78.

———. *Transformations: Growth and change in adult life.* New York: Simon & Schuster, 1978.

GRATCH, G., APPEL, K. J., EVANS, W. F., LECOMPTE, G. K., AND WRIGHT, N. A. Piaget's Stage IV object concept error: Evidence of forgetting or object conception? *Child Development,* 1974, *15,* 71–77.

GRAY, S. W., RAMSEY, B. K., AND KLAUS, R. A. From 3 to 20: The Early Training Project. Baltimore, Md.: University Park Press, 1982.

GRAY, W. M., AND HUDSON, L. M. Formal operations and the imaginary audience. *Developmental Psychology,* 1984, *20,* 619–627.

GREEN, E. H. Group play and quarrelling among preschool children. *Child Development,* 1933, *4,* 302–307.

GREEN, R. *The "Sissy boy syndrome" and the development of homosexuality.* New Haven, Conn.: Yale University Press, 1987.

———, NEUBERG, D. S., AND FINCH, S. J. Sex-typed motor behaviors of "feminine" boys, conventionally masculine boys, and conventionally feminine girls. *Sex Roles,* 1983, *9,* 571–579.

GREENBERG, D. J. Accelerating visual complexity levels in the human infant. *Child Development,* 1971, *42,* 905–918.

GREENBERG, M., AND MORRIS, N. Engrossment: The newborn's impact upon the father. *American Journal of Orthopsychiatry,* 1974, *44,* 520–531.

———, ROSENBERG, L., AND LIND, J. First mothers rooming-in with their newborns: Its impact upon the mother. *American Journal of Orthopsychiatry,* 1973, *43,* 783–788.

GREENBERGER, E. *Over-time data on the psychosocial maturity inventory: The South Carolina study.* Working papers (mimeo). Baltimore, Md.: Center for the Social Organization of Schools, The Johns Hopkins University, December 1974.

———. *Two three-year longitudinal studies of growth in psychosocial maturity* (mimeo). Baltimore, Md: Center for the Social Organization of Schools, The Johns Hopkins University, 1975.

———. Education and the acquisition of psychosocial maturity. In D. C. McClelland (Ed.), *The development of social maturity.* New York: Irvington, 1982.

———. Defining psychosocial maturity in adolescence. In P. Karoly and J. J. Steffen (Eds.), *Adolescent behavior disorders: Foundations and applications.* Lexington, Mass.: D. C. Heath, 1983. (a)

———. A researcher in the policy arena: The case of child labor. *American Psychologist,* 1983, *38,* 104–110. (b)

———, AND STEINBERG, L. *When teenagers work: The psychological and social costs of adolescent employment.* New York: Basic Books, 1986.

GREENFIELD, P. M. *Mind and media: The effects of T.V., video games, and computers.* Cambridge, Mass.: Harvard University Press, 1984.

———, AND SMITH, J. H. *The structure of communication in early language development.* New York: Academic Press, 1976.

GREENWOOD, K. *The development of communication with mother, father, and sibling.* Unpublished doctoral thesis, Cambridge University, 1983.

GREIF, E. B. *Sex differences in parent–child conversation: Who interrupts who?* Paper presented at the biennial meeting of the Society for Research in Child Development, San Francisco, March 1979.

———, AND ULMAN, K. J. The psychological impact of menarche on early adolescent females: A review of the literature. *Child Development,* 1982, *53,* 1413–1430.

GRENVILLE, T. N. E. *U.S. Decennial life tables for 1969–1971.* Washington, D.C.: U.S. Public Health Service, 1976.

GROSSMAN, F. K., EICHLER, L. S., AND WINICKOFF, S. A. *Pregnancy, birth, and parenthood.* San Francisco: Jossey-Bass, 1980.

GROSSMANN, K., THANE, K., AND GROSSMANN, K. E. Maternal tactual contact of the newborn after various postpartum conditions of mother–infant contact. *Developmental Psychology.* 1981, *17,* 159–169.

GRUNEBAUM, H., WEISS, H. J., COHLER, B., HARTMAN, C., AND GALLANT, D. *Mentally ill mothers and their children.* Chicago: University of Chicago Press, 1975.

GUERIN, D., AND GOTTFRIED, A. W. Infant temperament as a predictor of preschool behavior problems. *Infant Behavior and Development,* 1986, *9* (Special Issue: Abstracts of papers presented at the Fifth International Conference on Infant Studies), 152.

GUIGOZ, Y., AND MUNRO, H. N. Nutrition and aging. In C. E. Finch and E. L. Schneider (Eds.), *Handbook of the biology of aging.* New York: Van Nostrand Reinhold, 1985.

GUMP, P. Educational environments. In D. Stokols and I. Altman (Eds.), *Handbook of environmental psychology.* New York: Wiley, 1987.

GUMPERZ, J. J., AND TANNEN, D. Individual and social differences in language use. In C. J. Filmore, D. Kempler, and W. S. Y. Wang (Eds.), *Individual differences in language ability and language behavior.* New York: Academic Press, 1979.

GUNNAR, M. R., AND DONAHUE, M. Sex differences in social responsiveness between six months and twelve months. *Child Development,* 1980, *51,* 262–265.

———, AND STONE, C. The effects of positive maternal affect on infant responses to pleasant, ambiguous, and fear-provoking toys. *Child Development,* 1984, *55,* 1231–1236.

GUSTAFSON, G. E. Effects of the ability to locomote on infants' social and exploratory behaviors: An experimental study. *Developmental Psychology,* 1984, *20,* 397–405.

HAAN, N. Common dimensions of personality development: Early adolescence to middle life. In D. H. Eichorn, J. A. Clausen, N. Haan, M. P. Honzik, and P. H. Mussen (Eds.), *Present and past in middle life.* New York: Academic Press, 1981, pp. 117–153.

______. Processes of moral development: Cognitive or social disequilibrium? *Development Psychology*, 1985, *21*, 996–1006.

______, AERTS, E., AND COOPER, B. A. B. *On moral grounds: The search for practical morality*. New York: New York University Press, 1985.

______, SMITH, M. B., AND BLOCK, J. Moral reasoning of young adults: Political-social behavior, family background, and personality correlates. *Journal of Personality and Social Psychology*, 1968, *10*, 183–201.

HAAS, A. *Teenage sexuality: A survey of teenage sexual behavior*. New York: Macmillan, 1979.

HACKER, H. M. Blabbermouths and clams: Sex differences in self-disclosure in same-sex and cross-sex friendship dyads. *Psychology of Women Quarterly*, 1981, *5*, 385–401.

HAGEN, J. W., AND HALE, G. H. The development of attention in children. In A. D. Pick (Ed.), *Minnesota Symposia on Child Psychology*, Volume 7. Minneapolis: University of Minnesota Press, 1973.

HAGESTAD, G. O. Parent and child: Generations in the family. In T. M. Field, A. Huston, H. C. Quay, L. Troll, and G. E. Finley (Eds.), *Review of human development*. New York: Wiley-Interscience, 1982, pp. 485–499.

HAINS, A. A., AND RYAN, E. B. The development of social cognitive processes among juvenile delinquents and nondelinquent peers. *Child Development*, 1983, *54*, 1536–1544.

HAITH, M. M. The response of the human newborn to visual movement. *Journal of Experimental Child Psychology*, 1966, *3*, 112–117.

______, BERGMAN, T., AND MOORE, M. J. Eye contact and face scanning in early infancy. *Science*, 1977, *198*, 853–855.

HALES, D. J., LOZOFF, B., SOSA, R., AND KENNELL, J. Defining the limits of the maternal sensitive period. *Developmental Medicine and Child Neurology*, 1977, *19*, 454–461.

HALL, G. S. The contents of children's minds. *Princeton Review*, 1883, *59*, 38–43.

______. *Adolescence*, Volume 1. New York: D. Appleton, 1904.

______. *Senscence: The last half of life*. New York: Appleton, 1922.

HALL, W. M. *Observational and interactive determinants of aggressive behavior in boys*. Unpublished doctoral dissertation, Indiana University, 1974.

HALL, W. S., NAGY, W. E., AND LINN, R. *Spoken words: Effects of situation and social group on oral word usage and frequency*. Hillsdale, N.J.: Lawrence Erlbaum, 1984.

HALLINAN, M. T. Recent advances in sociometry. In S. R. Asher and J. M. Gottman (Eds.), *The development of children's friendships*. New York: Cambridge University Press, 1981.

HAMBURG, B. Early adolescence: A specific and stressful stage of the life cycle. In G. V. Coelho, D. A. Hamburg, and J. E. Adams (Eds.), *Coping and adaptation*. New York: Basic Books, 1974.

HAMMOND, C. B., AND MAXON, W. S. Current status of estrogen therapy for the menopause. *Fertility and Sterility*, 1982, *37*, 5–25.

HAMMOND, P. E. Aging and the ministry. In M. W. Riley, J. W. Riley, Jr., and M. E. Johnson (Eds.), *Aging and society*, Volume 2. *Aging and the professions*. New York: Russell Sage Foundation, 1969, pp. 293–323.

HANEY, D. Q. Boys found more likely to fantasize than girls. *Los Angeles Times*, October 28, 1984, part 8, p. 10.

HARBERT, A. S., AND GINSBERG, L. H. *Human services for older adults: Concepts and skills*. Belmont, Calif.: Wadsworth, 1979.

HARDING, P. G. R. The metabolism of brown and white adipose tissue in the fetus and newborn. *Clinical Obstetrics and Gynecology*, 1971, *14*, 685–709.

HARDWICK, D., MCINTYRE, A., AND PICK, A. The content and manipulation of cognitive maps in children and adults. *Monographs of the Society for Research in Child Development*, 1976, *41* (3, Serial No. 166).

HARDY, J. B., AND MELLITS, E. D. Does maternal smoking during pregnancy have a long-term effect on the child? *Lancet*, 1972, *2*, 1332–1336.

HARDY-BROWN, K., AND PLOMIN, R., AND DEFRIES, J. C. Genetic and environmental influences on the rate of communicative development in the first year of life. *Developmental Psychology*, 1981, *17*, 704–717.

HARKNESS, S., AND SUPER, C. M. The cultural context of gender segregation in children's peer groups. *Child Development*, 1985, *56*, 219–224.

HARLOW, H. F., AND GRIFFIN, G. Induced mental and social deficits in rhesus monkeys. In S. F. Osler and R. E. Cooke (Eds.), *The biosocial bases of mental retardation*. Baltimore, Md.: Johns Hopkins University Press, 1965.

______, AND ZIMMERMAN, R. R. Affectional responses in the infant monkey. *Science*, 1959, *130*, 431–432.

HARMAN, D. Free radical theory of aging: Effect of free radical reaction inhibitors on the mortality rate of male LAF mice. *Journal of Gerontology*, 1968, *23*, 476–482.

HARPER, J. F., AND COLLINS, J. K. The effects of early or late maturation on the prestige of the adolescent girl. *Australian and New Zealand Journal of Sociology*, 1972, *8*, 83–88.

HARRIS, B. Whatever happened to little Albert? *American Psychologist*, 1979, *34*, 151–160.

HARRIS, L., AND ASSOCIATES. *The myth and reality of aging in America*. Washington, D.C.: National Council on the Aging, 1975.

______. *Aging in the eighties: America in transition*. Washington, D.C.: National Council on the Aging, 1981.

HARRIS, P. L. Infant cognition. In P. H. Mussen (Ed.), *Handbook of child psychology*, Volume 2, New York: Wiley, 1983.

HARTER, S. *Children's understanding of multiple emotions: A cognitive-developmental approach*. Address given at the Ninth Annual Piaget Society meeting, Philadelphia, 1979.

______. Developmental perspectives on the self-system. In P. H. Mussen (Ed.), *Handbook of the child psychology*, Volume 4. New York: Wiley, 1983.

______, AND CONNELL, J. P. A comparison of alternative models of the relationships between academic achievement, and children's perceptions of competence, control, and motivational orientation. In J. Nicholls (Ed.), *The development of achievement-related cognitions and behaviors*. Greenwich, Conn.: JAI Press, 1982.

HARTLEY, A. A., AND HARTLEY, J. T. Performance changes in champion swimmers aged 30 to 84 years. *Experimental Aging Research*, *10*, 3, 1984. (a)

______, AND ______. In response to Stones and Kozma. *Experimental Aging Research*, *10*, 3, 1984. (b)

HARTSHORNE, H., AND MAY, M. A. *Studies in the nature of character*, Volume 1. *Studies in deceit*. New York: Macmillan, 1928.

______, ______, AND MALLER, J. B. *Studies in the nature of character*, Volume 2, *Studies in self-control*. New York: Macmillan, 1929.

______, ______, AND SHUTTLEWORTH, F. K. *Studies in the nature of character*, Volume 3, *Studies in the organization of character*. New York: Macmillan, 1930.

HARTUP, W. W. Peer interaction and social organization. In P. H. Mussen (Ed.), *Carmichael's manual of child psychology*, Volume 2. New York: Wiley, 1970.

______. Aggression in childhood: Developmental perspectives. *American Psychologist*, 1974, *29*, 336–341.

______. Peer relations. In P. H. Mussen (Ed.), *Handbook of child psychology*, Volume 4. New York: Wiley, 1983.

HASHER, L., AND CLIFTON, D. A. A developmental study of attribute encoding in free recall. *Journal of Experimental Child Psychology*, 1974, *7*, 332–346.

HAUSER, P. M. Aging and world-wide population change. In R. H. Binstock and E. Shanas (Eds.), *Handbook of aging and the social sciences*. New York: Van Nostrand Reinhold, 1976, pp. 58–116.

HAWLEY, I., AND KELLY, F. Formal operations as a function of age, education, and fluid and crystallized intelligence. Paper presented at the Annual Meeting of the Gerontological Society. Miami, Florida, November 1973.

HAWKINS, J., PEA, R. D., GLICK, J., AND SCRIBNER, S. "Merds that laugh don't like mushrooms'': Evidence for deductive reasoning by preschoolers. *Developmental Psychology*, 1984, *20*, 584–594.

HAY, D. F., NASH, A., AND PEDERSON, J. Responses of six-month-olds to the distress of their peers. *Child Development*, 1981, *52*, 1071–1075.

———, AND RHEINGOLD, H. L. The early appearance of some valued social behaviors. In D. L. Bridgeman (Ed.), *The nature of prosocial development: Interdisciplinary theories and strategies*. New York: Academic Press, 1983.

———, AND ROSS, H. S. The social nature of early conflict. *Child Development*, 1982, *53*, 105–113.

HAYFLICK, L. The cellular basis for biological aging. In C. E. Finch and L. Hayflick (Eds.), *Handbook of the biology of aging*. New York: Van Nostrand Reinhold, 1977, pp. 159–186.

HEBB, D. O. On watching myself get old. *Psychology Today*, November 1978, *12*, 15–23.

HEIDER, E. R. ''Focal'' color areas and the development of color names. *Developmental Psychology*, 1971, *4*, 447–455.

HEINICKE, C., DISKIN, S., RAMSEY-KLEE, D., AND OATES, D. *Prebirth parent characteristics and family development in the first two years of life*. Unpublished manuscript, University of California at Los Angeles, Department of Psychiatry, 1983.

HEINICKE, C. M., AND WESTHEIMER, I. *Brief separations*. New York: International Universities Press, 1966.

HEINONEN, O. P., SLONE, D., AND SHAPIRO, S. *Birth defects and drugs in pregnancy*. Littleton, Mass.: Publishing Sciences Group, 1976.

HEISEL, M. A., AND FAULKNER, A. O. Religiosity in an older black population. *Gerontologist*, 1982, *22*, 354–358.

HELMRATH, T. A., AND STERNITZ, E. M. Death of an infant: Parental grieving and the failure of social support. *Journal of Family Practice*, 1978, *6*, 785–790.

HELSON, R., AND CRUTCHFIELD, R. S. Mathematicians: The creative researcher and the average Ph.D. *Journal of Consulting and Clinical Psychology*, 1970, *34*, 250–257.

HENKER, B., AND WHALEN, C. K. The changing faces of hyperactivity: Retrospect and prospect. In C. K. Whalen and B. Henker (Eds.), *Hyperactive children: The social ecology of identification and treatment*. New York: Academic Press, 1980.

HERMAN, J. F., AND ROTH, S. F. Children's incidental memory for spatial locations in a large-scale environment: Taking a tour down memory lane. *Merrill-Palmer Quarterly*, 1984, *30*, 87–102.

HERSHENSON, M. Visual discrimination in the human newborn. *Journal of Comparative Physiological Psychology*, 1964, *58*, 270–276.

HESS, B. Friendship. In M. W. Riley, M. Johnson, and A. Foner *(Eds.)*, *Aging and Society*, Vol. 3, *A sociology of age stratification*. New York: Russell Sage Foundation, 1972, pp. 357–393.

HESS, R. D., HOLLOWAY, S. D., DICKSON, W. P., AND PRICE, G. G. Maternal variables as predictors of children's school readiness and later achievement in vocabulary and mathematics in sixth grade. *Child Development*, 1984, *55*, 1902–1912.

———, AND MCDEVITT, T. M. Some cognitive consequences of maternal intervention techniques: A longitudinal study. *Child Development*, 1984, *55*, 2017–2030.

HESTON, L. L. The genetics of schizophrenic and schizoid disease. *Science*, 1970, *167*, 249–256.

HESTON, L. L., AND MASTRI, A. R. The genetics of Alzheimer's disease: Associations with hemalologic malignancy and Down's syndrome. *Archives of General Psychiatry*, 1977, *34*, 976–981.

HETHERINGTON, E. M., COX, M., AND COX, R. *Aftermath of divorce*. Paper presented at the meeting of the American Psychological Association, Washington, D.C., September 1976. (a)

———, ———, AND ———. Divorced fathers. *Family Coordinator*, 1976, *25*, 417–428. (b)

———, ———, AND ———. *Family interaction and the social, emotional, and cognitive development of children following divorce*. Paper presented at the symposium on ''The family: Setting priorities.'' Sponsored by the Institute for Pediatric Service of the Johnson and Johnson Baby Company. Washington, D.C., May 17–20, 1978.

———, ———, AND ———. Effects of divorce on parents and children. In M. Lamb (Ed.), *Nontraditional families*. Hillsdale, N.J.: Lawrence Erlbaum, 1982.

HICKEY, T. *Health and aging*. Monterey, Calif.: Brooks/Cole, 1980.

HICKS, D. J. Imitation and retention of film-mediated aggressive peer and adult models. *Journal of Personality and Social Psychology*, 1965, *2*, 97–100.

HILL, C. R., AND STAFFORD, F. P. *Parental care of children: Time diary estimates of quantity predictability and variety*. Working paper series, Institute for Social Research, University of Michigan, Ann Arbor, 1978.

HILL, J. P. AND PALMQUIST, W. J. Social cognition and social relations in early adolescence. *International Journal of Behavioral Development*, 1978, *1*, 1–36.

HILL, K., AND SARASON, S. The relation of test anxiety and defensiveness to test and school performance over the elementary school years: A further longitudinal study. *Monographs of the Society for Research in Child Development*, 1966, *31* (2, Serial No. 104).

HILL, R. Drugs ingested by pregnant women. *Clinical Pharmacology Therapeutics*, 1973, *14*, 654–659.

———, FOOTE, N., ALDOUS, J., CARLSON, R., AND MACDONALD, R. *Family development in three generations*. Cambridge, Mass.: Schenkman, 1970.

———, AND MATTESSICH, P. Family devélopment theory and life-span development. In P. B. Baltes and O. G. Brim, Jr. (Eds.), *Life-span development and behavior*, Volume 2. New York: Academic Press, 1979, pp. 161–204.

HILLINGER, C. Help on way for expectant fathers. *Los Angeles Times*, June 17, 1984, Part 8, p. 14.

HINES, M., AND SHIPLEY, C. Prenatal exposure to diethylstilbesterol (DES) and the development of sexually dimorphic cognitive abilities and cerebral lateralization. *Development Psychology*, 1984, *20*, 81–94.

HINGSON, R., ALPERT, J. J., DAY, N., DOOLING, E., KAYNE, H., MORELOCK, S., OPPENHEIMER, E., AND ZUCKERMAN, B. Effects of maternal drinking and marijuana use on fetal growth and development. *Pediatrics*, 1982, *70*, 539–547.

HINTON, J. *Dying*. Harmondsworth, Eng.: Penguin, 1967.

HIRSCH, J. Cell number and size as a determinant of subsequent obesity. In M. Winick (Ed.), *Childhood obesity*. New York: Wiley, 1975.

HIRSH-PASEK, K., GOLINKOFF, R. M., FLETCHER, A., BEAUBEIN, F., AND CAULEY, K. *In the beginning: One-word speakers comprehend word order*. Paper presented at the Boston Language Conference, October 1985.

———, NELSON, D. G., JUSCZYK, P. W., AND WRIGHT, K. *A moment of silence: How the prosodic cues in motherese might assist language learning*. Paper presented at the International Conference on Infant Studies, Los Angeles, April 1986.

———, AND TREIMAN, R. DOGGEREL: Motherese in a new context. *Journal of Child Language*, 1982, *9*, 229–237.

———, AND SCHNEIDERMAN, M. Brown and Hanlon revisited: Mothers' sensitivity to ungrammatical forms. *Journal of Child Language*, 1984, *11*, 81–88.

HOCK, E. Working and nonworking mothers and their infants: A comparative study of maternal caregiving characteristics and infant social behavior. *Merrill-Palmer Quarterly*, 1980, *26*, 79–101.

HOFF-GINSBERG, E. Function and structure in maternal speech: Their relation to the child's development of syntax. *Developmental Psychology*, 1986, *22*, 155–163.

HOFFMAN, L. W. Changes in family roles, socialization and sex differences. *American Psychologist*, 1977, *32*, 644–657.

———. Maternal employment: 1979. *American Psychologist*, 1979, *34*, 859–865.

———. The effects of maternal employment on the academic attitudes and performance of school-aged children. *School Psychology Review*, 1980, *9*, 319–336.

———. Maternal employment and the young child. In M. Perlmutter (Ed.), *Parent–child interaction and parent–child relations in child development. Minnesota Symposia on Child Psychology*, Volume 17. Hillsdale, N.J.: Lawrence Erlbaum, 1984.

______. Moral development. In P. H. Mussen (Ed.), *Carmichael's manual of child psychology*, Volume 2. New York: Wiley, 1970.

______. Altruistic behavior and the parent–child relationship. *Journal of Personality and Social Psychology*, 1975, *31*, 937–943.

______. Sex differences in empathy and related behaviors. *Psychological Bulletin*, 1977, *84*, 712–722.

HOFFMAN-PLOTKIN, D., AND TWENTYMAN, C. T. A multimodal assessment of behavioral and cognitive deficits in abused and neglected preschoolers. *Child Development*, 1984, *55*, 794–802.

HOFSTAETTER, P. R. The changing composition of intelligence: A study of the t-technique. *Journal of Genetic Psychology*, 1954, *85*, 159–164.

HOGUE, C. C. Injury in late life: Prevention. *Journal of the American Geriatric Society*, 1982, *30*, 276–280. (a)

______. Injury in late life: Epidemiology. *Journal of the American Geriatric Society*. 1982, 30, 183–190. (b)

HOLLAND, J. *Making vocational choice: A theory of careers*. Englewood Cliffs, N.J.: Prentice-Hall, 1973.

HOLLINGWORTH, H. L. *Mental growth and decline: A survey of developmental psychology. New York: Appleton, 1927.*

HOLLOS, M. Comprehension and use of social rules in pronoun selection by Hungarian children. In S. Ervin-Tripp and C. Mitchell-Kernan (Eds.), *Child discourse*. New York: Academic Press, 1977.

HOLMES, D. L., RUBLE, N., KOWALSKI, J., AND LAUESEN, B. *Predicting quality of attachment at one year from neonatal characteristics*. Paper presented at the International Conference on Infant Studies, New York, April 1984.

HOLT, J. H. *How children fail*. New York: Dell, 1964.

HOOD, B., AND WILLATTS, P. Reaching in the dark: Object permanence in five-month-old infants. *Infant Behavior and Development*, 1986, *9* (Special Issue: Abstracts of papers presented at the Fifth International Conference on Infant Studies), 173a.

HORWITZ, R. A. Psychological effects of the "open classroom." *Review of Educational Research*, 1979, *49*, 71–86.

HOUSEHOLDER, J., HATCHER, R., BURNS, W. J., AND CHASNOFF, I. Infants born to narcotic-addicted mothers. *Psychological Bulletin*, 1982, *92*, 453–468.

HORN, J. Regriefing: A way to end pathological mourning. *Psychology Today*, May 1974, *7*, 104.

HORN, J. L. The theory of fluid and crystallized intelligence in relation to concepts of cognitive psychology and aging in adulthood. In F. I. M. Craik and S. Trehub (Eds.), *Aging and cognitive processes*. New York: Plenum, 1982, pp. 237–278.

HORNBLUM, J., AND OVERTON, W. Area and volume conservation among the elderly: Assessment and training. *Developmental Psychology*, 1976, *12*, 68.

HOWARD, J. Counseling: A developmental approach. In E. E. Bleck and D. A. Nagel (Eds.), *Physically handicapped children: A medical atlas for teachers*. New York: Grune and Stratton, 1982.

HOWARD, L., AND POLICH, J. P300 latency and memory span development. *Developmental Psychology*, 1985, *21*, 283–289.

HOYER, W. J. Aging and the development of expert cognition. In T. M. Schlechter and M. P. Taglia (Eds.), *New directions in cognitive science*. Norwood, N.J.: Ablex, 1984, pp. 69–87.

HUBBARD, L. In search of 40 winks. *Modern Maturity*, April/May, 1982, 72–74.

HUDSON, J., AND NELSON, K. Effects of script structure on children's story recall. *Developmental Psychology*, 1983, *19*, 625–635.

HUESMANN, L. R., ERON, L. D., LEFKOWITZ, M. M., AND WALDER, L. O. Stability of aggression over time and generations. *Developmental Psychology*. 1984, *20*, 1120–1134.

______, LAGERSPETZ, K., AND ERON, L. D. Intervening variables in the TV violence-aggression relation: Evidence from two countries. *Developmental Psychology*, 1984, *20*, 746–775.

HUNT, C. C. Interviewed by Peter Gorner, "The sleeping world of quiet mysteries and silent killers." *Chicago Tribune*, May 3, 1982, Section 3, p. 1.

HUNTER, F. T. Socializing procedures in parent–child and friendship relations during adolescence. *Developmental Psychology*, 1984, *201*, 1092–1099.

______, AND YOUNISS, J. Changes in functions of three relations during adolescence. *Developmental Psychology*, 1982, *18*, 806–811.

HUNTER, S., WOLF, T., SKLOV, M., WEBBER, L., AND BERENSON, G. *A-B coronary-prone behavior pattern and cardiovascular risk factor variables in children and adolescents: The Bogalusa Heart Study*. Paper presented at the American College of Cardiology, San Francisco, March 1981.

HUTTENLOCHER, J. The origins of language comprehension. In R. L. Solso (Ed.), *Theories in cognitive psychology*. Hillsdale, N.J.: Lawrence Erlbaum, 1974.

______, AND PRESSON, C. B. Mental rotation and the perspective problem. *Cognitive Psychology*, 1973, *4*, 277–299.

HUTTENLOCHER, P. Press release, The University of Chicago, July 1979.

HUYCK, M. H. From gregariousness to intimacy: Marriage and friendship over the adult years. In T. M. Field, A. Huston, H. C. Quay, L. Troll, and G. E. Finley (Eds.), *Review of human development*. New York: Wiley-Interscience, 1982, pp. 471–484.

HYDE, J. S. How large are gender differences in aggression? A developmental meta-analysis. *Developmental Psychology*, 1984, *20*, 722–736.

IANNOTTI, R. J. Naturalistic and structured assessments of prosocial behavior in preschool children: The influence of empathy and perspective taking. *Developmental Psychology*, 1985, *21*, 46–55.

ICKES, W., AND EARNES, R. D. Boys and girls together—and alienated: On enacting stereotyped sex roles in mixed-sex dyads. *Journal of Personality and Social Psychology*, 1978, *36*, 669–683.

INGE, G. Life and death in Swedish folk customs. Presented at International Work Group on Death and Dying, Rosenheim, Sweden, 1982.

INHELDER, B., AND PIAGET, J. *The growth of logical thinking from childhood to adolescence*. New York: Basic Books, 1958.

ISEN, A. M., HORN, N., AND ROSENHAN, D. L. Effects of success and failure on children's generosity. *Journal of Personality and Social Psychology*, 1973, *27*, 239–247.

IVERSON, L. L. The chemistry of the grain. *Scientific American*, 1979, *241*, 134–149.

IVEY, M. E., AND BARDWICK, J. M. Patterns of affective fluctuation in the menstrual cycle. *Psychosomatic Medicine*, 1968, *30*, 336–345.

IZARD, C. E. *The face of emotions*. New York: Appleton, 1971.

______. *Human emotions*. New York: Plenum, 1977.

______, HUEBNER, R. R., RISSER, D., MCGINNES, G. C., AND DOUGHERTY, L. M. The young infant's ability to produce discrete emotion expressions. *Developmental Psychology*, 1980, *16*, 132–141.

JACKSON, D., AND BURCH, P. R. J. Dental caries as a degenerative disease. *Gerontologia*, 1969, *15*, 203–216.

JACKSON, D. L., AND YOUNGER, S. Patient autonomy and "Death with Dignity." Some clinical caveats. *New England Journal of Medicine*, 1979, *301*, 404–408.

JACOBS, J. *Adolescent suicide*. New York: Wiley, 1971.

JACOBSON, J. L., AND WILLIE, D. E. The influence of attachment pattern on developmental changes in peer interaction from the toddler to the preschool period. *Child Development*, 1986, *57*, 338–347.

______ AND ______. *The influence of attachment pattern on peer interaction at 2 and 3 years*. Paper presented at the International Conference on Infant Studies, New York, April 1986.

JAQUISH, G. A., AND SAVIN-WILLIAMS, R. C. Biological and ecological factors in the expression of adolescent self-esteem. *Journal of Youth and Adolescence*, 1981, *10*, 473–485.

JARVIK, L. F. Discussion: Patterns of intellectual functioning in the later years. In L. F. Jarvik, C. Eisdorfer, and J. E. Blum (Eds.), *Intellectual functioning in adults: Psychological and biological influences*. New York: Springer, 1973, pp. 65–67.

JEFFKO, W. G. Redefining death. *Commonweal*, July 6, 1979, 394–397.

JEFFREY, W. The developing brain and child development. In M. Whittrock (Ed.), *The brain and psychology*. New York: Academic Press, 1980.

JELLIFFE, D. B., AND JELLIFFE, E. F. P. Breast-feeding is desirable—and vital for the poor. *Chicago Tribune*, May 5/6, 1982, Section 7, p. 13.

JENCKS, C., SMITH, M., ACLAND, H., BANE, M. J., COHEN, D., GINTIS, H., HEYNS, B., AND MICHAELSON, S. *Inequality: A reassessment of the effect of family and schooling in America*. New York: Harper & Row, 1972.

JENSEN, K. Differential reactions to taste and temperature stimuli in newborn infants. *Genetic Psychology Monographs*, 1932, *12*, 363–479.

JERNIGAN, T. L., ZATZ, L. M., FEINBERG, I., AND FEIN, G. The measurement of cerebral atrophy in the aged by computed tomography. In L. W. Poon (Ed.), *Aging in the 1980s*. Washington, D.C.: American Psychological Association, 1980, pp. 86–94.

JERSILD, A. T., AND HOLMES, F. B. *Children's fears*. Child Development Monograph 20. New York: Teachers College Press, Columbia University, 1935.

______, AND MARKEY, F. V. *Conflicts between preschool children*. Child Development Monograph 21. New York: Teachers College Press, Columbia University, 1935.

JERUCHIMOWICZ, J., AND HANS, S. L. Behavior of neonates exposed in utero to methadone as assessed on the Brazelton Scale. *Infant Behavior and Development*, 1985, *8*, 323–336.

JESSOR, R., COSTA, F., JESSOR, L., AND DONOVAN, J. E. Time of first intercourse: A prospective study. *Journal of Personality and Social Psychology*, 1983, *44*, 608–626.

______, AND JESSOR, S. L. The transition from virginity to nonvirginity among youth: A social-psychological study over time. *Developmental Psychology*, 1975, *11*, 473–484.

JOFFE, J. M. *Prenatal determinants of behavior*. Oxford: Pergamon, 1969.

JOHNSON, D. B. Self-recognition in infants. *Infant Behavior and Development*, 1983, *6*, 211–222.

JOHNSON, E., AND BURSK, B. Relationships between the elderly and their adult children. *Gerontologist*, 1977, *17*, 90–96.

JOHNSON, S. M., WAHL, G., MARTIN, S., AND JOHANSSEN, S. How deviant is the normal child: A behavioral analysis of the preschool child and his family. In R. D. Rubin, J. P. Brady, and J. D. Henderson (Eds.), *Advances in behavior therapy*, Volume 4. New York: Academic Press, 1973.

JOHNSTON, L. D., O'MALLEY, P. M., AND BACHMAN, J. G. *Use of licit and illicit drugs by American high school students*, 1975–1984. National Institute of Drug Abuse, United States Department of Health and Human Services. Washington, D.C., 1985.

JONES, H. B. A special consideration of the aging process, disease and life expectancy. In J. H. Lawrence and J. G. Hamilton (Eds.), *Advances in biological and medical physics*, Volume 4. New York: Academic Press, 1956, pp. 281–337.

JONES, K. L., SMITH, D. W., ULLELAND, C. N., AND STREISSGUTH, A. P. Pattern of malformation in offspring of chronic alcoholic mothers. *Lancet*, 1973, *1*, 1267–1271.

JONES, M. C. The later careers of boys who were early or late maturing. *Child Development*, 1957, *28*, 113–128.

______. Personality correlates and antecedents of drinking patterns in adult males. *Journal of Consulting and Clinical Psychology*, 1968, *32*, 2–12.

______, AND BAYLEY, N. Physical maturing among boys as related to behavior. *Journal of Educational Psychology*, 1950, *41*, 129–148.

JONES, S. J., AND MOSS, H. A. Age, state, and maternal behavior associated with infant vocalizations. *Child Development*, 1971, *42*, 1039–1051.

JUSCZYK, P. W. Perception of syllable-final stop consonants by 2-month-old infants. *Perception and Psychophysics*, 1977, *21*, 450–454.

______, ROSNER, B. S., CUTTING, J. E., FOARD, F., AND SMITH, L. B. Categorical perception of non-speech sounds by 2-month-old infants. *Perceptions and Psychophysics*, 1977, *21*, 50–54.

JUSTICE, E. M. Categorization as a preferred memory strategy: Developmental changes during elementary school. *Developmental Psychology*, 1985, *21*, 1105–1110.

KAGAN, D. M., AND SQUIRES, R. L. Eating disorders among adolescents: patterns and prevalence. *Adolescence*, 1984, *19*, 15–29.

KAGAN, J. *Change and continuity in infancy*. New York: Wiley, 1971.

______, AND MOSS, H. A. *Birth to maturity*. New York: Wiley, 1962.

______, AND TULKIN, S. R. Social class differences in child rearing during the first year. In H. R. Schaffer (Ed.), *The origins of human social relations*. London: Academic Press, 1971.

KAHANA, B., AND KAHANA, E. Grandparenthood from the perspective of the developing grandchild. *Developmental Psychology*, 1970, *3*, 98–105.

KAIL, R., AND NIPPOLD, M. A. Unconstrained retrieval from semantic memory. *Child Development*, 1984, *55*, 944–951.

KALES, A., JACOBSON, A., PAULSON, M., KALES, J., AND WALTER, R. Somnambulism: Psychophysiological correlates, I. *Archives of General Psychiatry*, 1966, *14*, 586–594.

KALISH, R. A. Death in a social context. In R. H. Binstock and E. Shanas (Eds.), *Handbook of aging and the social sciences*. New York: Van Nostrand Reinhold, 1976, pp. 483–507.

______, AND REYNOLDS, D. K. *Death and ethnicity: A psycho-cultural study*. Los Angeles: University of Southern California Press, 1976; reprinted, Farmingdale, N.Y.: Baywood Publishing Co., 1981.

KANNER, L. Problems of nosology and psychodynamics of early infantile autism. *American Journal of Orthopsychiatry*, 1949, *19*, 416–426.

KARACAN, I., ASLAN, C., AND HERSHKOWITZ, M. Erectile mechanisms in man. *Science*, 1983, *220*, 1080–1082.

KASTENBAUM, R. Time and death in adolescence. In H. Feifel (Ed.), *The meaning of death*. New York: McGraw-Hill, 1959, pp. 99–113.

______. The dimensions of future time perspective, an experimental analysis. *Journal of Genetic Psychology*, 1961, *65*, 203–218.

______. Is death a life crisis? On the confrontation with death in theory and practice. In N. Datan and L. H. Ginsberg (Eds.), *Life-span developmental psychology: Normative life crises*. New York: Academic Press, 1975, pp. 19–50.

______, AND AISENBERG, R. B. *The psychology of death* (rev. ed.). New York: Springer, 1976.

______. Death and dying: A lifespan approach. In J. E. Birren and K. W. Schaie (Eds.), *Handbook of the psychology of aging* (2nd ed.). New York: Van Nostrand Reinhold, 1985.

______, KASTENBAUM, B. K., AND MORRIS, J. Strengths and preferences of the terminally ill: Data from the National Hospice Demonstration Study. In preparation.

______, AND WEISMAN, A. D. The psychological autopsy as a research procedure in gerontology. In D. P. Dent, R. Kastenbaum, and S. Sherwood (Eds.), *Research planning and action for the elderly*. New York: Behavioral Publications, 1972.

KASWORM, C. E. The older student as an undergraduate. *Adult Education*, 1980, *31*, 30–47.

KAVANAUGH, R. D. Observations on the role of logically constrained sentences in the comprehension of "before" and "after." *Journal of Child Language*, 1979, *6*, 353–357.

______, AND JIRKOVSKY, A. M. Parental speech to young children: A longitudinal analysis. *Merrill-Palmer Quarterly*, 1982, *28*, 297–311.

KAYE, K. Toward the origin of dialogue. In H. R. Schaffer (Ed.), *Studies in mother-infant interaction*. London: Academic Press, 1977.

______. *The mental and social life of babies: How parents create persons*. Chicago: University of Chicago Press, 1982.

KEARSLEY, R. The newborn's response to auditory stimulation: A demonstration of orienting and defensive behavior. *Child Development*, 1973, *44*, 582–590.

KEATING, D. P., AND BOBBITT, B. L. Individual and developmental differences in cognitive-processing components of mental ability. *Child Development*, 1978, *49*, 155–167.

KEENAN, E. O. Making it last: Repetition in children's discourse. In S. Ervin-Tripp and C. Mitchell-Kernan (Eds.), *Child discourse*. New York: Academic Press, 1977.

KEENEY, T. J., CANIZZO, S. R., AND FLAVELL, J. H. Spontaneous and induced verbal rehearsal in a recall task. *Child Development*, 1967, *38*, 953–966.

KELLER, A., FORD, L. H., AND MEACHAM, J. A. Dimensions of self-concept in preschool children. *Developmental Psychology*, 1978, *14*, 483–489.

KELLOGG, R. *Analyzing children's art*. Palo Alto, Calif.: Mayfield, 1970.

KELLY, J. Cosmetic lib for men. *The New York Times Magazine*, September 25, 1977.

KEMPE, H. C., AND HELFER, R. E. (Eds.) *Helping the battered child and his family*. Philadelphia: Lippincott, 1972.

KENDRICK, C., AND DUNN, J. Caring for a second child: Effects on the interaction between mother and first-born. *Developmental Psychology*, 1980, *16*, 303–311.

KENNELL, J. H., JERAULD, R., WOLFE, H., CHESLER, D., KREGER, N. C., MCALPINE, W., STEFFA, M., AND KLAUS, M. H. Maternal behavior one year after early and extended postpartum contact. *Developmental Medicine and Child Neurology*, 1974, *16*, 172–179.

KERASOTES, D., AND WALKER, C. E. Hyperactive behavior in children. In C. E. Walker and M. C. Roberts (Eds.), *Handbook of clinical child psychology*. New York: Wiley, 1983.

KESSEN, W. Research design in the study of developmental problems. In P. H. Mussen (Ed.), *Handbook of research methods in child development*. New York: Wiley, 1960.

______. *The child*. New York: Wiley, 1965.

______ (Ed.). *Childhood in China*. New Haven, Conn.: Yale University Press, 1975.

KEYE, E. *The family guide to children's television*. New York: Random House, 1974.

KHAN, A. U., AND CATAIS, J. *Men and women in biological perspective: A review of the literature*. New York: Praeger, 1984.

KIDD, R. F., AND BERKOWITZ, L. Dissonance, self-concept, and helplessness. *Journal of Personality and Social Psychology*, 1976, *33*, 613–622.

KIMBLE, G. A., AND PENNYPACKER, H. W. Eyelid conditioning in young and aged subjects. *Journal of Genetic Psychology*, 1963, *103*, 283–289.

KIMMEL, D. C., PRICE, K. F., AND WALKER, J. W. Retirement choice and retirement satisfaction. *Journal of Gerontology*, 1978, *33*, 575–585.

KIMURA, D. The asymmetry of the human brain. *Scientific American*, 1975, 70–78.

KINLOCH, G. C. Parent-youth conflict at home: An investigation among university freshmen. *Journal of Orthopsychiatry*, 1970, *40*, 658–664.

KINSBOURNE, M. Minimal brain dysfunction as a neurodevelopmental lag. *Annals of the New York Academy of Sciences*, 1973, *205, 268–273.*

KINSEY, A. C., POMEROY, W. B., MARTIN, C. E., AND GEBHARD, P. H. *Sexual behavior in the human female*. Philadelphia: Saunders, 1953.

KIVNICK, H. Q. Grandparenthood: An overview of meaning and mental health. *Gerontologist*, 1982, *22*, 59–66.

KLAUS, M. H., AND KENNELL, J. H. Maternal-infant bonding. St. Louis: C. V. Mosby, 1976.

______, AND ______. *Hunting and gathering societies: An empirical basis for exploring biobehavioral processes in mothers and infants*. Paper presented at the International Conference on Infant Studies, New York, 1984.

______, ______, PLUMB, N., AND ZUEHLKE, S. Human maternal behavior at first contact with her young. *Pediatrics*, 1970, *46*, 187–192.

KLAUS, R., AND GRAY, S. The Early Training Project for Disadvantaged Children: A report after five years. *Monographs of the Society for Research in Child Development*, 1968, *33* (4, Serial No. 120).

KLECK, R. E., RICHARDSON, S. A., AND RONALD, L. Physical appearance cues and interpersonal attraction in children. *Child Development*, 1974, *45*, 305–310.

KLEEMEIER, R. W. Intellectual change in the senium. *Proceedings of the Social Statistics Section of the American Statistical Association*, 1962, pp. 290–295.

KLEIMAN, G. M. *Brave new schools: How computers can change education*. Reston, Va.: Reston, 1984.

KLEIN, R. E., HABICHT, J. P., AND YARBROUGH, C. Effects of protein-calorie malnutrition on mental development. *Advances in Pediatrics*, 1971, *18*, 75–91.

KLIMA, E., AND BELLUGI, U. Syntactic regularities—the speech of children. In J. Lyons and R. J. Wales (Eds.), *Psycholinguistic papers*. Edinburgh: Edinburgh University Press, 1966.

KLOCKE, R. A. Influence of aging on the lung. In C. E. Finch and L. Hayflick (Eds.), *Handbook of the biology of aging*. New York: Van Nostrand Reinhold, 1977, pp. 432–444.

KNOBLOCH, H., AND PASAMANICK, B. Prospective studies on the epidemiology of reproductive casualty: Methods, findings and some implications. *Merrill-Palmer Quarterly*, 1966, *12*, 27–43.

KOGAN, N. Creativity and cognitive style: A life-span perspective. In P. B. Baltes and K. W. Schaie (Eds.), *Life-span developmental psychology: Personality and socialization*. New York: Academic Press, 1973, pp. 145–178.

______. Cognitive styles in older adults. In T. M. Field, A. Huston, H. C. Quay, L. Troll, and G. E. Finley (Eds.), *Review of human development*. New York: Wiley-Interscience, 1982, pp. 586–601.

KOEPKE, J. E., HAMM, M., LEGERSTEE, M., AND RUSSELL, M. Neonatal imitation: Two failures to replicate. *Infant Behavior and Development*, 1983, *6*, 97–102.

KOHLBERG, L. Development of moral character and moral ideology. In L. W. Hoffman and M. L. Hoffman (Eds.), *Review of child development research*, Volume 1. New York: Russell Sage Foundation, 1964.

______. Stage and sequence: The cognitive-developmental approach to socialization. In D. A. Goslin (Ed.), *Handbook of socialization theory and research*. Chicago: Rand McNally, 1969.

______. Continuities in childhood and adult moral development revisited. In P. B. Baltes and K. W. Schaie (Eds.), *Life-span developmental psychology: Personality and socialization*. New York: Academic Press, 1973, pp. 179–204.

______. Moral stages and moralization: Cognitive-developmental approach. In T. Lickona (Ed.), *Moral development and behavior: Theory, research, and social issues*. New York: Holt, Rinehart & Winston, 1976.

______, COLBY, A., GIBBS, J., AND SPEICHER-DUBIN, B. Standard form scoring manual. Cambridge, Mass.: Center for Moral Education, Harvard Graduate School of Education, 1978.

______, AND GILLIGAN, C. The adolescent as a philosopher: The discovery of the self in a postconventional world. *Daedalus*, 1971, *100*, 1051–1086.

KOHN, M. L. Job complexity and adult personality. In N. J. Smelser and E. H. Erikson (Eds.), *Themes of work and love in adulthood*. Cambridge, Mass.: Harvard University Press, 1980, pp. 193–210.

KOHN, R. R. *Principles of mammalian aging*. Englewood Cliffs, N.J.: Prentice-Hall, 1971.

______. Heart and cardiovascular system. In C. E. Finch and L. Hayflick (Eds.), *Handbook of the biology of aging*. New York: Van Nostrand Reinhold, 1977, pp. 281–317.

KOLATA, G. Lowered cholesterol decreases heart disease. *Science*, 1984, *223*, 381–382.

———. Studying learning in the womb. *Science*, 1984, *225*, 302–303.

———. Obese children: A growing problem. *Science*, April 1986, *232*, 20–21.

KONOPKA, G. *Young girls: A portrait of adolescence*. Englewood Cliffs, N.J.: Prentice-Hall, 1976.

KOPP, C. B. Risk factors in development. In P. H. Mussen (Ed.), *Handbook of child psychology*, Volume 2. New York: Wiley, 1983.

———, AND PARMELEE, A. H. Prenatal and perinatal influences on infant behavior. In J. D. Osofsky (Ed.), *Handbook of infant development*. New York: Wiley, 1979.

KORN, S. J. Continuities and discontinuities in difficult/easy temperament: Infancy to young adulthood. *Merrill-Palmer Quarterly*, 1984, *30*, 189–199.

KORNER, A. F. Visual alertness in neonates: Individual differences and their correlates. *Perceptual and Motor Skills*, 1970, *31*, 499–509.

———. State as variable, as obstacle and as mediator of stimulation in infant research. *Merrill-Palmer Quarterly*, 1972, *18*, 77–94.

KORNER, A. F., AND THOMAN, E. G. Relative efficacy of contact and vestibular stimulation on soothing neonates. *Child Development*, 1972, *10*, 67–78.

KOTELCHUCK, M. The infant's relationship to the father: Experimental evidence. In M. E. Lamb (Ed.), *The role of the father in child development*. New York: Wiley, 1976.

KOTULAK, R. Baby removed from womb, returned in new surgery: *Chicago Tribune*, November 15, 1981, Section 1, pp. 1–12.

KOZMA, A., AND STONES, M. J. Predictors of happiness. *Journal of Gerontology*, 1983, *38*, 626–628.

KREBS, R. L. *Some relations between moral judgment, attention, and resistance to temptation*. Unpublished doctoral dissertation, University of Chicago, 1967.

———, AND KOHLBERG, L. Moral judgment and ego controls as determinants of resistance to cheating. Unpublished manuscript, Center for Moral Education, Harvard University, 1973.

KREPPNER, K., PAULSEN, S. AND SCHUETZE, Y. Infant and family development: From triads to tetrads. *Human Development*, 1982, *25*, 373–391.

KREPS, J. M. The economy and the aged. In R. H. Binstock and E. Shanas (Eds.), *Handbook of aging and the social sciences*. New York: Van Nostrand Reinhold, 1976, pp. 272–285.

KREUZ, I. E., AND ROSE, R. M. Assessment of aggressive behavior and plasma testosterone in a young criminal population. *Psychosomatic Medicine*, 1972, *34*, 321–332.

KROSNICK, J. A., AND JUDD, C. M. Transitions in social influence at adolescence: Who induces cigarette smoking? *Developmental Psychology*, 1982, *18*, 359–368.

KRUGMAN, S., WARD, R., AND KATZ, S. *Infectious diseases of children*. St. Louis: C. V. Mosby, 1977.

KÜBLER-ROSS, E. *On death and dying*. New York: Macmillan, 1969.

———, interviewed by D. Goleman. The child will always be there: Real love doesn't die. *Psychology Today*, September 1976, *10*, 48–52.

KUHL, P. K. Discrimination of speech by nonhuman animals: Basic auditory sensitivities conducive to the perception of speech-sound categories. *Journal of the Acoustical Society of America*, 1981, *70*, 340–349.

———. Perception of auditory equivalence classes for speech in early infancy. *Infant Behavior and Development*, 1983, *6*, 263–285.

KUHN, D., NASH, S. C., AND BRUCKEN, L. Sex-role concepts of two- and three-year-olds. *Child Development*, 1978, *49*, 445–451.

LABOUVIE-VIEF, G. V. Discontinuities in development from childhood to adulthood: A cognitive-developmental view. In T. M. Field, A. Huston, H. C. Quay, L. Troll, and G. E. Finley (Eds.), *Review of human development*. New York: Wiley-Interscience, 1982, pp. 447–455.

———, AND CHANDLER, M. J. Cognitive development and life-span developmental theory: Idealistic versus contextual perspectives. In P. B. Baltes (Ed.), *Life-span development and behavior*, Volume 1. New York: Academic Press, 1978, pp. 182–210.

———, AND GONDA, J. N. Cognitive strategy training and intellectual performance in the elderly. *Journal of Gerontology*, 1976, *31*, 327–331.

LACHMAN, J. L., AND LACHMAN, R. Age and the actualization of world knowledge. In L. W. Poon, J. L. Fozard, L. S. Cermak, D. Arenberg, and L. W. Thompson (Eds.), *New directions in memory and aging*. Hillsdale, N.J.: Lawrence Erlbaum, 1980, pp. 285–311.

LACHMAN, M., AND MCARTHUR, L. Adulthood age differences in causal attributions for cognitive, physical, and social performance. *Psychology and Aging*, 1986 *1*, 2, 127–132.

LADD, G. W. Effectiveness of a social learning method for enhancing children's social interaction and peer acceptance. *Child Development*, 1981, *52*, 171–178.

LAFRENIERE, P. J., AND SROUFE, L. A. Profiles of peer competence in the preschool: Interrelations among measures, influence of social ecology, and relation to attachment history. *Developmental Psychology*, 1985, *21*, 56–69.

LAFRENIERE, P. J., STRAYER, F. F., AND GAUTHIER, R. The emergence of same-sex affiliative preferences among preschool peers: A developmental ethological perspective. *Child Development*, 1984, *55*, 1958–1965.

LAIR, C. V., MOON, W. H., AND KAUSLER, D. H. Associative Interference in the paired-associate learning of middle-aged and old subjects. *Developmental Psychology*, 1969, *1*, 548–552.

LAMB, M. E. Parent-infant interaction in 8-month-olds. *Child Psychiatry and Human Development*, 1976, *7*, 56–63.(a)

———. Twelve-month-olds and their parents: Interaction in a laboratory playroom. *Developmental Psychology*, 1976, *12*, 237–244.(b)

———. Father-infant and mother-infant interaction in the first year of life. *Child Development*, 1977, *48*, 167–181.

———, AND EASTERBROOKS, M. A. Individual differences in parental sensitivity: Origins, components, and consequences. In M. E. Lamb and L. R. Sherrod (Eds.), *Infant social cognition*. Hillsdale, N.J.: Lawrence Erlbaum, 1981, pp. 127–154.

———, FRODI, A. M., HWANG, C. P., FRODI, M., AND STEINBERG, J. Mother- and father-infant interaction involving play and holding in traditional and nontraditional Swedish families. *Developmental Psychology*, 1982, *18*, 215–221.

———, THOMPSON, R. A., GARDNER, W., AND CHARNOV, E. L. *Infant-mother attachment: The origins and developmental significance of individual differences in strange situation behavior*. Hillsdale, N.J.: Lawrence Erlbaum, 1985.

LAMMER, E. J., CHEN, D. T., HOAR, R. M., AGNISH, N. D., BENKE, P. J., BRAUN, J. T., CURRY, C. J., FERNHOFF, P. M., GRIX, A. W., LOTT, I. T., RICHARD, J. M., AND SUN, S. C. Retinoic acid embryopathy. *The New England Journal of Medicine*, October 3, 1985, *313*, 837–841.

LANETTO, R. *Children's conceptions of death*. New York: Springer, 1980.

LANGLOIS, J. H., AND DOWNS, A. C. Mothers, fathers and peers as socialization agents of sex-typed play behavior in young children. *Child Development*, 1980, *51*, 1227–1247.

———, AND STEPHAN, C. The effects of physical attractiveness and ethnicity on children's behavioral attributions and peer preferences. *Child Development*, 1977, *48*, 1694–1698.

LAPORTE, R. E., BLACK-SANDLER, R., CAULEY, J. A., LINK, M., BAYLES, C., AND MARKS, B. The assessment of physical activity in older women: Analysis of the interrelationship and reliability of activity monitoring, activity surveys, and caloric intake. *Journal of Gerontology*, 1983, *38*, 385–393.

LAPSLEY, D. K., AND MURPHY, M. N. Another look at the theoretical assumptions of adolescent egocentrism. *Developmental Review*, 1985, *5*, 201–217.

LARSON, R. Thirty years of research on the subjective well-being of older Americans. *Journal of Gerontology*, 1978, *33*, 109–125.

———, CSIKSZENTMIHALYI, M., AND GRAEF, R. Mood variability and the psychosocial adjustment of adolescents. *Journal of Youth and Adolescence*, 1980, *9*, 469–490.

———, SYRDAL-LASKY, A., AND KLEIN, R. E. VOT discrimination by four- and six-and-a-half-month-old infants from Spanish environments. *Journal of Experimental Child Psychology*, 1975, *20*, 215–225.

LAURENCE, M. W., AND ARROWOOD, A. J. Classification style difference in the elderly. In F. I. M. Craik and S. Trehub (Eds.), *Aging and cognitive processes*. New York: Plenum, 1982, pp. 213–220.

LAZAR, I., DARLINGTON, R. B., MURRAY, H., ROYCE, J., AND SNIPPER, A. Lasting effects of early education. *Monographs of the Society for Research in Child Development*, 1982, *47* (2–3, Serial No. 195).

LEBOYER, F. *Birth without violence*. New York: Knopf, 1975.

LECOURS, A. R. Myelogenetic correlates of the development of speech and language. In E. H. Lenneberg and E. Lenneberg (Eds.), *Foundations of language development*. New York: Academic Press, 1975.

LEECH, S., AND WITTE, K. L. Paired associate learning in elderly adults as related to pacing and incentive conditions, *Developmental Psychology*, 1971, *5*, 180.

LEFKOWITZ, M. M. Smoking during pregnancy: Long-term effect on offspring. *Developmental Psychology*, 1981, *17*, 192–194.

______, ERON, L. D., WALDER, L. O., AND HUESMANN, L. R. *Growing up to be violent: A longitudinal study of the development of aggression*. Elmsford, N.Y.: Pergamon, 1977.

LEGG, C., SHERICK, I., AND WADLAND, W. Reaction of preschool children to the birth of a sibling. *Child Psychiatry and Human Development*, 1974, *5*, 3–39.

LEIFER, M. Psychological changes accompanying pregnancy and motherhood. *Genetic Psychology Monographs*, 1977, *95*, 55–96.

LEMPERS, J. D., FLAVELL, E. R., AND FLAVELL, J. H. The development in very young children of tacit knowledge concerning visual perception. *Genetic Psychology Monographs*, 1977, *95*, 3–53.

LENNEBERG, E. H. *Biological foundations of language*. New York: Wiley, 1967.

LENNEY, L. M. *Mother–daughter communication about maturation and sexuality*. Unpublished master's thesis, University of California, Irvine, 1985.

LEON, G. R., GILLUM, B., GILLUM, R., AND GOUZE, M. Personality stability and change over a 30-year period—Middle age to old age. *Journal of Consulting and Clinical Psychology* [in press].

LEPMAN, J. (ED.). *How children see our world*. New York: Avon, 1971.

LEPPER, M. R. Micro computers in education: Motivational and social issues. *American Psychologist*, 1985, *40*, 1–18.

______, AND GREENE, D. Turning play into work: Effects of adult surveillance and extrinsic rewards on children's intrinsic motivation. *Journal of Personality and Social Psychology*, 1975, *31*, 479–486.

______, AND GREENE, D. Overjustification research and beyond: Toward a means-ends analysis of intrinsic and extrinsic motivation. In M. R. Lepper and D. Greene (Eds.), *The hidden costs of reward*. Hillsdale, N.J.: Lawrence Erlbaum, 1978.

LERNER, J. V. The import of temperament for psychosocial functioning: Tests of a goodness of fit model. *Merrill-Palmer Quarterly*, 1984, *30*, 177–188.

LERNER, M. When, why, and where people die. In O. G. Brim, Jr., H. E. Freeman, S. Levine, and N. A. Scotch (Eds.), *The dying patient*. New York: Russell Sage Foundation, 1970.

LERNER, R. M., KARSON, M., MEISELS, M., AND KNAPP, J. R. Actual and perceived attitudes of late adolescents and their parents: The phenomenon of the generation gap. *Journal of Genetic Psychology*, 1975, *126*, 195–207.

______, AND LERNER, J. V. Effects of age, sex, and physical attractiveness on child-peer relations, academic performance, and elementary school adjustment. *Developmental Psychology*, 1977, *13*, 585 590.

LESSER, G. S., FIFER, G., AND CLARK, D. H. Mental abilities of children from different social-class and cultural groups. *Monographs of the Society for Research in Child Development*, 1965, *30* (4, Serial No. 102).

LESGOLD, A. M. *Expert systems*. Paper presented at the Cognitive Science 5 Meetings, Rochester, N.Y. May, 1983.

LESSERM, R., AND PAISNER, M. Magical thinking in formal operational adults. *Human Development*, 1985, *28*, 57–70.

LESTER, B. M., ALS, H., AND BRAZELTON, T. B. Regional obstetric anesthesia and newborn behavior: A reanalysis toward synergistic effects. *Child Development*, 1982, *53*, 687–692.

LEVER, J. Sex differences in games children play *Social Problems*, 1976, *23*, 478–487.

LEVIN, J. A., BORUTA, M. J., AND VASCONCELLOS, M. T. Microcomputer-based environments for writing. In A. C. Wilkinson (Ed.), *Classroom computers and cognitive science*. New York: Academic Press, 1983.

LEVIN, J. R. The mnemonic '80s: *Keywords in the classroom*. Theoretical paper No. 86, Wisconsin Research and Development Center for Individualized Schooling, Madison, 1980.

LEVINE, M. H., AND SUTTON-SMITH, B. Effects of age, sex and task on visual behaviour during dyadic interaction. *Developmental Psychology*, 1973, *9*, 400–405.

LEVINE, S. C. Hemispheric specialization and functional plasticity during development. *Journal of Children in Contemporary Society*, 1983, *16*, 77–98.

LEVINSON, D. J., DARROW, C. N. KLEIN, E. B., LEVINSON, M. H., AND MCKEE, B. *The seasons of a man's life*, New York: Knopf, 1978.

LEVITAN, M., AND MONTAGU, A. *Textbook of human genetics*. London: Oxford University Press, 1971.

LEVY, J. *Nature*, 1969, *224*, 614.

LEVY, R. *Tahitians*. Chicago: University of Chicago Press, 1973.

LEVY, R. I., AND MOSKOWITZ, J. Cardiovascular research: Decades of progress, a decade of promise.'' *Science*, 1982, *217*, 121–129.

LEWIS, C. C. The effects of parental firm control: A reinterpretation of findings. *Psychological Bulletin*, 1981, *90*, 547–563.

LEWIS, M. Individual differences in the measurement of early cognitive growth. In J. Hellmuth (Ed.), *Exceptional infant*, Volume 2. *Studies in abnormalities*. New York: Brunner/Mazel, 1971.

______, AND BROOKS, J. Self, other and fear: Infants' reactions to people. In M. Lewis and L. A. Rosenblum (Eds.), *The origins of fear*: New York: Wiley, 1974.

______, FEIRING, C., MCGUFFOG, C., AND JASKIR, J. Predicting psychopathology in six-year-olds from early social relations. *Child Development*, 1984, *55*, 123–136.

______, AND FREEDLE, R. Mother-infant dyad: The cradle of meaning. In P. Pliner, L. Krames, and T. Alloway (Eds.), *Communication and affect: Language and thought*. New York: Academic Press, 1973.

______, AND JASKIR, J. Infant intelligence and its relation to birth order and birth spacing. *Infant behavior and development*, 1983, *6*, 117–120.

______, AND MICHALSON, L. The socialization of emotions. In T. Field and A. Fogel (Eds.), *Emotion and early interaction*. Hillsdale, N.J.: Lawrence Erlbaum, 1982.

______, AND WEINRAUB, M. Sex of parent x sex of child: Socioemotional development. In R. C. Friedman, R. M. Richart, and R. L. Van de Wiele (Eds.), *Sex differences in behavior*. New York: Wiley, 1974.

______YOUNG, G., BROOKS, J., AND MICHALSON, L. The beginning of friendship. In M. Lewis and L. A. Rosenblum (Eds.), *Friendship and peer relations*. New York: Wiley-Interscience, 1975.

LEWIS, R. A. Parents and peers: Socialization agents in the coital behavior of young adults. *The Journal of Sex Research*, 1973, *9*, 156–170.

LEY, R. G., AND KOEPKE, J. E. *Sex and age differences in the departures of young children from their mothers*. Paper presented at the meeting of the Society for Research in Child Development, Denver, April 1975.

______, AND KOEPKE, J. E. Attachment behavior outdoors: Naturalistic observations of sex and age differences in the separation behavior of young children. *Infant Behavior and Development*, 1982, *5*, 195–201.

LIBEN, L. S. Perspective-taking skills in young children: Seeing the world through rose-colored glasses. *Developmental Psychology*, 1978, *14*, 87–92.

LIBERMAN, I. Y., SHANKWEILER, D., FISCHER, F. W., AND CARTER, B. Explicit syllable and phoneme segmentation in the young child. *Journal of Experimental Child Psychology*, 1974, *18*, 201–212.

LICHT, B. G., AND DWECK, C. S. Determinants of academic achievement: The interaction of children's achievement orientations with skill area. *Developmental Psychology*, 1984, *20*, 628–636.

LIDZ, T. The adolescent and his family. In H. Caplan and F. Lebovici (Eds.), *Adolescence: Psychosocial perspectives*. New York: Basic Books, 1969.

LIEBERMAN, M. A., AND TOBIN, S. *The experience of old age*. New York: Basic Books, 1983.

LIEBERMAN, P. *Intonation, perception, and language*. Cambridge, Mass.: MIT Press, 1967.

LIEBERT, R. M., AND POULOS, R. W. Television and personality development: The socializing effects of an entertainment medium. In A. Davids (Ed.), *Child personality and psychopathology: Current topics*, Volume 2. New York: Wiley, 1975.

LIGHT, P., AND NIX, C. "Own view" versus "good view" in a perspective-taking task. *Child Development*, 1983, *54*, 480–483.

LINDGREN, G. Height, weight and menarche in Swedish urban school children in relation to socioeconomic and regional factors. *Annals of Human Biology*, 1976, *3*, 510–528.

LINDEMANN, E. Symptomatology and management of acute grief. *American Journal of Psychiatry*, 1944, *101*, 141–148.

LIPSITT, L. P., ENGEN, T., AND KAYE, H. Developmental changes in the olfactory threshold of the neonate. *Child Development*, 1963, *34*, 371–376.

———, REILLY, B. M., BUTCHER, M. J., AND GREENWOOD, M. M. The stability and interrelationships of newborn sucking and heart rate. *Developmental Psychobiology*, 1976, *9*, 305–310.

LIST, J. A., COLLINS, W. A., AND WESTBY, S. D. Comprehension and inferences from traditional and nontraditional sex-role portrayals on television. *Child Development*, 1983, *54*, 1579–1587.

LITOVSKY, V. G., AND DUSEK, J. B. Perceptions of child rearing and self-concept development during the early adolescent years. *Journal of Youth and Adolescence*, 1985, *14*, 373–387.

LITTLE, L., AND HAAR, C. Terms of enlargement. *Family Weekly*, March 24, 1985, p. 15.

LITWAK, E. Reference group theory, bureaucratic career and neighborhood primary group cohesion. *Sociometry*, 1960, *23*, 72–84.

LIVESLEY, W. J., AND BROMLEY, D. B. *Person perception in childhood and adolescence*. London: Wiley, 1973.

LIVSON, F. B. Paths to psychological health in the middle years: Sex differences. In D. H. Eichorn, J. A. Clausen, N. Haan, M. P. Honzik, and P. H.Mussen (Eds.), *Present and past in middle life*. New York: Academic Press, 1981, pp. 195–222.

LOCKYER, L., AND RUTTER, M. A five to fifteen year follow-up study of infantile psychosis: IV. Patterns of cognitive ability. *British Journal of Social and Clinical Psychology*, 1970, *9*, 152–163.

LONDERVILLE, S., AND MAIN, M. Security of attachment, compliance and maternal training methods in the second year of life. *Developmental Psychology*, 1981, *17*, 289–299.

LONEY, J., LANGHORNE, J. E., AND PATERNITE, C. E. An empirical basis for subgrouping the hyperkinetic/minimal brain dysfunction syndrome. *Journal of Abnormal Psychology*, 1978, *87*, 431–441.

LONGO, V. G. *Pharmacological Review*, 18 (1966), 965.

LOPATA, H. Z. Widowhood: Societal factors in life-span disruption and alternatives. In N. Datan and L. H. Ginsberg (Eds.), *Life-span developmental psychology: Normative life crises*. New York: Academic Press, 1975, pp. 217–234.

———. The widowed family member. In N. Datan and N. Lohmann (Eds.), *Transitions of aging*. New York: Academic Press, 1980, pp. 93–118.

LOPICCOLO, J. Mothers and daughters: Perceived and real differences in sexual values. *The Journal of Sex Research*, 1973, *9*, 171–177.

LORENZ, K. *Studies in animal and human behavior*, Volume 1. Cambridge, Mass.: Harvard University Press, 1971.

LOUGEE, M. D., GRUENEICH, R., AND HARTUP, W. W. Social interaction in same- and mixed-age dyads of preschool children. *Child Development*, 1977, *48*, 1353–1361.

LOVENTHAL, M. F. AND CHIRIBOGA D. Transition to the empty nest: Crisis, challenge, or relief? *Archives of General Psychiatry*, *26*, 1972, 8–14.

———, THURNHER, M. AND CHIRIBOGA, D. *Four stages of life*. San Francisco: Jossey-Bass, 1975.

LOWREY, G. H. *Growth and development of children* (6th ed; 7th ed.). Chicago: Yearbook Medical Publishers, 1973, 1978.

LUCARIELLO, J., AND NELSON, K. Slot-filler categories as memory organizers for young children. *Developmental Psychology*, 1985, *21*, 272–282.

LUCAS, T. C., AND UZGIRIS, I. C. Spatial factors in the development of the object concept. *Developmental Psychology*, 1977, *13*, 492–500.

LUEPTOW, L. B. *Adolescent sex role and social change*. New York: Columbia University Press, 1984.

LUKASEVITCH, A., AND GRAY, R. F. Open space, open education and pupil performance. *Elementary School Journal*, 1978, *79*, 108–114.

LURIA, A. R. *The role of speech in the regulation of normal and abnormal behavior*. London: Pergamon, 1961.

———. Speech development and the formation of mental processes. In M. Cole and I. Maltzman (Eds.), *Handbook of contemporary Soviet psychology*. New York: Basic Books, 1969.

LURIA, Z., FRIEDMAN, S., AND ROSE, M. *Human sexuality*. New York: Wiley, 1987.

LUTKENHAUS, P. *Mother-infant attachment at 12 months and its relations to 3 year old's readiness to build up a new relationship*. Paper presented at the International Conference on Infant Studies, New York, April 1984.

LYND, R. S., AND LYND, H. M. *Middletown*, New York: Harcourt, Brace, & Co., 1929.

LYTTON, H. The socialization of two-year-old boys: Ecological findings. *Journal of Child Psychology and Psychiatry*, 1976, *17*, 287–304.

MACCOBY, E. E. Selective auditory attention in children. In L. P. Lipsitt and C. C. Spiker (Eds.), *Advances in child development and behavior*, Volume 3. New York: Academic Press, 1967.

———. Commentary and reply. In G. R. Patterson, Mothers: The unacknowledged victims, *Monographs of the Society for Research in Child Development*, 1980, *45*, 5, Whole No. 186, 56–63.

———. Social groupings in childhood: Their relationship to prosocial and antisocial behavior in boys and girls. In D. Olweus, J. Block, and M. Radke-Yarrow (Eds.), *Development of antisocial and prosocial behavior*. Orlando, Fla.: Academic Press, 1986.

———, DOERING, C. H., JACKLIN, C. N., AND KRAMER, H. Concentrations of sex hormones in umbilical-cord blood: Their relation to sex and birth order of infants. *Child Development*, 1979, *50*, 632–642.

———, AND FELDMAN, S. S. Mother-attachment and stranger-reactions in the third year of life. *Monographs of the Society for Research in Child Development*, 1972, *37* (1, Serial No. 146).

———, AND JACKLIN, C. N. *The psychology of sex of differences*. Stanford, Calif.: Stanford University Press, 1974.

———, AND ———. Sex differences in aggression: A rejoinder and reprise. *Child Development*, 1980, *51*, 964–980.

———, AND MARTIN, J. A. Socialization in the context of the family: Parent-child interaction. In P. H. Mussen (Ed.), *Handbook of child psychology*, Volume 4. New York: Wiley, 1983.

———, SNOW, M. E., AND JACKLIN, C. N. Children's dispositions and mother-child interaction at 12 and 18 months: A short-term longitudinal study. *Developmental Psychology*, 1984, *20*, 459–472. (a)

———, ———, AND ———. Continuities and discontinuities in early mother-child interaction: A longitudinal study at 12 and 18 months. In M. E. Lamb and A. L. Brown (Eds.), *Advances in developmental psychology*, Volume 3. Hillsdale, N.J.: Lawrence Erlbaum, 1984.

MACFARLANE, A. *The psychology of childbirth*. Cambridge, Mass.: Harvard University Press, 1977.

MACFARLANE, J. W., ALLEN, L., AND HONZIK, M. P. *A developmental study of the behavior problems of normal children between twenty-one months and fourteen years*. Berkeley: University of California Press, 1954.

MACKENZIE, B. Explaining race differences in IQ: The logics, the methodology, and the evidence. *American Psychologist*, 1984, *39*, 1214–1233.

MACKINODAN, T. Immunity and aging. In C. E. Finch and L. Hayflick (Eds.), *Handbook of the biology of aging*. New York: Van Nostrand Reinhold, 1977, pp. 379–408.

MADDEN, D. J. Aging and distraction by highly familiar stimuli during visual search. *Developmental Psychology*, 1983, *19*, 499–507.

MADDEN, J., O'HARA, J., AND LEVENSTEIN, P. 1984, Home again: Effects of the mother-child home program on mother and child. *Child Development*, 1984, *55*, 636–647.

MADDISON, D. C., AND WALKER, W. L. Factors affecting the outcome of conjugal bereavement. *British Journal of Psychiatry*, 1967, 113, 1057.

MADDOX, G. L. Persistence of life style among the elderly. In E. Palmore (Ed.), *Normal aging*. Durham, N.C.: Duke University Press, 1970, pp. 329–331.

______, AND DOUGLAS, E. B. Self-assessment of health. In E. Palmore (Ed.), *Normal aging II*. Durham, N.C.: Duke University Press, 1974, pp. 49–54.

MADSEN, M. C. Developmental and cross-cultural differences in the cooperative and competitive behavior of young children. *Journal of Cross-Cultural Psychology*, 1971, 2, 365–371.

MAGENIS, R. E., OVERTON, K. M., CHAMBERLIN, J., BRADY, T., AND LOVRIEN, E. Parental origin of the extra chromosome in Down's syndrome. *Human Genetics*, 1977, *37*, 7–16.

MAGRAB, P. R., AND CALCAGNO, P. L. Psychological impact of chronic pediatric conditions. In P. R. Magrab (Ed.)., *Psychological management of pediatric problems: Early Life conditions and chronic diseases*, Volume 1. Baltimore, Md.: University Park Press, 1978.

MAHLER, M. S., PINE, F., AND BERGMAN, A. *The psychological birth of the human infant*, New York: Basic Books, 1975.

MAIN, M., AND GEORGE, C. Responses of abused and disadvantaged toddlers to distress in agemates: A study in the day care setting. *Developmental Psychology*, 1985, *21*, 407–412.

______, AND GOLDWYN, R. Predicting rejection of her infant from mother's representation of her own experience: Implications for the abused-abusing intergenerational cycle. *Child Abuse & Neglect, The International Journal*, 1984, *8*, 203–217.

______, AND WESTON, D. R. The quality of the toddler's relationship to mother and to father: Related to conflict behavior and the readiness to establish new relationships. *Child Development*, 1981, *52*, 932–940.

MALATESTA, C. Z., AND HAVILAND, J. M. Learning display rules: The socialization of emotion expression in infancy. *Child Development*, 1982, *53*, 991–1003.

MALL, J. A study of U.S. teen pregnancy rate. *Los Angeles Times*, March 17, 1985, Part 7, p.27.

MALLICK, S. K., AND MCCANDLESS, B. R. A study of catharsis of aggression. *Journal of Personality and Social Psychology*, 1966, *4*, 590–596.

MANDLER, J. M. Representation and recall in infancy. In M. Moscovitch (Ed.), *Infant memory*. New York: Plenum, 1981.

MANOSEVITZ, M., PRENTICE, N. M., AND WILSON, F. Individual and family correlates of imaginary companions in preschool children. *Developmental Psychology*, 1973, *8*, 72–79.

MARANTZ, S. A., AND MANSFIELD, A. F. Maternal employment and the development of sex-role stereotyping in five- to eleven-year-old girls. *Child Development*, 1977, *48*, 668–673.

MARATSOS, M. Some current issues in the study of the acquisition of grammar. In P. H. Mussen (Ed.), *Handbook of child psychology*, Volume 3. New York: Wiley, 1983.

MARCIA, J. E. *Determination and construct validity of ego identity status*. Unpublished doctoral dissertation, Ohio State University, 1964.

______. Development and validation of ego identity status. *Journal of Personality and Social Psychology*, 1966, *3*, 551–558.

______. Identity in adolescence. In J. Adelson (Ed.), *Handbook of adolescent psychology*. New York: Wiley, 1980.

MARINI, M., AND GREENBERGER, E. Sex differences in occupational aspirations and expectations. *Sociology of Work and Occupation*, 1978, *5*, 147–178.

MARK, R. Memory and nerve cell connections. Oxford: Clarendon Press, 1974.

MARIS, R. W. *Pathways to suicide*. Baltimore: Johns Hopkins University Press, 1981.

MARKIDES, K. S. Aging, religiosity, and adjustment: A longitudinal analysis. *Journal of Gerontology*, 1983, *38*, 621–625.

MARKMAN, E. M. Realizing that you don't understand: Elementary school children's awareness of inconsistencies. *Child Development*, 1979, *50*, 643–655.

______. Comprehension monitoring. In W. P. Dickson (Ed.), *Children's oral communication skills*. New York: Academic Press, 1981.

MARKSON, E. W., AND HAND, J. Referral for death: Low status of the aged and referral for psychiatric hospitalization. *International Journal of Aging and Human Development*, 1970, *15*, 261–272.

MARSH, G. R., AND THOMPSON, L. W. Psychophysiology of aging. In J. E. Birren and K. W. Schaie (Eds.), *Handbook of mental health and aging*. New York: Van Nostrand Reinhold, 1977, pp. 219–248.

MARSHALL, H., AND ODEN, M. H. The status of the mature gifted individual as a basis for evaluation of the aging process. *Gerontologist*, 1962, 2, 301–306.

MARSHALL, J. C. On the biology of language acquisition. In D. Caplan (Ed.), *Biological studies of mental processes*. Cambridge, Mass.: MIT Press, 1980.

MARSHALL, V. W. Age and awareness of finitude in developmental gerontology. *Omega*, 1975, *6*, 113–129.

______. *Last chapters: A sociology of aging and dying*. Monterey, Calif.: Brooks/Cole, 1980.

MARTIN, J. A. A longitudinal study of the consequences of early mother–infant interaction: A microanalytic approach. *Monographs of the Society for Research in Child Development*, 1981, *46* (3, Serial No. 190).

MARTIN, R. M. Effects of familiar and complex stimuli on infant attention. *Developmental Psychology*, 1975, *11*, 178–185.

MARTIN, W. E. Singularity and stability of profiles of social behavior. In C. B. Stendler (Ed.), *Readings in child behavior and development*. New York: Harcourt, Brace & World, 1964.

MARTINEZ, C., AND CHAVEZ, A. Nutrition and development in infants of poor rural areas: I. Consumption of mother's milk by infants. *Nutrition Reports International*, 1971, *4*, 139–149.

MARVIN, R. S. An ethological-cognitive model of the attenuation of mother–child attachment behavior. In T. Alloway, P. Pliner, and L. Krames (Eds.), *Attachment behavior: Advances in the study of communication and affect*, Volume 3. New York: Plenum, 1977.

______, MOSSLER, D. G., AND GREENBERG, M. The development of conceptual perspective-taking in preschool children in a "secret game." Paper presented at the biennial meeting of the Society for Research in Child Development, Denver, March 1975.

MARX, J. L. Hormones and their effects in the aging body. *Science*, 1979, *206*, 805–806.

MASLIN, C. A., AND BATES, J. E. *Precursors of anxious and secure attachments: A multivariate model at age 6 months*. Paper presented at the biennial meeting of the Society for Research in Child Development, Detroit, 1983.

______, BRETHERTON, I., AND MORGAN, G. The influence of attachment security and maternal scaffolding on toddler mastery motivation. *Infant Behavior and Development*, 1986, *9* (Special Issue: Abstracts of papers presented at the Fifth International Conference on Infant Studies), 244.

MASUR, E. F., MCINTYRE, C. W., AND FLAVELL, J. H. Developmental changes in apportionment of study time among items in a multitrial free-recall task. *Journal of Experimental Child Psychology*, 1973, *15*, 237–246.

MATARAZZO, J. D. *Wechsler's measurement and appraisal of adult intelligence* (5th ed.). Baltimore, Md.: Williams & Wilkins, 1972.

MATAS, L., AREND, R. A., AND SROUFE, L. A. Continuity of adaptation in the second year: The relationship between quality of attachment and later competence. *Child Development*, 1978, *49*, 547–556.

MATHENY, A. P., JR., AND DOLAN, A. B. Persons, situations and time: A genetic view of behavioral change in children. *Journal of Personality and Social Psychology*, 1975, *32*, 1106–1110.

MATSUYAMA, S. S., AND JARVIK, L. F. Genetics and mental functioning in senescence. In J. E. Birren and R. B. Sloane (Eds.), *Handbook of mental health and aging*. Englewood Cliffs, N.J.: Prentice-Hall, 1980, pp. 134–148.

MATTHEWS, S. H., AND SPREY, J. The impact of divorce on grandparenthood: An exploratory study. *Gerontologist*, *24*, 1984, 41–47.

MATTSON, A. Long-term physical illness in childhood: A challenge to psychosocial adaptation. *Pediatrics*, 1972, *50*, 801–811.

MAURER, D., AND BARRERA, M. Infants' perception of natural and distorted arrangements of a schematic face. *Child Development*, 1981, *52*, 196–202.

______, AND YOUNG, R E. Newborn's following of natural and distorted arrangements of a facial features. *Infant Behavior and Development*, 1983, *6*, 127–131.

MAZESS, R. B. AND FORMAN, S.H. Longevity and age exaggeration in Vilacabamba, Ecuador. *Journal of Gerontology*, 1979, *34*, 94–98.

MCANARNEY, E., AND GREYDANUS, D. Adolescent pregnancy—a multifaceted problem. *Pediatrics in Review*, 1979, *1*, 123–126.

MCCABE, M. P. Toward a theory of adolescent dating. *Adolescence*, 1984, *19*, 150–170.

______, AND COLLINS, J. K. Dating desires and experiences: A new approach to an old question. *Australian Journal of Sex, Marriage, and the Family*, 1981, *2*, 165–173.

MCCABE, V. Abstract perceptual information for age level: A risk factor for maltreatment? *Child Development*, 1984, *55*, 267–276.

MCCALL, R. B. Challenges to a science of developmental psychology. *Child Development*, 1977, *48*, 333–344.

______. Environmental effects on intelligence: The forgotten realm of discontinuous nonshared within-family factors. *Child Development*, 1983, *54*, 408–415.

______. Underachiever—wasted talent. *The Brown University Newsletter*. February 1986, *2*, 1-3.

______. HOGARTY, P.S., AND HURLBURT, N. Transitions in infant sensorimotor development and the prediction of childhood IQ. *American Psychologist*, 1972, *27*, 728–741.

MCCARTHY, J. D., AND HOGE, D.R. Analysis of age effects in longitudinal studies of adolescent self-esteem. *Developmental Psychology*, 1982, *18*, 372–379.

______, AND HOGE, D.R. The analysis of self-esteem and delinquency. *American Journal of Sociology*, 1985, *90*, 396–410.

MCCLEARN, G. E. Genetic influences on behavior and development. In P.H. Mussen (Ed.), *Carmichael's manual of child psychology*, Volume 1. New York: Wiley, 1970.

MCCLELLAND, D. C. Testing for competence rather than for intelligence. *American Psychologist*, 1973, *28*, 1–14.

______, ATKINSON, J. R., CLARK, R. A., AND LOWELL, E. O. *The achievement motive*. New York: Appleton-Century-Crofts, 1953.

MCCLOSKY, M., CARAMAZZA, A., AND GREEN, B. Curvilinear motion in the absence of external forces: Naive beliefs about the motion of objects. *Science*, 1980, *210*, 1139–1141.

MCCONNEL, C. E., AND DELJAVAN, F. Consumption patterns of the retired household. *Journal of Gerontology*, 1983, 38, 480–490.

MCCONNELL, S. R. Retirement and employment. In D. S. Woodruff and J. E. Birren (Eds.), *Aging: Scientific perspectives and social issues* (2nd ed.) Monterey, Calif.: Brooks/Cole, 1983, pp. 333–350.

MCCOY, C. L., AND MASTERS, J. C. The development of children's strategies for the social control of emotion. *Child Development*, 1985, *56*, 1214–1222.

MCCRAE, R. R., BARTONE, P. T., AND COSTA, P. T., JR. Age, personality, and self-reported health. *International Journal of Aging and Human Development*. 1976, *6*, 49–58.

______, AND COSTA, P. T., JR. Aging, the life course, and models of personality. In T. M. Field, A. Huston, H. C. Quay, L. Troll, and G. E. Finley (Eds.), *Review of human development*. New York: Wiley-Interscience, 1982, pp. 602–613.

MCDEVITT, S. C. *A longitudinal assessment of longitudinal stability in temperamental characteristics from infancy to early childhood*. Unpublished doctoral dissertation, Temple University, 1976.

MCGHEE, P. E., AND GRANDALL, V. C. Beliefs in internal-external control of reinforcement and academic performance. *Child Development*, 1968, *39*, 91–102.

MCGRAW, M. B. *Growth: A study of Johnny and Jimmy*. New York: Appleton-Century-Crofts, 1935.

______. Swimming behavior in the human infant. In Y. Brackbill and G. G. Thompson (Eds.), *Behavior in infancy and early childhood*, St. Louis: C. V. Mosby, 1939/1967.

MCGREW, W. C. Aspects of social development in nursery school children, with emphasis on introduction to the group. In N. Blurton Jones (Ed.), *Ethological studies of child behavior*. London: Cambridge University Press, 1972.

MCINTIRE, M. L. The acquisition of American sign language hand configurations. *Sign Language Studies*, 1977, *16*, 247–266.

MCKAIN, W. C., JR. *Retirement marriage*. Storrs, Conn.: University of Connecticut Agriculture Experiment Station, 1969.

MCKEOWON T., AND BROWN R. G. Medical evidence related to English population changes in the eighteenth century. *Population Studies*, 1955, *9*.

MCLOUGHLIN, D., AND WHITFIELD, R. Adolescents and their experience of parental divorce. *Journal of Adolescence*, 1984, *7*, 155–170.

MCNEILL, D. Developmental psycholinguistics. In F. Smith and G. A. Miller (Eds.), *The genesis of language: A psycholinguistic approach*. Cambridge: Mass. MIT Press, 1966.

MEACHAM, J. A. The development of memory abilities in the individual and in society. *Human Development*, 1972, *15*, 205–228.

MEAD, M. Coming of age in Samoa. In *From the South Seas: Studies of adolescence and sex in primitive societies*. New York: William Morrow, 1939. (Originally published 1928)

______. Adolescence in primitive and modern society. In V. F. Calverton and S. D. Schmalhausen (Eds.), *The new generation*. New York: Arno Press and The New York Times, 1971. (Originally published 1930)

MECHANIC, D. The influence of mothers on their children's health attitudes and behavior. *Pediatrics*, 1964, *33*, 444–453.

MECKLENBURG, M. E., AND THOMPSON, P. G. The adolescent family life program as a prevention measure. *Public Health Reports*, 1983, *98*, 21–29.

MEHLER, J., BERTONCINI, J., BARRIERE, M., AND JASSIK-GERSCHENFELD, D. Infant recognition of mother's voice. *Perception*, 1978, *7*, 491–497.

MELTON, G. B. Toward "personhood" for adolescents: Autonomy and privacy as values in public policy. *American Psychologist*, 1983, *38*, 99–103.

MELTZOFF, A. N., AND MOORE, M. K. Imitation of facial and manual gestures by human neonates. *Science*, 1977, *198*, 75–78.

______, AND MOORE, M. K. Newborn infants imitate adult facial gestures. *Child Development*, 1983, *54*, 702–709.

MENG, Z., HENDERSON, C., CAMPOS, J., AND EMDE, R. N. *The effects of background emotional elicitation on subsequent problem solving in the toddler*. Unpublished manuscript, University of Denver, 1983.

MERRICK, F. Personal communication to Jean Sorrells-Jones, 1978.

MERVIS, C. B. On the existence of prelinguistic categories: A case study. *Infant Behavior and Development*, 1985, *8*, 293–300.

______, AND MERVIS, C. A. Leopards are kitty-cats: Object labeling by mothers for their thirteen-month-olds. *Child Development*, 1982, *53*, 267–273.

MERVIS, J. Adolescent behavior: What we think we know. *APA Monitor*, April 1984, *15*, 24–25.

MILBRATH, L. W., AND GOEL, M. L. *Political participation* (2nd ed.). Chicago: Rand McNally, 1977.

MILGRAM, R. M., MILGRAM, N. A., ROSENBLUM, G., AND RABKIN, L. Quantity and quality of creative thinking in children and adolescents. *Child Development*, 1978, *49*, 385–388.

MILLER, B. D., AND OLSEN, D. Typology of marital interaction and contextual characteristics: Cluster analysis of the I.M.C. Unpublished paper available from D. Olsen, Minnesota Family Study Center, University of Minnesota, 1978.

MILLER, L. C. Fears and anxiety in children. In C. E. Walker and M. C. Roberts (Eds.), *Handbook of clinical child psychology*. New York: Wiley, 1983.

MILLER, M. B. *The interdisciplinary role of the nursing home medical director*. Wakefield, Mass.: Contemporary Publishing, 1976.

MILLER, P. Y., AND SIMON, W. Adolescent sexual behavior: Context and change. *Social Problems*. 1974, *22*, 58–76.

______, AND ______. The development of sexuality in adolescence. In J. Adelson (Ed.), *Handbook of adolescent psychology*. New York: Wiley, 1980.

MILLER, S. Child rearing in the kibbutz. In J. G. Howells (Ed.), *Modern perspectives in international child psychiatry*. Edinburgh: Oliver and Boyd, 1969.

MILLS, M., AND FUNNELL, E. Experience and cognitive processing. In S. Meadows (Ed.), *Developing thinking: Approaches to children's cognitive development*. New York: Methuen, 1983.

MILNE, A. A. *Now we are six*. New York: E. P. Dutton, 1927.

MILNE, C., SEEFELDT, V., AND REUSCHLEIN, P. Relationship between grade, sex, race, and motor performance in young children. *Research Quarterly*, 1976, *47*, 726–730.

MILSTEIN, R. M. *Visual and taste responsiveness in obese-tending infants*. Unpublished paper, Yale University, 1978.

MINNETT, A. M., VANDELL, D. L., AND SANTROCK, J. W. The effects of sibling status on sibling interaction: Influence of birth order, age spacing, sex of child, and sex of sibling. *Child Development*, 1983, *54*, 1064–1072.

MINTON, J. H. The impact of *Sesame Street* on readiness. *Sociology of Education*, 1972, *48*, 141–151.

MINUCHIN, P. Sex-role concepts and sex typing in childhood as a function of school and home environments. *Child Development*, 1965, *36*, 1033–1048.

MOBERG, D. O. Religiosity in old age. In B. L. Neugarten (Ed.), *Middle age and aging*. Chicago: University of Chicago Press, 1968, *Modern maturity*. They Show How to Live on Less, 26 (June–July 1983), 94.

MOERK, E. L. *The mother of Eve: As a first language teacher*. Norwood, N.J.: Ablex, 1983.

MOLFESE, D. Neural mechanisms underlying the process of speech information in infants and adults: Suggestions of differences in development and structure from electrophysiological research. In V. Kirk (Ed.), *Neuropsychology of language, reading, and spelling*. New York: Academic Press, 1983.

MOLFESE, D. L., AND MOLFESE, V. J. Electrophysiological indices of auditory discrimination in newborn infants: The bases for predicting later language development? *Infant Behavior and Development*, 1985, *8*, 197–211.

MONEY, J., AND EHRHARDT, A. A. *Man and woman. Boy and girl*. Baltimore, Md.: Johns Hopkins University Press, 1972.

______, HAMPSON, J. G., AND HAMPSON, J. L. Imprinting and the establishment of gender role. *AMA Archives of Neurological Psychiatry*, 1957, *77*, 333–336.

MONGE, R., AND HULTSCH, D. Paired associate learning as a function of adult age and the length of anticipation and inspection intervals. *Journal of Gerontology*, 1971, *26*, 157–162.

MONTAGU, A. *Prenatal influences*. Springfield, Ill.: Charles C Thomas, 1962.

MONTEMAYOR, R. The relationship between parent–adolescent conflict and the amount of time adolescents spend alone and with parents and peers. *Child Development*, 1982, *53*, 1512–1519.

______. Parents and adolescents in conflict: All families some of the time and some families most of the time. *Journal of Early Adolescence*, 1983, *3*, 83–103.

______, AND EISEN, M. The development of self-conceptions from childhood to adolescence. *Developmental Psychology*, 1977, *13*, 314–319.

MONTHLY VITAL STATISTICS REPORT. *Final mortality statistics (1977), Nos. 79–1120*. Hyattsville, Md.: National Center for Health Statistics. Health Education, and Welfare, 1979.

MOORE, B. S., UNDERWOOD, B., AND ROSENHAN, D. L. Affect and altruism. *Developmental Psychology*, 1973, *8*, 99–104.

MOORE, W. M. The secular trend in physical growth of urban North American Negro school children. *Monographs of the Society for Research in Child Development*, 1970, *35* (7, Serial No. 140).

MORA, J. Q., AMEZQUITA, A., CASTRO, L., CHRISTIANSEN, N., CLEMENT-MURPHY, J., COBOS, L. F., CREMER, H. D., DRAGASTIN, S., ELIAS, M. F., FRANKLIN, D., HERRERA, M. G., ORTIZ, N., PARDO, F., DE TIANSEN, B., WAGNER, M., AND STARE, F. J. Nutrition, health and social factors related to intellectual performance. *World Review of Nutrition and Dietetics*, 1974, *19*, 205–236.

MORRISON, F. J., HOLMES, D. L., AND HAITH, M. M. A developmental study of the effects of familiarity on short term visual memory. *Journal of Experimental Child Psychology*, 1974, *18*, 412–425.

______, AND LORD, C. Age differences in recall of categorized material: Organization or retrieval? *Journal of Genetic Psychology*, 1982, *141*, 233–241.

MORRISON, H., AND KUHN, D. Cognitive aspects of preschoolers' peer imitation in a play situation. *Child Development*, 1983, *54*, 1054–1063.

MORRISON, J. R., AND STEWART, M. A. A family study of the hyperactive child syndrome. *Biological Psychiatry*, 1971, *3*, 189–195.

______, AND ______. The psychiatric status of legal families of adopted hyperactive children. *Archives of General Psychiatry*, 1973, *28*, 888–891.

MORTIMER, J. T., FINCH, M. D., AND KUMKA, D. Persistence and change in development: The multidimensional self-concept. In P. B. Baltes and O. G. Brim, Jr. (Eds.), *Life-span development and behavior*, Volume 4. New York: Academic Press, 1982.

MOSCOVITCH, M. *Infant memory*. New York: Plenum, 1981.

MOSS, H. A., AND SUSMAN, E. J. Longitudinal study of personality development. In O. G. Brim and J. Kagan (Eds.), *Constancy and change in human development*. Cambridge, Mass.: Harvard University Press, 1980.

MOSS, M. S., AND MOSS, S. Z. The impact of parental death on middle-aged children. Presented at the 38th Annual Meeting of the American Association of Marriage and Family Therapy, Toronto, Nov. 8, 1980.

MOSSLER, D. G., MARVIN, R. S., AND GREENBERG, M. Conceptual perspective taking in two- to six-year-old children. *Developmental Psychology*, 1976, *12*, 85–86.

MUCH, N. C., AND SHWEDER, R. A. Speaking of rules: The analysis of culture in the breach. In W. Damon (Ed.), *New directions for child development: Moral development*. San Francisco: Jossey-Bass, 1978.

MUELLER, E. The maintenance of verbal exchanges between young children. *Child Development*, 1972, *43*, 930–938.

______, BLEIER, M., KRAKOW, J., HAGEDUS, K., AND COURNOYER, P. The development of peer verbal interaction among two-year-old boys. *Child Development*, 1977, *48*, 284–287.

______, AND BRENNER, J. The origins of social skills and interaction among playgroup toddlers. *Child Development*, 1977, *48*, 854–861.

______, AND LUCAS, T. A developmental analysis of peer interaction among toddlers. In M. Lewis and L. A. Rosenblum (Eds.), *Friendship and peer relations*. New York: Wiley-Interscience, 1975.

______, AND VANDELL, D. Infant-infant interaction. In J. D. Osofsky (Ed.), *Handbook of infant development*. New York: Wiley, 1979.

MUIR, D., AND FIELD, J. Newborn infants' orientation to sound. *Child Development*, 1979, *50*, 431–436.

MUMFORD, M. D., AND OWENS, W. A. Individuality in a developmental context. *Human Development*, 1984, *27*, 84–108.

MUNROE, R. L., AND MUNROE, R. H. Male pregnancy symptoms and cross-sex identity in three societies. *Journal of Social Psychology*, 1971, *84*, 11–25.

MURPHY, L. B. *The widening world of childhood: Paths toward mastery.* New York: Basic Books, 1962.

MURRAY, D. M., JOHNSON, C. A., LUEPKER, R., AND MITTELMARK, M. The prevention of cigarette smoking in children: A comparison of four strategies. *Journal of Applied Social Psychology*, 1984, *14*, 274–288.

MURSTEIN, B. I. Mate selection in the 1970s. *Journal of Marriage and the Family*, 1980, *42*, 777–792.

MUSSEN, P. H., AND EISENBERG-BERG, N. *Roots of caring, sharing, and helping*. San Francisco: W. H. Freeman, 1977.

———, AND HAAN, N. A longitudinal study of patterns of personality and political ideologies. In D. H. Eichorn, J. A. Clausen, N. Haan, M. P. Honzik, and P. Mussen (Eds.), *Present and past in middle life*. New York: Academic Press, 1981.

———, AND JONES, M. C. Self conceptions, motivations, and interpersonal attitudes of late and early maturing boys. *Child Development*, 1957, *28*, 243–256.

MUSTE, M., AND SHARPE, D. Some influential factors in the determination of aggressive behavior in preschool children. *Child Development*, 1947, *18*, 11–28.

MYERS, N. A., AND PERLMUTTER, M. Memory in the years from two to five. In P. A. Ornstein (Ed.), *Memory development in children*. Hillsdale, N.J.: Lawrence Erlbaum, 1978.

MYLES-WORSLEY, M., CROMER, C. C., AND DODD, D. H. Children's preschool script reconstruction: Reliance on general knowledge as memory fades. *Developmental Psychology*, 1986, *22*, 22–30.

NADELMAN, L. Sex identity in American children: Memory, knowledge, and preference tests. *Developmental Psychology*, 1974, *10*, 413–417.

NAEYE, R. L. Relationship of cigarette smoking to congenital anomalies and perinatal death. *American Journal of Pathology*, 1978, *90*, 289–294.

———, AND PETERS, E. C. Working during pregnancy: Effects on the fetus. *Pediatrics*, 1982, *69*, 724–727.

NAGY, M. H. The child's theories concerning death. *Journal of Genetic Psychology*, 1948, *73*, 3–27.

NATIONAL CENTER FOR EDUCATIONAL STATISTICS (NCES). A 1985 study cited in C. Cordes. Test Tilt: Boys outscore girls on both parts of SAT. APA *Monitor*, June 1986, *17*, 30–31.

NATIONAL CENTER FOR HEALTH STATISTICS. *Anthropometric and clinical findings: Preliminary findings of the first Health and Nutrition Examiantion Survey. United States, 1971–1972*. Washington, D.C.: DHEW, 1975.

NATIONAL COMMISSION ON EXCELLENCE IN EDUCATION. *A nation at risk: The imperative for educational reform*. Washington, D.C.: U.S. Department of Education, 1983.

NEIMARK, E. D. Cognitive development in adulthood: Using what you've got. In T. M. Field, A. Huston, H. C. Quay, L. Troll, and G. E. Finley (Eds.), *Review of Human Development*. New York: Wiley-Interscience, 1982.

———. Longitudinal development of formal operational thought. *Genetic Psychology Monographs*, 1975, *91*, 171–225.

———. Adolescent thought: Transition to formal operations. In B. B. Wolman and G. Stricker (Eds.), *Handbook of developmental psychology*. Englewood Cliffs, N.J.: Prentice-Hall, 1982.

NEISSER, U. The control of information pickup in selective looking. In A. D. Pick (Ed.), *Perception and its development*. Hillsdale, N.J.: Lawrence Erlbaum, 1979.

NELSON, K. Structure and strategy in learning to talk. *Monographs of the Society for Research in Child Development*, 1973, *38* (1–2, Serial No. 149).

———. Concept, word, and sentence: Interrelations in acquisition and development. *Psychological Review*, 1974, *81*, 267–285.

———, AND GRUENDEL, J. At morning it's lunchtime: A scriptal view of children's dialogues. *Discourse Processes*, 1979, *2*, 73–94.

NELSON, K. E. Toward a rare event: Cognitive comparison theory and syntax acquisition. In P. Dale and D. Ingram (Eds.), *Children's language: An international perspective*. Baltimore, Md.: University Park Press, 1980.

———. Experimental gambits in the service of language acquisition theory. In S. Kuczaj (Ed.), *Language development: Syntax and semantics*. Hillsdale, N.J.: Lawrence Erlbaum, 1981.

———, DENNINGER, M. M., BONVILLIAN, J. D., KAPLAN, B. J., AND BAKER, N. Maternal input adjustments and non-adjustments as related to children's linguistic advances and to language acquisition theories. In A. D. Pellegrini and T. D. Yawkey (Eds.), *The development of oral and written languages: Readings in developmental and applied linguistics*. Norwood, N.J.: Ablex, 1983.

———, AND KOSSLYN, S. M. Recognition of previously labeled or unlabeled pictures by 5-year-olds and adults. *Journal of Experimental Child Psychology*, 1976, *21*, 40–45.

NESSELROADE, J. R., AND BALTES, P. B. Adolescent personality development and historical change: 1970–1972. *Monographs of the Society for Research in Child Development*, 1974, *39* (1, Serial No. 154).

NEUGARTEN, B. L. Adaptation and the life cycle. *Journal of Geriatric Psychiatry*, 1970, *4*, 71–87.

———, CROTTY, W. AND TOBIN, S. Personality types in an aged population. In B. L. Neugarten (Ed.), *Personality in middle and late life: Empirical studies*. New York: Atherton, 1964

———, AND WEINSTEIN, K. K. The changing American grandparent. *Journal of Marriage and the Family*, 1964, *26*, 199–204.

NEUMANN, C. G., AND ALPAUGH, M. Birthweight doubling time: A fresh look. *Courrier du Center Interantional de l'Enfance*, 1976, *26*, 507.

NEWCOMB, A. F., AND BRADY, J. E. Mutuality in boys' friendship relations. *Child Development*, 1982, *53*, 392–395.

NEWCOMBE, N., AND BANDURA, M. M. Effect of age at puberty on spatial ability in girls: A question of mechanism. *Developmental Psychology*, 1983, *19*, 215–224.

NEWPORT, E. L. Motherese: The speech of mothers to young children. In J. J. Castellan, D. B. Pisoni, and G. R. Potts (Eds.), *Cognitive theory*, Volume 3. Hillsdale, N.J.: Lawrence Erlbaum, 1976.

NEWSON, J. An intersubjective approach to the systematic description of mother-infant interaction. In H. R. Schaffer (Ed.), *Studies in mother-infant interaction*. London: Academic Press, 1977.

———, AND NEWSON, E. *Four years old in an urban community*. Harmondsworth, England: Pelican Books, 1968.

———, AND NEWSON, E. *Seven years in the home environment*. New York: Wiley, 1976.

NEWTON, N., AND MODAHL, C. Pregnancy: The closest human relationship. *Human Nature*, March 1978, *1*, 40–49.

NICHOLLS, J. G. Causal attributions and other achievement-related cognitions: Effects of task outcomes, attainment values, and sex. *Journal of Personality and Social Psychology*, 1975, *31*, 379–389.

NICHOLS, M., AND LEIBLUM, S. R. Lesbianism as personal identity and social role: Conceptual and clinical issues. Unpublished article. Rutgers University, 1983.

NILSEN, D. M. The youngest workers: 14- and 15-year-olds. *Journal of Early Adolescence*, 1984, *4*, 189–197.

NORTHERN, J., AND DOWNS, M. *Hearing in children*. Baltimore, Md.: Williams & Wilkins, 1974.

NOTTELMAN, E., AND SUSMAN, E. *Passage through puberty*. Paper presented at the annual meeting of the American Association for the Advancement of Science, Los Angeles, 1985.

NUCCI, L. P., AND TURIEL, E. Social interactions and the development of social concepts in pre-school children. *Child Development*, 1978, *49*, 400–408.

O'BRIEN, M., AND HUSTON, A. C. Development of sex-typed play behavior in toddlers. *Developmental Psychology*, 1985, *21*, 866–871.

O'BRYAN, K. G., AND BOERSMA, F. J. Eye movements, perceptual activity, and conservation development. *Journal of Experimental Child Psychology*, 1971, *12*, 157–169.

______, AND MACARTHUR, R. S. Reversibility, intelligence, and creativity in nine-year-old boys. *Child Development*, 1969, *40*, 33–45.

ODEN, S., AND ASHER, S. R. Coaching children in social skills for friendship making. *Child Development*, 1977, *48*, 495–506.

ODOM, R. D. A perceptual salience account of decalage relations and developmental change. In L. S. Siegel and C. J. Brainerd (Eds.), *Alternatives to Piaget: Critical essays on the theory*. New York: Academic Press, 1978.

______. Lane and Pearson's inattention to relevant information: A need for the theoretical specification of task information in developmental research. *Merrill-Palmer Quarterly*, 1982, *28*, 339–345.

OFFENBACH, S. I. A developmental study of hypothesis testing and cue selection strategies. *Developmental Psychology*, 1974, *10*, 484–490.

OFFER, D. *The psychological world of the teen-ager: A study of normal adolescent boys*. New York: Basic Books, 1969.

______. *The adolescent*. New York: Basic Books, 1981.

______, OSTROV, E., AND HOWARD, K. J. *The adolescent: A psychological self-portrait*. New York: Basic Books, 1981.

OHLSSON, M. Information processing related to physical fitness in elderly people. *Reports from the Institute of Applied Psychology*, 1976, *71*, 1–12.

O'KEEFE, E. S. C., AND HYDE, J. S. The development of occupational sex-role stereotypes: The effects of gender stability and age. *Sex Roles*, 1983, *9*, 481–492.

OKUDAIRA, N., FUKUDA, H., NISHIHARA, K., OHTANI, K., ENDO, S. AND TORII, S. Sleep apnea and nocturnal myoclonus in elderly persons in Vilcabamba, Ecuador. *Journal of Gerontology*, 1983, *38*, 436–438.

OLIVER, C. M., AND OLIVER, G. M. Gentle birth: Its safety and its effect on neonatal behavior. *Journal of Obstetrical, Gynecological and Neonatal Nursing*, 1978.

OLSON, G. M., AND SHERMAN, T. Attention, learning, and memory in infants. In P. H. Mussen (Ed.), *Handbook of child psychology*. New York: Wiley, 1983.

______, AND STRAUS, M. S. The development of infant memory. In M. Moscovitch (Ed.), *Infant Memory: Its relation to normal and pathological memory in humans and other animals*. New York: Plenum, 1981.

OLTON, D. S., AND FEUSTLE, W. A. *Experimental Brain Research*, 1981, *41*, 380.

OLWEUS, D. The stability of aggressive reaction patterns in human males: A review. *Psychological Bulletin*, 1979, *85*, 852–875.

______. Familial and temperamental determinants of aggressive behavior in adolescent boys: A causal analysis. *Developmental Psychology*, 1980, *16*, 644–660.

O'MALLEY, P. M., AND BACHMAN, J. G. Self-esteem: Change and stability between ages 13 and 23. *Developmental Psychology*, 1983, *19*, 257–268.

OPPENHEIMER, V. K. The changing nature of life-cycle squeezes: Implications for the socioeconomic position of the elderly. In R. W. Fogel, E. Hatfield, S. B. Kiesler, and E. Shanas (Eds.), *Aging: Stability and change in the family*. New York: Academic Press, 1981.

OREN, D. L. Cognitive advantages of bilingual children related to labeling ability. *Journal of Educational Research*, 1981, *74*, 164–169.

ORNSTEIN, P. A., AND NAUS, M. J. Rehearsal processes in children's memory. In P. A. Ornstein (Ed.), *Memory development in children*. Hillsdale, N.J.: Lawrence Erlbaum, 1978.

______, ______, AND LIBERTY, C. Rehearsal and organizational processes in children's memory. *Child Development*, 1975, *26*, 818–830.

______, STONE, B. P., MEDLIN, R. G., AND NAUS, M. J. Retrieving for rehearsal: An analysis of active rehearsal in children's memory. *Developmental Psychology*, 1985, *21*, 633–641.

ORNSTEIN, R. E. The split and whole brain. *Human Nature*, May, 1978, *1*, 76–83.

OSHERSON, D. N., AND MARKMAN, E. Language and the ability to evaluate contradictions and tautologies. *Cognition*, 1975, *3*, 213–226.

OSTREA, E. M., JR., AND CHAVEZ, C. J. Perinatal problems (excluding neonatal withdrawal) in maternal drug addition: A study of 830 cases. *The Journal of Pediatrics*, 1979, *94*, 292–295.

O'TOOLE, P. Casualties in the classroom. *The New York Times Magazine*, December 10, 1978, pp. 59–90.

OVERCAST, T. D., MURPHY, M. D., SMILEY, S. S., AND BROWN, A. L. The effects of instruction on recall and recognition of categorized lists in the elderly. *Bulletin of the Psychonomic Society*, 1975, *5*, 339–341.

OVERTON, W. F., AND CLAYTON, V. The role of formal operational thought in the aging process. Unpublished manuscript. State University of New York, Buffalo, 1976.

OWEN, G., FULTON, R., AND MARKUSEN, E. Death at a distance: A study of family survivors. *Omega*, 1983, *13*, 191–226.

OWENS, W. A. Age and mental abilities: A second adult follow-up. *Journal of Educational Psychology*, 1966, *57*, 311–325.

PALARDY, J. N. What teachers believe, what children achieve. *Elementary School Journal*, 1969, *69* 370–374.

PALEY, V. *Boys and girls: Superheroes in the doll corner*. Chicago: University of Chicago Press, 1984.

PALKOVITZ, R. Parental attitudes and fathers' interactions with their 5-month-old infants. *Developmental Psychology*, 1984, *20*, 1054–1060.

PALMORE, E. Health practices and illnesses. In E. Palmore (Ed.), *Normal aging II*. Durham, N.C.: Duke University Press, 1974.

______. Total chance of institutionalization among the aged. *Gerontologist*, 1976, *6*, 504–507.

______, GEORGE, L. K., AND FILLENBAUM, G. G. Predictors of retirement. *Journal of Gerontology*, 1982, *37*, 733–742.

PAPALIA, D. The status of several conservation abilities across the life span. *Human Development*, 1972, *15*, 229–243.

PAPOUSEK, H., AND PAPOUSEK, M. *Biological aspects of parent-infant communication in man*. Invited address at the International Conference on Infant Studies, Providence, R.I., March 1978.

PAPPAS, C. C. The relationship between language development and brain development. *Journal of Children in Contemporary Society*, 1983, *16*, 133–169.

PARIS, S. G., AND LINDAUER, B. K. The role of inference in children's comprehension and memory for sentences. *Cognitive Psychology*, 1976, *8*, 217–227.

______, AND OKA, E. R. Children's reading strategies, metacognition, and motivation. *Developmental Review*, 1986, *6*, 25–56.

PARKE, R. D. Punishment in children: Effects, side effects, and alternative strategies. In H. Hom and P. Robinson (Eds.) *Psychological processes in early education*. New York: Academic Press, 1977.

______. Children's home environments: Social and cognitive effects. In I. Altman and J. F. Wohlwill (Eds.), *Children and the environment*, Volume 3, *Human behavior and environment*. New York: Plenum, 1978.

______, AND O'LEARY, S. Family interaction in the newborn period: Some findings, some observations, and some unresolved issues. In K. Riegel and J. Meacham (Eds.), *The developing individual in a changing world*, Volume 2, *Social and environmental issues*. The Hague: Mouton, 1976.

______, AND SAWIN, D. *Infant characteristics and behavior as elicitors of maternal and paternal responsibility in the newborn period*. Paper presented at the meetings of the Society for Research in Child Development, Denver, Colo., April 1975.

PARKER, E. B., AND PAISLEY, W. J. *Patterns of adult information seeking*. Final Report on USOE Project No. 2583. Stanford, Calif.: Stanford University Press, 1966.

PARKES, C. M. *Bereavement: Studies of grief in adult life*. New York: International Universities Press, 1972.

______, AND WEISS, R. S. *Recovery from bereavement*. New York: Basic Books, 1983.

PARLEE, M. B. The friendship bond. *Psychology Today*, October 1979, *13*, 43–54+.

PARPAL, M., AND MACCOBY, E. E. Maternal responsiveness and subsequent child compliance. *Child Development*, 1985, *56*, 1326–1334.

PARSONS, J. E., ADLER, T. F., AND KACZALA, C. M. Socialization of achievement attitudes and beliefs: Parental influences. *Child Development*, 1982, *53*, 310–321.

———, KACZALA, C. M., AND MEECE, J. K. Socialization of achievement attitudes and beliefs: Classroom influences. *Child Development*, 1982, *53*, 322–339.

PARSONS, T., AND LIDZ, V. M. Death in American society. In E. S. Shneidman (Ed.), *Essays in self-destruction*. New York: Science House, 1967.

PASTOR, D. L. The quality of mother-infant attachment and its relationship to toddlers' initial sociability with peers. *Developmental Psychology*, 1981, *17*, 326–335.

PATTERSON, C. J., AND MASSAD, C. M. Facilitating referential communication among children: The listener as teacher. *Journal of Experimental Child Psychology*, 1980, *29*, 357–370.

PATTERSON, G. R. Mothers: The unacknowledged victims. *Monographs of the Society for Research in Child Development*, 1980, *45* (5, Serial No. 186).

———. *Coercive family process*. Eugene, Ore.: Castalia Press, 1982.

———, AND STOUTHAMER-LOEBER, M. The correlation of family management practices and delinquency. *Child Development*, 1984, *55*, 1299–1307.

PEAR, R. Census bureau finds turnout in federal elections is rising. *The New York Times*, November 22, 1983, A23.

———. Rise in poverty from '79 to '82 is found in U.S. *The New York Times*, February 24, 1984, A1+.

PEDERSEN, F. A., CAIN, R., ZASLOW, M., AND ANDERSON, B. Variation in infant experience associated with alternative family role organization. In L. Laesa and I. Sigel (Eds.), *Families as learning environments for children*. New York: Plenum, 1983.

PEERY, J. C., AND STERN, D. Gaze duration frequency distributions during mother-infant interaction. *Journal of Genetic Psychology*, 1976, *129*, 45–55.

PENNINGTON, B. F., AND SMITH, S. D. Genetic influences on learning disabilities and speech and language disorders. *Child Development*, 1983, *54*, 369–387.

———, WALLACH, L., AND WALLACH, M. A. Nonconservers' use and understanding of number and arithmetic. *Genetic Psychology Monographs*, 1980, *101*, 231–243.

PEPLAU, L. A. What homosexuals want in relationships. *Psychology Today*, March 1981, *15*, 28–38.

PEPLER, D. *Naturalistic observations of teaching and modeling between siblings*. Paper presented at the biennial meeting of the Society for Research in Child Development, Boston, April 1981.

———, CORTER, C., AND ABRAMOVITCH, R. Social relations among children: Siblings and peers. In K. Rubin and H. Ross (Eds.), *Peer relationships and social skills in childhood*. New York: Springer-Verlag, 1982.

PERLMUTTER, M. What is memory aging the aging of? *Developmental Psychology*, 1978, *14*, 330–345.

———. Age differences in adults' free recall, cued recall, and recognition. *Journal of Gerontology*, 1979, *34*, 533–539.

———. Development of memory in the preschool years. In R. Greene and T. D. Yawkey (Eds.), *Childhood development*. Westport, Conn.: Technomic Publishing Co., 1980.

———. Learning and memory through adulthood. In M. W. Riley, B. B. Hess, and K. Bond (Eds.), *Aging in society: Selected reviews of recent research*. Hillsdale, N.J.: Lawrence Erlbaum, 1983.

———, AND HALL, E. *Adult development and aging*. New York: Wiley, 1985.

———, AND LANGE, G. A developmental analysis of recall-recognition distinctions. In P. A. Ornstein (Ed.), *Memory development in children*. Hillsdale, N.J.: Lawrence Erlbaum, 1978.

———, AND LIST, J. A. Learning in later adulthood. In T. M. Field, A. Huston, H. C. Quay, L. Troll, and G. E. Finley (Eds.), *Review of human development*. New York: Wiley-Interscience, 1982.

———, AND MITCHELL, D. B. The appearance and disappearance of age differences in adult memory. In F. I. M. Craik and S. Trehub (Eds.), *Aging and cognitive processes*. New York: Plenum, 1982.

———, AND MYERS, N. A. Development of recall in 2- to 4-year-old children. *Developmental Psychology*, 1979, *15*, 73–83.

PERONE, M., AND BARON. A. Age-related effects of pacing on acquisition and performance of response sequences: An operant analysis. *Journal of Gerontology*, 1982, *37*, 443–449.

PESKIN, H. Pubertal, onset and ego functioning. *Journal of Abnormal Psychology*. 1967, *72*, 1–15.

PETTERSEN, L., YONAS, A., AND FISCH, R. O. The development of blinking in response to impending collision in preterm, full term, and post term infants. *Infant Behavior and Development*, 1980, *3*, 155–165.

PETZEL, S. R., AND CLINE, D. W. Adolescent suicide: Epidemiological and biological aspects. *Adolescent Psychiatry*, 1978, *6*, 239–266.

PEZDEK, K., AND HARTMAN, E. F. Children's television viewing: Attention and comprehension of auditory versus visual information. *Child Development*, 1983, *54*, 1015–1023.

PFEIFFER, E., AND DAVIS, G. C. Determinants of sexual behavior in middle and old age. *Journal of the American Geriatrics Society*, 1972, *20*, 151–158.

———, VERWOERDT, H., AND DAVIS, G. C. Sexual behavior in middle life. *American Journal of Psychiatry*, 1972, *128*, 1281–1287.

PHILLIPS, D. The illusion of incompetence among academically competent children. *Child Development*, 1984, *55*, 2000–2016.

PIAGET, J. The moral judgment of the child. Glencoe, Ill.: Free Press, 1932.

———. *The origins of intelligence in children*. New York: International Universities Press, 1952.

———. *The language and thought of the child*. New York: Meridian Books, 1955. (Originally published 1926)

———. *Play, dreams and imitation*. New York: Norton, 1962. (Originally published 1951)

———. *The child's conception of number*. New York: Norton, 1965.

———. *The child's conception of time*. New York: Basic Books, 1969.

———. Piaget's theory. In P. H. Mussen (Ed.), *Carmichael's manual of child psychology*, Volume 1. New York: Wiley, 1970.

———. Intellectual evolution from adolescence to adulthood. *Human Development*, 1972, *15*, 1–12.

———. *The child's conception of the world*. Totowa, N.J.: Littlefield, Adams, 1975. (Originally published 1929)

———, AND INHELDER, B. *The psychology of the child*. New York: Basic Books, 1969.

PICK, A. D., CHRISTY, M. D., AND FRANKEL, G. W. A developmental study of visual selective attention. *Journal of Experimental Child Psychology*, 1972, *14*, 165–175.

———, AND FRANKEL, G. W. A developmental study of strategies of visual selectivity. *Child Development*, 1974, *45*, 1162–1165.

PIERCE, J. E. *A study of 750 Portland, Oregon children during the first year*. Papers and Reports on Child Language Development, 8. Stanford, Calif.: Stanford University Press, 1974.

PIERS, M. W. *Infanticide*. New York: Norton, 1978.

PINE, V. R. *Caretaker of the dead: The American funeral director*. New York: Irvington, 1975.

PINES, M. Can a rock walk? *Psychology Today*, November 1983, 46–54.

———. Winning ways. *Psychology Today*, December 1984, 57–65.

PIOTRKOWSKI, C., AND STARK, E. Blue collar stress worse for boys. *Psychology Today*, June 1985, 15.

PITCHER, E. G., AND SCHULTZ, L. H. *Boys and girls at play—the development of sex roles*. New York: Praeger, 1984.

PITT, B. "Maternity blues." *British Journal of Psychiatry*, 1973, *122*, 431–433.

PLECK, J., AND RUSTAD, M. *Husbands' and wives' time in family and paid work in the 1975–1976 study of time use*. Unpublished manuscript, Wellesley College Center for Research on Women, 1980.

PLESS, I. B., AND ROGHMANN, K. J. Chronic illness and its consequences: Observations based on three epidemiologic surveys. *Journal of Pediatrics*. 1971, *79*, 351–359.

______, and Satterwhite, B. B. Chronic illness. In R. J. Haggerty, K. J. Roghmann, and I. B. Pless (Eds.), *Child health and the community*. New York: Wiley, 1975.

Plomin, R. Developmental behavioral genetics. *Child Development*, 1983, *54*, 253–259.

______, Loehlin, J. C. and DeFries, J. C. Genetic and environmental components of "environmental" influences. *Developmental Psychology*, 1985, *21*, 391–402.

Polani, P. E., Lessof, M. H., and Bishop, P. M. F. Colour-blindness in "ovarian agenesis" (gonadal dysplasia). *Lancet*, 1956, *2*, 118–119.

Pollock, L. *Forgotten children: Parent-child relations from 1500 to 1900*. Cambridge, England: Cambridge University Press, 1983.

Powell, G. F., Brasel, J. A., Raiti, S., and Blizzard, R. M. Emotional deprivation and growth retardation stimulating idiopathic hypopituitarism: II. Endocrinologic evaluation of the syndrome. *New England Journal of Medicine*, 1967, *276*, 1279–1283.

Power, T. G., and Parke, R. D. Patterns of mother and father play with their 8-month-old infant: A multiple analyses approach. *Infant Behavior and Development*, 1983, *6*, 453–459.

Price, G. G., Hess, R. D., and Dickson, W. P. Processes by which verbal-educational abilities are affected when mothers encourage preschool children to verbalize. *Developmental Psychology*, 1981, *17*, 554–564.

Price, J., and Feshbach, S. *Emotional adjustment correlates of television viewing in children*. Paper presented at the American Psychological Association, Washington, D.C., August 1982.

Prentice, N., and Fathman, R. Joking riddles: A developmental index of children's humor. *Developmental Psychology*, 1975, *2*, 210–216.

Pressey, S. L, Janney, J. E., and Kuhlen, R. G. *Life: A psychological survey*. New York: Harper, 1939.

Puntil, J. *Juvenile delinquency in Illinois: Highlights of the 1972 adolescent survey*. Chicago: Institute for Juvenile Research, Illinois Mental Health Institute, 1972.

Purpura, D. P. Morphogenesis of the visual cortex in preterm infants. In M. A. B. Brazier (Ed.), *Growth and brain development*. New York: Raven Press, 1975.

Putallaz, M. Predicting children's sociometric status from their behavior. *Child Development*, 1983, *54*, 1417–1426.

Querleu, D., and Renard, K. Les perceptions auditives du foetus humain. *Medicine et Hygiene*, 1981, *39*, 2102–2110.

Quetelet, A. *A treatise on man and the development of his faculties*. Edinburgh: William & Robert Chambers, 1842.

Quigley, M. E., Sheehan, K. L., Wilkes, M. M., and Yen, S. S. C. Effects of maternal smoking on circulating catecholomine levels and fetal heart rates. *American Journal of Obstetrics and Gynecology*, 1979, *133*, 685–690.

Quinn, J. F. *The early retirement decision: Evidence from the 1969 Retirement History Study*: U.S. Department of Health, Education, and Welfare, Social Security Administration, Office of Research and Statistics (Staff Paper No. 29). Washington, D.C.: U.S. Government Printing Office, 1978.

Quinn, P. K., and Resnikoff, M. The relationship between death anxiety and the subjective experience of time in the elderly. *Omega*, in press.

Rabbitt, P. An age-decrement in the ability to ignore irrelevant information. *Journal of Gerontology*, 1965, *20*, 233–238.

______. Changes in problem solving ability in old age. In J. E. Birren and K. W. Schaie (Eds.), *Handbook of the psychology of aging*. New York: Van Nostrand Reinhold, 1977

Rabinowitz, J. C., Ackerman, B. P., Craik, F. I. M. and Hinchley, J. L. Aging and metamemory: The roles of relatedness and imagery. *Journal of Gerontology*, 1982, *37*, 688–695.

Radke-Yarrow, M., Cummings, E. M., Kuczynski, L., and Chapman, M. Patterns of attachment in two- and three-year-olds in normal families and families with parental depression. *Child Development*, 1985, *56*, 884–893.

Ragozin, A. S., Basham, R. B., Crnic, K. A. Greenberg, M. T., and Robinson, N. M. Effects of maternal age on parenting role. *Developmental Psychology*, 1982, *18*, 627–634.

Ramey, C. T., and Campbell, F. A. Compensatory education for disadvantaged children. *School Review*, 1979, *87*, 171–189.

______, and Haskins, R. The modification of intelligence through early experience. *Intelligence*, 1981, *5*, 5–19.

Rao, S. *The effect of instruction on pupil reading strategies*. Unpublished dissertation, University of Reading, 1982.

Rapoport, R., and Rapoport, R. *Leisure and the family life cycle*. Boston: Routledge and Kegan Paul, 1976.

Ratner, H. H. Memory demands and the development of young children's memory. *Child Development*, 1984, *55*, 2173–2191.

Ratner, N. B., and Pye, C. Higher pitch in BT is not universal: Acoustic evidence from Quiche Mayan. *Journal of Child Language*, 1984, *11*, 515–522.

Rawat, A. Alcohol harms fetus, study finds. *Chicago Tribune*, April 19, 1982, section 1, p. 13.

Rebelsky, F., and Hanks, C. Fathers' verbal interaction with infants in the first three months of life. *Child Development*, 1971, *42*, 63–68.

Rebok, G. W., and Offermann, L. R. Behavioral competencies of older college students: A self-efficacy approach. *Gerontologist*, 1983, *23*, 428–432.

Reedy, M. N. Personality and aging. In D. S. Woodruff and J. E. Birren (Eds.), *Aging: Scientific perspectives and social Issues* (2nd ed.). Monterey, Calif.: Brooks/Cole, 1983.

______, Birren, J. E., and Schaie, K. W. Age and sex differences in satisfying love relationships across the adult life span. *Human Development*, 1981, *24*, 52–56.

Reese, H. W. Imagery and associative memory. In R. V. Kail, Jr., and J. W. Hagen (Eds.), *Perspectives in the development of memory and cognition*. Hillsdale, N.J.: Lawrence Erlbaum, 1977.

______, and Rodeheaver, D. Problem solving and complex decision making. In E. Birren and K. Schaie (Eds.), *The handbook of the psychology of aging*. New York: Van Nostrand Reinhold, 1985, p. 495.

Reinisch, J. M. Prenatal exposure to synthetic progestins increases potential for aggression in humans. *Science*, 1981, *211*, 1171–1173.

Renwick, P. A., and Lawler, E. E. What you really want from your job. *Psychology Today*, May 1978, *11*, 53–65+.

Rescorla, L. A. Category development in early language. *Journal of Child Language*, 1981, *8*, 225–238.

Rest, J. R. Morality. In P. H. Mussen (Ed.), *Handbook of child psychology*, Volume 3. New York: Wiley, 1983.

______, Davison, M. L., and Robbins, S. Age trends in judging moral issues: A review of cross-sectional, longitudinal, and sequential studies of the defining issues test. *Child Development*, 1978, *49*, 263–279.

______, Turiel, E., and Kohlberg, L. Level of moral development as a determinant of preference and comprehension of moral judgments made by others. *Journal of Personality*, 1969, *37*, 225–252.

Restak, R. Male, female brains: Are they different? *Boston Globe*, September 9, 1979, p. A1.

Rheingold, H. L. Independent behavior of the human infant. In A. D. Pick (Ed.), *Minnesota Symposia on Child Psychology*, Volume 7. Minneapolis: University of Minnesota Press, 1973.

______, and Cook, K. V. The contents of boys' and girls' rooms as an index of parents' behavior. *Child Development*, 1975, *46*, 459–463.

______, and Emery, G. N. The nurturant acts of very young children. In D. Olweus, J. Block, and M. Radke-Yarrow (Eds.), *Development of antisocial and prosocial behavior*. Orlando, Fla.: Academic Press, 1986.

———, Hay, D. F., and West, M. J. Sharing in the second year of life. *Child Development*, 1976, *47* 1148–1158.

Rholes, W. S., and Ruble, D. N. Children's understanding of dispositional characteristics of others. *Child Development*, 1984, *55*, 550–560.

Ricciuti, H. N. Fear and the development of social attachments in the first year of life. In M. Lewis and L. A. Rosenblum (Eds.), *The origins of fear*. New York: Wiley, 1974.

Rice, M. The role of television in language acquisition. *Developmental Review*, 1983, *3*, 211–224.

Richardson, S. A., Goodman, N., and Hastorf, A. H. Cultural uniformity in reaction to physical disabilites. *American Sociological Review*, 1961, *26*, 241–247.

Richman, N., Stevenson, J., and Graham, P. J. *Pre-school to school: A behavioural study*. London: Academic Press, 1982.

Richmond-Abbott, M. Sex-role attitudes of mothers and children in divorced, single-parent families. *Journal of Divorce*, 1984, *8*, 61–81.

Riegel, K. F. The dialectics of human development. *American Psychologist*, 1976, *31*, 689–699.

Riese, M. L. Temperament stability between the neonatal period and 24 months in full-term and preterm infants. *Infant Behavior and Development*, 1986, *9* (Special Issue: Abstracts of papers presented at the Fifth International Conference on Infant Studies), 305.

Riley, M. W., and Bond, K. Beyond ageism: Postponing the onset of disability. In M. W. Riley, B. B. Hess, and K. Bond (Eds.), *Aging in society: Selected reviews of recent research*. Hillsdale, N.J.: Lawrence Erlbaum, 1981.

———, and Foner, A. *Aging and society*, Volume 1, *An inventory of research findings*. New York: Russell Sage Foundation, 1968.

———, Johnson, M., and Foner, A. (Eds.). *Aging and society*, Volume 3, *A sociology of age stratification*. New York: Russell Sage Foundation, 1972.

Rinke, C. L., Williams, J. J., Lloyd, K. E., and Smith-Scott, W. The effects of prompting and reinforcement on self-bathing by elderly residents of a nursing home. *Behavior Therapy*, 1978, *9*, 873–881.

Rivlin, L., and Rothenberg, M. The use of space in open classrooms. In H. Proshansky, W. Ittelson, and L. Rivlin (Eds.), *Environmental psychology: People and their physical settings* (2nd ed.). New York: Holt, Rinehart & Winston, 1976.

Rizzato, G., and Marazzini, L. Thoracoabdominal mechanics in elderly men. *Journal of Applied Physiology*, 1970, *28*, 457–460.

Roach, M. Another name for madness. *The New York Times Magazine*, January 16, 1983, 22–31.

Roberts, C. J., and Lowe, C. R. Where have all the conceptions gone? *Lancet*, March 1, 1975, *7905*, 498–499.

Roberts, G. C., Block, J. H., and Block, J. Continuity and change in parents' child-rearing practices. *Child Development*, 1984, *55*, 586–597.

Roberts, R. J., and Patterson, C. J. Perspective taking and referential communication: The question of correspondence reconsidered. *Child Development*, 1983, *54*, 993–1004.

Robertson, J. Significance of grandparenthood: Perceptions of young adult-grandchildren. *Gerontologist*, 1976, *16*, 137–140.

Robinson, E. J. The child's understanding of inadequate messages and communication failure: A problem of ignorance or egocentrism. In W. P. Dickson (Ed.), *Children's oral communication skills*. New York: Academic Press, 1981.

———, and Robinson, W. P. Development in the understanding of causes of success and failure in verbal communication. *Cognition*, 1977, *5*, 363–378.

Robson, K. S., and Moss, H. A. Patterns and determinants of maternal attachment. *Journal of Pediatrics*, 1970, *77*, 976.

Rockstein, M., Chesky, J., and Sussman, Comparative biology and evolution of aging. In C. E. Finch and L. Hayflick (Eds.), *Handbook of the biology of aging*. New York: Van Nostrand Reinhold, 1977.

———, and Sussman, M. *Biology of aging*. Belmont, Calif.: Wadsworth, 1979.

Roe, K. V., and Roe, A. *Vocal stimulation early in life and infant vocal responsiveness to mother vs. stranger: A curvilinear relationship*. Paper presented at the International Conference on Infant Studies, New York, April 1984.

Rogers, C. R. Toward a theory of creativity. In H. H. Anderson (Ed.), *Creativity and its cultivation*. New York: Harper, 1959.

Rollins, B. C., and Galligan, R. The developing child and marital satisfaction of parents. In R. M. Lerner and G. B. Spanier (Eds.), *Child influences on marital and family interaction*. New York: Academic Press, 1978.

Romalis, C. Taking care of the little woman: Father-physician relations during pregnancy and childbirth. In S. Romalis (Ed.), *Childbirth: Alternatives to medical control*. Austin: University of Texas Press, 1981.

Roopnarine, J. L., and Field, T. M. Peer-directed behaviors of infants and toddlers during nursery school play. *Infant Behavior and Development*, 1983, *6*, 133–138.

———, and Johnson, J. E. Socialization in a mixed age experimental program. *Developmental Psychology*, 1984, *20*, 828–832.

Rosch, E. Cognitive representations of semantic categories. *Journal of Experimental Psychology*, 1975, *104*, 192–233.

———, Mervis, C. B., Gray, W. D., Johnson, D. M., and Boyes-Braem, P. Basic objects in natural categories. *Cognitive Psychology*, 1976, *8*, 382–439.

Rose, S. A., and Blank, M. The potency of context in children's cognition: An illustration through conservation. *Child Development*, 1969, *40*, 383–406.

Rose, W. L. *A documentary history of slavery in North America*. Oxford: Oxford University Press, 1976.

Rosemeier, H. P. and Minsel, W. R. Das kranke Kind und der Tod. [The sick child and death.] Presented at the Thanato-Psychologie Symposium, Vechta, West Germany, Nov. 5, 1982.

Rosen, B. Management perception of older employees. *Monthly Labor Review*, 1978, *101*, 33–35.

———, and D' Andrade, R. The psychosocial origins of achievement motivation. *Sociometry*, 1959, *22*, 185–218.

Rosenberg, M. *Society and the adolescent self-image*. Princeton, N.J.: Princeton University Press, 1965.

———, and Simmons, R. G. *Black and white self-esteem: The urban school child*. Washington, D.C.: American Sociological Association, 1971.

Rosenblatt, P. C., and Cunningham, M. R. Television watching and family tensions. *Journal of Marriage and the Family*, 1976, *38*, 105–110.

Rosenfeld, E. F. *The relationship of sex-typed toys to the development of competency and sex-role identification in children*. Paper presented at the meeting of the Society for Research in Child Development, Denver, 1975.

Rosenthal, D. *Genetic theory and abnormal behavior*. New York: McGraw-Hill, 1970.

Rosenthal, R. From unconscious experimenter bias to teacher expectancy effects. In J. B. Dusek (Ed.) *Teacher expectancies*. Hillsdale, N.J.: Lawrence Erlbaum, 1985.

———, Baratz, S. S., and Hall, C. M. Teacher behavior, teacher expectations, and gains in pupils' rated creativity. *Journal of Genetic Psychology*, 1974, *124*, 115–121.

Rosett, H. L., Weiner, L., Zuckerman, B., McKinlay, S., and Edelin, K. C. Reduction of alcohol consumption during pregnancy with benefits to the newborn. *Alcoholism: Clinical and Experimental Research*, 1980, *4*, 178–184.

Rosner, B. S., and Doherty, N. E. The response of neonates to intra-uterine sounds. *Developmental Medicine and Child Neurology*, 1979, *21*, 723–729.

Ross, D. M., and Ross, S. A. *Hyperactivity: Research, theory, and action* (1st ed.; 2nd ed.). New York: Wiley, 1976, 1982.

Ross, H. G., and Milgram, J. I. Important variables in adult sibling relationships: A qualitative study. In M. E. Lamb and B. Sutton-

Smith (Eds.), *Sibling relationships: Their nature and significance across the lifespan*. Hillsdale N.J.: Lawrence Erlbaum, 1982.

______, AND GOLDMAN, B. D. Establishing new social relations in infancy. In T. Alloway, P. Pliner, and L. Krames (Eds.), *Attachment behavior: Advances in the study of communication and affect*, Volume 3. New York: Plenum, 1977.

ROTH, C. Factors affecting developmental changes in the speed of processing. *Journal of Experimental Child Psychology*, 1983, 35, 509–528.

ROUTH, D. K., SCHROEDER, C. S., AND O'TUAMA, L. Development of activity level in children. *Developmental Psychology*, 1974, *10*, 163–168.

ROVEE-COLLIER, C. K., SULLIVAN, M. W., ENRIGHT, M. L., LUCAS, D., AND FAGEN, J. W. Reactivation of infant memory. *Science*, 1980, *208*, 1159–1161.

ROVET, J. Cognitive and neuropsychological test performance of persons with abnormalities of adolescent development: A test of Waber's hypothesis. *Child Development*, 1983, *54*, 941–950.

______, AND NETLEY, C. The triple X chromosome in childhood: Recent empirical findings. *Child Development*, 1983, *54*, 831–845.

ROWLEY, V., AND KELLER, E. D. Changes in children's verbal behavior as a function of social approval and manifest anxiety. *Journal of Abnormal Social Psychology*, 1962, *65*, 53–57.

RUBIN, J. Boosting baby's brain. *Psychology Today*, March 1986, 16.

RUBIN, J. Z., PROVENZANO, F. J., AND LURIA, Z. The eye of the beholder: Parents' views on sex of newborns. *American Journal of Orthopsychiatry*, 1974, *44*, 512–519.

RUBIN, K. H., AND DANIELS-BEIRNESS, T. Concurrent and predictive correlates of sociometric status in kindergarten and grade 1 children. *Merrill-Palmer Quarterly*, 1983, *29*, 337–351.

______, AND KRASNOR, L. R. Social-cognitive and social behavior perspectives on problem solving. In M. Perlmutter (Ed.), *Minnesota Symposia on Child Psychology*, Volume 18. Hillsdale, N.J.: Lawrence Erlbaum, 1986.

______, WATSON, K. S., AND JAMBOR, T. W. Free-play behaviors in preschool and kindergarten children. *Child Development*, 1978, *49*, 534–536.

RUBIN, L. B. The empty nest: Beginnings or ending? In L. A. Bond and J. C. Rosen (Eds.), *Competence and coping during adulthood*. Hanover, N.H.: University Press of New England, 1980.

RUBIN, Z. *Children's friendships*. Cambridge, Mass.: Harvard University Press, 1980.

RUBINSTEIN, R. L. Never married elderly as a social type. *The Gerontologist*, 1987, *27*, 108–113.

RUBLE, D. N., AND BROOKS, J. *Adolescent attitudes about menstruation*. Paper presented at the biennial meeting of the Society for Research in Child Development, New Orleans, March 1977.

______, AND BROOKS-GUNN, J. The experience of menarche. *Child Development*, 1982, *53*, 1557–1566.

RUFF, H. A. Components of attention during infants' manipulative exploration. *Child Development*, 1986, *57*, 105–114.

RUKE-DRAVINA, V. Modifications of speech addressed to young children in Latvian. In C. E. Snow and C. A. Ferguson (Eds.), *Talking to children: Language input and acquisition*. Cambridge, England: Cambridge University Press, 1977.

RUSHTON, J. P. Generosity in children: Immediate and long term effects of modeling, preaching, and moral judgment. *Journal of Personality and Social Psychology*, 1975 *31*, 459–466.

______. *Can genes help helping*? Paper presented at the annual convention of the American Psychological Association, Toronto, August 1984.

RUSSELL, D. E. The prevalence and seriousness of incestuous abuse: Stepfathers vs. biological fathers. *Child Abuse & Neglect, The International Journal*, 1984, *8*, 15–22.

RUSSELL, M. J. Human olfactory communication. *Nature*, 1976, *260*, 520–522.

RUTTER, M. On confusion in the diagnosis of autism. *Journal of Autism and Childhood Schizophrenia*, 1978, *8*, 137–161.

______. *Changing youth in a changing society: Patterns of development and disorder*. Cambridge, Mass.: Harvard University Press, 1980.

______, TIZARD, J., AND WHITMORE, R. *Education, health, and behavior*. New York: Wiley, 1970.

SAARIO, T., JACKLIN, C. N., AND TITTLE, C. K. Sex role stereotyping in the public schools. *Harvard Educational Review*, 1973, *43*, 386–404.

SAARNI, C *When not to show what you feel: Children's understanding of relations between emotional experience and expressive behavior*. Paper presented at the biennial meeting of the Society for Research in Child Development, San Francisco, March 1979.

SABATINI, P., AND LABOUVIE-VIEF, G. Age and professional specialization: Formal reasoning. Paper presented at the Annual Meeting of the Gerontological Society. Washington, D.C., November 1979.

SACHER, G. A. Life table modification and life prolongation. In C. E. Finch and L. Hayflick (eds.) *Handbook of the biology of aging*. New York: Van Nostrand Reinhold, 1977.

SACHS, J. The adaptive significance of linguistic input to prelinguistic infants. In C. E. Snow and C. A. Ferguson (Eds.), *Talking to children: Language input and acquisition*. Cambridge, England: Cambridge University Press, 1977.

______, AND JOHNSON, M. Language development in a hearing child of deaf parents. In W. Von Raffler Engel and Y. Le Brun (Eds.), *Baby talk and infant speech*. Neurolinguistics Series, Volume 5. Amsterdam: Swets and Zeitlinger, 1976.

SADKER, M., AND SADKER, D. Sexism in the schoolroom of the 80's. *Psychology Today*, March 1985, 54–57.

SALAPATEK, P. Pattern perception in early infancy. In L. B. Cohen and P. Salapatek (Eds.), *Infant perception: From sensation to cognition*, Volume 1, *Basic visual processes*. New York: Academic Press, 1975.

SALK, L., LIPSITT, L. P., STURNER, W. Q., REILLY, B. M., AND LEVAT, R. H. Relationship of maternal and perinatal conditions to eventual adolescent suicide. *Lancet*, March 16, 1985, 624–627.

SALTER, A. Birth without violence: A medical controversy. *Nursing Research*, 1978, *27*, 84–88.

SALTHOUSE, T. A. *Adult cognition*. New York: Springer-Verlag, 1982.

______. Why is typing rate unaffected by age? Paper presented at the Annual Scientific Meeting of the Gerontological Society. San Francisco, October 1983.

______, AND SOMBERG, B. L. Isolating the age deficit in speeded performance. *Journal of Gerontology*, 1982, *37*, 59–63.

SALTIN, B., BLOMQUIST, G., MITCHELL, J. H., JOHNSON, R. L., WILDENTHAL, K, AND CHAPMAN, C. B. Response to exercise after bed rest and after training. *American Heart Association Monograph*. No. 23. New York: American Heart Association, 1968.

SALTZ, E., CAMPBELL, S., AND SKOTKO, D. Verbal control of behavior: The effects of shouting. *Developmental Psychology*, 1983, *19*, 461–464.

SAMEROFF, A. J. The components of sucking in the human newborn. *Journal of Experimental Child Psychology*, 1968, *6*, 607–623.

______, AND CAVANAUGH, P. J. Learning in infancy: A developmental perspective. In J. D. Osofsky (Ed.), *Handbook of infant development*. New York: Wiley, 1979.

______, SIEFER, R., AND ELIAS, P. K. Sociocultural variability in infant temperament ratings. *Child Development*, 1982, *53*, 164–173.

______, AND ZAX, M. Early development of children at risk for emotional disorder. *Monographs of the Society for Research in Child Development*, 1982, *47* (7, Serial No. 199).

SAMUEL, J., AND BRYANT, P. Asking only one question in the conversation experiment. *Journal of Child Psychology and Psychiatry*, 1984, *25*, 315–318.

SAMUELS, H. R. The effect of an older sibling on infant locomotor exploration of a new enviornment. *Child Development*, 1980, *51*, 607–609.

SANADI, D. R. Metabolic changes and their significance in aging. In C. E. Finch and L. Hayflick (Eds.), *Handbook of the biology of aging*. New York: Van Nostrand Reinhold, 1977,

SANDER, L. Issues in early mother-infant interaction. *Journal of the American Academy of Clinical Psychiatry*, 1962, 141–166.

SANDERS, B., AND SOARES, M. P. Sexual maturation and spatial ability in college students. *Developmental Psychology*, 1986, *22*, 199–203.

SANDERS, C. M. A comparison of adult bereavement in the death of a spouse, child, and parent. *Omega*, 1979, *10*, 303–322.

SANFORD, E. C. Mental growth and decay. *Americal Journal of Psychology*, 1902, *13*, 426–449.

SANTROCK, J. W. *Adolescence: An introduction*. Dubuque, Iowa: W. C. Brown, 1984.

SAPIR, E. Language. New York: Harcourt, Brace, 1921. (Reprinted 1958.)

SARASON, S. B. Individual psychology: An obstacle to comprehending adulthood. In L. A. Bond and J. C. Rosen (Eds.), *Competence and coping during adulthood*. Hanover, N.H.: University Press of New England, 1980.

SAUL, S. *Aging: An album of people growing old* (2nd ed.). New York: Wiley, 1983.

SAVIN-WILLIAMS, R. C. *Dominance-submission behaviors and hierarchies in young adolescents at a summer camp: Predictors, styles and sex differences*. Unpublished doctoral dissertation, University of Chicago, 1977.

———, AND DEMO, D. H. Developmental change and stability in adolescent self-concept. *Developmental Psychology*, 1984, *20*, 1100–1110.

SAVITSKY, J. C., AND WATSON, M. J. Patterns of proxemic behavior among preschool children. *Representative Research in Social Psychology*, 1975, *6*, 109–113.

SAXBY, L., AND BRYDEN, M. P. Left-ear superiority in children for processing auditory emotional material. *Developmental Psychology*, 1984, *20*, 72–80.

SCAFIDI, F. A., FIELD, T. M., AND SCHANBERG, S. M. Effects of tactile/kinesthetic stimulation on the clinical course and sleep/wake behavior of preterm neonates. *Infant Behavior and Development*, 1986, *9*, 91–105.

SCANLON, J. *Self-reported health behavior and attitudes of youth 12–17 years*. Vital and Health Statistics, Series 11, 147. Washington, D.C.: U.S. Government Printing Office, 1975.

SCANLON, J. W., BROWN, W. V., WEISS, J. B., AND ALPER, M. H. Neurological responses of newborn infants after maternal epidural anesthesia. *Anesthesiology*, 1974, *40*, 121–128.

SCANZIONI, L., AND SCANZIONI, J. *Men, women and change: A sociology of marriage and the family*. New York: McGraw-Hill, 1976.

SCARR, S., AND MCCARTNEY, K. How people make their own environments: A theory of genotype–environment effects. *Child Development*, 1983, *54*, 424–435.

———, AND WEINBERG, R. A. The Minnesota Adoption Studies: Genetic differences and malleability. *Child Development*, 1983, *54*, 260–267.

SCARR-SALAPATEK, S. Genetics and the development of intelligence. In F. D. Horowitz (Ed.), *Review of child development research*, Volume 4. Chicago: University of Chicago Press, 1975.

SCHACTER, D. L., AND MOSCOVITCH, M. Infants, amnesics, and dissociable memory systems. In M. Moscovitch (Ed.), *Infant memory*. New York: Plenum, 1981.

SCHAEFER, E. S., AND BAYLEY, N. Maternal behavior, child behavior and their intercorrelations from infancy through adolescence. *Monographs of the Society for Research in Child Development*, 1963, *28* (3, Serial No. 87).

SCHAFFER, H. R. Acquiring the concept of the dialogue. In M. H. Bornstein and W. Kessen (Eds.), *Psychological development from infancy: Image to intention*. Hillsdale, N.J.: Lawrence Erlbaum 1978.

———, COLLIS, G. M. AND PARSONS, G. Vocal interchange and visual regard in verbal and preverbal children. In H. R. Schaffer (Ed.), *Studies in mother-infant interaction*. London: Academic Press, 1977.

———, AND CROOK, C. K. The role of the mother in early social development. In H. McGurk (Ed.), *Childhood social development*. London: Methuen, 1978.

———, AND EMERSON, P. E. The development of social attachments in infancy. *Monographs of the Society for Research in Child Development*, 1964, *29* (3, Serial No. 94).(a)

SCHAFFER, J. R., AND EMERSON, P. E. Patterns of response to physical contact in early human development. *Journal of Child Psychology and Psychiatry*, 1964, *5*, 1–13.(b)

SCHAIE, K. W. The primary mental abilities in adulthood: An exploration in the development of psychometric intelligence. In P. B. Baltes and O. G. Brim, Jr. (Eds.), *Life-span development and behavior*, Volume 2. New York: Academic Press, 1979,

———. Age changes in adult intelligence. In D. S. Woodruff and J. E. Birren (Eds.), *Aging: Scientific perspectives and social issues* (2nd ed.). Monterey, Calif.: Brooks/Cole, 1983.

———, AND HERTZOG, C. Fourteen-year cohort-sequential analyses of adult intellectual development. *Developmental Psychology*, 1983, *19*, 531–543.

———, AND PARHAM, I. A. Stability of adult personality: Fact or fable? *Journal of Personality and Social Psychology*, 1976, *36*, 146–158.

SCHAU, C. G., KAHN, L., DIEPOLD, J. H., AND CHERRY, F. The relationships of parental expectations and preschool children's verbal sex typing to their sex-typed toy play behavior. *Child Development*, 1980, *51*, 266–270.

SCHEINFELD, A. *Your heredity and environment*. Philadelphia/New York: Lippincott, 1965.

SCHERZ, R. G. Fatal motor vehicle accidents of child passengers from birth through 4 years of age in Washington State. *Pediatrics*, 1981, *68*, 572–576.

SCHIEFFELIN, B. B., AND OCHS, E. A cultural perspective on the transition from prelinguistic to linguistic communication. In R. M. Golinkoff (Ed.), *The transition from prelinguistic to linguistic communication*. Hillsdale, N.J.: Lawrence Erlbaum, 1983.

SCHMECK, H. M., JR. Fetal defects discovered early by new method. *New York Times*, October 18, 1983, p. C1.

SCHNEIDER-ROSEN, K., AND CICCHETTI, D. The relationship between affect and cognition in maltreated infants: Quality of attachment and the development of visual self-recognition. *Child Development*, 1984, *55*, 648–658.

SCHOLNICK, E. K. *New trends in conceptual representation: Challenges to Piaget's theory?* Hillsdale, N.J.: Lawrence Erlbaum, 1983.

SCHRAG, P., AND DIVOKY, D. *The myth of the hyperactive child*. New York: Pantheon, 1975.

SCHROEDER, C., TEPLIN, S., AND SCHROEDER, S. An overview of common medical problems encountered in schools. In C. R. Reynolds and T. B. Gutkin (Eds.), *The handbook of school psychology*. New York: Wiley, 1982.

SCHONFIELD, A. E. D. Learning, memory, and aging. In J. E. Birren and R. B. Sloane (Eds.), *Handbook of mental health and aging*. Englewood Cliffs, N.J.: Prentice-Hall, 1980.

SCHRAM, S. F., AND OSTEN, D. F. CETA and the aging. Aging and Work, 1978, *7*, 163–174.

SCHRANK, H. T., AND WARING, J. M. Aging and work organizations. In M. W. Riley, B. B. Hess, and K. Bond (Eds.), *Aging in society: Selected reviews of recent research*. Hillsdale, N.J.: Lawrence Erlbaum, 1983.

SCHUBERT, J. B., BRADLEY-JOHNSON, S., AND NUTTAL, J. Mother-infant communication and maternal employment. *Childhood Development*, 1980, *51*, 246–249.

SCHULZ, J. H. *The economics of aging* (2nd ed.). Belmont, Calif.: Wadsworth, 1980.

SCHULZ, R. *The psychology of death, dying, and bereavement*. Reading, Mass.: Addison-Wesley, 1978.

———. Emotionality and aging: A theoretical and empirical analysis. *Journal of Gerontology*, 1982, *37*, 42–51.

———, AND HANUSA, B. H. *Long-term effects of control and predic-*

tability enhancing interventions: Findings and ethical issues. San Francisco: Gerontological Society, 1977.

SCHWAB, D. P., AND HENEMAN, H. G., II. Effects of age and experience on productivity. *Industrial Gerontology*, 1977, *4*, 113–117.

SCHWARTZ, P. Length of day-care attendance and attachment behavior in eighteen-month-old infants. *Child Development*, 1983, *54*, 1073–1078.

SCHWARZ, J. C. *Young children's fears: Modeling or cognition?* Paper presented at the biennial meeting of the Society for Research in Child Development, San Francisco, March 1979.

______. *Effects of group day care in the first two years.* Paper presented at SRCD, Detroit, April 1983.

SCOLLEN, R. *One child's language from one to two: The origins of construction.* Unpublished doctoral dissertation, 1974.

SCOTT-MAXWELL, F. *The measure of my days.* New York: Penguin, 1979.

SEARS, R. R., RAU, L., AND ALPERT, R. *Identification and child rearing.* Stanford, Calif.: Stanford University Press, 1965.

______, AND WISE, G. W. Relation of cup feeding in infancy to thumbsucking and the oral drive. *American Journal of Orthopsychiatry*, 1950, *20*, 123–138.

SEAVEY, C. A., KATZ, P. A., AND ZALK, S. R. Baby X, the effect of gender labels on adult responses to infants. *Sex Roles*, 1975, *1*, 61–73.

SECORD, D., AND PEEVERS, B. The development and attribution of person concepts. In T. Mischel (Ed.), *Understanding other persons.* Oxford: Basil Blackwell, 1974.

SEIDEN, R. H. *Suicide among youth: A review of the literature*, 1900–1967. Chevy Chase, Md.: National Clearing House for Mental Health Information, 1969.

SELMAN, R. L. *The growth of interpersonal understanding: Developmental and clinical analyses.* New York: Academic Press, 1980.

______. The child as a friendship philosopher. In S. R. Asher and J. M. Gottman (Eds.), *The development of children's friendships.* New York: Cambridge University Press, 1981.

______, SCHORIN, M. Z., STONE, C. R., AND PHELPS, E. A naturalistic study of children's social understanding. *Developmental Psychology*, 1983, *19*, 82–102.

______, AND SELMAN, A. Children's ideas about friendship: A new theory. *Psychology Today*, October 1979, 70–80.

SELMANOWITZ, V. J., RIZER, R. I., AND ORENTREICH, N. Aging of the skin and its appendages. In C. E. Finch and L. Hayflick (Eds.), *Handbook of the biology of aging.* New York: Van Nostrand Reinhold, 1977.

SERBIN, L. A., AND CONNOR, J. M. Sex-typing of children's play preferences and patterns of cognitive performance *Journal of Genetic Psychology*, 1979, *134*, 315–316.

______, ______, AND DENIER, C. *Modification of sex typed activity and interactive play patterns in the preschool classroom: A replication and extension.* Paper presented at the Annual Meeting of the Association for the Advancement of Behavior Therapy, Chicago, 1978.

______, O'LEARY, K. D., KENT, R. N., AND TONICK, I. J. A comparison of teacher responses to the preacademic and problem behavior of boys and girls. *Child Development*, 1973, *44*, 796–804.

______, TONICK, I. J., AND STERNGLANZ, S. H. Shaping cooperative cross-sex play. *Child Development*, 1977, *48*, 924–929.

SEXTON, M. A., AND GEFFEN, G. Development of three strategies of attention in dichotic monitoring. *Developmental Psychology*, 1979, *15*, 299–310.

SHANTZ, C. U. Children's understanding of social rules and the social context. In F. C. Serafica (Ed.), *Social-cognitive development in context.* New York: Guilford Press, 1982.

SHAPIRO, B. J., AND O'BRIEN, T. C. Logical thinking in children ages six through thirteen. *Child Development*, 1970, *41*, 823–829.

SHATZ, M. *The comprehension of indirect directives: Can you shut the door?* Paper presented at the summer meeting of the Linguistic Society of America, Amherst, Mass., July 1974.

______. Children's comprehension of their mother's question-directives. *Journal of Child Language*, 1978, *5*, 39–46.

SHAYER, M., KUCHEMAN, D. E., AND WYLAM, H. The distribution of Piagetian stages of thinking in British middle and secondary school children. *British Journal of Educational Psychology*, 1976, *46*, 164–173.

______, AND WYLAM, H. The distribution of Piagetian stages of thinking in British middle and secondary school children: II. 14 to 16 years old and sex differentials. *British Journal of Educational Psychology*, 1978, *48*, 62–70.

SHEPHERD-LOOK, D. L. Sex differentiation and the development of sex roles. In B. B. Wolman and G. Stricker (Eds.), *Handbook of developmental psychology.* Englewood Cliffs, N.J.: Prentice-Hall, 1982.

SHERESHEFSKY, P. M., AND YARROW, L. J. *Psychological aspects of a first pregnancy and early postnatal adaption.* New York: Raven Press, 1973.

SHERIF, M., HARVEY, O. J., WHITE, B. J., HOOD, W. R., AND SHERIF, C. W. *Intergroup conflict and cooperation: The robbers' cave experiment.* Norman: University of Oklahoma Press, 1961.

______, AND SHERIF, C. W. *Reference groups.* New York: Harper & Row, 1964.

SHINN, M. Father absence and children's cognitive development. *Psychology Bulletin*, 1978, *85*, 295–324.

SHIPLEY, E. F., SMITH, C. S., AND GLEITMAN, L. R. A study in the acquisition of language: Free responses to commands. *Language*, 1969, *45*, 322–342.

SHIRLEY, M. M. *The first two years: A study of twenty-five babies.* Minneapolis: University of Minnesota Press, 1933.

SHNEIDMAN, E. S. You and death, *Psychology Today*, June 1971, *5*, 43–45 +.

______. *Death: Current perspectives* (2nd ed.). Palo Alto, Calif.: Mayfield, 1980.

SHOCK, N. W. Systems integration. In C. E. Finch and L. Hayflick (eds.), *Handbook of the biology of aging.* New York: Van Nostrand Reinhold, 1977, pp. 639–665.

SIEGAL, L. S. Children's and adolescents' reactions to the association of Martin Luther King; A study of political socialization. *Developmental Psychology*, 1977, *13*, 284–285.

SIEGAL, M., AND BARCLAY, M. S. Children's evaluations of fathers' socialization behavior. *Developmental Psychology*, 1985, *21*, 1090–1096.

______, AND COWEN, J. Appraisals of intervention: The mother's versus the culprit's behavior as determinants of children's evaluations of discipline techniques. *Child Development*, 1984, *55*, 1760–1766.

______, AND RABLIN, J. Moral development as reflected by young children's evaluation of maternal discipline. *Merrill-Palmer Quarterly*, 1982, *28*, 499–509.

SIEGEL, B. Doubts haunt town in sex abuse case. *Los Angeles Times*, December 29, 1984, pp. 1–28.

SIEGEL, E., BAUMAN, K., SCHAEFER, E., SANDERS, M., AND INGRAM, D. Hospital and home support during infancy: Impact on maternal attachment, child abuse and neglect, and health care utilization. *Pediatrics*, 1980, *66*, 183–190.

SIEGEL, L. S. The relationship of language and thought in the preoperational child: A reconsideration of non-verbal alternatives to Piagetian tasks. In L. S. Siegel and C. J. Brainerd (Eds.), *Alternatives to Piaget: Critical essays on the theory.* New York: Academic Press, 1978.

SIEGEL, R. K. The psychology of life after death. *American Psychologist*, 1980, *35*, 911–931.

SIEGLER, I. C., AND COSTA, P. T., JR. Health behavior relationships. In J. E. Birren and K. W. Schaie (Eds.), *Handbook of the psychology of aging* (2nd ed.). New York: Van Nostrand Reinhold [in press].

______, GEORGE, L. K., AND OKUN, M. A. A cross-sequential analysis of adult personality. *Developmental Psychology*, 1979, *15*, 350–351.

SIEGLER, R. S. Information processing approaches to development. In P. H. Mussen (Ed.), *Handbook of child psychology*, Volume 1. New York: Wiley, 1983. (a)

———. Five generalizations about cognitive development. *American Psychologist*, 1983, *38*, 263–277. (b)

———. *Children's thinking*. Englewood Cliffs, N.J.: Prentice-Hall, 1986.

SILVERMAN, I. W., AND STONE, J. M. Modifying cognitive functioning through participation in a problem solving group. *Journal of Educational Psychology*, 1972, *63*, 603–608.

SILVERMAN, P. R., MACKENZIE, D., PETTIPAS, M., AND WILSON, E. *Helping each other in widowhood*. New York: Health Sciences, 1974.

SIMAN, M. L. Application of a new model of peer group influence to naturally existing adolescent friendship groups. *Child Development*, 1977, *48*, 270–274.

SIMON, E. W., DIXON, R. A., NOWAK, C. A., AND HULTSCH, D. F. Orienting task effects on text-recall in adulthood. *Journal of Gerontology*, 1982, *37*, 575–580.

SINEX, F. M. The molecular genetics of aging. In C. E. Finch and L. Hayflick (Eds.), *Handbook of the biology of aging*. New York: Van Nostrand Reinhold, 1977.

SIMMONS, R. G., ROSENBERG, F., AND ROSENBERG, M. Disturbance in the self-image of adolescence. *American Sociological Review*, 1973, *38*, 553–568.

SIMNER, M. L. Newborns' response to the cry of another infant. *Developmental Psychology*, 1971, *5*, 136–150.

SIMONTON, D. K. Genius, creativity, and leadership: Histriometric inquiries. Cambridge: Harvard University Press, 1984.

———. Creative productivity and age. *Developmental Review*, 1983, *3*, 97–111.

SIMPSON, E. Moral development research. A case study of scientific cultural bias. *Human Development*, 1974, *17*, 81–105.

SINCLAIR-DE ZWART, H. *Acquisition du language et dévelopment de la pensée*. Paris: Dunod, 1967.

SINGER, J. B., AND FLAVELL, J. H. Development of knowledge about communication: Children's evaluations of explicitly ambiguous messages. *Child Development*, 1981, *52*, 1211–1215.

SINGER, J. L., SINGER, D. G., AND SHERROD, L. R. *Prosocial programs in the context of children's total pattern of TV viewing*. Paper presented at the biennial meeting of the Society for Research in Child Development, San Francisco, March 1979.

SINGER, L. M., BRODZINSKY, D. M., RAMSAY, D., STEIR, M., AND WATERS, E. Mother-infant attachment in adoptive families. *Child Development*, 1985, *56*, 1543–1551.

SIQUELAND, E. R., AND LIPSITT, L. P. Conditioned head-turning in human newborns. *Journal of Experimental Child Psychology*, 1966, *3*, 356–376.

SIRIGNANO, S. W., AND LACHMAN, M. E. Personality change during the transition to parenthood: The role of perceived infant temperament. *Developmental Psychology*, 1985, *21*, 558–567.

SIZER, T. *Horace's compromise: The dilemma of the American high school*. Boston: Houghton Mifflin, 1985.

SKARIN, K. Cognitive and contextual determinants of stranger fear in six- and eleven-month-old infants. *Child Development*, 1977, *48*, 537–544.

SKINNER, B. G. *Verbal behavior*. New York: Appleton-Century-Crofts, 1957.

SKODAK, M., AND SKEELS, H. M. A final follow-up study of one hundred adopted children. *Journal of Genetic Psychology*, 1949, *75*, 85–125.

SKOLNICK, A. Married lives: Longitudinal perspectives on marriage. In D. H. Eichorn, J. A. Clausen, N. Haan, M. P. Honzik, and P. H. Mussen (Eds.), *Present and past in middle life*. New York: Academic Press, 1981.

SLATER, A. M. Visual perception at birth. *Infant Behavior and Development*, 1986, *9* (Special Issue: Abstracts of papers presented at the Fifth International Conference on Infant Studies), 346.

SLOANE, R. B. Organic brain syndrome. In J. E. Birren and R. B. Sloane (Eds.), *Handbook of mental health and aging*. Englewood Cliffs, N.J.: Prentice-Hall, 1980.

SLOBIN, D. I. Imitation and grammatical development in children. In N. S. Endler, L. R. Boulter, and H. Osser (Eds.), *Contemporary issues in developmental psychology*. New York: Holt, 1968.

———. *Psycholinguistics*. Glenview, Ill.: Scott, Foresman, 1971.

———. Cognitive prerequisites for the development of grammar. In C. A. Ferguson and D. I. Slobin (Eds.), *Studies of child language development*. New York: Holt, Rinehart & Winston, 1973.

———. On the nature of talk to children. In E. H. Lenneberg and E. Lenneberg (Eds.), *Foundations of language development*, Volume 1. New York: Academic Press, 1975.

———, AND WELSH, G. A. Elicited imitation as a research tool in developmental psycholinguistics. In C. A. Ferguson and D. I. Slobin (Eds.), *Studies of child language development*. New York: Holt, Rinehart & Winston, 1973.

SMILEY, S. S., AND BROWN, A. L. Conceptual preference for thematic and taxonomic relations: A nonmonotonic age trend from preschool to old age. *Journal of Experimental Child Psychology*, 1979, *28*, 249–257.

SMITH, C. B., ADAMSON, L. B., AND BAKEMAN, R. Interactional predictors of early language. *Infant Behavior and Development*, 1986, *9* (Special Issue: Abstracts of papers presented at the Fifth International Conference on Infant Studies), 347.

SMITH, H. K. The responses of good and poor readers when asked to read for different purposes. *Reading Research Quarterly*, 1967, *3*, 53–84.

SMITH, L. B., KEMLER, D. G., AND ARONFREED, J. Developmental trends in voluntary selective attention: Differential effects of source distinctiveness. *Journal of Experimental Child Psychology*, 1975, *20*, 352–365.

SMITH, P. K., AND CONNOLLY, K. Patterns of play and social interaction in preschool children. In N. Blurton Jones (Ed.), *Ethological studies of child behavior*. Cambridge, England: Cambridge University Press, 1972.

SNAREY, J. R., REIMER, J., AND KOHLBERG, L. Development of social-moral reasoning among kibbutz adolescents: A longitudinal cross-cultural study. *Developmental Psychology*, 1985, *21*, 3–17.

SNOW, C. E. Mother's speech to children learning language. *Child Development*, 1972, *43*, 549–564.

SOAR, R. S., AND SOAR, R. M. *An attempt to identify measures of teacher effectiveness from four studies*. Paper presented at the meetings of the American Educational Research Association, San Francisco, April 1976.

SOERGEL, K. H., ZBORALSKE, F. F., AND AMBERG, J. R. Presbyesophagus: Esophageal motility in nonagenarians. *Journal of Clinical Investigation*, 1964, *43*, 1472–1479.

SOLNICK, R. L., AND BIRREN, J. E. Age and male erectile responsiveness. *Archives of Sexual Behavior*, 1977, *6*, 1–9.

———, AND CORBY, N. Human sexuality and aging. In D. S. Woodruff and J. E. Birren (Eds.), *Aging: Scientific perspectives and social issues* (2nd ed.). Monterey, Calif.: Brooks/Cole, 1983

SONNENSCHEIN, S., AND WHITEHURST, C. J. Training referential communication skills: The limits of success. *Journal of Experimental Child Psychology*, 1983, *35*, 426–436.

SONTAG, L. W., AND NEWBERY, H. Normal variations of fetal heart rate during pregnancy. *American Journal of Obstetrics and Gynecology*, 1940, *40*, 449–452.

SOPHIAN, C. Perseveration and infants' search: A comparison of two- and three-location tasks. *Developmental Psychology*, 1985, *21*, 187–194.

SORCE, J., EMDE, R. N., CAMPOS, J. J., AND KLINNERT, M. *Maternal emotional signaling: Its effect on the visual cliff behavior of one-year-olds*. Paper presented at the meeting of the Society for Research in Child Development, Boston, April 1981.

SORCE, J. F., EMDE, R. N., CAMPOS, J., AND KLINNERT, M. D. Maternal emotional signaling: Its effect on the visual cliff behavior of 1-year-olds. *Developmental Psychology*, 1985, *21*, 195–200.

SORENSEN, R. C. *Adolescent sexuality in contemporary America*. New York: World Publishing, 1973.

SORRELLS-JONES, J. *A comparison of the effects of Leboyer delivery and modern "routine" childbirth in a randomized sample*. Unpublished doctoral dissertation, University of Chicago, 1983.

Sostek, A. M., Scanlon, J. W., and Abramson, D. C. Postpartum contact and maternal confidence and anxiety: A confirmation of short-term effects. *Infant Behavior and Development*, 1982, *5*, 323–329.

Soumerai, S. B., and Avorn, J. Perceived health, life satisfaction, and activity in the urban elderly: A controlled study of the impact of part-time work. *Journal of Gerontology*, 1983, *38*, 356–362.

Spanier, G. B. Formal and informal sex education as determinants of premarital sexual behavior. *Archives of Sexual Behavior*, 1976, *5*, 39–67.

______, and Lewis, R. A. Marital quality: A review of the seventies. *Journal of Marriage and the Family*, 1980, *42*, 825–839.

Sparrou, S., and Zigler, E. Evaluation of a patterning treatment for retarded children. *Pediatrics*, 1978, *62*, 137–149.

Spear, L. Treatment of grief explored. *The New York Times*, November 6, 1983, WC9.

Spearman, C. *The abilities of man*. New York: Macmillan, 1927.

Spelke, E. Infants' intermodal perception of events. *Cognitive Psychology*, 1976, *8*, 553–560.

______. *International exploration by 4-month-old infants: Perception and knowledge of auditory-visual events*. Unpublished doctoral dissertation, Cornell University, 1978.

______. Exploring audible and visible events in infancy. In A. D. Pick (Ed.), *Perception and its development: A tribute to Eleanor J. Gibson*. Hillsdale, N.J.: Lawrence Erlbaum, 1979.(a)

______. Perceiving bimodally specified events in infancy. *Developmental Psychology*, 1979, *15*, 626–636.(b)

Spelke, E. S. Perceptual knowledge of objects in infancy. In J. Mehler, M. Garrett, and E. Walker (Eds.), *Perspectives on mental representation*. Hillsdale, N.J.: Lawrence Erlbaum, 1982.

______, and Owsley, C. J. Intermodal exploration and knowledge in infancy. *Infant Behavior and Development*, 1979, *2*, 13–27.

Spelt, D. K. The conditioning of the human fetus in utero. *Journal of Experimental Psychology*, 1948, *38*, 338–346.

Spence, J. T., and Helmreich, R. L. *Masculinity and femininity: Their psychological dimensions, correlates and antecedents*. Austin: University of Texas Press, 1978.

Spence, M. J., and DeCasper, A. J. *Human fetuses perceive maternal speech*. Paper presented at the meeting of the International Conference on Infant Studies, Austin, Texas, March 1982.

Spielberger, C. D. The effects of anxiety on complex learning and academic achievement. In C. D. Spielberger (Ed.), *Anxiety and behavior*, New York: Academic Press, 1966.

Spilton, D., and Lee, L. C. Some determinants of effective communication in four-year-olds. *Child Development*, 1977, *48*, 968–977.

Spinelli, D. N., Jensen, F. E., and Viana Di Prisco, G. Early experience effect on dendritic branching in normally reared kittens. *Experimental Neurology*, 1980, *68*, 1–11.

Spivak, G., and Shure, M. B. *Social adjustment of young children: A cognitive approach to solving real-life problems*. San Francisco, CA.: Jossey-Bass, 1974.

Sprafkin, C., Serkin, L., Denier, C., and Connor, J. Sex differentiated play: Cognitive consequences and early interventions. In M. Liss (Ed.), *Social and cognitive skills: Sex roles and children's play*. New York: Academic Press, 1983.

Spring, C., and Capps, C. Encoding speed, rehearsal, and probed recall of dyslexic boys. *Journal of Educational Psychology*, 1974, *66*, 780–786.

Sroufe, L. A. Wariness of strangers and the study of infant development. *Child Development*, 1977, *48*, 731–746.

______. Infant-caregiver attachment and patterns of adaptation in preschool: The roots of maladaptation and competence. In M. Perlmutter (Ed.), *Development and policy concerning children with special needs. Minnesota Symposium in Child Psychology*, Volume 16. Hillsdale, N.J.: Lawrence Erlbaum, 1983.

______, Schork, E., Motti, E., Lawroski, N., and LaFreniere, P. The role of affect in emerging social competence. In C. Izard, J. Kagan, and R. Zajonc (Eds.), *Emotion, cognition and behavior*. New York: Plenum, 1984.

______, and Waters, E. Attachment as an organizational construct. *Child Development*, 1977, *48*, 1184–1199.

Staiano-Coico, L., Darzynkiewicz, Z., Hefton, J. M., Dutkowski, R., Darlington, G. J., and Weksler, M. E. Increased sensitivity of lymphocytes from people over 65 to cell cycle arrest and chromosomal damage. *Science*, 1983, *219*, 1335–1337.

Stark, E. Taking a beating. *Psychology Today*, April 1985, 16.

______. Mom and dad: The great American heroes. *Psychology Today*, May 1986, 12–13.

Staub, E. A. *The development of prosocial behavior in children*, Morristown, N.J.: General Learning Press, 1975.

Steele, B. F. *Working with abusive parents from a psychiatric point of view*. U.S. Department of Health, Education and Welfare Publication No. (OHD) 75-70. Washington, D.C.: U.S. Government Printing Office, 1975.

Stein, A. H. Imitation of resistance to temptation. *Child Development*, 1967, *38*, 159–169.

______, and Bailey, M. The socialization of achievement orientation in females. *Psychological Bulletin*, 1973, *80*, 345–366.

______, and Friedrich, L. K. Impact of television on children and youth. In E. M. Hetherington (Ed.), *Review of child development research*, Volume 5. Chicago: University of Chicago Press, 1975.

Stein, N., and Glenn, C. An analysis of story comprehension in elementary school children. In R. O. Freedle (Ed.), *New directions in discourse processing*. Norwood, N.J.: Albex, 1979.

Steinberg, L. D. *Changes in family relations at puberty*. Paper presented at the biennial meeting of the Society for Research in Child Development, San Francisco, March 1979.

______. Early temperamental antecedents of adult Type A behaviors. *Developmental Psychology*, 1985 *21*, 1171–1180.

Steinhausen, H.-C. Psychological evaluation of treatment in phenylketonuria: Intellectual, motor and social development. *Neuropaediatrie*, 1974, *5*, 146–156.

Stern, D. N. Mother and infant at play: The dyadic interaction involving facial, vocal and gaze behaviors. In M. Lewis and L. A. Rosenblum (Eds.), *The effect of the infant on its caregiver*. New York: Wiley, 1974.

______. *The first relationship: Infant and mother*. Cambridge, Mass.: Harvard University Press, 1977.

Stern, M., Northman, J. E., and Van Slyck, M. R. Father absence and adolescent "problem behaviors": Alcohol consumption, drug use and sexual activity. *Adolescence*, 1984, *19*, 301–312.

Sternberg, R. J. *Intelligence, information processing, and analogical reasoning: The componential analysis of human abilities*. Hillsdale, N.J.: Lawrence Erlbaum, 1977.

______. Stalking the IQ quark. *Psychology Today*, September 1979, 42–54.

______. *Beyond I.Q.* Cambridge, England: Cambridge University Press, 1985.

______, Conway, B. E., Ketron, J. L., and Bernstein, M. People's conceptions of intelligence. *Journal of Personality and Social Psychology: Attitudes and Social Cognition*, 1981, *41*, 37–55.

______, and Powell, J. S. The development of intelligence. In P. H. Mussen (Ed.), *Handbook of child psychology*, Volume 3. New York: Wiley, 1983.

Sternglanz, S. H., and Serbin, L. A. Sex role stereotyping in children's television programs. *Developmental Psychology*, 1974, *10*, 710–715.

Sterns, H., Barrett, G. V., and Alexander, R. A. Accidents and the aging individual. In J. E. Birren and K. W. Schaie (Eds.), *Handbook of the psychology of aging*. New York: Van Nostrand Reinhold, 1985, pp. 703–724.

Sterns, H. L., and Sanders, R. E. Training and education of the elderly. In R. R. Turner and H. W. Reese (Eds.), *Life-span developmental psychology: Intervention*. New York: Academic Press, 1980.

Stevenson-Hinde, J., Hinde, R. A., and Simpson, A. E. Behavior at home and friendly or hostile behavior in preschool. In D.

Olweus, J. Block, and M. Radke-Yarrow (Eds.), *Development of antisocial and prosocial behavior*. New York: Academic Press, 1986.

STEWART M. A. Is hyperactivity normal? and other unanswered questions. *School Review*, 1976, *85*, 31–42.

STEWART, R. B. Sibling attachment relationships: Child-infant interactions in the strange situation. *Developmental Psychology*, 1983, *19*, 192–199.

———, AND MARVIN, R. S. Sibling relations: The role of conceptual perspective-taking in the ontogeny of sibling caregiving. *Child Development*, 1984, *55*, 1322–1332.

STINNETT, N., COLLINS, J., AND MONTGOMERY, J. E. Marital need satisfaction of older husbands and wives. *Journal of Marriage and the Family*, 1970, *32*, 428–434.

STIRES, L. The effect of classroom seating location on student grades and attitudes: Environment or self-selection? *Environment and Behavior*, 1980, *12*, 241–254.

STOCKDALE, D. F., GALEJS, I., AND WOLINS, L. Cooperative-competitive preferences and behavioral correlates as a function of sex and age of school-age children. *Psychological Reports*, 1983, *53*, 739–750.

STODDARD, R., AND TURIEL, E. Children's concepts of cross-gender activities. *Child Development*, 1985, *56*, 1241–1252.

STOLZ, H. R., AND STOLZ, L. M. *Somatic development of adolescent boys*. New York: Macmillan, 1951.

STONES, M. J., AND KOZMA, A. In response to Hartley and Hartley. *Experimental Aging Research*, 1984, *10*, (3).

STORCK, P., LOOFT, W., AND HOOPER, F. H. Interrelationships among Piagetian tasks and traditional measures of cognitive abilities in mature and aged adults. *Journal of Gerontology*, 1972, *27*, 461–465.

STOUDT, H. W., DAMON, A., MCFARLAND, R. A., AND ROBERTS, J. *Weight, height and selected body Measurements of adults. United States, 1960–1962*. U.S. Public Health Service Publication No. 1000, Series 11, No. 8. Washington, D.C.: U.S. Government Printing Office, 1965.

STRAUSS, M. E., LESSEN-FIRESTONE, J., STARR, R., AND OSTREA, E. M., JR. Behavior of narcotics-addicted newborns. *Child Development*, 1975, *46*, 887–893.

STRAUSS, M. S., AND COHEN, L. B. *Infant immediate and delayed memory for perceptual dimensions*. Unpublished manuscript, University of Illinois, 1978.

STROUD, J. G. Women's careers: Work, family, and personality. In D. H. Eichorn, J. A. Clausen, N. Haan, M. P. Honzik, and P. H. Mussen (Eds.), *Present and past in middle life*. New York: Academic Press, 1981.

STRUCKEY, M. F., MCGHEE, P. E., AND BELL, N. J. Parent-child interaction: The influence of maternal employment. *Developmental Psychology*, 1982, *18*, 635–644.

SULLIVAN, J. W., AND HOROWITZ, F. D. The effects of intonation on infant attention: The role of the rising intonation contour. *Journal of Child Language*, 1983, *10*, 521–534.

SULLLIVAN, K., AND SULLIVAN, A. Adolescent-parent separation. *Developmental Psychology*, 1980, *10*, 93–99.

SUOMI, S. J. Social interactions of monkeys reared in a nuclear family environment versus monkeys reared with mothers and peers. *Primates*, 1974, *15*, 311–320.

———. Adult male-infant interactions among monkeys living in nuclear families. *Child Development*, 1977, *48*, 1255–1270.

SUPER, D. E. *The psychology of careers*. New York: Harper & Row, 1967.

———, KOWALSKI, R., AND GOTKIN, E. *Floundering and trial after high school*. Unpublished manuscript, Columbia University, 1967.

SURBER, C. F., AND GZESH, S. M. Reversible operations in the balance scale task. *Journal of Experimental Child Psychology*, 1984, *38*, 254–274.

SUSKIND, R. M. Characteristics and causation of protein-calorie malnutrition in the infant and preschool child. In L. S. Greene (Ed.), *Malnutrition, behavior and social organization*. New York: Academic Press, 1977.

SUSSMAN, M. B. The family life of older people. In R. H. Binstock and E. Shanas (Eds.), *Handbook of aging and the social sciences*. New York: Van Nostrand Reinhold, 1976.

SUSSMAN, R. P. *Effects of novelty and training on the curiosity and exploration of young children in day care centers*. Unpublished doctoral dissertation, University of Chicago, 1979.

SUTTON-SMITH, B., AND ROSENBERG, B. G. Sixty years of historical change in the game preferences of American children. In R. E. Herron and B. Sutton-Smith (Eds.), *Child's play*. New York: Wiley, 1971.

SVEJDA, M., PANNABECKER, B., AND EMDE, R. N. Parent-to-infant attachment: A critique of the early 'bonding' model. In R. N. Emde and R. J. Harmon (Eds.), *The development of attachment and affiliative systems: Psychological aspects*. New York: Plenum, 1982.

SWANSON, J. M., AND KINSBOURNE, M. Food dyes impair performance of hyperactive children in a laboratory learning test. *Science, 1980, 207*, 1485–1487.

SWITSKY, H. N. Exploration, curiosity, and play in young children: Effects of stimulus complexity. *Developmental Psychology*, 1973, *10*, 321–329.

———, HAYWOOD, H. C., AND ISETT, R. Exploration, curiosity and play in young children: Effects of stimulus complexity. *Developmental Psychology*, 1974, *10*, 321–329.

SYLVA, K., BRUNER, J. S., AND GENOVA, P. The role of play in problem-solving of children three to five years old. In J. S. Bruner, A. Jolly, and K. Sylva (Eds.), *Play: Its role in development and evolution*. London: Penguin, 1976.

SYLVA, K., ROY, C., AND PAINTER, M. *Childwatching at playgroup and nursery school*. London: Grant McIntyre Ltd., 1980.

TAKAHASHI, K. Examining the strange-situation procedure with Japanese mothers and 12-month-old infants. *Developmental Psychology*, 1986, *22*, 265–270.

TALBERT, G. B. Aging of the reproductive system. In C. E. Finch and L. Hayflick (eds.), *Handbook of the biology of aging*. New York: Van Nostrand Reinhold, 1977.

TANNER, J. M. *Growth at adolescence* (2nd ed.). Oxford: Blackwell Scientific Publications, 1962.

———. The trend towards earlier physical maturation. In J. E. Meade and A. S. Parkes (Eds.), *Biological aspects of social problems*. Edinburgh: Oliver and Boyd, 1965.

———. Physical growth. In P. H. Mussen (Ed.), *Carmichael's manual of child psychology*, Volume 1. New York: Wiley, 1970.

———. Variability of growth and maturity in newborn infants. In M. Lewis and L. A. Rosenblum (Eds.), *The effect of the infant on its caregiver*. New York: Wiley, 1974.

———. *Foetus into man: Physical growth from conception to maturity*. London: Open Books, 1978.

TANZ, C. *Studies in the acquisition of deictic terms*. Cambridge, England: Cambridge University Press, 1980.

TANZER, D., AND BLOCK, J. L. *Why natural childbirth?* New York: Schocken Books, 1976.

TAVORMINA, J. B., BOLL, H., DUNN, N. J., LUSCOMB, R. L., AND TAYLOR, J. R. *Psychosocial effects of raising a physically handicapped child on parents*. Paper presented at the meeting of the American Psychological Association Convention, San Francisco, September 1975.

TAYLOR, M. E. Sex role sterotypes in children's readers. *Elementary English*, 1973, *50*, 1061–1064.

TAYLOR, P. M., TAYLER, F. H., CAMPBELL, S. B. G., MALONI, J., AND DICKEY, D. *Effects of extra contact on early maternal attitudes, perceptions, and behaviors*. Paper presented at the meetings of the Society for Research in Child Development, San Francisco, March 1979.

TEC, N. Some aspects of high school status and differential involvement with marihuana: A study of suburban teenagers. *Adolescence*, 1972, *6*, 1–28.

TELLIS-NAYAK, V. The transcendent standard: The religious ethos of the rural elderly. *Gerontologist*, 1982, *22*, 359–363.

TENNES, E. R., KISLEY, A., AND METCALF, D. The stimulus barrier in early infancy: An exploration of some formulations of John Benjamin. In R. Holt and E. Peterfreund (Eds.), *Psychoanalysis and contemporary science*, Volume 1. New York: Macmillan, 1972.

TERMAN, L. M. *Genetic studies of genius*, Volume 1, *Mental and physical traits of a thousand gifted children*. Stanford, Calif.: Stanford University Press, 1925.

______, AND ODEN, M. H. *Genetic studies of genus*, Volume 4, *The gifted group at midlife*. Stanford, Calif.: Stanford University Press, 1959.

TERR, L. C. A family study of child abuse. *American Journal of Psychiatry*, 1970, *127*, 665–671.

TESCH, S. A. Review of friendship development across the life span. *Human Development*, 1983, *26*, 266–276.

TETENS, J. N. *Philosophische Versuche uber die menschliche Natur und ihre Entwicklung*. Leipzig: Weidmanns Erben & Reich, 1777.

THELEN, E., FISHER, D. M., AND RIDLEY-JOHNSON, R. The relationship between physical growth and a newborn reflex. *Infant Behavior and Development*, 1984, *7*, 479–493.

THIRER, J., AND WRIGHT, S. D. Sport and social status for adolescent males and females. *Sociology of Sport Journal*, 1985, *2*, 164–171.

THOMAE, H. Personality and adjustment to aging. In J. E. Birren and R. B. Sloane (Eds.), *Handbook of mental health and aging*. Englewood Cliffs, N.J.: Prentice-Hall, 1980.

THOMAS A., AND CHESS, S. *Temperament and development*. New York: Brunner Mazel, 1977.

______, AND ______. *Correlation of early temperament with later behavioral functioning*. Paper presented at CIBA Foundation temperament conference, London, September 1981.

______, ______, and Birch, H. G. *Temperament and behavior disorders in children*. New York: New York University Press, 1968.

THOMAS, J. Gender differences in satisfaction with grandparenting. *Psychology and Aging*, 1986, *1*, (3), 215–219.

THOMPSON, M. G. *Life adjustment of women with anorexia nervosa and anorexic-like behavior*. Unpublished doctoral dissertation, University of Chicago, 1979.

THOMPSON, R. A., AND LAMB, M. E. Security of attachment and stranger sociability in infancy. *Development Psychology*, 1983, *19*, 184–191.

THOMPSON, S. K. Gender labels and early sex role development. *Child Development*, 1975, *46*, 339–347.

______, AND BENTLER, P. M. The priority of cues in sex discrimination by children and adults. *Developmental Psychology*, 1971, *5*, 181–185.

THORESEN, C., EAGLESTON, J., KIRMIL-GRAY, K., AND BRACKE, P. *Type A children anxious, insecure*. Paper presented at the annual meeting of the American Psychological Association, Los Angeles, August 1985.

THORNBURG, D. H. Sources of sex education among early adolescents. *Journal of Early Adolescence*, 1981, *1*, 174.

TIEGER, T. On the biological basis of sex differences in aggression. *Child Development*, 1980, *51*, 943–963.

TIESZAN, H. R. Children's social behavior in a Korean preschool. *Journal of Korean Home Economics Association*, 1979, *17*, 71–84.

TIGER, L. *Men in groups*. New York: Random House, 1969.

TISHLER, C., MCKENRY, P. C., AND MORGAN, K. C. Adolescent suicide attempts: Some significant factors. *Suicide and life-threatening behavior*, 1981, *11*, 86–92.

TIZARD, B., CARMICHAEL, H., HUGHES, M., AND PINKERTON, G. Four year olds talking to mothers and teacher. In L. A. Hersoveval (Ed.), *Language and language disorders in childhood* (Supplement No. 2, *Journal of Child Psychology and Psychiatry*). London: Pergamon Press, 1980.

______, AND REES, J. A comparison of the effects of adoption, restoration to the natural mother, and continued institutionalization on the cognitive development of four-year-old children. *Child Development*, 1974, *45*, 92–99.

TOBIN, S. The earliest memory as data for research in aging. In D. P Kent, R. Kastenbaum. and S. Sherwood (Eds.), *Research, planning, and action for the elderly*. New York: Behavioral Publications, 1972.

TODD, C. M., AND PERLMUTTER, M. Reality recalled by preschool children. In M. Perlmutter (Ed.), *New directions in child development: Children's memory*. San Francisco: Jossey-Bass, 1980.

TONNA, E. A. Aging of skeletal-dental systems and supporting tissue. In C. E. Finch and L. Hayflick (eds.), *Handbook of the biology of aging*. New York: Van Nostrand Reinhold, 1977.

TORRES, A., FORREST, J. D., AND EISMAN, T. Telling parents: Clinic policies and adolescents' use of family planning and abortion services. *Family Planning Perspectives* 1980, *12*, 284–292.

TOYNBEE, A. *Man's concern with death*. New York: McGraw-Hill, 1968.

TRAUPMANN, J., ECKELS, E., AND HATFIELD, E. Intimacy in older women's lives. *Gerontologist*, 1982, *22*, 493–498.

TRAUSE, M. A., VOOS, D., RUDD, C., KLAUS, M., KENNELL, J., AND BOSLETT, M. Separation for childbirth: The effect on the sibling. *Child Psychiatry and Human Development*, 1981, *12*, 32–39.

TRAUTNER, H. M. [Relationships between parental style of education and parent orientation in 10-14 year old girls.] *Zeitschrift für Entwicklungpsychologie und Padagogische Psychologie*, 1972, 4 (3), 116–182.

TREAS, J. Aging and the family. In D. S. Woodruff and J. E. Birren (Eds.), *Aging: Scientific perspectives and social issues* (2nd ed.). Monterey, Calif.: Brooks/Cole, 1983.

TREHUB, S. E. Infants' sensitivity to vowel and tonal contrasts. *Developmental Psychology*, 1973, *9*, 91–96.

______, AND CURRAN, S Habituation of infants' cardiac response to speech stimuli. *Child Development*, 1979, *50*, 1247–1250.

TRELEASE, M. L. Dying among Alaskan Indians: A matter of choice. In E. Kübler-Ross (Ed.), *Death: The final stage of growth*. Englewood Cliffs, N.J.: Prentice-Hall, 1975.

TRETHOWAN, W. H., AND CONLON, M. F. The couvade syndrome. *British Journal of Psychiatry*, 1965, *111*, 57–66.

TREVARTHEN, C. Descriptive analyses of infant communicative behavior. In H. R. Schaffer (Ed.), *Studies in mother-infant interaction*. London: Academic Press, 1977.

______. Development of the cerebral mechanisms for language. In V. Kirk (Ed.), *Neuropsychology of language, reading and spelling*. New York: Academic Press, 1983.

TROLL, L. E. Grandparenting. In L. W. Poon (Ed.), *Aging in the 1980s: Psychological issues*. Washington, D.C.: American Psychological Association, 1980.(a)

______. Intergenerational relations in later life: A family system approach. In N. Datan and N. Lohmann (Eds.), *Transitions of aging*. New York: Academic Press, 1980.(b)

______, Miller, S. J., and Atchley, R. C., *Families in later life*. Belmont, Calif.: Wadsworth, 1979.

TUDIVER, J. *Parental influences on the sex role development of the preschool child*. Unpublished manuscript, University of Western Ontario, London, Ontario, 1979.

TURIEL, E. An experimental test of the sequentiality of developmental stages in the child's moral judgments. *Journal of Personality and Social Psychology*, 1966, *3*, 611–618.

______. The development of social concepts. In D. DePalma and J. Foley (Eds.), *Moral development*. Hillsdale, N.J.: Lawrence Erlbaum, 1975.

TURKINGTON, C. Pituitary defect seen in anorexia. APA *Monitor*, January 1984, *15*, 17.(a)

______. Parents found to ignore sex stereotypes. APA *Monitor*, April 1984, *15*, 12.(b)

TURNER, B. The self concept of older women. *Research on Aging*, 1979, *1*, 464–480.

TURNURE, C. Response to voice of mother and stranger by babies in the first year. *Developmental Psychology*, 1971, *4*, 182–190.

TURNURE, J., BUIUM, N., AND THURLOW, M. The effectiveness of interrogatives for promoting verbal elaboration productivity in children. *Child Development*, 1976, *47*, 851–855.

TVERSKY, A., AND KAHNEMAN, D. Judgment under uncertainty: Heuristics and biases. *Science*, 1974, *185*, 1124–1131.

———, AND ———. The framing of decisions and the psychology of choice. *Science*, 1981, *211*, 453–458.

UDRY, J. R. Changes in the frequency of marital intercourse from panel data. *Archives of Sexual Behavior*, 1980, *9*, 319–325.

ULLIAN, D. Z. The development of conceptions of masculinity and feminity. In B. Lloyd and J. Archer (Eds.), *Exploring sex differences*. London: Academic Press, 1976.

U.S. BUREAU OF THE CENSUS. *Statistical abstract of the United States, 1982–83*. 103d edition. Washington, D.C.: U.S. Government Printing Office, 1982.

———. *Estimates of the population of the United States, by age, sex, and race: 1980–82*, P-25, No. 929. Washington, D.C.: U.S. Government Printing Office, 1983.

U.S. DEPARTMENT OF COMMERCE. *Social indicators*, Volume 3. Washington, D.C.: Author, 1981.

U.S. DEPARTMENT OF HEALTH AND HUMAN SERVICES. *Health, United States*, 1980. DHHS Publication No. (PHS) 81-1232, 1980, p. 276.

———. *Health, United States*, 1982. DHHS Publication No. (PHS) 83-1232, 1982.

U.S. PUBLIC HEALTH SERVICE. *Alzheimer's disease: Q & A*. NIH Pub. No. 80-1646. Washington, D.C.: U.S. Government Printing Office, 1980.

———. *Healthy people: The Surgeon General's Report on Health Promotion and Disease Prevention*. DHEW Pub. No. 79-55071. Washington, D.C.: U.S. Government Printing Office, 1979.

———. *Alcohol and health: Second special report to the U.S. Congress*. Rockville, Md.: National Institute on Alcohol Abuse and Alcoholism, 1974.

UTECH, D. A., AND HOVING, K. L. Parents and peers as competing influences in the decisions of children of different ages. *Journal of Social Psychology*, 1969, *78*, 267–274.

VAILLANT, G. E. *Adaptation to life: How the best and brightest came of age. Boston*: Little, Brown, 1977.

VANDENBERG, S. G. The nature and nurture of intelligence. In D. C. Glass (Ed.), *Genetics*. New York: Rockefeller University Press/Russel Sage Foundation, 1968.

———, AND KUSE, A. R., Spatial ability: A critical review of the sex-linked major gene hypothesis. In M. A. Wittig and A. C. Peterson (Eds.), *Sex related differences in cognitive functioning: Developmental issues*. New York: Academic Press, 1979.

VAUGHN, B., JOFFE, L., EGELAND, B., DIENARD, A., AND WATERS, E. *Relationships between neonatal behavioral organization and infant-mother attachment in an economically disadvantaged sample*. Paper presented at the meeting of the Society for Research in Child Development, San Francisco, March 1979.

VAUGHN, B. E., AND LANGLOIS, I. H. Physical attractiveness as a correlate of peer status and social competence in preschool children. *Developmental Psychology*, 1983, *19*, 561–567.

VENTURA, S. J. Teenage childbearing: United States 1966–1975. *Monthly Vital Statistics Reports: National Center for Health Statistics, 1977, 26*, (5) (supp.).

VEROFF, J., AND FELD, S. *Marriage and work in America: A study of motives and roles*. New York: Van Nostrand Reinhold, 1970.

VURPILLOT, E. The development of scanning strategies and their relation to visual differentiation. *Journal of Experimental Child Psychology*, 1968, *6*, 632–650.

WABER, D. P. Sex differences in cognition: A function of maturation rate? *Science*, 1976, *192*, 572–573.

———, MANN, M. B., MEROLA, J., AND MOYLAN, P. M. Physical maturation rate and cognitive performance in early adolescence: A longitudinal examination. *Developmental Psychology*, 1985, *21*, 668–681.

WACHS, T. D., AND GANDOUR, M. J. Temperament, environment, and six-month cognitive-intellectual development: A test of the organismic specificity hypothesis. *International Journal of Behavioral Development*, 1983, *6*, 135–152.

———, AND GRUEN, C. E. *Early experience and human development*. New York: Plenum, 1982.

WADDINGTON, C. H. *The strategy of the genes*. London: Allen & Unwin, 1957.

WAKE, F. R. Attitudes of parents towards the premarital sex behavior of their children and themselves. *Journal of Sex Research*, 1969, *5*, 170–177.

WALDROP, M. F., AND HALVERSON, C. F., JR. Intensive and extensive peer behavior: Longitudinal and cross-sectional analyses. *Child Development*, 1975, *46*, 19–26.

WALFORD, R. L. *Maximum life span*. New York: Norton, 1983.

———, WEINDRUCH, R. H., GOTTESMAN, S. R. S., AND TAM, C. F. The immunopathology of aging. In C. Eisdorfer, B. Starr, and V. J. Cristofalo (Eds.), *Annual Review of Gerontology and Geriatrics*, Volume 2. New York: Springer, 1981.

WALKER, E., AND EMORY, E. Commentary: Interpretive bias and behavioral genetic research. *Child Development*, 1985, *56*, 775–778.

WALKER, J. A., AND KERSHMAN, S. M. *The deaf-blind in social interaction*. Paper presented at the meeting of the Society for Research in Child Development, Boston, March 1981.

WALKER, L. J. The sequentiality of Kohlberg's stages of moral development. *Child Development*, 1982, *53*, 1330–1336.

WALLERSTEIN, J. S., AND KELLY, J. B. The effects of parental divorce: The adolescent experience. In E. J. Anthony and C. Koopernick (Eds.), *The child in his family: Children at psychiatric risk*, Volume 3. New York: Wiley, 1974.

———, AND ———. *Surviving the breakup*. New York: Basic Books, 1980.

WALSH, R. P. Age differences in learning and memory. In D. S. Woodruff and J. E. Birren (Eds.), *Aging: Scientific perspectives and social issues* (2nd ed.). Monterey, Calif.: Brooks/Cole, 1983.

WALSTER, E., ARONSON, V., ABRAHAM, D., AND ROLTMAN, L. Importance of physical attractiveness in dating behavior. *Journal of Personality and Social Psychology* 1966, *4*, 508–516.

WALTERS, J., PEARCE, D., AND DAHMS, L. Affectional and aggressive behavior of preschool children. *Child Development*, 1957, *28*, 15–26.

WANG, H. S., OBRIST, W. D., AND BUSSE, E. W. Neurophysiological correlates of the intellectual function of elderly persons living in the community. *American Journal of Psychiatry*, 1970, *126*, 1205–1212.

WANNER, R. A., AND MCDONALD, L. Ageism in the labor market: Estimating earnings discrimination against older workers. *Journal of Gerontology*, 1983, *38*, 738–745.

WANTZ, M. S., AND GAY, J. E. *The aging process: A health perspective*. Cambridge, Mass.: Winthrop, 1981.

WARD, I. L. Prenatal stress feminizes and demasculinizes the behavior of males. *Science*, 1972, *176*, 82–84.

WARSHAK, R., AND SANTROCK, J. W. *The effects of father and mother custody on children's social development*. Paper presented at the meetings of the Society for Research in Child Development, San Francisco, March 1979.

WASHBURN, S. L. Longevity in primates. In J. L. McGaugh and S. B. Kiesler (Eds.), *Aging: Biology and behavior*. New York: Academic Press, 1981.

WASSERMAN, G. A., ALLEN, R., AND SOLOMON, C. R. At-risk toddlers and their mothers: The special case of physical handicap. *Child Development*, 1985, *56*, 73–83.

WATERMAN, A. Identity development from adolescence to adulthood: An extension of theory and a review of research. *Developmental Psychology*, 1982, *18*, 341–358.

WATERMAN, A. S. Identity in the context of adolescent psychology. In A. S. Waterman (Ed.), *Identity in adolescence: Processes and contents*. New Directions for Child Development, No. 30. San Francisco: Jossey-Bass 1985.

WATERMAN, G., GEARY, P., AND WATERMAN C. Longitudinal study of changes in ego identity status from the freshman to the senior year at college. *Developmental Psychology*, 1974, *10*, 387–392.

Waters, E., and Deane, K. E. Defining and assessing individual differences in attachment relationships. *Monographs of the Society for Research in Child Development*, 1985, *50* (1–2, Serial No. 209), 41–45.

______, Hay, D., and Richters, J. Infant-parent attachment and the origins of prosocial and antisocial behavior. In D. Olweus, J. Block, and M. Radke-Yarrow (Ed.), *Development of antisocial and prosocial behavior*. New York: Academic Press, 1986.

______, Matas, L., and Sroufe, L. A. Infants' reactions to an approaching stranger: Description, validation, and functional significance of wariness. *Child Development*, 1975, *46*, 348–356.

______, Wippman, J., and Sroufe, L. A. Attachment, positive affect, and competence in the peer group: Two studies in construct validation. *Child Development*, 1979, *50*, 821–829.

Waterson, N. Growth of complexity in phonological development. In N. Waterson and C. E. Snow (Eds.), *The development of communication*. New York: Wiley, 1978.

Watson, J. B. *Psychological care of infant and child*. New York: Norton, 1928.

______, and Rayner, R. Conditioned emotional reactions. *Journal of Experimental Psychology*, 1920, *3*, 1–4.

Watson, M. W. The development of social roles: A sequence of social-cognitive development. *New Directions for Child Development*, 1981, *12*, 33–41.

Waugh, N. C., and Barr, R. A. Encoding deficits in aging. In F. I. M. Craik and S. Trehub (Eds.), *Aging and cognitive processes*. New York: Plenum, 1982, pp. 183–190.

Waxler, C. Z., and Radke-Yarrow, M. An observational study of maternal models. *Developmental Psychology*, 1975, *11*, 485–494.

Weatherly, D. Self-perceived rate of physical maturation and personality in late adolescence. *Child Development*, 1964, *35*, 1197–1210.

Webb, W. B. Sleep in older persons: Sleep structure of 50- to 60-year-old men and women. *Journal of Gerontology*, 1982, *37*, 581–586.

Weber, R. A., Levitt, M. J., and Clark, M. C. Individual variation in attachment security and strange situation behavior: The role of maternal and infant temperament. *Child Development*, 1986, *57*, 56–65.

Webster, R. L., Steinhardt, M. H., and Senter, M. G. Changes in infants' vocalizations as a function of differential acoustic stimulation. *Developmental Psychology*, 1972, *7*, 39–43.

Wechsler, D. *Manual for the Wechsler Intelligence Scale for children—Revised*. New York: Psychological Corporation, 1974.

Weg, R. B. Changing physiology of aging: Normal and pathological. In D. S. Woodruff and J. E. Birren (Eds.), *Aging: Scientific perspectives and social issues* (2nd ed.). Monterey, Calif.: Brooks/Cole, 1983.

Weil, W. B. Infantile obesity. In M. Winick (Ed.), *Childhood obesity*, New York: Wiley, 1975.

Weiman, L. A. Stress patterns of early child language. *Journal of Child Language*, 1976, *3*, 283–286.

Weiner, B., and Handel, S. J. A cognition-emotion-action sequence: Anticipated emotional consequences of causal attributions and reported communication strategy. *Developmental Psychology*, 1985, *21*, 102–107.

Weinraub, M., Clemens, L. P., Sockloff, A., Ethridge, T., Gracely, E., and Myers, B. The development of sex role stereotypes in the third year: Relationships to gender labeling, gender identity, sex-typed toy preference and family characteristics. *Child Development*, 1984, *55*, 1493–1503.

Weinraub, M., and Wolf, B. M. Effects of stress and social supports on mother-child interactions in single- and two-parent families. *Child Development*, 1983, *54*, 1297–1311.

Weir, C. Auditory frequency sensitivity of human newborns: Some data with improved acoustic and behavioral controls. *Perception and Psychophysics*, 1979, *26*, 287–294.

Weir, R. H. *Language in the crib*. The Hague: Mouton, 1962.

Weisfeld, G. E., Bloch, S. A., and Ivers, J. W. Possible determinants of social dominance among adolescent girls. *Journal of Genetic Psychology*, 1984, *144*, 115–129.

Weiss, L., and Lowenthal, M. F. Life course perspective on friendship. In M. F. Lowenthal, M. Thurnher, and D. Chiriboga (Eds.), *Four stages of life*. San Francisco: Jossey-Bass, 1975

Weiss, M. J., and Zelazo, P. R. *The cephalocaudal hypothesis: A comparison of infant leg kicks and arm flexions in water*. Paper presented at the International Conference on Infant Studies, New York, April 1984.

Weiss, R. S. *Going it alone: The family life and social situation of the single parent*. New York: Basic Books, 1979.

Weithorn, L., and Campbell, S. Competency of minors to make crucial decisions reported in S. Cunningham. Court fails to settle abortion rights of minors. *APA Monitor*, March 1984, *15*, 7.

Welford, A. T. *Ageing and human skill*. London: Oxford University Press, 1958.

Wellman, H. M. The early development of intentional memory behavior. *Human Development*, 1977, *20*, 86–101.(a)

______. Preschoolers understanding of memory-relevant variables. *Child Development*, 1977, *48*, 1720–1723. (b)

______, and Lempers, J. The naturalistic communication ability of two-year-olds. *Child Development*, 1977, *48*, 1052–1057.

______, Ritter, K., and Flavell J. H. Deliberate memory behavior in the delayed reactions of very young children. *Developmental Psychology*, 1975, *11*, 780–787.

Wells, G., and Raban, B. *Children learning to read*. SSRC Final Report No. HR 397/1. School of Education, University of Bristol, 1978.

Wells, R. V. Demographic change and the life cycle of American families. In T. K. Rabb and R. I. Rotberg (Eds.), *The family in history*. New York: Harper & Row, 1973, pp. 85–94.

Werner, J. S., and Perlmutter, M. Development of visual memory in infants. In H. W. Reese and L. P. Lipsitt (Eds.), *Advances in child development and behavior*, Volume 14. New York: Academic Press, 1980.

______, and Siqueland, E. R. Visual recognition memory in the preterm infant. *Infant Behavior and Development*, 1978, *1*, 79–94.

Werner, E. E., and Smith, R. S. *Vulnerable but invincible: A longitudinal study of resilient children and youth*. New York: McGraw-Hill, 1982.

Werts, C. Paternal influence on career choice. *Journal of Counseling Psychology*, 1968, *15*, 48–52.

West, R. L., Odom, R. D., and Aschkenasy, J. R. Perceptual sensitivity and conceptual coordination in children and younger and older adults. *Human Development*, 1978, *21*, 334–345.

Westinghouse and Ohio University. The impact of Head Start: An evaluation of the effects of Head Start on children's cognitive and affective development. In J. L. Frost (Ed.), *Revisiting early childhood education: Readings*. New York: Holt, Rinehart & Winston, 1973.

Weston, D. R., and Richardson, E. *Children's world views: Working models and quality of attachment*. Paper presented at the biennial meeting of the Society for Research in Child Development, Toronto, Canada, April 1985.

Weston, D. R., and Turiel, E. *Act-role relations: children's concepts of social roles*. Unpublished manuscript, University of California at Berkeley, 1979.

Whalen, C. K., and Henker, B. Hyperactivity and the attention deficit disorders: Expanding frontiers. *Pediatric Clinics of North America*, 1984, *31*, 397–427.

______, Henker, B., and Dotemoto, S. Teachers response to the methylphenidate (Ritalin) versus placebo status of hyperactive boys in the classroom. *Child Development*, 1981, *52*, 1005–1114.

Whisnant, L., and Zegans, L. A study of attitudes toward menarche among white middle-class American adolescent girls. *American Journal of Psychiatry*, 1975, *132*, 809–814.

Whitaker, J. M., M.D. *Reversing heart disease*. New York: Warner, 1985.

Whitbourne, S. K., and Waterman, A. S. Psychosocial development during the adult years: Age and cohort comparison. *Developmental Psychology*, 1979, *15*, 373–378.

White, B. L. An experimental approach to the effects of experience on early human behavior. In J. P. Hill (Ed.), *Minnesota Symposia on Child Psychology*, Volume 1. Minneapolis: University of Minnesota Press, 1967.

White, S. E., and Reamy, K. Sexuality and Pregnancy: A review. *Archives of Sexual Behavior*, 1982, *11*, 429–443.

Whitehurst, G., and Vasta, R. Is language acquired through imitation? *Journal of Psycholinguistic Research*, 1975, *4*, 37–59.

Whiten, A. Assessing the effects of perinatal events on the success of the mother-infant relationship. In H. R. Schaffer (Ed.), *Studies in mother-infant interaction*, London: Academic Press, 1977.

Whiting, B. B., and Pope-Edwards, C. (Eds.) *The effects of age, sex, and modernity on the behavior of mothers and children*. Report to the Ford Foundation, January 1977.

Wideman, M. V., and Singer, J. E. The role of psychological mechanisms in preparation for childbirth. *American Psychologist*, 1984, *39*, 1357–1371.

Wiener, G. Psychologic correlates of premature birth: A review. *Journal of Nervous and Mental Diseases*, 1962, *134*, 129–144.

Wiersma, W., and Klausmeier, H. J. The effect of age upon speed of concept attainment. *Journal of Gerontology*, 1965, *20*, 398–400.

Wiesenfeld, A. R., and Klorman, R. The mother's psychophysiological reactions to contrasting affective expressions by her own and an unfamiliar infant. *Developmental Psychology*, 1978, *14*, 294–304.

Wilen, J. B., and Petersen, A. C. *Young adolescents' responses to the timing of pubertal changes*. Paper presented at the annual meeting of the American Psychological Association, Montreal, September 1980.

Wilkening, F. Integrating velocity, time, and distance information: A developmental study. *Cognitive Psychology*, 1981, *13*, 231–247.

Wilkie, F., and Eisdorfer, C. Intelligence and blood pressure. In E. Palmore (Ed.), *Normal aging: II*. Durham, N.C.: Duke University Press, 1974, pp. 87–94. (a)

_____, and _____. Terminal changes in intelligence. In E. Palmore (Ed.), *Normal aging: II*. Durham, N.C.: Duke University Press, 1974, pp. 103–115. (b)

Willatts, P. Stages in the development of intentional search by young infants. *Developmental Psychology*, 1984, *20*, 389–396.

Williams, A. F. Children killed in falls from motor vehicles. *Pediatrics*, 1981, *68*, 576–578.

Williams, R. L. Black pride, academic relevance and individual achievement. *Counseling Psychologist*, 1970, *2*, 321–325.

Williams, T. M. *The impact of television: A study of three Canadian communities* ERIC Document ED 171 401. Vancouver: University of British Columbia, 1977.

Williams, T. P., and Lillis, R. P. Changes in alcohol consumption by 18-year-olds following an increase in New York State's purchase age to 19. *Journal of Studies on Alcohol*, 1986, *47*, 290–296.

Wilmore, J. H., and McNamara, J. J. Prevalence of coronary heart disease risk factors in boys 8 to 12 years of age. *Journal of Pediatrics*, 1974, *84*, 527–533.

Wilson, J. G. Current status of teratology. In J. G. Wilson and F. C. Fraser (Eds.) *Handbook of teratology*. New York: Plenum, 1977.

_____, and Fraser, F. C. (Eds.)., *Handbook of teratology*. New York: Plenum, 1977.

Wilson, R. S. The Louisville Twin Study: Developmental synchronies in behavior. *Child Development*, 1983, *54*, 298–316.

_____. Risk and resilience in early mental development. *Developmental Psychology*, 1985, *21*, 795–805.

_____, and Harpring, E. B. Mental and motor development in infant twins. *Developmental Psychology*, 1972, *7*, 277–287.

Windle, M., and Lerner, R. M. The "goodness of fit" model of temperament-context relations: Interaction or correlation? In J. V. Lerner and R. M. Lerner (Eds.), *New directions for child development: Temperament and social interaction during infancy and childhood*. San Francisco: Jossey-Bass, 1986.

Winterbottem, M. R. The relation of need for achievement to learning experiences in independence and mastery. In J. W. Atkinson (Ed.), *Motives in fantasy, action, and society*. New York: Van Nostrand, 1958.

Wish, M., Deutsch, M., and Kaplan, S. J. Perceived dimensions of interpersonal relations. *Journal of Personality and Social Psychology*, 1976, *33*, 409–420.

Wiswell, R. A. Relaxation, exercise, and aging. In J. E. Birren and R. B. Sloane (Eds.), *Handbook of mental health and aging*. Englewood Cliffs, N.J.: Prentice-Hall, 1980, pp. 943–958.

Witelson, S. F. Sex and the single hemisphere: Specialization of the right hemisphere for spatial processing. *Science*, 1976, *193*, 425–426.

Wolf, T. M., Sklov, M. C., Wenzl, P. A., Hunter, S. M., and Berenson, G. S. Validation of a measure of Type A behavior pattern in children: Bogalusa Heart Study. *Child Development*, 1982, *53*, 126–135.

Wolff, P. H. Observations on the early development of smiling. In B. M. Foss (Ed.), *Determinants of infant behavior*, Volume 2. New York: Wiley, 1963.

Wolfinger, R. E., and Rosenstone, S. J. *Who votes?* New Haven, Conn.: Yale University Press, 1980.

Wolstenholme, G. E. W., and O'Connor, M. *Endocrinology of the testis*. Boston: Little, Brown, 1967.

Wood, V., and Robertson, J. The significance of grandparenthood. In J. F. Gubrium (Ed.), *Time, roles, and self in old age*. New York: Human Sciences Press, 1976.

Woodruff, D. S. and Birren, J. E. Age changes and cohort difference in personality. *Developmental Psychology*, 1972, *6*, 252–259.

Worobey, J. *Temperament ratings in infancy: The salience of perceived difficulty*. Paper presented at the International Conference on Infant Studies. New York, April, 1984.

Wright, P. H. Men's friendships, women's friendships and the alleged inferiority of the latter. *Sex Roles*, 1982, *8*, 1–20.

Wyatt, R. J., interviewed by R. Young. A conversation with Richard Jed Wyatt. *Psychology Today*, August 1983, *17*, 30–41.

Wynne, E. A. Behind the discipline problem: Youth suicide as a measure of alienation. *Phi Delta Kappan*, 1978, *59*, 307–315.

Yakovlev, P. I., and Lecaurs, A. R. The myelogenetic cycles of regional maturation of the brain. In A. Minkowski (Ed.), *Regional development of the brain in early life*. Oxford: Basil Blackwell, 1967.

Yalom, I. D., Lunde, D. T., Moos, R. H., and Hamburg, D. A. "Postpartum blues" syndrome: A description and related variables. *Archives of General Psychiatry*, 1968, *18*, 16.

Yankelovich, D. *New rules: Searching for self-fulfillment in a world turned upside down*. New York: Random House, 1981.

Yarrow, L. J., Rubenstein, J. L., and Pedersen, F. A. *Infant and environment: Early cognitive and motivational development*. Washington, D.C.: Hemisphere, 1975.

Yeates, K. O., MacPhee, D., Campbell, F. A., and Ramey, C. T. Maternal IQ and home environment as determinants of early childhood intellectual competence: A developmental analysis. *Developmental Psychology*, 1983, *19*, 731–739.

Yerkes, R. M. Psychological examining in the United States army. *Memoirs of the National Academy of Sciences*, 1921, *15*, 1–890.

Yesavage, J. A., Rose, T. L. and Bower, G. H. Interactive imagery and affective judgments improve face-name learning in the elderly. *Journal of Gerontology*, 1983, *38*, 197–203.

Yogman, M. W. *The goals and structure of face-to-face interaction between infants and fathers*. Paper presented at the biennial meeting of the Society for Research in Child Development, New Orleans, March 1977.

Yonas, A., and Pettersen, L. *Responsiveness in newborns to optical information for collision*. Paper presented at the biennal meeting of the Society for Research in Child Development, San Francisco, March 1979.

YOUNISS, J., AND VOLPE, J. A relational analysis of children's friendship. In W. Damon (Ed.), *New directions in child development: Social cognition*. San Francisco: Jossey-Bass, 1978.

YUSSEN, S. R., AND LEVY, U. Developmental changes in conscious knowledge about different retrieval problems. *Developmental Psychology*, 1977, *13*, 114–120.

ZACKS, R. T. Encoding strategies used by young and elderly adults in a keeping track task. *Journal of Gerontology* 1982, *37*, 203–211.

ZAHN-WAXLER, C., FRIEDMAN, S. L., AND CUMMINGS, E. M. Children's emotions and behaviors in response to infants' cries. *Child Development*, 1983, *54*, 1522–1528.

______, IANNOTTI, R., AND CHAPMAN, M. Peers and prosocial development. In K. H. Rubin and H. S. Ross (Eds.), *Peer relationships and social skills in childhood*. New York: Springer-Verlag, 1982.

______, AND RADKE-YARROW, M. *A developmental analysis of children's responses to emotions in others*. Paper presented at the biennial meeting of the Society for Research in Child Development, San Francisco, March 1979.

______, ______, AND BRADY-SMITH, J. Perspective-taking and prosocial behavior. *Developmental Psychology*, 1977, *13*, 87–88.

______, ______, AND KING, R. A. Child rearing and children's prosocial initiations toward victims of distress. *Child Development*, 1979, *50*, 319–330.

______, ______, AND ______. Early altruism and guilt. *Academic Psychology Bulletin*, 1983, *5*, 247–259.

ZAKS, P. M., AND LABOUVIE-VIEF, G. Spatial perspective taking and referential communication skills in the elderly: A training study. *Journal of Gerontology*, 1980, *35*, 217–224.

ZARBATANY, L., HARTMANN, D. P., AND GELFAND, D. M. Why does children's generosity increase with age: Susceptibility to experimenter influence or altruism? *Child Development*, 1985, *56*, 746–756.

ZELAZO, P. R. From reflexive to instrumental behavior. In L. P. Lipsitt (Ed.), *Developmental psychobiology: The significance of infancy*. Hillsdale, N.J.: Lawrence Erlbaum, 1976.

______. The development of walking: New findings and old assumptions. *Journal of Motor Behavior*, 1983, *15*, 99–137.

______, ZELAZO, N. A., AND KOLB, S. Walking in the newborn. *Science*, 1972, *176*, 314–315.

ZELNICK, M., AND KANTNER, J. F. Sexual activity, contraceptive use and pregnancy among metropolitan-area teenagers, 1971–1979. *Family Planning Perspectives*, 1978, *12*, 230–237.

______, AND KIM, Y. J. Sex education and its association with teenage sexual activity, pregnancy, and contraception use. *Family Planning Perspectives*, 1982, *14*, 117–126.

ZIGLER, E., LEVINE, J., AND GOULD, L. Cognitive challenge as a factor in children's human appreciation. *Journal of Personality and Social Psychology*, 1967, *6*, 332–336.

ZIVIAN, M. T., AND DARIES, R. W. Free recall by in-school and out-of-school adults: Performance and metamemory. *Developmental Psychology*, 1983, *19*, 513–520.

ZUCKER, K. J. *The development of search for mother during brief separation*. Unpublished doctoral dissertation, University of Toronto, 1982.

ZUCKER, R. A. *Sex role identity patterns and drinking behavior in adolescents*. Center for Alcohol Studies, Rutgers University, 1967.

ZUCKERMAN, D. Mom's jobs, kid's careers. *Psychology Today*, February 1985, 6.

ZUSSMAN, J. V. Situational determinants of parental behavior: Effects of competing cognitive activity. *Child Development*, 1980, *51*, 792–800.

Photo Credits

Introduction Page ii: Thomas Hopker/Woodfin Camp.

Chapter 1 Opener: Courtesy N. and S. Davidson. Pages 7, 8, and 11: The Bettmann Archive. Page 15: Courtesy Albert Bandura. Page 16: Jim Erikson. Page 18: Yves DeBrane/Black Star.

Chapter 2 Opener: Herb Gehr/Life Magazine, Time Inc. Pages 31 and 32: Courtesy Marcia Skaarup. Page 35: Nina Leen/Life Magazine, Time Inc. Page 37: (left) Ulrike Welsch; (right) Bohdan Hrynewych/Stock, Boston. Page 41: (right) Cary Wolinsky/Stock, Boston; (left) Tom Ballard/EKM-Nepenthe. Page 43: Courtesy Dr. George Engel.

Chapter 3 Opener: Peter Simon/Stock, Boston. Page 58: Paul Fortin/Stock, Boston. Page 60: Sarah Putnam/The Picture Cube. Page 62: Alice Kandell/Photo Researchers. Page 67: Newsday. Page 68: Evan Johnson/Jeroboam. Page 71: Leonard McCombe/Time Magazine, Time Inc.

Chapter 4 Opener: Transworld Feature/Black Star. Page 79: Lennart Nilsson, from *Behold man*. Boston: Little Brown, 1978. Page 80: Philippe Ledru/Sygma. Page 83: (top left) Transworld Feature/Black Star; Claude Edelmann/Petit Format et Guignoz/Black Star. Page 87: Alice Kandell/Photo Researchers. Page 93: Sipa Press/Black Star. Page 95: (left) Alice Kandell/Photo Researchers; (right) From J. W. Hanson et al., Fetal alcohol syndrome experiment with 41 patients, *Journal of the American Medical Association*, 1976, *235*, 1458–1460. By permission of The American Medical Association. Page 98: (top left) Suzanne Szasz/Photo Researchers; (right) Mariette Pathy Allen; (bottom) Abigail Heyman/Archive Pictures. Page 100: David Powers/Stock, Boston. Page 101: Suzanne Arms/Jeroboam.

Chapter 5 Opener: Erika Stone/Peter Arnold. Page 110: Suzanne Arms/Jeroboam. Page 112: Dr. Kathryne Jacobs. Page 113: Omikron/Photo Researchers. Page 114: Alan Carey/The Image Works. Page 118: (top left) Jason Lauré/Woodfin Camp; (top right and center left) Susan Berkowitz/Taurus Photos; (center right) Linda Ferrer/Woodfin Camp; (bottom) DPI. Page 120: Stella Kupferberg. Page 123: Gerry Cranham/Photo Researchers. Page 125: (left) Erika Stone; (center) Suzanne Arms/Jeroboam; (right) Erika Stone/Peter Arnold. Page 126: Courtesy Major Arthur Ginsburg, Aviation Vision Laboratory, Wright Patterson Air Force Base. Page 131: Courtesy Dr. Joseph Campos. Page 133: George Bellerose/Stock, Boston. Page 135: Elizabeth Crews.

Chapter 6 Opener: Shirley Zeiberg/Taurus Photos. Page 143: Dr. Lewis Lipsitt. Page 145: Jason Lauré/Woodfin Camp. Page 147: James R. Holland/Stock, Boston. Page 148: Erika Stone/Peter Arnold. Page 150: James R. Holland/Stock, Boston. Page 151: Gale Zucker/Stock, Boston. Page 152: Mimi Cotter/International Stock Photo. Page 157: Judith Sedwick/The Picture Cube. Page 159: Elizabeth Crews/Stock, Boston. Page 161: George Bellerose/Stock, Boston. Page 164: (top) Peter Simon/Stock, Boston; (bottom) Suzanne Szasz; (right) Gary Renand/Magnum. Page 165: Elizabeth Crews.

Chapter 7 Opener: Erika Stone. Page 171: Olof Källström/Jeroboam. Page 173: Mariette Pathy Allen. Page 175: Hufnagle/Monkmeyer. Page 176: Horst Schäfer/Peter Arnold. Page 178: Erika Stone/Peter Arnold. Page 180: B. Kliewe/Jeroboam. Page 182: Suzanne Szasz/Photo Researchers. Page 185: (left) Suzanne Szasz; (right) Wayne Miller/Magnum. Page 189: Dr. Mary Ainsworth. Page 191: Alison Clarke-Stewart. Page 195: Terry Evans/Magnum.

Chapter 8 Opener: Shirley Zieberg/Taurus Photos. Page 207: University of California, Institute of Human Development. Page 209: (left) Science Photo Library; (right) Taurus Photos. Page 213: Elizabeth Crews. Page 215: James R. Holland/Stock, Boston.

Chapter 9 Opener: Jean-Claude Lejeune/Stock, Boston. Page 220: (left) Burk Uzzle/Magnum Photos; (right) Peter Vandermark/Stock, Boston. Page 223: Graham Bell. Page 226: Michael Weisbrot and Family. Page 229: Burt Glinn/Magnum Photos. Page 243: Peter Vandermark/Stock, Boston. Page 247: (left) Ann L. Reed/Taurus Photos; (right) Elizabeth Crews. Page 249: (top) Erika Stone/Jeroboam; (bottom) Elizabeth Hamlin/Stock, Boston.

Chapter 10 Opener: Tequila Minsky. Page 256: Peter Vandermark/Stock, Boston. Page 257: Erika Stone/Peter Arnold. Page 259: (left) Bill Stanton/Magnum; (right) Mary Stuart Lang. Page 262: Elizabeth Crews. Page 266: (left) Tequila Minsky; (right) Deborah Kahn/Stock, Boston. Page 268: (left) Tequila Minsky; (right) Shirley Zeiberg/Taurus Photos. Page 271: Alison Clarke-Stewart. Page 273: David M. Grossman. Page 274: (left) Elizabeth Crews; (right) Shirley Zeiberg/Taurus Photos. Page 279: (left) Barbara Rios/Photo Researchers; (right) Shirley Zeiberg/Taurus Photos. Page 282: (left) Bettye Lane; (right) David S. Strickler/The Picture Cube.

Chapter 11 Opener: Tequila Minsky. Page 293: David Strickler/Monkmeyer. Page 294: Alice Kandell/Photo Researchers. Page 298: David M. Grossman/Photo Researchers.

Chapter 12 Opener: Elizabeth Crews. Page 306: Barbara Rios. Page 313: Deborah Kahn/Stock, Boston. Page 317: Charles Gatewood/Stock, Boston. Page 324: (left) Laimute Druskis/Taurus Photos; (right) John Eastcott, Yva Monatiuk/The Image Works. Page 329: Robert V. Eckert, Jr./EKM-Nepenthe. Page 331: Elizabeth Crews.

Chapter 13 Opener: Peter Simon/Stock, Boston. Page 341: (left) John Eastcott, Yva Monatiuk/The Image Works; (right) Steve and Mary Skjold/The Image Works. Page 345: Alan Carey/The Image Works. Page 348: Ulrike Welsch. Page 357: Jean-Claude Lejeune/Stock, Boston.

Chapter 14 Opener: Steve Baratz/The Picture Cube. Page 370: Donald C. Deitz/Stock, Boston. Page 371: Stan Goldblatt/Photo Researchers. Page 376: Erika Stone. Page 378: Ron Byers. Page 380: Gale Zucker/Stock, Boston. Page 384: Mitchell Phyne/Jeroboam. Page 386: Susan Rosenberg/Photo Researchers.

Chapter 15 Opener: Paul Conklin/Monkmeyer. Page 393: Erika Stone. Page 390: Paul Conklin/Monkmeyer. Page 396: Ann Chwatsky. Page 398: David M. Grossman. Page 402: (top) Alan Carey/The Image Works; (bottom) Barbara Alper/Stock, Boston.

Chapter 16 Opener: Derek Berg. Page 408: Ed Lettau/Photo Researchers. Page 412: Peter Southwick/Stock, Boston. Page 417: Elizabeth Crews. Page 419: Paul Conklin/Monkmeyer. Page 421: Tequila Minsky. Page 423: Robert Kalman/The Image Works. Page 427: Tequila Minsky. Page 428: Alan Carey/The Image Works. Page 433: Barbara Alper/Stock, Boston.

Chapter 17 Opener: Van Bucher/Photo Researchers. Page 442: Peter Southwick/Stock, Boston. Page 443: Ulrike Welsch. Page 446: Barbara Alper/Stock, Boston. Page 448: Stephen Chidester/The Image Works. Page 452: Peter Menzel. Page 455: Arthur Tress/Photo Researchers. Page 456: Tequila Minsky.

Chapter 18 Opener: Charles Gupton/Stock, Boston. Page 465: Laimute Druskis/Stock, Boston. Page 463: Hazel Hankin. Page 464: Sepp Seitz/Woodfin Camp. Page 469: Ulrike Welsch. Page 471: Billy E. Barnes/Stock, Boston. Page 472: Lynne Jaeger Weinstein/Woodfin Camp. Page 474: (left) Mark Antman/The Image Works; (right) Ulrike Welsch. Page 475: Jill A. Cannefax/EKM-Nepenthe.

Chapter 19 Opener: Hazel Hankin. Pages 480 and 483: Ulrike Welsch. Page 486: (left) Erika Stone/Photo Researchers; (center) Alan Carey/The Image Works; (right) Cathy Cheney/EKM-Nepenthe. Page 491: John Coletti/Stock, Boston. Page 494: Ann Chwatsky/Phototake. Page 502: Ulrike Welsch. Page 503: David M. Grossman. Page 504: Stu Rosner/Stock, Boston. Page 498: (left) Ulrike Welsch; (center) Lynne Jaegar Weinstein/Woodfin Camp; (right) Beryl Goldberg.

Chapter 20 Opener: Marc and Evelyne Bernheim/Woodfin Camp. Page 514: Mark Antman/The Image Works. Page 516: Alan Carey/The Image Works. Page 519: Ulrike Welsch. Page 521: David M. Grossman. Page 532: Ulrike Welsch. Page 536: George W. Gardner/The Image Works. Page 535: Ulrike Welsch/The Boston Globe.

Chapter 21 Opener: Ulrike Welsch. Page 546: Alan Carey/The Image Works. page 551: Watriss/Baldwin/Woodfin Camp. Page 553: Robert A. Isaacs/Photo Researchers. Page 554: (left) George W. Gardner; (right) Peter Menzel. Page 558: Bob Sacha. Page 561: UPI/Bettmann Newsphotos. Page 563: Audrey Topping/Rapho/Photo Researchers.

Chapter 22 Opener: Thomas Hopker/Woodfin Camp. Page 569: (left) Tequila Minsky; (right) Ida Wyman. Page 571: Tequila Minsky. Page 574: Eric Kamp/Phototake. Page 576: Ulrike Welsch. Page 578: Ida Wyman. Page 579: (left) Watriss/Baldwin/Woodfin/Camp; (right) Hazel Hankin. Page 582: George Bellerose/Stock, Boston. Page 583: Timothy Eagan/Woodfin Camp. Page 584: Owen Franken/Stock, Boston.

Chapter 23 Opener: Dan Chidest/The Image Works. Page 591: Ulrike Welsch. Page 596: Michael Heron/Woodfin Camp. Page 597: Ray Ellis/Photo Researchers. Page 599: Gale Zucker/Stock, Boston. Page 602: Robert V. Eckert Jr./EKM-Nepenthe. Page 605: Bill Strode/Woodfin Camp. Page 606: Mike Douglas/The Image Works.

Epilogue Page 610: Craig Aurness/Woodfin Camp.

Source Notes

Chapter 1

Page 9, Figure 1.1 W. K. Frankenberg and J. B. Dodds. The Denver Development Screening Test. *Journal of Pediatrics*, 1967, *71*, 181–191.

Chapter 3

Page 55, Figure 3.1 Adapted from I. I. Gottesman. Genetic aspects of intelligent behavior. In N. Ellis (Ed.), *The handbook of mental deficiency: Psychological theory and research*. New York: McGraw-Hill, 1963.

Page 57, Figure 3.3 Adapted from C. H. Waddington. *New patterns in genetics and development*. Copyright © 1962 Columbia University Press. By permission.

Page 66, Figure 3.4 Adapted from L. Erlenmeyer-Kimling and L. F. Jarvik. Genetics and intelligence: A review. *Science*, 1963, *142*, 1477–1479. Copyright © 1963, by the American Association for the Advancement of Science.

Chapter 4

Page 77, Figure 4.1 C. Grobstein. External human fertilization. *Scientific American*, 1979, *240(6)*, 57–68. Copyright © 1979, by Scientific American, Inc. All rights reserved.

Page 97, Figure 4.3 K. L. Moore. *The developing human: Clinically oriented embryology* (3rd ed.). Philadelphia: W. B. Saunders, 1982.

Chapter 5

Page 120, Figure 5.2 W. K. Berg, C. D. Adkinson, and B. L. Strock. Duration and periods of alertness in neonates. *Developmental Psychology*, 1973, *9*, 434. Copyright © 1973, by the American Psychological Association. Adapted by permission of the authors.

Page 128, Figure 5.4 R. L. Fantz. Pattern discrimination and selective attention as determinants of perceptual development from birth. In A. H. Kidd and J. L. Rivoire (Eds.), *Perceptual development in children*. Madison, CT: International Universities Press, 1966.

Page 129, Figure 5.6 M. S. Banks and P. Salapatek. Infant visual perception. In P. Mussen (Ed.), *Handbook of child psychology*, Vol. 2. New York: Wiley, 1983. Copyright © 1983, by John Wiley and Sons.

Page 130, Figure 5.7 P. Salapatek. Infant perception: From sensation to cognition. In L. B. Cohen and P. Salapatek (Eds.), *Basic visual processes*, Vol. 1. Orlando, FL: Academic Press, 1975.

Page 136, Figure 5.8 B. I. Bertenthal, J. J. Campos, and M. M. Haith. Development of visual organization: The perception of subjective contours. *Child Development*, 1980, *51*, 1072–1080. Copyright the Society for Research in Child Development, Inc.

Chapter 7

Page 175, Figure 7.1 L. A. Sroufe. Socioemotional development. In J. D. Osofsky (Ed.), *Handbook of infant development*. New York: Wiley, 1979. Copyright © 1979, by John Wiley and Sons.

Page 177, Figure 7.2 E. Waters. L. Matas, and L. A. Sroufe. Infants' reactions to an approaching stranger: Description, validation, and functional significance of wariness. *Child Development*, 1975, *46*, 348–356. Copyright The Society for Research in Child Development.

Chapter 8

Page 205, Figure 8.1 J. M. Tanner. *Foetus into man*. London: Open Books Publishing, Ltd., 1978.

Page 206, Figure 8.2 C. Golomb. *Young children's sculpture and drawing: A study in representation development*. Cambridge, Mass.: Harvard University Press, 1974. Reprinted by permission.

Page 207, Table 8.1 G. H. Lowrey. *Growth and development of children*, 7th ed. Chicago: Yearbook Medical Publishers, 1978. Reproduced with permission.

Chapter 9

Page 232, Figure 9.4 R. Gelman. Conservation acquisition. *Journal of Experimental Child Psychology*, 1969, *7*, 174.

Page 250, Figure 9.6 S. W. Gray, B. K. Ramsey, and R. A. Klaus. *From 3 to 20: The early training project*. Baltimore, Md.: University Park Press, 1982.

Chapter 11

Page 299, Figure 11.1 S. Farnham-Diggory. *Learning disabilities*. Cambridge, Mass.: Harvard University Press, 1978.

Chapter 12

Page 309, Table 12.1 L. Kohlberg. Stage and sequence: The cognitive-developmental approach to socialization. In D. A. Goslin (Ed.), *Handbook of socialization theory and research*, Table 6.2. Boston. Houghton Mifflin, 1969.

Page 310, Figure 12.1 A. Colby, L. Kohlberg, J. Gibbs, and M. Lieberman. A longitudinal study of moral development. *Monographs of the Society for Research in Child Development*, 1983, *48* (1, Serial No. 200). Copyright the Society for Research in Child Development.

Page 320, Figure 12.3 Copyright © 1949, 1974 by The Psychological Corporation. Reproduced by permission. All rights reserved.

Page 321, Figure 12.4 From the Raven Standard Progressive Matrices, by permission of J. C. Raven, Ltd.

Page 323, Figure 12.5 J. W. Getzels and P. W. Jackson. *Creativity and intelligence*. New York: Wiley, 1962. Copyright © 1962, by John Wiley and Sons.

Chapter 13

Page 338, Figure 13.1 S. Ellis, B. Rogoff, and C. Cromer. Age segregation in children's social interactions. *Developmental Psychology*, 1981, *17*, 399–407. Copyright © 1981, by the American Psychological Association. Adapted by permission of the authors.

Page 359, Table 13.1 A. Thomas, S. Chess, and H. G. Birch. *Temperament and behavior disorders in children*. Copyright © 1968 by New York University Press.

Chapter 14

Page 368, Figure 14.2 J. M. Tanner and R. H. Whitehouse. Clinical longitudinal standards for height, weight, height velocity, weight velocity, and the stages of puberty. *Archives of Disease in Childhood*, 1976, *51*, 170–179. J. M. Tanner. *Fetus into man: Physical growth from conception to maturity.* Cambridge, Mass.: Harvard University Press, 1968.

Page 373, Figure 14.3 E. Koff, J. Rierdan, and E. Silverstein. Changes in the representation of body image as a function of menarcheal status. *Developmental Psychology*, 1978, *14*, 635–642. Copyright © 1978, by the American Psychological Association. Reprinted by permission of the authors.

Page 378, Figure 14.4 C. A. Darling, D. J. Kallen, and J. E. Van Dusen. Sex in transition, 1900-1980. *Journal of Youth and Adolescence*, 1984, *13*, 385–399.

Chapter 16

Page 411, Figure 16.1 R. G. Simmons, E. F. Van Cleave, D. A. Blyth, and D. M. Bush. Entry into early adolescence: The impact of school structure, puberty, and early dating in self-esteem. *American Sociological Review*, 1979, *44*, 956.

Page 424, Figure 16.2 T. J. Berndt. Developmental changes in conformity to peers and parents. *Developmental Psychology*, 1979, *15*, 608–616. Copyright © 1979, by the American Psychological Association. Adapted by permission of the author. P. R. Costanzo and M. E. Shaw.

Conformity as a function of age level. *Child Development*, 1966, *37*, 967–975. Copyright the Society for Research in Child Development, Inc.

Chapter 17

Page 446, Figure 17.1 G. A. Kimble, N. Garmezy, and E. Zigler. *Principles of Psychology*, 6th ed. New York: Wiley, 1984, p. 106. Copyright © 1984, by John Wiley and Sons.

Pages 447–450, Figures 17.2, 17.3, 17.4 G. E. Nelson. *Biological Principles with Human Perspectives*, 2nd ed. New York: Wiley, 1984. Copyright © 1980, 1984, by John Wiley and Sons.

Page 457, Figure 17.5 A Kinsey et al. *Sexual Behavior in the Human Female*. Philadelphia: W. B. Saunders, 1953, Figure 102, p. 519, Figure 99, p. 514. A. Kinsey et al., *Sexual Behavior in the Human Male*. Philadelphia: W. B. Saunders, 1948, Table 44, p. 220, Table 45, p. 226. Reprinted by permission of the Kinsey Institute for Research in Sex, Gender, and Reproduction, Inc.

Chapter 18

Page 472, Figure 18.2 J. L. Horn. Organization of data on lifespan development. In Goulet and P. Baltes, *Life-span developmental psychology: Research and theory*. Orlando, FL: Academic Press, 1970.

Chapter 19

Page 491, Figure 19.1 From *The seasons of a man's life*, by Daniel Levinson et al. Copyright © 1978, by Daniel J. Levinson and the Sterling Lord Agency. Reprinted by permission of Alfred A. Knopf, Inc.

Page 497, Table 19.3 H. M. Hacker. "Blabbermouths and clams: Sex differences in same-sex and cross-sex friendship dyads." *Psychology of Women Quarterly*, 1981, *5*, 393.

Chapter 20

Pages 523, 526, and 527, Figures 20.1, 20.2, and 20.3 G. E. Nelson. *Biological principles with human perspectives*, 2nd ed. New York: Wiley, 1984, pp. 144, 147, 170–171. Copyright © 1980, 1984, by John Wiley and Sons.

Chapter 21

Page 550, Figure 21.1 K. W. Schaie and C. R. Stother. A cross-sequential study of age changes in cognitive behavior. *Psychological Bulletin*, 1968, *70*, 671–680. Copyright © 1968 by the American Psychological Association. Reprinted by permission of the authors.

Selected Name Index

Index